Progress in Tryptophan and Serotonin Research

Progress in Tryptophan and Serotonin Research

Proceedings
Fourth Meeting of the
International Study Group for
Tryptophan Research ISTRY
Martinsried, Federal Republic of Germany
April 19 – 22, 1983

Editors
H. G. Schlossberger · W. Kochen
B. Linzen · H. Steinhart

W DE G

Walter de Gruyter · Berlin · New York 1984

Editors

Hans Georg Schlossberger, Dr. rer. nat.
Max-Planck-Institut für Biochemie
D-8033 Martinsried
Federal Republic of Germany

Walter Kochen, Professor, Dr. rer. nat.
Universitäts-Kinderklinik
Isotopenlaboratorium
Neuenheimer Feld 150
D-6900 Heidelberg
Federal Republic of Germany

Bernt Linzen, Professor, Dr. rer. nat.
Universität München
Zoologisches Institut
Luisenstraße 14
D-8000 München 2
Federal Republic of Germany

Hans Steinhart, Professor, Dr. rer. nat., Dr. agr. habil
Technische Universität München
Institut für Ernährungsphysiologie
D-8050 Freising-Weihenstephan
Federal Republic of Germany

CIP-Kurztitelaufnahme der Deutschen Bibliothek

Progress in tryptophan and serotonin research:
proceedings, Martinsried, Fed. Republic of Germany, April 19 – 22, 1983 /
ed. H. G. Schlossberger . . . – Berlin; New York: de Gruyter, 1984.
(. . . Meeting of the International Study Group for Tryptophan
Research, ISTRY; 4)
ISBN 3-11-009760-5

NE: Schlossberger, Hans G. [Hrsg.]; International Study Group for
Tryptophan Research: . . . Meeting of the . . .

Library of Congress Cataloging in Publication Data

International Study Group for Tryptophan Research.
Meeting (4th: 1983: Martinsried, Germany)
Progress in tryptophan and serotonin research.
„In honor of Professor Adolf Butenandt"--Pref.
Bibliography: p.
Includes indexes.
1. Tryptophan--Congresses. 2. Tryptophan--Metabolism--Congresses.
3. Serotonin--Congresses. 4. Butenandt, Adolf, 1903– . I. Schlossberger,
H. G. (Hans Georg), 1920– . II. Butenandt, Adolf, 1903– . III. Title. [DNLM:
1. Tryptophan--Congresses. 2. Receptors, Serotonin--Physiology--Congresses.
W3 IN907 4th 1983p / QU 60 1596 1983p]
QP562.T7 I58 1983 574.19'245 84-1719
ISBN 3-11-009760-5

Printing: Gerike GmbH, Berlin. – Binding: Lüderitz & Bauer Buchgewerbe GmbH, Berlin.
Printed in Germany.

Preface

The Fourth International Meeting of ISTRY was held at Martinsried in the outskirts of Munich. The Contents of these Proceedings show at a glance what was also felt by all participants: Tryptophan research is still going strong. Comparison with previous meetings reveals that new research areas are developing and that emphasis in traditional areas is changing. In the Martinsried conference, strong attention was paid to the serotonin pathway and the relation of tryptophan supply to brain function. A special workshop was organized (by Ernest Hartmann) to discuss the clinical and behavioural effects of treatment with tryptophan. Although a wealth of analytical methods for determination of tryptophan and its metabolites can be found in the existing literature, the impact of recent high performance separation techniques and of extremely sensitive physical detection methods led to a lively discussion of their applications to tryptophan and its metabolites (this has been reported by Mendel Friedman). Classical fields of tryptophan research: enzymology, nutritional aspects, clinical significance of tryptophan metabolism, metabolism in mammals and insects are also included in this volume along with recent developments in plant and microbial tryptophan metabolism. To the existing list of biologically active or toxic compounds derived from tryptophan, e.g. nicotinamide, melatonin, indoleacetic acid or the Amanita toxins, new ones have been added, such as blepharismone, a mating hormone in ciliates, or the teleocidin-type tumor promoters. Proper consideration has been given to tryptophan in proteins and peptides, and the opportunities for cleavage, and the pitfalls encountered by the protein chemist.
The meeting was held in honour of Professor Adolf Butenandt at the occasion of the 80th anniversary of his birthday. This book is also dedicated to him in recognition of his pioneering work in tryptophan metabolism.

The meeting was sponsored by the Deutsche Forschungsgemeinschaft and by the Max-Planck-Gesellschaft. We gratefully acknowledge their support. We would also like to thank all those persons who helped in the preparation and during the meeting.

In editing of this book, we had to achieve a compromise between detailed editing and a reasonable publication time. Hence we restricted corrections to a minimum and did not enforce uniformity in nomenclature or in the abbrevations. Furthermore, as in any international meeting the English was not always flawless. Finally, several controversial issues have been included among the contributions since we felt that the flair of scientific debate should also be reflected in the Proceedings of a meeting. It must also be mentioned that this volume contains a few contributions which could not be presented during the meeting due to serious impediments to travel.

In the preparation of the final manuscript and compilation of the index, the unflagging help of Mrs. Christiane Birnstiel, Mrs. Uta Schimanko, and Dr. Petra Stützle is gratefully acknowledged. Thanks go also to the Publishers for their helpful cooperation.

H.G. Schlossberger
W. Kochen
B. Linzen
H. Steinhart

CONTENTS

OPENING SESSION

QUALITATIVE AND QUANTITATIVE ANALYSIS OF TRYPTOPHAN AND ITS METABOLITES

WORKSHOP: ANALYSIS OF TRYPTOPHAN

TRYPTOPHAN AND SEROTONIN: NEUROBIOLOGY

PATHOLOGY, CLINICAL ASPECTS

Nutritional Aspects

Tryptophan Metabolism in Mammals

TRYPTOPHAN METABOLISM IN INSECTS

Tryptophan Metabolism in Plants and Microorganisms

TRYPTOPHAN IN PROTEINS AND PEPTIDES

Opening Session

ADOLF BUTENANDT'S WORK ON THE METABOLISM OF TRYPTOPHAN

E. Biekert

Knoll AG
D-6700 Ludwigshafen

I have the privilege and honor to open the scientific part of the ISTRY conference although I am far from being a famous researcher in the field of tryptophan metabolism. As a matter of fact I have not been concerned with tryptophan since those happy years from 1950 - 1961, when I had the opportunity of working together with A. Butenandt on the subject of chemical aspects based on ideas and conceptions of this great teacher. I later entered the research branch of the pharmaceutical industry where quite a few of Butenandt's co-workers had already become so successful that the Dahlem-Tübingen-School had come to be regarded as a special talent school for industrial research.

If my best friend Ulrich Schiedt, one of Butenandt's most gifted and successful students, had not met his tragic death so soon, it would have surely been his place to stand here today looking back upon the tryptophan history of the Kaiser-Wilhelm-Institute and the Max-Planck-Institute for Biochemistry. Ladies and gentlemen of the ISTRY circle and dear friends and admirers of Adolf Butenandt: I trust you will forgive me if I begin my short talk on tryptophan with a personal foreword. A foreword addressed toward our esteemed teacher and friend which will give those of you, who were not lucky enough to experience it themselfes, a small impression of the Tübingen workshop, where the biochemical aspects of tryptophan metabolism were one of the main subjects.

The years following the second world war, the post-war generation and the scientific revival have been talked about very often, almost too often.

Progress in Tryptophan and Serotonin Research

The fact that the University of Tübingen had not been destroyed was a stroke of luck and a particular blessing to those who returned from the war. An illustrious group of scientists gathered here who would have otherwise surely never have come to this, in those times, rather provincial university.
Foremost of all Adolf Butenandt, the biochemist who brought the aura of the Nobel Prize-winner to dreamy Tübingen, together with his whole institute: Schramm, Friedrich-Freksa, Dannenberg, Ruhenstroth, Schmidt-Thomé and Karlson, just to mention a few. Also Alfred Kühn, who brought new glamour into the somewhat stuffy zoology department and who, together with Butenandt, started on the subject to which today's tryptophan lectures will be dedicated, at least in part. W. Kossel in the physics department, G. Wittig in the chemistry department and W. Hückel, the great theoretical chemist, settled in the pharmacy department. The list of names could be continued and supplemented by others from the fields of medicine and the liberal arts, who were equally well known (E. Kretschmer, E. Spranger, R. Guardini, H. Thielicke), and with whom a close exchange of views existed.
You, dear Prof. Butenandt, were the star among the great scientists of Tübingen for us students in many ways: you were cosmopolitan and known all over the world; your visits to the USA had helped to quickly renew your close ties across the Atlantic. In this context I remember in particular Mr. Haas' visit (Roehm + Haas, USA) who gave the first Perkin-Elmer IR spectograph to the institute, which considerably encouraged the tryptophan ommochrome work group.
Your appearance was youthful and athletic so that you could hardly be distinguished from a young student when you arrived at the institute before 8 o'clock in the morning; the rumour that you had been horseback riding for one hour before your lecture contributed much to your legendary reputation.
Scientifically inspiring and always "one step ahead", you were considered by everyone to be the type of scientist who could bring our science, the Kaiser-Wilhelm-Gesellschaft and the universities back up to international standards. And at last but not least your lectures: these were the highlights of scientific life in the Tübingen of the late forties and the

early fifties and are still praised as the best an academic teacher can offer. Your students, diploma and doctoral candidates, scientific assistants from the chemical, biological or medical departments, had the chance to attend the "Kinderstunde", a meeting which you held once a week to help them "cram" a subject. No one who had the privilege to participate will ever forget these sessions.

Any academic teacher may be recommended to do the same, however, I fear that this manner of caring for those young persons entrusted to you, will remain unique. In the Tübingen of those days you ruled the Kaiser-Wilhelm-Institute, later Max-Planck-Institute for Biochemistry and the Physiological-Chemical-Institute of the University with complete sovereignty. And many of your students who may sometimes have felt your strict and uncompromising way of teaching as a burden are indebted to you for having experienced a very special manner of scientific thinking and working which can only be described by one expression "the Butenandt family".

The scientific work at the Tübingen institute - besides virology which, at that time, had already become almost independent - concentrated on the search for the pupation hormone (Ecdyson, Karlson), on the work on insect attractants (Bombycol, Hecker) and the question as to how genetic information in the cell nucleus is expressed.

You had personally chosen these 3 topics whose progress you supervised in almost daily discussions; the laboratories in which the work was being carried out were your favourite haunts. Coming to tryptophan: In close cooperation with Alfred Kühn you focussed on the question of how genes influence the characteristics of an organism ascribed to them. This question was raised by Alfred Kühn's observations on various strains of the Mediterranean flour moth. Before going into details, I would like to mention the result of this work, and here I am quoting you, dear Prof. Butenandt: " Our mutual work resulted in the knowledge that certain genetic factors enable the cell to synthesize certain biocatalysts, so-called enzymes or ferments. In general terms this means that certain genes enable the cell to produce certain specific proteins: this discovery opened up the path to the full understanding of the mechanism of action of the genetic factors in individual development, as it is nowadays described in molecular genetics". (A. Butenandt, Orden "Pour le Mérite", Reden und

Gedenkworte 9/1968/69).

What was the historical background of this work and what made you choose this subject, dear Prof. Butenandt?

Mendel's laws - long forgotten - were rediscovered around the turn of the century. Shortly afterwards, the first communications on biochemical effects attributable to genetic processes appeared.

Bateson (Cambridge) found that alcaptonuria, a disease known since as early as 1685, is recessively hereditary according to Mendel's laws. In this disease, homogentisic acid, responsible for blackening of urine, is enriched. This homogentisic acid had been isolated by Bodecker from the urine of patients as early as 1858. This disease is described among

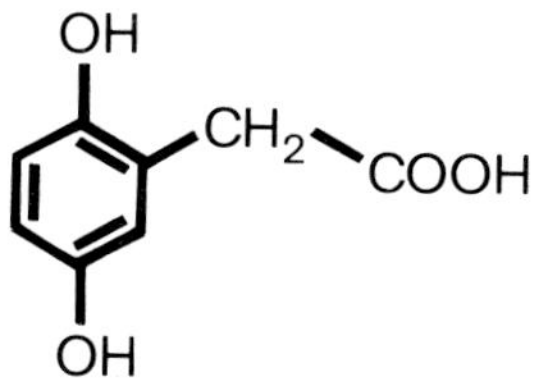

Homogentisic acid

other hereditary diseases in Garrod's book (1909) "Inborn errors of Metabolism". Then in 1914 Gross isolated the enzyme which degrades homogentisic acid. This enzyme is absent in the urine of alcaptonuria patients. This unambiguous finding was not appropriately appreciated by the geneticists and biochemists. In his speech on the occasion of being awarded the Nobel Prize Beadle said that unfortunately he had only read Garrod's book after he had completed his work on the pigments in the eyes of insects. The work of Butenandt-Kühn, Beadle-Tatum and others on insect pigments are, therefore, the actual beginning of biochemical genetics. The starting point was Kühn's research on the inheritance of eye colour in the Mediterranean flour moth (Ephestia 1930). Analogous work was performed shortly afterwards by Ephrussi and Beadle on the fruit fly (Drosophila).

Between 1930 and 1940 numerous mutants of Ephestia and Drosophila were bred in which pigmentation had been changed by a modification of an allele. These experiments resulted in the substantial information that deficient mutants can regain the ability to form pigment by tissue grafting or by injection of extracts from wild forms.

Becker gave a very detailed description of the pigments themselves as early as 1937, and it was he who coined the term ommochromes which is still in use today. He also found in 1937 that the processes in the Mediterranean flour moth and in Drosophila are identical The a^+ substance of the flour moth and the v^+ substance of Drosophila were the same.

Both working groups (Butenandt-Kühn, Beadle-Tatum) enriched extracts containing this active substance. In analogy to Butenandt's experiments on sexual hormones, where metabolic deficiencies could be repaired by tissue grafting or injection of extracts, the existence of some type of gene hormone was presumed.

To prepare the extracts Butenandt used Calliphora which is easy to breed; Tatum found that bacteria are able to produce an active substance from tryptophan and this finding was confirmed by Butenandt with Corynebacteria. In this context it is particularly noteworthy that experiments with the Corynebacterium mediolanum on steroids, which were performed at the former Kaiser-Wilhelm-Institute, probably represented the first attempts to convert chemical substances by means of microorganisms using biotechnological methods which are presently so succesful.

At this point it was quite clear that the substance in question must be some type of amino acid, but investigations of various amino acids (including tryptophan) had failed. The indication that bacteria can produce active substances from tryptophan caused Butenandt to examine tryptophan derivatives and this finally lead to success.

Kynurenine, an amino acid isolated from rabbit urine in 1935 by Kotake - which forms in rabbits when tryptophan is added to a rice diet - had all properties of the active substance. For example: it also develops a distinct odour of jasmine as had been observed with Calliphora extracts, when alkali is added. Tatum, who had found an easy way to produce active extracts with his bacteria cultures, could not identify the active substance since he had isolated a kynurenine complex with sucrose from his bacteria cultures.

At this stage the two working groups started to go separate ways. Butenandt steadfastly followed the tryptophan-kynurenine-ommochrome idea and finally succeeded in demonstrating the whole pathway and in elucidating the complex structure of the ommochromes.

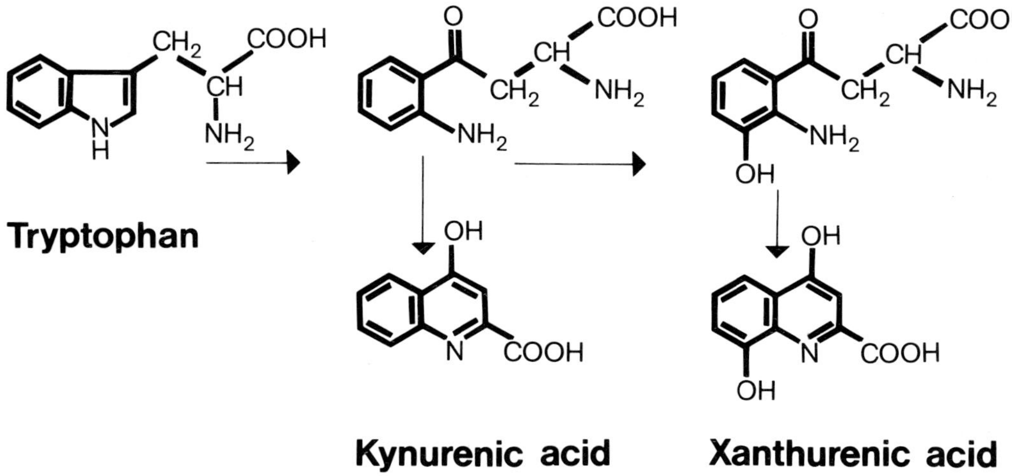

Beadle and Tatum turned to the microorganisms and,using the yeast Neurospora which is particularly suited for this type of work, made further major contributions to the biochemistry of genes.
Here again the "Butenandt School" proved its worth. The chemical-synthetic and analytical approach to the problem was a pragmatic one profiting from Butenandt's enormous experience in the field of organic synthesis.
Kynurenine was synthesized and proved unambiguously that this is the active substance responsible for pigment formation in deficient mutants, at which time the structural formula first presented by Kotake could be revised.
A further misconception became evident later on. 2-Hydroxytryptophan had been described as an active substance. Wieland had isolated this compound by hydrolysis of phalloidine from the death cup mushroom (Amanita phalloides). 2-Hydroxytryptophan is not, however, part of the tryptophan-kynurenine pathway. The activity found is due to the degradation of dissolved 2-hydroxytryptophan to kynurenine upon standing.
Even at this stage it became evident: kynurenine is not a "gene hormone". The effect of kynurenine on pigment formation was tested quantitatively and showed definitely that kynurenine is an intermediate for the synthesis of ommochrome. Since, however, kynurenine is formed by enzymatic conversion the following conclusion was apparent: genes provide enzymes responsible for the reaction pathways. Later more precisely stated as: one gene one enzyme.
Now the future activities were predestinated. To search for the next step

in the tryptophan-kynurenine-ommochrome pathway. Research work conducted at different places suggested that the substance in question must be closely related to kynurenine and is likely to be formed by oxidation.
It was also known that in mammalian metabolism not only is kynurenic acid formed from tryptophan via kynurenine but also xanthurenic acid, which would call for the presence of a 3-hydroxykynurenine. Since isolation attempts had failed for a long time, 3-hydroxykynurenine was synthesized. It turned out to be the active substance sought after.
Already during the work on 3-hydroxykynurenine the idea was put forward that this compound, in analogy to the formation of melanin from dihydroxyphenylalanine (dopa), might undergo conversion by oxidation into a quinoid system. The further studies to elucidate the structure of ommochromes confirmed this idea.

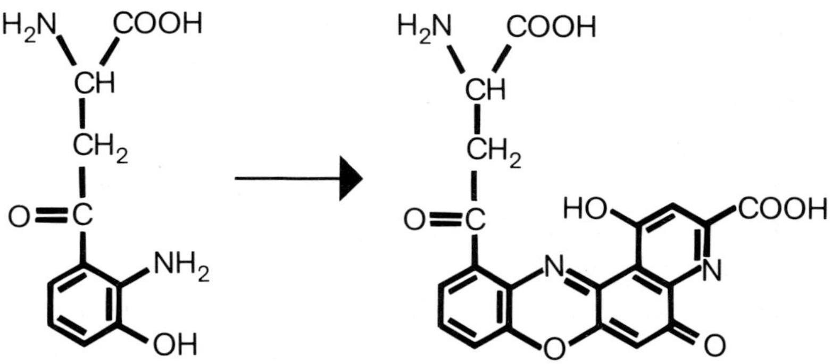

3-Hydroxykynurenine **Xanthommatin**

New organisms were used for the investigation of the structure of ommochromes. Becker had already investigated the secretions of the butterfly Vanessa urticae produced for bursting the chrysalis and had classified them as ommochromes. The methods he employed at that time were, however, not sophisticated enough to allow purification and structure elucidation. Separation of the dyestuffs proved to be extremely difficult, they could not be crystallized and their determination was rather global and on an intuitive basis.
The break-through was finally achieved with the introduction of new methods of separation and analysis, in particular paper chromatography, which for the first time was used for preparative separation purposes.

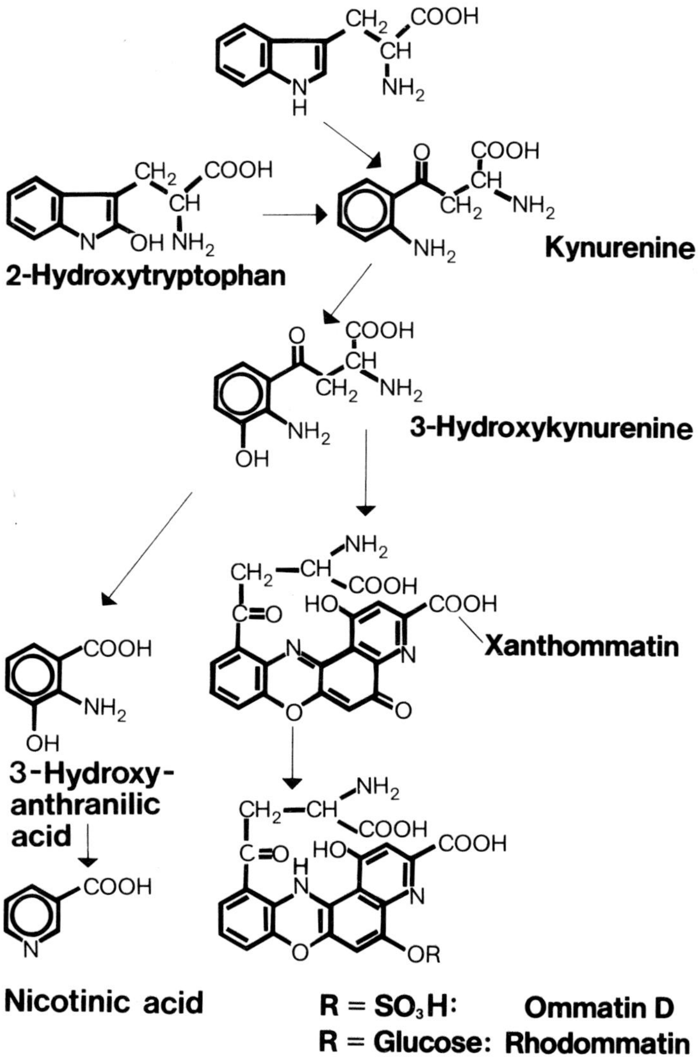

Structure elucidation following separation was facilitated by the aforementioned spectrographs, above all the IR spectrometer, which had been donated by Roehm and Haas. The purified xanthommatin was degraded and xanthurenic acid was formed which indicated that 3-hydroxykynurenine must be a structural element of ommochromes.

Model experiments and systematic comparative spectroscopy revealed a phenoxazone structure. The final result was the elucidation of the ommochrome structure for the closely related compounds xanthommatin, rhodommatin and ommatin D.

The structure of ommin, which belongs to a different class of substances, has not yet been definitely elucidated; ommin is also a final product of tryptophan metabolism and probably the most common insect pigment. At present only certain structural elements are known although diligent students of Butenandt have been working hard on this vexing natural dyestuff. A challenge for young scientists engaged in natural product chemistry and an idea for Butenandt's 85th birthday: the ommin formula as a present.
This would then complete the major part of Butenandt's tryptophan work. It has been a long way (30 years) from Kühn's first mutation studies to the synthesis of ommochromes. When admiring today's knowledge of the structure and action of genes and the enormous success of gene technology one should keep in mind that the studies on tryptophan metabolism in connection with the synthesis of ommochromes were the first steps in this direction.

TRYPTOPHAN, THE HEART OF THE TOXIC PEPTIDES FROM AMANITA MUSHROOMS

Th. Wieland
Max-Planck-Institut für medizinische Forschung, Jahnstr. 29, D-6900 Heidelberg

Introduction

Nature makes use of the indole nucleus for multiple purposes, tryptophan being one of the most important ingredients. Without this amino acid life of animals would not be possible but it is also a component of seemingly outsider peptides such as the toxins of the poisonous Amanita fungi.[1)] Within the many years of employment with the toxic compounds from A.phalloides and A.virosa we have isolated more than 20 toxic peptides the center of which is the indole moiety of tryptophan.

The Amanita toxins

In all cases the indole ring is substituted in 2-position by a sulfur residue. This S-atom is part of a cysteine originally present in the peptide which has been crosslinked to the indole by a biochemical oxidation. The conjugate so formed was called tryptathionine. Tryptathionine is the bridge in the bicyclic phallotoxins and amatoxins. At the latter ones bio-oxidation has proceeded to produce a (R)-sulfoxide. In most members of this series in the 6-position of indole an additional OH-group is present.

PHALLOIDIN

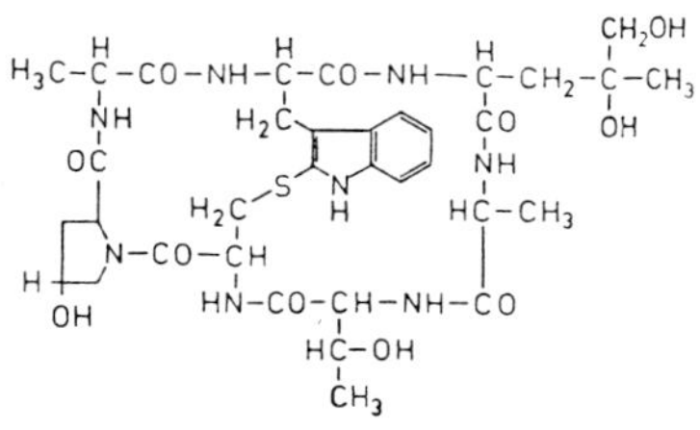

α-AMANITIN

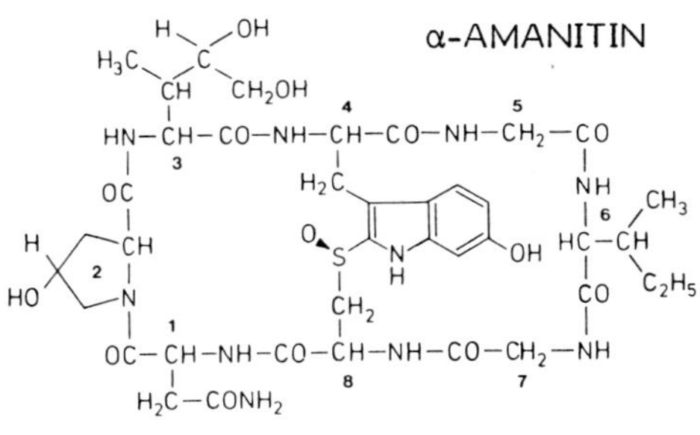

Progress in Tryptophan and Serotonin Research

In the family of the virotoxins one finds the sulfur in 2 -position of the tryptophan moiety oxidized even to the sulfone state. They posess a methyl sulfonyl group, CH_3SO_2, two more OH-groups than the phallotoxins and are monocyclic heptapeptides. Since their amino acid composition and sequence resembles closely that of the phallotoxins a biochemical connection is very probable (see later).

VIROTOXINS

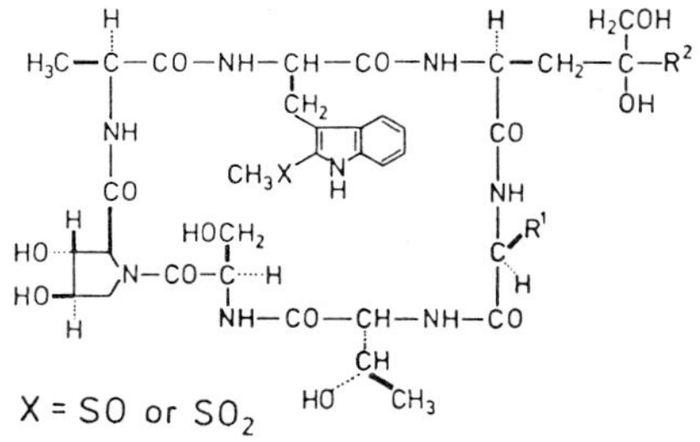

X = SO or SO_2

In the following we will neglect the peripheral amino acids of the cyclic toxins but consider only relations between structural changes of the tryptathionine moiety and toxicological activity.

Tryptophan in the Amatoxins

The amatoxins are (R)-sulfoxides. The OH group in position 6 of the indole ring is not essential for their toxicity since the main poison of A.virosa, amanine amide is equally toxic as α-amanitin although it doesn't contain this OH-group. The toxicity is due to a very strong inhibition ($K_i = 10^{-8}$-10^{-9}M) of the DNA dependent RNA polymerase II (or B) in the nucleus of eukaryotic cells (liver and kidney). The depletion of m-RNA so caused leads to a stop of protein synthesis with its fatal consequences. According to A.C. Vaisius[2)] it is not the formation of the phosphodiester bond between the 3'-OH of the growing RNA chain and the next nucleotide triphosphate which is blocked rather than the translocation of the RNA under formation. A ternary complex consisting of template DNA, polymerase, and the nascent RNA stabilized by α-amanitin can be isolated, so to speak, as a snapshot.

State of sulfur and conformation of amatoxins

In order to see if the R-sulfoxide moiety is responsible for the strong binding of the amatoxins to RNA polymerase II, we removed the oxygen from sulfur by reducing agents.[3] The thioether so obtained is as toxic as the R-sulfoxide. It was re-oxidized by hydrogen peroxide yielding besides the original (R)-sulfoxide the diastereomeric (S)-sulfoxide and the sulfone. The (S)-sulfoxide is about 100 times less toxic than the (R) isomer, but the sulfone, surprisingly, is about as toxic as the (R)-sulfoxide although bearing one oxygen atom at the same side as the (S)-sulfoxide.

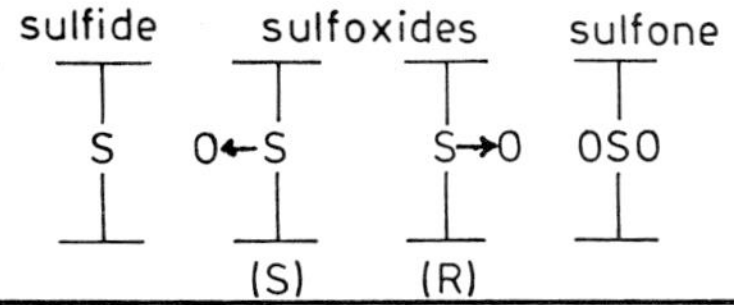

	sulfide	(S)	(R)	sulfone	
K_i	2.3	20	2.6	5.0	nM
LD_{50}	0.25	20	0.3	0.5	mg/kg

The striking difference of toxicities of the (S)-sulfoxide and the other amatoxins formerly was ascribed to differences of conformations. Therefore comparing structure analyses were conducted by ^{1}H-NMR analysis and by X-ray crystallographic studies.[4] The result was that the 3-dimensional structures of all of the derivatives are practically identic. Among other considerations it is suggested that a reason for the differing affinities for the polymerase of (R)- and (S)-sulfoxides can be the opposite direction of the strong SO-dipoles by which the negatively charged oxygen atom is or is not rejected by a respective negative charge on the protein.

Tryptophan in the Phallotoxins

The phallotoxins, represented by their best known member phalloidin contain the chromophoric system of tryptathionine. In the rigid molecule the 2-indolylthioether moiety is fixed in a helical conformation.

This inherent dissymmetric chromophore is responsible for the positive Cotton effects in the CD spectrum around 300 nm, the region of the long wave absorption bands in uv(fig. 1). The question whether the system has positive or negative helicity could be answered by x-ray diffraction analysis of a crystalline model peptide with

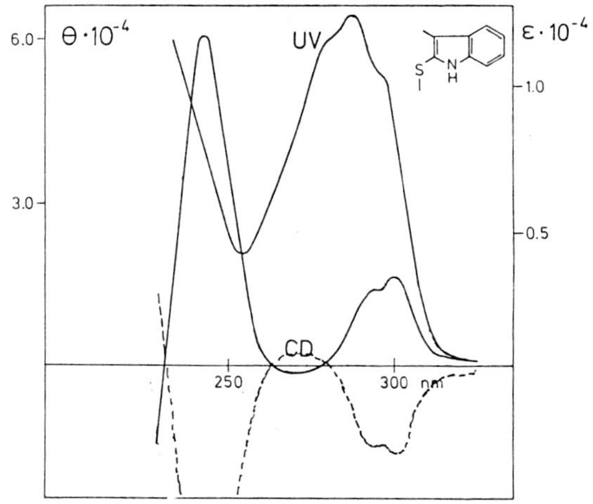

Figure 1. UV spectrum (λ_{max} 292 nm) and CD spectrum (positive Cotton effects around 300 nm) of phalloidin in water. Broken line (negative Cotton effects): CD spectrum of model thioether.

an analogous thioether structure but exhibiting negative Cotton effects around 300 nm. Since the x-ray analysis revealed negative helicity for the model compound the phallotoxins exhibiting positive cotton effects must have a positive helical chromophore (fig. 2).

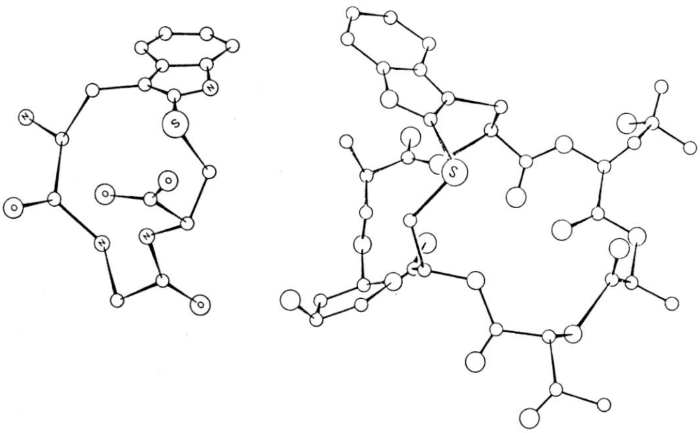

Figure 2. Spatial formulae of model thioether (left, negative helical) and phalloin (right, positive helical).

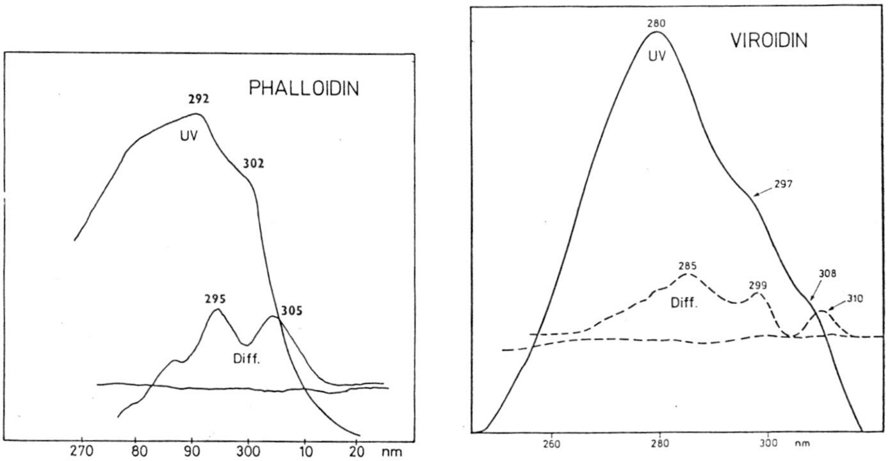

Figure 3. UV spectra (in water) of phalloidin (left) and viroidin, and respective differences of absorbance between the peptides alone and in the presence of F-actin. The chromophores are embedded in a lipophilic region.

Interaction with F-actin

Phallotoxins bind strongly to F-actin. The equilibrium G-actin F-actin is shifted extremely to the F-actin side because the filaments are stabilized effectively by phallotoxins.[5)] It can be demonstrated by uv-difference spectroscopy that the indole part of the toxin participates in the strong association of toxin with the protein.

The interaction of phallotoxins with actin in liver cells is causatively connected with their toxicity. This can be shown by comparing the association constants of derivatives and analogs to the respective lethal doses. As an example a product of oxidation of phalloidin with peroxyacetic acid, the (S)-sulfoxide is not toxic with doses up to 20 mg per kg (white mouse) and has minimal affinity for F-actin, whereas phalloidin which binds to F-actin with an association constant of about 2×10^7 M has LD_{50} (lethal dose for 50%) of 2 mg per kg (white mouse).

The oxidation of phalloidin mentioned above beside the (S)-sulfoxide yields an (R)-sulfoxide and a sulfone. Most interestingly, as in the case of amatoxins, also the (R)-sulfoxide and the sulfone are toxic, and bind strongly to F-actin (K_{ass} 0.5×10^7). For explanation also here the opposite directions of both the S-O dipoles can be offered. When the 3-dimensional structure of actin perhaps should be solved in due time the possibility would exist to prove the correctness of the dipole hypothesis.

Phallotoxins with affinity labels

For affinity labelling various derivatives of phalloidin were prepared. An amino function was introduced by 2 routes[6,7] which allowed to fix a (radioactive) iodoacetyl residue reacting with thiol groups close to the binding site. A novel diazirino-benzoic acid[8] served as a photolabel coupled via a ^{3}H- -alanine spacer to aminophalloin.

$$\text{PHN-NH}_2 \begin{array}{l} \nearrow \text{PHN-NHCOCH}_2\text{I} \\ \searrow \text{PHN-NHCOCH}_2\text{CH}_2\text{NHCOC}_6\text{H}_4\text{C(N}_2\text{)CF}_3 \end{array}$$

On irradiation in the presence of F-actin the photolabelled phalloin via a carbene will covalently bind to nucleophiles at the binding site.

Phallotoxin carriers

Phallotoxins are rapidly taken up by the liver. In order to detect putative carriers in the cell membrane isolated rat hepatocytes were irradiated while incubated with the radioactive diazirin-labelled phalloin. After that by gel electrophoresis 3 proteins covalently linked to the label were traced with molecular masses of 48, 52, and 95 kD, beside, of course, actin (42 kD). The three carrier membrane proteins are the same which bind also antamanide[9], the cyclic decapeptide preventing phalloidin death of experimental animals by inhibiting the entering of the toxin into the cell.[10] It is interesting to note that the same carrier proteins are also responsible for the uptake of bile salts by the liver as reported from another laboratory.[11]

Tryptophan in the Virotoxins

The virotoxins are found only in the white Amanita species A.virosa. Their mode of toxic activity is analog to that of phallotoxins, i.e. they cause swelling of the liver and death after 2-5 hours by binding to and stabilizing the F-actin of liver cells. The peptide ring of the virotoxins is monocyclic, not crosslinked, the tryptophan side chain is free.[12] However, also here the 2-position is occupied by a methylsulfenyl, CH_3SO, or methylsulfonyl group, CH_3SO_2. The L-cysteine moiety of the tryptathionine of the phallotoxins is replaced by a D-serine. This additional OH-group seems to be necessary for binding to F-actin. Since the virotoxins have no rigid conformation for their adaption to the protein additional binding groups are required. One of them could be that of D-serine, but strong binding as with the phallotoxins occurs only by a third attaching unit. This is the additional OH group in the hydroxyproline moiety: the virotoxins contain all-trans-3,4-dihydroxy-L-proline.

The supposed biochemical relationship of virotoxins and phallotoxins is represented by the following sequence of reactions (fig. 4); with the last oxidation step also the 3'-OH group into L-hydroxyproline may be introduced.

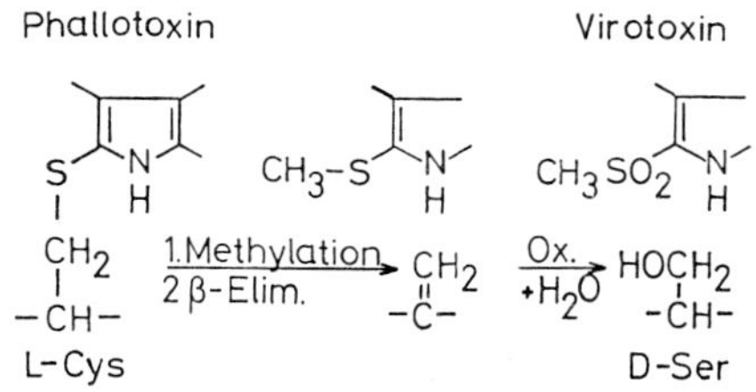

Figure 4. Suggested biochemical transformation of the tryptathionine moiety of the phallotoxins to that of the virotoxins.

Conclusion

The existence of two different F-actin stabilizing drugs, the phallotoxins and the virotoxins in two different Amanita species points to a biological function unrevealed so far. Likewise the amatoxins, other than only luxurious products, could play an essential role in the life of their producer organisms.

References

1. Wieland, Th., Faulstich, H., Crit. Rev. Biochem. 5, 185-260 (1978).
2. Vaisius, A.C., Wieland, Th., Biochemistry 21, 3097-3101 (1982).
3. Buku, A., Altmann, R., Wieland, Th., Liebigs Ann. Chem. 1974, 1580-1586.
4. Wieland, Th., Götzendörfer, Ch., Dabrowski, J., Lipscomb W. Shoham, G., Biochemistry 22, (1983) in press.
5. Wieland, Th. Naturwiss. 64, 303-309 (1977).
6. Wieland, Th., Deboben, A., Faulstich, H., Liebigs Ann. 1980, 416-424.
7. Wieland, Th., Hollosi, M., Nassal, M., Liebigs Ann. 1983, in press.
8. Nassal, M., Liebigs Ann. Chem. 1983, in press.
9. Publication in preparation.
10. Wieland, Th.: Chemistry and Biology of Peptides, Proc. of the 3rd Amer. Peptide Symp., J. Meienhofer ed., Ann Arbor Science Publ. Inc., p. 377-390, 1972.
11. Kramer, W., Bickel, U., Buscher, H.P., Gerok, W. Kurz, G. Eur. J. Biochem. 129, 13-24 (1982).
12. Faulstich, H., Buku, A., Bodenmüller, H., Wieland, Th. Biochemistry 19, 3334-43 (1980).

FORTY YEARS OF TRYPTO-FUN

Bernhard Witkop

National Institutes of Health, National Institute of Arthritis, Diabetes, and Digestive and Kidney Diseases, Laboratory of Chemistry, Bethesda Maryland 20205 USA

Butenandt's investigations on the sequence of events that convert tryptophan (2) to ocular insect pigments[1] prompted the revision (1943) of the hitherto accepted structure (1) for kynurenine[2] and its replacement by the correct o-amino-phenone expression (3).[3] For indole alkaloids[4], such

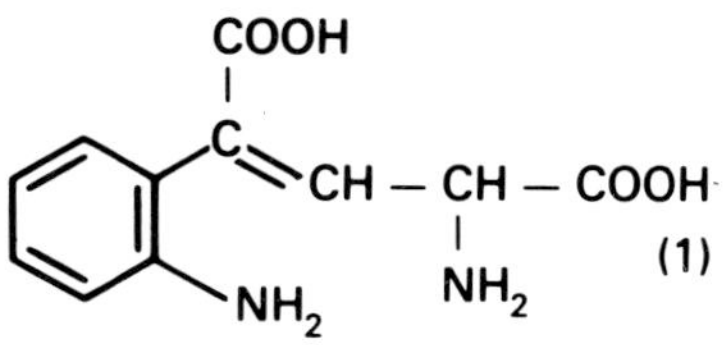

(1)

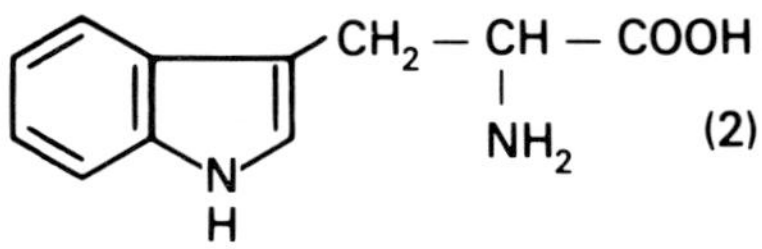

(2)

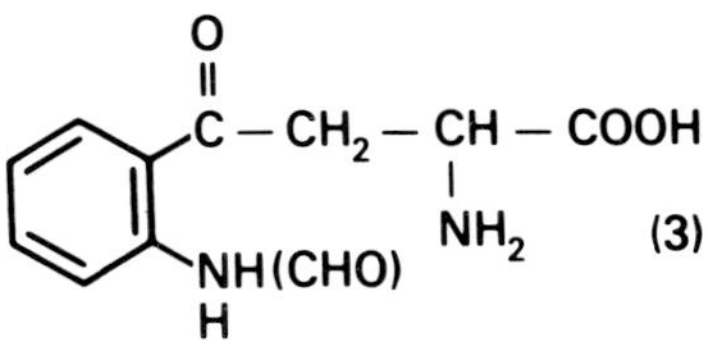

(3)

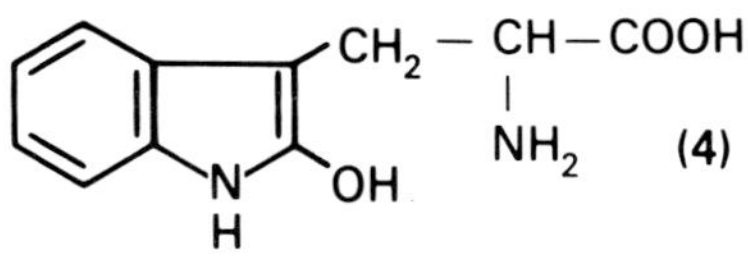

(4)

as yohimbine[5], the easy attack of ozone on the 2,3-double bond of the pyrrole part of indole with oxidative cleavage[6] set an important precedent. The application of this oxidation to tryptophan (2) constituted a one-step synthesis of (formyl)kynurenine, which appeared in the Annalen in 1944 at the height of total war.[7] Not long thereafter neither the famous Munich laboratory, nor the printing presses for the Annalen in Leipzig nor my apartment in Schwabing existed any longer. In wise anticipation Heinrich Wieland had moved me to Freising-Weihenstephan where, as a guest in his former student, Fritz Reindel's[8] laboratory, another conversion was practiced, namely, direct oxidation of tryptophan to

Progress in Tryptophan and Serotonin Research

α-hydroxytryptophan (4).[9] When I sent Professor Butenandt a note on this for publication in the "Naturwissenschaften" together with the last Annalen reprint on the kynurenine paper, still printed in Leipzig, he graciously sent the letter (following page) which I had kept until today. Although the winds of war howled fiercely, the house of science was still standing, at least as a concept.

The most artful tale raises little curiosity when it is known to be false: neither the α-hydroxytryptophan, isolated from the hydrolysate of phalloidin, the toxin of the death-cap,[10] is real,[11] nor its putative role as a prokynurenine in the pathway to ommochromes,[12] nor the simplistic picture I had when I studied the action of peroxyacids on indoles.[9]

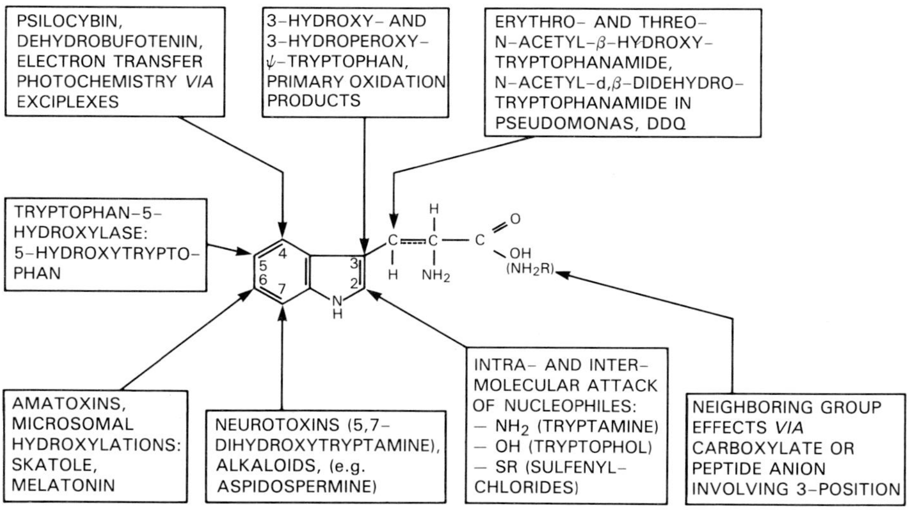

There are in the molecule of tryptophan seven positions capable of being hydroxylated. Forty years of tryptophan research can now be summarized by the presentation above. In the course of these 30 years numerous discussions with Osamu Hayaishi helped to refine the various oxidation mechanisms involved and their importance for biochemistry and metabolism.

PROF. A. BUTENANDT
KAISER WILHELM-INSTITUT
FÜR BIOCHEMIE

(1) BERLIN-DAHLEM, DEN
THIEL-ALLEE 69-73 (Bahnstation Lichterfelde-West)
Fernsprecher: 76-25-56
Zweigstelle Tübingen, d. 26. 1. 1945.
(14) Tübingen, Silcherstr. 7

Herrn Dr. B. W i t k o p

(13 b) Freising Obb.
Chemisches Institut
Weihenstephan.

Sehr geehrter Herr Kollege!

Ihr Schreiben vom 8. Jan. ist mir nach hier nachgereist. Ich danke Ihnen sehr herzlich für die Übersendung des Sonderdrucks Ihrer schönen Arbeit, die schon bei ihrem Erscheinen bei meinen Mitarbeitern und mir große Freude und Begeisterung ausgelöst hat. Ihre für die "Naturwissenschaften" bestimmte Notiz läßt neue schöne Erfolge ahnen, denn die Darstellung von Oxytryptophan aus Tryptophan war ein lohnendes Problem, dem auch wir lange, aber leider vergeblich, nachstrebten. Ich beglückwünsche Sie sehr zu Ihren Erfolgen.

Gern will ich mich bei Professor Süffert für das möglichst schnelle Erscheinen Ihrer Notiz in den "Naturwissenschaften" einsetzen, doch machen die Zeitverhältnisse es wohl unmöglich, irgend eine Aussage darüber zu machen, in welcher Schnelligkeit neue Hefte erscheinen werden.

Mit besten Grüßen

Ihr Butenandt

O/0615

My first (Norwegian) doctoral student at Harvard, Arvid Ek, synthesized 5- (4') and 7-hydroxytryptophan (6).[13] 5-Hydroxytryptophan (4')

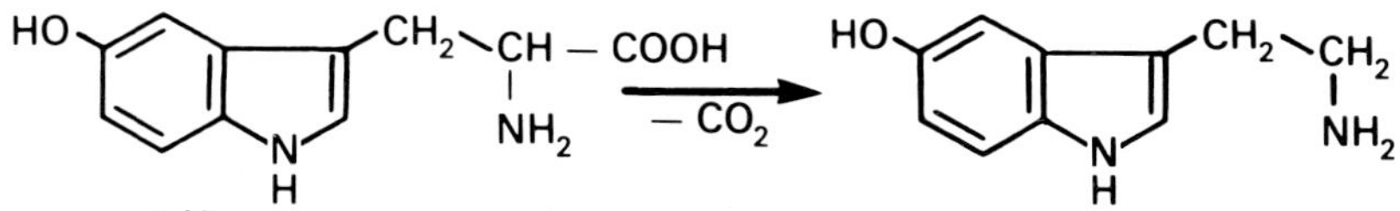

(4') (5)

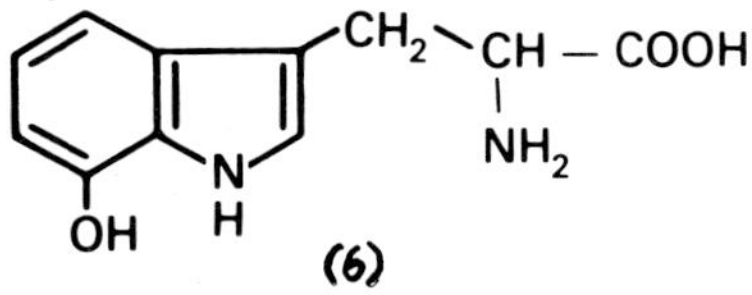

(6)

is a new naturally occurring amino acid, the precursor of serotonin, whose importance as a neurotransmitter, a precursor of melatonin, a regulator of sleep and affective disorders is still increasing.[14] A metabolite of 5-hydroxytryptophan in rabbit small intestine is 5-hydroxykynurenine[15] which was first synthesized by Butenandt and his group.[16]

It was my good fortune to have Sidney Udenfriend next door at NIH who wasted no time to show in 1953 that L-5-hydroxytryptophan was the natural substrate of aromatic amino acid decarboxylase.[17] He also aided in the development of a rapid spectrophotometric assay for L-amino acid oxidase based on L-kynurenine.[18]

Again, in collaboration with S. Udenfriend, a much more surprising phenomenon was discovered in 1966, namely, the migration of deuterium or tritium in $5\text{-}^{2}H$- of $5\text{-}^{3}H$-tryptophan (7) to 5-hydroxy-$4\text{-}^{2}H$- or $\text{-}4\text{-}^{3}H$-tryptophan (8) with up to 90% retention of label.[19] This migration of a substituent in the process of enzymatic hydroxylation of aromatic substrates was termed "NIH-Shift"[20] and served as the starting point of worldwide investigations on the significance or danger of arene oxide intermediates, either in drug metabolism, long-range toxicity or carcinogenesis.[21-23] Here we observe "wie ein Tritt tausend Fäden regt" ("one step--and thousand threads arise . . ."). Selective exchange of nuclear protons in hydroxyindoles was just a parenthetic observation, mentioned here only to show the impact of a method still

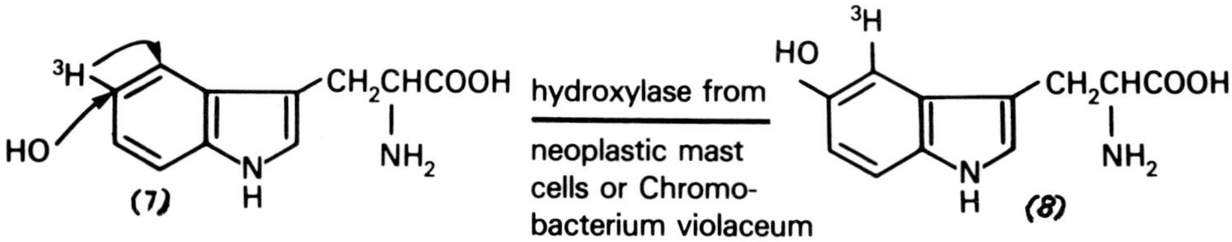

comparatively new at the time, viz., NMR spectroscopy.[24]

Although Arvid Ek did prepare 5,6-dibenzyloxytryptamine as early as 1953, it was not until Hans Schlossberger knew how to prepare and handle free 5,6-dihydroxytryptamine (9) that this and the 5,7-isomere (10) were discovered as serotinergic neurotoxins[25] in the same way as 6-hydroxy-dopamine (11) was first brought to life in a metabolic context, by my

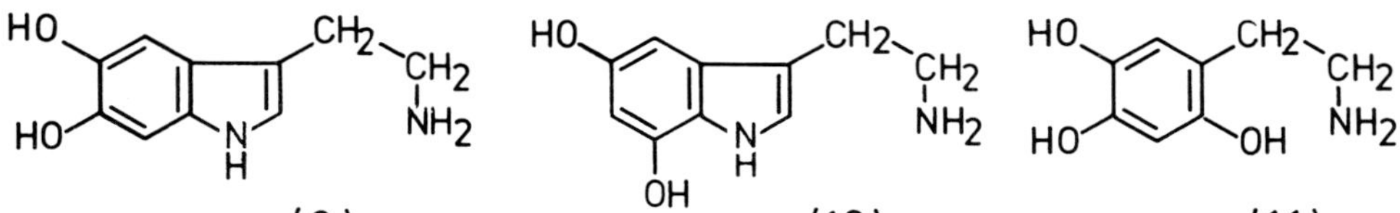

(9) (10) (11)

first scientific collaborator from Japan, Siro Senoh, when it turned out later to be an agent for chemical sympathectomy by mechanisms probably both involving related quinoid intermediates capable of addition of nucleophilic SH groups from adjacent (receptor) proteins (reviewed in ref. 25).

Both photoreduction[26] as well as reduction in the ground state[27] provided partially hydrogenated tryptophans, such as 2,3-dihydro- or ditritio-L-tryptophan (12) convertible to L-[2-^{3}H]tryptophan (13) by liver microsomes. The N,N'-dibenzoyl derivative (14) proved useful for intramolecular acylation involving the 4-position (15), a kind of functionalization with many uses, e.g., for synthetic access to lysergic

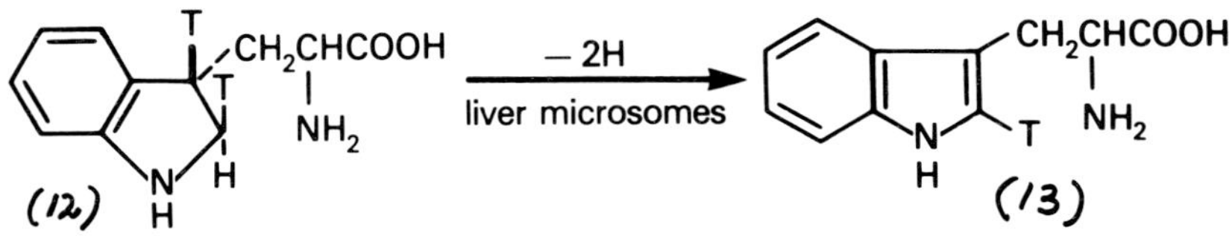

cis-[2,3-^{3}H] dihydro-L-tryptophan L-[2-^{3}H] tryptophan

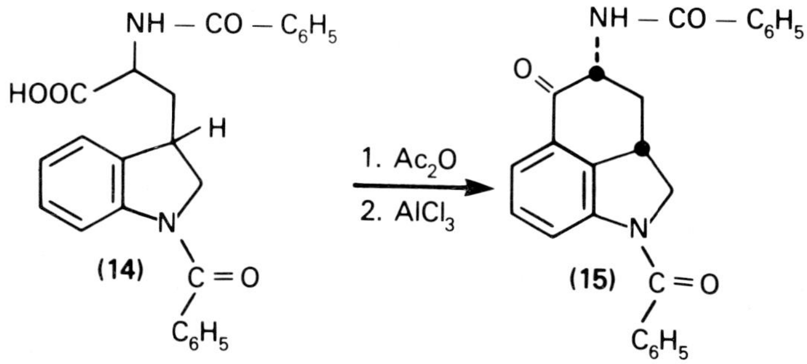

acid derivatives (in ten steps from tryptophan)[28] or to unusual indole alkaloids, such as the regulovasines.[29] A new type of photocyclization provides another approach to tricyclic indoles (18-20) involving the 4- (and 2-) position.[23] The starting materials are simple N-chloroacetyl-tryptophans (16) or -tryptamines (17), e.g. 5-methoxytryptamine.[30] My colleague Osamu Yonemitsu has studied these competitive cyclizations into the 4 and 2 positions in considerable and sophisticated mechanistic detail.[31, 32]

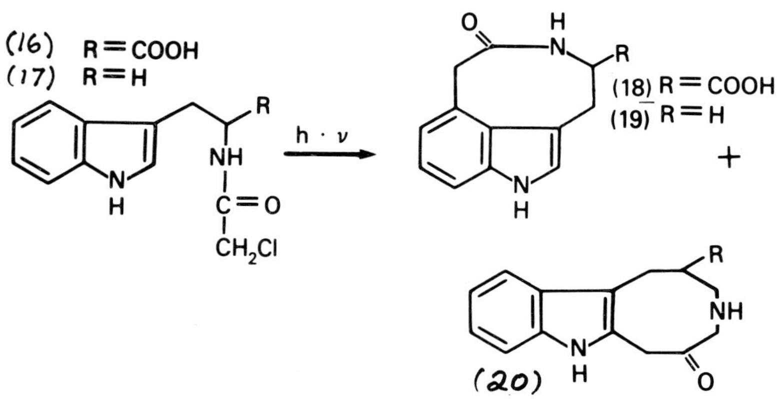

The method has been expanded to photochemical modification of proteins in the presence of chloroacetamide[33] and to synthetic applications in the alkaloid field.[34,35] Dehydrobufotenin (22) should arise in the toad from bufotenin (21).

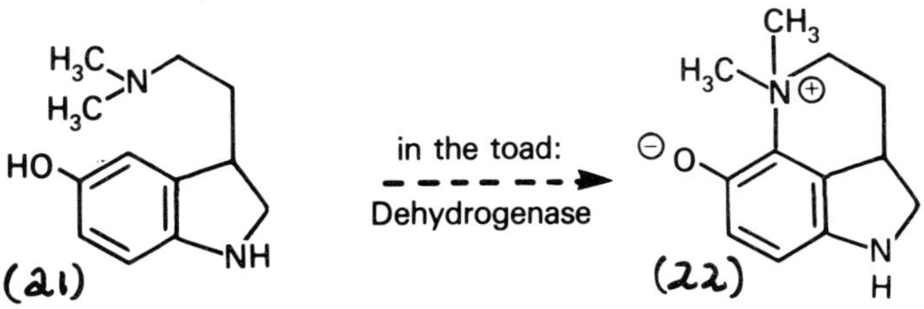

Attempts in this direction were started by Siro Senoh almost 20 years ago, but did not succeed and should be resumed by younger powers.

Tryptophan (2), as such, in peptides or bound in proteins, reacts so fast (23→24→25) with positive halogen, e.g., N-bromosuccinimide, tribomocresole or 2-nitrophenyl-sulfonyl chlorides (NPS-Cl, DNPS-Cl, Scoffone, Fontana, 1966) that these reactions permit an easy titration of bound tryptophan.[36,37] Selective cleavage of tryptophyl peptide

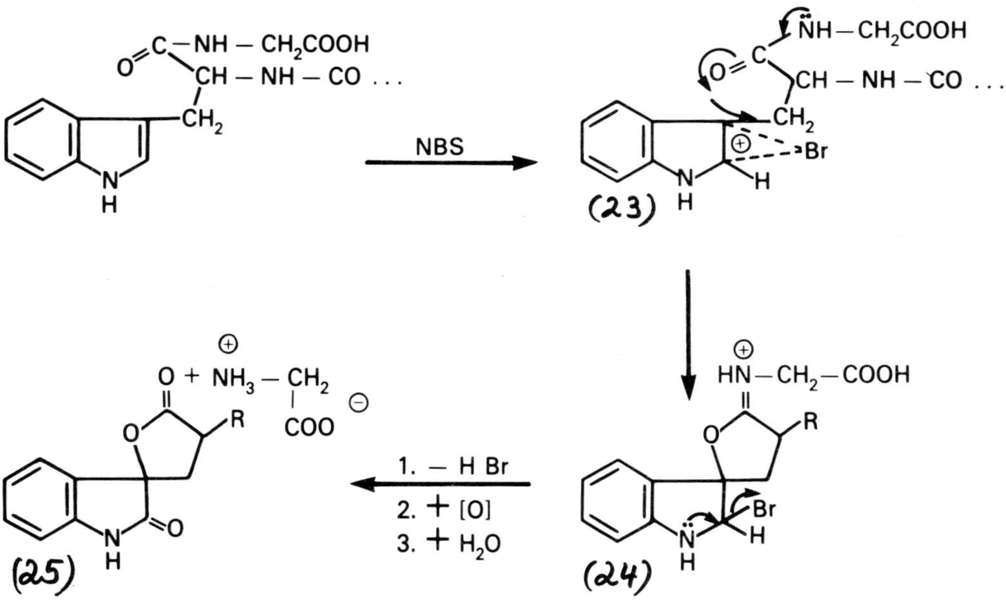

bonds, under appropriate conditions, has been utilized for sequencing of peptides and proteins, but not as systematically and extensively as the cyanogen bromide cleavage, discovered with the late Erhard Gross in 1961.[38]

The classical study on kynurenine by Butenandt[3] reports the formation of an (N,N'-diacetyl)lactone from dihydrokynurenine. Almost one generation later this reaction found a new use. Ozonolysis of bound tryptophan (26), reduction (28) of bound kynurenine (27) so formed and utilization of the new neighboring γ-hydroxyl group, in the hands of Fumio

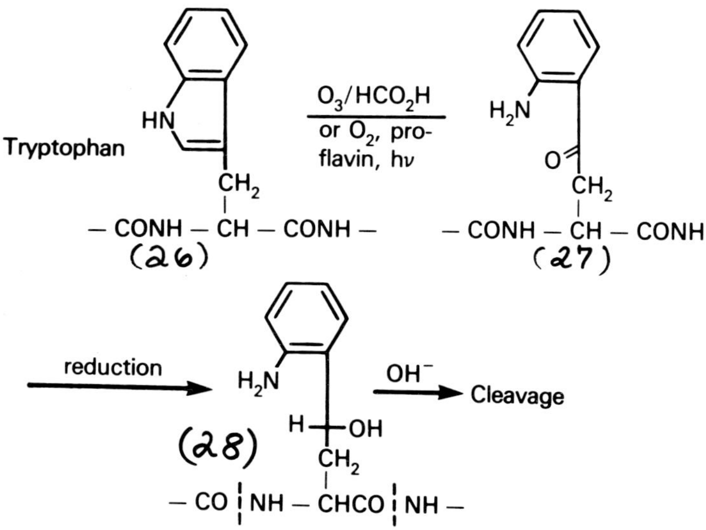

Sakiyama and his group, has made possible useful cleavages, probably more preferential than selective.[37] M. Ozaki helped to develop tetrahydroharmanes as inhibitors of monamine oxidase.[39] Very recently such condensation products (30) of tryptophan have attracted attention as--probably only exogenous--inhibitors of benzodiazepine receptors,[40] suitable as models for anxiety in primates,[41] and as (-)-trypargine (29), a serotinergic principle, isolated from the African frog *Kassina senegalensis*.[42,43]

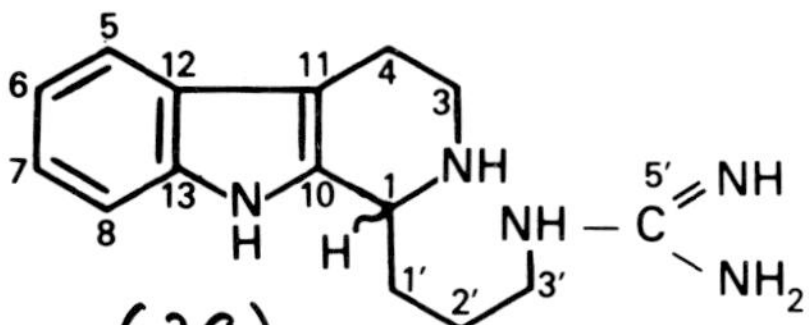

(29)

(—) - trypargine: C (1) - αH

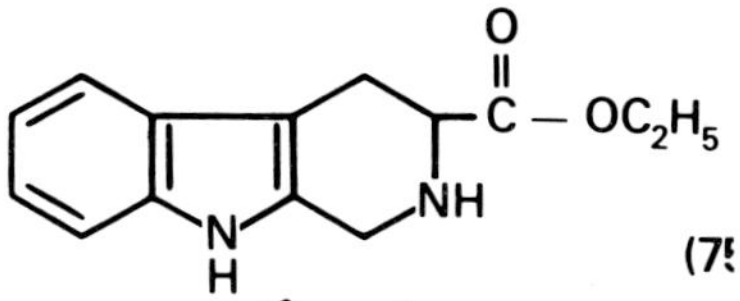

(30)

Epilog

Through my teacher, Heinrich Wieland,[44] and the doctoral dissertation on the deathcap, Amanita phalloides,[10] my scientific path crossed ever so slightly the path of Adolf Butenandt whose life work encompasses a plethora of scientific fields. There is another less well known crossroad of interest. In November 1929 H. Wieland reported: For two years we have been busy to understand the chemical nature of the female sex hormones. We have been concentrating the active substance and, in one case, arrived at a crystalline preparation.[45] Butenandt, then in his twenties, reported the crystallization of estrone[46] followed by structural elucidation not long thereafter.

References

1. Butenandt, A.: Das Werk eines Lebens, herausg. v.d. Max-Planck-Ges., Band I/3, pp. 1 ff., Vandenhoeck & Ruprecht. Göttingen und Zürich 1981.
2. Kotake, Y.: Ergebnisse der Physiologie 37, 245 (1935).
3. Butenandt, A., Weidel, W., Becker, E., Derjugin, V.: Hoppe-Seyler's Z. für physiol. Chemie 279, 27 (1943).
4. Witkop, B.: Die Chemie 56, 265 (1943).
5. Witkop, B.: Ann. 554, 83 (1943).
6. Witkop, B.: Goodwin, S.: J. Amer. Chem. Soc. 75, 3371 (1953).
7. Witkop, B.: Ann. 556, 103 (1944).

8. Fritz Reindel was with H. Wieland at the T. H. 1921; cf. Ber. Dtsch. Chem. Ges. 55, 3313 (1922).

9. Witkop, B.: Ann. 558, 98 (1947).

10. Wieland, H., Witkop, B.: Ann. 543, 171 (1940).

11. Wieland, Th.: Chemie in unserer Zeit 13, 56 (1979).

12. Butenandt, A., Weidel, W., Becker, E.: Naturwiss. 28, 447 (1940).

13. Ek, A., Witkop, B.: J. Amer. Chem. Soc. 75, 500 (1953).

14. Bender, D. A.: Molecular Aspects of Medicine, Biochemistry of Tryptophan in Health and Disease 6, 101-197 (1983).

15. Fujiwara, M., Shibata, M. Nomiyama, Y., Sugimoto, T., Hirata, F., Tokuyama, T., Senoh, S., Hayaishi, O.: Proc. Natl. Acad. Sci. USA 76, 1145 (1979).

16. Butenandt, A., Schulz, G., Hanser, G.: Hoppe-Seylers Z. Physiol. Chem. 295, 404 (1953).

17. Clark, C. T., Weissbach, H., Udenfriend, S.: J. Biol. Chem. 210, 139 (1954).

18. Weissbach, H. H., Robertson, A. V., Witkop, B., Udenfriend, S.: Anal. Biochem. 1, 286 (1960).

19. Renson, J., Daly, J., Weissbach, H., Witkop, B., Udenfriend, S.: Biochem. Biophys. Res. Commun. 5, 504 (1966).

20. Guroff, G., Daly, J. W., Jerina, D., Rensen, J., Udenfriend, S., Witkop, B.: Science 158, 1524 (1967).

21. Witkop, B.: The NIH-Shift, Kagaku no Ryoiki 23, 37 (1967); in Japanese.

22. Witkop, B.: Experientia 27, 1121 (1921).

23. Witkop, B.: The NIH Shift and Its Implications for the Mechanism of Biological Oxidations, Current Topics in Biochemistry, 109-133, Academic Press. New York 1973.

24. Daly, J. W., Witkop, B.: J. Amer. Chem. Soc. 89, 1032 (1967).

25. Baumgarten, H. G., Jenner, S., Bjorklund, A., Klemm, H. P., Schlossberger, H.G.: Biology of Serotonergic Transmission. Osborne, N. N. (Ed.), pp. 249, John Wiley. New York 1982.

26. Yonemitsu, O., Cerutti, P., Witkop, B.: J. Amer. Chem. Soc. 88, 3941 (1966).

27. Daly, J. W., Mauger, A. B., Yonemitsu, O., Antonov, V. K., Takase, K., Witkop, B.: Biochemistry 6, 648 (1967).

28. Rebek, J., Tai, D. F.: Tetrahedron Letters 24, 859-860 (1983).

29. Rebek, J., Shue, Y. K.: J. Amer. Chem. Soc. 102, 5426 (1980).

30. Kobayashi, T., Spande, T. F., Aoyagi, T. F., Witkop, B.: J. Med. Chem. 12, 636 (1969).

31. Yonemitsu, O.: Electron Transfer Photochemistry, Yakugaku Zasshi 102, 716-734 (1982).

32. Narito, S. and Yonemitsu, O.: Tetrahedron Letters, 3399 (1975); Yonemitsu, O.: Yakugaku Zasshi 102, 716-734 (1982).

33. Hamada, T., Yonemitsu, O.: Chem. Pharm. Bull. 25, 271 (1977).

34. Sundberg, R. J., Smith, F., X.: J. Org. Chem. 40, 2613 (1975).

35. Wu, A., Snieckus, V., Tetrahedron Letters, 2060 (1975).

36. Witkop, B.: Nonenzymatic Methods for the Preferential and Selective Cleavage and Modification of Proteins, Adv. Protein Chemistry 16, 221 (1961).

37. Spande, T. F., Witkop, B., Degani, Y., Patchornik, A.: Adv. Protein Chemistry 24, 98 (1970).

38. Gross, E., Witkop, B.: J. Amer. Chem. Soc. 83, 1510 (1961).

39. Ozaki, M., Weissbach, H., Ozaki, A., Witkop, B., Udenfriend, S.: J. Med. Pharm. Chem. 2, 591 (1960).

40. Braestrup, C., Nielson, M., Olsen, C. E.: Proc. Natl. Acad. Sci. USA 77, 2288 (1980).

41. Ninan, P., Insel, T., Cook, J., Skolnick, P., Paul, S.: Science 218, 1333 (1982).

42. Synthesis: Shimizu, M., Ishikawa, M., Komoda, Y., Nahajima, T.: Chem. Pharm. Bull. (Japan) 30, 909 (1982).

43. Witkop, B.: Amphibian Venoms, The Alkaloids. Brossi, A. (Ed.). Vol. XXI, pp. 139-253, Academic Press. New York 1983.

44. Witkop, B.: Heinrich Wieland hundert Jahre: Sein Werk und Vermächtnis heute. Angew. Chemie 89, 575 (1977); Internat. Ed. 16, 559 (1977).

45. Wieland, H., Straub, W., Dorfmüller, Th.: Hoppe-Seylers Z. für physiol. Chemie 186, 97 (1930); this work was started in 1927.

46. Butenandt, A.: Naturwiss 17, 879 (1929).

INDOLEAMINE DIOXYGENASE —— A POSSIBLE BIOLOGICAL FUNCTION ——

Osamu Hayaishi, Ryotaro Yoshida, Osamu Takikawa
and Hiroaki Yasui

Department of Medical Chemistry
Kyoto University Faculty of Medicine
Yoshida, Sakyo-ku, Kyoto 606
Japan

Prof. Butenandt, Mr. Chairman, Ladies and Gentlemen. I would like to join the other speakers in expressing hearty congratulations to Prof. Butenandt on the occasion of his 80th birthday and also my profound admiration of his pioneering work in the field of tryptophan research. I consider it a great honour to have been invited to this meeting and to talk about our recent work on indoleamine dioxygenase.

This photograph was taken when Prof. Butenandt came to Japan and visited Prof. Yashiro Kotake at Wakayama Medical College on April 19, 1955, exactly 28 years ago. In 1927, Kotake fed rabbits a large amount of tryptophan and isolated and crystallized a new compound from the urine. He named it "kynurenine" assuming it to be a precursor of kynurenic acid (1). In 1943, the correct structure of this compound was established by

Progress in Tryptophan and Serotonin Research

Prof. Butenandt (2) and kynurenine turned out to be the key intermediate in tryptophan metabolism, as we all know today.

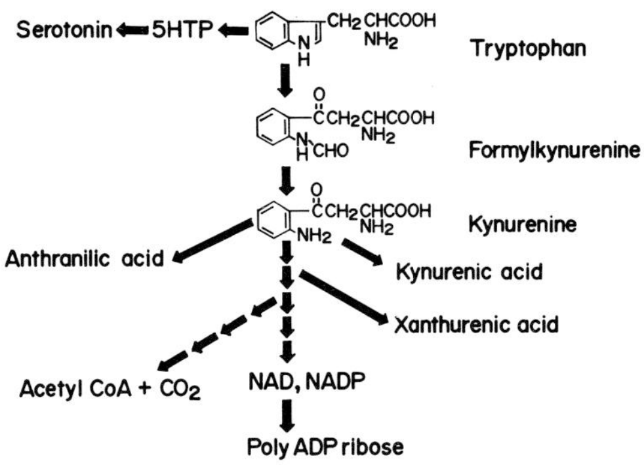

Fig. 1 Major metabolic pathway of tryptophan

The major metabolic pathways of tryptophan in mammals are shown in Fig. 1. It is now well-known that kynurenine is not only the precursor of anthranilic acid, kynurenic acid and xanthurenic acid but is also converted to 3-hydroxykynurenine and other intermediates and is ultimately metabolized to acetyl CoA and CO_2 leading to the complete oxidation of the tryptophan molecule, or to the biosynthesis of important coenzymes, NAD and NADP and then to poly(ADP-ribose). Poly(ADP-ribose) is a homopolymer of ADP-ribose derived from NAD and plays a crucial role in the repair of damaged DNA, and hence in oncogenesis, differentiation and mutation (3). In 1936, Kotake and Masayama isolated an enzyme from liver that catalyzes the conversion of tryptophan to formylkynurenine (4). Because this enzyme cleaves the pyrrole moiety of tryptophan, Kotake termed it "tryptophan pyrrolase". In 1957, we demonstrated that molecular oxygen was incorporated into formylkynurenine and thereafter the enzyme was renamed tryptophan 2,3-dioxygenase, abbreviated as TPO (5). It has been tacitly assumed that this enzyme is solely responsible for the formation of formylkynurenine from tryptophan in mammals and that it is the rate-limiting step of the kynurenine pathway. Actually when a large amount of tryptophan is given to experimental animals or glucocorticoid is injected to stimulate gluconeogenesis from amino acids, this enzyme is induced in the liver, as shown in Fig. 2.

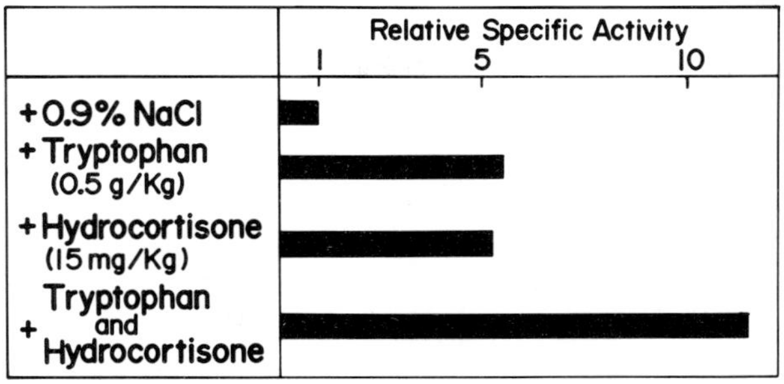

Fig. 2 Effects of tryptophan and hydrocortisone on TPO in adrenalectomized rats

Civen and Knox (6) reported that when either tryptophan or hydrocortisone was given to adrenalectomized rats, the tryptophan dioxygenase activity in the liver increased about 5-fold, as compared with the control. When both were given simultaneously, the increment was even greater. These results clearly indicate that tryptophan dioxygenase plays a crucial role in the degradation of tryptophan in the liver under physiological conditions. It has been known for a long time that various tryptophan metabolites such as kynurenine, 3-hydroxykynurenine, 3-hydroxyanthranilic acid, and so forth are excreted in elevated amounts in the urine of patients under various pathological conditions. For example, in a large percentage of patients with rheumatoid arthritis, tuberculosis, leukemia, Hodgkin's disease, bladder tumor or prostate disorders, excess amounts of tryptophan metabolites such as kynurenine, 3-hydroxykynurenine and so forth are excreted into the urine. However, under these pathological conditions, the induction or elevation of the tryptophan oxygenase activity in the liver was never demonstrated.

Fig. 3 shows the lack of correlation between tryptophan oxygenase activity and urinary excretion of kynurenine in patients with bladder cancer, according to Gailani and coworkers (7). The tryptophan dioxygenase activity was determined with surgical and needle biopsy liver specimens. These results indicate that abnormal increase of kynurenine in the urine is not due to the increased activity of tryptophan oxygenase in the liver.

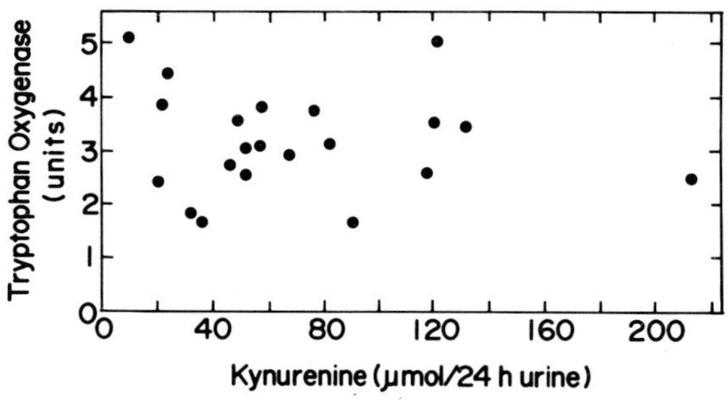

Fig. 3 Lack of correlation between TPO activity and urinary excretion of kynurenine

Therefore, the excretion of large amounts of kynurenine and its metabolites could not be explained on the basis of excess production of kynurenine in the liver.

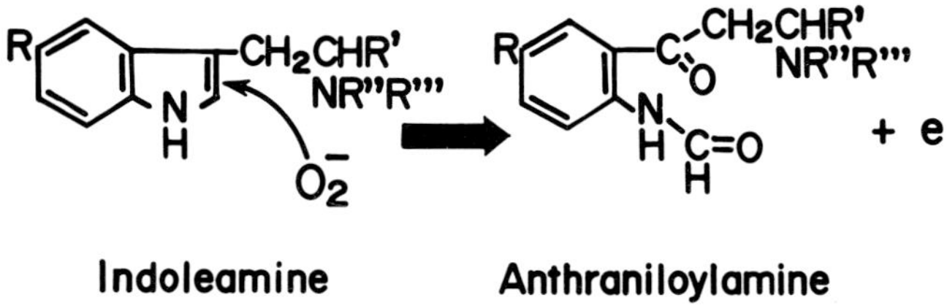

Fig. 4 Indoleamine 2,3-dioxygenase

In 1963 we discovered in the rabbit intestine another enzyme that also catalyzes the formation of kynurenine from tryptophan (8). In contrast to tryptophan dioxygenase, this enzyme exhibits broad substrate specificity and acts upon various indoleamine derivatives, including tryptophan. Therefore, it was termed indoleamine 2,3-dioxygenase, abbreviated as IDO. As shown in Fig. 4, indoleamine dioxygenase utilizes superoxide anion, a univalently reduced molecular oxygen for activity. Distribution and properties of these two enzymes are different in many respects. For example, tryptophan dioxygenase is found only in the liver. It acts on only tryptophan, whereas indoleamine dioxygenase is widely distributed in almost all tissues and organs except in the liver. It has a broad substrate specificity, and acts upon not only tryptophan but also 5-hydroxytryptophan, tryptamine, serotonin, etc. Tryptophan dioxygenase utilizes molecular oxygen, but indoleamine dioxygenase required superoxide anion and molecular oxygen for activity. The molecular and catalytic properties of the purified enzyme preparations of these two oxygenases are also different as regard to the molecular weight, subunit structure, carbohydrate and heme contents and the turn-over number. They are also different in terms of immunological properties. Furthermore, unlike tryptophan dioxygenase, indoleamine dioxygenase is not induced by either tryptophan or glucocorticoid. These results clearly indicate that there are two entirely different enzymes both catalyzing essentially identical reactions, namely the formation of formylkynurenine from tryptophan. What is then the physiological function of indoleamine dioxygenase? The superoxide anion is an active form of oxygen and is produced in leukocytes when bacteria are ingested. During phagocytosis, NADPH oxidase on the membrane of phagosomes is activated and molecular oxygen is reduced by one electron to produce superoxide anion. The superoxide anion alone, or together with H_2O_2 which is generated from the superoxide, kills the ingested bacteria. Therefore, the superoxide anion is implicated in the defence mechanism against bacterial infection. Because indoleamine dioxygenase utilizes the superoxide anion and the lung is an aerobic organ and probably generates a fair amount of superoxide anion, we subjected mice to viral infection and determined the indoleamine

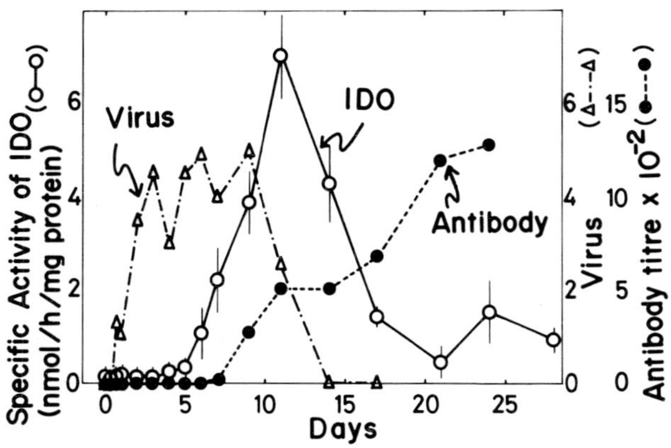

Fig. 5 Induction of IDO by virus infection

dioxygenase activity in the lung. When we exposed mice to influenza virus, the virus replication in the lung began within 24 hours after the infection, reached a peak by the 3rd day and persisted until the 9th day, causing viral pneumonia. Thereafter the virus was rapidly eliminated from the lung and by the 14th day almost completely disappeared. The indoleamine dioxygenase activity in the lung increased almost linearly from the 5th day after infection. More than 120-fold increase was observed on the 11th day after which the activity gradually decreased to normal values over a two week period (9). The serum antibody did not appear until the 9th day after infection when the virus had begun to disappear and gradually increased (Fig. 5). These results indicate that, with viral infection, the indoleamine dioxygenase activity increases tremendously and the breakdown of tryptophan and other indoleamine derivatives in the lung must be accelerated to a great extent. Since this experimental system is somewhat complicated, we then used bacterial endotoxin instead of influenza virus to simplify the experimental conditions.

Bacterial endotoxin or lipopolysaccharide, abbreviated as LPS, of gram negative bacteria is an inflammatory agent and induces a nonspecific immune response. It is a large molecular compound consisting of 3 different portions, namely lipid A, core polysaccharide and O-antigen specific polysaccharide. When the lipopolysaccharide fraction of E coli was given intraperitoneally to mice, again we observed a remarkable increase in the indoleamine dioxygenase activity with a concomitant increase in the serum kynurenine level, which is shown in Fig. 6A. As can be seen in the lower panel, induction of indoleamine dioxygenase in the mouse lung was observed almost immediately after intraperitoneal administration of LPS. The specific activity in the high speed supernatant of the lung increased almost immediately, reached a plateau after about 24 hours, then gradually decreased and reached a normal value within seven days. A single injection of about 5 µg/mouse of LPS was sufficient to induce a 30-50 fold, sometimes 100-fold increase in this enzyme activity. The plasma kynurenine level also increased and gradually decreased, as shown in the upper panel. The time course of the increase and decrease in plasma kynurenine level is essentially identical to that of IDO activity in the lung. Under exactly the same conditions, tryptophan dioxygenase activity in the liver decreased to less than 50% of the normal activity, and there was no correlation with the kynurenine level in the plasma (Fig. 6B).

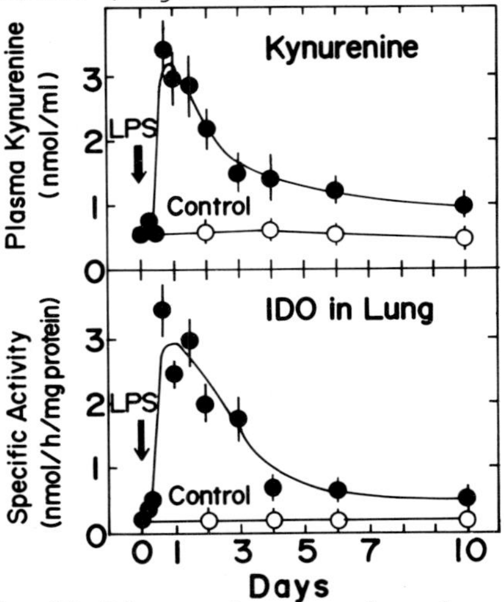

Fig.6A Plasma kynurenine level and IDO activity in lung after LPS treatment

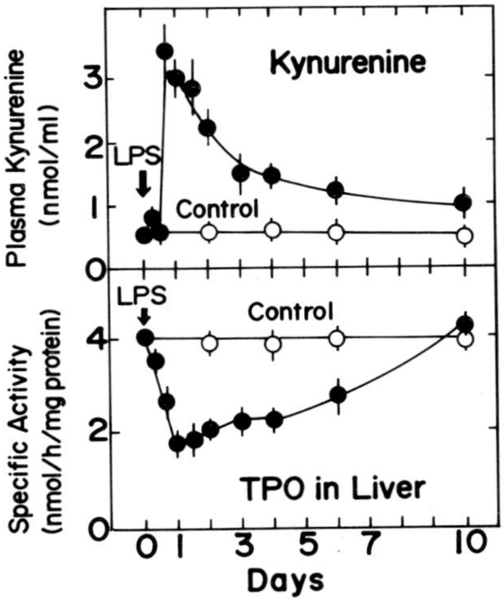

Fig.6B Plasma kynurenine level and TPO activity in liver after LPS treatment

These results clearly indicate that the induction of indoleamine dioxygenase is related to the increase in the plasma kynurenine level and that tryptophan is probably degraded actively in the lung and other tissues in the presence of elevated levels of indoleamine dioxygenase.

The increase in the enzyme activity under these conditions appears to be due to the net synthesis of enzyme protein rather than to the removal of inhibitors, formation of activators or other factors, because the concomitant administration of an inhibitor of RNA or protein synthesis such as actinomycin D or cycloheximide abolished the increase in the enzyme activity produced by LPS. Immunological analyses using antibody against IDO are in line with this conclusion because the increment in enzyme activity was mainly due to an increase in the amount of immunologically reactive protein. The induction of IDO activity by LPS is most pronounced in the lung. However, the enzyme activity increased at least several fold in many other organs and tissues, including heart, spleen, colon and so forth.

To determine whether or not the induction was specific for the indoleamine dioxygenase activity, various other enzyme activities in the lung were determined. For example, enzymes involved in inflammation such as monoamine oxidase, prostaglandin cyclooxygenase, superoxide dismutase, lipoxygenase and so forth and typical lysosomal enzymes such as acid phosphatase were not induced to any significant extent.

To find out about the exact location of indoleamine dioxygenase activity, the lung tissue was digested by trypsin and collagenase, fractionated and the cellular distribution determined. The indoleamine dioxygenase activity was found exclusively in the alveolar interstitial cells. Immunohistochemical studies with anti-IDO serum also support this conclusion.

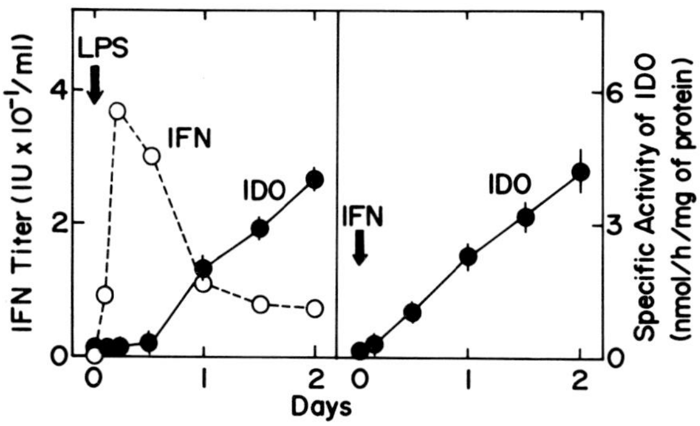

Fig. 7 Induction of IDO by LPS and interferon in mouse lung slices

In an attempt to determine the mechanism of enzyme induction, we then used lung slices and exposed the slices to LPS in vitro. As can be seen on the left-hand side of Fig. 7, when LPS was given at zero time, indoleamine dioxygenase was gradually induced. LPS and virus are known to be inducers of interferon and the interferon titre increased almost instantaneously after LPS administration. When interferon itself was administered at zero time, there was no lag in the induction of indoleamine dioxygenase and interferon appears to be an even more effective inducer of this enzyme. These time courses indicate that when LPS enters the cells, interferon is produced and triggers the induction of indoleamine dioxygenase. This induction of indoleamine dioxygenase by interferon is inhibited by a number of antiinflammatory agents such as indomethacin, aspirin and so forth, all known to inhibit the biosyntheses of prostaglandins. The inhibition is dose dependent and 50% inhibition is observed with about 20 μ M (10). These results are compatible with the possible sequence of events shown in Fig. 8.

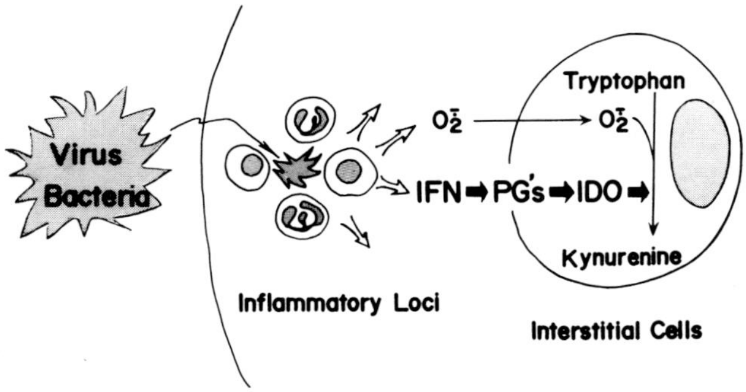

Fig. 8 Biological significance of IDO induction

In Fig. 8 is shown our current working hypothesis concerning the biological significance of indoleamine dioxygenase induction. When some tissues are invaded by virus, bacteria, or bacterial endotoxin, leukocytes and lymphocytes may accumulate in the inflammatory loci and the interferon so produced migrates and interacts with the surface of the interstitial cells and stimulates the intracellular synthesis of prostaglandins which trigger indoleamine dioxygenase induction. As a consequence of inflammation, superoxide anion is liberated and accerelates the conversion of tryptophan to kynurenine catalyzed by indoleamine dioxygenase. Whether or not indoleamine dioxygenase plays a role in the antiviral state caused by interferon remains to be elucidated, and is the subject of ongoing research in our laboratory.

References

1. Matsuoka, Z., and Yoshimatsu, N.: Hoppe-Seyler's Zeitshrift für Physiologische Chemie. 143, 206-210 (1925).
2. Butenandt, A., Weidel, W., and Becker, E.: Naturwissenschaften 28, 63-64 (1940).
3. Hayaishi, O., and Ueda, K.: ADP-Ribosylation Reactions —— Biology and Medicine ——. Academic Press, 3-16 (1982).
4. Kotake, Y., and Masayama, T.: Hoppe-Seyler's Zeitshrift für Physiologische Chemie. 243, 237-244 (1982).
5. Hayaishi, O., Rothberg, S., Mehler, A. H., and Saito, Y.: J. Biol. Chem. 229, 889 (1957).
6. Civen, M., and Knox, W. E.: J. Biol. Chem. 234, 1787 (1959).
7. Gailani, S., Murphy, G., Kenny, G., Nussbaum, A., and Silvernail, P.: Cancer Research 33, 1071-1077 (1973).
8. Higuchi, K., Kuno, S., and Hayaishi, O.: Federation Proc. 22, 242 (1963).
9. Yoshida, R., Urade, Y., Tokuda, M., and Hayaishi, O.: Proc. Natl. Acad. Sci. USA 76, 4084-4086 (1979).
10. Sayama, S., Yoshida, R., Oku, T., Imanishi, J., Kishida, T. and Hayaishi, O.: Proc. Natl. Acad. Sci. USA 78, 7327-7330 (1981).

Qualitative and Quantitative Analysis of Tryptophan and its Metabolites

LIQUID CHROMATOGRAPHY/ELECTROCHEMISTRY DETERMINATION OF BIOGENIC AMINES: IMPROVED PERFORMANCE USING LOW DEAD VOLUME MULTIPLE ELECTRODE TRANSDUCERS

Peter T. Kissinger

Department of Chemistry, Purdue University, and Bioanalytical Systems, W. Lafayette, IN 47906, U.S.A.

Introduction

The first practical liquid chromatography/electrochemistry (LCEC) experiments were carried out in early 1972. The technological developments followed the need to solve an important problem in neuropharmacology. Determination of catecholamine and serotonin neurotransmitters in brain tissue using such diverse techniques as fluorescence, gas chromatography/mass spectrometry, and various radiochemical techniques left much to be desired. LCEC appeared to be a good solution to some of the problems and after a decade of development, nearly a thousand publications have appeared. The primary attributes of the technique are its good selectivity, low detection limits, wide applicability (especially vs. fluorescence), and low cost (especially vs. GCMS and radiochemical procedures).

In recent years a number of new applications (enzyme activity measurements, GABA, acetylcholine, NADH, pterins, etc.) have been developed. Multiple-electrode LCEC detection systems are now available which permit significantly improved performance. The new transducers are compatible with short, high speed columns and longer microbore columns. With multiple electrode LCEC the identification of individual

Progress in Tryptophan and Serotonin Research

substances can be confirmed and the selectivity can be much improved for very complex samples such as brain tissue and body fluids. In addition, better detection limits can often be achieved for some compounds. The latest developments in LCEC technology will be briefly described with respect to 1) multiple working electrodes, 2) "high speed" columns, 3) pre-column derivatization, and 4) post-column reactions.

It is not my intention to review earlier applications of the LCEC technique in any detail. An extensive bibliography is available which is frequently updated (1). A "minireview" emphasized applications in the neurosciences (2) and an extensive text details the principles and most popular applications (3).

Multiple Electrode LCEC

Electrochemistry in thin layers of solution is a very highly developed field of electroanalytical chemistry. The thin-layer geometry is ideal for LCEC in that it provides a very low volume transducer which can faithfully reproduce the shape of concentration profiles ("peaks") eluting from very efficient LC columns. Figure 1 illustrates the most popular LCEC detector cell. The thin-layer channel is defined by a gasket held between the upper block (a stainless steel auxiliary electrode with low dead volume fittings) and the lower block (an inert polymeric material containing one or more working electrodes at which the reactions of interest occur). The effective dead volume can be made less than 1 μL, a very difficult task for optical detectors. Routine determination of 1 pmole of an analyte is generally very straightforward and often 0.1 pmole or less can be quantitated in optimized procedures.

The simultaneous use of two working electrodes greatly improves both the qualitative and quantitative aspects of an

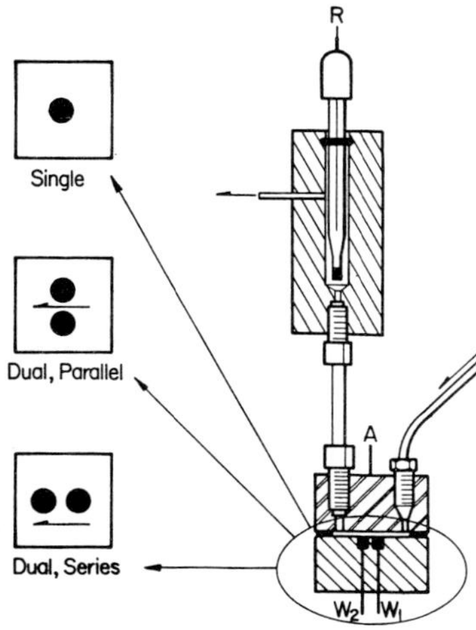

Figure 1. Thin-Layer Amperometric Detector Cell.

LCEC experiment. Electrodes of the same or different materials may be used and the electrode potentials may be independently controlled. In the "parallel mode" the compounds eluting from the column pass over each electrode at the same time. The following applications are quite useful:

1. The ratio of currents monitored at each electrode can provide confirmation of peak identity and purity.

2. Oxidations and reductions can be carried out simultaneously. This saves time and enhances selectivity. This can be ideal for compounds present in several different redox states (e.g. pterins).

3. Signals from low and high potential reactions can be recorded simultaneously, providing both greater selectivity and wider applicability in a single experiment.

4. A difference signal can be plotted to subtract out "common mode" information while enhancing detection of the desired compound.

In the "series mode" the lower block is rotated 90° in relation to the flow stream. Products of the upstream electrode reaction can be detected downstream. If an oxidation is carried out upstream a reduction is accomplished downstream and vice versa. The following applications are popular:

1. The ratio of currents monitored at each electrode can provide confirmation of peak identity and purity.
2. Selectivity is enhanced at the downstream electrode because compounds with chemically irreversible reactions upstream are discriminated against.
3. The upstream electrode can "derivatize" compounds to enhance detectability at the downstream electrode. Overall selectivity and detection limits can be greatly improved.
4. Dissolved oxygen can be discriminated against, simplifying LCEC of compounds that ordinarily would require mobile phase deoxygenation (e.g. nitro compounds).
5. "Common mode" currents can be discriminated against by taking the difference between the two signals.

Both series and parallel dual electrode LCEC provide many opportunities for study of tryptophan metabolism because of the wide range of redox properties involved, including some metabolites that are relatively difficult to oxidize (e.g. tryptophan, kynurenine) and some that oxidize extremely easily (e.g. serotonin and hydroxykynurenines).

"High Speed" Columns for LCEC

Short LC columns (typically 3-10 cm) packed with small particles (typically 3 μm) can provide very high resolution separations in a few minutes. For example, Figure 2 illustrates the separation of 17 neurochemicals in under 8 minutes using an experimental 10 cm reverse phase column (4). This technology requires modification of a conventional LCEC system to minimize dead volume, but these modifications are quite straightforward and present no unusual problems (5). This development presents a great opportunity for solving certain types of problems, but by no means replaces LCEC experiments using more conventional 25-30 cm columns. In natural products and neurochemical research it is often desirable to simultaneously quantitate more metabolites in a given pathway, where previously many separate assays were necessary. These very high resolution systems will definitely make this work easier.

Pre-column Derivatization for LCEC

Almost all tyrosine and tryptophan metabolites of neurochemical interest are electrochemically reactive and can be

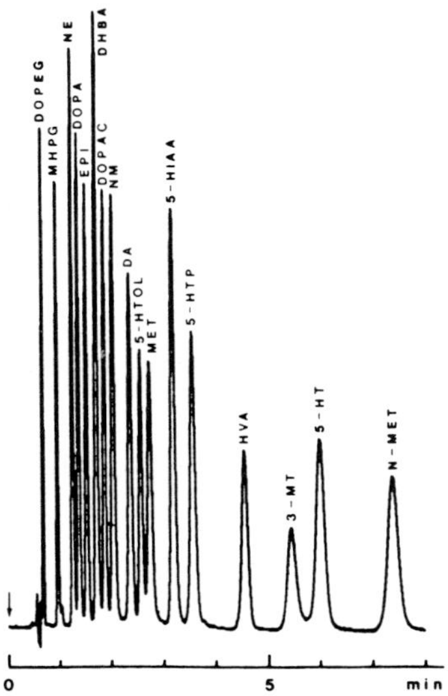

Figure 2. Separation of neurochemical standards.

directly determined by LCEC. Without doubt, this fact is the primary reason the technique has become so popular in the last few years, in contrast to other methodologies which require pre-column derivatization. Simple assays (few steps) always are the most reliable. Nevertheless, there are some problems in which derivatization appears to be the only practical possibility. Amino acids, with few exceptions, are universally determined by derivatization whether the technique be optical absorbance, fluorescence, gas chromatography, liquid chromatography or thin-layer chromatography. Fatty acids (e.g. prostaglandins), alcohols, alkylamines, and many thiols also present a challenge for most analytical methods. This challenge is often met by derivatization.

A large number of classical (e.g. 2,4-dinitrophenylhydrazine) and more recent (e.g. 2,4,6-trinitrobenzenesulfonic acid) derivatizing reagents are based on the excellent optical properties of the aromatic nitro group. Fortunately, this same group is ideal as a derivatization approach for LCEC. The nitro group is easily reduced to a hydroxylamine:

$$\phi NO_2 + 4e + 4H^+ \longrightarrow \phi NHOH + H_2O$$

Using a dual series detector the latter can be oxidized to the nitroso derivative at very low positive potentials:

$$\phi NHOH \longrightarrow \phi NO + 2e + 2H^+$$

Because the nitro group is quite unusual in biology, it provides a very selective handle for LCEC of many nonelectroactive compounds. Derivatives with more than one nitro group add sensitivity (more electrons transferred per molecule). Several good examples have appeared in the literature (6-8), including determination of GABA in brain tissue (9, 10).

Recently Joseph and Davies discovered that the o-phthalaldehyde/mercaptoethanol (OPT) derivatives of amines, commonly used for LCF of amino acids, are ideal for LCEC as well (11). This discovery greatly expands the potential role of LCEC in biomedical research. There is a strong possibility that OPT derivatives will be widely studied by LCEC. The detector response is more uniform than for the LCF technique, where for certain compounds detection limits suffer due to a poor fluorescence quantum yield.

Post-Column Reactions in LCEC

Post-column chemical reactions coupled to electrode reactions are also expanding the range of LCEC applications. Figure 3 illustrates four configurations. In **A** a reagent is added, mixed, and reacted in a delay line followed by electrochemical detection. A superb example of this is the determination of acetylcholine in brain tissue by reverse-phase LCEC (12). Acetylcholine esterase and choline oxidase are mixed in and the detection process proceeds as follows:

$$\text{acetylcholine} \longrightarrow \text{choline} + \text{acetic acid}$$

$$\text{choline} \longrightarrow \text{betaine} + H_2O + 2H_2O_2$$

The peroxide is detected electrochemically at a platinum electrode.

In **B**, a catalyst is immobilized and a cofactor is detected. For example, using a dehydrogenase enzyme an alcohol can be detected indirectly by monitoring the turnover of NAD to NADH, the latter being ideal for electrochemical detection. In C an upstream electrode generates a reagent (e.g. Br_2 from Br^- in the mobile phase) which reacts with a nonelectroactive compound (e.g. an unsaturated fatty acid) and the decrease in reagent concentration is monitored downstream. In **D**, the analytes of interest are converted at an upstream electrode into a product which is more selectively detected downstream. Examples include reduction of a nitro compound to a hydroxylamine, reduction of a disulfide to a thiol, and reduction of a nitrate ester to generate the nitrite ion.

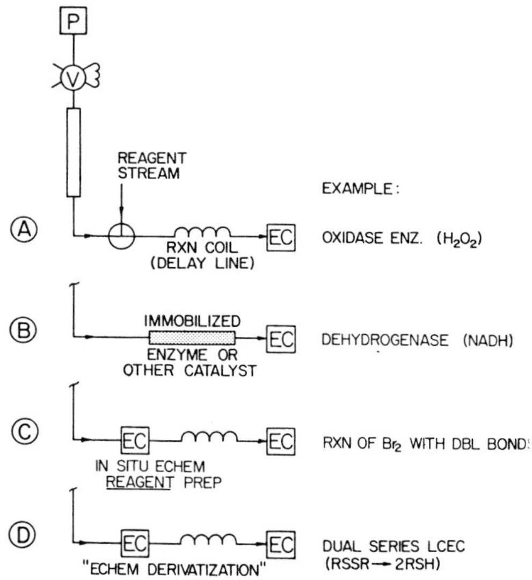

Figure 3. Post-column strategies for LCEC.

It was the intention of this brief review to indicate that LCEC technology is advancing rapidly. The technique is far more reliable and now exhibits significantly better sensitivity and detection limits. When coupled with pre- and post-column reactions the range of applicable compounds has been dramatically increased.

Acknowledgement

The author wishes to thank Dr. Ron Shoup and his crew at Bioanalytical Systems R&D for their continuous input of new ideas. Prof. LeRoy Blank of the University of Oklahoma is thanked for his contribution of Figure 2 and the innovative work that made it possible.

References

1. Shoup, R.E., Ed., Recent Reports on Liquid Chromatography/Electrochemistry, BAS Press, W. Lafayette (1982).

2. Kissinger, P.T., Bruntlett, C.S., Shoup, R.E.: Life Sciences 28, 455-465 (1981).

3. Kissinger, P.T., Ed., Liquid Chromatography/Electrochemistry: Principles and Applications, BAS Press, W. Lafayette (1983), in press.

4. Lin, P.Y.T., Blank, C.L.: Current Separations 5(1), 3-6 (1983).

5. Shoup, R.E.: Current Separations 5(1), 7-10 (1983).

6. Kissinger, P.T., Bratin, K., Davis, G.C., Pachla, L.A.: J. Chromatog. Sci. 17, 137-141 (1979).

7. Shoup, R.E.: Current Separations 4(4), 66-67 (1982).

8. Jacobs, W.A., Kissinger, P.T.: J. Liq. Chromatog. 5, 881-895 (1982).

9. Caudill, W.L., Houck, G.P., Wightman, R.M.: J. Chromatog. Biol. Appl. 227, 331(1982).

10. Caudill, W.L., Papach, L.A., Wightman, R.M.: Current Separations 4(4), 59-61.

11. Joseph, M.H., Davies, P.: Current Separations 4(4), 62-65 (1982).

12. Potter, P.E., Meek, J.L., Neff, N.H., 1982 Society for Neuroscience Abstracts, No. 143.8.

GLC AND HPLC ANALYSIS OF TRYPTOPHAN, KYNURENINE AND RELATED METABOLITES

Michael H. Joseph

Division of Psychiatry, MRC Clinical Research Centre, Watford Road, Harrow, U.K.

Determination of the pyrrolase pathway metabolites of tryptophan in urine has traditionally been carried out following ion-exchange fractionation. The different metabolites are then determined by native fluorescence or by colourimetry following condensation (1). The marked native fluorescence of tryptophan which is so useful in its determination is lost when the pyrrole ring is opened; fluorescence analysis is therefore not available for most of the pyrrolase pathway metabolites, with the exception of the bicyclic kynurenine and xanthurenic acids (1).

A desire to extend studies on pyrrolase pathway metabolites to unloaded subjects, and to tissues and physiological fluids other than urine has led to a search for more sensitive, and if possible, more convenient methods, especially for the "main line" metabolites kynurenine (K), 3-hydroxy-kynurenine (3HK) and 3-hydroxyanthranilic acid (3HA). Kynurenine is of special interest since it is the first detectable metabolite on the pathway, and is not preceeded by any branch points. In addition it is the effective product (since formyl-kynurenine is present in vanishingly small concentrations) of the first and rate limiting enzyme of the pathway, tryptophan pyrrolase, and as such is measured in studies of pyrrolase enzymology.

GLC Methods

In the late sixties and early seventies GLC was the method of choice for high sensitivity. A number of studies on GLC separations of a series of pure metabolite standards using silylation or other esterification of substituents (2, 3, 4) appeared. Other method papers dealt with physiological samples, but usually only one metabolite was measured after

Progress in Tryptophan and Serotonin Research

separation; urinary 3HA (5, 6) or urinary K (7) for example. The use of trifluoroacetic anhydride for derivatisation by the last mentioned authors permitted the use of the more sensitive electron capture detector.
In enzymological studies on tryptophan pyrrolase, knurenine is usually measured by its absorption at 365 nm. For more specific measurement, a useful aspect of the chemistry of K (and of 3HK) is its cleavage to the more volatile o-aminoacetophenone (3-hydroxy, 2-aminoacetophenone) by heating in strong alkali. This reaction has been made the basis of a number of methods for K, the o-aminoacetophenone (OAAP) being separated by steam distillation or solvent extraction, and determined by colourimetry or GLC with ECD (7-10). In our hands differential extraction before and after heating and aqueous phase trifluoroacetylation coupled with GLC-ECD results in a simple method which is sensitive enough to determine K in plasma and in brain and even in human CSF, where the baseline K level is 5 - 10 ng/ml (10, 11, Fig.1,2).

Fig.1. GLC records from kynurenine analysis. (a) Blank, (b) 150 ng, (c) 300 ng, (d) 600 ng added kynurenine standard, (e) autopsied human brain cortex. Injection is indicated by arrow; the first major peak is α,α'dichloro-p-xylene internal standard; the second is the OAAP derivative.

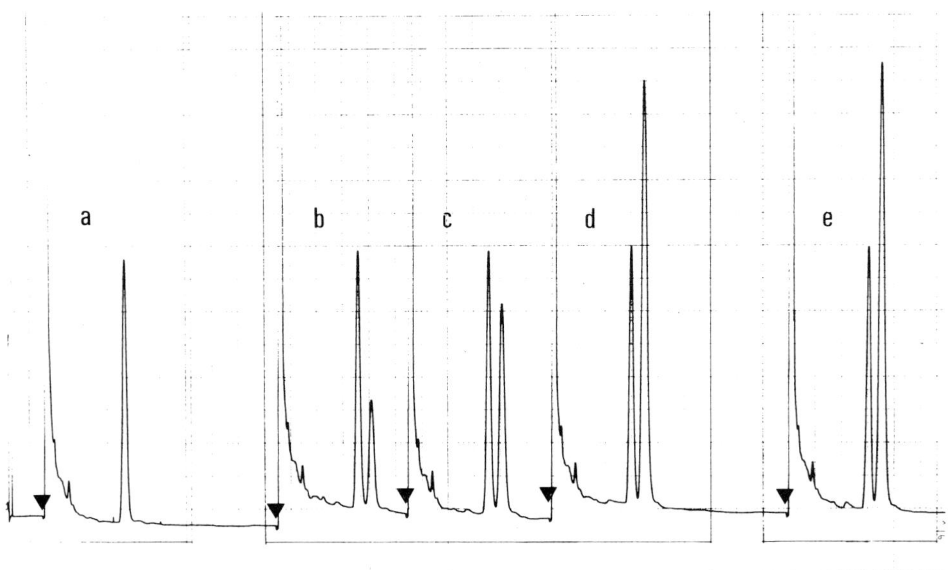

Fig.2. Summary of procedure for GLC analysis of kynurenine.

A) Homogenise brain (or plasma) in 5 volumes (10 volumes) acid butanol; add heptane to supernatant and back extract into 0.1N HCl.

B) 0.1N HCl phase from A, or untreated urine or CSF : add $\frac{1}{3}$ volume of 10N NaOH and Tiron (final concentration 5 mM; to suppress tryptophan interference at next stage) and extract with butyl acetate.

C) Heat alkaline aqueous phase in boiling water bath for 20 minutes to cleave

Kynurenine (K) to o-amino acetophenone (OAAP)

D) Cool and extract into butyl acetate containing 200 ng/ml dichloro-p-xylene (DCPX) as internal standard.

E) Add NaOH/Borate buffer and mix briefly. After settling add trifluoroacetic anhydride (TFAA) and mix again immediately.

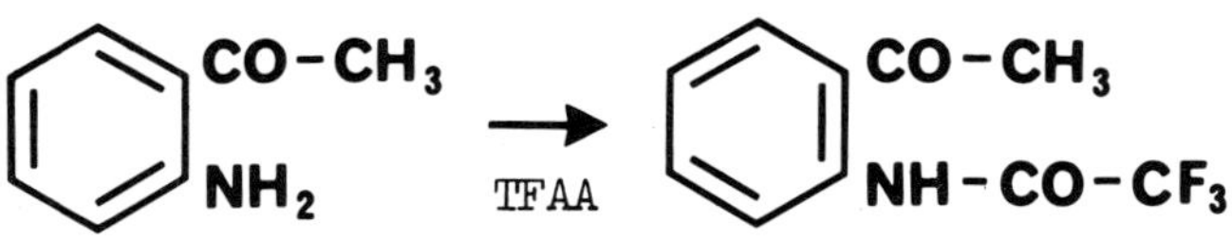

F) Centrifuge and remove upper (organic) phase - injected 1µl into GLC.

Since the GLC traces obtained are so clean, further improvements in sensitivity can be obtained by using internal standards further separated from the OAAP derivative (α,α' dibromo-*p*-xylene or α,α' dichloro-*o*-xylene) and a higher column temperature (12, 13). The GLC method could also be useful in monitoring the activity of tryptophan pyrrolase, especially during purification, since it does not distinguish between formylkynurenine and kynurenine.

The only other method for kynurenine which has been applied to brain, as far as I am aware, is that of Gal and Sherman (14). This more complex procedure can determine 3HK and kynuramine as well as K, tryptophan and other related metabolites. These are determined by GLC-ECD of HFB derivatives of separate fractions obtained by ion exchange chromatography. However, I am not aware of any data in the literature on the validity or specificity of this method, which makes it difficult to evaluate discrepancies. One example is that these authors report the presence of kynuramine in rat brain tissue, which we were unable to detect (10). Another is that Allenmark et al (15) using an HPLC-EC method find less than one tenth the concentration of 3HK in rat plasma reported by Gal and Sherman (14).

HPLC Methods

Given the mixed hydrophobic/hydrophilic nature of the pyrrolase pathway metabolites, HPLC is an attractive alternative to GLC for analysis, as has proved the case over the last few years for the 5HT pathway metabolites and the catecholamine pathway metabolites. K and 3HK retain, in common with tryptophan, the α-amino acid structure, and thus are in principle analysable by methods used for amino acids. Amino acid analysers use ion exchange HPLC, and Møller (16) has reported the use of an analyser with ninhydrin detection to determine kynurenine, along with the other amino acids in plasma. The major problem is the low level of K in relation to other plasma amino acids.

The aromatic amino acids run well also on reversed phase packings, now the most commonly used type of HPLC columns. A number of studies on the separation of tryptophan and its pyrrolase pathway metabolites have appeared (17-21) but had not yet reached this stage of routine methods for determining the metabolites at their endogenous levels in physiological samples (at least, prior to the publication of the present volume!) Tryptophan itself can be detected either by its native fluorescence (for example at 254/330nm), or by its electrochemical activity. Electrochemical (EC) detection depends on oxidation of the species of interest at relatively modest potentials at a working electrode. 5-hydroxyindoles (and catechols) are readily oxidised at 0.4 - 0.5 V and the resulting

Fig.3. HPLC analysis of cerebrospinal fluid. Arrows indicate time of injection on left (fluorescent) trace and right (electrochemical) trace. Unretained peaks have been truncated for clarity.

(A) f.s.d. = 50mV fluorescent, 50nA EC; V = 1.0
(B) f.s.d. = 100mV fluorescent, 20nA EC; V = 0.65

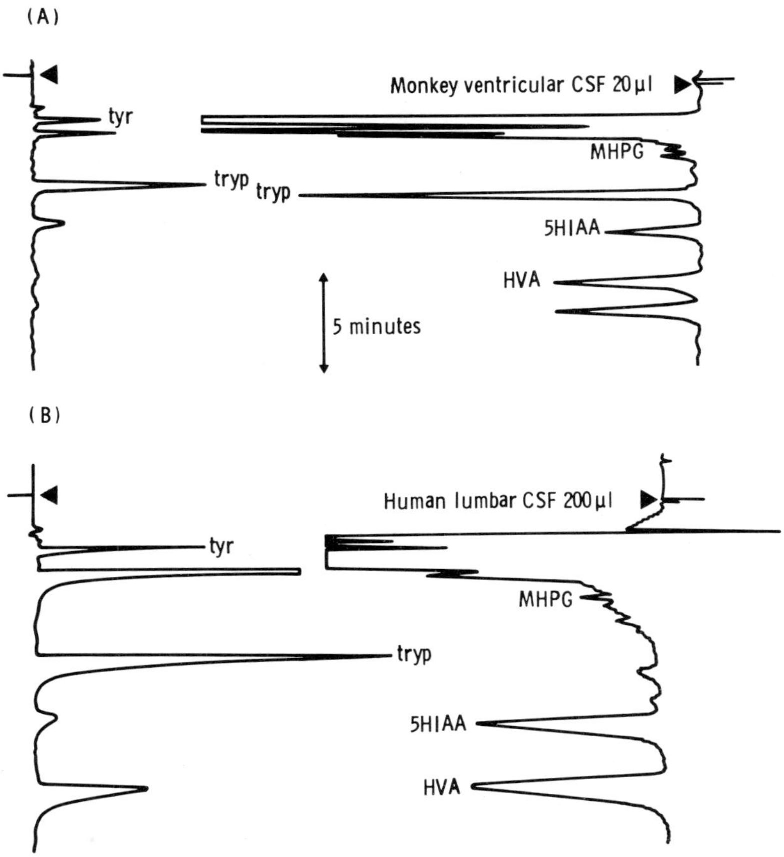

current, of the order of 10 nA is detected and displayed on a recorder to produce the chromatographic trace. Methylation of (one of) the hydroxyl group(s) results in a higher potential being required for oxidation. In the absence of (one of) the hydroxyl group(s) (as in tryptophan (tyrosine)), oxidation still occurs, but at a higher potential; 0.8 - 0.9 V. Thus tryptophan can be readily quantitated by running untreated CSF or deprotenised plasma on a C_{18} reversed phase column and using either fluorescence detection or EC detection at 0.8 - 0.9 V (22-24, Fig.3).

These findings naturally raise the question of whether the pyrrolase pathway metabolites are electroactive also. In the case of K the answer is a qualified yes; it is electroactive at 1.1 V. This voltage is rather high for practical purposes since noise is relatively high. Moreover any species active at a lower potential will also be seen, so that the chances of interference are increased. However we have been able to measure kynurenine in plasma following a tryptophan load. Direct EC detection at 1.1 V following separation on Hypersil ODS 5 was used, and the results agreed well with those obtained by our GLC method. In the case of 3HK and 3HA the answer is an unqualified yes since the vicinal hydroxy and amino groups on the ring appear to be as active as the catechol grouping. Thus 3HK and 3HA can be detected at 0.4 - 0.5 V. We have used this approach to determine 3HK and 3HA binding to plasma protein in buffer solution (25). Allenmark et al (15) are publishing an HPLC method for the determination of 3HK in plasma and CSF using EC detection.

Another approach to the HPLC analysis of amino acids is precolumn derivatisation with o-phthalaldehyde (OPT)/mercaptoethanol reagent. This results in the formation of stable isoindoles (Fig.4) which chromatograph well on reversed phase and can be detected fluorimetrically (26, 24). We have recently established (27) that these derivatives are also electro-active at voltages of 0.5 V and above. The electroactivity profile depends upon the structure of the amino acid involved, and it appears that some amino acids including those which are intrinsically electroactive are also active at lower potentials. Thus when a mixture of OPT-amino acids are separated by a methanol gradient applied to a reversed phase column, fluorescent peaks are seen for each one including tryptophan, but not for cysteine, K and 3HK. These amino acids do form derivatives however since

Fig.4. Reaction of OPT with an amino acid and mercaptoethanol

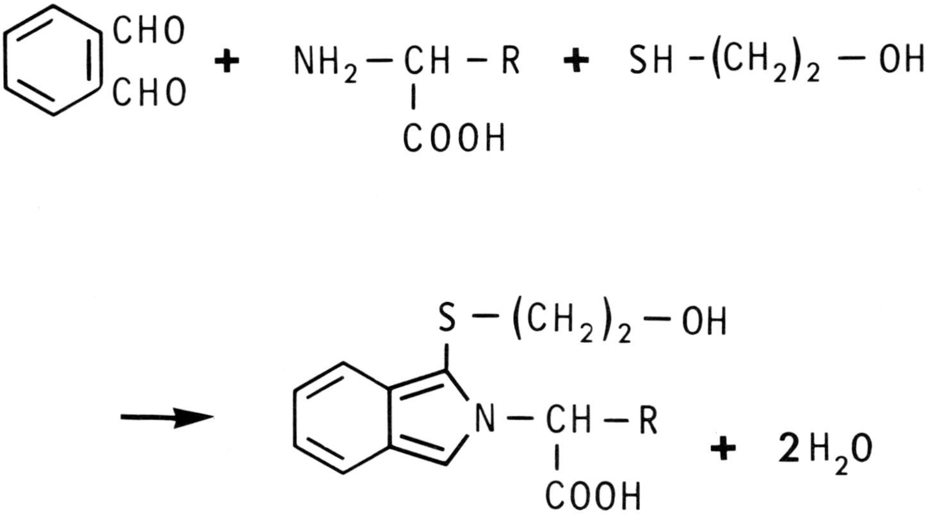

they are detectable electrochemically. With the EC detector set at 0.4 V or less peaks are seen only for tryptophan, 3HK, and a few other basic amino acids. This may reasonably provide a method for the determination of K and especially of 3HK as OPT-amino acids, in spite of the presence of much larger amounts of the common amino acids.

In conclusion HPLC-EC shows great promise for the analysis of pyrrolase pathway metabolites. For the analysis of kynurenine, however, the GLC-ECD method offers the highest sensitivity at present.

References

1. Price, J.M., Brown, R.R., Yess, N.: Adv. Metab. Disorders. 2, 159-225 (1965).

2. Dalgliesh, C.E., Horning, E.C., Horning, M.G., Knox, K.L., Yarger, K.: Biochem. J. 101, 792 (1966).

3. Noguchi, T., Kaseda, H., Konishi, N., Kido, R.: J. Chromatogr. 55, 291-295 (1971).

4. Altschuler, H., Gold, M.: J. Chromatogr. Sci. 9, 377-381 (1971).

5. Rose, D.P., Toseland, P.A.: Clin. Chim. Acta. 17, 235-238 (1967).

6. Hirano, K., Mori, K., Tsuboi, N., Kawai, S., Ohno, T.: Chem. Pharm. Bull. 20, 1412-1416 (1972).

7. Naruse, M., Hirano, K., Kawai, S., Ohno, T., Masada, Y., Hashimoto, K.: J. Chromatogr. 82, 331-336 (1973).

8. Tompsett, S.L.: Clin. Chim. Acta 4, 411-419 (1959).

9. Joseph, M.H., Risby, D.: Clin. Chim. Acta 63, 197-204 (1975).

10. Joseph, M.H. J. Chromatogr. 146, 33-41 (1978).

11. Joseph, M.H., Baker, H.F., Crow, T.J., Riley, G.J., Risby, D.: Psychopharmacology 62, 279-285 (1979).

12. Kennett, G.A., Joseph, M.H.: Neuropharmacology 20, 39-43 (1981).

13. Young, S.N., Joseph, M.H., Gautier, S.: J. Neurol. Neurosurg. Psychiat. submitted for publication (1983).

14. Gal, E.M., Sherman, A.D.: J. Neurochem. 30, 607-613 (1978).

15. Allenmark, S., Bengtsson, M.B., Nordström, K., Hallert, C.: Microchemical Journal. in press (1983).

16. Møller, S.E.: Anal. Biochem. 79, 590-593 (1977).

17. Grushka, E., Kikta, E.J., Naylor, E.W.: J. Chromatogr. 143, 51-56 (1977).

18. Krstulovic, A.M., Matzura, C.: J. Chromatogr. 163, 72-76 (1979).

19. Yong, S., Lau, S.: J. Chromatogr. 175, 343-346 (1979).

20. Geeraerts, F., Schimpfessel, L., Crokaert, R.: Clin. Chim. Acta. 102, 247-251 (1980).

21. Otsuka, H., Ugata, M., Nomura, S., Shibata, Y.: J. Aichi. Med. Univ. Assn. 8, 221-224 (1980).

22. Anderson, G.M., Young, J.G., Cohen, D.J.: J. Chromatogr. 164, 501-505 (1979).

23. Joseph, M.H., Baker, H.F., Ridley, R.M.: in Central Neurotransmitter Turnover (Pycock, C.J. and Taberner, P.V., eds.), Croom Helm, London, pp. 162-167 1981.

24. Joseph, M.H., Johnson, J.A., Kennett, G.A.: This volume.

25. Joseph, M.H., Kadam, B.V. Unpublished results.

26. Lindroth, P., Mopper, K. Anal. Chem. 51, 1667-1674 (1979).

27. Joseph, M.H., Davies, P.: J. Chromatogr. in press (1983).

MONITORING OF TRYPTOPHAN LEVELS IN BIOLOGICAL FLUIDS BY HPLC: AN IMPROVED METHOD.

Antonio Bettero, Carlo Costa and Graziella Allegri

Istituto di Chimica Farmaceutica, Università di Padova-Centro Studio Chim.Farmaco e Prodotti Biol.Attivi, CNR Padova-Padova Italy.

Summary

A direct determination of Tryptophan(Trp) in biological fluids was performed by measuring its native fluorescence in a fully automated reversed phase HPLC system. The method involves the retention of protein fraction in a suitable precolumn by a combined action of acetonitrile and phosphate buffer at pH 3.5 used as mobile phase. Good correlation of the standard addition method (r=0.98) and high reproducibility of fifty determinations(± 2% RSD) before changing the fitting material of the precolumn was obtained. Determination requires less than 10 minutes and the detection limit was found at levels of picomoles. No interfering substance was observed.

Introduction

The important role of Trp and its metabolites in both normal and pathological states is under investigation(1-4) since decades and therefore many methods were developed to analyze them(5-7). An improved analytical method is here presented for the direct determination of Trp: it is based on HPLC analysis of biological fluids directly and without use of deproteinizing agents through native fluorescence detection.

Materials and Methods

A Perkin Elmer series 3B liquid chromatograph equipped with a Rheodyne 7125 injector valve, ASV1 automatic switching valve

Progress in Tryptophan and Serotonin Research

and 650-10 LC fluorimetric detector set at excitation and emission wavelengths of 282 nm and 370 nm respectively was emploved.All detections were carried out by a RP-8 (5 µm) Merck column with a precolumn fitting with lichroprep RP-8 (20-40 µm) using a mobile phase of acetonitrile and phosphate buffer 0.004 M at pH 3.5 (15-85 %) at flow rate of 1.0 ml/min. A Sigma 15 data station with a standard addition method was employed. An Amicon model 12 ultrafiltration cell and Diaflo XM 50 were used for free Trp analysis. Standard solutions of Trp (0.002-20 γ/ml in water) and human fluids, obtained from healthy volunteers, was filtered through millipore 0.45 µm before use. 10 µl of each sample were injected into the chromatographic system. Reagents grade chemical were used.

Results and Discussion

The schematic steps in pretreatment of biological fluids and HPLC system employed are shown in Fig.1.
10 µl of each sample were injected directly or by automatic sampling device.

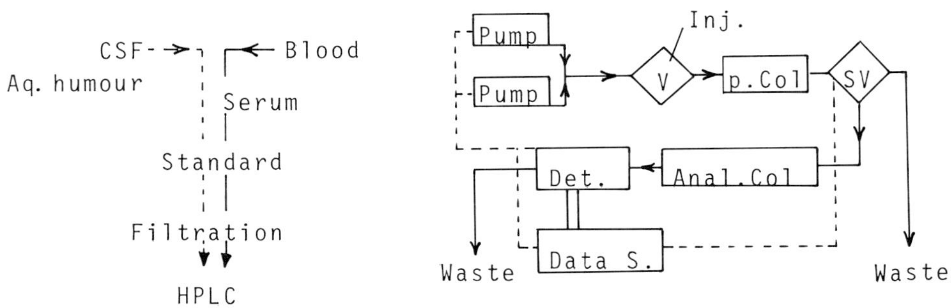

Fig.1- Steps of biological fluids and HPLC system.

Typical chromatograms of Trp in human serum, cerebrospinal fluid (CSF) and aqueous humour are reported in Fig.2. For the analysis of the free Trp, the albumin bound fraction was separated by ultrafiltration (8). Peaks identification was done by standard addition technique and by fluorescence caracterization (excitation and emission spectra) with a stop flow method at the same retention time of standard. Quantitative determinations were made from calibration curve obtained by standard addition method.

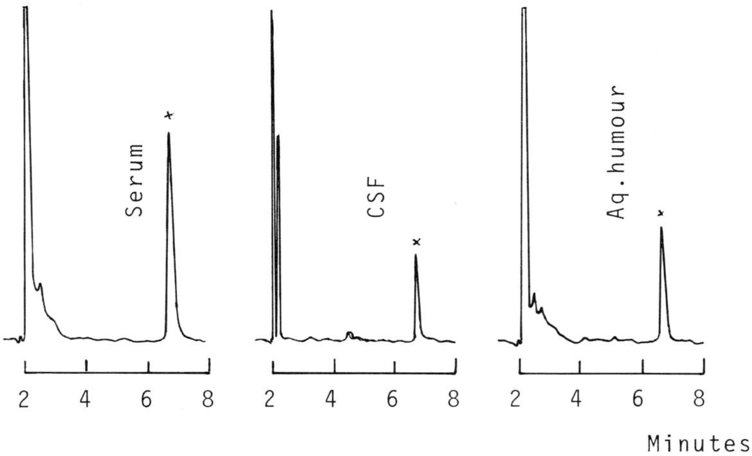

Fig.2- Typical chromatograms of Trp*

Fig.3 reports the slope and correlation curves from a human serum samples achieved directly and after the classical samples pretreatment.

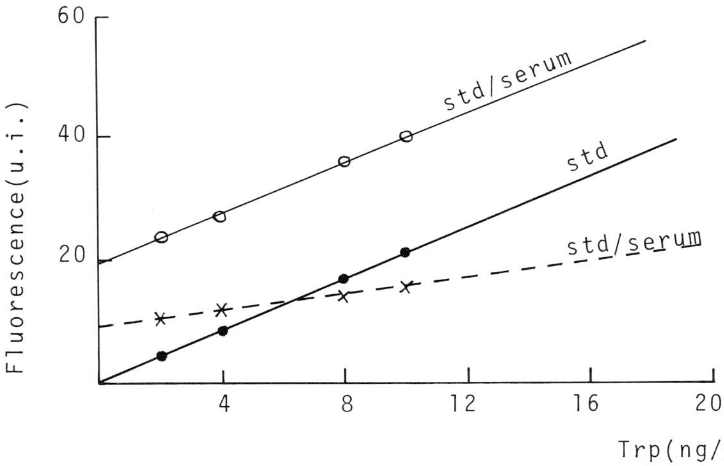

Fig.3-Calibration curves of Trp: —●— standard(std)in water, —o— std in serum sample,-x-std in serum sample with deproteinizing agents.

The results listed in table I confirmed the usefulness of the HPLC method proposed.

Table I-Analytical and chromatographic parameters

Data	std/water	std/serum s.
Inj. volume (µl)	10	10
Retention time (min.)	6.7	6.7 ±1% RSD
Capacity factor	3.05	3.05
Recovery % (n=5)	----	97 ± 2% RSD
Detection limit(pmoles)	20	20
Reproducibility (n=50)	----	± 2% RSD
Calibration curve	y=2.21x -0.49	y=1.99x +19.52
Correlation(r)	r=0.99	r=0.98

The influence of the precolumn in the capacity factor was negligible.The packing material retains the protein fraction of the biological sample by the combined action of the acetonitrile and pH of the mobile phase without memory effects. Recovery and reproducibility were found excellent for detection at levels of picomoles.Automatic ASV1 switching valve provide to control of precolumn by washing in backflush mode. The packing material was replaced when the resolution decreases,on average after fifty determinations,in dependence of the samples used.Moreover,the method of standard addition is an extremely valuable tool for the fluorimetric measurements of Trp in biological fluids:the presence of interference can be observed by a different slope of the sample and standard calibration.Finally,the HPLC method described has been shown to be a selective and sensitive procedure for the detection of Trp and its indole metabolites in biological fluids,suitable for clinical investigations.

References

1.Musajo,L.,Benassi,C.A.:Adv.Clin.Chem.7,63(1964).
2.Price,J.M.,Brown,R.R.,Yess,N.:Advanc.Metab.Disord.2,159(1965).
3.Barchas,J.and Usdin,E.:Serotonin and behavior,Academic Press New York.1973.
4.Wolf,H.:Scand.J.Clin.Lab.Invest.33(Suppl.136),1(1974).
5.Yong,S.and Lau,S.:J.Chromatogr.175,343(1979).
6.Anderson,G.M.and Purdy,W.C.:Anal.Chem.51,283(1979).
7.Richards,D.A.:J.Chromatogr.175,293(1979).
8.Eccleston,E.G.:Clin.Chem.Acta 48,269(1973).

RAPID DETERMINATION OF TRYPTOPHAN AND ITS METABOLITES ALONG THE KYNURENINE PATWAY BY HPLC.

Carlo Costa, Antonio Bettero, Graziella Allegri

Istituto di Chimica Farmaceutica - Centro di Studio sulla Chimica del Farmaco e dei Prodotti Biologicamente Attivi del CNR - Università di Padova - Italy.

Introduction

Several methods have been employed to determine the tryptophan metabolites in the physiological fluids. But, the relatively low sensitivity of these analytical methods has required the use of loading doses of tryptophan, which can reduce availability of vitamin B6. Recently, some authors (1-4) have developed HPLC procedures for the separation of some tryptophan metabolites in body fluids. Here we report a rapid and selective reversed phase HPLC method for the estimation of tryptophan metabolites along the kynurenine pathway in urine or serum directly injected into the chromatograph without sample pre-treatment steps. Tryptophan (TRP), kynurenine (KYN), 3-hydroxykynurenine (3-OHKYN), kynurenic acid (KA), xanthurenic acid (XA), 3-hydroxyanthranilic acid (3-OHAA) and anthranilic acid (AA) were quantitatively separated in approximately 15 minutes at level of picomoles.

Materials and methods

A Perkin Elmer series 3B Liquid Chromatograph with LC-85 UV and LS-4 fluorimetric detectors collected through ASV 1 automatic switching valve to ISS-100 Intelligent Sampling System was employed. The chromatographic separations were achieved by RP 18 (10 μm) Merck column and precolumn

Progress in Tryptophan and Serotonin Research

packed with RP-8 (20-40 μm) LiChroprep at flow rate of 2 ml/min, using a mobile phase of acetonitrile (A) and phosphate buffer (0.004M) at pH 3.5 by a linear gradient elution in the fully automated series 3B gradient prophyles (Tl= 17 min; A=5-35%; Teq.=5 min; A= 5%).
The fluorimetric detector was set at wavelengths of excitation and emission depending on the native fluorescence of each metabolite under investigation, according to the relative retention times throughout the analysis. For detection of 3-OHKYN and KYN the UV detector was set at 226 nm. 20 μl of each standard solution containing mixture of TRP metabolites and biological fluids were injected directly into the chromatographic system after filtration with millipore 0.45 μm. A Perkin Elmer Sigma 15 Data Station with a modified external Standard method was employed for quantitative detection. Grade chemical reagents were used.

Results and discussion

Fig. 1 reports the separation of mixtures of 7 compounds

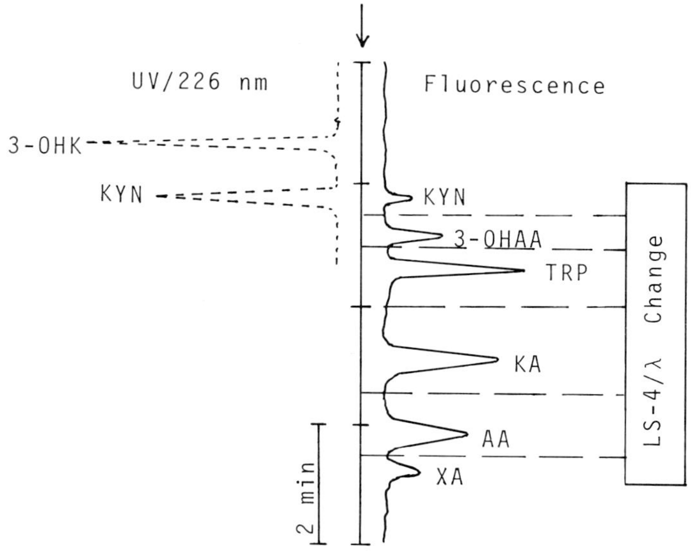

Fig. 1. Example of automatic steps of standard solutions of metabolites.

determined by automatic UV and fluorimetric detections in the HPLC system. A typical chromatogram of a urine sample in the conditions above described is shown in Fig. 2.

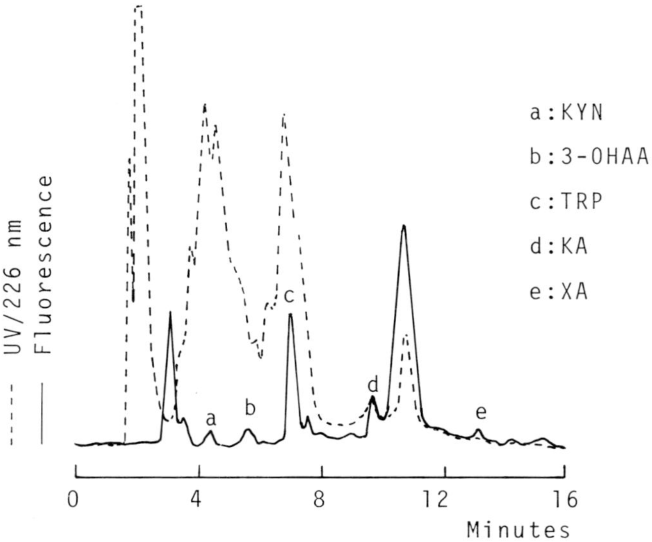

Fig. 2. Chromatogram of a urine sample obtained by fluorescence changes and UV at 226 nm.

The chromatographic parameters, average of 5 analyses of standard and urine sample solutions, are summarized in Table 1. It is shown that the reversed phase HPLC method described can be a powerful tool in the detection of TRP metabolites. The use of the automatic multi-wavelength fluorimetric detector permits a great selectivity in the analysis of biological fluids. Only 3-OHKYN does not fluoresce naturally and its UV detection, low in sensitivity, can cause artefacts. A suitable derivatization procedure for KYN and 3-OHKYN is under consideration. The identification of peaks eluted was achieved by standard addition method and stop flow technique. Recovery and detection limits at levels of picomoles show that the HPLC method proposed is a powerful technique for investigation of TRP metabolites in a wide variety of diseases without loading doses of TRP. Moreover, the physiological samples were analyzed directly without deproteinizing agents since

Table 1. HPLC parameters for TRP and its metabolites.

Metabolites		nm	K	Calibration	r	D.L.
3-OHKYN	UV	226	0.35	F=0.015x-0.31	0.99	16.5
	F	-	-	-	-	-
KYN	UV	226	1.18	F=0.018x+2.90	0.99	4.8
	F	400/481	1.18	F=0.004x-0.34	0.99	68.0
3-OHAA	F	305/415	1.80	F=0.11 x+0.09	0.99	3.3
TRP	F	285/320	2.40	F=0.21 x-0.32	0.99	1.2
KA	F	315/390	3.95	F=0.02 x+6.38	0.99	9.0
AA	F	288/398	5.15	F=0.31 x-4.20	0.99	1.8
XA	F	330/456	5.70	F=0.017x-0.39	0.99	15.5

D.L. (detection limit) = picomoles/µl; reproducibility less than 2% RSD

the protein fraction is retained in the precolumn system by effect of the mobile phase and pH used.

Reference

1. Grushka, E., Kikta, E.J.,Jr.: J.Chromatogr. 143, 51 (1977).

2. Krstulovic, A.M., Matzura, C.: J.Chromatogr. 163, 72 (1979).

3. Yong, S., Lau, S.: J.Chromatogr. 175, 343 (1979).

4. Ghebregzabher, M., Rufini, S., Castellucci, M.G., Lato, M.: J.Chromatogr. 222, 191 (1981).

HPLC-ANALYSIS OF XANTHURENIC ACID AND ITS 8-METHYL ETHER

Jaap Schrijver, Frank van Schaik, Jan van der Greef,
Klaas D. Bos and Wil H.P. Schreurs
CIVO-Institutes TNO, P.O. Box 360, 3700 AJ Zeist,
The Netherlands

Introduction

High-performance liquid chromatographic (HPLC) methods are described for the analysis of XA and of its 8-methyl ether (8-MeXA) in small urine samples. The specificity of both HPLC methods in determining XA and 8-MeXA respectively have been substantiated by field desorption-mass spectrometric (FD-MS) analyses.

Materials and methods

(1) XA (Sigma) was dissolved, using a small amount of NaOH, in water (100 mg/l) and stored at 4 °C. Dilutions were made with water. (2) Mobile phase I: 0.09 M KH_2PO_4 and 9 % (v/v) methanol, pH 2.25 with H_3PO_4. Mobile phase II: 0.04 M KH_2PO_4 and 16.7 % (v/v) methanol, pH 4.6. (3) 8-MeXA was synthesized according to Furst and Olson (1).

HPLC procedures

1. <u>HPLC analysis of XA with electrochemical detection (ECD)</u>

   A Knauer stainless steel column (80x4.6 mm i.d.) was filled with ODS-Hypersil, 3 µm (Shandon). A Rheodyne injection valve was used for the injection of 200 µl samples. The column was eluted with mobile phase I at a

Progress in Tryptophan and Serotonin Research

flow rate of 1.0 ml/min with a Knauer HPLC pump. The effluent of the column was monitored with a BAS amperometric detector (LC-4A control unit and a TL-5A amperometric flow cell: glassy carbon electrodes and Ag/AgCl as the references; detection at +700 mV). Elution profiles were displayed on a Kipp BD 41 recorder (1 V full scale).

2. HPLC analysis of 8-MeXA

Pump, injection valve (for the injection of 20 µl samples) and recorder (set at 10 mV full scale) used were as described under 1. An HPLC column (250x4.6 mm i.d.), filled with ODS-Hypersil, 5 µm, was eluted at 1.0 ml/min with mobile phase II. The fluorescence of 8-MeXA was measured in a Perkin Elmer model 204A spectrofluorometer, equipped with a flow cell of 100 µl volume, at 450 nm, with excitation set at 350 nm.

Other procedures

Field desorption (FD) measurements were performed with a Varian MAT 731 mass spectrometer (MS) equipped with a combined EI/FI/FD source.

Urine samples, collected overnight , were obtained from volunteers before and after an oral dose of 2 g L-tryptophan (2). Samples were stored at -20 °C. For the analysis of 8-MeXA, after thawing the samples were centrifuged 10 min at 2000 g and 4 °C to remove precipitates.

For the analysis of XA, urine samples were thawed and centrifuged as described above and acidified with 0.25 ml 1 N HCl/ml urine. Prepurification of these samples was performed on disposable extraction columns (aromatic sulfonic acid silane, Baker, cat.no. 7090). These columns were pretreated by flushing with 5 ml 5 N HCl and 10 ml distilled water. Thereafter, 1 ml sample was brought into the column and flushed with 4 ml 0.2 N HCl. XA was eluted from the column with 5 ml of distilled water.

Results

Typical elution profiles of XA are shown in Figure 1. XA eluted from the column after about 15 min. The ECD response was linear up to 100 μg XA/ml prepurified urine. The detection limit of the HPLC-ECD method for the analysis of urinary XA was as low as 15 ng/ml urine. The specificity of the HPLC method for XA analysis was substantiated through identification and quantitation of XA by FD-MS analysis.

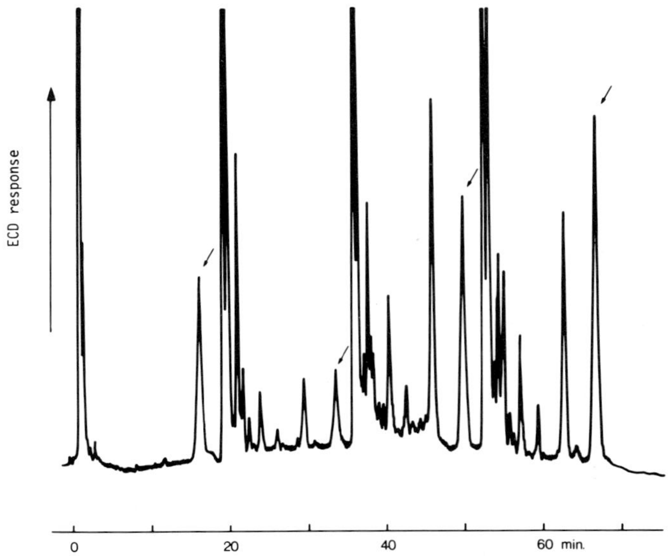

Figure 1. Elution profiles of XA from standard (20 ng) and 3 urine samples in HPLC-ECD analysis. Arrows indicate XA from different samples.

The reliability of the HPLC-ECD analysis of XA is given in Table I.

Excretion of XA with urine has been determined in some urine samples before and after an oral dose of 2 g L-tryptophan (2). The results are given in Table II. The urinary excretion of XA is increased by the use of OAC. Furthermore, a stronger increase in XA excretion after an oral tryptophan load was observed in the female using OAC, containing oestrogens.

Table I

RELIABILITY OF HPLC-ECD ANALYSIS OF URINARY XA

Samples*	A	B	C
mean (μg/ml)	0.145	1.172	7.200
sd (μg/ml)	0.008	0.092	0.614
CV (%)	5.5	7.9	8.5

*Three different 8 h urine samples (A, B and C) were analysed for XA on six consecutive days (between-assay-precision).

Table II

URINARY EXCRETION OF XA AND 8 MeXA BEFORE AND AFTER TRYPTOHAN LOADING

Subject	Before tryptophan		After tryptophan	
	μg/8 h urine			
	XA	8-MeXA	XA	8-MeXA
male	100	30	1110	290
female	110	30	1940	530
female (OAC)*	460	290	6640	4500

* OAC: oral anticonceptives containing oestrogens

Typical elution profiles of 8-MeXA from the synthesized standard, from the XA standard and from urine samples are shown in Figure 2. The XA standard contained 8-MeXA in spite of the supplier stating that this standard is substiantially free from 8-MeXA. In addition to the HPLC results, 8-MeXA was identified and quantitated by FD-MS analysis. It appeared that 8-MeXA was present in the XA standard in an amount of 0.96 % (w/w).

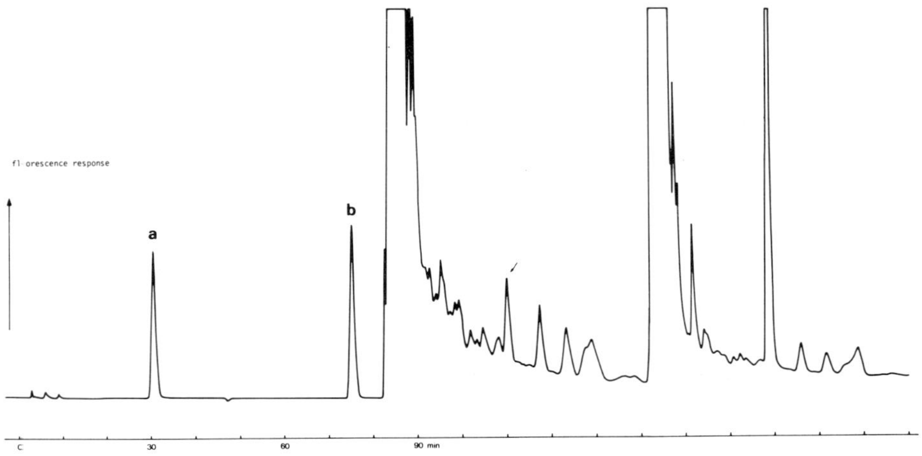

Figure 2. Elution profiles of 8-MeXA from standard (b: 1.1 ng), XA-standard (a: 100 ng) and urine samples in HPLC-fluorometric analysis. Arrows indicate 8-MeXA from different samples.

The amounts of 8-MeXA excreted with the urine before and after an oral load of 2 g L-tryptophan are also given in Table II. The proportional increase in the 8-MeXA excretion after an oral tryptophan load is comparable to that of XA.

References

1. Furst, A., Olson, J.: J. Org. Chem. 16, 412 (1951)
2. Coursin, D.B.: Am. J. Clin. Nutr. 14, 56 (1964)

FREE AND TOTAL TRYPTOPHAN MICRODETERMINATION BY AN AUTOMATED FLUOROMETRIC METHOD

Jean-Claude Bigot, Marie-Pierre Moineau, Jean-François Ménez,
Louis-Georges Bardou.

Laboratoire de Biochimie A, Centre Hospitalier Régional A. Morvan,
5 avenue Foch, 29279 Brest Cédex, France.

Introduction

The interest of total tryptophan (T.T.) and free tryptophan (F.T.) measurement in plasma is well known (1). Moreover the tryptophan content in urine and in spinal fluid also present some interest (2,3). Fluorometric technics have been described for the determination of this compound: they are based on norharman formation with formaldehyde by heating in acidic medium (3, 5) and oxidative amplification (1, 4). In plasma, T.T. is determined after deproteinization with trichloracetic acid (1) or after dilution (4). For delicate F.T. determination, several procedures have been reported: most workers prefer to measure F.T. in ultrafiltrates of plasma (1, 3) or after equilibrium dialysis (6). However, such methods are always time-consuming, so Mason (4) proposed a continuous flow dynamic dialysis technique to determine F.T. But such a procedure requires 4 ml of sample which is unsuitable for microsamples, particularly in pediatrics.
We propose an automated continuous flow method to determine both T.T. and F.T. in microsamples (1 ml), without preliminary steps. Accordingly, it was necessary to optimize the fluorometric conditions.

Material and Methods

1°) Samples preparation. Blood (1 ml) was collected into heparinized tube (1.43 UI/ml) and centrifuged at 3000 g for 10 mn at 4°C. All samples were analyzed within 24 H or frozen at -20°C. Spinal fluid and urine were treated in the same conditions. T.T. was determined after deproteinization of 100 µl of plasma with 400 µl of trichloracetic acid 3.5 % (w/v) in phos-

Progress in Tryptophan and Serotonin Research

phate buffer 0.125 M (pH 7.4). F.T. and spinal fluid tryptophan were determined without any treatment. Urine tryptophan required a previous dilution in water (1/10).

2°) Reagents for automated method (justified in Fig. 1).

* Trichloracetic acid (TCA) 3.5 % (w/v) in distilled water.
* Formaldehyde 9 % obtained after dilution 1/4 of formaldehyde solution 35 %.
* Hydrogen peroxide obtained after dilution 3 % of hydrogen peroxide (30 vol. per litre).
* Disodium citrate 0.116 M.

All these reagents contained 1 ml of "Brij 35" per litre.

* Wash solution : phosphate buffer 1.32 mM, NaCl 145 mM, pH 7.4.
* Tryptophan standard solutions were prepared to cover the range 2-20 μM by dilution of 2 mM solution in deproteinizating reagent containing sodium azide 3 mM.

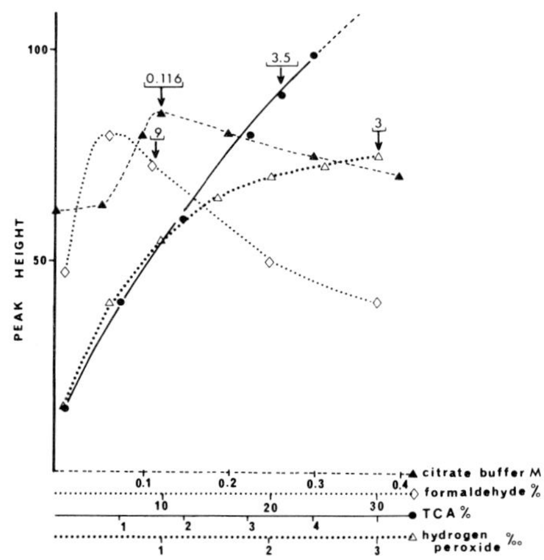

Figure 1

3°) Automated assay.

The flow diagram is shown in fig. 2. The system was assembled with Technicon components (Domont, F).

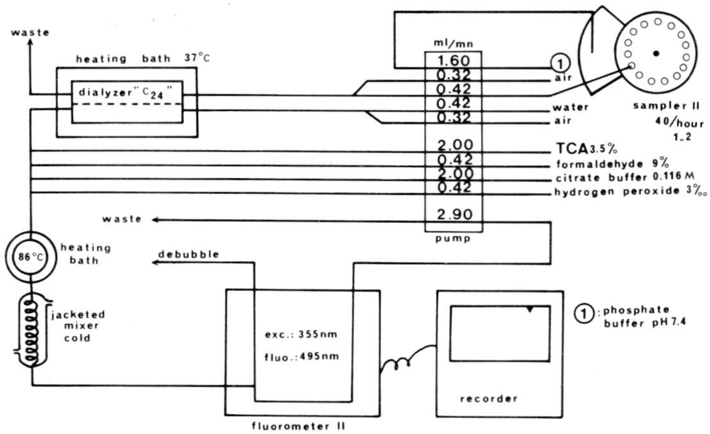

Figure 2

- SAMPLER II

The sampleswere aspirated at the rate of 40/h with a sample to wash ratio of 2 : 1.

The wash solution is phosphate buffer 1,32 mM, Na cl 145 mM ph7,4.

- PROPORTIONING PUMP II

- (1) Phosphate buffer ph 7,4	rate 1,60 ml/mm (blue)
- Air	0,32 ml/mn (black)
- Sample	0,42 ml/mn (orange)
- Water	0,42 ml/mn
- Air	0,32 ml/mn
- T C A 3,5 %	2,00 ml/mn (green)
- formaldehyde 9 %	0,42 ml/mn
- citrate buffer 0,116 M	2,00 ml/mn
- hydrogen peroxide 3 ‰	0,42 ml/mn

Pump tubing are Technicon designations.

- DIALYZER "C 24"

Samples are dialyzed against water for removal of protein in plasma.

The dialyzer "type C24" Techicon with Pre-mount Dialysis Membrane is thermostated at 37°C.

- NORHARMAN FORMATION

  The dialyzate containing free tryptophan is mixed with TCA 3,5%, with formaldehyde 9 %, with citrate buffer 0,116 M and at last with hydrogen peroxide 3 ‰ and is passed through a glass coil heated to 86°C in a circulating oil bath Technicon.

- DETECTION

  After cooling the fluorescence of Norharman is measured in a fluoronephelometer (Technicon) with primary (excitation 355nm) and secondary (fluorescence 495 nm) filters and with a recorder (Technicon).

Results and discussion

The results are obtained 10 minutes after the sample pumping. A typical strip chart recording is shown on fig. 3.

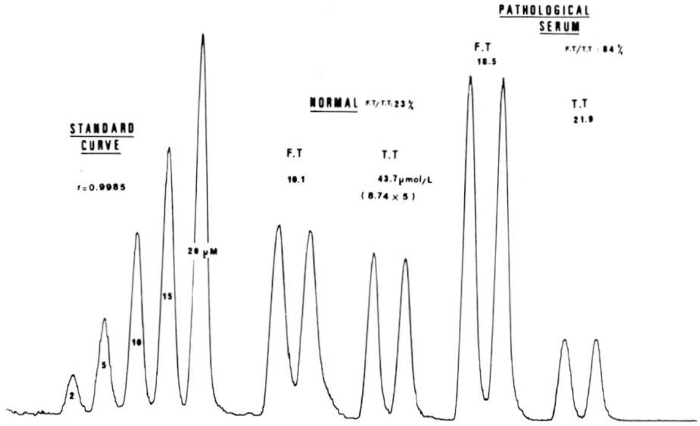

Figure 3

The heating at 37°C compared to 25°C improves the dialysis for standard solution; it was also chosen for simulating physiological conditions.Higher concentration of hydrogen peroxide and trichloracetic acid also produce carbon dioxide which causes an erratic flow rate. The formaldehyde 9 % concentration is justified by the fact that with the formaldehyde 4.5 %

we observed some important differences in tryptophan determination if this latter compound is in water, buffer or acidic medium. This observation is not true with formaldehyde 9 %. In our system the norharman fluorescence is maximum at pH 2.5, which is realised by the addition of citrate solution Analytical criteria were : within assay precision (standard solution : n = 30, $\bar{x}$ = 20 μmol/l, CV = 1.17 % ; F.T. : n = 30, $\bar{x}$ = 16.5 μmol/l, CV = 3 % ; T.T. : n = 30, $\bar{x}$ = 55.9 μmol/l, CV = 2.17 % ; between assay precision (standard solution : n = 21 days, $\bar{x}$ = 20 μmol/l, CV = 1.5 %) ; quantitative detection limit (0.6 μmol/l) ; overload analytical recovery was found within 97 and 103 % for a standard solution. Linear response was obtained in the range 2-20 μM. Some deproteinizing reagents have been tested (sulfosalicylic acid, picric acid, p-toluenesulfonic acid), they produce an interference. Moreover we noted an interference with acetylsalicylic acid. The specificity of such a method has been discussed (3). The mean values for an normal unselected population (20-45 years) were :

- total plasmatic tryptophan n = 18, $\bar{x}$ = 59.8 ± 4.8 μmol/l ($\bar{x}$ ± SEM)
- free plasmatic tryptophan n = 18, $\bar{x}$ = 8.9 ± 0.6 μmol/l
- F.T. : T.T. plasmatic ratio n = 18, $\bar{x}$ = 14.8 ± 0.8 %
- urine tryptophan n = 16, $\bar{x}$ = 89.9 ± 21.4 μmol/24 H
- spinal fluid n = 36, $\bar{x}$ = 1.76 ± 0.16 μmol/l

Our results are comparable with those described in literature and summarized in (5) so this method seems to us to be suitable for routine determination. It allows us to assess 40 samples per hour with 300 μl of plasma for both F.T. and T.T., and consequently, pediatric samples can be achieved.

References

1. Escande C., Bousquet B., Dreux C. : Ann. Biol. Clin., 35, 387-395 (1977)
2. Lehmann M.D. : Scand. J. Clin. Lab. Invest., 22, 72-78 (1968).
3. Flentge F., Venema K., Korf J. : Biochem. Med., II, 234-241 (1974).
4. Mason F.A., Diez J.A. et al : Anal. Biochem., 84, 231-239 (1978).
5. Escande C., Bousquet B., Dreux C. : Path. Biol., 26, (7), 453-457 (1978).
6. Mc Menamy R.H., Lund C.C. et al : Arch. Biochem. Biophys., 93, 135-139 (1961).

COMPARISON OF HPLC AND GC-MS METHODS FOR THE DETERMINATION OF TRYPTOPHAN AND ITS INDOLE METABOLITES IN BIOLOGICAL SAMPLES

Emili Martinez, Albert Adell, Cristina Suñol, Francesc Artigas, Josep Mª Tusell, Emili Gelpí
Analytical Neurochemistry Unit, Instituto de Química Bio-Orgánica, C.S.I.C., Barcelona-34, Spain

Introduction

The determination of tryptophan (TP) and its metabolites serotonin (5HT), 5-hydroxyindoleacetic acid (5HIAA) has been a matter of interest in neurochemical studies since 5HT is a well established neurotransmitter in CNS (1, 2). Tryptamine (T), the non-hydroxylated analog of 5HT, whose exact role is far from being defined (3), has received comparatively less attention mainly due to lack of reliable analytical methods for the assay of T and IAA (indoleacetic acid) in brain tissue.

Aside from the classical fluorimetric methods for TP, 5HT and 5HIAA there are two choice techniques for the determination of these compounds in biological samples: GC-MS (for a review see ref. 4) and HPLC (5, 6), where the introduction of efficient flow-through microcells for fluorimetric and electrochemical detection together with the advent of microparticulate reversed-phase columns have provided a turning point in this application. Being both chromatographic methods they offer the potential of obtaining metabolic profiles, thus facilitating the evaluation of metabolite interrelationships in normal as well as in pharmacologically or pathologically altered conditions.

In this report we will attempt to summarize and evaluate, according to our own experience in the field, the analytical features of these chromatographic techniques in the study of tryptophan and its indole metabolites.

Progress in Tryptophan and Serotonin Research

Methods

The analytical methods have been described in detail elsewhere. Briefly, they can be summarized as follows:

GC-MS conditions (Ref. 7): OV-101 capillary column, 25 m long, 0.3 mm i.d. Carrier gas: Helium. Column temperature from 180 to 220ºC at 10ºC. Injection mode: splitless, 60ºC. GC-MS interface: open split type. Electron impact, 70 eV. Apparatus: Hewlett Packard 5992.

HPLC methods: μBondapak columns with guard columns were used in all HPLC procedures. The solvent delivery pumps (M6000) and programmer were from Waters Ass. Detectors; fluorimeter (Perkin Elmer 650 10S) set at 280/340 nm for excitation and emission wavelengths, respectively, or an amperometric detector (BAS) with glassy carbon electrode set at +0.7V.

Method A (8) Solvent 1: 5 mmol/L of 1-pentanesulfonic acid at pH 3.0. Solvent 2: 7/3 methanol/water. Linearly programmed elution from 20 % to 100 % of solvent 2 in 20 min. Flow: 1.0 mL/min. Fluorimetric detection as in methods B and C.

Method B (9). Isocratic elution with a mixture of 60% 6 mmol/L sodium acetate at pH 4.0 and 40% methanol. Flow: 1.0 mL/min.

Method C (10). Isocratic elution with a mixture of 50% 5 mmol/L of 1-pentanesulfonic acid and 50% methanol. Flow: 1.0 mL/min.

Method D (11). Isocratic elution with a mixture of 94% phosphate-citrate buffer (0.03 mol/L citric acid, 0.06 mol/L sodium hydrogen phosphate) and 6% methanol. Flow: 1.5 mL/min. Amperometric detection.

Results and discussion

GC-MS: Our major contribution to the study of the TP indole pathway has been the concurrent determination of TP, 5HT, T, 5HIAA and IAA in the same sample of brain tissue (12), urine or CSF. The method relies on the single step XAD-2 adsorption of the indoles contained in samples of brain homogenates or physiological fluids and further analysis by GC-MS-Selected ion monitoring (SIM) of the methanolic eluate. The mean concentration values found in rat brain (n=6) were: TP 4.16 μg/g, 5HT 526 ng/g, 5HIAA 442 ng/g, T $<$ 400 pg/g and IAA 13 ng/g. Consequently, this method allows the study of the two major indole pathways and clearly shows that T levels are minimal relative to other classical central neurotransmitters. Also, the acid/amine ratios (0.84 for 5HIAA/5HT, > 35 for IAA/T suggest that T is unprotected from the action of MAO (monoamine oxidase), probably due to poor neuronal storage mechanisms, even though its subcellular distribution resembles that of 5HT (13). Figure 1 illustrates the determination of endogenous T in adult rat brain which we have recently achieved by GC-MS with glass capillary colums. The levels found in this case were 450 ± 178 pg/g ($\bar{X}$±SD) (n=6), confirming earlier results by different groups (12, 14, 15). Unequivocal identification of endogenous T was performed by monitoring m/z values 276 and 289. Accordingly, GC-MS-SIM offers the unique capability of structural confirmation of this analyte, even though its concentration may be as low as 0.5 ppb in a complex sample matrix such as brain tissue. This, plus the possibility of quantitation by isotopic dilution using stable la-

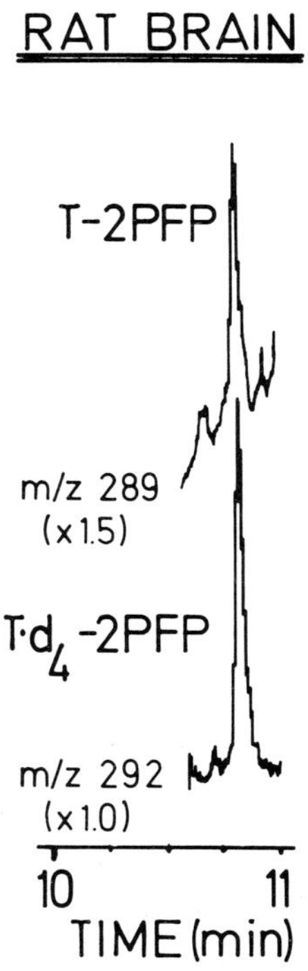

Figure 1. SIM profiles of perfluoroacylated T. Top; response of endogenous at m/z 289. Bottom; response of deuterated internal standard. Conditions as described in the text.

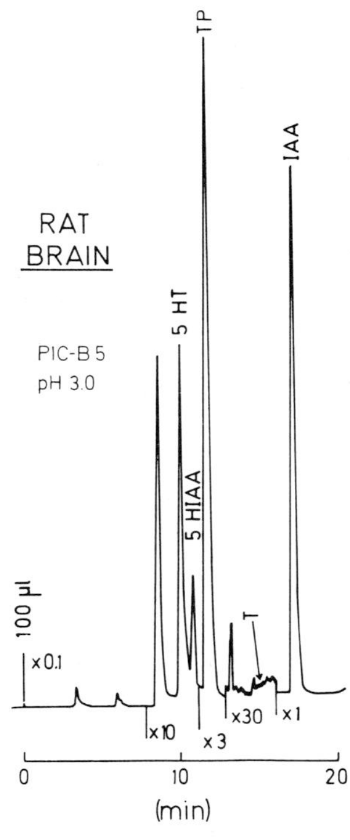

Figure 2. HPLC profile obtained from the brain homogenate of a rat treated i.p. with 150 mg/kg of T. (Reprinted with permission of J. Liq. Chrom. Ref. 8).

belled analogs make it a reference analytical technique for TP and its indole metabolites.

HPLC: We have approached the analysis of the indoles using two different detection systems: a) fluorimetric (F) and b) electrochemical (ECD). Direct fluorimetric detection is readily achieved in this case due to the intense natural fluorescence of the indole ring. Figure 2 shows the profile obtained with method A from the brain of a rat treated i.p. with T. The tissue was homogenized according to described procedures (12) and the supernatant injected directly into the HPLC system. 5HT, 5HIAA, TP and IAA are clearly resolved and identified, the latter compound being about 1000 times greater than in control rats. This is because extracerebral T crosses the blood brain barrier and is converted intracerebrally into IAA. The basal levels of this acid are below the detection limit of this procedure so that its determination would require a concentration step in ethyl acetate. Basal IAA in control rat brains can be determined using HPLC method B. The levels in this case are 16 ng/g, in agreement with values reported using GC-MS techniques (see above).

HPLC coupled to fluorimetric detection is also a good procedure for the determination of IAA in plasma. After a single deproteinization step, human or rat plasma is injected directly into the HPLC column (method C). The mean value in samples from 15 healthy individuals is 269 ng/mL (1.54 ± 0.59 µmol/L, $\bar{X} \pm SD$) (10). On the other hand, 5HT and 5HIAA can be best monitored by amperometric detection. Figure 3a shows the profile of an injection equivalent to 2 mg of cortex+striatum of rat brain (11).

The method allows the additional concurrent determination of dopaminergic (DA) and noradrenergic (NA) pathways, as can also be seen in Fig. 3b in which MHPG, the main cerebral metabolite of NA is shown (CSF sample). Both chromatograms have been obtained by HPLC method D. With this method, we have been able to determine the low levels of 5HIAA in human plasma (51.8 nmol/L (9.9 ng/mL)$\pm$20.1nmol/L, n=16). These results show the high sensitivity of ECD for easily oxidizable compounds, such as the 5-hydroxyindoles and catecholamines.

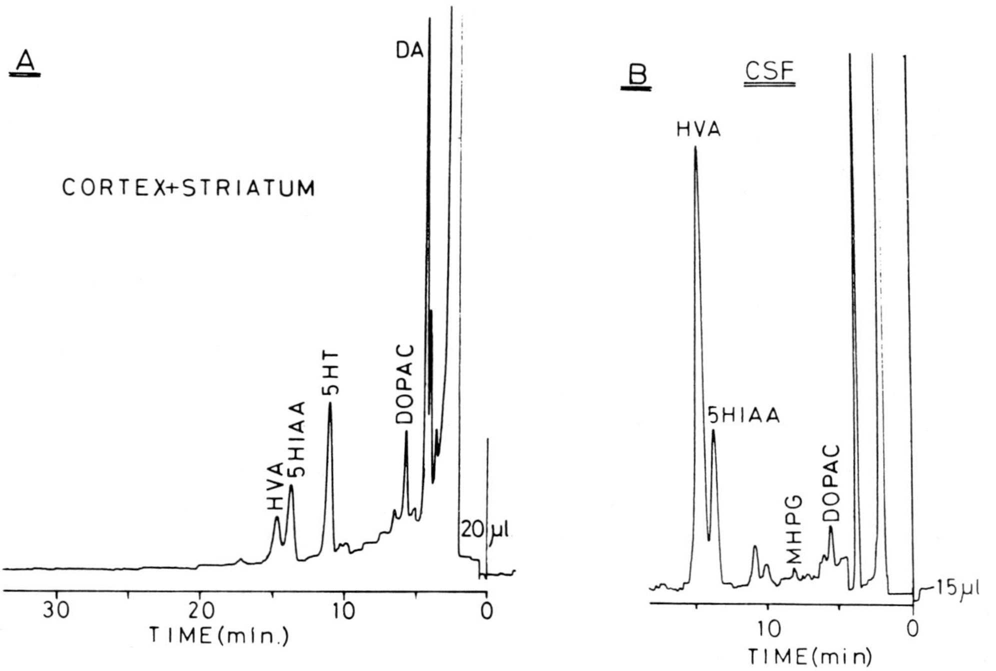

Figure 3. HPLC profiles from rat brain and human CSF samples monitored with an ECD at +0.7V. See method D. (Reprinted with permission of Chromatographia. See ref. 11)

Comparison of HPLC and GC-MS techniques. Qualitative aspects: According to our own experience in the use of both techniques, some important comparative features of GC-MS vs HPLC could be underlined: 1) Price. Obviously this is a conclusive factor and does not need any further comment. 2) Instrumental breakdown time. In general, GC-MS equipments are prone to more instrumental failures than HPLC systems. 3) Possibility of fast routine analysis. Most HPLC analysis of indoles in biological samples do not require any kind of prepurification or concentration steps (except brain homogenization or plasma deproteinization), resulting in high sample throughput, specially with an automatic injector. All of these points are favourable to HPLC, although the coelution of the analyte(s) and unknown compounds could be a disavantage. Nevertheless, this can be overcome by using several eluent systems, applying pharmacological manipulations, etc.

Clearly, the best system would be to check the HPLC method vs an analytical reference technique, such as GC-MS.

Quantitative aspects: Table I summarizes the detection limits for TP and its indole metabolites using GC-MS-SIM and HPLC, either with fluorimetric or electrochemical detection (+0.7V).

Table I

	ABSOLUTE DETECTION LIMITS (pg)		
	GC-MS-SIM(16)	HPLC-ECD(11)	HPLC-F(8)
TP	10	-- (**)	180
5HT	20	5	20
5HIAA	10	10	120
T	10	-- (**)	30
IAA	10(*)	-- (**)	80

(*) 5 pg can be easily achieved using capillary columns.

(**) Not easily detectable in biological systems. Interferences due to the higher ECD voltage required.

A quantitative evaluation of the HPLC-F and GC-MS-SIM methods for IAA was carried out in rat brain. The accuracy of the HPLC procedure (checked vs GC-MS-SIM) is very good. The ordinate at the origin was only 0.8 ng/g and the slope of the regression straight line (defined as SIM value = a + b HPLC value) is 0.995, both parameters indicating that the two methods are actually quantitating the same compound. A coefficient of correlation of 0.991 was obtained. The endogenous concentrations of IAA in rat brain determined by GC-MS-SIM and HPLC-F were 18 and 16 ng/g, respectively.

As a final conclusion, it could be stated that HPLC methods are the best alternative to analyze TP indole metabolites (with the only exception of brain T in samples of brain tissue and physiological fluids). Moreover, ECD is the detection method of choice for 5HT and 5HIAA. However, HPLC-F would be extremely useful in metabolic studies since it allows the direct determination of the TP⟶5HT⟶5HIAA pathway, as well as for the quantitative estimation of IAA in brain and physiological fluids. GC-MS remains as a reference analytical technique, its use preferably being restricted to those cases in which HPLC fails to determine correctly any of the TP indole metabolites.

It should be evident, however, that any recommended method along the lines of this discussion would necessarily depend on the availability of the equipment, the nature of the sample, the concentration of the analyte(s) and the number of components one wishes to concurrently determine. Taking into account the analytical features of each technique, table II condenses our opinion as to which have proven to be in practice the most suitable methods for determinations of these indole TP metabolites.

Table II

BRAIN AND PHYSIOLOGICAL FLUIDS

COMPOUND	BEST METHOD	ALTERNATIVE METHOD
TP	HPLC-F	GC-MS
5HT	HPLC-ECD	HPLC-F
5HIAA	HPLC-ECD	HPLC-F
T	GC-MS	---
IAA	GC-MS	HPLC-F

Nevertheless, these should be considered as general statements that are not implied to be true in an absolute sense (that is, for each compound in any type of sample). Thus, for instance, the determination of TP, 5HT and 5HIAA in brain tissue can be performed with a single HPLC run, using a fluorimetric detector, whereas the use of an ECD would be valid if one is only interested in 5HT and 5HIAA. Furthermore, the low concentration of 5HIAA in human plasma precludes its estimation by HPLC-F, requiring the use of an ECD.

Acknowledgments

This work has been carried out under research grants of the Spanish CSIC and the Fondo de Investigaciones Sanitarias de la Seguridad Social.

References

1. Costa, E., Greengard, P, (Eds): Advances in Biochemical Psychopharmacology. Vols. 10, 11. Raven Press, New York 1974.
2. Boullin, D.J. (Ed): Serotonin in Mental Abnormalities. John Wiley & Sons, New York 1978.
3. Boulton, A.A.: Trace Amines and the Brain. pp. 21. E. Usdin, M. Sandler, Eds. Marcel Dekker, New York 1976.
4. Gelpí, E.: Techniques and Instrumentation in Analytical Chemistry. Vol. 4. Part A. Analysis of Biogenic Amines, G.B. Baker, R.T. Coutts, Eds. Elsevier, Amsterdam 1982.
5. Kissinger, P.T., Bountlett, C.S., Shoup, R.E.: Life Sci. 28, 455 (1981).
6. Anderson, G.M., Young, J.G.: Life Sci. 28, 507 (1981).
7. Suñol, C., Artigas, F., Martínez, E., Tusell, J.M., Gelpí, E. (Submitted).
8. Adell, A., Tusell, J.M., Artigas, F., Martínez, E., Suñol, C., Gelpí,E. J. Liq. Chrom. (in press).
9. Tusell, J.M., Suñol, C., Martínez, E., Artigas, F., Gelpí, E. (Submitted).
10. Martínez, E., Artigas, F., Suñol, C., Tusell, J.M., Gelpí, E. Clin. Chem. (in press).
11. Tusell, J.M., Suñol, C., Artigas, F., Martínez, E., Gelpí, E. Chromatographia 16, 112 (1982).
12. Artigas, F., Gelpí, E.: Anal. Biochem. 92, 233 (1979).
13. Boulton, A.A., Baker, G.B.: J. Neurochem. 25, 477 (1975).
14. Warsch, J.J., Godse, D.D., Stancer, H.C., Chan, P.W., Coscina, D.V.: Biochem. Med. 18, 10 (1977).
15. Philips, S.R., Durden, D.A., Boulton, A.A.: Can. J. Biochem. 52, 447 (1974).
16. Artigas, F., Martinez, E., Gelpí, E.: J. Chrom. Sci. 20, 75 (1982).

PINEAL INDOLES: THEIR DETERMINATION BY HIGH PERFORMANCE LIQUID CHROMATOGRAPHY AND THE INFLUENCE OF DIURNAL VARIATION AND TRYPTOPHAN ADMINISTRATION ON THEIR LEVEL IN THE RAT

Simon N. Young

Department of Psychiatry, McGill University, 1033 Pine Avenue West, Montreal, Quebec, Canada, H3A 1A1

George M. Anderson

Department of Laboratory Medicine and the Child Study Center, Yale University, New Haven, Connecticut, U.S.A.

Introduction

The pineal is a rich source of tryptophan metabolites but until recently the lack of convenient analytical methodology has hindered their study. We have been using high performance liquid chromatography (HPLC) for the determination of indoles in brain (1). Because this method involves only homogenization of the brain and direct injection of the supernatant into the HPLC, and because indoles are detected by the intense fluorescence of the indole ring, we felt that it might be a suitable approach for the quantitation of pineal indoles.

Materials and Methods

1) Samples and sample preparation. For animal experiments, male Sprague-Dawley rats (100-125 g) were kept in a light-dark 12 h/12 h cycle for one week before use. Human pineals were obtained at autopsy at 8-24 h after the death of the patients, most of whom died of cancer. Rat pineals were sonicated in 200 µl of ice-cold 0.1 M perchloric acid containing 0.1% ascorbic acid. After centrifugation 10-100 µl of the protein-free supernatant was injected directly into the chromatograph. Human pineals were sonicated in 0.5 ml of perchloric acid but otherwise

Progress in Tryptophan and Serotonin Research

the procedure was the same. These methods have been described in greater detail (2).

2) Chromatography. Samples were chromatographed on a μBondapak C_{18} reversed-phase column (Waters) and detection was using a modified (1) Aminco Fluoromonitor. Sometimes an electrochemical detector (LC-4 from BAS) was run in series with the Fluoromonitor. The carbon-paste working electrode was set at +0.8 V versus the Ag/AgCl reference electrode. Isocratic systems were used with buffers of 0.01 M sodium acetate at a pH of 4.25 or 4.50 with 8-35% methanol, depending on the compound to be measured.

Results

1) Compounds detected. Using this system we were able to quantitate various indoles at detection limits of 10-40 pg. Standard addition studies for the compounds investigated gave recoveries of greater than 95%. The following compounds were measurable in both rat and human pineals; tryptophan, 5-hydroxytryptamine (5HT), 5-hydroxyindoleacetic acid (5HIAA), 5-hydroxytryptophol (5HTOL), N-acetyl-5-hydroxytryptamine (NAc-5HT) and melatonin. 5-Methoxytryptophol and 5-methoxyindoleacetic acid were not detected although the detection limits (0.1 ng/rat pineal) were much lower than the levels at which these compounds have been reported previously using less specific methodology (3, 4, 5). The identity of 5HTOL, measured for the first time in single rat pineals, was based on chromatography in 3 different systems and on the close aggrement (r=0.996) between fluorometric and electrochemical measurements. Human pineals, unlike rat pineals, showed a variety of unidentified peaks on HPLC (2), suggesting that much remains to be learned about indole metabolism in human pineal.

2) Human pineals. The concentrations of compounds in human pineals given as mean of 26 ± SD in ng per pineal are 5HT, 1620 ± 1500; 5HIAA, 327 ± 234; 5HTOL, 56.9 ± 50.0; melatonin, 22.8 ± 20.5. There were no apparent relationships between clinical variables and the concentrations of the various compounds.

3) Rat pineals. All six compounds measured in rat pineals showed a signi-

ficant diurnal variation. Melatonin and NAc-5HT were high during the night and low during the day, while tryptophan, 5HT, 5HIAA and 5HTOL showed the opposite rhythm. The concentrations of 5HT, 5HIAA and melatonin that we found were similar to those in the literature. We found lower levels of tryptophan than reported previously in a study which used a less specific fluorometric method (6). The concentration of 5HTOL, which had not been reported previously, was from about 0.5 ng/pineal at night to 3.5 ng/pineal during the day. Thus, the peak pineal content of 5HTOL was higher than the nighttime peak melatonin content of 2 ng/pineal.

Tryptophan administration (100 mg/kg, intraperitoneally) increased all six indoles in rat pineal one hour later. This was true during both day and night. 5HT and 5HTOL increased 2 to 3-fold while melatonin went up only about 50%.

Discussion

Our study suggests that HPLC with fluorometric detection is a simple and convenient way of measuring pineal indoles. One interesting result from our study is that the concentration of 5HTOL can be higher than that of melatonin in both human and rat pineals. Whether 5HTOL has any function is uncertain at the moment but it is known to have an antigonadotrophic action similar to that of melatonin (7). Unlike melatonin, 5HTOL is known to elicit cerebroarterial contractions by activating tryptaminergic receptors and it also interferes with the action of 5HT on these receptors (8). When given to animals, 5HTOL induces sleep (9). Thus, the hypnotic action of tryptophan in humans (10) may be mediated in part by an increase in 5HTOL. The hormonal and behavioural influences of 5HTOL deserve further consideration especially because the concentration of 5HTOL in human CSF is at least 20-fold that of melatonin (11, 12).

The change in pineal indoles on tryptophan administration suggests that variations in pineal tryptophan brought about by dietary variations may influence pineal metabolism. Thus, possible dietary influences on pineal metabolism and function are worth studying.

Acknowledgements

We thank P.A. Martiquet for excellent technical assistance. This work was supported by the Medical Research Council of Canada, by MHCRC grant 30929. and by the Soloman R. and Rebecca D. Baker Foundation, Inc.

References

1. Anderson, G.M., Young, J.G., Batter, D.K., Young, S.N., Cohen, D.J., Shaywitz, B.A.: J. Chromatog. 223, 315-320 (1981).
2. Anderson, G.M., Young, J.G., Cohen, D.J., Young, S.N.: J. Chromatog. 228, 155-163 (1982).
3. Quay, W.B.: Pineal Chemistry in Cellular and Physiological Mechanisms, Thomas, Springfield, pp. 137-200, 1974.
4. Wilson, B.W., Lynch, H.J., Ozaki, Y.: Life Sci. 23, 1019-1024 (1978).
5. Smith, I., Francis, P., Leone, R.M., Muller, P.E.: Biochem. J. 185, 537-540 (1980).
6. Sugden, D.: J. Neurochem. 33, 811-813 (1979).
7. Reiter, R.J., Vaughan, M.K.: Life Sci. 21, 159-172 (1977).
8. Fu, L.H.W., Hayashi, S., Toda, N.: Brit. J. Pharmacol. 68, 17-18 (1980)
9. Feldstein, A., Chang, F.H., Kucharski, J.M.: Life Sci. 9, 323-329 (1970).
10. Hartmann, E.: Amer. J. Psychiatry 134, 366-370 (1977).
11. Beck, O., Borg, S., Holmstedt, B., Stribler, H.: Biochem. Pharmacol. 29, 693-696 (1980).
12. Brown, G.M., Young, S.N., Gauthier, S., Tsui, H., Grota, L.J.: Life Sci. 25, 929-936 (1975).

HIGHER-ORDER DERIVATIVE SPECTROPHOTOMETRY FOR THE QUANTITATIVE DETERMINATION OF TRYPTOPHAN, TYROSINE AND PHENYLALANINE IN TWO- AND THREE-COMPONENT MIXTURES, PEPTIDES AND PROTEINS

Gerhard Talsky, Sibylle Tichy

Institut für Technische Chemie, Lehrstuhl I, der Technischen Universität München
D-8046 Garching

Introduction

Structure and functional behaviour of proteins have been studied by a variety of spectroscopic techniques, among which UV/VIS spectrophotometry is probably most frequently employed.

In addition to cysteine which absorbs weakly in the ultraviolet, phenylalanine (Phe), tyrosine (Tyr) and tryptophan (Trp) absorb actively between 200 and 300 nm. While individual chromophores can easily be determined, due to the overlapping of the spectra, this cannot be done at all or only unsatisfactorily for mixtures containing two or three of these amino acids (Fig. 1). Other difficulties which have to be overcome are the vastly differing absorption coefficients of the three amino acids (ε_{257}^{Phe} $1.90 \cdot 10^2$, ε_{275}^{Tyr} $1.4 \cdot 10^3$, ε_{279}^{Trp} $5.55 \cdot 10^3$ l/mol·cm).

It is occasionally possible to determine the composition of a binary mixture from the ratio of the absorption coefficients at two wavelengths. However, the problem can be solved far better by computer analysis of the superposed curves. In order to get satisfactory results in this, one must know the number of components as well as any disturbing substances and their spectra which must be of about the same absorption strength.

Progress in Tryptophan and Serotonin Research

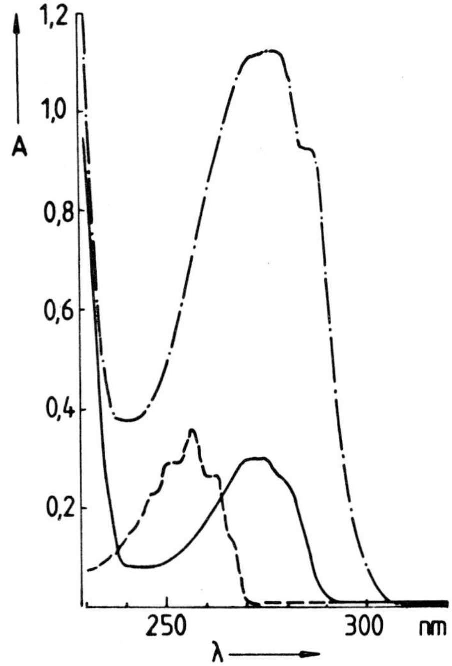

Fig. 1. UV spectra of Tyr (——) conc. $2\cdot10^{-4}$ $mol\cdot l^{-1}$, Trp (-·-·-), conc. $2\cdot10^{-4}$ $mol\cdot l^{-1}$ and Phe (---) conc. $2\cdot10^{-3}$ $mol\cdot l^{-1}$; pH 7.

Another way to solve the problem is to generate derivatives of the fundamental spectra. MATSUSHIMA et al. (1) used the first derivative (d^1) for the quantitative estimation of Phe in the presence of Tyr and Trp. ICHIKAWA et al. (2) showed that the second derivative (d^2) of the spectra of Tyr and Trp does not disturb the characteristic bands in the spectrum of Phe in proteins.

In 1978 we reported about a new analog computer which made it possible for the first time to generate on-line and immediately low-noise derivatives up to the ninth order (3). Thus it was possible to study systematically whether and to what extent higher derivatives ($n > 2$) in particular, improve the information obtainable from UV spectra of mixtures of the three amino acids and of proteins.

Results

1) Free amino acids. In two-component mixtures the relative molar ratio of the amino acids is decisive for the accuracy (3). Trp and Phe could be estimated nearly undisturbed between a molar ratio of 4:1 and 1:4. In the presence of an excess of Tyr the values for Trp rise over 100 %. Phe and Tyr could be recovered well (Tab. 1).

Tab. 1. Two-component mixtures of Trp, Tyr and Phe.
d derivative order; the λ-range refers to the peaks which were evaluated; P-P peak-peak evaluation; Z-P zero-peak evaluation.

mixture		pH	d	λ-range	eval.	mol. ratio	recovery
Trp + Phe							
Trp		7	d^2	284 - 289	P-P	4:1 - 1:7	100.0 - 101.0
	Phe	7	d^4	249 - 252	P-P	4:1 - 1:4	100.0 - 92.5
Trp + Tyr							
Trp		10	d^1	289	Z-P	4:1 - 1:4	101.7 - 118.5
	Tyr	10	d^1	248 - 266	P-P	4:1 - 1:7	86.4 - 97.6
Trp		7	d^2	290	Z-P	10:1 - 1:4	101.6 - 114.3
Phe + Tyr							
	Tyr	7	d^2	277 - 283	P-P	7:1 - 1:4	104.6 - 100.0
	Tyr	7	d^3	280 - 286	P-P	4:1 - 1:4	100.0 - 99.2
	Tyr	7	d^4	280 - 286	P-P	4:1 - 1:4	92.9 - 94.9
Phe		7	d^4	249 - 252	P-P	4:1 - 1:4	98.1 - 103.1

To find the optimal conditions for the simultaneous estimation of the three amino acids in three-component mixtures, 54 various samples (permutation of 3 volume elements from 1 to 8; a + b + c = 10) of standard solutions with different molar ratios were mixed and 24 standard solutions of the pure amino acids were prepared; then the derivatives of all samples up to the sixth order were evaluated. The concentrations of Phe were varied between $5.0\cdot10^{-4}$ and $40.0\cdot10^{-4}$ $mol\cdot l^{-1}$, those of Tyr and Trp between $0.5\cdot10^{-4}$ and $4.0\cdot10^{-4}$ $mol\cdot l^{-1}$. For instance to prepare a mixture with a molar ratio of 10:4:4 to 250 ul Phe (standard solution $5\cdot10^{-3}$ $mol\cdot l^{-1}$), 1000 µl Tyr ($5\cdot10^{-4}$ mol l^{-1}) and 1000 µl Trp ($5\cdot10^{-4}$ $mol\cdot l^{-1}$) were added and diluted with 250 µl buffer to an end volume of 2500 µl.

For the quantitative determination of Trp and Phe the 4th derivatives of the pH 7-spectra were used (Fig. 2a). The peaks of Trp between 282 - 286 nm are not disturbed by those of Phe at 270 to 234 nm. The situation is different with Tyr, which in higher molar concentrations more or less overlaps the spectra of Trp and Phe. The estimation of Tyr succeeded at 244 nm, pH 10 and d^4. It is not affected by Trp and only in high concentration by Phe (Fig. 2b).

Better results are obtained if an optimized condition for each substance is used for the 3 amino acids instead of only a single computer program. Thus it is possible to estimate small amounts of Phe in the presence of Tyr and Trp although ε of Phe is much smaller in comparison to Tyr and Trp. Beyond it the standard addition method eliminates the disturbance (see chapter "proteins").

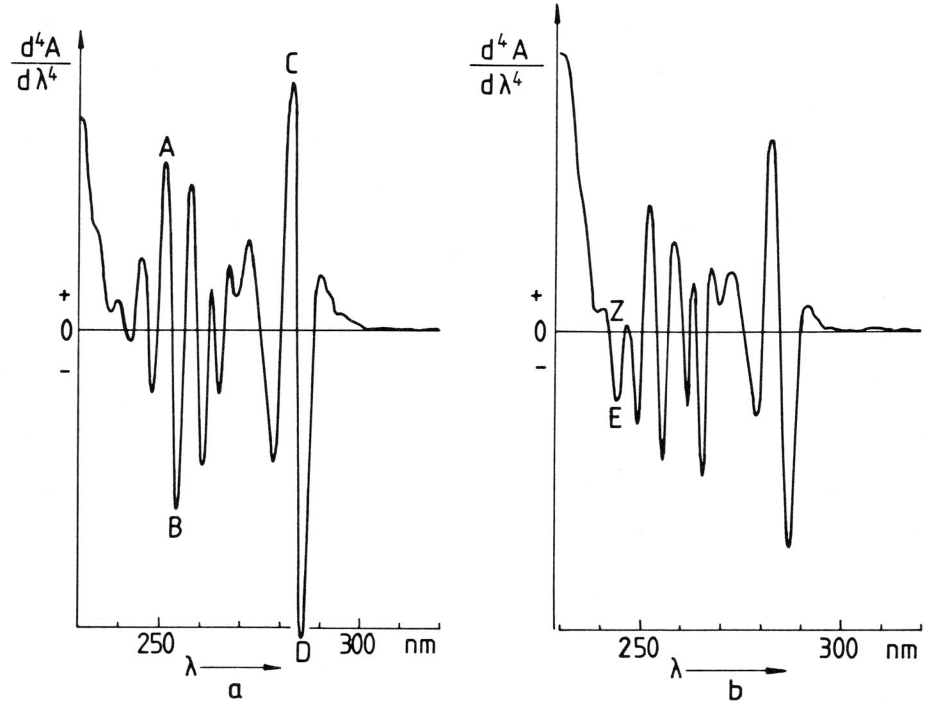

Fig. 2. 4th derivative of a three-component mixture of Phe ($1,5 \cdot 10^{-3}$ $mol \cdot l^{-1}$), Tyr ($1,5 \cdot 10^{-4}$ $mol \cdot l^{-1}$), and Trp ($1,5 \cdot 10^{-4}$ $mol \cdot l^{-1}$).

a) pH 7; $\overline{AB}$ is equivalent to the concentration of Phe, $\overline{CD}$ is equivalent to Trp (P-P evaluation). LTH 1 cm, slit 1 nm, scan rate 2 nm/s

b) pH 10; $\overline{ZE}$ is equivalent to the concentration of Tyr (Z-P evaluation).

Same conditions as given in a).

2) Peptides. Contrary to the free amino acids the quantitative determination of peptides affords an additional term, i.e. whether the UV-active Phe, Tyr and Trp is or is not at the N-terminus, C-terminus or inside the sequence. When the various spectra of the free amino acids are compared, their N-acetyl ethylesters, N-acetyl amids and the tripeptides Gly-X-Gly (X = Trp, Tyr, Phe) more or less differ. Thus the standard solutions may not always be representative for the unknown substances. This may effect an eror of about 10 to 15 %. Current investigations deal with this problem.

3) Proteins. MATUSHIMA et al.(1) applied the first order derivative absorption spectrophotometry as a tool to analyse aromatic amino acid residues in native proteins and BALESTRIERI et al. (4) used the derivatives of the second order for this purpose.

Our first studies of the possibility of estimating Phe, Tyr and Trp by higher-order spectrophotometry (HODS) were promising (3). In most cases further investigations gave good results.

A solution of the unknown protein ($1 - 3 \cdot 10^{-4}$ $mol \cdot l^{-1}$ in buffer pH 7 or pH 10) was diluted to give an absorption of about 1 A (end conc. about $0,2$ to $0,9 \cdot 10^{-4}$). Then the representative peaks in the derivatives were optimized for each amino acid and standard solutions of the amino acid esters or peptides Gly-X-Gly (X = Trp, Tyr, Phe) in various concentrations were added (Fig. 3). In a diagram the increase of the relevant peak caused by addition of the standard was plotted vs. the standard concentration.

For the estimation of Trp in a phosphate buffer pH 7 the evaluation of the d^2 leads to the best results. On the other hand d^4 was most suitable for the estimation of Tyr in H_3BO_3/KCl-NaOH buffer pH 10, d^4 or d^5 were suitable for Phe in phosphate buffer pH 7 (Tab. 2).

Recently, FELL (5) reported on his studies about extending the fourth and sixth derivatives on aromatic amino acids, enzymes and other biological materials.

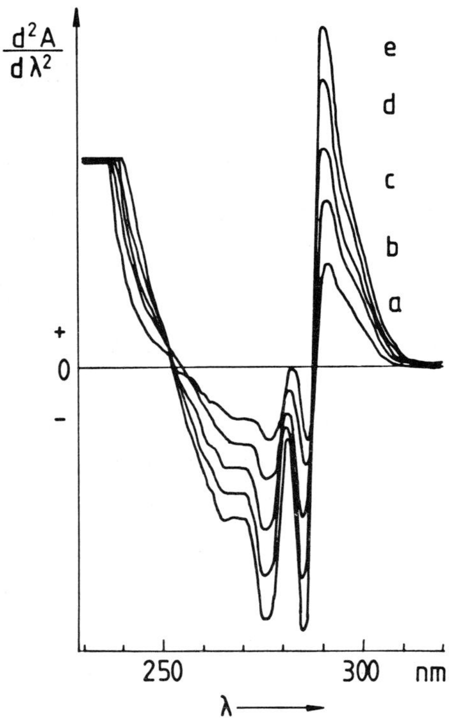

Fig. 3 d^2 spectra of chymotrypsin (a $2 \cdot 10^{-5}$ mol·l^{-1}) in the presence of various amounts of N-acetyl tryptophan ethylester, (b $0{,}5 \cdot 10^{-5}$, c $1 \cdot 10^{-5}$, d $1{,}5 \cdot 10^{-5}$, e $2 \cdot 10^{-5}$ mol·l^{-1})

Tab. 2. Amino acid estimation in proteins.

protein	d	Trp eval.	nm	exp.	th.	Phe d	eval.	λ[nm]	exp.	th.	Tyr d	eval.	nm	exp.	th.	St.
Chymotrypsin	d^2	Z-P	291	7.5	8	d^4	P-P	252/54	5.8	6	d^4	P-P	246/52	3.0	4	E
		Z-P	291	8.3	8		P-P	252/54	6.55	6		P-P	246/52	3.1	4	P
Papain	d^2	Z-P	290	5.3	5	d^4	P-P	251/54	3.2	4	d^4	Z-P	267	17,8	17-19	E
RNase	d^2	-	-	-	0	d^5	P-P	255/59	2.9	3	d^4	Z-P	264	5.6	6	P
Subtilisin	d^2	Z-P	290	3.3	3	d^5	P-P	251/54	3.0	3	d^4	Z-P	267	9.0	10	E

d derivative order — eval. evaluation — λ characteristic wave length
exp. experimentally — th. theoretically — St. standard solution
E amino acid esters — Gly-X-Gly (X Trp, Phe, Tyr)

All spectra were taken with a spectrophotometer SP8 - 100 (Philips Pye Unicam, Cambridge, UK.) and Differentiator TLB 6000 (Lucius and Baer, Geretsried, FRG).

Acknowledgement. These investigations were supported by the "Deutsche Forschungsgemeinschaft" and the "Freunde der Technischen Universität München e.V."; the authors are also obliged to Miss Franziska Reisinger for her technical help.

References

1. Matsushima, A., Inoue, Y., Shibata, K.: Anal. Biochm. 65, 362-368 (1975).
2. Ichikawa, T., Tereda, H.: Biochim. Biophys. Acta 580, 120-128 (1979)
3. Talsky, G., Mayring, L., Kreuzer H.: Angew. Chem. Int. Ed. Engl. 17, 785-799 (1978)
4. Balestrieri, C., Colonna, G., Giovane, A., Irace, G., Servillo, L.: Europ. J. Biochm. 90, 433-440 (1978).
5. Fell, A.: UV Spectrum. Group Bull. 8, 5-24 (1979)

THE APPLICATION OF A SPECTROFLUOROMETER - OPTICAL MULTICHANNEL ANALYZER SYSTEM TO THE DETERMINATION OF TRYPTOPHAN

Hans Steinhart, Petra Stützle

Institut für Ernährungsphysiologie der Technischen Universität München
D-8050 Freising-Weihenstephan

Introduction

The quantitative determination of tryptophan (Trp) especially in biological material has not been considered satisfactory until now because of the sensitivity of Trp to oxidation and photooxidation (1). This sensitivity causes problems both with the hydrolysis of biological samples and with detection methods. It has been shown that losses of Trp are dependent upon the various hydrolysis procedures and hydrolysis times normally used in the preparation of foods, feedstuffs and organs for analysis (2,3). There are also problems with the sensitivity of the commonly used methods. The aim of this work was to overcome some of the difficulties by using a spectrofluorometer - Optical Multichannel Analyzer (OMA) system. The system which we first used had some drawbacks with regard to operation mode and the influence of temperature (4). An improved system is presented below.

Instrumental

The principle of the measurement is that a simultaneous spectrum can be made within 10 - 70 ms. An EG & G Princeton Applied Research OMA-2 was adapted to a Kratos spectrofluorometer RRS 1000 for these experiments. The scheme of the system is shown in Figure 1. A double monochromator is used in the spectrofluorometer to suppress stray light. This results in a higher monochromacy of the excitation radiation which gives a more correct fluorescence spectrum. It is also important that the emission monochromator

Progress in Tryptophan and Serotonin Research

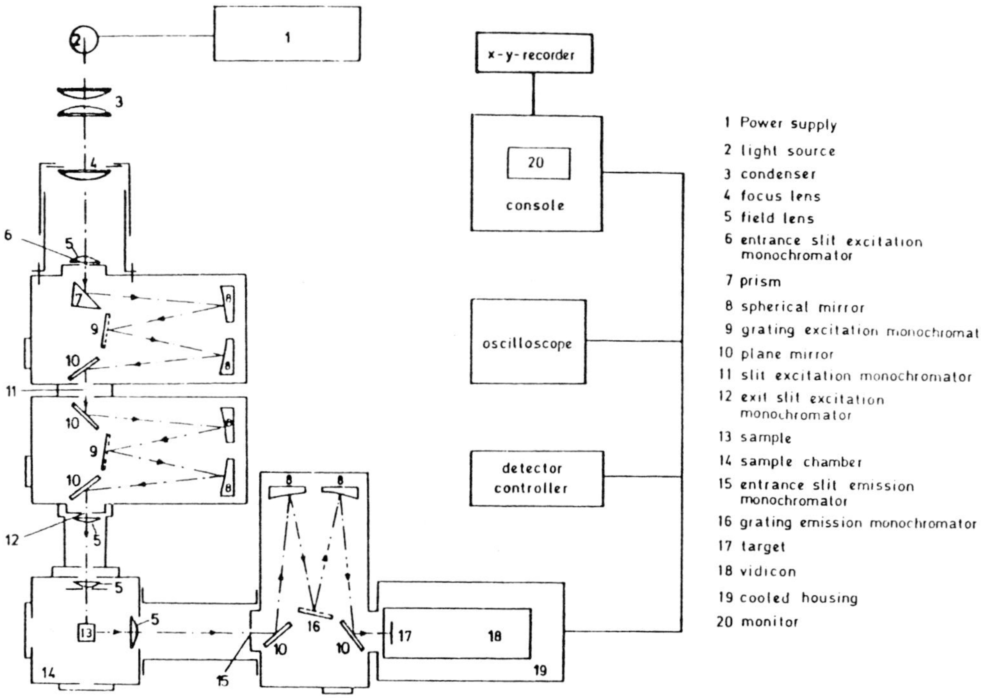

Figure 1: Scheme of the System

grating can be changed rapidly, for the number of the grooves of the grating has a strong influence on the spectral region of the detected spectrum. Contrary to a standard spectrofluorometer, this system has no exit slit at the emission monochromator, because the fluorescence signal is detected as a spectrum at the so-called target of the vidicon. This vidicon is used instead of a photomultiplier. In order to get a correct spectrum, the focal plane of the signal, which is determined by the second spherical concave mirror of the emission monochromator, has to lie on the target. The technical problem of this adaptation was solved by constructing a special metal adapter. The target, which has a size of 12.5 x 12.5 mm, is electronically divided into 500 detection channels. The arriving signals can be stored on a floppy disc and can be viewed on a monitor. In order to keep the dark current constant and low, the vidicon was kept in a cooled housing. A general report on fluorescence spectrometry with optoelectronic image detectors is given by Talmi et al. (5).

Results and Discussion

The experiments for testing the system were made with the Trp derivative norharman which has a λ_{ex}= 373 nm and a λ_{em}= 456 nm. Fluorescence peaks are normally broad peaks. Therefore it is important to get a large spectral window on the target. The spectral region depends on the number of grooves of the emission grating. If a grating with few grooves is chosen, e.g. 73 grooves/mm, a spectrum of 550 nm can be detected, and with 2360 grooves/mm, the spectrum comprises only 17 nm. For practical purposes, a grating with 147 grooves/mm was used which gave a spectrum of 275 nm. The resolution of a spectrum depends, however, not only on the emission grating, but also on the entrance slitwidth of the emission monochromator. A large spectral window shows additional peaks but with a lower resolution. In some hydrolysates, this may result in the appearance of fluorescence peaks which can influence the major peak of interest. In contrast, a small spectral window result in high resolution. This is important for the detection and recognition of peaks very close to the major peak of interest. These shoulder peaks could be readily separated using derivative spectroscopic methods. Additionally it is possible, independently of the emission grating used, to detect every shift in the spectrum. The system allows for the determination of underground compensated spectra by subtracting dark current and solvent influences from the fluorescence spectrum.

There are three operation modes with the spectrofluorometer - OMA system:

a. Real Time Mode: This mode can be compared best with the way a photomultiplier tube (PMT) works. In this mode, each target channel is addressed every 10 to 70 ms. Contrary to the PMT, where only a part of the spectrum can be seen at once, the vidicon detects the whole spectrum which is asserted by the emission grating. The wavelength axis is not, as at the PMT, a time axis.

b. Accumulation Mode: This means that the target is scanned several times, and the information of each scan is accumulated in the memory. The number of accumulation cycles is dependent on the dynamic range of the system (storage capacity). The advantage of this mode is that the signal-to-noise ratio is improved because the signal is directly proportional to the square root of the number of cycles accumulated. An improved signal-to-noise ratio increases the sensitivity of the measurement. The drawback of this mode is that it is somewhat more time-consuming than the real time mode.

c. On-Target Signal Integration: The arriving signal is integrated on the target over a period of time that can be varied according to requirements. At the end of this time, the signal is read out from the target. The signal-to-noise ratio is further increased compared to the accumulation mode, the reason being that the thermal dark charge generation is greatly reduced at low remperatures (-50°C), and the noise of the preamplifier and the electron readout beam are very small compared to the integrated signal. As a consequence, very weak signals can be recovered. The on-target signal integration is, however, more time consuming than the accumulation mode.

The following conclusions can be drawn for the quantitative determination of Trp.

a. Fluorescence spectra of Trp can be derived within a few ms. Consequently shifts and disturbing peaks can be detected immediately. Application of derivative spectroscopy would be very helpful in this regard.

b. Time dependent changes of the spectrum can be detected. Thus the system is suitable for using kinetic methods for the evaluation of Trp in alkaline or in enzymatic hydrolysates of biological material (6, 7).

c. Flexibility and high sensitivity recommend this system as suitable for the general determination of Trp.

References

1. Steinhart, H.: Landwirtsch. Forsch. 32, 63-73 (1989).
2. Steinhart, H.: Arch. Tierernährung 29, 211-220 (1979).
3. Stützle, P.: Thesis, Technische Universität München (1983).
4. Steinhart, H., Sandmann, J.: Anal. Chem. 49, 950-953 (1977).
5. Talmi, Y., Baker, D.C., Jadamec, J.R., Saner, W.A.: Anal. Chem. 50, 936A-952A (1978).
6. Steinhart, H.: Z. Tierphysiol., Tierernähr. u. Futtermittelkde. 40, 332-343 (1978).
7. Steinhart, H.: Anal. Chem. 51, 1012-1016 (1979).

PICOMOLE DETERMINATION OF TRYPTOPHAN BY DERIVATIZATION TO 9-HYDROXYMETHYL-β-CARBOLINE

Katsuji Takai, Shintaro Inoue

Department of Medical Chemistry, Kyoto University Faculty of Medicine, Kyoto 606, Japan

Introduction

The classical Denkla and Dewey method or its modification (1,2) used in the majority of laboratories is specific for tryptophan but not acceptable in sensitivity and reproducibility on assessments of plasma free tryptophan or the regional distribution of this neurotransmitter precursor with reasonable amounts of samples. In the present study, we reexamine the Pictet-Spengler reaction and developed a new method, whereby a single product, 9-hydroxymethyl-β-carboline, is quantitatively obtained with sub-picomole quantities of tryptophan. The method thus makes possible the quantification of free tryptophan with 1 μl or less of plasma ultrafiltrates or the brain tissue weighing less than 1 mg (3). Under our conditions, 9-hydroxymethyl-β-carboline-3-carbonyl peptides are formed from peptides with N-terminal tryptophan such as delta sleep inducing peptide.

Results

When tryptophan was incubated under conditions prescribed by Denkla and Dewey, the products showed heterogeneity as judged by an HPLC system developed for β-carbolines (column μ-Bondapak C_{18} (3.9 X 300 mm); solvent system, DMF/8.5% NaPB, pH 1.5/H_2O (20/20/55); monitor, fluorescence at 455 nm on excitation at 375 nm. In addition to Peak II (β-carboline), Peak I and several other peaks were eluted (Fig. 1A). On raising HCHO concentration (Fig. 1B) Peak I increased extensively and under the conditions elaborated by us (Fig. 1C), Peak I was obtained as a single product with the

Progress in Tryptophan and Serotonin Research

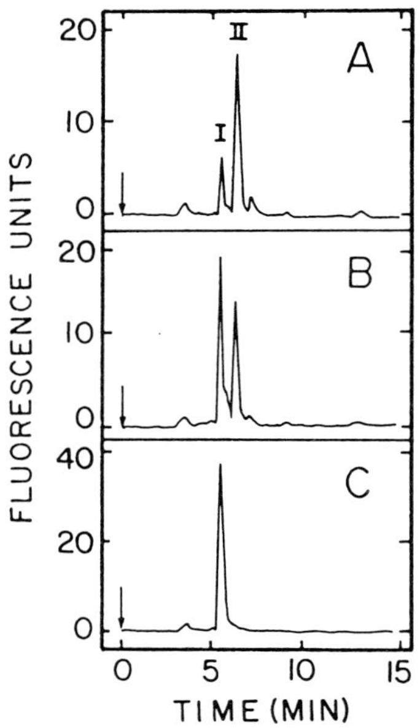

Fig. 1. HPLC of products from tryptophan obtained by Denkla/Dewey method (A), the same method except that HCHO concentration was raised (B), and by our method (C).

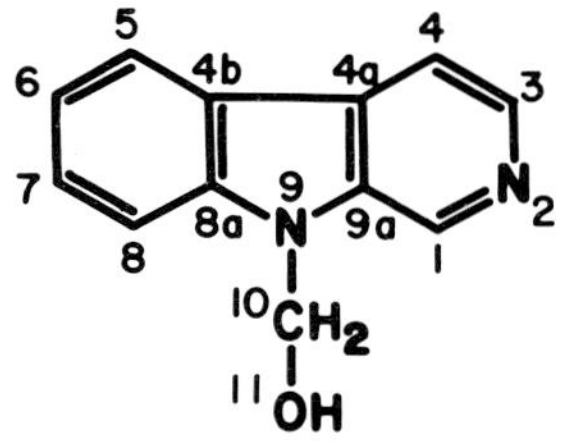

Fig. 2. Structure of 9-hydroxymethyl-β-carboline

highest yield. The crystalline Peak I was identified as 9-hydroxymethyl-β-carboline (Fig. 2) by C-13 NMR. Outline of the new method established by us is shown in Fig. 3. The raised HCHO concentration, reduced acidity, and the use of ferricyanide, extensively increased the rate of reaction (15') and, the homogeneity as well as the yield of the product. The product was assayed either by the HPLC system in the absence or presence of internal standard (5-methyl-DL-trypto-

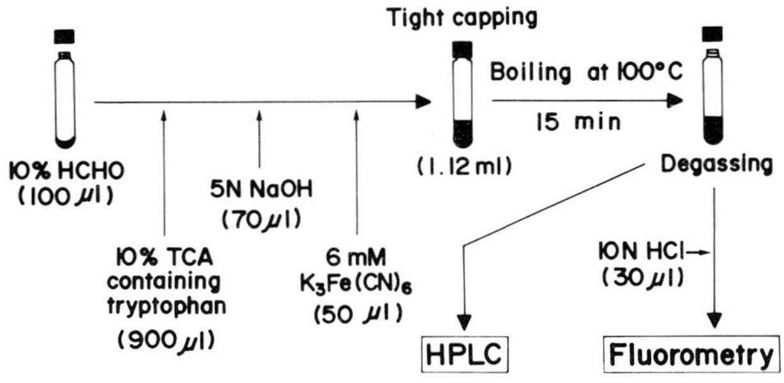

Fig. 3. Outline of the new method for the determination of tryptophan. A 200 µl aliquot or less is injected on HPLC.

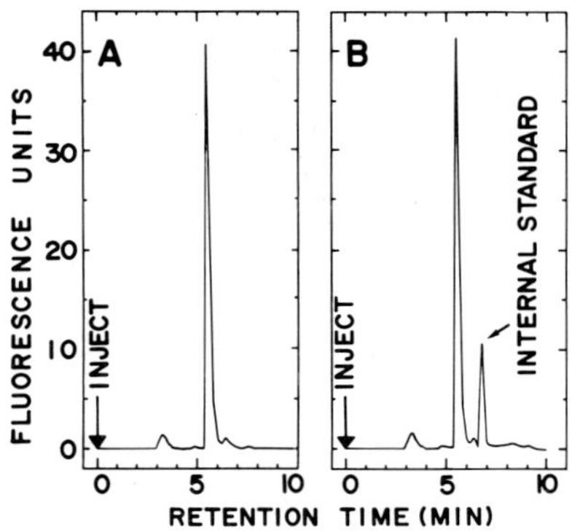

Fig. 4. HPLC patterns of the product obtained by the new method in the absence (A) or presence (B) of internal standard.

phan) (Fig. 4), or direct spectrofluorometry after acidification. Fig. 5 shows the standard curve for each of 3 methods. On HPLC assay, the curves were linear with minimum deviations down to 3 picomole/ml TCA, the amount corresponding to 0.5 picomole on injection. Thus, our method is two orders of magnitude more sensitive than Denkla and Dewey method. The use of internal standard was verified and, even on direct spectrofluorometry, 10 picomole/ml TCA was quantified with acceptable reproducibility.

The new method was successfully applied to biological samples; a minute volume of plasma ultrafiltrates gave accurate and reproducible results (Fig. 6) and the values of 29 $\pm$ 3 nmole/g were obtained with the brain homogenate corresponding to 1 mg of tissue. Foregoing results fully justified the use of our method for determination of tryptophan. The differential determination of tryptophan, tryptamine, β-carbolines, and peptides

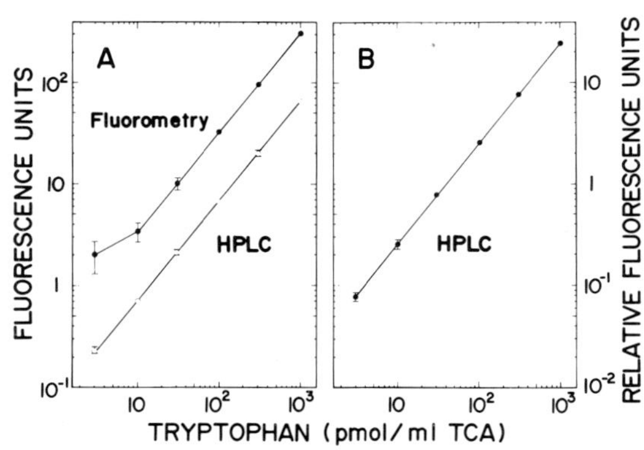

Fig. 5. Standard curves for quantification of tryptophan in the absence (A) or presence (B) of internal standard.

Sample	Total Trp (μM)	Filtrate (μl)	Free Trp (μM)	%Free Trp	Mean ± SD Free Trp (μM)	Mean ± SD %Free Trp
KT	65.4	4.0	11.0	16.8	10.8 ± 0.21	16.6 ± 0.31
		6.0	11.1	17.0		
		6.0	10.8	16.5		
		7.0	10.8	16.5		
		13.0	10.5	16.1		
RY	56.7	4.0	8.88	15.7	8.99 ± 0.28	15.9 ± 0.50
		4.0	9.14	16.1		
		7.0	9.36	16.5		
		8.0	9.05	16.0		
		8.0	8.52	15.0		
YU	77.1	3.0	12.2	15.8	10.8 ± 1.28	14.0 ± 1.65
		5.0	9.87	12.8		
		6.0	10.7	13.9		
		8.0	8.97	11.6		
		10.0	12.2	15.8		

Fig. 6. Application of the new method to human plasma

with N-terminal tryptophan that is converted to the corresponding β-carbonyl peptides (Fig. 7) under our conditions, is now feasible.

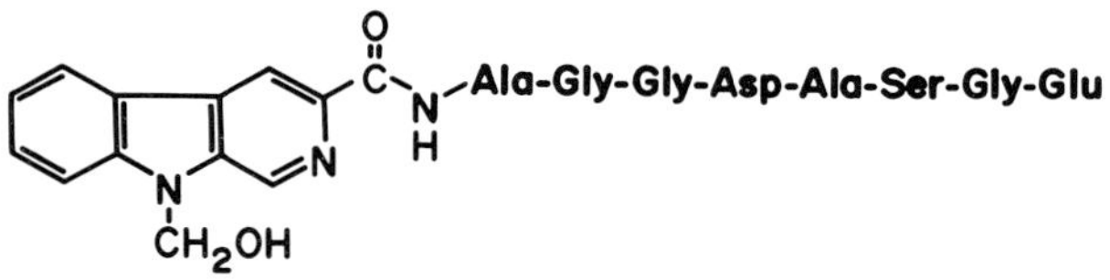

Fig. 7. Structure of the product from delta sleep inducing peptide (DSIP); 9-hydroxymethyl-β-carboline-3-carbonyl DSIP

References

1. Denkla, D. W., Dewey, H. K.: J. Lab. Clin. Med. 69, 160-169 (1967).

2. Bloxam, D. L., Warren, W. H.: Anal. Biochem. 60, 621-625 (1974).

3. Inoue, S., Tokuyama, T., Takai, K.: Anal. Biochem. in press (1983).

DETERMINATION OF TRYPTOPHAN IN FOOD AFTER ALKALINE HYDROLYSIS: AN EVALUATION

Henrik K. Nielsen and Richard F. Hurrell
Nestlé Products Technical Assistance Co. Ltd.
CH-1814 La Tour-de-Peilz, Switzerland

Introduction

Alkaline hydrolysis followed by column chromatography seems to be the most promising method for tryptophan analysis of foods. Based on the conditions of Hugli and Moore (1) and using goat's casein ('cold' or labelled with tritiated tryptophan) we have investigated the influence of several hydrolysis parameters on the recovery of tryptophan. Reverse phase HPLC with a phosphate buffer - acetonitrile gradient was used to measure tryptophan and internal standards.

Experimental

Hydrolysis. Approximately 50 mg of goat's casein ('cold' or labelled with ^{3}H-tryptophan) was weighed into a polypropylene liner in a glass tube. Alkali and internal standards (4 μmoles) were added to give a final volume of 4 ml and an alkali concentration of 4.2 M. One drop of 1-octanol was added to prevent foaming. Partially hydrolyzed potato starch (0.17 g) (1) or thiodiglycol (0.12 g) were added and a vacuum was drawn. For the high vacuum (pressure 0.02 mbar) an oil pump was used; for lower vacuums (pressure 15 - 550 mbar) a tap-water pump. The tubes were sealed and hydrolysis was carried out in an oven at 110°C for 20 h.

HPLC. Tryptophan and internal standards were measured by

Progress in Tryptophan and Serotonin Research

reverse phase high performance liquid chromatography using a C18 column and fluorimetric detection with excitation at 285 nm and emission measured above 320 nm (cut-off filter). Solvent A was 0.02 M phosphate buffer pH 3.25, and solvent B was 30% acetonitrile in solvent A. The flow rate was 1.5 ml/min with a gradient of 15% to 59% solvent B in 28 min.

Results and Discussion

Choice of alkali. Hydrolyzing food proteins in LiOH was recently reported to give higher tryptophan values than hydrolyzing in NaOH (2). Our results however (Table 1) show no significant differences in the recoveries of protein-bound ^{3}H-tryptophan or of the internal standards between the two alkalis and we see no advantage in changing from the normally used NaOH.

Choice of protective agent. Conflicting results exist in the literature (1, 3, 4) as to which protective agent best reduces the degradation of tryptophan during alkaline hydrolysis. We found (Table 1) that the addition of partially

Table 1. Recoveries of protein-bound ^{3}H-tryptophan, 5-methyltryptophan and α-methyltryptophan after hydrolysis in NaOH or LiOH with addition of partially hydrolyzed starch or thiodiglycol.

PROTECTIVE AGENT	Trp/int. std.	NaOH	LiOH
Partially hydrolyzed starch	^{3}H-Trp	93.0	93.7
	5-Me-Trp	92.1	90.3
	α-Me-Trp	97.9	97.6
Thiodiglycol	^{3}H-Trp	89.2	89.6
	5-Me-Trp	89.4	88.1
	α-Me-Trp	94.2	95.1

All values are an average of 3 determinations made using the high vacuum system.

hydrolyzed starch, in all instances, had a larger protective effect than did thiodiglycol.

Choice of internal standard. 5-methyltryptophan was introduced as an internal standard (4) under the assumption that its degradation was similar to that of protein-bound tryptophan. In Table 1, the recovery of protein-bound tryptophan (from goat's casein labelled with tritiated tryptophan) can be seen to be on average 1.4% higher than the recovery of 5-methyltryptophan and 4.8% lower than the recovery of α-methyltryptophan. Both these differences were significant ($P < 0.05$). The results in Fig. 1 on hydrolyses made under different levels of vacuum also show that 5-methyltryptophan will tend to slightly overcorrect for tryptophan losses, but indicate that it will satisfactorily and, on the whole, better correct for tryptophan losses than α-methyltryptophan.

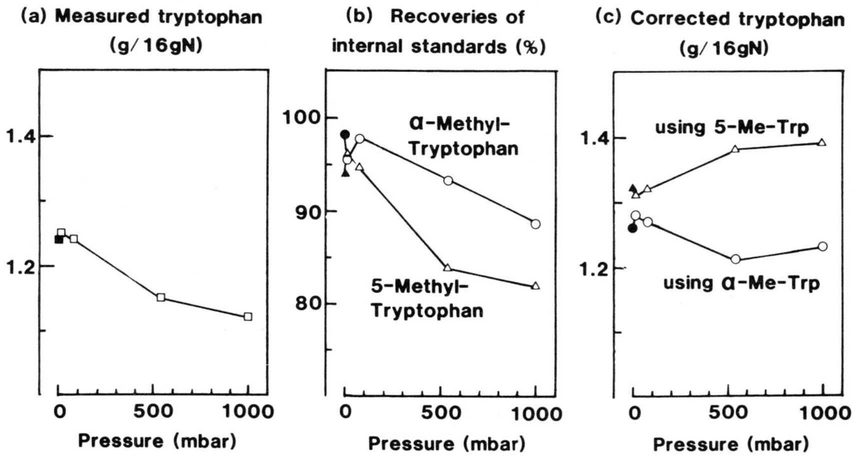

Fig 1. Influence of oxygen on tryptophan values and internal standard recoveries. Filled symbols represent high vacuum system and open symbols various levels of tap-water pump vacuum. Hydrolyses were made in NaOH with added partially hydrolyzed starch.

Influence of oxygen. The procedure of Hugli and Moore (1) includes the drawing of a high vacuum (0.07 mbar) before hydrolysis. This is time-consuming and a tap-water pump vacuum would be more convenient. Fig. 1a shows the measured tryptophan level in goat's casein at high vacuum (pressure 0.02 mbar), at different levels of water pump vacuum (pressure 15 - 550 mbar) and with no vacuum. Fig. 1b shows the corresponding recoveries of the internal standards and Fig. 1c the corrected tryptophan values. There were no significant differences between values obtained at 0.02 mbar and 15 mbar. The high vacuum system can therefore be replaced by a tap-water pump vacuum system.

Use of a 'scavenger'. It has been shown that adding relatively large amounts of substances with a chemical structure similar to tryptophan 'dilute' the losses of tryptophan during acid hydrolysis (5). However, we found that indolyl-3-propionic acid (20 or 100 µmole) or 5-methyltryptophan (40 or 200 µmole) added as 'scavengers' prior to alkaline hydrolysis caused no improvement in the tryptophan recovery.

Degradation products. Chromatograms of tritiated goat's casein hydrolysates showed that many tryptophan degradation products were formed in small amounts during alkaline hydrolysis.

References

1. Hugli, T.E., Moore, S.: J. Biol. Chem. 247, 2828-2834 (1972).
2. Lucas, B., Sotelo, A.: Anal. Biochem. 109, 192-197 (1980).
3. Oelshlegel, F.J., Schroeder, J.R., Stahlmann, M.A.: Anal. Biochem. 34, 331-337 (1970).
4. Wilkinson, M., Iacobucci, G.A., Myers, D.V.: Anal. Biochem. 70, 470-478 (1976).
5. Gruen, L.C., Nicholls, P.W.: Anal. Biochem. 47, 348-355 (1972).

COLORIMETRIC DETERMINATION OF TRYPTOPHAN IN CORN-COB-MIX (CCM)

Erika Votisky
Research Center of Animal Production, Research Institute for Animal Nutrition Herceghalom, Hungary

Determination of tryptophan in proteins is very problematical because a loss of tryptophan is observed during hydrolysis of the protein. There are many publications about the hydrolysis (1,2,3,4,5,6,7) where the authors reported on the reduction or the increase of the decomposition. Several steps have to be carried out in these methods and they are time consuming, too. Tryptophan determination in unhydrolysed protein appears to be more accurate (8,9,10). This method has also been applied to meat samples (10).

Experiment

CCM was collected at the time of making silage and after a period of conservation. Samples were predried at 65°C and ground to dust (size of the particles below 0,5 mm). This fine dust was used for colour reaction.

Composition of CCM:

Dry Matter	90,0 - 92,0 %
Crude Protein	7,0 - 8,0 %
Crude fiber	4,8 - 5,0 %
Ash	1,4 - 1,7 %
Lysine	0,22- 0,28 %

Reagents: p-dimethylamino - benzaldehyde(DAB), $NaNO_2$, 19 n H_2SO_4

Progress in Tryptophan and Serotonin Research

Method

Tryptophan reacts with DAB to produce a condensation product which develops a well defined blue colour with $NaNO_2$.
A quantity to 40-50 mg of CCM was given into 2 test tubes (thick walled). Then 20 ml of the reagent (300 mg DAB solved in 100 ml 19 n H_2SO_4) freshly prepared, were added to tube "a" and 20 ml 19 n H_2SO_4 to tube "b". Both tubes were kept in dark for 16 hours for colour development. Then 0,1 ml of $NaNO_2$ solution (0,045 % in water) was added to "a" and the tube kept for half an hour in dark again.
After repeated shaking the mixture was filtered over glass filterand, using blind test, extinction values (E) were measured at 590 µm. The content of tube "b" was also filtered and measured in the same manner.
A calibration curve was developed with L tryptophan.
The content of tryptophan was calculated by using extinction values ($E_{"a"} - E_{"b"} = E$) on the basis of the curve.
For comparison tryptophan content of all the samples was determined after hydrolysis of the protein (400 mg) according to Eggum (4).

Results

The Table shows the mean values obtained by direct measuring the tryptophan content of the samples according to the method described in this paper (I) and after hydrolyses (II) according to Eggum (4).

Tryptophan content of CMM samples [%±SD]

N	I	II
30	0.111(±0.012)	0.099 (±0.007)

The difference proved to be statistically significant at 1 % level. Tendency of the values was the same. This proves that the P=1 % level existing between the data pairs (r=0,79) was significant.

Summary

We determined the tryptophan content of the CCM containing a small amount of tryptophan. We found reliable results with our method that is based on a direct colour reaction measurement without the hydrolysis of the protein. On the basis of our results which are in a good agreement with those obtained with hydrolysis (and using amino acid analyzer) this method can be applied for rapid determination of tryptophan in corn.

References

1. Spies, J.R., Chambers, D.C.: Anal. Chem. 21, 1249 (1949).
2. Penke, B., Ferenczi, R., Kovács, K.K.: Anal. Biochem. 60, 45-50 (1974).
3. Miller, E.L.: J.Sci. Fd.Agric. 18, 381 (1967).
4. Bjorn, O. Eggum: Acta Agriculture Scandinavica, 18, 127-131 (1968).
5. Spies, J.R.: Analyt. Chem. 39, 1422 (1967).
6. Votisky, E.: Proceeding Symposium on amino acids, Brno, Czehoslovaki. D-7 1-7 (1973).
7. Tyczkowska, K., Slawinski, P.: Tagungsberichten Akad. Landw. DDR, Berlin, (1974).
8. Concon, J.M.: Anal. Biochem., 67, 206 (1975).
9. Spies, J.R., Chambers, D.C.: Anal. Chem. 20, 30 (1948).
10. Rékásiné Szombati É., Orlando Presa Caballero, Mihályiné. Kengyel V., Dr. Körmendy, L.: Élelmiszervizsgálati Közlemények, XXIII.k. 17-26 (1978).

COMPARISON OF TRYPTOPHAN ASSAYS FOR FOOD PROTEINS

Mendel Friedman, Carol E. Levin, Amy T. Noma, William C. Montague, Jr. and James C. Zahnley

Western Regional Research Center, Agricultural Research Service, U.S. Department of Agriculture, Berkeley, California 94710, U.S.A.

The amino acid tryptophan is of paramount importance in nutrition, physiology, and food safety. It not only is nutritionally essential for normal growth, but also serves as the in vivo precursor for vitamin B_6, serotonin, and numerous other metabolites with presently unknown functions (1). In vitro, tryptophan can be nitrosated to N-nitrosotryptophan (2) and transformed to a carcinogenic indole derivative (3). The D-isomer of tryptophan and an oxidation product, N-formylkynurenine, are potential non-nutritive sweetening agents (4, 5). Because of these and other functions, analytical methods for tryptophan have been extensively studied, but without complete success, primarily because of interference by carbohydrates and because the indole ring of tryptophan is degraded under acid conditions used for protein hydrolysis (6,8). Since tryptophan cannot be analyzed along with the other amino acids by standard amino acid analysis techniques, numerous attempts have been made to devise modified hydrolytic methods that do not destroy tryptophan. Both hydrolysis by HCl in the presence of thioglycollic acid as well as organic acid hydrolysis by mercaptoethanesulfonic, methanesulfonic, and p-toluenesulfonic acids and basic hydrolysis by NaOH and $Ba(OH)_2$ have been proposed to minimize loss of trytophan (9-19). To assess the utility of the reported methods, we evaluated six hydrolytic and one nonhydrolytic, acid-ninhydrin method with

Progress in Tryptophan and Serotonin Research

three proteins and a number of carbohydrate-containing food products.

For the acid-ninhydrin method, a standard curve for lysozyme was used as a reference to calculate the tryptophan content of the other materials tested. The results (Tables 1 and 2) show that for carbohydrate-containing foods: (a) the thioglycollic and organic acid methods appear ineffective; (b) basic hydrolysis by NaOH and $Ba(OH)_2$ appears preferable over hydrolysis in organic acids; and (c) the acid-ninhydrin method merits further study to assess its general applicability to complex foods because it gave good reproducibility with a wide range of commonly used food products.

Table 1. Tryptophan content of proteins and foods. (Values in g/16 g N).[a,b]

Analytical method	Lysozyme	Casein	Milk powder	Soybean protein	Soybean flour
Thioglycollic acid (9)	5.84	1.27	Not detected	0.33	Not detected
Mercaptoethansulfonic acid (10)	6.69	1.26	0.09	1.24	0.31
Methanesulfonic acid (11)	5.72	0.71	Not detected	0.65	Not detected
p-Toluenesulfonic acid (12)	6.64	1.07	Not detected	1.01	Not detected
NaOH (13)	5.65	1.10	1.26	1.01	1.07
$Ba(OH)_2$, amino acid analysis (15, 16)		1.30	1.25	1.81	0.81
$Ba(OH)_2$, spectrophotometric (16)	6.54	1.25	1.03	0.75	1.12
Acid ninhydrin (17-19)[c]	7.66	1.70	1.73	1.36	1.91

[a]Values are averages from three separate determinations except for a single assay by the acid ninhydrin procedure.

[b]Nitrogen content (%): Lysozyme, 16.2; casein, 13.6; milk powder, 5.32; soybean protein, 14.0; soybean flour, 8.32.

[c]The tryptophan content of lysozyme was calculated to be 7.66g/16g N from its known amino acid sequence and determined nitrogen content.

Observed values were corrected for contributions of tyrosine to the absorbance at 385 nm by the following formula (Zahnley and Davis, 1973):

$$\frac{\text{TRP (observed)}}{\text{TRP (expected)}} = 1.013 + 0.0335 \left(\frac{\text{TYR}}{\text{TRP}}\right)$$

Table 2. Tryptophan content of flours. (Values in g/16 g nitrogen)

PROTEIN SOURCE	N (%)	ACID NINHYDRIN METHOD[a] FLOUR	ACID NINHYDRIN METHOD[a] PROTEIN EXTRACT OF FLOUR[b]	$Ba(OH)_2$ HYDROLYSIS[c]
Barley	1.26	1.553 ± 0.1145 (5)[d]	1.271 ± 0.0476 (3)	1.69; 1.83
Lima beans	3.32	1.422 ± 0.0666 (3)	1.305 ± 0.0627 (4)	1.62; 1.69
Beef, minced	13.57	1.250 ± 0.0401 (3)	1.402 ± 0.0918 (4)	1.58; 1.61
Corn, high lysine	1.59	1.795 ± 0.2416 (8)	1.537 ± 0.0339 (3)	1.46; 1.44
Corn	1.46	1.853 ± 0.1714 (7)	1.277 ± 0.0784 (4)	1.12; 0.98
Cottonseed	10.01	1.373 ± 0.0760 (4)	- -	1.45; 1.56
Rice	0.98	1.724 ± 0.1087 (5)	1.371 ± 0.0645 (3)	1.73; 1.82
Wheat	2.22	1.557 ± 0.1412 (5)	1.115 ± 0.0155 (2)	1.54; 1.58
Soybeans	8.32	1.430 ± 0.0682 (5)	1.425 ± 0.1024 (4)	1.12; 1.13
Oats	2.56	1.682 ± 0.0228 (3)	1.329 ± 0.1994 (3)	

[a] Determined by a modified acid ninhydrin methd.

[b] Proteins were extracted by the procedure described in: Concon, J.C., Chemical determination of critical amino acids in cereal grains and other foodstuffs. In: "Protein Nutritional Quality of Foods and Feeds", M. Friedman, Ed., Marcel Dekker, New York, 1975. Part 1, pp. 311-379.

[c] The basic hydrolyzate was treated with p-dimethylaminobenzaldehyde and the derivatized tryptophan was determined spectrophotometrically (Miller, 1967).

[d] Values in parentheses are number of separate determinations.

[e] Values from duplicate experiments.

Listed values are preliminary and are not corrected for tyrosine absorption. Values for protein extracts are not corrected for incomplete extraction of all of the proteins.

REFERENCES

1. Brown, R. R. The tryptophan load test as an index of vitamin-B_6 nutrition. In "Methods in Vitamin-B_6 Nutrition", J. E. Leklem and R. D. Reynolds, Eds., Plenum Press, New York, 1981, pp. 321-340.

2. Nakai, H., Cassens, R. G., Greaser, M. L. and Woolford, G. Significance of the reaction of nitrite with tryptophan. J. Food Sci. 43: 1857 (1978).

3. Hosaka, S., Matsushima, T., Hirono, I. and Sugimura, T. Carcinogenic activity of a pyrolysis product of tryptophan. Cancer Lett. 13: 23 (1981).

4. Yamada, S., Yamamoto, M., Hong, C. and Chibata, I. Preparation of optically active 6-chlorotryptophan and tryptophan. J. Agric. Food Chem. 23: 653 (1975).

5. Finley, J. W. and Friedman, M. New sweetening agents: N-formyl- and N-acetylkynurenines. J. Agric. Food Chem. 21: 33 (1973).

6. Friedman, M. and Finley, J. W. Methods of tryptophan analysis. J. Agric. Food Chem. 19: 626 (1971).

7. Finley, J. W., Johnston, P. H. and Friedman, M. Improved method for tryptophan analysis of proteins in the presence of carbohydrates. In "Protein Nutritional Quality of Foods and Feeds", M. Friedman, Ed., Marcel Dekker, New York, 1975. Part 1, pp. 553-562.

8. Sarwar, G., Christensen, D. A., Finlayson, A. J., Friedman M., Hackler, L. R., Mackenzie, S. L., Pellett, P. L. and Tkachuk, R. Inter- and intra-laboratory variation in amino acid analysis of food proteins. J. Food Science, 48: 526-531 (1983).

9. Matsubara, H. and Sasaki, R. M. High recovery of tryptophan from acid hydrolysates of proteins. Biochem. Biophys. Res. Commun. 35: 175 (1969).

10. Penke, B., Ferenczi, and Kovacs, K. A new acid hydrolysis method for determining tryptophan in peptides and proteins. Anal. Biochem. 60: 45 (1974).

11. Simpson, R. J., Neuberger, M. R. and Liu, T. Y. Complete amino acid analysis of proteins from a single hydrolysate. J. Biol. Chem. 251: 1936 (1976).

12. Liu, T. Y. and Chang, Y. H. Hydrolysis of proteins with toluenesulfonic acid. J. Biol. Chem. 246: 2842 (1971).

13. Hugli, T. E. and Moore, S. Determination of the tryptophan content of proteins by ion-exchange chromatography of alkaline hydrolysates. J. Biol. Chem. 247: 2828 (1972).

14. Lucas, B. and Sotelo, A. Effect of different alkalis, temperature, and hydrolysis time on tryptophan determination of pure proteins and of foods. Anal. Biochemistry, 109: 192-197 (1980).

15. Knox, R., Kohler, G. O., Palter, R. and Walker, H. G. Determination of tryptophan in feeds. Anal. Biochem. 36: 136 (1970).

16. Miller, E. L. Determination of the tryptophan content of feedingstuffs with particular reference to cereals. J. Sci. Food Agric. 18: 381 (1967).

17. Gaitonde, M. K. and Dovey, T. A rapid and direct method for the quantitative determination of trytophan in the intact protein. Biochem. J. 117: 907 (1970).

18. Zahnley, J. C. and Davis, J. G. Effect of high tyrosine content on the determination of tryptophan in proteins by the acidic ninhydrin method. Biochem. J. 135: 59 (1973).

19. Sodek, L., Veccia, P.T.D. and Lima, M.L.G.P. Rapid determination of tryptophan in beans by the acid ninhydrin method. J. Agric. Food Chem. 23: 1147 (1975).

Workshop:
Analysis of Tryptophan

DISCUSSION AT THE WORKSHOP ON "ANALYSIS OF TRYPTOPHAN"

Coordinated by Mendel Friedman

United States Department of Agriculture, Agricultural Research Service,
Western Regional Research Center
Berkeley, California 94710 USA

I. ANALYSIS OF TRYPTOPHAN IN PROTEINS AND PEPTIDES

H. K. Nielsen (Nestle, Switzerland). Dr. Friedman, I have several questions about your presentation. First, your acid ninhydrin method seems very interesting and I would like to ask what is the specificity of the method for tryptophan in foods? Do tryptophan derivatives and degradation products also give color formation? Do you measure absorbance on a solution, a filtrate, or a suspension? What were the recoveries of the alkaline protein extraction procedure which you described?

M. Friedman (USDA, Berkeley) In response to Dr. Nielsen's comments: (1) First, let me say that we did not originate the acid ninhydrin method for tryptophan (see our paper in this volume for appropriate references on this point). Our main objective is to assess the utility of the method for complex foods. (2) We do not know whether the method is specific for tryptophan. (3) We measure absorbance of a clear solution after centrifugation and filtration. (4) We estimate, on the basis of nitrogen analyses, that the recovery in the alkaline protein extraction is nearly quantitative.

H. K. Nielsen. In our opinion, determination after alkaline hydrolysis should be done by ion-exchange chromatography or reverse phase HPLC, in order to assure that only tryptophan is determined. This would also

Progress in Tryptophan and Serotonin Research

permit the use of an internal standard, e.g. 5-methyltryptophan, whose recovery during alkaline hydrolysis we have shown to agree with the recovery of protein-bound tryptophan (see this volume and comment by Dr. Baumann below). We were not able to obtain reproducible recoveries of added tryptophan when $Ba(OH)_2$ was used to hydrolyze the proteins, presumably due to adsorption of tryptophan to the precipitated $BaSO_4$. It is also noteworthy that we found many tryptophan degradation products in small amounts during NaOH hydrolysis of goat's casein containing tritium-labeled tryptophan. I would also like to point out that if an attempt is made to transform tryptophan to an acid-stable derivative, the specificity of the reaction as well as the presence of the derivative prior to analysis will have to be investigated.

Finally, I would like to ask Dr. Friedman whether he can explain the observed differences in tryptophan recovery following NaOH and $Ba(OH)_2$ hydrolysis, respectively.

M. Friedman. Very recently we found that the results are more comparable if we analyze for tryptophan immediately after hydrolysis by NaOH. Evidently, if this is not done, some of the tryptophan is degraded on standing, as was already alluded to by Dr. Nielsen.

H. Steinhart (Munich). We find that the recovery of tryptophan during barium hydroxide hydrolysis of feedstuffs is only 85 to 95% of the theoretical value. Since the recovery differs with different materials, this is not a reliable method. We have to find a new hydrolysis procedure.

S. Bech-Andersen (Copenhagen). In our experience, estimation of tryptophan by fluorescence spectroscopy after HPLC chromatography is a useful way for measuring tryptophan in feedstuffs after alkaline hydrolysis. Digesting feeds containing heavy cell-walls would probably lead to results somewhere between total and available tryptophan.

F. Gelpi (Barcelona). Our experience with non-tryptophan-containing peptides leads me to suggest that separation and collection of tryptophan-containing peptides by HPLC followed by radioimmunoassay might be the basis for a tryptophan assay.

M. Friedman. Would you need a different antibody for each peptide?

E. Gelpi. Not necessarily, because immunochemical cross-reaction might be an advantage here.

H. Steinhart. A key question is, do we need complete enzymatic digests or does a peptide digest followed by fluorescence or derivative spectroscopy, described by Dr. Talsky (see below), offer the potential for useful analytical methodology.

The use of fluorescence to measure tryptophan directly has at least two disadvantages: (a) photo-oxidation of tryptophan that occurs under the impact of light affects the results; (b) the method is difficult to apply to insoluble proteins present in feedstuffs, organs, etc. To avoid alkaline hydrolysis, it may be useful to prepare an enzymatic peptide digest followed by fluorescence assay of the tryptophan in the peptides.

M. Friedman: Would the fluorescence intensities of the peptides in the digest differ from each other and from the corresponding value for free tryptophan?

A. Fontana (Padua). In addition to ultraviolet and derivative spectroscopy, fluorescence methods can be used to determine tryptophan in intact proteins. The intrinsic fluorescence emmision of native proteins is due to tyrosine and tryptophan. Upon excitation at 295 nm, where tyrosine does not absorb, the protein fluorescence emission is due only to tryptophan with emission at 330-340 nm. Unfortunately, this fluorescence cannot be used to quantify tryptophan content because fluorescence properties of native proteins (quantum yield, maximum of emission) depend on the microenvironment of tryptophan residues within the protein. For example, buried residues upon excitation at 295 nm emit at 330-340 nm and exposed residues at ~ 350 nm. One way to circumvent this problem would be to measure the fluorescence emission intensity of tryptophan-containing proteins under strong denaturing conditions, such as solutions containing 6M guanidine hydrochloride. Since, under these solvent conditions, proteins are fully unfolded, all tryptophan residues are in full contact with the aqueous solvent medium and are, therefore, expected to fluoresce "normally," as the model compound N-acetyltryptophanamide.

J. J. Toulme (Paris). I would like to comment on the analysis of the tryptophan content of proteins using enzymatic digestion followed by fluorometric measurements. I have used such a method to determine the tryptophan content of gene 32 protein from bacteriophage T4. The protein was extensively digested by a cocktail of proteases (trypsins, α-chymotrypsin, pronase). The fluorescence of the digests was measured under denaturing conditions (8M urea). Increasing amounts of tryptophan were then added to the sample and the fluorescence measured after each addition according to the method of Sasaki et al. (Anal. Biochem. 65, 396, 1975).

Extrapolation of the linear plot of fluorescence intensity versus tryptophan concentration makes it possible to estimate the tryptophan concentration in the protein digests, as illustrated in Figure 1.

The major problem with this method lies in the choice of an internal standard since the fluorescence of the indole ring depends on its environment. The fluorescence emission of tryptophan is slightly different from that of N-acetyltryptophanamide in 8M urea. This means that a free residue has a different emission from the corresponding residue in a peptide or protein chain. For instance, the peptide lysyl-tryptophyl-α-lysine has a lower fluorescence quantum yield than free tryptophan in 8M urea. Therefore, tryptophan may be used as a standard only when digestion is complete. Moreover, partial digestion can lead to peptides that still have secondary structure even in 8M urea. In the case of the gene 32 protein, this leads to an error of about 15% of the actual tryptophan concentration. The accuracy of the method will probably vary from protein to protein and depend on the nature of the neighboring amino acid residues near tryptophan in the primary sequence. (B. Milligan and L. A. Holt offer a relevant discussion of the factors which influence the complete enzymatic hydrolysis of proteins in:
Protein Crosslinking: Nutritional and Medical Consequences, M. Friedman (Ed.), Plenum Press, New York, 1977, pp. 277-284).

M. H. Joseph (Harrow). In the method for analysis for kynurenine that I reported on, we find additional kynurenine appears to be formed from free tryptophan during alkaline cleavage. This may be a source of tryptophan loss also during barium hydroxyide hydrolysis of proteins. However, the formation of additional kynurenine, and presumably the loss of tryptophan

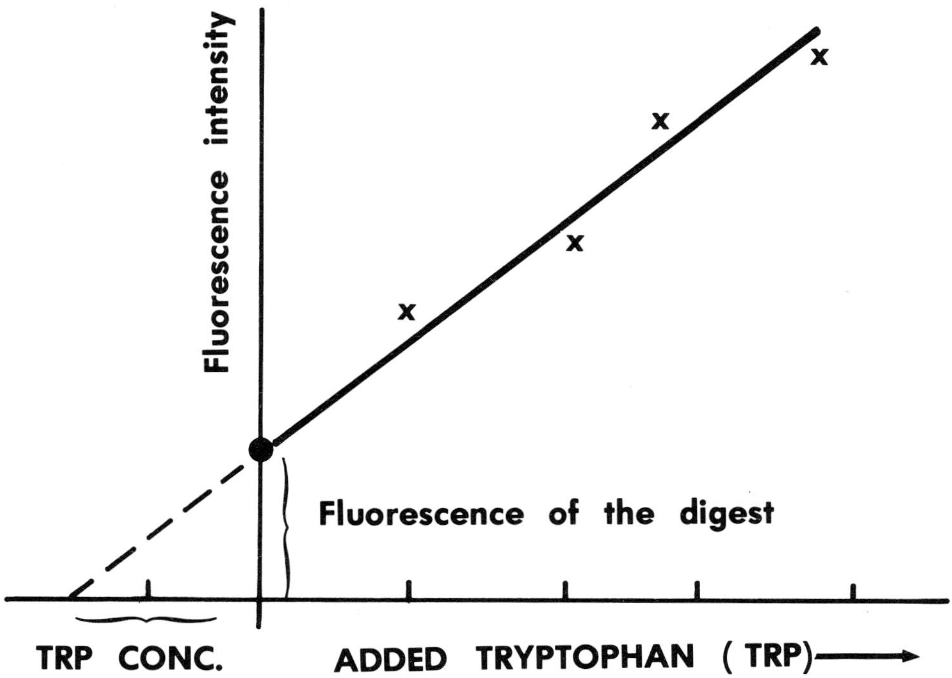

Figure 1. See text.

can be minimized by adding the chelating agent-"Tiron" (1,2 dihydroxy benzene-3,4 di-sulfonic acid) - to the reaction mixture (Joseph, J. Chromatog. 146, 33-41 (1978). It might be useful to include this during peptide hydrolysis.

In view of the low levels of tryptophan encountered, it might also be useful to determine tryptophan separately from the other amino acids. It can be simply determined by direct fluorscence, after removal of Ba ions as carbonate (Duggan and Udenfriend, JBC 223, 313 (1965)). Alternatively, norharman, formed by the action of formaldehyde on tryptophan by heating under acid conditions in the presence of an oxidizing agent (Denckla and Dewey, J. Lab. Clin. Med. 69, 160 (1967), might be a useful, acid-stable derivative for analytical purposes. It should be possible to analyse norhaman rapidly by either fluorescence or electrochemistry. (See also comment below by Dr. Baumann about analysis for norharman).

As to whether peptides could also be measured by the technique I had described for amino acids, we have limited experience with measuring orthopthaladehyde (OPT) derivatives of peptides electrochemically. We find that N-terminal tryptophyl peptide derivatives are less electrochemically active than the free tryptophan derivative. However, electrochemistry seems to have a major advantage over fluorescence for these peptide derivatives: fluorescence intensity is rapidly lost as chain length increases from di- to tri- and higher peptides. However, in electrochemistry, the intensity does not appear to fall off so rapidly with increase in peptide length.

II. ACID DEGRADATION PRODUCTS OF TRYPTOPHAN

M. Friedman. Can anyone comment about the nature of the acid-degradation product(s) of tryptophan?

A. Fontana. I think it is relevant to comment on the fate of tryptophan during the usual conditions of acid hydrolysis of tryptophan-containing proteins and peptides (6N HCl, 110°C, 22 hrs). Thus, it is commonly stated that tryptophan is "destroyed" under these conditions. However, it is also generally known that an "unknown" compound derived from tryptophan is eluted just prior to tryptophan from the short column of the automatic clinical two-column amino acid analyzer, (Liu, T-Y. and Chang, Y-H. (1971) J. Biol Chem. 246, 2842-2848; Liu, T-Y. (1972) Methods Enzymol. 25. 44-55). This compound has been clearly identified as oxindolylalanine (Figure 2). Its formation results from the interaction of sulfur amino acids with HCl. Methionine sulfoxide can participate in a series of equilibrium reactions in the presence of concentrated aqueous HCl, leading to the formation of methionine sulfonium chloride (R_2S^+Cl). This compound acts as an oxidizing/halogenating agent for the indole nucleus of tryptophan, similar to the dimethyl sulfoxide/HCl system (Savige, W.E. and Fontana, A. (1977) Methods Enzymol. 47, 442-453 and (1980) Int. J. Peptide Protein Res. 16, 285-287). Since methionine residues can be easily oxidized by traces of oxygen present in the hydrolysis tube, this reaction probably represents a main pathway of degradation (i.e. oxidation) of tryptophan during acid hydrolysis of a peptide or protein containing both tryptophan and methionine.

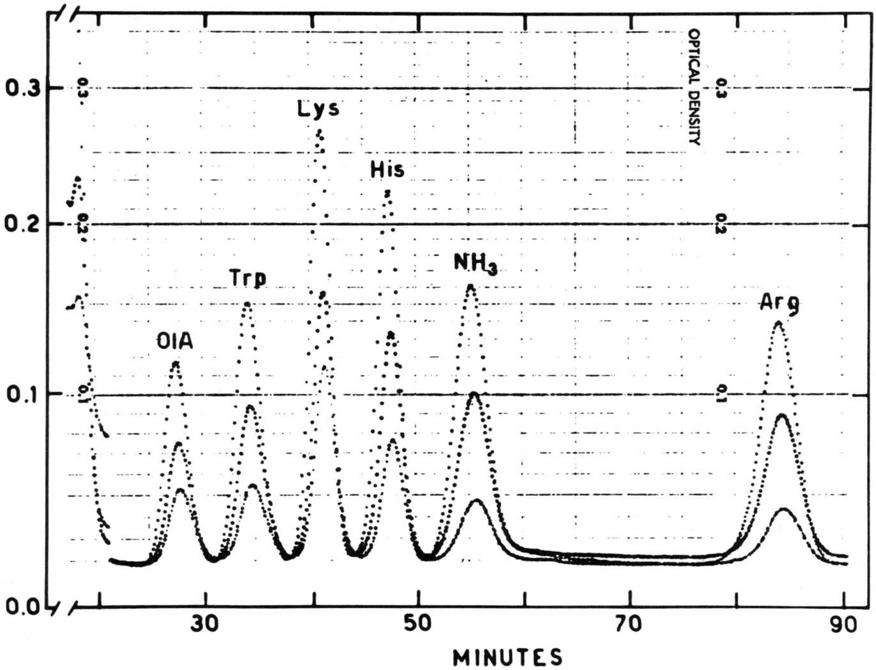

Figure 2. Elution pattern of oxindolylalanine (Oia) (0.2 μmol), tryptophan (0.1 μmol) and the basic amino acids (0.1 μmol each) from the short column of the amino acid analyzer (standard two column system). The column (0.9 x 12 m) was eluted with citrate buffer, pH 5.23, accordingly to Spackman, D. H., Stein, W. H. and Moore, S. (1958) Anal Chem. 30, A. 1190-1205). Taken with permission from Savige, W. E. and Fontana, (1977) Methods Enzymol. 47, 442-453.

A second pathway of oxidation results from the interaction of cystine with concentrated aqueous HCl (Nakai, T. and Ohta, T. (1976) Biochim. Biophys. Acta 420, 258-264; Ohta, T. and Nakai, T. (1978) Biochim. Biophys. Acta 533, 440-445). The haloacid leads to disulfide fission with concomitant formation of a reactive sulfenyl chloride (RSSR + HCl → RSCl + RSH). The sulfenyl halide then interacts at the 2-position of the indole nucleus of tryptophan (Scoffone, E., Fontana, A. and Rocchi, E. (1968) Biochemistry 7, 971-979). This leads to a thioether derivative in which cysteine is convalently linked to tryptophan (the well known "double" amino acid, tryptathionine). The tryptathionine moiety hydrolyzes in aqueous acid to the acid-stable oxindolylalanine (Wieland, Th., Weiberg, O., Dilger, W. and Fischer, E. (1955) Liebigs Ann. Chem. 582, 68-80; Veronese, F.M., Fontana, A., Boccu, E. and Benassi, C.A. (1967) Z. Naturforsch. 23b, 1219-1225).

The reactive species (methionine sulfonium chloride and cysteine sulfenyl chloride) discussed earlier do not form in the presence of sulfuric acid or p-toluenesulfonic acid. This presents a sound chemical basis for the almost quantitative recovery of tryptophan when proteins are hydrolyzed with p-toluenesulfonic acid, methanesulfonic acid, or mercaptoethanesulfonic acids. Perhaps it is more appropriate to state that tryptophan is "oxidized to oxindolylalanine" instead of being "destroyed".

The two reactions discussed above apply to proteins containing sulfur amino acids; however, carbohydrates are also detrimental to tryptophan recovery after acid hydrolysis.

M. Friedman. I would like to comment about the use of the three sulfonic acids mentioned by Prof. Fontana for hydrolyzing proteins. We have reported at this conference (see this volume) that these sulfonic acids are not suitable media for the hydrolysis of tryptophan-containing food proteins in the presence of carbohydrates.

III. ANALYSIS OF FREE AND BOUND TRYPTOPHAN IN PLASMA

P. Baumann (Lausanne). Besides the problem of the analysis of tryptophan by fluorometric or other methods, the determination of free, i.e. diffusible tryptophan has always been a challenge for biochemists. Several methods have been proposed: equilibrium dialysis, continuous dialysis, ultracentrifugation, ultrafiltration, and ultrafiltration by centrifugation. In a recent symposium (1), Fernstrom (2) and Curzon (3) thoroughly discussed the conditions needed to separate diffusible tryptophan from plasma protein-bound tryptophan. In several studies (4,5), the optimal conditions were determined experimentally. Thus, it has been shown that equilibrium dialysis yields 60% higher values for free tryptophan than ultrafiltration. It seems clear that ultrafiltration by centrifugation is not a suitable method, as pH and temperature cannot be controlled. Nevertheless, a mathematical model has been developed (6) to correct values obtained by ultrafiltration, continuous flow dialysis, or equilibrium dialysis.

Investigation of the fate of free tryptophan in blood is still important. Collaborative studies should be organized to standardize the analysis of tryptophan in blood by chemical methods and methods of separating free from total tryptophan.

Concerning the remarks of Dr. Nielsen about the use of 5-methyltryptophan as an internal standard in the determination of Trp by HPLC:

we described a method using the same internal standard when Trp is derivatized to norharman and subsequently analysed by GC or GC-MS (5).

References (to Dr. Baumann's comments)

1. Journal of Neural Transmission (Supplementum 15) "Transport Mechanism of Tryptophan in Blood Cells, Nerve Cells, and at the Blood-Brain Barrier. (Proceedings of the International Symposium, Philly-Lausanne, Switzerland, July 6-7, 1978). P. Baumann, Ed. Springer-Verlag, Wien, New York, (1979).

2. Fernstrom, J.D. - Comment on the methodology for separating and determining free and albumin-bound tryptophan levels in blood samples. In: 1, (pp. 217-219).

3. Curzon, G. - Methodological problems in the determination of total and free plasma tryptophan. In: 1, (pp. 221-226).

4. Baumann, P. and Perey, M. - The analysis of free tryptophan in human blood with the ultrafiltrator: a comparison with other methods. Clinica Chimica Acta, 76, p. 223-231 (1977).

5. Baumann, P., Rivier, L. and Perey, M. - Assay of tryptophan and related compounds such as norharman derivatives by gas chromatography - mass spectrometry in plasma. In: A. Frigerio (Ed.), Recent Developments in Mass Spectrometry in Biochemistry Medicine and Environmental Research, 8, p. 7-18 (1983).

6. Garnier, J.P., Bousquet, B., Dreux, C. - Séparation du tryptophanne libre et lié à l'albumine plasmatique: Modéles mathématiques d'étude de la liaison. Thérapie 33, p. 423-433 (1978).

H. Steinhart. It is impossible to determine free and albumin-bound tryptophan in plasma correctly for several reasons. Since the equilibrium is so sensitive, it is readily changed by manipulation, including centrifugation which changes the volume of the solution. A related point is that although it has been reported that the pH of blood changes on storage, we found this not to be the case with the aid of the Ingold electrode for blood from broiler chickens, even after three to four hours.

F. Artigas (Barcelona). There is an excellent paper by Bloxam and Curzon (Anal. Biochem. 83:130-142 (1977)) dealing with these topics (sample pH and volume, centrifugation time, etc.) I agree that determining free and albumin-bound tryptophan is quite a difficult matter. However, we should try to standardize methods so that results from different laboratories can be compared.

H. Steinhart. I think this is a good idea. One of the major tasks of ISTRY should be to recommend a reference method.

R. Oliver (Salford). I would point out that there are a number of problems in exchanging plasma samples. What we need is to exchange primary materials of tryptophan metabolites. I would also like to suggest that one of the major roles of ISTRY should be to recommend suitable analytical methods for tryptophan and tryptophan metabolites. The proliferation of a large number of analytical methods for tryptophan metabolites is not a desirable goal.

M. H. Joseph. If one were to exchange samples, it would be nice if one could agree on a method. I would like to support the suggestion that ISTRY might take a lead in recommending methods. A session on the determination of tryptophan in bound and free forms would be useful at the next ISTRY meeting.

IV. HIGHER ORDER DERIVATIVE SPECTROPHOTOMETRY OF TRYPTOPHAN AND METABOLITES.

W. Kochen (Heidelberg).* I would like to ask Dr. Talsky what are the advantages of the UV-derivative spectroscopic estimation of Trp?

G. Talsky (Garching). It is a direct quantitative method without preliminary treatment of the samples and is very sensitive. The oxidation of Trp leading to lower values does not occur. Beyond that, it is possible to estimate simultaneously Phe and Tyr at pH 7 and pH 10, respectively. The repeated differentiation causes a decrease of the half-width of the peaks and makes it possible to separate the overlapping signals. The second-order derivative gives best results for Trp estimation. On the other hand, Phe and Tyr could be optimally detected in the fourth or fifth derivatives. An additional signal which is undisturbed occurs at pH 10 in the Tyr-spectrum.

W. Kochen. Up to what order is the differentiation significant?

G. Talsky. The order is well chosen if all monopolar signals are converted to bipolar ones and all shoulders and inflection points dissolve to sharp peaks. In most cases this is true if the third to fifth derivatives are taken. The sixth order can be helpful in some special questions.

---

*Although the discussion between Drs. Kochen and Talsky did not take place during the workshop it is included here because it contributes to the theme of the workshop.

W. Kochen. Is it possible to make use of the higher-order derivative spectrophotometry (HODS) in clinical problems?

G. Talsky. Yes. For example, we made use of the HODS to direct routine investigations of sera, hemoglobins and estimation of pharmaceuticals and metabolites in biological fluids. The evaluation of unwanted background or turbidity of such samples by differentiation was very helpful.

W. Kochen. What is the simplest procedure to take derivative spectra?

G. Talsky. The simplest device for differentiation is an analog computer which is connected to the output of the photometer and the input of the recorder. But it is very important to separate the disturbing noise from the effective electrical signal. Otherwise the signal-to-noise ratio is worse and one cannot obtain quantitative data.

W. Kochen. Do you prefer analog or digital computing of derivatives?

G. Talsky. To generate first or second derivatives both methods are equivalent. For higher-order derivatives ($n > 2$), digital modules are more complicated and also more expensive if one compares spectra of identical quality.

W. Kochen. Is it possible to apply HODS also to chromatography?

G. Talsky. Yes. On the one hand, one can use the HODS for fine-resolving of elution diagrams at constant wave length. On the other hand, derivatives are taken of defined points of the elution diagram between λ_1 and λ_2 to identify the fractions. One has to take care that the scanning velocity is much higher than the elution velocity.

References (to Dr. Talsky's comments)

1. Talsky, G., Mayring L. and Kreuzer, H. -High-Resolution, Higher-Order UV/VIS-Derivative Spectrophotometry. Agnew. Chem. Int. Ed. Engl. 17. p. 785 (1978).

2. Talsky, G., Götz-Maler, S., and Betz, H. -High-Resolution, Higer-Order UV/VIS Derivative Sepctrophotometry in Micro-analytical Chemistry. Microchim. Acta. (Wien) II, pp. 1-9 (1981).

3. Talsky, G., Dostal, J. Glasbrenner, M. and Götz-Maler, S. - Higher-Order Derivative Spectrophotometry for Studying and Characterization of Proteins and other Native and Synthetic Macromolecules. Angew. Macromol. Chem. 105, pp. 49-59 (1982).

4. Talsky, G. - Higer-Order Derivative Spectrophotometry in Environmental Analytical Chemistry. Intern. Environ. Anal. Chem. 14, pp. 81-91 (1983).

M. Friedman. Let me conclude this session by stating that it was a fruitful and productive experience which produced a number of stimulating ideas for future research. It is becoming evident that we need interaction among different yet related disciplines, embracing all aspects of tryptophan in medicine and nutrition, to solve many of the stated problems.

Tryptophan and Serotonin: Neurobiology

TRYPTOPHAN TRANSPORT TO THE BRAIN: NEWER FINDINGS AND OLDER ONES RECONSIDERED.

G.Curzon and G.S.Sarna

Department of Neurochemistry, Institute of Neurology Laboratories, 33 John's Mews, London WC1N 2NS, U.K.

Introduction

How tryptophan is supplied to the brain has been a focus of much interest (reviews, 1,2,3) partly because tryptophan supply influences 5-hydroxytryptamine (5HT) synthesis and partly because of controversy about the relative importance of various influences on brain tryptophan concentration. Early work showed that brain tryptophan and 5HT synthesis were increased after i.p. injection of tryptophan (4) even in amounts as small as 12 mg/kg (5). Also, rather remarkably, brain tryptophan was found to be increased in two circumstances in which it might have been guessed that a decrease would occur i.e. following the intake of a tryptophan-free high carbohydrate meal (6) or following 24 h food deprivation (7). The effect of the high carbohydrate meal was explained in terms of insulin secretion causing decreased plasma concentrations of a group of large neutral amino acids (principally phenylalanine, tyrosine, leucine, isoleucine, valine) which compete with tryptophan for transport to the brain. The effect of food deprivation was related to the resultant increased plasma unesterified fatty acid concentration. These substances bind to plasma albumin, thus weakening the binding of tryptophan and making the free tryptophan available to the brain. The two mechanisms can be combined in a single scheme (1).

Progress in Tryptophan and Serotonin Research

The role of free tryptophan

Although numerous findings were consistent with free tryptophan being a determinant of brain tryptophan concentration (reviewed, 1) its role was disputed in a number of papers published in the mid 70s and it is perhaps now appropriate to reassess some of this data. For example, work on tryptophan uptake by brain (8) was interpreted to indicate that its binding to albumin has little effect on its availability to the brain. In these investigations, the uptake of tryptophan by brain was measured using a bolus of radioactive tryptophan in buffer injected into the carotid. As adding bovine albumin to the bolus had little effect on uptake it was deduced that albumin binds tryptophan so weakly that the brain readily strips it off so that bound tryptophan is almost as available to the brain as is free. These experiments however are not easy to interpret, firstly, because of the use of bovine albumin in the rat and secondly, because recent work shows that the binding characteristics of albumin are critically dependent on purification procedures and plasma ions (9,10) so that different preparations may have very different properties. It is hardly surprising therefore that in similar experiments but with a bolus of the rats own plasma instead of buffer we came to a different conclusion as altering free tryptophan in the bolus when total tryptophan was constant considerably increased brain tryptophan but altering total tryptophan with free tryptophan constant had no effect (11). Nevertheless, it may be that in some species the brain can readily strip tryptophan off from albumin as indicated in the rabbit (12) but not in the dog (13).

Other work (14) which discounted the role of free tryptophan was based on dietary experiments which were taken to show that it was without value as a predictor of brain tryptophan but that total tryptophan was of value (in particular, when divided by the sum of the concentrations of the competing amino acids). We have criticised these findings on methodological grounds as some of the diets used had high fat content and we have shown (15) that plasma collected from rats fed such diets shows active *in vitro* lipolysis, the increased unesterified fatty acid leading to high but artefactual plasma free tryptophan values unless precautions are taken to avoid *in vitro* changes. If the dietary experiments (14) are reconsidered in the light of these findings, it can be calculated that when the results on animals on high fat diets are eliminated, then the correlation between brain tryptophan and the serum free tryptophan/competers ratio becomes significant ($r = 0.83$, $p < 0.01$). A comparable correlation coefficient was reported (14) when total tryptophan/competers was used ($r = 0.79$, $p < 0.01$) but the distribution of values was less indicative of a linear relationship (see Fig.1, filled in circles). These results suggest that in this specific acute feeding situation, when plasma free tryptophan and plasma competers are both varying over a 3-4 fold range, then they both have appreciable influences on brain tryptophan values.

Unfortunately, these experiments have often been used to justify investigations in which free tryptophan was ignored and only total tryptophan/competers was taken as a predictor of brain tryptophan. On the other hand, in some other studies the competers have been ignored and only plasma free and total tryptophan determined.

The role of competing amino acids

How important are physiological alterations of plasma concentrations of the competers ? Fig.1. summarizes data from three published papers (14,16,17). Each point indicates data for a group of rats given a particular diet. When brain tryptophan is plotted against the ratio, plasma tryptophan/competers a rather striking relationship with brain tryptophan is suggested. However, a ratio can be altered by changing either the denominator (competing amino acids) or the numerator (tryptophan). Fig.1. reveals that the very steep relationship shown by some of the results from ref.14 is completely explicable in terms of plasma tryptophan alone and that this also could explain part of the relationship shown by the results from ref.17. The residual findings from ref.16 indicate that a 4-fold range of competer concentrations appears to lead to a 1.5 fold range of brain tryptophan values while the results of the other two papers suggest rather smaller responses.

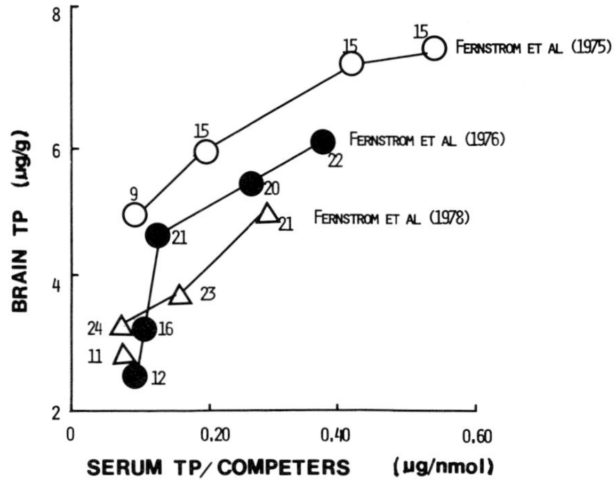

Figure 1. Relationships between brain tryptophan and plasma tryptophan/competers. Data from refs. 14,16,17. Each symbol indicates the mean value for a group of rats given one meal of a particular composition/24h. Nos. against symbols indicate mean plasma total tryptophan concentrations.

These results may be compared with those calculated below from our own work in which tryptophan influx was determined using a bolus of radioactive tryptophan in buffer injected into the carotid (18). Values were obtained at different tryptophan concentrations and the results used to determine Vmax ($0.0312 \pm 0.0040 \mu mol.g^{-1}.min^{-1}$) and Km ($0.252 \pm 0.041$ mM) for the saturable component of the influx and Kd ($0.0094 \pm$

0.0010 $ml.g^{-1}.min^{-1}$) the diffusional constant for the non-saturable component of the influx. From these values and kinetic data for the competers (19) the inhibitory effects of different plasma competer concentrations can be calculated as shown in Fig.2 which gives data from both our own results (18) on Sprague-Dawley rats and from those obtained by others on Long-Evans animals (20).

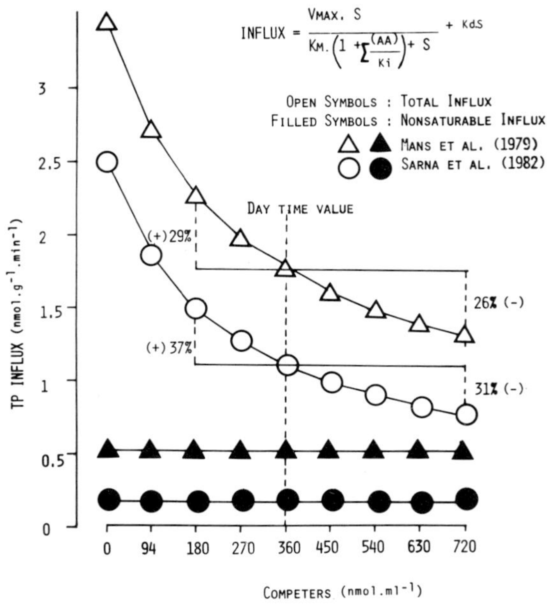

Figure 2. Calculated effects on tryptophan influx into brain of amino acids competing with it for transport to brain. Values obtained using results from ref. 18 (Sprague-Dawley rats), and ref. 20 (Long-Evans rats). Tryptophan concentration, 20 $nmol.ml^{-1}$. Results are comparable over tryptophan range 2-200 $nmol.ml^{-1}$.

It can be seen that a typical day time plasma competer level markedly inhibits influx in comparison with the non-physiological situation of complete absence of competers. However, the effects of physiologically likely competer changes are much smaller. Thus halving the found competer concentration increases influx by 29-31% while doubling it decreases influx by 26-31%, i.e. a 4 fold overall change of competer concentrations causes about a 2 fold change in brain tryptophan influx. This is reasonably in accord with the effects of competers on brain tryptophan concentration indicated in Fig.1. Note also that similar increments of plasma competer concentration cause progressively smaller % changes of tryptophan influx as competer concentration rises.

The extent to which these findings can be extrapolated to man is debatable. It is worth mentioning that one group (21) found that human subjects fed single high carbohydrate or high protein meals, which animal work (14,16,17) suggests might lead to very different plasma tryptophan/competer ratios, in fact had similar ratios. Also, while the laboratory responsible for the latter study found 2 fold differences in the ratio between subjects fed a protein-free or high protein meal (22), our own work on patients undergoing psychosurgery did not suggest that 1.5 fold differences of plasma competer concentration influenced brain tryptophan values (23).

Conditions under which brain tryptophan reflects differences of plasma free tryptophan concentration but does not reflect differences of plasma competing amino acid concentration.

In most of the above work, differences of plasma tryptophan binding is not taken into account. This is, of course, unnecessary in circumstances in which binding does not vary appreciably. In other situations, it can be the only useful predictor of brain tryptophan concentration. This is illustrated in Table 1.

Table 1. Correlation coefficients of brain tryptophan (TP) with plasma free TP, total TP and competing amino acid (COMP) concentrations.

	BRAIN TP vs PLASMA FREE TP	BRAIN TP vs PLASMA FREE TP/COMP	BRAIN TP vs PLASMA TOTAL TP	BRAIN TP vs PLASMA TOTAL TP/COMP
Expt 1	0.70*	0.61*	0.13(ns)	0.03(ns)
Expt 2	0.97*	0.56(ns)	0.58(ns)	-0.86(ns)

Expt 1. 29 sham operated rats (liver shunt expt.) (24). Expt 2. 5 patients undergoing brain surgery (23). In Expt 1, plasma TP values were divided by the sum of the plasma competer values but each of these was first divided by its uptake Km (19). *$p < 0.001$

In these studies, significant correlations were obtained between brain tryptophan and plasma free tryptophan and these became <u>less</u> significant when competing amino acids were taken into account. It should be noted that in both investigations, plasma competing amino acid values varied over relatively small ranges and that in the human subjects they tended to parallel plasma TP values. Values for the rats were rather higher (c.500 nmoles/ml) than in the influx experiment (Fig. 2). It is striking that neither of the correlations with plasma total tryptophan was significantly positive and in the human subjects, allowing for the competers gave a coefficient approaching a significant <u>negative</u> value.

Somewhat similar results were obtained in part of another experiment (25) in which we kept groups of rats for various periods on diets which had different TP contents but which were otherwise identical (Table 2). The rats on the 0.3% tryptophan diet showed a rather small variation with time of plasma total tryptophan, clearly insufficient to explain the considerably larger brain tryptophan differences. Plasma competers showed a much larger variation with time and correlated positively and significantly with brain tryptophan. As a consequence of this, the ratio of plasma total to competers correlated negatively with brain tryptophan, the value (as in the human study - Table 1) approaching significance. Animals on the 0.08% tryptophan diet did show a significant positive correlation between the plasma total tryptophan/competers ratio and brain tryptophan but this relationship was largely explicable in terms of plasma total tryptophan alone.

Table 2. Correlation coefficients of brain tryptophan (TP) with plasma free TP, total TP and competing amino acid (COMP) concentrations: rats on diets of different TP contents.

DIETARY TP	BRAIN TP vs PLASMA FREE TP	BRAIN TP vs PLASMA TOTAL TP	BRAIN TP vs COMP	BRAIN TP vs PLASMA TOTAL TP/COMP
0.3%	0.44(ns)	0.69(ns)	0.86*	-0.77(ns)
0.08%	0.37(ns)	0.77(ns)	0.03(ns)	0.81*

Rats killed between 10.00 and 12.00 h after 6 days on 0.3% TP diet and 2,4,7,14,21 and 26 subsequent days on either 0.3% or 0.08% TP diet. Concentrations of the variables are given in ref.25. *$p < 0.05$.

The plasma free tryptophan findings in this experiment were intruiging as values plotted against brain tryptophan gave distinct regression lines for rats killed before and after 7 days on the diet (25). This difference cannot be explained in terms of differences of plasma competing amino acid concentrations and is consistent with an increase of effectiveness of brain tryptophan uptake occurring during the study. As subsequent work points to decreased effectiveness of uptake with increasing age (see below), these results may reflect some environmental change in this particular experiment. Whatever their explanation, they illustrate that differences not only of plasma composition but also of the effectiveness of tryptophan uptake by the brain are important influences on brain tryptophan concentration.

Dependence of brain tryptophan concentration on age: influences of plasma free tryptophan and tryptophan influx.

The importance of tryptophan influx mechanisms in the control of brain tryptophan concentration was indicated by work on rats of various ages between 8 days and 16 months (18).

Figure 3 shows the age-dependencies of plasma free tryptophan, of plasma competers, of the brain uptake index for tryptophan as determined by the Oldendorf method and also of brain tryptophan values. There is an overall qualitative correspondence between plasma free and brain tryptophan concentrations but quantitative parallelism is poor, their ratio rising sharply during the preweaning period and then gradually falling. The rise very probably reflects the large decrease, between 8 and 19 days, in plasma concentrations of competing amino acids. The subsequent fall of the ratio appears to be explicable by the progressive decrease of brain uptake index (and hence fractional extraction) for tryptophan between 19 days and 16 months. Thus, free plasma and brain tryptophan only correlate significantly during this period if the plasma values are multipled by the brain uptake indices for tryptophan to give a measure of brain unidirectional influx. Corresponding correlations using plasma total tryptophan were not significant.

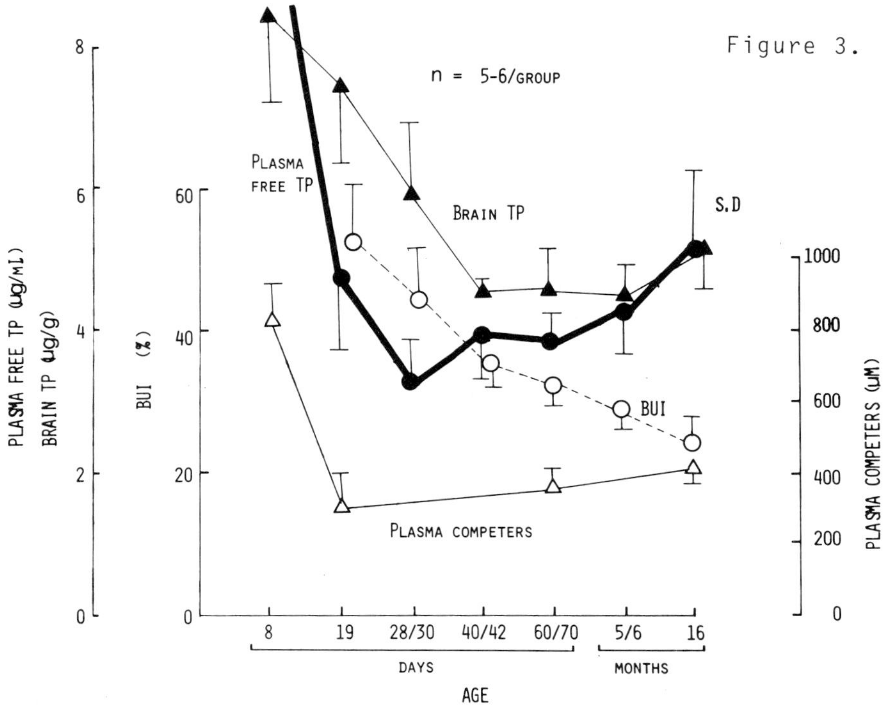

Table 3. Correlation coefficients of brain tryptophan (TP) with plasma free TP, total TP and brain uptake index for TP: rats aged between 19 days and 16 months.

BRAIN TP VS				
PLASMA FREE TP	PLASMA FREE TP x BRAIN UPTAKE INDEX	PLASMA TOTAL TP	PLASMA TOTAL TP x BRAIN UPTAKE INDEX
0.29 (ns)	0.90 $p < 0.02$	-0.69 (ns)	0.50 (ns)

Values derived from data shown in Fig.3.

The correlation coefficient of brain tryptophan with the product of plasma free tryptophan and uptake index indicates that 81% of the variance of brain tryptophan with age between 19 days and 16 months is explicable by these variables. Attempts to correct the plasma values for competitive effects by dividing them by the sum of the competing amino acids did not help to explain but instead obscured relations between plasma and brain values (18).

Acute dietary effects: influences of plasma competing amino acids and free tryptophan on brain tryptophan.

Many of the above studies thus suggest that plasma competing amino acid concentrations are in some circumstances a hindrance rather than a help when trying to predict brain tryptophan values. However, brain tryptophan can be appreciably affected by large changes of plasma competer concentration such as can occur following their administration (26) or in metabolic disease (27) or in certain feeding experiments (14,16,17).

For example, when we exposed rats to a high carbohydrate diet during the light period using a procedure similar to that of Fernstrom et al (14,16,17) in which standard chow was allowed ad lib until 18 h previously, then results essentially agreed with those previously reported. Thus, brain tryptophan was raised at the end of the 2 h feeding period. This occurred even though plasma free tryptophan remained constant and total tryptophan fell somewhat. There was a concurrent large fall of competing amino acids. Our brain influx experiments (Fig. 2) predict this would lead to a 57% increase of brain tryptophan (actual rise = 63%). It is also apparent that the time course of brain tryptophan change almost exactly paralleled that of the plasma free tryptophan/competers ratio (Fig.4). Thus, after a high carbohydrate diet is made available to the

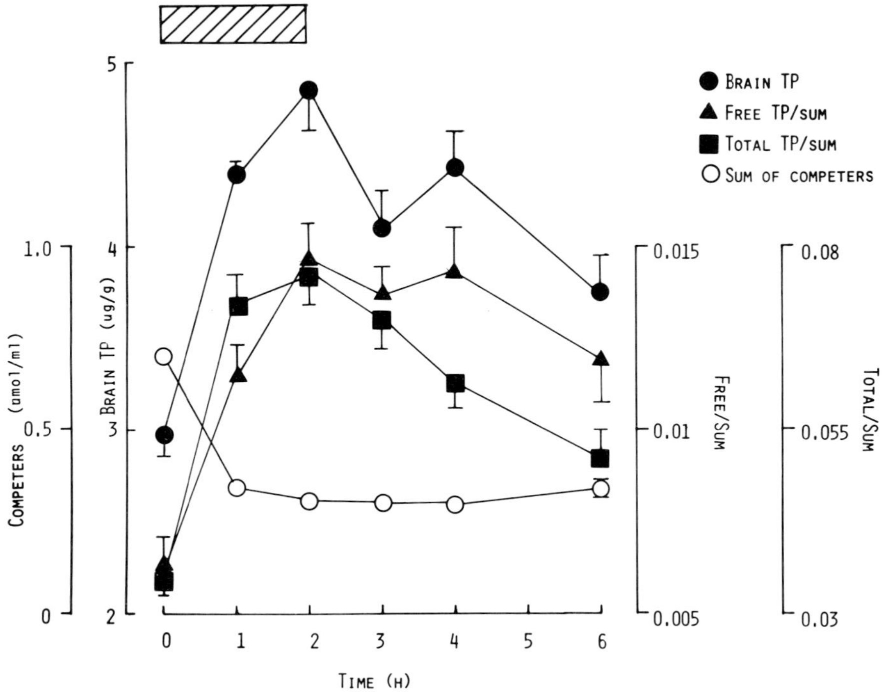

Figure 4. Effect of a high carbohydrate meal at 11.30 h - 13.30 h (06.00 h - 18.00 h white light) on rat plasma and brain tryptophan and on plasma competing amino acids (Sarna and Curzon, unpublished).

rat for 2 h/day there is a large fall of plasma competers and a resultant rise of brain tryptophan.

In another acute feeding experiment the animals were trained to eat a standard diet (41B, Labsure, Poole, Dorset) for 4 h during the dark period (when they normally eat). The subsequent fall of brain tryptophan agrees with earlier work (28). It may be considered as the converse of the rise of brain tryptophan on food deprivation which is associated with an increase of plasma free tryptophan due to increased unesterified fatty acid (7).

Table 4. Effect of a standard chow meal at 10.00-14.00 h (red light, 06.00-18.00 h). % Changes of rat plasma and brain tryptophan (TP) etc. (Sarna and Curzon, unpublished).

		% Change
BRAIN	TP	-23
	Free TP	-27
	Total TP	+31
PLASMA	Unesterified fatty acid	-52
	Competers	+30
	Free TP/competers	-48
	Total TP/competers	+1

All the above values, except that of total TP/competers are derived from significant changes of concentration ($p < 0.01$).

In the present experiment, we see falls of plasma unesterified fatty acid and free tryptophan and a fall of brain tryptophan. There is also a rise of plasma competers. The brain change may reflect that of plasma free tryptophan or of its ratio to that of the competers but it cannot reflect plasma total tryptophan as this rises or the ratio of total plasma tryptophan to the competers as this remains constant.

Tryptophan transport beyond the blood-brain barrier.

Recently, we have investigated whether there are physiologically important barriers for tryptophan transport within the brain (i.e. beyond the blood brain barrier) as *in vitro* results (29,30) suggest that such barriers may exist and influence 5HT synthesis.

For this work, a new method was developed to investigate transport *in vivo* from the CSF space to brain tissue using lateral ventricular injection of ^{3}H-labelled substrates + a ^{14}C reference marker (31). The method was authenticated using ^{3}H-5HT as regional differences of its uptake paralleled the distribution of 5HT terminals and uptake was reduced by destroying 5HT neurones with 5,7-DHT. Uptake of tryptophan (unlike transport from blood to brain) appeared to be entirely passive (i.e. unsaturable). Destroying 5HT neurones did not alter tryptophan uptake as indicated by the *in vivo* method or an *in vitro* synaptosome method. Also, the relationship between brain tryptophan concentration and 5HT synthesis was not affected by the amino acid valine at a pharmacological dosage (200 mg/kg i.p.) which considerably decreased brain tryptophan concentration. This result suggests that although competing amino acids can impair the transport of tryptophan into the brain, they do not effectively impair its transport at the 5HT neuronal membrane *in vivo*.

References

1. Curzon, G.: J.Neural Trans. (Suppl.) 15, 81-92 (1979).
2. Curzon, G.: in Biochemical and Medical Aspects of Tryptophan Metabolism (eds. O.Hayaishi et al.) pp. 267-279 Elsevier (1980).
3. Wurtman, R.J.: in Biochemical and Medical Aspects of Tryptophan Metabolism (eds. O.Hayaishi et al.) pp. 31-46 Elsevier (1980).
4. Eccleston, D., Ashcroft, G.W., Crawford, T.B.B.: J.Neurochem. 12, 493-503 (1965).
5. Fernstrom, J.D., Wurtman, R.J.: Science 173, 149-152 (1971).
6. Fernstrom, J.D., Wurtman, R.J.: Science 174, 1023-1025 (1971).
7. Knott, P.J., Curzon, G.: Nature 239, 452-453 (1972).
8. Yuwiler, W., Oldendorf, W.H., Geller, E., Braun, L.: J.Neurochem. 28, 1015-1023 (1977).
9. Bruderlein, H., Bernstein, J.: Biochem.Pharmacol. 29, 1915-1918 (1980).
10. Droge, J.H.M., Wilting, J., Janssen, L.H.M.: Biochem. Pharmacol. 31, 3781-3786 (1982).
11. Bloxam, D.L., Tricklebank, M.D., Patel, A.J., Curzon, G.: J.Neurochem. 34, 43-49 (1980).
12. Daniel, P.M., Love, E.R., Moorhouse, S.R., Pratt, O.E.: J.Physiol. 312, 551-562 (1981).
13. Huet, P.M., Pomier-Layrargues, G., Duguay, L., DuSouich, P.: Amer.J.Physiol. 241, G163-G169 (1981).
14. Fernstrom, J.D., Hirsch, M.J., Faller, D.V.: Biochem.J. 160, 589-595 (1976).
15. Hutson, P.H., Knott, P.J., Curzon, G.: Nature 262, 142-143 (1976).
16. Fernstrom, J.D., Faller, D.V.: J.Neurochem. 30, 1531-1538 (1978).
17. Fernstrom, J.D., Faller, D.V., Shabshelowitz, H.: J.Neural Trans. 36, 113-121 (1975).
18. Sarna, G.S., Tricklebank, M.D., Kantamaneni, B.D., Hunt, A., Patel, A.J., Curzon, G.: J.Neurochem. 39, 1280-1283 (1982).
19. Pardridge, W.M., Oldendorf, W.H.: Biochim.Biophys.Acta 382, 377-392 (1975).
20. Mans, A.M., Biebuyck, J.F., Saunders, S.J., Kirsch, R.E., Hawkins, R.A.: J.Neurochem. 33, 409-418 (1979).

21. Ashley, D.V., Barclay, D.V., Chauffard, F.A., Moennoz, D., Leathwood, P.D.: Amer.J.Clin.Nutr. 36, 143-153 (1982).

22. Fernstrom, J.D., Wurtman, R.J., Hammarstrom-Wiklund, B., Rand, W.M., Munro, H.N., Davidson, C.S.: Amer.J.Clin.Nutr. 32, 1912-1922 (1979).

23. Gillman, P.K., Bartlett, J.R., Bridges, P.K., Hunt, A., Patel, A.J., Kantameneni, B.D., Curzon, G.: J.Neurochem. 37, 410-417 (1981).

24. Bloxam, D.L., Curzon, G.: J.Neurochem. 31, 1255-1263 (1978).

25. Ashley, D.V.M., Curzon, G.: J.Neurochem. 37, 1385-1393 (1981).

26. Carlsson, A., Lindqvist, M.: Naunyn-Schmiedeberg's Arch. Pharmacol. 303, 157-164 (1978).

27. Daniel, P.M., Moorhouse, S.R., Pratt, O.E.: Psychol.Med. 6, 277-286 (1976).

28. Perez-Cruet, J., Tagliamonte, A., Tagliamonte, P., Gessa, G.L.: Life Sci., 11 Part 2, 31-39 (1972).

29. Swann, A.C., Heninger, G.R., Roth, R.H., Maas, J.W.: Life Sci. 28, 347-354 (1981).

30. Denizeau, F., Sourkes, T.L.: J.Neurochem. 28, 951-959 (1977).

31. Curzon, G., Green, P., Sarna, G.S., Tricklebank, M.D.: J.Physiol. (1983) in press.

TRYPTOPHAN AVAILABILITY AND SEROTONIN SYNTHESIS IN RAT BRAIN: EFFECTS OF EXPERIMENTAL DIABETES

John D. Fernstrom, Ph.D.

Departments of Psychiatry and Pharmacology
University of Pittsburgh School of Medicine
Pittsburgh, Pennsylvania 15213

Introduction

The rate of serotonin (5HT) synthesis in brain is influenced to a considerable extent by variations in the availability of tryptophan (TRP) to brain. This relationship follows from the fact that tryptophan hydroxylase is not fully saturated with substrate at normal brain TRP concentrations [1]. As a consequence, variations in brain TRP concentrations can directly influence the rate of TRP hydroxylation, and thereby the overall rate of 5HT production [1,2]. The physiologic and pharmacologic importance of this precursor-product relationship has repeatedly been demonstrated under conditions in which tryptophan levels in brain have been modified acutely (e.g., following the injection of tryptophan [2] or insulin [3], after the ingestion of single meals [3,4], following stress [5]), or chronically (e.g., following the chronic ingestion of tryptophan-poor [6] or -enriched [7,8] diets). It has also been shown to hold under certain pathophysiologic conditions, such as experimental phenylketonuria [9,10] and cirrhosis [11,12]. Recently, a number of reports have appeared suggesting that the relationship may also be relevant to another pathophysiologic state, diabetes. Specifically, in diabetes, and the absence of insulin secretion (or of its effects on target tissues), brain levels of TRP and of one or more 5-hydroxyindoles are abnormally low [13-15]. This review summarizes the evidence showing that the reduced brain levels of TRP in diabetes probably (1) cause a diminution in the rate of 5HT formation in brain, and (2) follow from reduced tryptophan uptake from blood, secondary to increased competition from the branched-chain amino acids (BCAA). In addition, as discussed below, the increased blood levels of the BCAA in diabetes appear to cause predictable changes in the competitive uptake of not just TRP, but also of the other aromatic amino acids [16,17].

Progress in Tryptophan and Serotonin Research

Effects of Experimental Diabetes on 5HT formation in Rat Brain

A number of neuropharmacologists have observed brain tryptophan and 5-hydroxyindoleacetic acid (5HIAA) levels to be below normal in rats made diabetic with streptozotocin [13-15]. Associated with this reduction are decreases in the brain levels of both 5HT and 5HIAA (though the reduction in 5HT is usually quite small; see **Table 1**). These changes appear to follow from the absence of normal insulin secretion, inasmuch as they disappear when diabetic animals receive daily injections of insulin [15]. The fact that 5-hydroxyindole levels are low in diabetes suggests reduced synthesis rate, but other mechanisms are always possible. We therefore measured the *in vivo* rate of 5HT synthesis, using two different methods. The *first* technique employed was that of Carlsson et al. [18], in which the *in vivo* rate of TRP hydroxylation is estimated by measuring the linear accumulation of 5-hydroxytryptophan (5HTP) following inhibition of aromatic L-amino acid decarboxylase (AAAD). Since TRP hydroxylase catalyzes the rate-limiting step in 5HT synthesis [19], estimation of its *in vivo* activity presumably provides a good index of the overall rate of 5HT synthesis. As indicated in **figure 1**, the accumulation of 5HTP in brain following injection of the AAAD inhibitor was slower in diabetic than in normal rats [20]. The *second* technique used was that of Tozer et al. [21], in which 5HT synthesis rate is inferred from the measured rate of decline in 5HIAA levels following administration of a monoamine oxidase (MAO) inhibitor. The method *requires* steady-state conditions (satisfied in these studies) to enable synthesis rate to be derived from metabolite clearance rate (5HIAA is the major metabolite of 5HT in brain). **Table 2** summarizes the results of one such study, and shows that indeed, 5HT synthesis rate is below normal in diabetic rat brain [20].

The reduced 5HT synthesis rate observed in the diabetic rat brain probably does follow indirectly from the decreased brain levels of TRP, rather than from direct inhibition of TRP hydroxylase, since (1) injections of TRP can increase rapidly 5HT and 5HIAA levels in the diabetic rat brain [22], and (2) the *in vitro* activity of tryptophan hydroxylase, obtained from diabetic rat brain, is *not* below normal [23]. (In fact, the V_{max} of tryptophan hydroxylase is increased in the diabetic brain [23]. This finding led to the speculation that 5HT levels are normal in the diabetic brain, despite below normal levels of tryptophan, because tryptophan hydroxylase activity [and presumably 5HT synthesis rate] was above normal. This latter speculation now

TABLE 1. Effects of glucose intubation on brain indoles in normal and diabetic rats.

Parameter	Time	Normal	Diabetic
	(hr)		
Serum Tryptophan	0	102 $\pm$ 4	86 $\pm$ 5[a]
(nmol/ml)	2	90 $\pm$ 3	96 $\pm$ 8
Serum Tryptophan	0	0.16 $\pm$ 0.01	0.09 $\pm$ 0.01[a]
Ratio	2	0.30 $\pm$ 0.02[b]	0.13 $\pm$ 0.04[a]
Brain Tryptophan	0	26.5 $\pm$ 2.0	13.7 $\pm$ 1.5[a]
(nmol/g)	2	36.3 $\pm$ 0.5[b]	15.7 $\pm$ 2.0[a]
Brain 5HT	0	4.11 $\pm$ 0.07	4.16 $\pm$ 0.18
(nmol/g)	2	4.48 $\pm$ 0.05[b]	4.30 $\pm$ 0.07
Brain 5HIAA	0	3.47 $\pm$ 0.14	2.95 $\pm$ 0.13[a]
(nmol/g)	2	4.38 $\pm$ 0.18[b]	3.25 $\pm$ 0.15[a]

Rats were made diabetic with streptozotocin (65 mg/ kg, intracardiac) 10 days prior to use. Groups of 6 overnight-fasted normal and diabetic rats were intubated with glucose (1 ml/100 g of a 2 M solution) and killed 2 hours later. In normal animals, serum insulin rose to 78 $\pm$ 9 μU/ml 30 minutes post-intubation, while in diabetics, serum insulin did not exceed 15 μU/ml at any time. [a] $P < 0.05$ compared to normal values at the same time; [b] $P < 0.05$ compared to 0-hour values. [From 15]

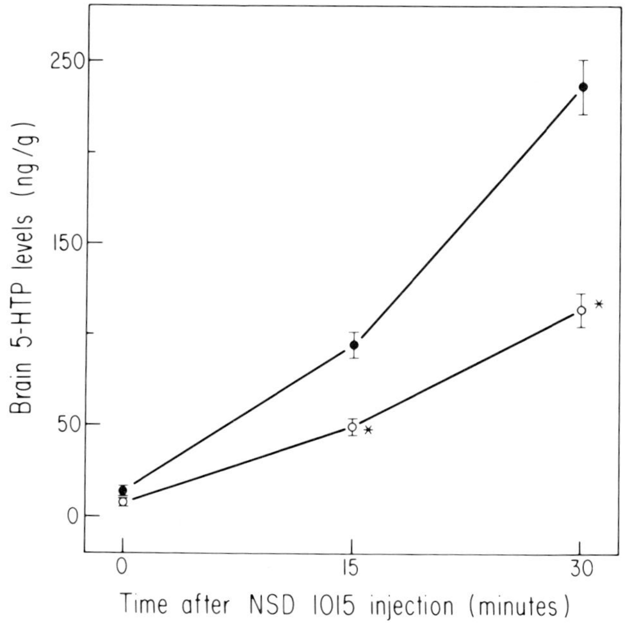

Figure 1. 5-Hydroxytryptophan accumulation in normal and diabetic rats following decarboxylase inhibition.

Groups of six male normal or diabetic rats, fasted overnight, received injections of NSD-1015 (100 mg/kg, ip, an inhibitor of AAAD) and were killed 15 or 30 minutes later. Closed circles, normal rats; open circles, diabetic rats. Data points represent means $\pm$ standard errors. The data were analyzed by two-way analysis of variance. Significant F values ($P < 0.001$) were obtained for time after NSD-1015 injection, diabetic vs. normal, and the interaction between the two factors. Differences between group means were then determined using Student's t-test: * diabetic value differs significantly from normal value at the same timepoint, $P < 0.001$. [From 20]

TABLE 2. Serotonin synthesis rates in brains of normal and diabetic rats

Group	Time	Brain 5HIAA	k	5HT Synthesis Rate
	(min)	(ng/g)	(1/hr)	(g/g/hr)
Normal	0	491 ± 09	0.52 ± 0.02	0.26 ± 0.01
	30	343 ± 13		
	60	278 ± 07		
Diabetic	0	392 ± 17*	0.30 ± 0.03*	0.12 ± 0.01*
	30	332 ± 12		
	60	285 ± 21		

Groups of 6 male rats, fasted overnight, received pargyline (75 mg/kg, i.p.) and were killed at the times indicated. The rate of change in brain 5HIAA level after pargyline treatment was used to calculate k [21]. The k values were used to determine serotonin synthesis rates. Data are expressed as means ± standard errors. * different from normal (P < 0.001, Student's t-test). [From 20] appears to be incorrect, because direct measurements of hydroxylation rate in vivo show 5HT synthesis to be reduced.)

Effects of Experimental Diabetes on Tryptophan Uptake and Levels in Rat Brain

Published data showing reduced brain levels of TRP in the diabetic rat suggested that TRP uptake into brain might be reduced. The reduction of TRP levels in the diabetic brain could presumably result from at least three changes: (1) the net uptake of TRP into brain might be reduced; (2) the balance between TRP addition to and removal from the brain protein pool might be shifted so as to produce a net increase in the removal of TRP from the free brain pool for deposition into the protein pool; and (3) the loss of TRP to other metabolic pathways (e.g., hydroxylation, oxidation, decarboxylation) might be increased. The second possibility can probably be discounted - if protein synthesis were increased, then one might expect the free pools of most amino acids to be abnormally low in the diabetic brain. Such is not the case ([16,17], and **Table 3**), and indeed, the brain concentrations of some amino acids are above normal in the diabetic brain. The third possibility can also be discarded, at least for hydroxylation (**Table 2**; **Figure 1**). Moreover, since the relative rates of TRP utilization in other known pathways in brain are probably small, relative to the size of the brain TRP pool [see 24], such other pathways can probably also be discounted as significant vehicles for the reduction in brain TRP levels in diabetes. The first possibility thus remains, and is at present the most attractive possibility.

Several lines of evidence support reduced TRP uptake into the diabetic brain. One early report was that of MacKenzie and Trulson [22], who observed that above-normal increments in blood TRP were required to produce the same elevations in brain TRP seen in normal rats following smaller increments in blood TRP concentration. These data lend themselves to two interpretations (at least), one that diabetes somehow directly retards TRP transport into brain, possibly by altering the carrier itself, and the other that the disease indirectly inhibits TRP uptake into brain, possibly via increasing blood levels of the BCAA (see below). The first possibility can be ruled out. For unlike cirrhosis, in which chronically altered brain levels of TRP appear to follow from a direct change in the kinetics of the carrier system that transports this amino acid into brain [25,26], no such direct effect on transport kinetics is noted in

Table 3. Twenty-four hour variations in serum and brain aromatic amino acids and valine in normal and diabetic rats

Amino Acid	Time of Day					
	10 am	2 pm	6 pm	10 pm	2 am	6 am
Serum Tryptophan (nmol/ml)*						
Normal**	160±11	156±7	136±13	184±5	181±11	192±5
Diabetic	133±17	126±3	143±14	130±21	139±13	158±8
Brain Tryptophan (nmol/g)*						
Normal**	28±1	32±2	31±2	33±1	32±1	34±2
Diabetic	20±1	18±2	24±2	21±2	22±1	23±1
Serum Phenylalanine (nmol/ml)*						
Normal**	64±2	52±2	51±5	74±3	70±4	67±2
Diabetic**	68±4	63±4	60±2	86±2	64±4	62±3
Brain Phenylalanine (nmol/g)*						
Normal	31±2	32±1	33±2	29±1	34±2	30±2
Diabetic	23±1	18±1	23±1	21±2	20±2	21±1
Serum Valine (nmol/ml)*						
Normal**	228±27	165±9	147±21	271±5	232±11	214±3
Diabetic**	421±51	665±103	686±57	1069±161	822±66	585±55
Brain Valine (nmol/g)*						
Normal	48±4	53±2	41±7	46±3	58±2	54±6
Diabetic	77±7	97±12	91±10	95±9	98±14	100±9

Groups of 6 normal and diabetic male rats were killed at the indicated times of day. Data are means ± standard errors. By analysis of variance, significant effects of group (normal vs. diabetic; * $P < 0.05$) and time-of-day (** $P < 0.05$) were noted. [From 16]

diabetes [27]. (Of course, the data reported here were all obtained in rats that had been diabetic for relatively short periods; i.e., days to weeks. It is possible that the transport kinetics might change in the brains of animals diabetic for much longer. In prolonged diabetes, deterioration of competitive transport might accompany degeneration of cerebral capillaries, since the relevant transport systems reside in capillary endothelial cells. However, such studies have yet to be performed.) The second possibility remains, and receives support from several directions. The hypothesis is based on the fact that TRP is transported into brain via a competitive carrier it shares with several other large neutral amino acids (LNAA) [28,29]. Because of the competitive nature of the carrier, brain uptake of TRP can be reduced (1) by lowering blood TRP levels, and/or (2) by increasing blood concentrations of one or more of its competitors. In the case of uncontrolled diabetes, blood TRP levels are sometimes reduced (e.g., a small reduction in Table 1; a larger reduction in Table 3, though at some times of day, no reduction at all is apparent). However, blood levels of the BCAA are always elevated (e.g., see valine in Table 3). Such increments in blood BCAA levels could well account for the consistent reduction seen in brain TRP levels in diabetes. They could also, as noted above, account for the findings of MacKenzie and Trulson [22].

Other data support indirectly the view that TRP uptake into the diabetic brain is below normal, secondary to increased competition by the BCAA. For example, if brain TRP is reduced because of the increased competition for brain uptake offered by the high circulating levels of the BCAA, then the brain levels of the other aromatic amino acids should also be reduced. They share the competitive LNAA transport carrier, and their blood levels are about normal in the diabetic rat (Table 3). Such is indeed the case, as is indicated in table 3 for phenylalanine (the same effect is also seen for tyrosine in brain [16]). Additionally, it is interesting to note that the brain level of each BCAA is increased in the diabetic brain (see valine data in table 3) [16,17], consistent with the view that, given both the competitive nature of the LNAA carrier, and the blood LNAA changes in diabetes, the brain uptakes of the BCAA should be increased, and those of the aromatic amino acids decreased. A similar finding has been made in retina [30], suggesting that the same competitive transport mechanism may exist at the blood-retinal barrier as occurs at the blood-brain barrier.

Another indirect form of evidence follows from studies of the acute effects of ingesting carbohydrates on blood LNAA pattern and brain TRP. In normal rats, the ingestion of carbohydrates leads within 1-2 hours to reductions in the blood levels of the BCAA, tyrosine, and phenylalanine, and to an increase in brain TRP concentration [3]. (Because the duration of the study is so short, the rise in brain TRP level is interpreted as suggesting a net increase in the rate of TRP uptake into brain.) The rise in brain TRP follows mainly from the reduced levels of its competitors in blood (especially the BCAA), inasmuch as the carbohydrate-induced rise in brain TRP can be blocked by co-ingestion of a mixture of BCAA, tyrosine, and phenylalanine [31]. If (1) the elevated blood levels of the BCAA are contributing to the reduction in brain TRP level seen in diabetic rats, because of increased competition with TRP for transport carriers, and (2) the rise in brain TRP that rapidly follows carbohydrate ingestion by normal rats is due mainly to the insulin-induced reduction in blood levels of the BCAA (thereby reducing competition with TRP for brain uptake), then one should not only find diminished steady-state levels of TRP in the diabetic brain, but also that the ingestion of carbohydrates fails both to lower blood BCAA levels and to raise brain TRP levels. As indicated in **table 1**, such indeed is the case: no rise in brain TRP follows glucose ingestion by diabetic rats, though an increase does occur in normal rats. (In this table, the results for the BCAA levels in blood are reflected in those for the serum TRP ratio. This ratio is simply the ratio of the serum concentrations of (total) TRP to the sum of those of the BCAA, tyrosine, and phenylalanine. Since under the experimental conditions employed here, the changes in the blood concentrations of the BCAA are always the largest, the ratio is a good indirect indicator of the changes in blood BCAA levels. Since the ratio did not rise appreciably in diabetic rats intubated with glucose, blood levels of the BCAA were not altered by this treatment [See reference 15 for the individual values of each BCAA]. More generally, the serum TRP ratio is used as an index of the competitive position of TRP in blood relative to the other LNAA [4]. Note in table 1 that carbohydrate ingestion by normal but not diabetic rats raised the serum TRP ratio and brain TRP, a result again consistent with the view that the change (or absence of change) in brain TRP follows indirectly from competition-related alterations in brain TRP uptake.)

It should again be emphasized that all these results support indirectly the hypothesis that TRP transport into brain is below normal in diabetes. Single-pass techniques [e.g., 28,29] for studying unidirectional uptake provide important information about carrier kinetic constants, but they too are indirect, in that uptake is estimated from a bolus of injected fluid, not from blood directly. Though arterio-venous differences have been used [17] to infer changes in amino acid uptake across the diabetic rat brain, and give results consistent with the arguments in the above discussion, a more direct, kinetic study is needed (e.g., using the indicator-dilution method) to evaluate the brain uptake of TRP and of the other LNAA under the conditions prevailing *in vivo* in the bloods of normal and diabetic rats.

Finally, if an alteration in the competitive LNAA transport of tryptophan into brain provides an attractive hypothesis for explaining the reduction in brain TRP levels in diabetes, a change in the binding of TRP to serum albumin does *not*. Though detailed studies remain to be performed, the fact that serum non-esterified fatty acid (NEFA) levels are often elevated in diabetes should predict an *increase* in serum free TRP levels, since NEFA displaces TRP from albumin binding sites [32]. If this is the case, then such data will tend to support the notion that variations in the size of the free TRP pool do not bear on the access of TRP to brain in diabetes. Limited data concerning this question are already available [13]. However, they summarize experiments in which the animals were studied just 72 hr post-streptozotocin treatment, and plasma NEFA levels had not changed significantly. Nonetheless, serum free TRP levels *had* increased and brain TRP decreased, suggesting the likely outcome of further experimentation (i.e., to dissociate these two variables).

Conclusions

The results described in this short review support the notions that 5HT synthesis is reduced in the diabetic brain, secondary to reduced brain TRP levels (and as a consequence, TRP hydroxylation rate). The decline in brain TRP seems likely to follow from diminished uptake from blood via the competitive LNAA transport carrier, owing to the sizeable increases in blood levels of its competitors, the BCAA. Such appears also to hold true for the other aromatic amino acids, tyrosine and phenylalanine, whose brain levels are also reduced in diabetic rats.

If 5HT synthesis rate and 5HIAA levels are decreased in the diabetic brain, but 5HT concentration is only slightly modified, perhaps these effects together indicate reduced 5HT release from brain neurons. If so, then brain functions controlled in part by neuronal circuits containing 5HT neurons may be altered. Candidate brain functions would include regulation of appetite and nutrient intake, regulation of blood pressure, and control of pituitary hormone secretion, among others. It remains to be determined if such functions are predictably modified in diabetes, and if so, if the changes can be linked experimentally to a deficiency in 5HT release by brain neurons. Hopefully, such studies will be forthcoming. If they show that specific disturbances in brain tryptophan, and in the synthesis and turnover of 5HT, do induce predictable changes in brain functions, they would potentially be of considerable therapeutic significance.

Acknowledgements

The studies described in this review were supported in part by a grant from the Kroc Foundation.

References

1. Carlsson, A., and Lindqvist, M.: Naunyn-Schmiedeberg's Arch. Pharmacol. 303, 157-164 (1978).
2. Fernstrom, J.D., and Wurtman, R.J.: Science 173, 149-152 (1971).
3. Fernstrom, J.D., and Wurtman, R.J.: Science 174, 1023-1025 (1971).
4. Fernstrom, J.D., and Wurtman, R.J.: Science 178, 414-416 (1972).
5. Knott, P.J., and Curzon, G. Nature 239, 452-453 (1972).
6. Hirsch, M.J. and Fernstrom, J.D.: J. Neurochem. 28, 877-879 (1977).
7. Green, H., Greenberg, S.M., Erickson, R.W., Sawyer, J.L., and Ellison, T.: J. Pharmacol. Exptl. Therap. 136, 174-178 (1962).
8. Wang, H.L., Harwalkar, V.H., and Waisman, H.A.: Arch. Biochem. Biophys. 96, 181-184 (1962).
9. Boggs, D.E., and Waisman, H.A. Arch. Biochem. Biophys. 106, 307-311 (1964).
10. Yuwiler, A., Geller, E., and Slater, G.G.: J. Biol. Chem. 240, 1170-1174 (1965).

11. Baldessarini, R.J., and Fischer, J.E.: Nature New Biology 245, 25-27 (1973).
12. Cummings, M.G., James, J.H., Soeters, P.B., Keane, J.M., Foster, J., and Fischer, J.E.: J. Neurochem. 27, 741-746 (1976).
13. Curzon, G., and Fernando, J.C.R.: Brit. J. Pharmacol. 60, 401-408 (1977).
14. MacKenzie, R.G., and Trulson, M.E.: J. Neurochem. 30, 205-211 (1978).
15. Crandall, E.A., and Fernstrom, J.D.: Diabetes 29, 460-466 (1980).
16. Crandall, E.A., and Fernstrom, J.D.: Diabetes 32, 222-230 (1983).
17. Brosnan, J.T., Man, K.C., Hall, D.E., Colbourne, S.A., and Brosnan, M.E.: Amer. J. Physiol. 244 (Endocrinol. Metab. 7), E151-E158 (1983).
18. Carlsson, A., Kehr, W., Lindqvist, M., Magnussen, T., and Atack, C.V.: Pharmacol. Rev. 24, 371-384 (1972).
19. Kaufman, S. In: AROMATIC AMINO ACIDS IN THE BRAIN, edited by G.E.W. Wolstenholme and D.W. Fitzsimons. Elsevier: North Holland, pp. 85-108 (1974).
20. Crandall, E.A., Gillis, M.A., and Fernstrom, J.D.: Endocrinology 109, 310-312 (1981).
21. Tozer, T.N., Neff, N., and Brodie, B.: J. Pharmacol. Exptl. Therap. 153, 177-182 (1966).
22. MacKenzie, R.G., and Trulson, M.E.: J. Neurochem. 31, 157-160 (1978).
23. Trulson, M.E., and MacKenzie, R.G.: J. Pharmacol. Exptl. Therap. 212, 269-273 (1980).
24. Fernstrom, J.D. In: Serotonin in Health and Disease, Volume 3: The Central Nervous System, edited by W.B. Essman. New York: Spectrum, pp. 1-49 (1978).
25. James, J.H., Escourrou, J., and Fischer, J.E.: Science 200, 1395-1397 (1978).
26. Zanchin, G., Rigotti, P., Bettineschi, F., Vasanelli, P., and Battistin, L.: J. Neurosci. Res. 4, 291-299 (1979).
27. McCall, A.L., Millington, W.R., and Wurtman, R.J.: Proc. Natl. Acad. Sci. (US), in press.
28. Oldendorf, W.H.: Amer. J. Physiol. 221, 1629-1639 (1971).
29. Pardridge, W.M., and Oldendorf, W.H.: Biochim. Biophys. Acta 401, 128-136 (1975).
30. Frayser, R., and Buse, M.G.: Diabetologia 14, 171-176 (1978).
31. Fernstrom, J.D., Faller, D.V., and Shabshelowitz, H.: J. Neural Transm. 36, 113-121 (1975).
32. McMenamy, R.H., and Oncley, J.L.: J. Biol. Chem. 233, 1436-1447 (1958).

ANATOMY AND FUNCTION OF CENTRAL SEROTONERGIC NEURONS

Hans Georg Baumgarten, Hans Georg Schlossberger

Dept. Neuroanatomy, Freie Universität Berlin-West,
Max-Planck-Institut für Biochemie, Martinsried, GFR

Recent studies on the developing and adult rat CNS have shown that the classification of serotonergic neurons introduced by Dahlström and Fuxe in 1964 (B_1 - B_9) can be replaced by the more conventional nomenclature used for subdividing reticular nuclei in the brainstem which is based on classical methods. Table 1 delineates our concept on the localization of serotonergic cells throughout the rat brainstem. This concept is mainly derived from material labelled with ^{14}C-5,7-dihydroxytryptamine, a serotonin analogue known to be accumulated in serotonergic structures with a remarkable degree of selectivity provided uptake into catecholaminergic neurons is prevented (cf. Baumgarten et al. 1982).

The groups of serotonin neurons found in the adult rat CNS develop from two separate primordia, a mesencephalic and metencephalic primordium, with a gap of cells at the isthmus rhombencephali (Lidov and Molliver 1982). The ultimate nuclear configuration is attained by an interaction of the following mechanisms: migration of neuroblasts, peri- and postnatal death of some neurons, fusion of the brainstem along the midline resulting in the formation of a glial raphé, and displacement of subgroups of serotonin neurons by the development of non-serotonin nuclei and pathways.

The serotonergic nuclei in the adult rat, cat and in primates contain non-serotonin neurons, the percentage of which varies in each nucleus. According to Léger and Wiklund (1982), approximately 67% of the medium-sized cells in the nucleus dorsalis raphé, 55% in the nucleus raphé pallidus and 35% of cells in the nucleus centralis superior are serotonergic neurons; for nucleus raphé pontis and magnus, Léger and Wiklund have calculated that only 9% and 15% of the neurons are indoleamine-containing. These figures are important when serotonin deprivation is studied by unspecific lesioning of the raphé nuclei; such experiments are of limited value because

Progress in Tryptophan and Serotonin Research

T a b l e 1

Serotonin cell groups in the rat brainstem

1) Nucleus raphê dorsalis, substantia grisea centralis of mesencephalon and paramedian subependymal gray of pons extending caudally up to the rhombencephalic isthmus (corresponding to group B7 and B6); dorsolaterally, serotonin cells extend through the pedunculus cerebellaris superior into the cuneiform nucleus of the midbrain tegmentum.

2) Nucleus linearis intermedius and centralis superior (collectively called medianus raphê); the cells in the ventral portion of the nucleus centralis superior are continuous with serotonin cells in a) the dorsal and lateral aspects of the nucleus interpeduncularis and within the ventromedian midbrain tegmentum including the adjacent substantia nigra, and b) the nucleus reticularis tegmenti pontis of Bechterew (B8).

3) A small group of serotonin neurons within the ventral part of the nucleus locus coeruleus and subcoeruleus area and a periventricular group of lower pontine/medullary serotonin neurons scattered in the subependymal neuropil underneath the floor of the IVth. ventricle extending from the n. locus coeruleus to the vestibular comlex (B4).
This group of cells supplements the pontine serotonin neurons within the periventricular gray (B6) which belongs to the dorsal raphê nucleus (B7 and B6). The dorsomedian serotonin cells of groups B7, B6 and B4 may be considered as part of a superior brainstem pericentricular neuron system which may have similar projections.

4) Nucleus raphê pontis (B5) and nucleus reticularis tegmenti pontis. It represents a continuation of the n. medianus raphê (B8).

5) Nucleus raphê magnus (B3), cells within the nucleus reticularis gigantocellularis and paragigantocellularis lateralis.

6) Nucleus raphê obscurus (B2) and scattered neurons within the nucleus reticularis gigantocellularis and reticularis ventralis.

7) Nucleus raphê pallidus (B1 medial) and neurons in the gigantocellular and ventral reticular nucleus and, a ventrolateral group, located mainly ventral to the dorsal and medial accessory olivary nuclei (B1 lateral); according to Andrezik et al (1981), this may be a caudal portion of the nucleus paragigantocellularis lateralis. Scattered neurons encroach upon the decussating fibres of the pyramidal tract in the rostralmost spinal cord.

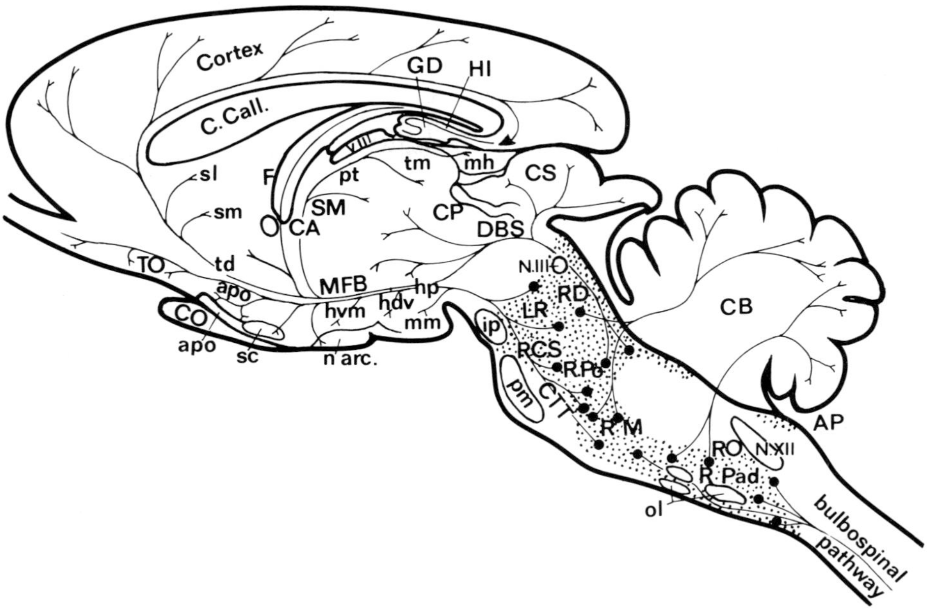

Fig.1: Paramedian-sagittal view of rat brain depicting the nuclei of origin of the central serotonin system (stippled area in brainstem) and their ascending and descending projections. AP, Area postrema; apo, area preoptica; CB, cerebellum; C.Call., Corpus callosum; CA, Commissura anterior. CO, Chiasma opticum; CP, Commissura posterior; CS, Colliculus superior; CTT, Central tegmental tract; DBS, Dorsal bundle of Schütz; F, Fornix; GD, Gyrus dentatus; HI, Hippocampus; hdv, n.hypothalami dorsalis, ventral part; hvm, n. hypothalami ventromedialis; hp, area hypothalamica posterior; ip, n.interpeduncularis; LR, n.linearis raphé; mh, medial habenula; mm, n.mammillaris medialis; MFB, Median forebrain bundle; n arc, n.arcuatus hypothalami; N.III, n.oculomotorius; NXII, n.hypoglossi; ol, nuclei olivares; pm, medial pontine nuclei;pt, n.parataenialis thalami; RCS, n.raphé centralis superior; RD, n.raphé dorsalis; RM, n.raphé magnus; R Pad, n.raphé pallidus; RPo, n. reticularis pontis; RO, n.raphé obscurus; sc, n.suprachiasmatic hypothalami; sm, n.septi medialis; sl, n.septi lateralis; SM, Stria medullaris; To, Tuberculum olfactorium; td,n.tractus diagonalis; tm, n.thalami medialis; VIII, third ventricle.

non-serotonergic elements are ablated together with serotonin neurons. Even if lesioning of serotonergic neurons is rendered specific by neurotoxin administration, the resulting functional deficits may not only be attributed to disruption of serotonergic transmission since some serotonin neurons in the lower brainstem contain peptide transmitter candidates such as substance P and TRF (cf. Hökfelt et al. 1978; Gilbert 1981).

Figure 1 portrays the main projections orginating in the paramedian serotonergic nuclei. The central superior and dorsal raphê nucleus give rise to a large telodiencephalic pathway which penetrates the ventral tegmentum to join the medial forebrain bundle. At various levels of the forebrain, groups of serotonergic fibres leave the MFB and become associated with known myelinated trajectories through which they reach target fields, such as the ansa peduncularis-ventral amygdaloid bundle (innervating the amygdaloid complex, hippocampal formation, lateral and posterior cortex, ventral thalamus and corpus striatum).

At the telodiencephalic junction, fibres are seen to enter the stria medullaris, terminalis, the fornix and cingulum (to reach the amygdala, thalamus, hippocampal formation and medial and dorsal cortex). More rostral fibre components innervate the tuberculum olfactorium, septum, nucleus accumbens, transition cortex and main olfactory bulb.

Additional fibres from the mesencephalic cell groups ascend together with the bundle of Schütz (to reach the periventricular thalamic and hypothalamic areas); this periventricular projection system is supplemented by axons from serotonin neurons scattered in the lower pontine/upper medullary subependymal gray; other fibres reach the cerebellum via the superior cerebellar peduncle. It is important to note that almost all ascending fibres to the forebrain project to ipsilateral target fields. Projections from the two major mesencephalic serotonergic nuclei (centralis superior and dorsalis raphê) overlap to a large extent; thus, they need not be considered separately. Both nuclei appear to have a minor descending projection to the brainstem and spinal cord which largely overlaps with the major descending bulbospinal pathways which originate in the nucleus raphê magnus, obscurus and pallidus. By means of short collaterals, the latter nuclei and the rostral groups of serotonin neurons innervate many brainstem centres, among those the substantia nigra, the interpeduncular nucleus, the motor nuclei of the cranial nerves, the inferior olivary complex, locus coeruleus, periventricular gray and the raphê nuclei. Within the spinal cord, descending serotonergic axons are contained in the lateral and anterior funiculus. These axons supply the gray matter of all segments of the spinal cord. A rich plexus of terminals is found in lamina 1 and 2 of the dorsal column (Ruda et al. 1981) and in the intermediolateral column. Significant contacts to α- and γ-motoneurons have also been demonstrated.

A characteristic feature of many serotonergic neurons is the high degree of axonal collateralization, i.e. one neuron innervates many targets along the neuraxis suggesting that changes in the firing rate of these neurons may alter the functional state of non-serotonin neurons in many circuits throughout the CNS. However, as shown recently, not all of the serotonin neurons conform to this collateralized type of reticular neuron (Fallon and Loughlin 1982).

The relationship of serotonin axons to target cells differs from that of other neurotransmitter systems in several respects: in most CNS regions, the releasing-points, i.e. the axonal varicosities, do not (or only to a minor degree) form true synaptic contacts. The preponderance of passing-by synaptoid contacts has lead to the concept of diffuse, neurohormonal actions of serotonin on its potentially variable targets; according to this hypothesis, serotonergic stimuli would only indirectly affect transmission in target neurons by altering their responsiveness to other afferents. There is physiological evidence to support this concept, e.g. serotonergic facilitation of glutamate-mediated excitation of motoneurons in the absence of direct effects of serotonin on these cells (McCall and Aghajanian 1979). Anatomically, such synaptoid contacts are mainly of the axodendritic Gray I type, and less frequently axosomatic ones. So far, the substantia gelatinosa and cerebellar mossy fibre rosettes (provided they contain 5-HT) represent the only target regions in which all established contacts are junctional (Chan-Palay 1977; Ruda et al. 1981). Another unique type of relationship concerns ependymal structures; within the ventricular cavities, varicosities of serotonergic axons form abundant non-junctional contacts to the ependymal cells, often showing small desmosome-like attachements (Richards et al. 1980, 1981). When released from these axons into the CSF, serotonin may interact with receptors located at the ventral aspect of the lower brainstem thereby influencing central regulation of respiration and vasomotion. Non-junctional axo-ependymal (and axo-axonic) contacts seem also to occur in the median eminence (Baumgarten and Lachenmayer 1974; Calas 1977). By contrast, all contacts

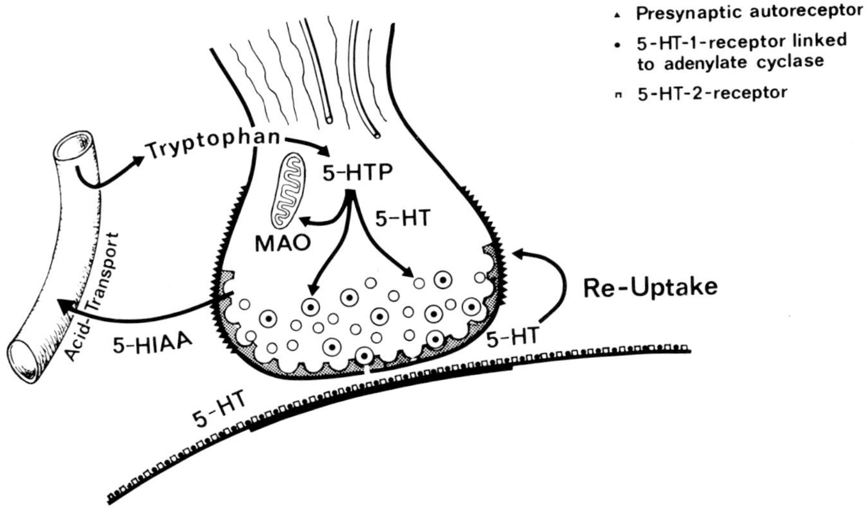

Fig. 2:
Schematic portray of a serotonergic synapse, consisting of a presynaptic nerve ending and postsynaptic membrane. Tryptophan uptake appears to be a carrier mediated process which can be inhibited by neutral amino acids. Tryptophan is converted to 5-hydroxytryptophan (5-HTP) by tryptophan-5-hydroxylase, an enzyme which requires oxygen and biopterin as cofactor. This enzyme is selectively localized in serotonergic neurons. 5-HTP is decarboxylated by unspecific aromatic L-amino acid decarboxylase which has a near ubiquitous distribution. The resulting 5-HT is either translocated to the storage vesicles or metabolized to 5-hydroxyindole acetic acid (5-HIAA) by mitochondrial monoamine oxidase (mainly the A-form). Released 5-HT may interact with any of the pre- and postsynaptic receptors and is mainly inactivated by re-uptake into the nerve ending and subsequent metabolism by MAO or re-incorporation into the synaptic vesicles. The presynaptic autoreceptor when stimulated by released 5-HT inhibits serotonin release and serotonin formation (autocontrol). The ultimate metabolite 5-HIAA leaves the brain by an active transport mechanism located at the capillary endothelium and choroid plexus epithelium. Part of the 5-HIAA uses the CSF as vehicle and can be recovered in lumbar taps.

established by serotonergic axons to the basal processes of the specialized, secretory ependyma in the subcommissural organ of the rat (Baumgarten et al. 1978; Møllgard and Wiklund 1979) are junctional contacts. Circumstantial evidence suggests dendro-dendritic contacts within the serotonergic nuclei that are responsible for auto-inhibition of firing (autoregula-

Table 2

Proposed classification of central 5-HT receptors

Presynaptic receptor	autoreceptor on 5-HT neurons mediating inhibition of firing (S_2-receptor of Aghajanian, 1981), LSD and various indoleamine hallucinogens (psilocin, N,N-dimethyltryptamine and 5-methoxy-N,N-dimethyltryptamine) are powerful agonists as is the non-hallucinogenic ergot derivative lisuride. "Classical" 5-HT antagonists (see below) are ineffective. Resembles the presynaptic inhibitory α_2 (auto)-receptor on adrenergic neurons.
5-HT-1 receptor	high-affinity (postsynaptic) receptor which binds ^{3}H-5-HT and ^{3}H-LSD at nanomolar concentrations (S_3-receptor of Aghajanian 1982); binding to this receptor is regulated by guanine nucleotides and a 5-HT-sensitive adenylate cyclase has been found in the immature and adult rat brain (v.Hungen et al. 1978; Enjalbert et al. 1978a,b; Fillion et al. 1979a,b). The 5-HT-1 receptor suppresses neuronal activity and is widely distributed in the forebrain. "Classical" 5-HT antagonists are ineffective with the exception of methergoline and lisuride which bind to this receptor at nanomolar range (Peroutka and Snyder 1982). The "synaptic" connections of 5-HT axons which operate through this type of receptor remain to be elucidated (see text for possible relationship of serotonin axons to targets).
5-HT-2 receptor	postsynaptic receptor which binds ^{3}H-5-HT, ^{3}H-LSD and ^{3}H-spiroperidol at micromolar concentrations (Leysen et al. 1978, 1982) and mediates excitation in the periphery and in the CNS. Sensitive to blockade by "classical" 5-HT antagonists such as methysergide, cinanserin or cyproheptadine. In brain, such receptors have been found on motoneurons and on reticular neurons. They are involved in elicitation of the "serotonin behavioural syndrome" consisting of head-waving, hind-limb abduction, tremor, forepaw padding and myoclonic fits. Several neuroleptic drugs have affinity to this receptor and block various aspects of this syndrome by a) interference with 5-HT transmission at 5-HT-2 receptors and b) interference with dopaminergic transmission at postsynaptic DA receptors (see text). The number of 5-HT-2-binding sites is decreased following chronic antidepressant drug administration (Peroutka and Snyder 1980).

tory control mediated by presynaptic inhibitory receptors; cf. Aghajanian 1981). Facilitatory actions of 5-HT (as those exerted in motor neurons of the brainstem and spinal cord and in reticular neurons) are mediated by receptors that are sensitive to blockade by the well known serotonin antagonists (e.g. cyproheptadine, methysergide, cinanserin; cf. Table 2 for receptor classification); these excitatory 5-HT-2-receptors may be directly localized on the target neurons and, therefore, not on interneurons). Inhibitory effects of serotonergic neurons exerted in many forebrain areas appear to be mediated by a different type of serotonin receptor for which selective blockers are not yet available (cf. Aghajanian 1981); these 5-HT-1 receptors, that are probably coupled to an adenylate cyclase, may be located on inhibitory interneurons (and would have to be considered as excitatory receptors), on excitatory afferents to target neurons (and would thus presynaptically inhibit an excitatory transmitter) or on the target neurons themselves where they should raise the threshold for excitatory inputs. Future studies are required to clarify this issue. To understand the action of drugs on serotonergic mechanisms, the properties of serotonin nerve terminals need to be considered (cf. Fig. 2). The serotonergic varicosity has the entire enzymatic machinery necessary for serotonin formation; maintenance of this capacity requires intact axoplasmic transport mechanism. The rate-limiting enzyme tryptophan-5-hydroxylase requires oxygen and tetrahydrobiopterin (THB_4) for optimum function; the cofactor regenerating enzyme appears to be present within the serotonergic neurons (Levine et al. 1979). Cofactor deficiencies impair serotonin synthesis and may be involved in serious mental disturbances, e.g. in patients with atypical phenylketonuria (cf. Niederwieser et al. 1983). Under fasting conditions, tryptophan hydroxylase is not saturated with its substrate; increases in serum free and, subsequently, brain tryptophan concentrations e.g. following raises in unesterified fatty acids or displacement of albumin-bound tryptophan by drugs (e.g. salicylate) or ingestion of large amounts of tryptophan (only 1-3% reach the brain) result in increases of serotonin synthesis (Fernstrom 1981). This does not necessarily imply enhanced serotonergic transmission in brain since the additionally formed 5-HT may be translocated to storage vesicles or may be metabolized by MAO or, when released, may stimulate inhibitory autoreceptors. Therefore, the serotonergic axonal varicosities have a rigidly inbuilt homeostatic economy (autoregulatory

Table 3

Drugs with known effects on central serotonergic mechanisms

Serotonin-depleting compounds

p-Chloroamphetamine: depletion of cerebral 5-HT by a) acute release of 5-HT from vesicular stores, inhibition of 5-HT re-uptake and interference with MAO metabolism, and b) destruction of axon terminals in some cell-body-distant target fields. The retarded neurotoxic action appears to depend on the formation of an active metabolite formed by interaction with microsomal hydroxylase(s). Only used in animal experiments.

Fenfluramine: serotonin-releasing drug similar to p-chloroamphetamine; since it is less potent a neurotoxin in comparison with chloroamphetamine, it can be administered to humans. By indirectly stimulating serotonin receptors in the hypothalamus it reduces preferentially non-protein caloric intake and serves as an appetite suppressant.

Drugs which interfere with the inactivation of serotonin by selectively blocking its reuptake

Zimelidine: selective serotonin uptake blocker which by liver metabolism is transformed into norzimelidine. Norzimelidine has an even longer half life than its parent compound and inhibits 5-HT transport even more effectively. It potentiates the endocrine and motor effects of 5-HTP, is non-sedative, has no effect on convulsion threshold, enhances morphine analgesia and fails to potentiate alcohol or barbiturate sedation in contrast to many of the non-selective tricyclic antidepressants (e.g. imipramine, chloroimipramine, amitriptyline).

Precursors in the biosynthetic pathway

L-Tryptophan: only cells which have tryptophan-5-hydroxylase are able to convert tryptophan to 5-HTP and thus to serotonin. Cells which lack this enzyme and take up and decarboxylate tryptophan may form tryptamine, an intermediate which has so far not been shown to be a separate transmitter. Only 1-3% of an oral dose of L-tryptophan will be available for serotonin synthesis in brain, thus it is questionable whether it is possible to enhance central serotonergic functions by moderate doses of tryptophan (e.g. 1g/dose given to humans). The dioxygenase pathways and the transamination and methylation pathways give rise to a large number of metabolites which may cause undesired side effects. Severe insomnia and depression are not ameliorated by even high doses of tryptophan. In mild depression and insomnia, tryptophan is not clearly superior to placebo.

L-5-Hydroxytryptophan: Immediate precursor in the biosynthesis of 5-HT which can be taken up and decarboxylated in everey peripheral and central cell that possess decarboxylase activity. Its oral administration in humans results in gastrointestinal side effects such as nausea and diarrhoea. This drug should, therefore, only be administered in conjunction with a peripheral decarboxylase inhibitor (e.g.carbidopa). A reasonable indication for 5-HTP is posthypoxic myoclonus in patients with lowered CSF 5-HIAA levels.

control) which also operates at the level of the cell bodies where released 5-HT causes inhibition of firing by the same type of autoreceptor (possibly via dendro-dendritic contacts). Released 5-HT is mainly inactivated by the extremely efficient high-affinity uptake mechanism (the ^{3}H-imipramine binding site). Blockade of this mechanism enhances serotonin transmission; however, upon chronic administration of uptake blockers (e.g. of Zimelidine), the serotonergic neuron and its target adapt to the increased availability of 5-HT in the synaptic space by a) reduction in 5-HT-turnover and b) changes in postsynaptic receptors sensitivity (decrease in 5-HT-2-binding sites, appeareance of a low-affinity binding site at the 5-HT-1-receptor, cf. Fuxe et al. 1983), and c) by decreasing the number of 5-HT-transport sites (imipramine-binding site). This compensatory down-regulation (to achieve homeostasis of the system) results in attenuation of serotonin-mediated (excitatory) effects, e.g. of tonic facilitation of prolactin and LH release (Fuxe et al. 1983). However, not all serotonin-modulated functions are altered, e.g. the inhibitory actions of 5-HT on TSH and growth hormone release are apparently uneffected indicating regional specificity of the adaptive long-term attenuation of serotonergic transmission (Fuxe et al. 1983). The functional consequences of this adaptive process are as yet poorly understood but should have an impact on noradrenergic and dopaminergic mechanisms since these are under inhibitory serotonergic control. Table 3 lists drugs with known effects on central serotonergic mechanisms.

Serotonergic modulation of state-dependent functions is well documented in endocrine regulation and sleep mechanisms though many details are still controversial. Tonic serotonergic stimuli are mediators of a central clock governing endogenous rhythms which result in chracteristic diurnal pattern of the temporal spacing and amplitude of the secretory profile of several hypothalamic and hypophysial hormones (CRF/ACTH; LRF/LH; TRF/TSH) that are harmonized with retino-hypothalamic afferences and other exogenous influences at the level of the suprachiasmatic nucleus (Kordon et al. 1981; Baumgarten et al. 1978). The endocrine state by itself appears to have profound effects on the physiology of serotonergic transmission (Bilgon et al. 1980) and the idea has been proposed that abnormal mental conditions such as depression -secondary to disturbances in sex hormones balance in females - might be susceptible to treatment with drugs improving serotonin transmission, e.g. serotonin reuptake inhibitors (Ladisich 1977). This concept has been expanded to a general hypothesis of depression which postulates alterations in serotonin receptor sensitivity/serotonin transport capacity/serotonin release and metabolism as a major pathophysiological mechanism of disturbed mood (Coppen 1967) supplementing previous concepts on the role of altered noradrenergic transmission in depression and mania (Schildkraut 1973). While the etiological role of 5-HT in depression is not settled, its importance in treating depressive disease by monoamine oxidase inhibitors and, more recently, by selective serotonin reuptake inhibitors (such as zimelidine and norzimelidine) is beyond dispute (see Aberg and Holmberg 1979; Coppen et al. 1979; Gershon et al. 1982).

The involvement of serotonin in sleep is still a matter of debate. Recent electrophysiological studies in freely moving cats suggest a role for serotonin in motor activity during REM sleep and in phasic and tonic arousal. Thus, SW sleep seems to correlate with and depend on a reduction of serotonin transmission (see Trulson and Jacobs 1981). At this point it seems relevant to mention that improvement of serotonergic transmission by reuptake blockers that are devoid of sedating side effects such as zimelidine cause an increase in alpha-activity of the EEG in humans accompanied by increased levels of attention and improvement in mental capacity and performance (cf. Schenk et al. 1980).

If serotonergic sensitization of motor neuron reactivity is essential for rapid initiation of motor performance, failure of serotonergic transmission might well be responsible for states of atonia in e.g. narcolepsy (cf. Lachenmayer et al. 1980 for references). This possibility should be investigated since there are various therapeutic ways for stabilizing serotonergic transmission (MAO-inhibitors and selective 5-HT uptake inhibitors which may be combined with low doses of tryptophan).

As recordings in awake cats were mainly performed on dorsal raphê neurons (Trulson and Jacobs 1981), it cannot be excluded that other serotonin neurons which are endowed with the task of inhibiting arousing sensory stimuli and mediate habituation to repeated acoustic, tactile and visual stimuli at the level of integrating relay nuclei in central sensory pathways may facilitate entry into SW sleep (see Vogt 1982 for details). Furthermore, the interaction between noradrenergic and serotonergic mechanisms in

the various sleep stages need to be studied in more detail. Apart from the role of serotonin in circadian hormone fluctuations, there is also evidence for paticipation of 5-HT in short-term, reflex-elicited endocrine functions. The best-documented example is the obligatory role of serotonergic mechanisms in suckling-induced prolactin surges (Kordon et al. 1973, Mena et al. 1976). Furthermore, serotonin augments the hypoglycemia-induced rise in plasma growth hormone levels and attenuates the stress-induced rise in ACTH (Müller et al. 1976); the role of 5-HT in these neuroendocrine reflexes is only permissive. Basal levels of prolactin secretion seem to be positively influenced by serotonin neurons and drugs which stimulate serotonergic transmission have been shown to enhance prolactin release from the pituitary (Fuller 1980).
Animals with lesions of the central serotonergic fibre systems show exaggerated responses to sensory stimuli, increased (intra- and interspecific) aggressiveness and, eventually, episodes of bizarre social behaviour (Baumgarten and Lachenmayer 1972) and increased sexual activity (da Prada et al. 1972); similar findings have been reported in para-chlorophenylalanine-treated cats (Holyland et al. 1970) indicating that serotonin "protects the organism from damaging responses of environmental stimuli" (Vogt,1982).

That serotonin is involved in antinociception has been documented in many different experiments but the exact site(s) and mode of interaction with nociceptive pathways throughout the brain remain to be elucidated. As mentioned before, the extremely dense serotonergic innervation of neurons in lamina I and II (Ruda et al. 1981) suggests that an important site of interference with pain information processing is the spinal cord. The effects of morphine are attenuated (but not abolished) by destruction of the bulbospinal serotonergic projections using 5,6-DHT (Vogt 1982) indicating that morphine interacts with pain processing also at supraspinal levels of the CNS (see Snelgar and Vogt 1981 and Vasko and Vogt 1981). It remains to be shown whether peptides co-stored with serotonin in part of the lower brainstem 5-HT neurons are also involved in pain transmission and antinociception.

As outlined above, the electrical and glutamate-mediated excitability of brainstem and spinal cord motoneurons is greatly enhanced in the presence of serotonin (McCall and Aghajanian 1979; White and Neuman 1980). Serotonin also facilitates the excitation of motoneurons produced by stimulation of trigeminal afferents and the red nucleus (Aghajanian 1981). It is thus conceivable that serotonin when released in excess at brainstem and spinal cord motoneurons could initiate localized or widespread, disseminated myoclonic jerking similar to those produced by 5-HTP in rats with supersensitive postsynaptic 5-HT receptors (Stewart et al. 1976a, b) or in guinea-pigs with a high dose of 5-HTP or in rats or mice treated with either tryptophan or 5-HTP following prior MAO-inhibition (Grahame-Smith 1971a; Modigh 1972), by acutely serotonin-releasing drugs such as fenfluramine and p-chloroamphetamine (Trulson and Jacobs 1976, Green and Kelly 1976) and by drugs acting directly on postsynaptic serotonin and/or tryptamine receptors (e.g. 5-methoxy-N,N-dimethyltryptamine or tryptamine in high doses; cf. Grahame-Smith 1971b and Marsden and Curzon 1978). While it appears that brainstem and spinal cord serotonergic mechanism are certainly involved in parts of the behavioural syndrome (Jacobs and Klemfuss 1975), most aspects of the syndrome require also intact dopaminergic transmission (as evidenced by the inhibitory effects of haloperidol and prior lesions of the DA-systems of 6-OH-DA, cf. Green 1981, Foldes and Costa 1975) and manifestation of the symptomatology is also due to interaction of 5-HT and/or tryptamine with tryptamine receptors in addition to serotonin receptors (Quock and Weick 1978; Marsden and Curzon 1978). Accordingly, it is not surprising that some 5-HT-2-receptor blockers, e.g.cyproheptadine, cinanserin and mianserin, fail to antagonize the major abnormalities provoked by 5-HTP or 5-methoxy-N,N-demethyltryptamine (see Green 1981 for details). It is possible that this syndrome has a human equivalent since certain forms of posthypoxic myoclonus are aggravated by 5-HTP whereas other forms respond favourably to 5-HTP (cf. Baumgarten et al. 1982 for details and for references). At higher levels of the CNS, serotonergic mechanisms appear to be involved in inhibition of motor activities rather than stimulation; this is expressed by elevation of the seizure threshold (for ECT, pentetrazol and acoustic stimuli) by raised 5-HT levels in the CNS (Kilian and Frey 1973). Therefore, subsystems of the central serotonin neurons may subserve antagonistic roles in brain function.

Further progress in our understanding of the role of serotonin in brain functions depends on more detailed information about the microcircuitry of

5-HT axon terminals in different brain areas, about the mechanisms of transmission at different synapses and on the development of antagonists that block the action of serotonin at (the still hypothetical) postsynaptic receptors that mediate suppression of neuronal activity (S_3 receptors of Aghajanian 1981; 5-HT_1 binding sites of Peroutka and Snyder 1979). According to Leysen (this volume), this binding site does not fulfill the criteria of a "physiological" receptor.

References

Åberg, A., Holmberg, G.: Acta psychiat. scand. 59, 45-58 (1979).

Aghajanian, G.K.: The MIT Press, Cambridge, Mass., London, England (1981).

Andrezik, J.A., Chan-Palay, V., Chan-Palay, S.L.: Anat. Embryol.161,355-371 (1981).

Baumgarten, H.G., Lachenmayer, L.: Z.Zellforsch.135, 399-414 (1972).

Baumgarten, H.G., Lachenmayer, L.: Z.Zellforsch.147, 285-292 (1974).

Baumgarten, H.G., Klemm, H.P., Lachenmayer, L., Björklund, A., Lovenberg,W., Schlossberger, H.G.: Ann. N.Y.acad. Sci. 305, 3-24 (1978).

Baumgarten, H.G., Jenner, S., Björklund, A., Klemm, H.P., Schlossberger, H. G.: Biology of Serotonergic Transmission, edited by N.N. Osborne, New York: John Wiley & Sons, 249-277 (1982).

Bilgon, A., Bercovitz, H., Samuel, D.: Brain Res. 187, 221-225 (1980).

Calas, A.: Advances in Biochem. psychopharmacology, 16, 79-88 (1977).

Chan-Palay, V.: Cerebellar dentate nucleus, Springer-Verlag (1977).

Coppen, A.: Br. J. Psychiat. 113, 1237-1264 (1967).

Coppen, A., Rama Rao, V.A., Swade, C., Wood, K.: Psychopharmacology,63, 199-202 (1979).

Dahlström, A., Fuxe, K.: Acta physiol.scand.62, 1-55 (1964).

Enjalbert, A., Bourgoin, S., Hamon, M., Adrien, J., Bockaert , J.: Mol. Pharmacol.14, 2-10 (1978a).

Enjalbert, A., Hamon, M., Bourgoin, S., Bockaert, J.: Mol. Pharmacol.14, 11-23 (1978b).

Fallon, J.H., Loughlin, S.E.: Brain Res. Bull.9, 295-307 (1982).

Fernstrom, J.D.: The MIT Press, Cambridge, Mass., London, England (1981).

Fillion, G., Rousselle, J.C., Beaudoin, D., Pradelles, P., Goiny, M., Dray, F., Jacob, J.: Life Sci, 24, 1813-1822 (1979a).

Fillion, G., Beaudoin, D., Rousselle, J.C., Denion, J.M., Fillion, M.P., Pray, F., Jacob, J.: Neurochem. 33, 567-570 (1979b).

Foldes, A., Costa, E.: Biochem. Pharmac. 24, 1617-1621 (1975).

Fuller, R.W., Ann. Rev. Pharmacol. Toxicol.20, 111-27 (1980).

Fuxe, K., Ögren, S.O., Aguati, L.F., Benfenati, F., Fredholm, B., Anderson, K., Zini, I., Eneroth, P.:Neuropharmacology, 22, 389-400 (1983).

Gershon, S., Bush, D.F., Georgotas, A.: The British Journal of Clinical Practice, 54-63 (1982).

Gilbert, R.F.T.: Thesis, Cambridge, U.K. (1981).

Grahame-Smith, D.G.: Br. J. Pharmac. 43, 856-864 (1971a).

Grahame-Smith, D. : J. Neurochem. 18, 1053-1066 (1971b).

Green, A.R., Kelly, P.H.: Br. J. Pharmac. 57, 141-147 (1976).

Green, A.R.: J. Physiol., Paris, 77, 437-447 (1981).

Hökfelt, T., Ljungdahl, A., Steinbusch, H., Verhofstad, A., Nilsson, G., Brodin, E., Pernow, B., Goldstein, M.: Neurosc. 3, 517-538 (1978).

Holyland, V.J., Shillito, E. E., Vogt, M.: Br. J. Pharmac. 40, 659-667(1970).

von Hungen, K., Roberts, S., Hill, D.F.: Neurosc.3, 517-538 (1978).

Jacobs, B.L., Klemfuss, H.: Brain Res. 100, 450-457 (1975).

Kilian, M., Frey, H.-H.: Neuropharmacology, 12, 681-692 (1973).

Kordon, C., Blake, C.A., Terkel, J., Sawjer, C.: Neuroendocrinology 13, 213-223 (1973).

Kordon, C., Hery, A., Szafarczyk, A., Ixert, G., Assenmacher, J.: J.Physiol. Paris, 77, 489-496 (1981).

Lachenmayer, L., Baumgarten, H.G., Wuttke, W.: Springer-Verlag,Berlin Heidelberg; Path. Erregbarkeit d.Nervensystems u.ihre Behandlung (1980).

Ladisich, W.: Psychoneuroendocrinology, 2, 203-314 (1977).

Lêger, L. Wiklund, L.: Brain Res. Bull.9, 245-252 (1982).

Levine, R.A., Kuhn, D.M., Lovenberg, W.: Neurochem. 32, 1575-1578 (1979).

Leysen, J.E., Sommeren, W., Laduron, P.: Arch. Int. Physiol. Biochim.86, 874-875 (1978).

Leysen, J.E., Tollenaere, J.P.: Biochem. Models f. Serotonin Receptors in Annual Rep. in Medicinal Chemistry, 17, 1-11 (1982).

Lidov, H.G.W., Molliver, M.E.: Brain Res. Bull.9, 559-604 (1982).

Marsden, C.A., Curzon, G.: Psychopharmacology, 57, 71-76 (1978).

McCall, R.B., Aghajanian, G.K.: Brain Res. 169, 11-27 (1979).

Mena, F., Enjalbert, A., Carbonell, L., Priam, M., Kordon, C.: Endocrinology 99, 445-452 (1976).

Modigh, K.: Psychopharmacologia, 23, 48-54 (1972).

Møllgard, K., Wiklund, L.; J. Neurocytol. 8, 445-467 (1979).

Müller, E.E., Udeschini, G., Tecci, C., Zambatti, F., Panerai, A., Vicenti, L., Cocola, F., Mantegazza, P.: Acta Endocr.82, 71-91 (1976).

Niederwieser, A., Curtius, H.-Ch., Levine, R.A., Lovenberg, W., Woggon, B., Angst, J., Müldner, H., Leupold, D.: This volume, 437.

Peroutka, S.J., Snyder, S.H.: Mol. Pharmacol. 16, 687-699 (1979).

Peroutka, S.J., Snyder, S.H.: Science 210, 88-90 (1980).

Peroutka, S.J., Snyder, S.H.: Biology of Serotonergic Transmission, edited by N.N.Osborne, John Wiley & Sons Ltd., 279-298 (1982).

Da Prada, M., Carruba, R.A.,O'Brian, A., Sauer, A., Pletscher, A.: Eur.J. Pharmacol. 19, 288-290 (1972).

Quock, R.M., Weick, B.G.: J. Pharm. Pharmac. 30, 280-283 (1978).

Richards, J.G., Lorez, H.P., Colombo, V.E., Guggenheim, R., Kiss, D., Wu, J. Y.: Neuroscience, 5, 1489-1502 (1980).

Richards, J.G. Lorez, H.P., Colombo, V.E., Guggenheim, R., Kiss, D.: J. Physiol. Paris, 77, 219-224 (1981).

Ruda, M., Allen, B., Gobel, St.: J. Physiol., Paris, 77, 205-209 (1981).

Schenk, G.K., Filler, W., Rauft, W., Zerbin, D.: Acta Psychiatrica Scandinavica 63, 303-313 (1980).

Schildkraut, J.J.: Am.J. Psychiatry, 130, 695-699 (1973).

Snelgar, R.S., Vogt, M.: J. Physiol. 314, 395-410 (1981).

Stewart, R.M., Growdon, J.H., Cancian, D., Baldessarini, R.J.: Neurology 26, 690-692 (1976a).

Stewart, R.M., Growdon, J.H., Cancian, D., Baldessarini, R.J.: Neuropharmacology, 15, 449-455 (1976b).

Trulson, M.E., Jacobs, B.L.: Eur. J. Pharmac. 36, 149-154 (1976).

Trulson, M.E., Jacobs, B.L.: The MIT Press, Cambridge, Mass., London, England (1981).

Vasko, M.R., Vogt, M.: Br. J. Pharmac. 73, 245P-246P (1981).

Vogt, M.: Biology of Serotonergic Transmission, edited by N.N. Osborne, John Wiley & Sons Ltd. 299-314 (1982).

White, S.R., Neumann, R.S.: Brain Res. 188, 127 (1980).

SEROTONIN RECEPTOR BINDING SITES: LOCALIZATION AND ROLE IN THE BRAIN AND PERIPHERAL TISSUES

Josée E. Leysen

Department of Biochemical Pharmacology, Janssen Pharmaceutica,
B-2340 Beerse, Belgium.

Introduction

Different hypotheses on distinct types of serotonin receptors have accrued from research using various pharmacological or biochemical approaches. In early pharmacological studies on isolated tissues, the existence of 'M'- and 'D'-type serotonergic receptors on gastrointestinal smooth muscle was postulated (1, 2). More recently, a different type of serotonin receptor was detected on vascular and bronchial smooth muscle (3, 4). In *in vivo* pharmacological studies, various behavioural models measuring serotonergic effects were described: the serotonin syndrome in rats and mice (5); tryptamine-induced clonic seizures in rats (6); mescaline or 5-hydroxytryptophan-induced head twitches in rats and mice (7, 8). Electrophysiological research in rat brain suggested the occurrence of postsynaptic inhibitory and excitatory receptors (9) and of serotonin autoreceptors (10). Finally from biochemical investigations, three distinct models of putative serotonin receptor sites were put forward: the serotonin stimulated adenylate cyclase (11), the serotonin-S_1 binding sites labelled with ^{3}H-serotonin (12) and the serotonin-S_2 binding sites, labelled with potent serotonin antagonists (13, 14, 15).
In this review the features and localization of the biochemically detected putative serotonin receptor sites and their relation to various pharmacological serotonin receptor models will be discussed.

Serotonin stimulated adenylate cyclase. Von Hungen *et al.* (11) detected in *in vitro* preparations of the colliculi an adenylate cyclase showing enhanced 3',5'-adenosine monophosphate (c-AMP) production in the presence of μmolar concentrations of serotonin or of its structural congeners. The enzyme has a very limited occurrence (16, 17, 18, 19) and is shown to be unrelated to serotonin-S_1 and serotonin-S_2 binding sites (20).

Progress in Tryptophan and Serotonin Research

IC_{50}-values (concentration inhibiting 50 % of the stimulated enzyme activity) of serotonin antagonists are presented in Table 1.

Serotonin-S_1 binding sites. Binding sites labelled by nanomolar concentrations of ^3H-serotonin in membrane preparations of rat brain (12) were termed 5-hydroxytryptamine$_1$ (5-HT_1) or serotonin-1 (S_1) binding sites (21, 22). The binding affinity of ^3H-serotonin was found to be enhanced by mmolar concentrations of Ca-ions (23) and is decreased by guanyl nucleotides (18, 24). For a series of compounds, the binding affinities, expressed as K_i-values (equilibrium inhibition constants, indicating the concentration required for half-maximal occupation of the binding sites), for serotonin-S_1 binding sites are presented in Table 1. Structural congeners of serotonin reveal high binding affinity, but structurally different serotonin agonists (quipazine, mescaline) bind very weakly to the sites, and so do various neurotransmitters. Amongst various serotonin antagonists, the ergoline derivatives, possessing an indole moiety, and the tricyclic derivative, metitepine bind rather potently to the sites with K_i-values between 10 to 100 nM. Serotonin antagonists of other chemical classes have very weak binding affinity for the serotonin-S_1 binding sites. In rat brains serotonin-S_1 binding sites are mostly concentrated in the hippocampus and the striatum (maximal number of binding sites 25 pmoles/g tissue). The relative distribution in various brain areas is shown in Table 2.
In studies using ^3H-lysergic acid diethylamide (^3H-LSD) as the radioligand and hippocampal membrane preparations, the labelled sites showed the features of serotonin-S_1 binding sites (25). However, in brain areas containing high levels of both serotonin-S_1 and serotonin-S_2 binding sites (see below), ^3H-LSD labels both sites concomitantly (21).
High and low affinity ^3H-serotonin binding sites were purportedly detected in human platelets (27). However, it can be calculated that the number of high affinity binding sites reported in that study (1.5 fmoles or 33 DPM per 10^8 platelets) falls within the experimental variations of the radioactivity counting.

Serotonin-S_2 binding sites. In 1977, ^3H-spiperone was found to label with high affinity sites in rat frontal cortical membrane preparations for which various serotonin antagonists showed high binding affinity. The

Table 1. Potencies of various compounds in biochemical models of serotonin receptors

Compounds, grouped by type*		1. Adenyl Cyclase Colliculus (19) IC_{50}, nM	2. S_1-Sites Hippocampus (25, 26) K_i, nM	3. S_2-Sites Frontal Cortex (15) K_i, nM
(i)	pirenperone	-	>> 1 000	0.28
	ketanserin	-	>> 1 000	0.39
(ii)	pizotifen	68 300	1 500	0.28
	metitepine	11 900	62	0.39
	cyproheptadine	11 500	700	0.44
	mianserin	27 600	1 100	1.4
	chlorpromazine	-	3 100	3.3
	amitryptiline	-	1 200	4.2
	clomipramine	-	>> 1 000	9.8
	imipramine	-	>> 1 000	37
	desimipramine	-	>> 1 000	78
(iii)	cinanserin	21 500	3 500	2.0
(iv)	spiperone	3 800	160	0.53
	pipamperone	-	5 000	0.78
	haloperidol	-	>> 1 000	22
(v)	metergoline	18 100	20	0.28
	methysergide	-	99	0.94
	LSD	-	20	2.5
(vi)	bufotenin	-	62	118
	serotonin	-	7.8	296
	tryptamine	-	197	1 500
(vii)	quipazine	no activation	1 600	235
(viii)	mescaline	-	70 000	5 900
(ix)	dopamine	-	20 000	93 500
	epinephrine	-	-	47 000
	norepinephrine	-	200 000	>>100 000
	histamine	-	-	>>100 000
	acetylcholine	-	>>100 000	120 000

Spearman rank correlation:	1 versus 2 (n=7)+ r_s=0.571, N.S.	
(N.S. = not significant, Student's t-test)	1 versus 3 (n=7)+ r_s=0.09, N.S.	2 versus 3 (n=23)+ r_s=0.073, N.S.

* Compound types: (i) 4-(p-fluorobenzoyl)piperidines; (ii) tri- and tetracyclic compounds related to cyproheptadine; (iii) cinanserin; (iv) butyrophenones; (v) ergolines; (vi) indole derivatives; (vii) quinoline derivative; (viii) phenylethylamine derivative; (ix) neurotransmitters. Groups (i) through (iv) are antagonists; group (v) includes putative mixed agonist-antagonists; groups (vi) through (viii) are agonists.

+ Calculated for compounds in groups (i) through (viii).

Table 2. Proportional distribution in rat brain subcortical areas of serotonin-S_2 and serotonin-S_1 binding sites and serotonin uptake

	S_2	S_1	Uptake
prefrontal cortex	100	100	100
tuberculum olfactorium	56	67	156
nucleus accumbens	40	-	92
striatum	36	150	93
substantia nigra	15	-	131
thalamus	10	82	132
hypothalamus	9	-	109
medulla oblongata	10	-	87
hippocampus	12	198	67
spinal cord	1	-	48
cerebellum	0	0	17

IC_{50} values of the antagonists correlated with their potencies to antagonize tryptamine-induced clonic seizures in rats. It was concluded that these were serotonergic receptor sites which were distinct from dopaminergic receptor sites in the striatum, also labelled by ^{3}H-spiperone (13, 14). Later on, the ^{3}H-spiperone labelled serotonergic receptor sites were designated as 5-HT_2 or serotonin-S_2 receptor sites (21, 22). Similar sites were reported to be detectable by using ^{3}H-mianserin, however, this ligand proved neither to be selective, since it concomitantly interacts with serotonin-S_2 and histamine-H_1 binding sites (28). The first selective radioligand for serotonin-S_2 binding sites was the novel serotonin antagonist ^{3}H-ketanserin (15, 26). Specific binding of ^{3}H-ketanserin defined by inhibition with methysergide, apparently represents binding to a single class of sites and is of nanomolar affinity. The equilibrium dissociation constant (K_D) at 37° C was found to be 0.4-1.1 nM depending on the ionic composition of the assay buffer (15). Data in Table 1 illustrate that various potent serotonin antagonists reveal high binding affinity for serotonin-S_2 binding sites. High binding affinity is also noted for certain of the tricyclic serotonin reuptake blockers. Selective serotonin reuptake blockers such as zimelidine and citalopram do not bind to the serotonin-S_2 sites. The most potent serotonin agonists: serotonin, bufotenin and the non-indole-like agonist quipazine, bind at concentrations between 100 and 500 nM, less potent agonists:tryptamine and mescaline, bind at micromolar concentrations. Various neurotransmitters are virtually inactive (see Table 1).

Serotonin-S_2 binding sites were found to be similarly distributed in brains of rats, guinea pigs, rabbits, dogs and humans (29, 30, 31). The highest density occurs in the prefrontal cortex (in rats: B_{max} = 33 fmoles/mg tissue). In cortical areas the density gradually decreases from the frontal towards the occipital part. The distribution in the subcortical areas, proportional to the density in the prefrontal cortex, of serotonin-S_1 and serotonin-S_2 binding sites and of specific serotonin uptake is shown in Table 2. No parallelism in distribution between either binding site and the uptake is observed.

Recently, using ^{3}H-ketanserin, the existence of serotonin-S_2 binding sites on cat platelet membranes was demonstrated (32, 33). The density amounts to 86 fmoles/10^9 platelets. The binding affinities of drugs for serotonin-S_2 binding sites in rat frontal cortical and cat platelet membranes are highly significantly correlated (32, 33).

Destruction of noradrenergic and dopaminergic neurones of the medial forebrain bundle by 6-hydroxydopamine does not alter the number of either serotonin-S_1 or serotonin-S_2 binding sites in the cortical brain areas or the striatum, indicating that the sites are not localized on dopaminergic or noradrenergic nerve endings (29). Neither site seems to be localized on serotonergic neurones originating in the raphe, since no decrease in binding sites was observed after 5,6- or 5,7-dihydroxytryptamine lesions of these neurones (12,30). Occasionally reported increases in serotonin-S_1 (23) or serotonin-S_2 binding sites (34) following serotonergic neurone denervation has not consistently been confirmed (25, 30). Frontal cortical bisection was found to reduce serotonin-S_2 binding sites in the prefrontal cortex by 30 % (29). Injection of rather high dosages of kainic acid (3-4 µg) in the frontal cortex reduced the number of serotonin-S_2 binding sites in that area by 60 to 70 % (30). Glutamic acid decarboxylase activity was concomitantly decreased whereas choline acetyl transferase was unaffected. It was tentatively suggested that serotonin-S_2 binding sites could be localized on glutaminergic neurones in the frontal cortex (30).

Role of the binding sites in pharmacological processes. Serotonin-antagonistic potencies of compounds in *in vivo* pharmacological tests: tryptamine-induced clonic seizures in rats (14, 15, 20), mescaline-induced

head twitches in rats (15, 20), 5-hydroxytryptophan-induced head twitches in rats (22), the 5-hydroxytryptophan-induced serotonin syndrome in rats (35) and the serotonin-induced paw edema (35), correlate highly significantly with the binding affinities of the compounds for the serotonin-S_2 binding sites. This suggests a role for serotonin-S_2 type receptors in serotonin-induced behavioural excitation and inflammation. No significant correlation exists between potencies of compounds in the above _in vivo_ tests and binding affinities for serotonin-S_1 sites or serotonin stimulated adenylate cyclase (20).
The _in vitro_ binding data were similarly compared to potencies of compounds to antagonize serotonin-induced contractions in isolated smooth muscle preparations. No relationship was apparent between any of the three receptor binding models and the pharmacological 'D'-receptor, measured in the rat fundus (15, 20) or with the pharmacological 'M'-receptor (26). Serotonin-S_2 type receptors were shown to mediate serotonin-induced bronchial smooth muscle contractions (4) and serotonin-induced contractions in rat caudal arteries (3, 15, 20, 25) and rabbit femoral artery (3). In these tissues no role was apparent for serotonin-S_1 sites or serotonin stimulated adenylate cyclase (20). Recently, Peroutka suggested that serotonin-induced contractions of the canine basilar artery would be mediated by serotonin-S_1 type receptor sites (36). However, the evidence can be questioned because of an unusual analysis of the _in vivo_ pharmacological data (Van Nueten _et al._, in preparation).
Other different purported roles for serotonin-S_1 binding sites were refuted (20): i.e. the suggestion by Ennis (37) that serotonin-S_1 sites would mediate the inhibition by serotonin of K^+-evoked dopamine release from striatal slices and the unsubstantiated supposition that serotonin-S_1 sites could be related to neuronal inhibition by serotonin (22). Moreover, the argument that the nanomolar binding affinity of serotonin itself for the binding sites is an indication for the specificity or physiological relevance of the sites is a misconception. This is not in agreement with the high concentrations of endogenous serotonin which are expected to be reached in synaptic clefts. Nanomolar binding affinities of neurotransmitters for their receptors, localized in synaptic clefts, would result in a constant overstimulation of the receptors. Moreover,

it would be impossible to antagonize the effects which result from such high affinity neurotransmitter binding with systemically administered compounds, unless the antagonists were to have binding affinities of picomolar order.

Thus far, only the serotonin-S_2 type binding sites, have been shown to have roles in various pharmacological processes. Besides their involvement in processes described above, evidence was recently provided that the serotonin-S_2 binding sites detected on cat platelet membranes have a functional role in serotonin-induced platelet aggregation (De Clerck et al., this volume).

Finally, therapeutic actions of more selective serotonin antagonists, such as ketanserin, in human pathologies involving impaired microcirculation, is currently being found. This can be reconciled with the role of serotonin-S_2 receptor sites in vasoconstriction and platelet aggregation. Indications for selective serotonin antagonists in diseases of the central nervous system are to be expected and await investigation.

References

1. Gaddum, J.H., Picarelli, Z.P.: Brit. J. Pharmacol. 12, 323-328 (1957).

2. Vane, J.R.: Brit. J. Pharmacol. 12, 344-349 (1957).

3. Van Nueten, J.M., Janssen, P.A.J., Van Beek, J., Xhonneux, R., Verbeuren, T.J., Vanhoutte, P.M.: J. Pharmacol. Exp. Ther. 218, 217-230 (1981).

4. Van Nueten, J.M., Leysen, J.E., Vanhoutte, P.M., Janssen, P.A.J.: Arch. Int. Pharmacodyn. Ther. 256, 331-334 (1982).

5. Jacobs, B.L.: Life Sci. 19, 777-786 (1976).

6. Niemegeers, C.J.E., Lenaerts, F.M., Artois, K.S.K., Janssen, P.A.J.: Arch. Int. Pharmacodyn. Ther. 227, 238-253 (1977).

7. Corne, S.J., Pickering, R.W.: Psychopharmacologia 11, 65-78 (1967).

8. Niemegeers, C.J.E, Colpaert, F.C., Leysen, J.E., Awouters, F., Janssen, P.A.J.: Drug Devel. Res. (in press) (1983).

9. Aghajanian, G.K.: Serotonin Neurotransmission and Behavior, Eds. B.L. Jacobs and A. Gelperin, MIT Press, Cambridge, M.A., 156-185 (1981).

10. Rogawski, M.A., Aghajanian, G.K.: J. Neurosci. 1, 1148-1154 (1981).

11. Von Hungen, K., Roberts, S., Hill, D.F.: Brain Res. 84, 257-267 (1975).

12. Bennett, J.P., Snyder, S.H.: Mol. Pharmacol. 12, 373-389 (1976).

13. Leysen, J.E., Laduron, P.M.: Arch. Int. Pharmacodyn. Ther. 230, 337-339 (1977).

14. Leysen, J.E., Niemegeers, C.J.E., Tollenaere, J.P., Laduron, P.M.: Nature 272, 168-171 (1978).

15. Leysen, J.E., Niemegeers, C.J.E., Van Nueten, J.M., Laduron, P.M.: Mol. Pharmacol. 21, 301-314 (1982).

16. Enjalbert, A., Bourgoin, S., Hamon, M., Adrien, J., Bockaert, J.: Mol. Pharmacol. 14, 2-10 (1978).

17. Enjalbert, A., Hamon, M., Bourgoin, S., Bockaert, J.: Mol. Pharmacol. 14, 11-23 (1978).

18. Nelson, D.L., Herbet, A., Enjalbert, A., Bockaert, J., Hamon, M.: Biochem. Pharmacol. 29, 2445-2453 (1980).

19. Fillion, G., Beaudoin, D., Rousselle, J.C., Jacob, J.: Brain Res. 198, 361-374 (1980).

20. Leysen, J.E., Tollenaere, J.P.: Ann. Rep. Med. Chem. 17, 1-10 (1982).

21. Peroutka, S.J., Snyder, S.H.: Mol. Pharmacol. 16, 687-699 (1979).

22. Peroutka, S.J., Lebovitz, R.M., Snyder, S.H.: Science 212, 827-829 (1981).

23. Nelson, D.L., Herbet, A., Bourgoin, S., Glowinski, J., Hamon, M.: Mol. Pharmacol. 14, 983-995 (1978).

24. Peroutka, S.J., Lebovitz, R.M., Snyder, S.H.: Mol. Pharmacol. 16, 700-708 (1979).

25. Leysen, J.E.: J. Physiol. 77, 351-362 (1981).

26. Leysen, J.E., Awouters, F., Kennis, L., Laduron, P.M., Vandenberk, J., Janssen, P.A.J.: Life Sci. 28, 1015-1022 (1981).

27. Peters, J.R., Grahame-Smith, D.G.: Eur. J. Pharmacol. 68, 243-256 (1980).

28. Peroutka, S.J., Snyder, S.H.: J. Pharmacol. Exp. Ther. 216, 142-148 (1981).

29. Leysen, J.E., Geerts, R., Gommeren, W., Verwimp, M., Van Gompel, P.: Arch. Int. Pharmacodyn. Ther. 256, 301-305 (1982).

30. Leysen, J.E., Van Gompel, P., Verwimp, M., Niemegeers, C.J.E.: CNS Receptors - From Molecular Pharmacology to Behaviour, Eds. P. Mandel and F.V. DeFeudis, Raven Press . New York (in press).

31. Schotte, A., Maloteaux, J.M., Laduron, P.M.: Brain Res. (in press).

32. Leysen, J.E., Gommeren, W., De Clerck, F.: Eur. J. Pharmacol. (in press) (1983).

33. Leysen, J.E., Gommeren, W., De Clerck, F.: Arch. Int. Pharmacodyn. Ther. (in press) (1982).

34. Brunello, N., Chuang, D.M., Costa, E.: Science 215, 1112-1115 (1982).

35. Ortmann, R., Bischoff, S., Radeke, E., Buech, O., Delini-Stula, A.: Naunyn-Schmiedeberg's Arch. Pharmacol. 321, 265-270 (1982).

36. Peroutka, S.J., Noguchi, M., Tolner, D.J., Allen, G.: Brain Res. 259, 327-330 (1983).

37. Ennis, C., Kemp, J.D., Cox, B.: J. Neurochemistry 36, 1515-1520 (1981).

INVOLVEMENT OF $5\text{-}HT_2$ (S_2)-RECEPTOR SITES IN THE SEROTONIN-INDUCED ACTIVATION OF CAT PLATELETS

Fred De Clerck, Benoit Xhonneux, Walter Gommeren and Josée Leysen

Laboratory of Haematology and Department of Biochemical Pharmacology, Janssen Pharmaceutica Research Laboratories, B-2340 Beerse, Belgium

Introduction

5-Hydroxytryptamine (5-HT) activates platelets by interaction with specific surface membrane receptors which are distinct from those involved in the cellular 5-HT uptake mechanism (1,2). The present study reports on the possible relationship between these functional platelet receptors for 5-HT and the $5\text{-}HT_2$ (S_2) receptor binding sites characterized by ligand binding experiments in brain tissues (2,3).

Methods

Platelet function experiments. Physiological responses to 5-HT were assessed by measurements of the initial slope of aggregation of cat platelets induced by 5-HT (5×10^{-6} M) or ADP (1×10^{-5} M), and of the prostaglandin production induced by arachidonic acid (1×10^{-3} M). The effect of drugs on the uptake of 5-HT by platelets was measured using ^{14}C-labelled amine at 5×10^{-6} M (60 sec at 37° C); (4).

Membrane ligand binding experiments. Inhibition of the [^3H] ketanserin binding, as a marker for $5\text{-}HT_2$ (S_2) receptor sites (2), to isolated cat platelet-, rat striatum- and rat pre-frontal cortex membrane preparations was assessed as described earlier (5).

Progress in Tryptophan and Serotonin Research

Results

Platelet function. The slope of aggregation of cat platelets induced by 5-HT (5 x 10^{-6} M) was inhibited in a concentration-dependent way by the various subclasses of compounds, IC_{50}-values (nM) showing a rank order of potency of pipamperone (4.57) > ketanserin (11.5) > spiperone (12.6) > methysergide (25) > cyproheptadine (35.5) > cinanserin (300) > domperidone (700) > amitryptiline (891) > clomipramine (3890) > haloperidol (4200) > pyrilamine (9100) > desimipramine (17 400) > prazosin (19 000) > citalopram (37 000).
In contrast to appropriate reference compounds, the five most potent 5-HT receptor antagonists did not significantly affect platelet reactions non-related to 5-HT-induced receptor activation and only slightly reduced the active uptake of 5-HT at high concentrations.

Membrane ligand binding. Specific [^{3}H] ketanserin binding to cat platelet membranes was inhibited differentially, K_i-values (nM) for the compounds, in order of IC_{50} ranking for 5-HT antagonism, being 4, 0.6, 1.1, 7.1, 2, 14, 54, 21, 47, 32, 1140, 210, 850, 2500 respectively (5).

Correlations ligand binding/physiological response. Linear regression analysis (Pearson correlation coefficient) revealed a strong correlation between the log IC_{50} for inhibition of 5-HT-induced cat platelet activation and the log K_i for inhibition of specific [^{3}H] ketanserin binding to cat platelet membranes (Fig. 1).

Conclusions

Originally it was assumed that receptors responsible for the 5-HT-induced platelet activation resemble the pharmacologically defined D-type receptor of intestinal smooth muscle (6), which are different from the 5-HT_1 (S_1) and 5-HT_2 (S_2) receptor binding sites characterized by ligand binding in brain tissues (3).
Using [^{3}H] ketanserin as a selective label for 5-HT_2 (S_2) receptor binding sites (2), the presence of the latter subtype of receptors on cat

Fig. 1

CORRELATION BETWEEN INHIBITION OF SPECIFIC [^{3}H]-KETANSERIN BINDING TO CAT PLATELET MEMBRANE PREPARATIONS AND INHIBITION OF 5-HYDROXYTRYPTAMINE-INDUCED AGGREGATION OF CAT PLATELETS

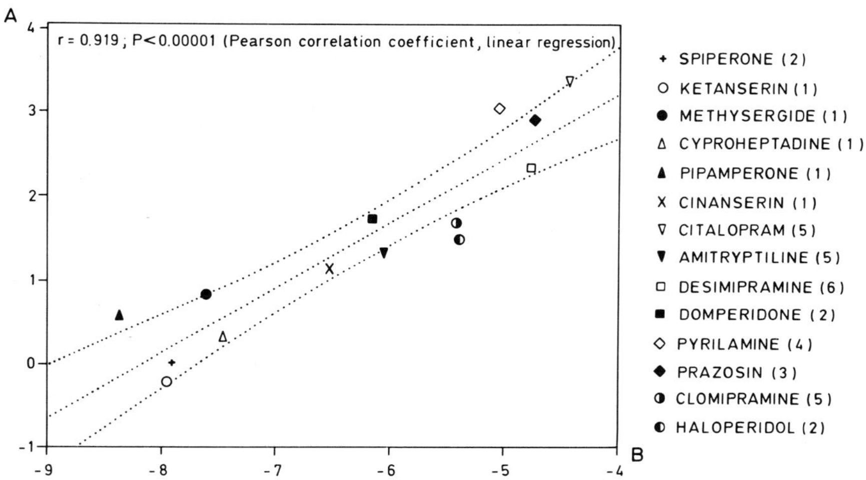

A = log Ki (nM) for inhibition of specific [^{3}H]-ketanserin binding to cat platelet membrane preparations
B = log IC_{50} (M) for inhibition of 5-HT-induced aggregation (initial slope) of cat platelets in P.R.P.
(1) Serotonin antagonist, (2) Dopamine antagonist, (3) α_1-Adrenergic antagonist, (4) H_1-Histamine antagonist, (5) Serotonin uptake blockers, (6) Noradrenaline uptake blockers.

platelet membrane preparations was recently demonstrated (5). The present study shows a strong inhibition of 5-HT-induced platelet activation by serotonin antagonists through a specific effect at the level of receptor-agonist interaction. The strong correlation between ligand binding- and physiological response data suggests that the platelet receptors responsible for the 5-HT-induced platelet activation are identical to the 5-HT_2 (S_2) receptor sites radiolabelled with [^{3}H] ketanserin in platelet- and brain tissue membrane preparations. These findings therefore demonstrate a functional role for the 5-HT_2 (S_2) receptor binding sites in the activation of platelets by 5-HT.

References

1. Peters, J.R., Grahame-Smith, D.G.: Eur. J. Pharmacol. 68, 243-256 (1980).

2. Leysen, J.E., Niemegeers, C.J.E., Van Nueten, J.M., Laduron, P.M.: Mol. Pharmacol. 21, 301-314 (1982).

3. Leysen, J.E., Awouters, F., Kennis, L., Laduron, P.M., Vandenberk, J., Janssen, P.A.J.: Life Sci. 28, 1015-1022 (1981).

4. De Clerck, F., David, J.L., Janssen, P.A.J.: Agents Actions 12, 388-397 (1982).

5. Leysen, J.E., Gommeren, W., De Clerck, F.: Eur. J. Pharmacol., in press (1983).

6. Michal, F.: Nature 221, 1253-1254 (1969).

THE REGULATION OF 5-HYDROXYTRYPTAMINE RELEASE FROM SUPERFUSED SYNAPTOSOMES BY TRYPTOPHAN

Hugh A. Suter and Keith J. Collard
Department of Physiology, University College, Cardiff, CF1 1XL, Wales, U.K.

Introduction

It has recently been demonstrated that tryptophan (TRP) may be released from synaptosomes in a Ca^{2+}-dependent manner during depolarisation (1). In addition, a number of studies have failed to demonstrate a significant correlation between the high affinity uptake of TRP into nerve terminals and the distribution of 5-HT terminals, 5-HT receptors or 5-HT uptake (2). Furthermore, TRP has certain actions in the central nervous system which cannot be ascribed to any of its known metabolites in the brain (3, 4). These findings taken together indicate that TRP may play a specific role in synaptic transmission in addition to its function as a precursor for protein and 5-HT synthesis. The finding that TRP may be released from nerve endings by depolarisation suggests that it may act as a neurotransmitter or neuromodulator. It is already well established that 5-HT is able to act on presynaptic receptors to modulate its own release (5). The aim of this study was to determine whether tryptophan is able to regulate 5-HT release in a similar manner.

Methods

Synaptosomes were prepared from rat forebrain by the method of Gray and Whittaker (6). The synaptosomal pellet was resuspended in 20 ml of Krebs solution and a 10 ml portion was incubated with 10^{-7} M [^{3}H]-5-HT (specific activity 11-14 Ci/m mol) for 10 min at 37°C. Beds of synaptosomes were prepared from 4.5 ml aliquots of the incubated suspension and placed in perfusion chambers as previously described (7). The beds of synaptosomes

Progress in Tryptophan and Serotonin Research

were perfused with oxygenated (95% O_2 + 5% CO_2) Krebs solution at 37°C and at a rate of 8 ml/min. Serial 4.0 ml samples of effluent perfusate were collected on ice in a fraction collector. After 8 minutes of perfusion, the fluid was switched to one containing either L- or D-TRP. The amino acid remained in the fluid for the remainder of the experiment. After 12 minutes of perfusion, a 38 second pulse of Krebs solution containing either 7.5 mM or 15 mM K^+ (replacing Na^+) was applied. The perfusion fluid was then returned to normal. 2.0 ml portions of each sample were taken for separation of [^{3}H]-5-HT by ion-exchange chromatography using 5 x 25 mm columns of Biorex 70. Separated [^{3}H]-5-HT was measured by liquid scintillation counting.

Results

The release of 5-HT is expressed as the percentage of tissue 5-HT released per fraction. The release in response to depolarisation was calculated as the total 5-HT released above the baseline during the application of the K^+ pulse.

The basal efflux of [^{3}H]-5-HT was unaffected by concentrations of TRP from 1-200 nM suggesting that there was little conversion of TRP to 5-HT with subsequent displacement of [^{3}H]-5-HT. This was also confirmed by chromatographic analysis which showed that the applied TRP was not appreciably converted to either 5-HTP, 5-HT, or tryptamine.

Preliminary studies examined the effect of 1-200 nM L-TRP on the release of [^{3}H]-5-HT evoked by the application of a pulse of Krebs solution containing 15 mM K^+. The results of this study are shown in Table 1. The higher concentrations of TRP studied (50-200 nM) had little effect on the K^+-evoked release of [^{3}H]-5-HT. In contrast, lower concentrations (1-10 nM) potentiated [^{3}H]-5-HT release reaching significance at a concentration of 10 nM.

Table 1

The effect of TRP on the K^+-evoked release of $[^3H]$-5-HT from superfused synaptosomes.

The release of 5-HT is expressed as the percentage of tissue 5-HT released per K^+ pulse. The results are expressed as the mean release ± S.E.M. The number of experiments is given in parentheses. Significant differences between the control response and the response in the presence of TRP are given by * P<0.05, ** P<0.02, † P<0.01, †† P<0.001. Students t test.

TRP conc. (nM)	L-TRP		D-TRP
	15 mM K^+	7.5 mM K^+	7.5 mM K^+
0	1.339 0.063 (26)	0.540 0.037 (16)	0.620 0.065 (10)
1	1.310 0.130 (3)	-	-
5	1.410 0.080 (9)	0.568 0.037 (7)	0.502 0.078 (5)
10	1.660 0.070 (7)*	0.762 0.063 (10)†	0.548 0.070 (5)
20	-	0.785 0.062 (6)††	0.605 0.070 (6)
35	-	0.678 0.053 (5)**	-
50	1.266 0.153 (5)	0.625 0.081 (6)	0.618 0.076 (6)
100	1.370 0.090 (5)		
200	1.320 0.080 (3)		

In order to examine this effect further, a second study examined the effect of 5-50 nM L-TRP on the release of $[^3H]$-5-HT evoked by a pulse of Krebs solution containing 7.5 mM K^+. Under these conditions, the K^+-evoked release of $[^3H]$-5-HT was increased by L-TRP at all concentrations studied and reached significance at 10, 20, and 35 nM (see Table 1).

A third study examined the effect of D-TRP (5-50 nM) on the release of $[^3H]$-5-HT evoked by 7.5 mM K^+. In contrast to the effects observed with L-TRP, D-TRP had no significant effects on the K^+-evoked release of $[^3H]$-5-HT. This result indicates some degree of stereospecificity of the effect of TRP on the K^+-evoked release of $[^3H]$-5-HT.

Discussion

The results of this study indicate that low concentrations of L-TRP can potentiate the K^+-evoked release of $[^3H]$-5-HT from superfused synaptosomes. The effect appears to be specific for the physiological L-isomer and suggests that low concentrations of TRP may act as a positive feedback regulator of 5-HT release.

These findings add support for the suggestion that TRP may act as a modulator of transmitter release in the mammalian brain. It remains to be determined whether these effects are exerted via specific presynaptic receptors.

References

1. Collard, K.J., Evans, T.N.W., Suter, H.A., Wilkinson, L.S.: J. Neural Transmission 53, 223-230 (1982).
2. Hockel, S.H.J., Muller, W.E., Wollert, U.: Neurosci. Lett. 8, 65-69 (1978).
3. Aghajanian, G.K.: Fed. Proc. 31, 91-96 (1972).
4. Modich, K.: Acta Physiol. Scand. Suppl. 403, 1-56 (1974).
5. Mounsey, I.M., Brady, K.A., Carroll, J., Fisher, R., Middlemiss, D.M.: Biochem. Pharmacol. 31, 49-53 (1982).
6. Gray, E.G., Whittaker, V.P.: J. Anat. (Lond) 96, 79-87 (1962).
7. Collard, K.J., Cassidy, D.M., Pye, M.A., Taylor, R.M.: J. Neurosci. Meth. 4, 163-179 (1981).

SELECTIVE INCREASES IN HIGH AFFINITY TRYPTOPHAN TRANSPORT IN SYNAPTOSOMES FROM DIFFERENT BRAIN AREAS INCUBATED IN SODIUM - FREE MEDIA.

Lawrence S. Wilkinson, Christopher J. Price and Keith J. Collard

Department of Physiology, University College, P.O. Box 78, Cardiff, CF1 1XL, Wales, U.K.

Introduction

A number of recent findings obtained in this laboratory considered in the light of previously observed anomalies between tryptophan (TRP) and 5-hydroxytryptamine (5-HT) (1, 2) has prompted a reappraisal of the role of TRP in neural function. These findings have suggested that TRP may have a specific synaptic function in addition to its well documented role as a precursor amino acid for 5-HT and protein synthesis (3). The aim of this study was to examine the regional distribution and effect of ionic manipulation on high affinity TRP uptake into rat brain synaptosomes as part of a re-evaluation of the relationship between this transport system and 5-HT neurones.

In character with a great deal of the literature concerning the properties of TRP transport into mammalian brain tissue, its ionic requirements are not established. Thus it would appear that the transport of TRP at the level of the neuronal membrane can be either Na^+-dependent (1, 4), Na^+-independent (5) or Na^+-inhibited (6). Previous work in this laboratory has indicated that TRP uptake by synaptosomes prepared from whole fore-brain is increased by incubation in Na^+-free medium. This study attempted to determine whether this effect was a generalised effect or whether regional differences occurred.

Progress in Tryptophan and Serotonin Research

Methods

Synaptosomes were prepared from brain areas known to vary in 5-HT activity. A dissection procedure was developed which enabled samples of cerebellum, cerebral cortex, midbrain, hippocampus, striatum and hypothalamus to be isolated quickly, accurately and reproducibly. Brain areas were pooled from six animals and synaptosomes were prepared according to the method of Gray and Whittaker (7). The final synaptosomal pellet was resuspended in either standard Krebs solution or in a Na^{+}-deficient buffer in which NaCl and $NaHCO_3$ had been isosmotically replaced with sucrose and Tris respectively. 1.125 ml aliquots of the synaptosomal suspension were gassed with 95% O_2 + 5% CO_2 and pre-incubated in a shaking water bath at 37°C for 4 mins. 0.125 ml of [^{3}H]-TRP (concentration range 0.5 - 10 μM, specific activity 0.14 Ci/mMol) was then added and the incubation continued for one minute. Preliminary studies had shown that uptake over one minute was linear in all brain regions studied and thus gave a good approximation of the initial rate of TRP uptake. The reaction was terminated using Millipore filtration (pore-size 0.65 μM) followed by rapid washing. The Millipore filters were placed in scintillation vials and dried at room temperature for 5 hr. Homogeneous counting conditions were obtained by the addition of 1 ml of Ethanol/Methanol (4:1) followed by 10 ml of scintillation cocktail. The radioactivity was measured by liquid scintillation counting. On the basis of the concentration of protein in each sample and the specific activity of the [^{3}H]-TRP, the active uptake of TRP was expressed as pmoles of TRP accumulated /mg protein /minute. That part of total label on the filter paper due to active uptake was ascertained by running appropriate synaptosomal (0°C) and filter blanks. Kinetic constants were determined by the Eadie - Hofstee linear transformation method.

Results and Discussion

The apparent Km values in each of the brain areas were unaffected by the removal of Na^+. Apparent Km values obtained in Na^+-rich conditions were similar for each brain area studied. This is in contrast to the results of other workers who observed marked regional variation in the affinity of the uptake system for TRP (1, 8). In the presence of Na^+, regional differences in the apparent Vmax were observed. Hypothalamic synaptosomes displayed the highest and **cerebellum** synaptosomes the lowest capacity for TRP accumulation. In the absence of Na^+ there was a significant and selective increase in the apparent Vmax values in cortical, hippocampal and striatal synaptosomes. Synaptosomes prepared from the hypothalamus and cerebellum showed no significant change in TRP uptake in response to this ionic manipulation whilst midbrain synaptosomes displayed a decrease.

Table 1

The apparent Vmax and Km values for TRP transport into synaptosomes prepared from various brain regions in the presence and absence of Na^+ as determined using the Eadie - Hofstee linear transformation method. Vmax is expresed as pmoles of TRP taken up /mg protein /minute. Km is expressed as μM. n = the number of separate experiments.

		Na^+-rich		Na^+-free	
Area	n	Km	Vmax	Km	Vmax
Cerebellum	4	4.00	192	3.20	163
Cortex	6	4.22	216	4.91	644
Midbrain	4	5.70	346	3.80	182
Hippocampus	4	3.48	220	4.46	451
Striatum	4	4.34	335	5.50	578
Hypothalamus	4	3.00	375	4.00	421

This selective increase in TRP uptake seen in the absence of Na^+ appears to reflect with one exception the degree of 5-HT activity (as exemplified by the density of 5-HT terminals, receptors and high affinity 5-HT uptake activity) present in each brain area. The cerebellum and midbrain regions being relatively low and hippocampal, striatal and cortical regions being relatively high in 5-HT activity. The results obtained with the hypothalamus are anomalous with regard to this logic.

It is suggested that the results presented here might be indicative of there being two mechanisms by which L-TRP can enter nerve terminals. One system, active in Na^+-rich incubation conditions and present in all brain regions studied, and another which becomes selectively manifest in the absence of Na^+-ions. Such an observation would lend support to the hypothesis being formed that TRP might play a dual role in neural function since such a role could involve the accumulation of TRP at functionally and perhaps spatially distinct sites within the brain.

References

1. Hockel, S.H.J., Muller, W.E., Wollert, U. :Neurosci. Lett. 8, 65-69 (1978).
2. Kuhar, M.J., Roth, R.H. Aghajanian, G.K. : J. Pharm. Exp. Ther. 181, 36-45 (1972).
3. Collard, K.J., Evans, T.N.W., Suter, H.A., Wilkinson, L.S. : J. Neural transmission 53, 223-230 (1982).
4. Korpi, E.R. : J. Neurochem. 40, 425-433 (1983).
5. Grahame-Smith, D.G., Parfitt, A.G. : J. Neurochem. 17, 1339-1353 (1970).
6. Bruinvels, J. : Nature 257, 606-607 (1975).
7. Gray, E.G., Whittaker, V.P. : J. Anat. (Lond) 96, 79-87 (1962).
8. Denizeau, F., Sourkes, T.L. : J. Neurochem. 28, 951-959 (1977).

FORMATION AND FUNCTION OF TETRAHYDRO-ß-CARBOLINES WITH SPECIAL REFERENCE TO THEIR ACTION ON [3H]-TRYPTAMINE BINDING SITES

H. Rommelspacher and G. Brüning
Institut für Neuropsychopharmakologie der Freien Universität Berlin
1000 Berlin 19, FRG

FORMATION OF TETRAHYDRO-ß-CARBOLINES (THBC's)

The THBC's are formed in vitro and in vivo by a reaction of indoleamines with aldehydes or ketones. The reaction of amines, like that of 5-hydroxytryptamine and tryptamine, has mostly been studied with formaldehyde and acetaldehyde.

Tetrahydronorharmane (THN) which is synonymous to tetrahydro-ß-carboline (THBC) and 6-hydroxy-THN are both formed by a reaction involving formaldehyde. Other compounds detected in mammalian tissue include tetrahydroharmane as well as harmane which are both derived from tryptamine and either acetaldehyde or possibly pyruvic acid. In contrast to the former substances harmane is a fully aromatic compound (1-methyl-ß-carboline, 1-Me-BC).

Several pathways were propagated by Brossi for the formation of 1-Me-tetrahydroisoquinolines (TIQ's). THBC's may be formed by reactions similar to those for TIQ's since both classes of compounds are derived from biogenic amines and aldehydes. The first pathway proceeds non-enzymatically with a Schiff's base as an intermediate product (Pictet-Spengler reaction). The second pathway requires pyruvic acid (Hahn and Stiehl 1936, Hahn and Rumpf 1938). In analogy tryptamine would be converted into 1-carboxyl-1-methyl-THBC, which would be oxidized and decarboxylated to a dihydro-ß-carboline (Fig. 1). A third

Progress in Tryptophan and Serotonin Research

FORMATION OF 1-METHYL-β-CARBOLINES

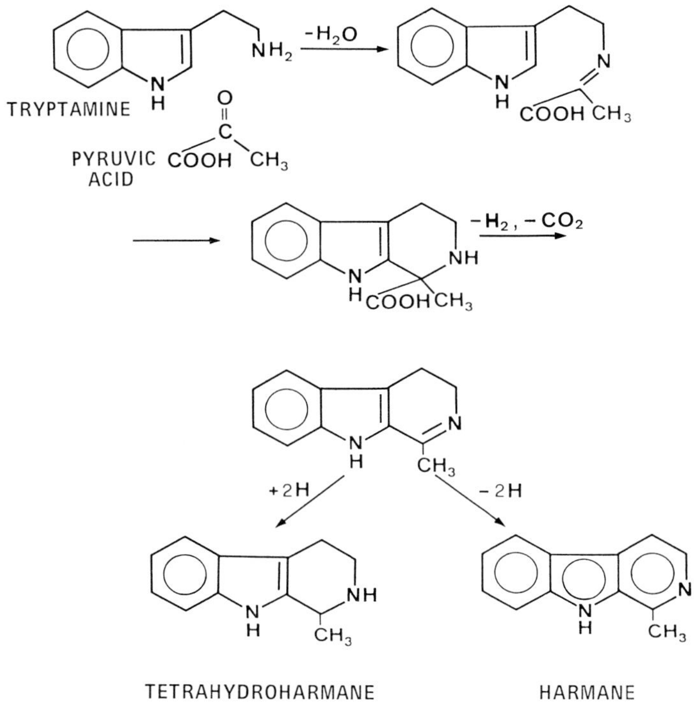

FIGURE 1 (See text for details.)

pathway for the formation of TIQ's has been described in plants in the biosynthesis of norlaudanosoline from dopamine and 3,4-dihydroxyphenylacetaldehyde. The enzyme, norlaudanosoline synthetase, mimics the Pictet-Spengler synthesis; it was found to be completely specific for phenylacetaldehydes. Phenylpyruvic acids were not utilized by this system. Indolacetaldehyde on the other hand was not tested (Rueffer et al. 1981).

The formation of the fully aromatic BC's remains to be elucidated. A possible reaction would proceed as described above with a carboxylic acid as intermediate. By oxidative decarboxylation the dihydro-BC is formed. This compound is converted either to 1-Me-THBC's or by an oxidative reaction to the fully aromatic BC's like harmane (Fig. 1). This pathway has recently been demonstrated in the brain of rats (Susilo and Rommelspacher, unpublished).

FUNCTION OF THBC'S

The biological activity of THBC's has been investigated in more detail. THN acts 5-HT-mimetic and dopamine - antagonistic (Pannier et al. 1981, Nielson et al. 1982). Low doses of THN and other THBC's and BC's augment the conflict behavior in rats and suppress the exploration of mice (Rommelspacher et al. 1982 and unpublished results). Both types of behavior are opposite to those elicited by anxiolytic drugs. However, the affinity of THBC's to various binding sites is relatively low with the exception of those of (^{3}H) - spiroperidol labelling dopamine receptors and (^{3}H)-5-HT (Table I).

Table 1

IC_{50}-values (µmol/l) of THN and harmane in various receptor binding assays*

	Harmane	THN
(^{3}H)-Flunitrazepam	7 ± 1	920 ± 5
(^{3}H)-Quinuclidinyl benzylate	24 ± 5	638 ± 128
(^{3}H)-Naloxone O mM Na+	2.8 ± 4	355 ± 70
(^{3}H)-Naloxone 150 mM Na+	42 ± 1	1050 ± 50
(^{3}H)-Spiroperidol	163 ± 22	26 ± 3
(^{3}H)-5-Hydroxytryptamine	101 ± 8	4 ± 1

*Data from Müller et al. (1981)

Several, but not all of the biological effects of THBC's may be elicited by the interaction with these binding sites. Therefore, the search for other receptors was commenced. Two considerations were pursued. Either the THBC's act via receptors of their own or via receptors of substances with similar chemical structure. Recently we detected high affinity binding sites for (^{3}H)-THN in rat brain membranes with a K_D of 2.7 nmol/l. Whether the labelled sites are activated by THN to elicite its biological effects remains to be established (Rommelspacher and Brüning 1982). The chemical similarity of THN and tryptamine suggested that both substances may utilize the same receptors. High affinity binding sites for (^{3}H)-tryptamine were described recently (Kellar and Cascio 1982; Rommelspacher and Brüning 1983). In the following section we report the kinetics, the specificity and the pharmacological characterization of the (^{3}H)-tryptamine binding sites on brain membranes and the interaction with THN and other ß-carbolines.

Methods

Brain of male rats (Wistar - 250-350 g) was removed from the skull and the cerebral cortex was dissected off. The white matter was excised. The remainder was homogenized in 25 volumes of 0.32 mol/l sucrose using a glass-teflon homogenizer. The homogenate was centrifuged at 1000 g for 10 minutes at 1°C. The resulting supernatant was centrifuged at 30000 g for 30 minutes. The pellet was washed once with a tenfold diluted incubation buffer which consisted of Tris-HCl (50 mmol/l) pH 7.4, 5 mmol/l ascorbic acid and 220 µmol/l nialamide. (^{3}H)-Tryptamine binding was performed in a manner similar to that described by Kellar and Cascio (1982). The homogenate was preincubated for five minutes at 37°C. Aliquots of 300 µl corresponding to 5 mg of wet tissue, were incubated with (^{3}H)-tryptamine (specific activity 41.5 Ci/mmol, NEN, Dreieich) and various concentrations of displacers for 50 minutes at 0°C in a final volume of 500 µl. Incubation was terminated by addition of 5 ml ice cold buffer and rapid filtration over Whatman GF/B filters, followed by two washes with the same volume of buffer. Samples were counted using 5 ml Biofluor (NEN) as scintillation fluid wih an efficiency of 39%. Unspecific binding was measured by incubation with 10 µmol/l tryptamine. The receptor binding experiments with (^{3}H)-5-HT as ligand were performed using the same membrane preparation of rat brain as outlined above with minor modifications according to a procedure published previously (Müller et al. 1981).

Kinetics of (^{3}H)-Tryptamine Binding

The binding characteristics of (^{3}H)-tryptamine to rat brain membranes have not been described yet in detail. The forward reaction between (^{3}H)-tryptamine and tryptamine binding sites was studied under pseudo-first-order conditions. (^{3}H)-Tryptamine (11 nmol/l) was incubated with brain membranes at 0°C and the rate of complex formation was measured. A typicl experiment is depicted in fig. 2. The binding reaction reaches a nearly steady-state level without a lag, within 6 minutes after the reaction was initiated. In two out of three experiments a decrease in specific binding was already observed after 60 minutes. From the least squares fit of the data an observed association

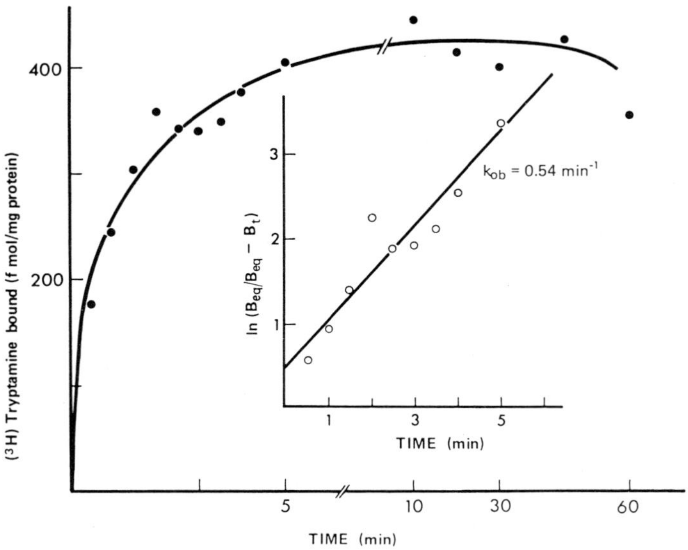

FIGURE 2

Association kinetics of (^{3}H)tryptamine specific binding to membranes from rat cerebral cortex. Insert: pseudo-first order plot, Beq is (^{3}H)tryptamine bound at equilibrium (after 50 minutes and Bt bound at time t.)

FIGURE 3

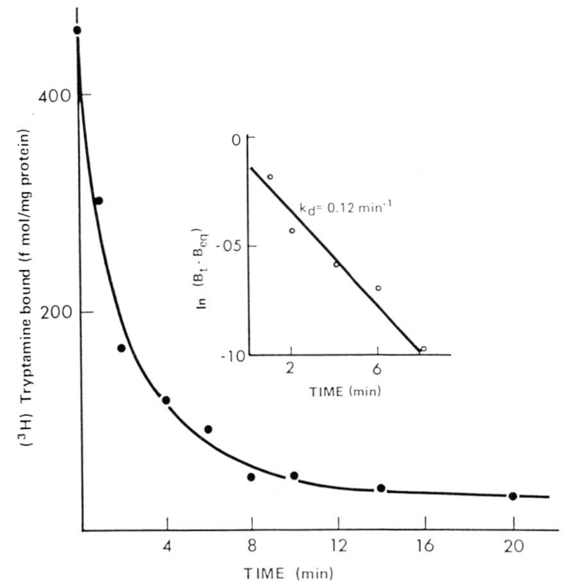

Dissociation kinetics, initiated by combined isotope/volume dilution after 50 minutes of incubation.
Insert: First order plot. For details see legend Figure 2.

constant k_{ob} = 0.54 min^{-1} resulted. The second order rate constant k_a was calculated from $k_a = (k_{ob} - k_d)/\ T$ = 0.039 +/- 0.012 $nmol^{-1}\ min^{-1}$ where k_d is the dissociation rate constant and T is the concentration of (^{3}H)-tryptamine in the reaction mixture.

The dissociation was started after equilibrium binding (50 minutes of incubation at 0°C) by the addition of 5 ml of 10 µmol/l tryptamine to the reaction mixture. From the first order plot of dissociation a rate constant k_d = 0.93 +/- 0.067 min^{-1} was obtained (fig. 3). The ratio k_d/k_a gives a kinetically determined equilibrium dissociation constant of 2.4 nmol/l.

Effect of THN on (^{3}H)-Tryptamine and (^{3}H)-5-Hydroxytryptamine Binding

After characterization of (^{3}H)-tryptamine binding we investigated whether THN displaces (^{3}H)-tryptamine from its binding sites. As discussed above, in this case some of the pharmacological effects of THN might be explained by such a mechanism. In figure 4A saturation experiments under equilibrium conditions are illustrated. By Scatchard analysis a K_D-value of 3.7 nmol/l was calculated for

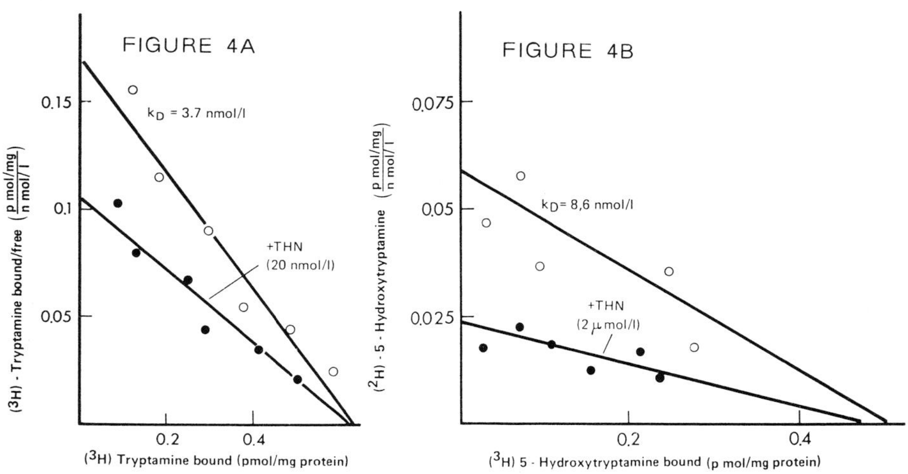

Scatchard graphs of saturation analysis of (^{3}H)tryptamine and (^{3}H) 5-Hydroxytryptamine specific binding in the absence (open) and presence (filled symbols) of 20 nmol-l or 2 µmol-l THN.

(^{3}H)-tryptamine binding which is in good agreement with the data published by Kellar and Cascio (1982). THN displaced (^{3}H)-tryptamine at a concentration of 20 nmol/l apparently in a competitive manner.

Parallel experiments were conducted with (^{3}H)-5-hydroxytryptamine as a ligand. The equilibrium dissociation content was determined to be 8.6 nmol/l. THN inhibited (^{3}H)-5-hydroxytryptamine binding at a concentration which exceeds approximately onehundredfold the concentration which displaces (^{3}H)-tryptamine (Fig. 4B).

Characterization of (^{3}H)-Tryptamine and of (^{3}H)-5-HT Binding Sites by ß-Carbolines

Comparing the various ß-carbolines tested, THN is amoung the most effective to both binding sites (tab. 2). This has been found by Cascio and Kellar (1982) as well. However, we observed harmaline (4.9-dihydro-7-methoxy-1-methyl-3H-pyrido [3,4-b] indole) to be the most potent inhibitor of (^{3}H)tryptamine binding. 6-OH-THN inhibits both (^{3}H)-tryptamine and (^{3}H)-5HT binding in the same concentration range, but in comparison to the other ß-carbolines it exerts low affinity to tryptamine and intermediate affinity to 5-HT sites. This result suggests a specificity of the tryptamine binding sites at least with respect to it hydroxylated derivative.

Table 2

Inhibition of (^{3}H)-Tryptamine (2.4 nmol/l) and (^{3}H)-5-Hydroxytryptamine (4.3 nmol/l) specific binding by ß-carbolines (Ki-values)

	(^{3}H)-Tryptamine (nmol/l)	(^{3}H)-5-Hydroxytryptamine (μmol/l)
Harmaline	21 ± 4	68 ± 15
THN	32 ± 5	2 ± 1
Norharmane	39 ± 9	61 ± 20
Harmine	63 ± 15	124 ± 50
Harmane	71 ± 14	45 ± 19
Harmalol	73 ± 16	83 ± 29
Harmol	245 ± 75	76 ± 13
6-OH-THN	6000 ± 550	11 ± 3

In conclusion, although previously reported results suggest the existence of THBC binding sites on brain membranes, the present findings are compatible with the hypothesis that THBC's act on tryptamine receptors. Whether both binding sites are two states of the same protein or different receptors remains to be elucidated.

Acknowledgement

This study was supported by the Deutsche Forschungsgemeinschaft.

References

Brossi, A.: In: Beta-Carbolines and Tetrahydroisoquinolines (Bloom F., Barchas J., Sandler M., Usdin E., eds.) (A.R. Liss, New York) (1982) pp 125-133.

Cascio, C.S., Kellar, K.J.: Neuropharmacology 21, (1982) 1219-1221.

Hahn, G., Stiehl, K.: Chem. Ber. 69, (1936) 2627-2654.

Hahn, G., Rumpf, F.: Chem. Ber. 71, (1938) 2141-2153.

Kellar, K.J., Cascio, C.S.: Eur. J. Pharmacol. 78, (1982) 475-478.

Müller, W.E., Fehske, K.J., Borbe, H.O., Wollert, K., Nanz, Ch., Rommelspacher, H.: Pharmacol. Biochem. Behav. 4, (1981) 693-699.

Nielsen, E.B., White, F.J., Holohean, A.M., Callahan, P.M., Appel, J.B.: Life Sciences 31, (1982) 2433-2439.

Pannier, L., Rommelspacher, H.: Neuropharmacology 20, (1981) 1-8.

Rommelspacher, H. Brüning G.: Hoppe-Seyler's Z. Physiol. Chem. 363, (1982) 1284.

Rommelspacher, H., Barbey, M. Strauss, S., Greiner, B., Fähndrich, E.: In: Beta-Carbolines and Tetrahydroisoquinolines (Bloom F., Barchas J., Sandler M., Usdin E., eds.) (A.R. Liss, New York) (1982) pp 41-55.

Rommelspacher, H., Brüning, G.: Naunyn-Schmiedeberg's Arch. Pharmacol. 322, (1983) R92.

Rueffer, M., El-Shagi, H., Nagakura, N., Zenk, M.H.: FEBS Lett. 129, (1981) 5-9.

REGULATIONS OF SEROTONERGIC NEURONS INNERVATING THE BASAL GANGLIA: IN VIVO RELEASE STUDIES

P. Soubrié and J. Glowinski
INSERM U 114, Collège de France
11, pl. M. Berthelot
75231 Paris cedex 05

Introduction

Functional changes in the activity of central neurons can be estimated using various biochemical and electrophysiological parameters. A crucial one deals with the in vivo release of the neurotransmitter from nerve terminals or dendrites. Hence, a methodology has been developed for studying serotonin (5-HT) release in distinct cat brain structures such as the substantia nigra or the caudate nucleus which both receive a substantial 5-HT projection from the dorsal and/or the median raphe nuclei (4,17,34,39).

This methodology combines the push-pull cannula technique with a radioisotopic assay (30). Halothane anaesthetized or "Encephale isolé" cats are stereotaxically implanted with push-pull cannula allowing a continuous superfusion of the intended brain area(s) with ^{3}H-tryptophan so that the ^{3}H-5-HT continuously formed from its ^{3}H-precursor and released can be collected and isolated by ion-exchange chromatography and then measured by liquid scintillation spectrophotometry.

The aim of this review is to present some basic information on what such a methodology is able to provide regarding the mechanisms involved in the control of 5-HT transmission in the brain, and more particularly in the basal ganglia. In fact, this review will be centered on some interactions

Progress in Tryptophan and Serotonin Research

between GABA (gamma-aminobutyric acid)-ergic and 5-HT neurons. Hence, we will envisage succesively local GABAergic influences upon nigral 5-HT nerve terminals and the role of GABAergic cells of the dorsal raphe in controlling 5-HT release in both caudate nuclei and substantia nigra under two conditions: benzodiazepine treatment and manipulation of the activity of habenulo-raphe pathways.

Local GABAergic control of nigral 5-HT release

The substantia nigra receives GABA afferent inputs from the caudate nucleus and the globus pallidus and also contains intrinsic as well as highly collateralized, efferent GABA neurons (7, 27). On account of these data we investigated whether nigral GABAergic transmission regulates 5-HT release in the substantia nigra. We observed that when applied into one substantia nigra, GABA (10^{-5}M) or muscimol (10^{-6}M) reduced the *in vivo* ^{3}H-5-HT release locally whereas the blockade of GABA transmission in one substantia nigra by local application of picrotoxin (10^{-5}M) induced an opposite effect (31).

Our results indicate that in addition to its influence upon dopaminergic neurons the nigral GABAergic transmission also controls 5-HT release in the substantia nigra. This control may be direct via an action on 5-HT nerve terminals, since Gale (9) found a significant population of GABA receptors on nigral 5-HT nerve endings, or indirect, by affecting the activity of other nigral transmitter systems. Indeed, GABA modulates the release of substance P or of dopamine in the substantia nigra, both transmitters being able to exert important influences over the *in vivo* nigral release of 5-HT (14,26).

Moreover, in these experiments, cats were implanted with one push-pull cannula in each substantia nigra and each caudate nucleus, so that it was possible to investigate whether manipulating GABAergic transmission in one substantia nigra could induce only local or also distal changes in 5-HT release. Thus altering nigral GABA transmission affected neither striatal nor contralateral ^{3}H-5-HT release (31). This type of control, referred as to local control, can also be observed during the alteration of glutamatergic transmission in one substantia nigra (26). Conversely, stimulating or interrupting dendritic release of dopamine by nigral application of dopamine or alpha -methyl paratyrosine not only modulated local 5-HT release but also affected 5-HT release in each caudate nucleus (14). Such a pattern, of distal changes in 5-HT release may involve neurons contributing to a feed- back pathway from the substantia nigra to the raphe nuclei (29) which could only be sensitive to a restricted group of nigral neurotransmitter systems. These distal changes also suggest the occurrence of phenomena of information transfer within the basal ganglia, the push-pull cannula methodology is particularly apt to reveal.

As suggested by electrophysiological and biochemical experiments (10, 28) GABAergic interneurons localized within the dorsal raphe (3), seem to play an inhibitory role upon 5-HT cells. Since the pushpull cannula methodology can be appropriate for investigating the privileged sites of action of drugs, we decided to study whether bezodiazepines (drugs known to facilitate GABAergic transmission, 12) could affect the activity of 5-HT neurons at the level of the dorsal raphe.

Benzodiazepines and 5-HT neurons

Evidence has been accumulated which indicates that one possible mechanism of the anti-anxiety activity of benzodiazepines lies in their ability to reduce 5-HT transmission within the brain (33). Thus, iontophoretic or systemic administrations of benzodiazepine are highly potent in depressing serotonergic cell firing or in potentiating the inhibitory action of GABA agonists (10,16,38). Consonant with these electrophysiological findings, biochemical data showed a decreased serotonin turnover in brain structures following benzodiazepine administration (18-28). In vitro studies have revealed that benzodiazepines are able to facilitate the potassium-evoked serotonin release from mid-brain raphe slices (6,37), but to depress or block the potassium-evoked serotonin release from terminal areas such as the hippocampus or the cerebral cortex (2,6). Therefore, experiments were undertaken in order to delineate more precisely whether cell bodies and/or dendrites versus terminals of serotonergic neurons may represent preferential target sites for the in vivo action of benzodiazepines.

In a first set of experiments, the administration of chlordiazepoxide (10 mg/kg ip) to "encephale isolé" cats was found to reduce significantly the nigral and striatal release of ^{3}H-5-HT whereas enhancing ^{3}H-5-HT release in the dorsal raphe. Several data suggest that these chlordiazepoxide-induced changes are initiated at the dorsal raphe level.

First, the delivery of chlordiazepoxide (10^{-5}M) to the dorsal raphe decreased nigral ^{3}H-5-HT release. More interestingly, the continuous perfusion of the dorsal raphe with Ro 15-1788 -a benzodiazepine receptor antagonist (16)- abolished all of the changes in ^{3}H-5-HT release induced by the peripheral administration of chlordiazepoxide (Fig. 1).

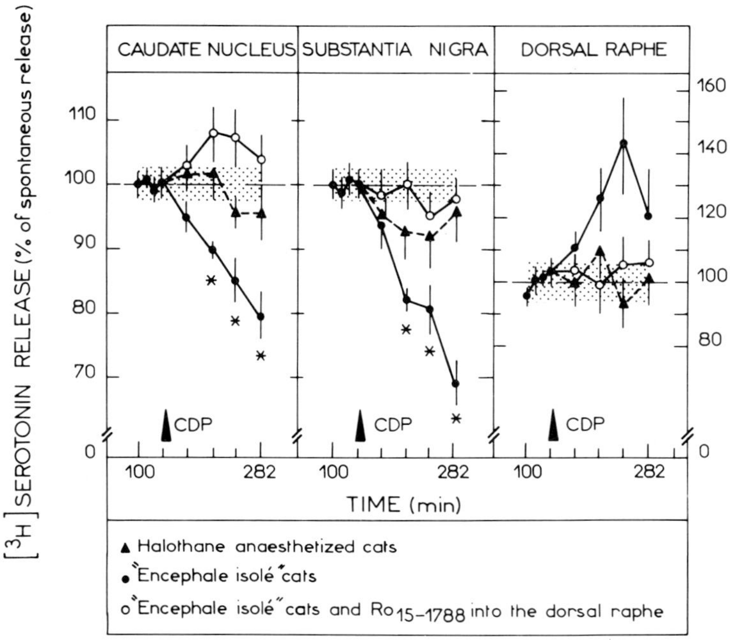

Figure 1 : Effects of an intraperitoneal injection of chlordiazepoxide (CDP) on ³H-5-HT release in the cat caudate nucleus, substantia nigra and dorsal raphe. Using push-pull cannula, each caudate nucleus, each substantia nigra and the dorsal raphe were continuously superfused with an artificial CSF medium containing L-³H-tryptophan. Chlordiazepoxide was injected intraperitoneally (10 mg/kg in a volume of 5 ml) 136 min after the onset of the superfusion with ³H-tryptophan. ³H-5-HT release was estimated in 12 min successive superfusate fractions and each fractional values was expressed as a percentage of the average release from the four 12 min fractions collected after a 90 min labelling period. During the 144 min period following chlordiazpoxide administration, an averaged value of ³H-5-HT release was calculated on blocks of 3 successive fractions (36 min). * $p<0.05$ when compared to corresponding control values (stippled area).

Second, the application of chlordiazepoxide (10^{-5}M) either to the substantia nigra, the caudate nucleus or the dorsal hippocampus never altered the local release of ³H-5-HT.

The unsensitivity of nigral serotonergic nerve terminals to local applications of chlordiazepoxide was unexpected since we have previously indicated that nigral application of GABA or picrotoxin locally altered ^{3}H-5-HT release. In fact, these data are in keeping with the notion that benzodiazepines only interact with certain sub-classes of GABA receptors (5). Taken together these data emphasize the significant role of the 5-HT cell bodies and/or dendrites (8) in mediating benzodiazepine-induced decrease in 5-HT transmission within the brain. In contrast to that it can be concluded from in vitro studies, our in vivo experiments are not in favor of the concept that nerve terminals are crucial target sites through which benzodiazepines influence 5-HT transmission in an intact brain.

In agreement with electrophysiological studies which have emphasized the inability of benzodiazepines to alter 5-HT cell activity in anaesthetized animals (10), none of the alterations in ^{3}H-5-HT release caused in "encephale isolé" cats by chlordiazepoxide could be reproduced in halothane-anaesthetized animals. These data indicate further that anaesthesia and/or the level of the animal arousal is a crucial factor in controlling the reactivity of 5-HT neurons to various treatments or neuronal inputs (15).

The last type of information provided by the push-pull methodology which we would like to consider deals with the involvement of GABAergic neurones in more complex regulatory mechanisms such as the habenulo-raphe feedback loops differentially implicated in the control of nigral versus striatal 5-HT release.

Habenulo-raphe neurons and nigral and striatal 5-HT release

Anatomical data obtained in the rat and the cat indicate that the dorsal raphe receives an important input from neurons originating or passing through the lateral habenula (1,13,29). It has been suggested that this neuronal input exerts a pronounced control over 5-HT neurons of the dorsal raphe. This hypothesis was proposed since electrical stimulation of the rat lateral habenula decreased the firing rate of 5-HT neurons (35,36,40). In addition, as shown in the cat, the unilateral application of the depolarizing agent, potassium chloride, to the lateral habenula altered the *in vivo* release of 5-HT in brain structures which receive a 5-HT innervation from the dorsal raphe (32).

Therefore, we investigated the effects of various treatments made at different levels of the habenula-raphe axis on the *in vivo* nigral and striatal ^{3}H-5-HT release. Changes induced by the unilateral electrical stimulation of the lateral habenula were investigated first. Since the cat lateral habenula receives an input from the entopeduncular nucleus (19,20) and since experiments performed in rodents (21) have suggested that this projection is GABAergic, we then examined the effects of picrotoxin infusion into the lateral habenula.

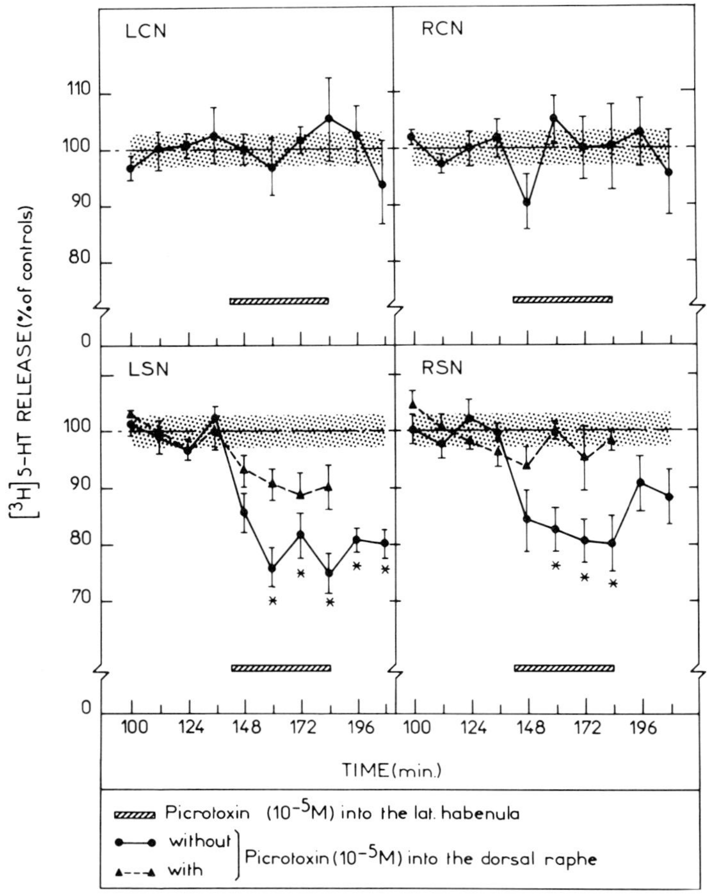

Figure 2 : Effects of unilateral application of picrotoxin to the left lateral habenula on ^3H-5-HT release in both caudate nuclei and substantia nigra. Picrotoxin (10^{-5}M) was applied for 48 min to the left lateral habenula (by way of a push-pull cannula) 136 min after the onset of the delivery of ^3H-tryptophan to each basal ganglia structure (both left and right caudate nuclei (LCN and RCN) and substantia nigra (LSN and RSN)). In another series of experiments, picrotoxin (10^{-5}M) was applied simultaneously to the left lateral habenula and dorsal raphe for 48 min, 136 min after the onset of the delivery of ^3H-tryptophan to each substantia nigra. * $p<0.05$ when compared with corresponding control values (stippled area).

In addition, since dopaminergic neurons originating in the VTA (ventral tegmental area) have been described in rodents to project to the habenula (23), the effect of habenular applications of dopamine were also investigated. Finally, because the inhibition of the dorsal raphe neurons induced by the electrical stimulation of the lateral habenula has been found to be reduced by intradorsal raphe application of picrotoxin (40), attempts were made to determine whether our two latter manipulations coud influence ^{3}H-5-HT release in the basal ganglia by way of GABA neurons impinging on 5-HT cells of the dorsal raphe.

We found that unilateral electrical stimulation of the cat lateral habenula reduced ^{3}H-5-HT release in both caudate nuclei and substantia nigra. Furthermore, blockade of the inhibitory GABAergic input by picrotoxin application into the lateral habenula induced a decrease in ^{3}H-5-HT release in both substantia nigra (Fig 2). Likewise, dopamine infusion into the habenula reduced nigral ^{3}H-5-HT release (Fig 3). The results obtained confirm the role of the lateral habenula in bilaterally controlling 5-HT transmission in structures of the basal ganglia and suggest that activation of neurons (and fibers) of the lateral habenula can inhibit the activity of 5-HT neurons innervating the basal ganglia. Since the simultaneous application of picrotoxin to the dorsal raphe prevented the inhibition of nigral ^{3}H-5-HT release induced by the delivery of the GABA antagonist or of dopamine to the lateral habenula, changes in GABA transmission within the dorsal raphe seem to play an important role in the alteration of ^{3}H-5-HT release observed in the basal ganglia following lateral habenula manipulations.

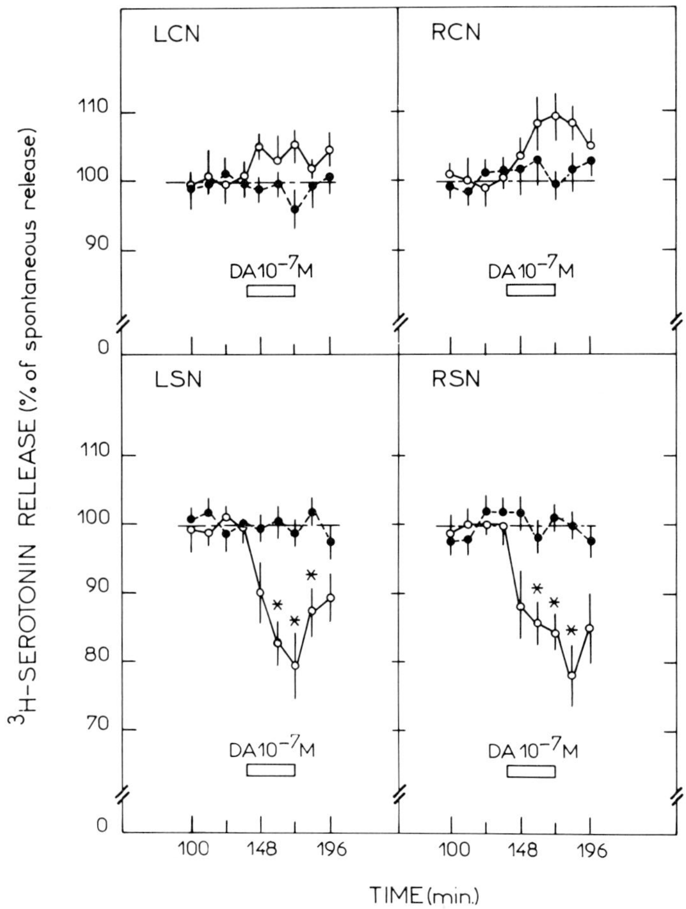

Figure 3 : Effects of unilateral application of dopamine (DA, 10^{-7}M for 2 fractions) to the left lateral habenula on ^{3}H-5-HT release in each caudate nucleus (LCN & RCN) and substantia nigra (LSN & RSN). See legend fig 2. * $p<0.05$ between control (open circle) and treated (closed circle) cats.

The presence of a direct GABAergic lateral habenula-dorsal raphe pathways proposed by Wang and Aghajanian (4) has been questioned by several authors. (3,11,22). Accordingly it has been proposed that non-GABAergic lateral habenula-dorsal raphe neurons can activate GABA interneurons in the dorsal raphe which could be involved in the control of 5-HT neurons (22). Such an anatomical organisation could explain the results of Wang and Aghajanian as well as our observations.

In the course of these experimentsan interesting fact emerged : the application of picrotoxin or of dopamine to the lateral habenula decreased ^{3}H-5-HT release in the substantia nigra but not in the caudate nucleus. A similar dissociation was previously observed following sensory stimuli which were found to enhance nigral but not striatal 5-HT release in "encephale isolé" cats (25). These results may be surprising in light of anatomical studies indicating that, in rats, the majority of dorsal raphe 5-HT cells which innervate the substantia nigra send collaterals to the ipsilateral striatum (39). Thus, our findings could suggest that the fine organization of 5-HT projections originating from the dorsal raphe is not entirely similar in the cat and the rat. Alternatively,it is conceivable that 5-HT neurons innerva ting the substantia nigra are, under certain conditions, regulated differently from those projecting to the caudate nucleus, the push-pull cannula methodology being particularly appropriate for detecting such fine regulatory processes.

In conclusion, all of our experiments using push-pull cannula for studying the in vivo release of ^{3}H-5-HT continuously formed from its ^{3}H-precursor underline the different types of control that GABAergic processes may exert upon serotonin transmission. A local inihibitory influence can be observed at the nigral level. An indirect inhibitory control appears to be provided by neurons of entopeduncular origin projecting to habenula cells which innervate the dorsal raphe. An important role of GABAergic interneurons of the dorsal raphe is suggested by experi- ments dealing with benzodiazepines and by the ability of intra raphe application of picrotoxin to counteract the consequences of changing the activity of habenula raphe cells. Finally, GABAergic neurons of the dorsal raphe may be involved (at least) partly in the differential reactivity of 5-HT neurons in "encephale isolé" versus halothane anaesthetized cats.

References

1. Akagi, K., Powell, E.W.: Exp. Neurol. 182, 263-27 (1968).

2. Balfour, D.J.K.: European. J. Pharmacol. 68, 11-16 (1980).

3. Belin, M.F., Aguara, M., Tappaz, M., Mc Rac-Deguerce, A., Bobillier, P., Pujol, J.F.: Brain Res., 170, 279- 297 (1979).

4. Bobillier, P., Seguin, S., Petitjean, F., Salvert, D., Touret, M., Jouvet, M.: Brain res., 113, 449-486 (1976).

5. Brennan, M.J.W.: J. Neurochem. 38, 264-266 (1982).

6. Collinge, J., Pycock, C.: European. J. Pharmacol. 85, 9-14 (1982).

7. Di Chiara, G., Morelli, M., Porceddu, M., Mulas, M., Del Fiacco, M.: Brain Res. 189, 193-199 (1980).

8. Felten, D.L., Harrigan, P.: Neurosci. Lett. 16, 275-280 (1980).

9. Gale, K.: Neurosci. Abstr. 9, 71 (1979).

10. Gallager, D.W.: European J. Pharmacol., 49, 133-143 (1978).

11. Gottesfeld, Z., Hoover, D.B., Muth, E.A., Jacobowitz, D.M.: Brain Res. 141, 2353-2356 (1978).

12. Haefely, W., Pieri, L., Polc, P., Schaffner, R.: Handbook of Experimental Pharmacology ed. by F. Hoffmeister and G. Stille, 55, 13-262 (1981).

13. Herkenham, M., Nauta, J.H.W., 187 , 19-48 (1979).

14. Hery, F., Soubrié, P., Bourgoin, S., Montastruc, J., Artaud, F., Glowinski, J.: Brain Res. 193, 143-151 (1980).

15. Heym, J., Trulson, M.E., Jacobs, B.L.: European J. Pharmacol. 74, 117-125 (1981).

16. Hunkeler, W., Möhler, H., Pieri, L., Polc, P., Bonetti, E.P., Cumin, R., Schaffner, R., Haefely, W.: Nature, (Lond) 290, 514-517 (1981).

17. Jacobs, B.L., Foots, S.L., Bloom, F.E.: Brain Res. 147, 149-153 (1978).

18. Jenner, P., Chadwick, D., Reynolds, E.H., Marsden, C.D.: J. Pharm. Pharmac. 27, 707-710 (1970).

19. Larsen, K.D., McBride, R.L.: J. Comp. Neurol. 184, 293-308 (1979).

20. McBride, R.L.: J. Comp. Neurol. 198, 89-99 (1981).

21. Nagy, J.I., Carter, D.A., Lehmann, J., Fibiger, H.C.: Brain Res. 145, 360-364 (1978).

22. Neckers, L.M., Schwartz, J.P., Wyatt, R.J., Speciale, S.G.: Exp. Brain Res. 37, 619-623 (1979).

23. Phillipson, O., Griffith, A.: Brain Res. 197, 213-218 (1980).

24. Reisine, T., Soubrié, P., Artaud, F., Glowinski, J.: J. Neurosci. 2, 1062-1071 (1982a).

25. Reisine, T., Soubrié, P., Artaud, F., Glowinski, J.: Brain Res. 232, 77-87 (1982b).

26. Reisine, T., Soubrié, P., Artaud, F., Glowinski, J.: Brain Res. 236, 317-327 (1982).

27. Ribak, C., Vaughn, J., Roberts, E.: Brain Res., 192, 413-423 (1980).

28. Saner, A., Pletscher, A.: European J. Pharmaco. 55, 315-318 (1979).

29. Sakai, K., Salvert, P., Touret, M., Jouvet, M.: Brain Res. 137, 11-35 (1977).

30. Soubrié, P.: J. Histochem. Cytochem. 30, 817-820 (1982).

31. Soubrié, P., Montastruc, J.L., Bourgoin, S., Reisine, T., Artaud, F., Glowinski, J.: European J. Pharmacol. 69, 483-488 (1981).

32. Soubrié, P., Reisine, T., Artaud, F., Glowinski, J.: Brain Res. 222, 192-197 (1981).

33. Stein, L., Belluzi, J.D., Wise, C.D., Amer. J. Psychiat. 174, 665-669 (1977).

34. Steinbusch, H.W.M.: Neuroscience 6, 557-618 (1981).

35. Stern, W.C., Johnson, A., Bronzino, J.D., Morgane, P.J.: Exp. Neurol. 65, 326-342 (1979).

36. Stern, W.C., Johnson, A., Bronzino, J.D., Morgane, P.J.: Neuropharmacology 20, 974-979 (1981).

37. Thiebot, M.H., Hamon, M., Soubrié, P.: Neuroscience 7, 2287-2294 (1982).

38. Trulson, M.E., Preussler, D.W., Howell, G.A., Frederickson, C.J.: Neuropharmacology 21, 1045-1050 (1982).

39. Van der Kooy, D., Hattori, T.: Brain Res. 186, 1-7 (1980).

40. Wang, R.Y., Aghajanian, G.K.: Science 197, 89-91 (1977a).

BLOOD PLATELETS AS MODELS FOR NEURONS: SIMILARITIES OF THE 5-HYDROXYTRYPTAMINE SYSTEMS

Alfred Pletscher, Hubert Affolter, Andrea M.Cesura, Paul Erne, Kurt Müller

Departement Forschung des Kantonsspitals Basel, Hebelstr. 20, CH-4031 Basel

Blood platelets of man and animals exhibit various subcellular elements typical for the 5-hydroxytryptamine (5HT) system in the central nervous system (CNS). These include a specific uptake mechanism for 5HT at the plasma membrane, intracellular storage organelles for biogenic amines and a 5HT-receptor at the cell surface (Fig. 1). However, in other respects there are differences. For instance, platelets in contrast to neurons are in general unable to synthesize 5HT (absence of tryptophan hydroxylase and decarboxylase) and have a very slow turnover of 5HT. Furthermore, human platelets contain monoaminoxydase (MAO)-B which in neurons is less specific for 5HT than MAO-A (1, 2). The physiological and pathophysiological role of the 5HT-system in platelets is only partly understood and will not be the topic of this presentation. We shall rather focus on some similarities between platelets and 5HT-neurons regarding uptake of 5HT, its intracellular storage and the membrane receptor for the amine.

Uptake at the plasma membrane

Platelets take up 5HT in at least two ways, i.e. by passive diffusion (unspecific uptake) and by a specific mechanism. At

Progress in Tryptophan and Serotonin Research

low external 5HT-concentrations the latter is much more efficient than the former. The specific uptake depends on Na^+ and Cl^-, shows the characteristics of a carrier-mediated transport (e.g. saturation kinetics) and needs metabolic energy. It has been demonstrated in intact normal and reserpinized platelets (in the latter the intracellular 5HT-storage is abolished) and also in isolated membrane vesicles of platelets. The Km of the uptake of 5HT ($\sim 10^{-7}$M) is much smaller than that of other biogenic amines, e.g. of dopamine (DA; > 10^{-5}M) and noradrenaline (NA; > 10^{-4}M), which are probably taken up to a small extent by co-transport with 5HT (1, 2).

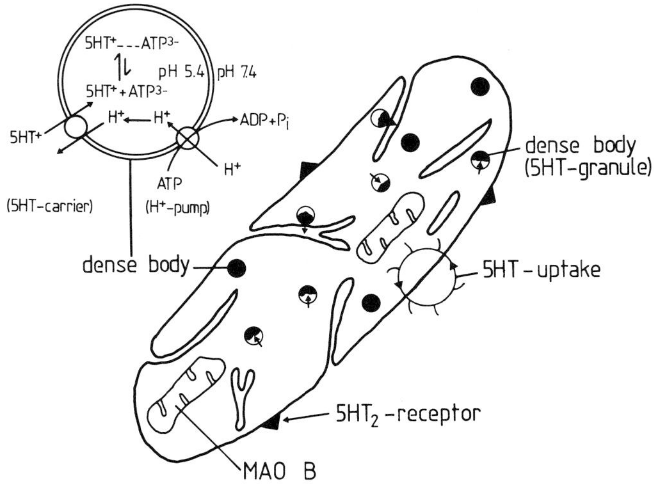

Fig. 1: Scheme of a blood platelet with the elements of the 5HT-system and of a 5HT-storage vesicle.

Various drugs inhibit the specific uptake of 5HT in platelets. Among the most potent compounds are antidepressants, e.g. imipramine (IC_{50} in the nanomolar range), whereas 5HT-receptor antagonists, e.g. ketanserin (IC_{50} > 100 μM), do not have a marked inhibitory action. In addition, radiolabelled imipramine shows specific and reversible binding to platelet membranes, the Kd being of the order of 10^{-9}M. Drugs which inhibit the 5HT-uptake by platelets also prevent the ^{3}H-imipramine binding and the order of potencies of the drugs is similar in both experimental systems (3). Therefore, ^{3}H-imipramine seems to be a marker for the 5HT-carrier. Whether the antidepressant binds to the same site of the carrier molecule as 5HT is not yet known.

The imipramine-binding protein has been solubilized with digitonin. It has a molecular weight of approximately 300,000 and still shows saturable high affinity binding of ^{3}H-imipramine (Kd = 4.6±1.2 nM). In the solubilized preparations as in intact membranes, antidepressants which inhibit 5HT-uptake are much more potent in preventing ^{3}H-imipramine binding than 5HT and 5HT-receptor antagonists (Table 1). Attempts to purify the protein are under way.

DRUG	INHIBITION OF ^{3}H-IMI-BINDING (IC_{50}, nM)
Imipramine (IMI)	7.9 ± 3.0
Amitryptiline	46.0 ± 7.5
Desmethyl-IMI	230 ± 31
5HT	7200 ± 1500
ketanserin	140000 ± 12000

Table 1: Inhibition of specific ^{3}H-imipramine binding to solubilized human platelet membranes by drugs. Each value is an average with s.e.m. of 3 independent experiments.

The 5HT-uptake system of brain synaptosomes shows several similarities to that of platelets. Thus, the Km-values for 5HT-uptake are of the same order in both systems. Furthermore, those drugs (e.g. antidepressants) which inhibit 5HT-uptake in platelets also interfere with the uptake of the amine in synaptosomes, and the order of potencies of the drugs is similar in both cell types (1, 2). In addition, ^{3}H-imipramine specifically binds to brain synaptosomes. The Kd of this binding is of the same order of magnitude in synaptosomes and in platelets, and in synaptosomes the order of potencies of drugs preventing imipramine binding is similar to that for inhibition of 5HT-uptake. Therefore, it is conceivable that in 5HT-neurons, too, imipramine interacts with the carrier for 5HT-uptake at the plasma membrane and that the carrier in synaptosomes shows similarities to that in platelets (3).

Intracellular storage

The bulk of the intracellular monoamines in platelets consists of 5HT, whereas other amines, especially catecholamines (4), generally occur only in trace amounts. The 5HT is localized in at least two pools, i.e. in specific electron-dense subcellular organelles (vesicles, granules, dense bodies) and at extragranular sites. The latter are not well understood and will not be dealt with in this presentation. The intragranular pool shows a very high concentration of the amine, which exeeds that of the extragranular pool by several orders of magnitude (1). Various experimental findings indicate that for the storage of 5HT (and other amines) in platelets a proton gradient across the granular membrane, created by an ATP-driven proton pump, is an important factor. The amine enters the granule passively along the electrochemical gradient, a carrier probably being involved (Fig. 1) (5). Recent experiments indicate that in cerebral neurons, too, intragranular 5HT-storage depends on a proton gradient (although the membrane potential also plays a role). For instance, ionophores which cause a collapse of H^+-gradients across the granular membranes (e.g. monensin and FCCP*) were found to induce a marked release of granular 5HT not only from platelets but also from brain synaptosomes (Fig.2), this being of a non-exocytotic nature (no concomitant

*Carbonyl cyanide p-(trifluoromethoxy)-phenylhydrazone

release of ATP from platelets, no dependence on Ca^{++} in synaptosomes) (6, 7).

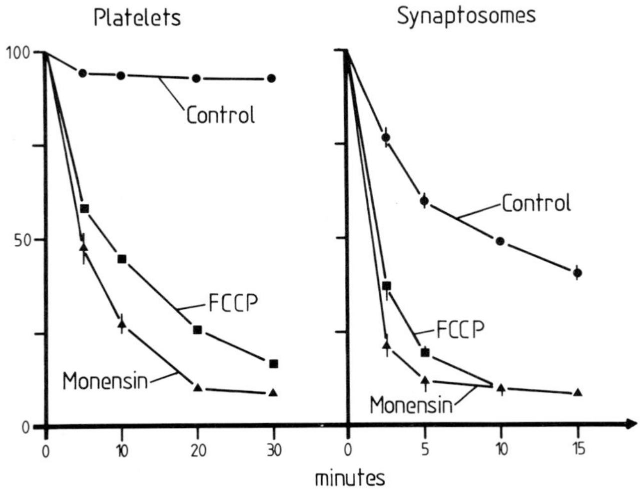

Fig. 2: Effect of monensin (15 and 4.5 µM respectively) and FCCP (16 µM) on the content of radiolabelled 5-hydroxytryptamine (5HT*) in human platelets and brain synaptosomes (Ca^{2+}-free medium) of guinea-pigs. Each point is an average with s.e.m. of at least 6 independent experiments (6, 7).

There are other mechanisms of release of granular 5HT which show similarities in platelets and neurons. In platelets there is a release by exocytosis (i.e. of unmetabolized 5HT) caused by agents like thrombin and collagen, intracellular Ca^{2+} functioning as a mediator. In 5HT-neurons, too, 5HT is liberated by Ca^{2+}-dependent exocytosis, however the stimuli for exocytosis (e.g. electrical and K^+-induced depolarisation) differ from those in platelets.

A non-exocytotic 5HT-release from granular stores is that induced by reserpine and related compounds. These drugs probably interfere with the 5HT-carrier of the granular membrane. They act similarly in platelets and neurons. For instance, in both cell types the 5HT-release is not stoichiometric, one molecule of reserpine probably liberating hundreds or thousands of 5HT-molecules. Furthermore, the 5HT leaves both platelets and neuronal cells partly metabolized, i.e. as 5-hydroxyindolacetic acid, or in the case of platelets, also as 5-hydroxytryptophol.

Finally, in both platelets and neurons intragranular 5HT is liberated by arylalkylamines such as amphetamine. This release probably occurs by intragranular displacement (possibly from a loose 5HT-ATP-complex). One molecule of an arylalkylamine has been estimated to displace at the most a few molecules of 5HT which leaves the cell mainly unchanged, since arylalkylamines inhibit MAO (1).

5HT-receptors

The 5HT-receptors of platelets have been characterized by binding experiments with radiolabelled 5HT and by a functional parameter, the shape change reaction. Binding experiments with 5HT are somewhat problematic, since unspecific binding is high and since platelets actively take up the amine. Nevertheless, specific, high affinity binding of ^{3}H-5HT to platelet membranes has been detected, two Kd-values being obtained. Experiments with 5HT-receptor antagonists and 5HT-uptake inhibitors indicate that the high and low affinity constants reflect binding to 5HT-receptors and 5HT-uptake sites respectively. With regard to the 5HT-receptor it is of interest that the value for the low Kd of platelets (4.7 ± 0.7 and 4.1 ± 0.7 nM in SIV and Lewis rats respectively) shows the same order of magnitude as that of neuronal membranes (6.3 ± 0.6 and 3.9 ± 0.9 nM) (1, 8).

The shape change reaction is characterized by a transition of the platelets from the normally occuring discoid shape into a spheroid form, whereby the light absorption of the platelet suspension increases. Various 5HT-receptor agonists, generally in micromolar concentrations, induce a shape change reaction.

In addition, typical 5HT-receptor antagonists counteract the 5HT-induced shape change reaction, the most potent compounds (e.g. spiroperidol, methysergide) acting in the nanomolar range. D-LSD and some tryptamine derivatives are mixed agonists-antagonists. Also, there is marked stereospecificity, D-LSD and D-butaclamol being highly potent in contrast to their stereoisomers (9). The 5HT-receptor of brain shows a similar pharmacological spectrum, since 5HT-agonists and -antagonists in micro- and nanomolar concentrations respectively prevent the ^{3}H-5HT-binding to brain membranes and since LSD and butaclamol exhibit marked stereoselectivity (10).

Based on these and other results, it can be assumed that the shape change reaction is a functional correlate of 5HT-receptor stimulation and that the 5HT-receptor of platelets resembles the $5HT_2$-receptor of neurons.

Recent experiments indicate that stimulation of $5HT_2$-receptors increases intracellular free Ca^{2+} in human platelets (measured by the fluorescent probe Quin-2) (Fig. 3) and that this rise may mediate functional effects, e.g. the shape change reaction. The latter as well as the Ca^{2+}-increase were counteracted by the $5HT_2$-antagonist ketanserin. In contrast, 5HT did not affect adenylate cyclase and cyclic AMP of platelets. In neurons, the $5HT_1$- but not the $5HT_2$-receptor has been shown to be connected with the adenylate cyclase system (11). It is therefore conceivable that in nerve cells, too, the functional effects of $5HT_2$-receptor stimulation are mediated by changes in intracellular Ca^{2+}.

Practical consequences

The 5HT-system of blood platelets can be considered to be a limited model for that of 5HT-neurons. The platelet model is suitable for the screening and characterisation of drugs and may be of interest for elucidating pathophysiological mechanisms, especially in the fields of neuropsychiatric and cardiovascular diseases. A potential use of platelets for the diagnosis and/or prognosis of such disorders can also be foreseen.

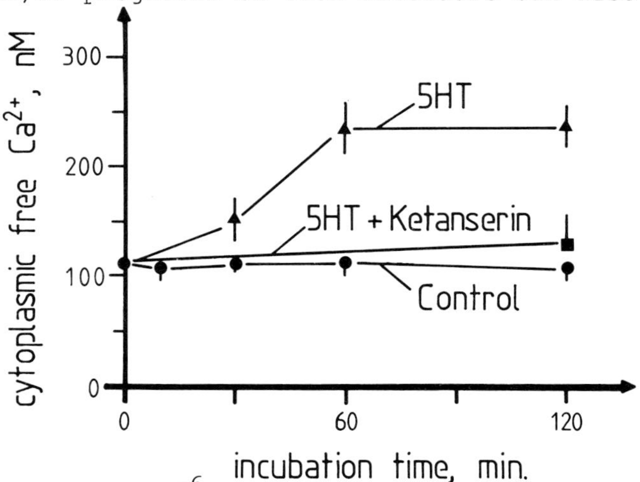

Fig. 3: Effect of 10^{-6}M 5HT and the $5HT_2$-blocker ketanserin (10^{-6}M) on intracellular free Ca^{2+}-levels. Each value is an average with s.e.m. of 3 independent experiments.

References

1. Pletscher, A.: Essays in Neurochemistry and Neuropharmacology, Vol. 3, Ed. M.B.H. Youdim, John Wiley and Sons Ltd., London 1978, pp 49-101.

2. Sneddon, J.M.: Progress in Neurobiology, Vol. 1, part 2, Ed. G.A. Kerkut and J.W. Phillis, Pergamon, Oxford 1973, pp 151-198.

3. Langer, S.Z., Briley, M.: Trends Neurosci. 4, 28-31 (1981).

4. Da Prada, M., Picotti, G.B.: Brit.J.Pharmacol. 65, 653-662 (1979).

5. Johnson, R.G., Scarpa, A., Salganicoff, L.: J.Biol.Chem. 253, 3638-3641 (1980).

6. Affolter, H., Pletscher, A.: Mol.Pharmacol. 22, 94-98 (1982).

7. Affolter, H., Peyer, M., Pletscher, A.: Brit.J.Pharmacol. 78, 111-116 (1983).

8. Von Hahn, H.P., Honegger, C.G., Pletscher, A.: Neurosci. Lett. 20, 319-322 (1980).

9. Graf, M., Pletscher, A.: Brit.J.Pharmacol. 65, 601-608 (1979).

10. Peroutka, S.J., Snyder, S.H.: Mol.Pharmacol. 16, 687-699 (1979).

11. Peroutka, S.J., Snyder, S.H.: Fed.Proc. 42, 213-217 (1983).

IN-VIVO-METABOLISM OF ^{14}C-5-HT, ^{14}C-5,6-DHT and ^{14}C-5,7-DHT BY MAO/COMT/ ALDEHYDE DEHYDROGENASE IN RAT BRAIN

Hans Georg Baumgarten, Hans Peter Klemm, Hans Georg Schlossberger

Dept. Neuroanatomy, Freie Universität Berlin (West);
Dept. Neurology, Universität Hamburg, GFR;
Max-Planck-Institut für Biochemie, Martinsried, GFR

Introduction

In vitro and in vivo studies by Creveling et al. (1975) failed to establish clear-cut relationships between the autoxidation rates of dihydroxytryptamines (DHT's) and their neurotoxic potency and in vitro studies by Rotman et al. (1976) established correlations between the degree of protein-binding of autoxidation products from 5,6-DHT and its capacity to cross-link proteins but failed to demonstrate similar relationships for 5,7-DHT. From the observation that MAO-inhibition counteracted the degeneration of adrenergic nerves in mouse heart, Creveling et al. (1975) concluded that MAO-metabolism might be the prerequisite for 5,7-DHT to develop toxicity. In vitro and in vivo studies by Baumgarten et al. (1975, 1977, 1978, 1982), Björklund et al. (1975) and by Klemm et al. (1979) on the mode and mechanism of action of a variety of DHT's and their monomethoxylated derivatives lead to the concept that MAO-metabolism plays no decisive role in the in vivo neurotoxicity of 5,6-DHT and a supporting role in the in vivo toxicity of 5,7-DHT. To obtain more direct evidence for the validity of this concept, the MAO-metabolism of 5,6-DHT and 5,7-DHT and, for comparison, of 5-HT was investigated in vivo and its role in the protein-binding of radioactivity was evaluated using inhibitors of MAO and of aldehyde dehydrogenase.

Progress in Tryptophan and Serotonin Research

Results

Time course of the action of MAO on ^{14}C-5,6-DHT, ^{14}C-5,7-DHT, and ^{14}C-5-HT after intraventricular administration in the rat brain

Fig.1a shows the time-dependent changes in the accumulation and disappearance of acid metabolites in the rat brain following intrathecal administration of ^{14}C-5,6-DHT, ^{14}C-5,7-DHT and ^{14}C-5-HT (0,2µmol). The deaminated material was collected with the "deaminated fraction" and identified by TLC. In the case of ^{14}C-5,6-DHT, the products formed by MAO represent mixtures of 5,6-DHIAA and its O-methylated derivatives. There is a continuous, nearly linear rise of ^{14}C-5-HIAA, ^{14}C-5,6-DHIAA and its methylated derivatives, and of ^{14}C-5,7-DHIAA until steady-state concentrations are reached within about 30 minutes. Conversion of the tryptamines by MAO is most efficient in the case of ^{14}C-5-HT. From the two dihydroxytryptamines, ^{14}C-5,6-DHT is oxidized more rapidly than ^{14}C-5,7-DHT. Whereas ^{14}C-5-HIAA is rapidly cleared from brain after reaching stady-state concentrations, clearance of ^{14}C-5,6-DHIAA and ^{14}C-5,7-DHIAA is retarded until about 60 minutes after injection. Part of the radioactivity recovered from brain extracts after intrathecal administration of the labelled amines was collected with the "non-deaminated fraction" and essentially consisted of the non-metabolized free amine as proved by TLC. Fig.2a shows the kinetics of disappearance of the amines from brain after this route of administration as a result of a complex action of MAO and other clearing mechanisms. Substantial differences between ^{14}C-5-HT and the dihydroxytryptamines were noted when the pca-insoluble fraction ("protein-bound fraction") was analyzed with respect to retention af radioactivity. Contrary to ^{14}C-5-HT, binding of radioactivity derived from ^{14}C-5,6-DHT and ^{14}C-5,7-DHT reaches a maximum after 1 h which remains high (Fig.3a) even when the free radioactivity has been eliminated from brain. Whereas binding of radioactivity derived from ^{14}C-HT seems to be accounted for by unspecific binding mechanisms which depend on the equilibrium between free and bound material, there is a clear increase of bound radioactivity in the face of diminishing concentrations of extractable, free radioactivity in the case of 5,7-DHT and, more impressively, of 5,6-DHT.
Selective inhibition of MAO-A by a single injection of clorgyline (2 mg/kg

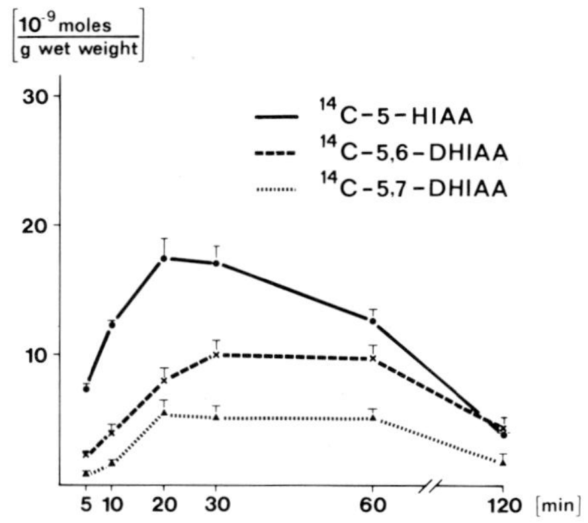

Fig.1a: Accumulation and disappearance of (14C)-indoleacetic acids in rat brain

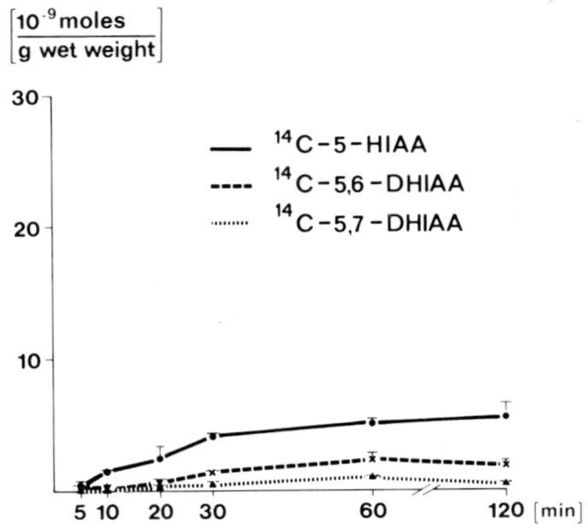

Fig.1b: Accumulation of (14C)-indoleacetic acids in rat brain after selective inhibition of MAO-A by clorgyline

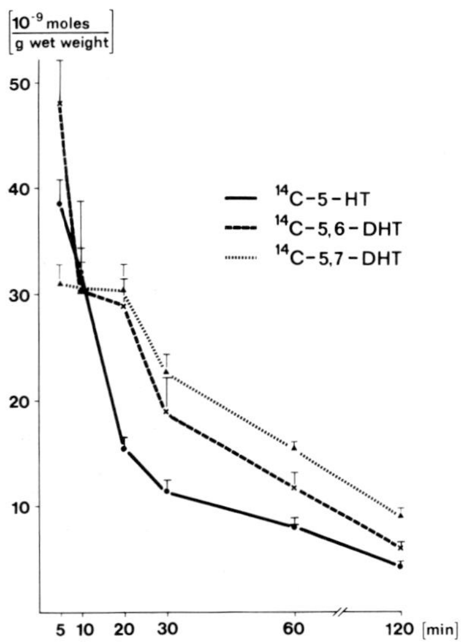

Fig.2a: Disappearance of (14C)-amines in rat brain

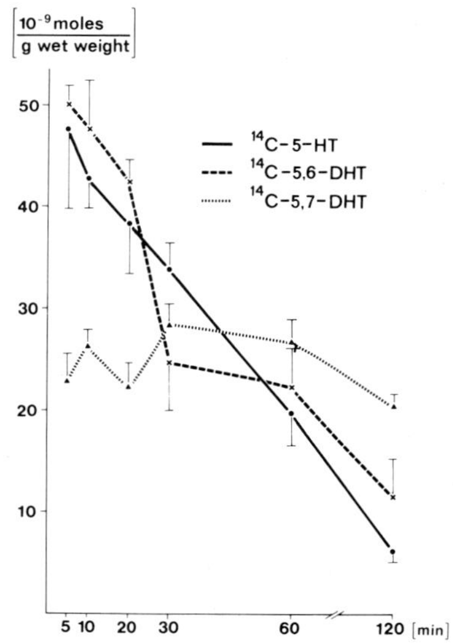

Fig.2b: Disappearance of (14C)-amines in rat brain after selective inhibition of MAO-A by clorgyline

i.p.). resulted in a nearly complete loss of activity of MAO-A (measured with ^{14}C-5-HT as substrate) to residual values 6 h after injection of clorgyline (6,7% of control). A single injection of pargyline (10mg/kg), a selective inhibitor of MAO-B at this dose range, caused loss of activity of MAO-B to residual values of 18,3 $\pm$ 3,7% of control whereas MAO-A activity remained at 73,5 $\pm$ 14,6% of corresponding control values. The effects of selective inhibition of MAO-A and -B on the conversion of ^{14}C-5,6-DHT, ^{14}C-5,6-DHT, ^{14}C-5,7-DHT and ^{14}C-5-HT were studied after pretreatment of the aminals with clorgyline or pargyline. Near-complete inhibition of MAO-A by clorgyline results in a substantial decrease of formation of the acid metabolites expecially in the case of the dihydroxytryptamines. On the other hand selective and extensive inhibition of MAO-B by pargyline fails to lower the formation of 5,6-DHIAA, 5,7-DHIAA or 5-HIAA to a significant extent. Fig. 1b demonstrates the effect of selective and nearly complete inhibiton of MAO-A on the time course of formation of the deamination products. Contrary to very low levels of MAO-A activity measured in vitro (6,7% of control) there is still an accumulation of acid material, especially 5-HIAA, pointing to a possible substituting role of MAO-B under conditions when MAO-A is inhibited, or pointing to other mechanisms that account for the oxidation process after elimination of MAO (see Discussion). Inhibition of MAO-A delays, but does not abolish, the disappearance of ^{14}C-5,6-DHT and ^{14}C-5-HT from brain (Fig. 2b) probably because of the lack of conversion to actively transported metabolites, such as the acids. Moreover in the case of ^{14}C-5,7-DHT the outward transport of radioactivity is almost completely blocked after inhibition of MAO-A by clorgyline. This suggest that 5,7-DHT is unable to penetrate the brain-blood-barrier and also unable to use the probenecid-sensitive transport mechanism that are responsible for elimination of 5-HT and 5,6-DHT metabolites from brain. Prior deamination and oxidation to the acid metabolite (5,7-DHIAA) seems to be a prerequisite for clearance of 5,7-DHT from brain.

Inhibition of MAO-A reduces the extent of binding of radioactivity to protein as revealed by analysis of the "protein-bound fraction" (Fig. 3b). The prevention of protein binding by clorgyline is evident at a very early stage (i.e. 5min after injection of the amines), i.e. before substantial deamination of the amines to the corresponding aldehydes has taken place in untreated controls (compare to Fig. 1a). Prevention of binding of radio-

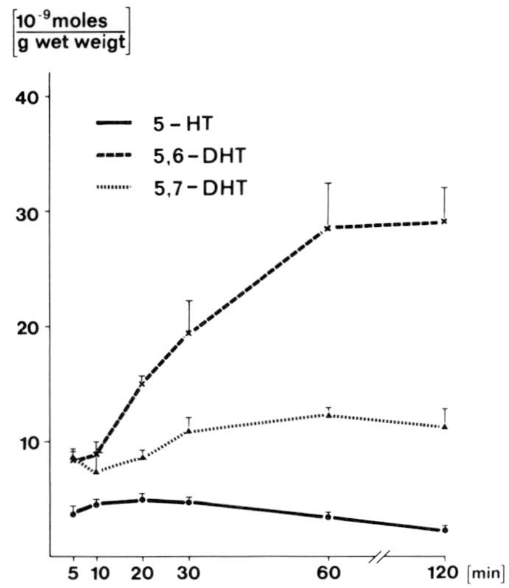

Fig.3a: Protein-bound radioactivity in rat brain after injection of (14C)-5-HT,(14C)-5,6DHT and (14C)-5,7-DHT

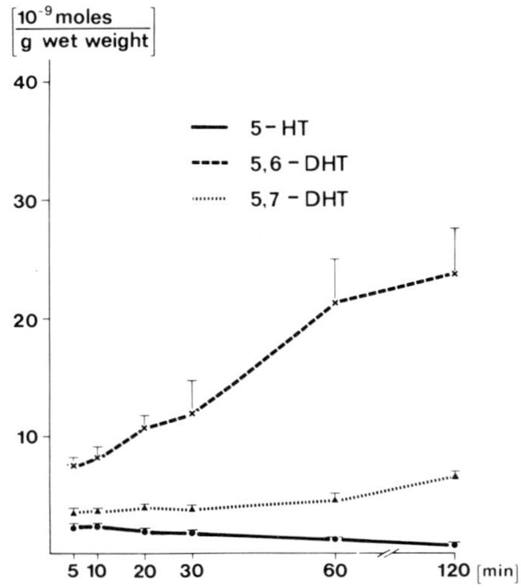

Fig.3b: Protein-bound radioactivity in rat brain after selective inhibition of MAO-A by clorgyline

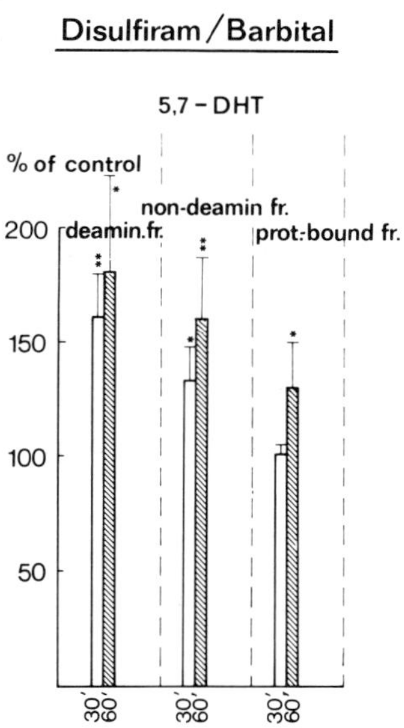

Fig.4: Metabolism of intraventricularly injected (14C)-5,7-DHT in rats treated with disulfiram and barbital

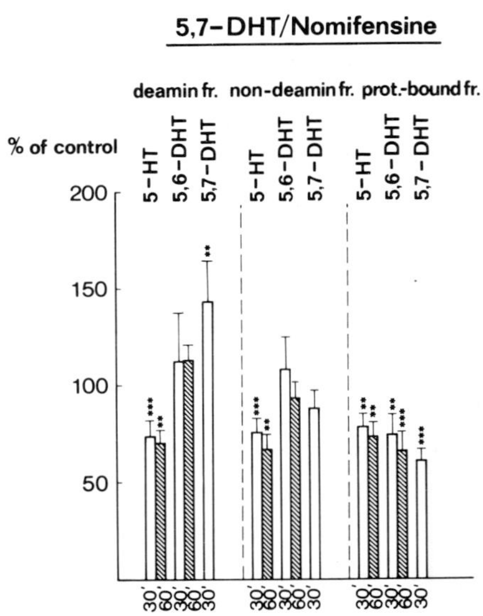

Fig.5: Metabolism of (14C)-5,6-DHT and (14C)-5,7-DHT in the brain of rats with selective and profound destruction of the serotonergic projections

activity due to inhibition of MAO activity is most pronounced during that period of time in which a nearly linear production of deaminated material is observed in the untreated animals. At later time points, binding of radioactivity derived from ^{14}C-5,6-DHT and ^{14}C-5,7-DHT tends to approximate the control values. In the case of ^{14}C-5,6-DHT, no significant difference between treated and untreated animals is seen at 120 min after injection.
To elucidate the possible involvement of the aldehyde intermediate in the binding process, we studied the effect of inhibition of aldehyde dehydrogenase by combined administration of disulfiram and barbital (Fukumori et al. 1979) on the metabolism of intraventricularly injected ^{14}C-5,7-DHT (Fig.4). 30 and 60 min after administration of ^{14}C-5,7-DHT, the levels of deaminated and non-deaminated material increase significantly in animals treated with disulfiram/barbital suggesting a decreased turnover of ^{14}C-5,7-DHT and an impaired elimination of its metabolite(s) from brain due to the blockade of aldehyde dehydrogenase. There is no significant increase in the amount of protein-bound radioactivity derived from ^{14}C-5,7-DHT 30 min after injection of this amine, and only a weak enhancement of binding observed 60 min after application. This points to a restricted participation of the aldehyde intermediate in the overall binding process.

Effects of impaired uptake and storage on the metabolism of intraventricularly administered ^{14}C-5,6-DHT, ^{14}C-5,7-DHT and ^{14}C-5-HT

Intraventricular injection of a high dose of 5,7-DHT (150 µg of the free base) five days before measurement of the monoamine uptake in crude synaptosomes results in a substantial loss of ^{3}H-5-HT uptake sites and moderate but distinct impairment of uptake capacity for ^{3}H-NA. Prior treatment of the animals with nomifensine, a NA and DA uptake inhibitor, protects the catecholamine uptake sites, but not the 5-HT uptake sites, from destruction by 5,7-DHT (see Table 1).
Fig.5 shows the effect of a selective destruction of the 5-HT uptake sites on the metabolism of ^{14}C-5,6-DHT, ^{14}C-5,7-DHT and ^{14}C-5-HT after intrathecal administration. There is a clear and substantial reduction in the binding of radioactivity derived from the three amines to the protein pellet

fraction, and a substantial decrease in the concentration of free (non-metabolized) and deaminated material in the case of 5-HT, which possibly reflects the accelerated turnover and elimination of exogeneously administered serotonin in the absence of uptake capacities that may prevent the amine from rapid catabolism. Interestingly, this effect of reduction of the 5-HT uptake sites on the turnover of the amines is not seen in the case of the DHT's, suggesting a restricted affinity to serotonin uptakes sites and a restricted involvement of serotoninergic neurons in the metabolism of these compounds when administered in high, neurotoxic doses.

Table 1

Effects of 5,7-DHT and of 5,7-DHT administered to nomifensine pretreated rats on the uptake of ^{3}H-5-HT and ^{3}H-NA into synaptosome-rich homogenates of rat brain. The results are expressed as % of control.

Control	^{3}H-5-HT	100 ± 5,4 %	
	^{3}H-NA	100 ± 6,9 %	
150 µg 5,7-DHT	^{3}H-5-HT	31,8 ± 3,0 %	xxx
	^{3}H-NA	75,1 ± 5,9 %	xx
150 µg 5,7-DHT given to nomifensine treated rats	^{3}H-5-HT	25,8 ± 4,4 %	xxx
	^{3}H-NA	95,8 ± 4,9 %	

Discussion

In vivo, 5,6- and 5,7-DHT are almost exclusively metabolized by MAO-A. By contrast, near-complete inactivation of MAO-A by clorgyline attenuates but does not abolish conversion of 5-HT to 5-HIAA suggesting that MAO-B may substitute for MAO-A under certain circumstances.In support of this, Mitra and Guha (1980) have shown that serotonin oxidation by mitochondria obtained from clorgyline-treated rats resembles B-type characteristics.

The postulated role of the aldehyde intermediates of 5,6- and 5,7-DHT in the process of protein-binding of reactive metabolites of either drug could not directly be verified as the labile aldehydes could not be isolated. However, indirect evidence in favour of a participation of the aldehyde intermediate in the irreversible binding of radioactivity to brain protein was obtained for 5,7-DHT by inhibition of MAO-A and of aldehyde dehydrogenase, respectively. While selective inhibition of MAO-A resulted in attenuation and retardation of binding of radioactivity to brain protein, inhibition of aldehyde dehydrogenase lead to a small increase in the amount of radioactivity bound to protein at 60 min after intraventricular injection of ^{14}C-5,7-DHT. In harmony with these observations are the findings by Baumgarten et al. (1979) that the extent of long-term 5-HT reductions is slightly attenuated in some brain regions by pretreatment of animals with pargyline. As these effects are only seen with 5,7-DHT and only of minor importance in the process that renders 5,7-DHT toxic, we conclude that mitochondria-promoted, accelerated oxidation of 5,6-DHT and 5,7-DHT to quinoidal intermediates and formation of secondary reaction products such as H_2O_2 and radicals are most significant in the in vivo toxicity of the DHT's (cf. Klemm et al. 1980, Baumgarten et al. 1982). Considering that protein binding is assumed to be an essential prerequisite of a portion of the neurotoxicity of the dihydroxytryptamines, prevention of protein binding after damaging (by 5,7-DHT) demonstrates convincingly that specific toxicity is evolved within monoaminergic structures by an irreversible attack of the neurotoxins to specific protein structures. Interestingly, both inhibition of catecholamine uptake by nomifensine (Baumgarten et al., 1982) or desipramine (Breese et al., 1975) and inhibition of MAO by pargyline (Breese et al., 1975) improves selectivity of 5,7-DHT for serotonin neurons and prevents catecholamine neurons from damage by this drug. However, the preventive effect of MAO-inhibition on the toxicity of 5,7-DHT exerted in NA neurons can also be explained by a) reduction in incorporation of 5,7-DHT into NA neurons by competition for the carrier of the increased NA and b) by the radical scavenging properties of NA. As shown in previous experiments (Klemm et al. 1979), spontaneous and mitochondria-promoted oxidation of 5,6-DHT and 5,7-DHT results in the production of highly reactive quinoidal products. These quinoidal species, formed within monoaminergic neurons through mitochondria-stimulated oxidation after accumulation of the amines by the high affinity uptake system, may be responsible for the toxic action of 5,6-DHT and 5,7-DHT in concert with radicals formed during the oxidation of the dihydroxytryptamines. The relative importance of the irreversible binding of quinoidal products to proteins and the damaging effects of the radicals formed during the oxidation - in particular of 5,7-DHT - needs to be studied in future.

References

Baumgarten, H.G., Victor, S.J., Lovenberg, W.: Psychopharmac.Commun.1, 75-88 (1975).

Baumgarten, H.G., Lachenmayer, L., Björklund, A.: Methods in Psychobiology, vol.3, edited by R.D.Myers. New York: Academic Press, 47-98 (1977).

Baumgarten, H.G., Klemm, H.P., Lachenmayer, L., Björklund, A., Lovenberg,W., Schlossberger, H.G.: Ann. N.Y. Acad. Sci. 305, 3-24 (1978).

Baumgarten, H.G., Jenner. S., Schlossberger, H.G.: Neurotoxins, Fundamental and Clinical Advances, I.W. Chubb, L.B. Geffen (eds.) 221-226. Adelaide University Union Press, Adelaide, South Australia (1979).

Baumgarten, H.G., Jenner, S., Björklund, A., Klemm, H.P., Schlossberger, H.G.: Biology of Serotonergic Transmission, edited by N.N. Osborne. New York: John Wiley & Sons, 249-277 (1982).

Breese, G.R., Cooper, B.R.: Brain Res. 98, 517-527 (1975).

Björklund, A., Baumgarten, H.G., Rensch, A.: J.Neurochem.24, 833-835 (1975).

Creveling, C.R., Lundström, J., McNeal, E.T., Tice, L., Daly, J.W.: Mol Pharmac. 11, 211-222 (1975).

Fukumori, R., Minegishi, A., Satoh, T., Kitagawa, H., Yanaura, S.: Life Sci-25, 123-130 (1979).

Klemm, H.P., Baumgarten, H.G., Schlossberger, H.G.: J.Neurochem.32, 111-120 (1979).

Klemm, H.P., Baumgarten, H.G., Schlossberger,H.G.: J. Neurochem.35, 1400-1408 (1980).

Mitra, Ch., Guha, S.R.: Biochem. Pharmac. 29, 1213-1216 (1980).

Rotman, A., Daly,J.W., Creveling, C.R.: Mol. Pharmacol. 12, 887-899 (1976).

TRYPTOPHAN METABOLISM IN THE PINEAL GLAND

Russel J. Reiter

Department of Anatomy, The University of Texas Health Science Center at San Antonio, San Antonio, TX 78284 USA

Introduction

The most thoroughly studied end product of tryptophan metabolism within the pineal is melatonin (1, 2). This compound was isolated from bovine pineal tissue over 20 years ago (3) and subsequently it has been the subject of extensive investigative efforts (4, 5). There is no longer any doubt that melatonin is a hormone of the mammalian pineal gland which is of paramount importance in influencing a number of functions of the organism (5, 6). The most thoroughly tested physiological actions of melatonin are those related to the reproductive system (7, 8). Besides its production in the pineal gland, melatonin also seems to be synthesized in the retina (9, 10) and the Harderian gland (11, 12). Its production at these sites should not detract from melatonin's role as a pineal hormone since the bulk of the melatonin in the blood seems to be derived from the pineal gland (13); additionally, the day:night rhythm in the blood levels of this constituent rely on its secretion by the pineal.

Other pineal tryptophan-derived products which deserve consideration as potentially active pineal substaces include 5-methoxytryptophol, 5-hydroxytryptophol, and 5-methoxytryptamine. 5-Methoxytryptophol is found in the blood of mammals where it exhibits a rhythm much like that for melatonin (14). In the proper experimental paradigm it inhibits reproductive physiology in a manner similar to melatonin (15). On the whole, however, the evidence supporting its candidacy as an active pineal secretory product is meager compared to that for melatonin (4, 5).

5-Hydroxytryptophol has never seriously been considered as a secretory product of the pineal although in isolated experiments its exogenous

Progress in Tryptophan and Serotonin Research

administration produces physiological consquences (15). Finally, 5-methoxytryptamine has recently received interest because of its alleged hormonal actions in reference to the reproductive system (16); again, however, the supportive data are generally weak.

In the final analysis, melatonin remains the only tryptophan-derived pineal indole that can be legitimately classified as a pineal hormone (4, 5). This does not exclude the possibility that the compounds mentioned above or other indoles or peptides (17) will be proven, at a later date, to be equally effective as melatonin in influencing the physiology of the organism.

Pineal Tryptophan Metabolism

1) Synthesis of serotonin. Tryptophan is taken up by the pinealocytes, the endocrine cells of the gland, from the peripheral circulation. Its conversion to serotonin involves two enzymes, tryptophan hydroxylase and aromatic amino acid decarboxylase. Both enzymes are found in tissues other than the pineal gland but the levels of tryptophan hydroxylase in the pineal are particularly high (18). The activity of this enzyme is considered to rate limit serotonin synthesis; the hydroxylation of tryptophan takes place in mitochondria (19). The activity of tryptophan hydroxylase requires a pterdine co-factor, probably tetrahydrobiopterin (20). At least in the rat pineal gland, tryptophan hydroxylase activity reportedly exhibits a 24 hour rhythm with highest activity being present during the normal period of darkness (21). The regulation of tryptophan hydroxylase activity in the pineal gland has not been completely defined. The administration of tryptophan to rats increases the amount of serotonin in the pineal gland (22).

   Aromatic L-amino acid decarboxylase is present in high quantities in the pineal and is found mainly localized in the cytosol fraction of the pinealocytes (23). Pyridoxal phosphate serves as a co-factor for the decarboxylation step in serotonin formation (24). Unlike the activity of several other enzymes in the rat pineal, aromatic L-amino acid decarboxylase is stimulated when animals are exposed to constant light for several days (25).

The serotonin content of the pineal gland is higher than virtually any other brain area (26); it is located primarily in the pinealocytes although a small percentage is also found in the sympathetic nerve endings in the gland (27). The serotonin concentration exhibits a 24 hour rhythm with highest levels being present during the day (28). At least three factors regulate the levels of serotonin within the pineal gland. Certainly, the availability of tryptophan is critical to the amount of serotonin within the pineal gland (22). A second important factor is the activity of monoamine oxidase (MAO) which oxidatively deaminates the monoamine. Pharmacologic inhibition of MAO causes an elevation in pineal serotonin content (29). Finally, the rate at which serotonin is N-acetylated may be influential in determining the amount of serotonin within the pineal (30). Usually, the acetylation of serotonin is greater at night than during the day (31).

2) Synthesis of melatonin. Especially during the night, serotonin is N-acetylated by the enzyme serotonin N-acetyltransferase (NAT). The activity of the enzyme exhibits a pronounced rhythm; for example, in the rat pineal gland the nocturnal increase in enzyme activity may be 30-70 fold (30). Compared to the liver and brain, there are distinct differences in the substrate specificity of NAT in the pineal gland (32). NAT is located exclusively in the cytoplasm and works by transfering the acetyl group from the co-factor acetyl CoA to acceptor amines (33). Although the primary physiologic acetyl acceptor in the pineal is serotonin, the NAT can also acetylate 5-methoxytryptamine.

NAT activity is regulated by an adrenergic-cyclic AMP mediated mechanism (34). Normally, during darkness norepinephrine is released from the sympathetic nerve endings within the gland; the neurotransmitter acts on β-adrenergic receptors on the pinealocyte resulting in the stimulation of adenylate cyclase activity. This eventually results in the synthesis of new RNA and protein (35) with the subsequent stimulation of NAT activity. Cyclic AMP within the pineal gland acts in two ways to influence NAT activity. First, it promotes the production of mRNA and protein which are essential for stimulating NAT activity (35); secondly, cyclic AMP has a stabilizing effect on the enzyme thereby maintaining it in an active form (36).

The high activity of NAT in the rat pineal at night results in an increased accumulation of N-acetylserotonin (37). Indeed, in this species there seems to be a close correlation between the activity of the acetylating enzyme and the level of N-acetylserotonin. Hence, drugs which interfere with the nocturnal rise in the activity of NAT also prevent the elevation in N-acetylserotonin; likewise, drugs which stimulate the enzyme result in increased levels of N-acetylserotonin (37).

The final enzyme in the melatonin pathway is hydroxyindole-O-methyltransferase (HIOMT). It acts to transfer a methyl group from the co-factor S-adenosylmethionine to acceptor molecules; in the pineal gland the acceptor molecules are 5-hydroxyindoles (38). HIOMT was initially thought to be confined to the pineal gland although more recent studies indicate its presence in retina and Harderian gland as well. Also, whereas early studies reported a daily rhythm in HIOMT activity with highest levels at night (39), recent investigations have failed to confirm this (35). In general, it seems that if there is a 24 hour cycle in the activity of HIOMT the rhythm is very weak.

In all species examined there is a marked daily rhythm in the melatonin content of the pineal gland (40-43). The production and secretion of melatonin by the pineal gland determines blood titers and urinary levels of the indole (44, 45). Melatonin levels in the pineal gland are high at night regardless of whether the species in question is diurnally or nocturnally active (46, 47). The acute exposure of domestic rodents to even low irradiance light during the normal dark period rapidly depresses pineal melatonin levels (48). However, in wild captured diurnal rodents much higher irradiances of light are required to reduce pineal melatonin levels (49).

3) Synthesis of 5-methoxytryptamine. 5-Methoxytryptamine has been promoted as a pineal hormone (16); this is based on the evidence that its injection into hamsters inhibits testicular growth in a manner similar to melatonin. However, considering that its activity is only about one-tenth that of melatonin (50) and the amount of data concerning its gonad-inhibiting is scanty, as of this writing it seems premature to classify 5-methoxytryptamine as a pineal hormone.

5-Methoxytryptamine is formed in the pineal gland by the methylation of serotonin by the enzyme HIOMT. Relatively little is known concerning the production of 5-methoxytryptamine within the pineal gland or its release into the peripheral circulation.

4) Synthesis of 5-methoxytryptophol. Very early a claim was made that 5-methoxytryptophol was actually more effective than melatonin in inhibiting ovarian growth in maturing rats (51). There have been few reports documenting this in recent years and, in fact, 5-methoxytryptophol is not often looked upon as a pineal secretory product.

Besides its conversion to N-acetylserotonin and 5-methoxytryptamine by NAT and HIOMT, respectively, serotonin can be oxidatively deaminated by MAO to 5-hydroxyindoleacetaldehyde which, in turn, is acted upon by aldehyde dehydrogenase to yield 5-hydroxyindoleacetic acid (52). Also, 5-hydroxyindoleacetaldehyde undergoes reduction to 5-hydroxytryptophol through the action of an enzyme aldehyde reductase (53); 5-hydroxytryptophol then undergoes methylation to 5-methoxytryptophol by HIOMT.

5-Methoxytryptophol is produced in greater quantities at night than during the day and is presumably secreted into the blood vascular system. The influence of the light:dark cycle as well as changes induced by different irradiances of light in terms of 5-methoxytryptophol production remain, for the most part, uninvestigated.

Concluding Remarks

Unquestionably, indole metabolism is a primary feature of the mammalian pineal gland. A great deal is known concerning tryptophan metabolism in this highly active organ of internal secretion. At least one of the tryptophan-derived factors, melatonin, is a pineal hormone. Other indoles may also be placed into the hormonal category in the years ahead.

References

1. Smith, J. A.: Adv. Biosci. 29, 135 (1981).
2. Wurtman, R. J., Ozaki, Y.: J. Neural Transmis., Suppl. 13, 59 (1978).
3. Lerner, A. B., Case, J. D., Takahashi, Y., Lee, T. H., Mori, W.: J. Am. chem. Soc. 80, 2587 (1958).
4. Reiter, R. J.: Endocr. Rev. 1, 109 (1980).
5. Cardinali, D. P.: Endocr. Rev. 2, 327 (1981).
6. Reiter, R. J.: Prog. Psychobiol. physiol. Psychol. 9, 323 (1980).
7. Reiter, R. J.: Am. J. Anat. 162, 287 (1981).
8. Reiter, R. J.: In Frontiers in Neuroendocrinology, Vol. 7, eds. Ganong, W. F., Martini, L., p. 287, Raven, New York 1982.
9. Gern, W. A., Ralph, C. L.: Science 204, 183 (1979).
10. Cardinali, D. P., Rosner, R. A.: Endocrinology 89, 301 (1971).
11. Bubenik, G. A., Brown, G. M., Grota, L. J.: J. Histochem. Cytochem. 24, 1173 (1976).
12. Reiter, R. J., Richardson, B. A., Matthews, S. A., Lane, S. J., Ferguson, B. N.: Life Sci. 32, 1229 (1983).
13. Lewy, A. J., Tetsuo, M., Markey, S. P., Goodwin, F. K., Kopin, I. J.: J. clin. Endocr. Metab. 50, 204 (1980).

14. Wilson, B. W., Lynch, H. J., Ozaki, Y.: Life Sci. 23, 1019 (1978).

15. Vaughan, M. K., Reiter, R. J., Vaughan, G. M., Bigelow, L., Altschule, M. D.: Gen. comp. Endocr. 18, 372 (1972).

16. Pevet, P., Haldar-Misra, C., Öcal, T.: J. neural Transmis. 51, 303 (1981).

17. Benson, B., Ebels, I.: In The Pineal Gland, Vol. I, Anatomy and Biochemistry, ed. Reiter, R. J., p. 165, CRC Press, Boca Raton 1981.

18. Lovenberg, W., Jequier, E., Sjoerdsma, A.: Science 155, 217 (1967).

19. Hori, S., Kuroda, Y., Saito, K. Ohotaui, S.: J. Neurochem. 27, 911 (1976).

20. Levine, R. A., Kuhn, D. M., Lovenberg, W.: J. Neurochem. 32, 1575 (1979).

21. Sugden, D.: J. Neurochem. 33, 811 (1979).

22. Deguchi, T., and Barchas, J.: Mol. Pharmacol. 8, 770 (1972).

23. Snyder, S. H., and Axelrod, J.: Biochem. Pharmacol. 13, 805 (1964).

24. Lovenberg, W., Weissbach, H., Udenfriend, S.: J. biol. Chem. 237, 89 (1962).

25. Synder, S. H., Axelrod, J., Wurtman, R. J., and Fischer, J. E.: J. Pharmacol. exp. Ther. 147, 371 (1965).

26. Saavedra, J. M., Brownstein, M., Axelrod, J.: J. Pharmacol. exp. Ther. 186, 508 (1973).

27. Neff, N. H., Barret, R. E., Costa, E.: Eur. J. Pharmacol. 5, 348 (1969).

28. Quay, W. B.: Proc. Soc. exp. Biol. Med. 115, 710 (1964).

29. Snyder, S. H., Zweig, M., Axelrod, J., Fischer, J. E.: Proc. nat. Acad. Sci., USA, 53, 301 (1965).

30. Klein, D. C., Weller, J. L.: Science 169, 1093 (1970).

31. Rudeen, P. K., Reiter, R. J., and Vaughan, M. K.: Neurosci. Let. 1, 225 (1975).

32. Yang, Y. H. T., Neff, N. H.: Neuropharmacology 15, 561 (1976).

33. Weissbach, H., Redfield, B. G., Axelrod, J.: Biochim. Biophys. Acta 54, 190 (1961).

34. Klein, D. C., Berg, G. R.: Adv. biochem. Psychopharmacol. 3, 241 (1970).

35. Klein, D. C.: In Endocrine Rhythms, ed. Krieger, D. T., p. 203, Raven, New York 1979.

36. Klein, D. C., Buda, M., Kopoor, C. L., Krishna, G.: Science 199, 309 (1978).

37. Brownstein, M., Saavedra, J. M., Axelrod, J.: Mol. Pharmacol. 9, 205 (1972).

38. Axelrod, J., Weissbach, H.: J. biol. Chem. 236, 211 (1961).

39. Wurtman, R. J., Axelrod, J., Phillips, L.: Science 142, 1071 (1963).

40. Panke, E. S., Rollag, M. D., Reiter, R. J.: Endocrinology 104, 194 (1979).

41. Lynch, H. J.: Life Sci. 10, 791 (1971).

42. Goldman, G., Hall, V., Hollister, C., Reppert, S., Roychoudhury, P., Yellon, S., Tamarkin, L.: Biol. Reprod. 24, 778 (1981).

43. Reiter, R. J., Craft, C. M., Johnson, J. E. jr., King, T. S., Richardson, B. A., Vaughan, G. M., Vaughan, M. K.: Endocrinology 109, 1295 (1981).

44. Wilkinson, M., Arendt, J., Bradtke, J., deZiegler, D.: J. Endocr. 72, 243 (1977).

45. Lynch, H. J., Wurtman, R. J., Moskovitz, M. A., Archer, M. C., Ho, M. H.: Science 187, 169 (1975).

46. Matthews, S. A., Evans, K. L., Morgan, W. W., Petterborg, L. J., Reiter, R. J.: Prog. clin. biol. Res. 92, 35 (1982).

47. Reiter, R.J., King, T. S., Richardson, B. A., Hurlbut, E. C., Karasek, M., Hansen, J. T.: Neuroendocr. Let. 4, 1 (1982).

48. Brainard, G. C., Richardson, B. A., Petterborg, L. J., Reiter, R. J.: Brain Res. 233, 75 (1982).

49. Reiter, R. J., King, T. S., Richardson, B. A., Hurlbut, E. C.: J. neural Transmis. 54, 275 (1982).

50. Rollag, M. D.: Life Sci. 31, 2699 (1982).

51. McIsaac, W. M., Taborsky, R. G., Farrell, G.: Science 145, 63 (1964).

52. Udenfriend, S., Titus, E., Wiessbach, H., Peterson, R. E.: J. biol. Chem. 219, 235 (1956).

53. Keglevic, D., Supek, Z., Kveder, S., Iskric, S., Keckes, S., Kisic, A.: Biochem. J. 73, 53 (1959).

SOME PHYSIOLOGICAL RESPONSES AND EFFECTS OF THE PINEAL INDOLEAMINE MELATONIN

Josephine Arendt, Department of Biochemistry, University of Surrey, Guildford, Surrey, GU2 5XH, U.K.

Introduction

Rene Descartes' (1596-1650) concept of the pineal gland as the seat of the soul, and particularly his intricate description of hypothetical links between the eyes, the pineal, the brain and the body, bear a curiously close relation to the current modern idea of the pineal as a neuro-endocrine transducer of environmental (particularly photic) information. Following the discovery and isolation of the indoleamine melatonin (5-methoxy-N-acetyltryptamine), the amphibian skin-lightening factor, from bovine pineal glands (1), 3 centuries after Descarte's death, and the description of its synthesis from tryptophan in the pineal gland (2), the major pioneering work reported as early as 1963 that the indoleamine metabolism of the rodent pineal did indeed relate to the eyes: that dark stimulated and light inhibited the synthesis of pineal melatonin (3). A great deal of subsequent work on the pineal, up to and including today, has related to observations of environmental effects. Physiological events related to photoperiod length are candidates for modulation by the pineal, and indeed the pineal is an essential component of such systems in some species (4). A multitude of biologically active compounds are now demonstrable in the pineals of different species (4) all of which could be said to be potential effectors of pineal-dependent events. The indole-amines are the most thoroughly studied, however, and melatonin itself is of probable physiological significance in the control of season-related activity in photoperiodic higher vertebrates. Until recently it was the only pineal compound for which the weight of evidence indicated a role of biological importance, notably in that timed administration of physiological amounts of it will induce major reproductive changes in hamsters (4, 5) and sheep (6, 7). Melatonin is also the best peripheral index of pineal function to date: pinealectomy greatly reduces its plasma levels in sheep (8) and rats (9) and its major metabolite in monkeys (10). In rats,

Progress in Tryptophan and Serotonin Research

plasma melatonin is demonstrably a good index of pineal synthesising activity (11).
The development of assay methods for melatonin (see 12 for a review) has provided a tremendous impetus to the study of pineal function. Some of the author's work on the application of melatonin assay will be reviewed here within the framework of the field in general.

Factors Controlling Melatonin Secretion:

1. Daily and seasonal variations in melatonin.
   In all species studied to date including man, melatonin is secreted during the dark phase of the day (see 13 for a review). Melatonin secretion may be suppressed by light, dependent on the intensity and quality of light used and the species studied (14). Initial studies seemed to indicate that human melatonin secretion was not suppressed by artificial light at night (12, 15), however this was later amended following the discovery that sufficiently intense light (equivalent to bright sunlight) would suppress melatonin secretion (16, 17). Thus the human pineal is likely to respond to changes in natural light.
   Seasonal variations in melatonin secretion are demonstrable in natural conditions in a number of species including man (18, 19, 20, 21). Variations in melatonin synthesis or secretion in different artificial photoperiod lengths have been described for sheep (19, 21, 22) and hamsters (23, 24, 25). In sheep, the duration of high plasma levels appears to monitor the dark phase. Variations of pineal melatonin content in the tortoise are found in constant photoperiod but different thermoperiod (26). Thus while the reported seasonal variations of melatonin in natural conditions probably depend largely on photoperiod, a thermoperiod component cannot be ruled out and certainly merits further investigation. The effects of seasonal variations in diet have not been thoroughly investigated, however, variations in plasma tryptophan do not appear to relate to melatonin secretion in the sheep (27). Short-term fasting (28) or short-term dietary changes, calculated to modify tryptophan availability, (Arendt and Hartman, unpublished) had no effect on plasma melatonin in man.

These observations on the environmental control of melatonin secretion have laid a solid foundation for the design of experiments concerned with definition of its physiological role. Experiments in sheep for example show that melatonin adminstered in physiological quantities and given in such a way as to mimic short-day variations in plasma levels in ambient long-day light, override the photoperiod and mimic short-day effects on reproductive function (6, 7) in artificial conditions. These effects on reproductive function are associated with marked changes in a circadian rhythm; that of prolactin secretion (29). It remains to be seen whether the same effects can be observed in natural environmental conditions.

2. Hormonal and other factors influencing melatonin secretion: Ovariectomy in the ewe markedly increases, in the short-term, the amount of circulating melatonin (30) suggesting a suppressive effect of the gonads on melatonin secretion. A report of low early morning melatonin at ovulation in the menstrual cycle (31) may relate to phase shifts rather than variation in total secretion. Total secretion, assessed by 6-hydroxymelatonin sulphate secretion, apparently does not change in the menstrual cycle (32). The rat (33) but not the hamster (34) appears to show oestrus cycle variations in melatonin content. A comprehensive survey of pubertal development as related to 24 hour variations in plasma melatonin is clearly difficult to accomplish, at least in man, but in view of controversial reports in the literature, would be of great interest. One may speculate that the mechanisms operative in the induction and repression of seasonal reproductive competence may well also apply to the sexual maturation process.

Several groups have shown a decline in the quantity of pineal, plasma or CSF melatonin with age (31, 35, 36). In the human neonate, melatonin is present in the urine in the first week of life and the 24h rhythm of production begins to develop during the first six weeks of life (37). The total amount excreted appears to increase during early life (38).

In experimental animals, and man (39, 18), the quantity of melatonin synthesised, or secreted, varies greatly from one individual to another. In the rat, at least, this quantity relates directly to the activity of the enzyme serotonin-N-acetyltransferase in groups of age- and weight-matched animals. In man, however, it is clearly difficult to establish such a relationship. Circulating melatonin (integrated 24h secretion or concentration at 2400 h) in man, related significantly to body weight and the obesity index in a group of 11 normal men and women (28) and a further group of 12 normal men (40). This observation has considerable importance in the interpretation of melatonin data wherever gross changes in weight are encountered, e.g. puberty, anorexia nervosa, depression, cachexia of old age. Subjects should be weight-matched for valid comparison, when possible.

3. Pharmacological manipulation of melatonin secretion:
Benserazide, (Ro4-4602) an inhibitor of aromatic amino acid decarboxylase, almost completely suppresses melatonin production in the rat pineal gland, in peripherally acting doses (41). A marked parellel decline in pineal, but not hypothalamic, serotonin suggests that serotonin availability is an important determinant of melatonin synthesis in vivo. Human pineal melatonin secretion, as in the rat, can be blocked by adrenergic β-receptor antagonists. A peripherally-acting β-blocker, atenolol, is effective, suggesting that in both man and rat the pineal is pharmacologically without the blood-brain barrier (42).
A number of anti-depressant drugs in pharmacological concentrations will stimulate the short-term production of pineal melatonin (43). If used in clinically effective doses however, no short-term effect is observed (13). It remains to be seen whether clinically effective doses of anti-depressants have an observable long-term effect on melatonin production in man.

Summary

The relative contribution of different inputs, environmental and hormonal in particular, in the control of pineal secretion is currently being assessed together with the relationship of pineal secretion to age, weight and pathological disturbances. Whilst we cannot as yet formulate a role for the pineal in human metabolism, melatonin measurement in particular has paved the way for convincing demonstrations of its essential function in the control of photoperiod-dependent reproductive functions. However, if melatonin can act in man as an organiser of circadian and seasonal rhythms as it appears to do in sheep, it may well prove of therapeutic value in pathological conditions associated with disturbed biological rhythms.

Acknowledgements

I thank the M.R.C. and the A.R.C. for financial support.

References

1. Lerner, A.B., Case, J.D., Takahashi, Y., Lee, T.H. and Mori, W.: J. Am. Chem. Soc. 80, 2587 (1958).
2. Axelrod, J.: Science 184, 1341-1348 (1974).
3. Wurtman, R.J., Axelrod, J. and Phillips, L.S.: Science 142, 1071-1073 (1963).
4. Reiter, R.J.: Endocrine Reviews 1, 109-131 (1980).
5. Tamarkin, L., Westrom, W.K., Hamill, A.I. and Goldman, B.D.: Endocrinology 99, 1534-1541 (1976).
6. Kennaway, D.J., Gilmore, T.A. and Seamark, R.F.: Endocrinology 110, 1766-1772 (1982).
7. Arendt, J., Symons, A.M., Laud, C.A. and Pryde, S.J.: J. Endocrinol in press (1983).
8. Arendt, J., Brown, W.B., Forbes, J.M. and Marston, A.: J. Endocr. 85, 1-2P (1980).
9. Lewy, A.J., Tetsuo, M., Markey, S.P., Goodwin, F.K. and Kopin, I.J.: J. Clin. Endocrinol. Metab., 50, 204-205 (1980b).
10. Tetsuo, M., Perlow, M.J., Mishkin, M. and Markey, S.P.: Endocrinology 110, 997-1003 (1982).

11. Wilkinson, M., Arendt, J., Bradtke, J. and Ziegler, D. de: J. Endocrinol. 72, 243-244 (1977).

12. Arendt, J.: J. Neural Transmission. Suppl. 13, 265-279 (1978).

13. Arendt, J.,: Photobiology-Biochronometry-Endocrinology, eds., Oksche, A. and Pevet, P. Elsevier-North Holland, pp. 247-260 (1981).

14. Reiter, R.J., King, T.S., Richardson, B.A., Hurlbut, E.C., Karasek, M.A. and Hansen, J.T.: Neuroendocrinol. Lett. 4, 1-5 (1982).

15. Lynch, H.J., Ozaki, Y. and Wurtman, R.J.: J. Neural. Transmission Suppl. 13, 251-265 (1978).

16. Wetterberg, L.: J. Neural Transmission Suppl. 13, 289-310 (1978).

17. Lewy, A.J., Wehr, T.A., Goodwin, F.K., Newsome, D.A. and Markey, S.P.: Science 210, 1267-1268 (1980a).

18. Arendt, J., The pineal gland of vertebrates including man, eds., Kappers, J.A. and Pevet, P.: Prog. Brain Research 52, 249-258 (1979).

19. Arendt, J., Symons, A.M. and Laud, C.: Experientia, 37, 584-584 (1981).

20. Vivien-Roels, B., Arendt, J. and Bradtke, J.: Gen. Comp. Endocr. 37, 197-210 (1979).

21. Rollag, M.D., O'Callaghan, P.L. and Niswender, G.D.: Biol. Reprod. 18, 279-285 (1978).

22. Lincoln, G., Almeida, O.F.X. and Arendt, J.: J. Reprod. Fert. Suppl. 30, 23-31 (1981).

23. Panke, E.S., Rollag, M.D. and Reiter, R.J.: Comp. Biochem. Physiol. 66A, 691-693 (1980).

24. Tamarkin, L., Reppert, S.M. and Klein, D.C.: Endocrinology 104, 385-389 (1978).

25. Hoffman, K., Illnerova, H. and Vanacek, J.: Biol. Reprod. 24, 551-556 (1981).

26. Vivien-Roels, B. and Arendt, J.: Adv. Biosciences 29, 401-406 (1981).

27. Kennaway, D.J., Porter, K.J. and Seamark, R.F.: Aust. J. Biol. Sci. 31, 49-52 (1978).

28. Arendt, J., Hampton, S., English, J., Kwasowski, P. and Marks, V.: Clin. Endocrinol. 16, 89-93 (1981).

29. Symons, A.M., Arendt, J. and Laud, C.A.: J. Endocr. submitted, 1983.

30. Arendt, J., Laud, C.A., Symons, A.M. and Pryde, S.J.: J. Reprod. Fert., in press, 1983.

31. Wetterberg, L., Arendt, J., Paunier, L., Sizonenko, P.C., Van-Donselaar, W. and Heyden, T.: J. Clin. Endocrinol. Metab. 98, 1418-1422 (1976).

32. Fellenberg,A.J., Phillipou, G. and Seamark, R.F.: Clinical Endocr. 17, 71-75 (1982).

33. Johnson, L.Y., Vaughan, M.K., Richardson, B.A., Petterborg, L.J. and Reiter, R.J.: Proc. Soc. Exp. Biol. Med. 169, 416-419 (1982).

34. Rollag, M.D., Chen, H.J., Ferguson, B.N. and Reiter, R.J.: Proc. Soc. Exp. Biol. Med. 162, 211-213 (1979).

35. Reiter, R.J., Richardson, B.A., Johnson, L.Y., Ferguson, B.N. and Dinh, D.T.: Science 210, 1372-1373 (1980).

36. Brown, G.M., Young, S.N., Gauthier, S., Tsui, H. and Grota, L.J.: Life Sci. 25, 929-936 (1979).

37. Marston, A., Arendt, J., Campbell, C., Armstrong-Esther, C., Hawkins, L. and Goodison, S.: Abstr. Neurosi. Letts. Suppl. 5, 403 (1980).

38. Lemaitre, B.J., Bouillie, J. and Hartman, L.: Clin. Chim. Acta 110, 77-83 (1981).

39. Lynch, H.J., Wurtman, R.J., Moskovitz, M.A., Archer, M.C. and Ho, M.H.: Science 187, 169-171 (1975).

40. Ferrier, I.N., Arendt, J., Johnstone, E.C. and Crow, T.J.: Clin. Endocrinol. (in press).

41. Arendt, J., Ho, A.K., Laud, C., Marston, A., Nohria, V., Smith, J.A. and Symons, A.M.: Brit. J. Pharmac. 72, 257-267 (1981).

42. Cowen, P.J., Fraser, S., Sammons, R. and Green, A.R.: Brit. Med. J. in press (1983).

43. Wirz-Justice, A., Arendt, J. and Marston, A.: Experientia. 36, 442-444 (1980).

PHARMACOLOGIC REGULATION OF SERUM N-ACETYLSEROTONIN

Gregory M. Brown, Timothy G. Burns, Stella Psarakis
McMaster University
Hamilton, Ontario

Lee J. Grota
Department of Psychiatry, University of Rochester
Rochester, New York

Introduction

N-Acetylserotonin (NAS) is derived from serotonin (5-HT) and is the precursor of melatonin in the pineal gland. The main source of serum melatonin is the pineal gland (1), but relatively little is known about the source of serum NAS. NAS has been regarded solely as an intermediate between pineal 5-HT and melatonin. However, serum NAS in rats has a 24 hour rhythm which can be dissociated from that of melatonin (2) and is significantly reduced but not eliminated by pinealectomy (3) suggesting that the pineal is not the only source of serum NAS. Recent findings from this laboratory using a more specific NAS assay support that concept (4). In addition, both serotonin N-acetyl transferase (NATase) activity and NAS content in the rat pineal exhibit peaks during the dark period of a light-dark cycle (5,6,7). Findings to date suggest the concept that serum NAS is regulated separately from melatonin and that it may not derive exclusively from the pineal gland. We have used pharmacological treatments to produce alterations in indolealkylamines. Systemic injections of 5-HTP (5-hydroxytryptophan) or 5-OHNAT (5-hydroxy-N-acetyltryptophan) were used. Inhibition of tryptophan hydroxylase by 6-fluorotryptophan was done. Pharmacological manipulation of the ß-adrenoceptor activated enzyme NATase in brain and pineal was done by administration of the ß-agonist, isoproterenol or the ß-blocker, propranolol. Morphine, which increases turnover of 5-HT in rat brain (8) was also investigated.

Progress in Tryptophan and Serotonin Research

Methods

Adult male Sprague-Dawley rats were used in tryptophan and 5-HTP loading experiments. Adult male Wistar rats were used in other studies. Animals were housed under a 12:12 light-dark schedule (0700-1900,lights on). Drugs were given intraperitoneally during the lights-on portion of the light-dark cycle. Nialamide (Sigma) was dissolved in 0.01 N HCl. Carbidopa (Merck, Sharp, and Dohme), and 6-fluoro-tryptophan (Sigma) were suspended in 1% methyl cellulose in saline. 5-OHNAT was synthesized by V. Snieckus (Waterloo) by acetylation of commercially available 5-HTP. 5-OHNAT and 5-HTP were dissolved in a small amount of 50:50 ethanol/water and then suspended in 1% methyl cellulose. Isoproterenol (Sigma), propranolol (Sigma), and morphine sulfate (Sternway) were dissolved in saline. Animals were killed by decapitation, trunk blood collected and centrifuged and serum stored at -20^{0}C until assay. 1) 5-HTP and 5-OHNAT loading; Animals were injected with carbidopa (100 mg/kg), followed 10 min later by nialamide (100 mg/kg). Sixty min later, these animals were injected with either 5-OHNAT (33 mg/kg), 5-HTP (28 mg/kg), or saline vehicle. 5-OHNAT and 5-HTP were given in equimolar dosages. Animals were killed one half hour following the last injection. 2) 6-Fluorotryptophan; Different groups of animals were injected with 6-fluorotryptophan (115 mg/kg) or saline vehicle. Animals were killed 3 hours post-injection. 3) Isoproterenol/propranolol; Animals were injected with isoproterenol (10 mg/kg) followed by saline one hour later. Control animals were given two injections of saline one hour apart. Propranolol control animals were given saline followed one hour later by propranolol (20 mg/kg). The fourth group of animals were given isoproterenol (10 mg/kg) followed by propranolol (20 mg/kg) one hour later. Animals were killed 15 min after the last injection. 4) Morphine; Animals were injected with morphine sulfate (5 mg/kg) or saline vehicle and killed 90 min later. Assay: Serum NAS was assayed by the method of Pang et al (10). Cross Reactivity: All drugs were tested for their ability to compete with ^{3}H-NAS for binding to the NAS specific antibody.

Results

None of the drugs used had significant crossreactivity. Carbidopa treatment caused a significant fall in serum NAS. 5-HTP loading led to a significant increase (p>.05), in serum NAS at 30 min as compared to the controls (1.38±0.40 vs 0.58±0.04: ng/ml ± S.E.M., n=6/group). NAS elevation was accompanied by profound behavioural changes characteristic of the serotonin syndrome: wet-dog shakes, forelimb tremor, and hindlimb abduction. 5-OHNAT administration had no significant effect on serum NAS as compared to controls (0.65±0.15 vs 0.58±0.04: ng/ml ± S.E.M., n=6/group). There was no significant difference between 6-fluorotryptophan treated animals and controls (3.44±0.61 vs 3.11±0.23: ng/ml ± S.E.M.,n=5/group). There was no significant difference between isoproterenol and saline treated animals (2.38±0.64 vs 2.39±0.51: ng/ml ± S.E.M., n=6/group). Likewise, propranolol was without effect on serum NAS. Finally, no difference in serum NAS levels was observed between morphine treated rats and saline controls (4.09±0.80 vs 3.66±0.20: ng/ml ± S.E.M., n=5/group).

Discussion

These findings indicate that circulating NAS is lowered by carbidopa suggesting that a peripheral 1-aromatic amino acid decarboxylase is involved in synthesis of serum NAS. Serum NAS elevation observed in 5-HTP treated animals suggests that decarboxylase inhibition is incomplete. 6-Fluorotryptophan treatment was without effect on serum NAS. NAS production is enzymatically removed by 2 steps from the hydroxylation of tryptophan. It may require a longer period of drug action or a larger dose before effects on NAS can be measured. Administration of isoproterenol, which causes a rise in circulating melatonin (1), did not result in a rise in serum NAS. Increased conversion of 5-HT to NAS in the pineal may simply have caused a subsequent increase in the production of melatonin. Another possibility is that conversion of pineal 5-HT to NAS did not appreciably add to the already relatively high NAS levels in serum. Morphine was without effect on serum NAS levels. However, morphine increases 5-HT turnover but does not affect its absolute levels (11). Elevation in serum NAS following 5-HTP loading was accompanied by the presence of gross behavioural changes, as previously reported (9). These behavioural changes are believed to be due to activation of central serotonergic synapses (12). Interestingly, 5-OHNAT treated animals as a group fail to develop the syndrome within 30 min and showed no elevation in serum NAS although individual animals which showed behavioural signs also had the highest serum NAS.

References

1. Lewy, A.J., Tetsuo, M., Markey, S.P., Goodwin, F.K., Kopin, I.J.: J. Clin. Endocrinol. Metab. 50, 204-205 (1980).
2. Grota, L.J., Brown, G.M., Pang, S.F. (Submitted).
3. Yu, H.S., Pang, S.F., Tang, P.L., Brown, G.M.: Neuroendocrinol. 21, 197-200 (1981).
4. Harvey, L., Brown, G.M., Grota, L.J.: (Submitted).
5. Klein, D.C., Weller, J.L.: Science, 169, 1093-1095 (1970).
6. Klein, D.C., Weller, J.L.: Fed. Proc. 31, 222 (1972).
7. Pang, S.F., Brown, G.M., Grota, L.J., Chambers, J.W., Rodman, R.L.: Neuroendocrinol. 23, 1-13 (1977).
8. Snelgar, R., Vogt, M.: J. Physiol. 284, 128P (1978).
9. Burns, T.G., Szechtman, H., Brown, G.M., Snieckus, V.: Progr. in Neuro-Psychopharmacol. and Biol. Psychiatry 6, 359-363 (1982).
10. Pang, S.F., Brown, G.M., Campbell, S.L., Snieckus, V., de Silva, S.O., Young, S.N., Grota, L.J.: J. Immunoassay,2, 263-276 (1981).
11. Haubrich, D.R., Blake, D.E.: Biochem. Pharm. 22, 2753-2759 (1973).
12. Jacobs, B.L.: Life Sci. 19, 777-786 (1976).

Acknowledgements

We acknowledge the expert technical assistance of L. Campbell and D. Waxman and support by the MRC (Canada).

N-ACETYL-TRYPTOPHAN: BIOCHEMICAL AND PHARMACOLOGICAL STUDIES

Jean-Pierre Garnier, Christine Escande, Bernard Bousquet,
Gilles Tran, Claude Dreux

Hôpital Saint-Louis, Laboratoire de Biochimie,
75475 Paris Cedex 10, France

Introduction

The neuroamine 5-hydroxytryptamine (5-HT) is known to fluctuate in the human brain in response to altered mental condition (1): its level decreases during depression (2) and rises in schizophrenia (3). The level of brain serotonin depends on the blood concentration of its precursor tryptophan (Trp), especially its plasma free level (4). For this reason, certain authors suggested the use of Trp (5) or of 5-hydroxytryptophan (5-HTP) as antidepressant (6) or as a neurosedative (7). It was, therefore, of interest to study the effects of Trp derivatives like N-acetyl-tryptophan (NAT) on Trp and hydroxyindole metabolism in blood and brain, including the mechanism of action of such derivatives, particularly their ability to cross the blood-brain-barrier, in order to envisage the possibility of a direct nervous central action.

Experimental protocol

1) Animal experiments: 265 male Sprague-Dawley rats were given doses ranging from 150 to 900 mg/kg of NAT salts containing Mg^{++}, Li^{+} or morpholine, either per os (PO, n = 85) or intraperitoneally (IP, n = 75). The remaining rats served as controls (T, n = 105). Rats were decapitated 0 to 180 minutes after NAT administration, the brains were dissected out of the

Progress in Tryptophan and Serotonin Research

skull and the blood collected on an ACD anticoagulant. Decapitation was performed between 10 and 11 hours a.m. to avoid diurnal changes in 5-HT (8, 9) and in Trp (10, 11). Dissected brains and platelet rich plasma are frozen at -20° C.
2) Clinical studies were carried out in 13 apparently healthy men, after oral doses of NAT ranging from 0.5 to 5.1 g. Thereafter, only the blood was analyzed.

Material and analytical methods

In order to separate free and albumin-bound fractions of Trp and NAT, we used ultrafiltration by 20 minutes centrifugation with Aminco PM 10 filters, after pH equilibrium with air containing 5 % CO_2. In rat brain we measured Trp, NAT, 5-HT and 5-HIAA (5-hydroxyindolylacetic acid) and in blood 5-HT, free and total NAT and Trp. The separation was performed by cation exchange high performance liquid chromatography and detection by spectrofluorometry. The conditions were: - partisil SCX 10 µm 25 cm column - pressure: 140 bars - mobile phase: pH 2.30 10^{-2}M $NH_4H_2PO_4$ - flow rate: 2 ml/min. - spectrofluorometric wavelengths: λE = 300 nm, λF = 340 nm - separation time: 15 min.

Results

1) Animal experiments: A preliminary study showed an increase in brain and blood hydroxyindoles after PO or IP NAT load. This confirms previous work (12). Ultrafiltration and HPLC allowed us to investigate the possible mechanism of action of NAT. Intraperitoneal NAT: Blood NAT shows a maximum 1 hour after injection and its plasma concentration is proportional to the injected dose. Neither Trp nor 5-HT show significative changes (statistical test of Kolmogorov-Smirnov with a 5 % risk). NAT poorly penetrates the BBB; thus brain concentrations are lower

than 1/2000 of plasma concentration. Trp (+240 %), 5-HT (+ 42%) and 5-HIAA (+ 50 to 500 %) show peaks by 1 hour after injection.

Per os NAT: Findings after peroral NAT strikingly differ from results after IP NAT. NAT is poorly absorbed when given per os. The plasma concentration peak is approximatively 33 % of that after IP injections (comparing the same dose); the peak level is reached by only one and a half hour after NAT ingestion. Trp, also at variance with the IP route, shows a peak (+ 50 %) at 1 1/2 h after ingestion. This seems to be due to a hepatic "first pass effect" (13) which is absent in IP injections. In brain, the results are similar after both routes of administration. We noticed that N-acetyl-serotonin (NAS) (< 0.1 nmol/g or /ml) does not accumulate in blood or brain.

2) Mechanism of action of NAT. After IP injection, NAT increases cerebral but not plasma Trp. An explanation may be the displacement of albumin-bound Trp by NAT, and an increase of plasma free Trp, and subsequently in cerebral Trp. On the contrary, after oral administration of NAT, brain and plasma Trp are elevated. NAT may be deacetylated in liver thus giving rise to Trp. Therefore we analyzed the free and albumin-bound fractions of NAT and Trp. First we were able to show in vitro that NAT is bound to albumin with a greater affinity (x2) than Trp. Secondly the kinetics show clearly a displacement of bound Trp (T_0 : 38 µmol/1, T_0 + 1 h : 3 µmol/1) by NAT (300 µmol/l after 1 h). This displacement involves a significant elevation of free Trp (T_0 : 15 µmol/1, T_0 + 1 h : 65 µmol/1) and of brain Trp (+ 80 %). On the contrary, after NAT oral administration, plasma free, bound, total and brain Trp fluctuate in parallel (peak at T_0 + 1h). This involves liver metabolism of NAT into Trp.

Clinical study

After oral administration of NAT, the results are equal in

man and rat. We observed no decrease of free Trp but increase of plasma free, bound and total Trp (+ 20 to 50 %).

Conclusion

After oral or IP administration, NAT increases brain Trp, 5-HT and 5-HIAA. It does not cross the blood-brain-barrier. For the IP route of administration the major mechanism seems to be a displacement of albumin-bound Trp. On the contrary, for the oral route of administration (in man and rat) the mechanism appears to be a liver deacetylation of NAT into Trp.

References

1. Lereboullet, J., Resa, A., Olivier-Martin, R.: Rev. Prat. (Paris) 25, 1235-1255 (1975).
2. Coppen, A.: Serotonin and Behavior, Barchas et Usdin ed., Acad. Press, New-York, 523-527 (1973).
3. Bender, D.A., Bamji, A.N.: J. Neurochem. 22, 805-809 (1974)
4. Deniker, P., Loo, H., Zarifian, E. et al.: Ann. Med. Psychol. (Paris) 136, 1, 169-175 (1978).
5. Prange, A.J., Sisk, J.C., Wilson, I.C. et al.: Serotonin and Behavior, Barchas and Usdin ed., Acad. Press, New-York, 539-548 (1973).
6. Richou, H., Hug, R.: Actual. Psych. 1, 87-93 (1979).
7. Greenwood, M.H., Friedel, J., Bond, A.J. et al.: Clin. Pharmacol. Ther. 16, 3, 455-464 (1974).
8. Carlsson, A., Svennerholm, L., Winblad, B.: Acta Psychiatr. Scand. 61, 75-85 (1980).
9. Mester, L., Szabados, L., Mester, M., Yadav, N.: Preg. Fd. Nutr. Sci. 4, 107, 1-20 (1980).
10. Wirtz-Justice, A., Kung, W., Menzi, R.: Experientia 32, 1210-1211 (1976).
11. Sugden, D.: J. Neurochem. 33, 811-813 (1979).
12. Green, A.R., Youdim, M.B.H.: Monoamine oxidase and its inhibition. Ciba Foundation Symposium 39, Elsevier, Amsterdam (1976).
13. Jacquot, C.: Therapie 33, 6, 683-697 (1978).

SEROTONIN, A MAJOR METABOLITE OF TRYPTOPHAN IN THE RABBIT RETINA

By: Neville N. Osborne, Saraswati Patel and David W. Beaton.

Nuffield Laboratory of Ophthalmology
University of Oxford
Walton Street
OXFORD. OX2 6AW. U.K.

Introduction

The vertebrate retina is derived embryologically from the brain and offers considerable advantages as a model preparation for integrating the morphology, physiology and biochemistry of specific neuronal systems. It would not therefore be illogical to assume that substances which are proposed chemical messengers in the brain have similar functions in the retina. The evidence for considering acetylcholine (1), dopamine (2) and γ-aminobutyric acid (3) as transmitters in the retina is reasonably convincing; all have similar functions in brain tissue as well (4). In contrast, it is debatable whether serotonin has a transmitter role in mammalian retinas, although evidence for such a function in non-mammalian retinas is good (5,6,7,8,9). The main argument against serotonin being a transmitter in mammalian retina is the inability to show that the amine is restricted to specific retinal neurones (5,9). In an attempt to substantiate the claim that serotonin is important in mammalian retinas we carried out a series of experiments on the rabbit retina. While it is true that we have not yet localised the endogenous serotonin in the rabbit retina, the combined data support the idea that serotonin is a neuromediator in this tissue.

Metabolism of ^{3}H-tryptophan

There is compelling evidence that bird and amphibian retinas possess the enzymes for the formation of melatonin from tryptophan (see 10,11). Incubation of chopped rabbit retinal pieces with ^{3}H-tryptophan for periods of

Progress in Tryptophan and Serotonin Research

up to 1½ hours at 37°C failed to reveal the formation of any melatonin or N-acetylserotonin. The major products from ^{3}H-tryptophan were 5-hydroxy-tryptophan (5-HTP), and serotonin (see Table 1), although a very small amount of 5-hydroxyindoleacetic acid (5-HIAA) was also detectable. p-Chlorophenylalanine did not seem to influence the ^{3}H-tryptophan levels of the tissues but drastically affected the metabolism of the amino acid (Table 1).

TABLE 1

Tissue (min.)	Condition	Tryptophan	5-HTP	Serotonin	5-HIAA
5	Saline	2000	600	700	n.d.
	Saline + p.Chlor.	1800	550	660	n.d.
15	Saline	6000	700	1000	n.d.
	Saline + p.Chlor.	5900	600	900	100
30	Saline	1100	840	1900	180
	Saline + p.Chlor.	1100	470	960	n.d.
60	Saline	1300	900	2400	250
	Saline + p.Chlor.	1400	430	920	100
90	Saline	1600	980	3000	380
	Saline + p.Chlor.	1400	440	1000	200

Results as expressed as incorporation of radioactivity in d.p.m./mg. retina. Tissues were incubated in physiological saline at 37°C containing 10^{-5}M ^{3}H-tryptophan. In some instances 10^{-4}M p-chlorophenylalanine (p-Chlor.) was added to the medium. The tissues were then subjected to thin layer chromatography on Silica gel precoated plates. Separation was achieved with n- butanol /pyridine /acetic acid /water (15:2:3:5 by volume). Results are mean values for 6 different experiments.

p-Chlorophenylalanine is known to be a potent inhibitor of the enzyme tryptophan-hydroxylase (12). We conclude from these experiments that a path way for the formation of serotonin from tryptophan exists in the rabbit retina. The inability to show the formation of melatonin from ^{3}H-tryptophan is slightly surprising in view of some evidence for the existence of melatonin and its enzymes in the rat retina (13,14,15), the only mammal thoroughly studied. It may well be necessary to use ^{3}H-serotonin as a substrate to demonstrate melatonin synthesis in the rabbit retina as carried out in studies on the trout retina (11).

Identification of indoleamines in the rabbit retina

Since it has been proposed that the true indoleamine of the vertebrate retina is not serotonin, but a substance closely related to the amine (16, 17) it was essential to use a procedure that could discriminate between serotonin and other likely indole derivatives. Fig.1. shows an electrochemical detection and complete HPLC separation of serotonin from a variety of other indole and catecholamine derivatives. Tryptamine and other non-hydroxyindoleamines are not electrochemically detectable. The only indole derivative definitely found in the rabbit retina was serotonin (Fig.2). Crude perchloric acid extracts of rabbit retina were subjected to chromatography with or without a variety of indole derivatives added to the extracts. As shown in the lower trace of Fig.2. where 5-HIAA and serotonin were added to the extract, serotonin could be identified but 5-HIAA acid was absent. Our HPLC data showed that small amounts of serotonin (18 ± 3 ng/g tissue n = 3 ± S.E.M.) definitely exist in the rabbit retina, but there was no evidence of other tryptophan metabolites.
In initial developmental studies we also analysed retinas from 6 day and 24 day old rabbits by HPLC. At the 6 day stage the eyes are still shut and we could not find any serotonin in the retina. At 24 days the eyes are open and small amounts of serotonin were detected. The level is lower than that in adult retinas, though a detailed comparison has not yet been made.

Identification of possible serotonergic neurones

In an attempt to localise the endogenous serotonin in the rabbit retinas we used an immunohistochemical procedure where a monoclonal antibody (YC5/45 HC) is employed (see 7). We observed serotonin-containing neurones in a variety of non-mammalian retinas (5,6,7,8) but never any positive staining in the rabbit or any other mammalian retinas. When rabbit retinas are first incubated with exogenous serotonin (10^{-7}M) at 37°C for 30 min. and then processed by immunohistochemistry to localise serotonin, specific

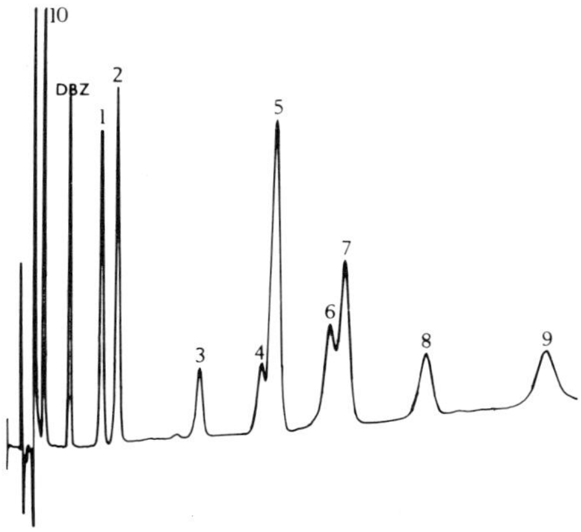

Fig.1.

HPLC separation of serotonin (7) from noradrenaline (10), dopamine (1), dihydroxyphenylacetic acid (2), 5,6-dihydroxytryptamine (3), methoxytyramine (4), 5-hydroxyindoleacetic acid (5), homovanillic acid (6), 6-hydroxytryptamine (8) and bufotenine (9). DBZ (dihydroxybenzylamine) is an internal standard (for details see reference 7).

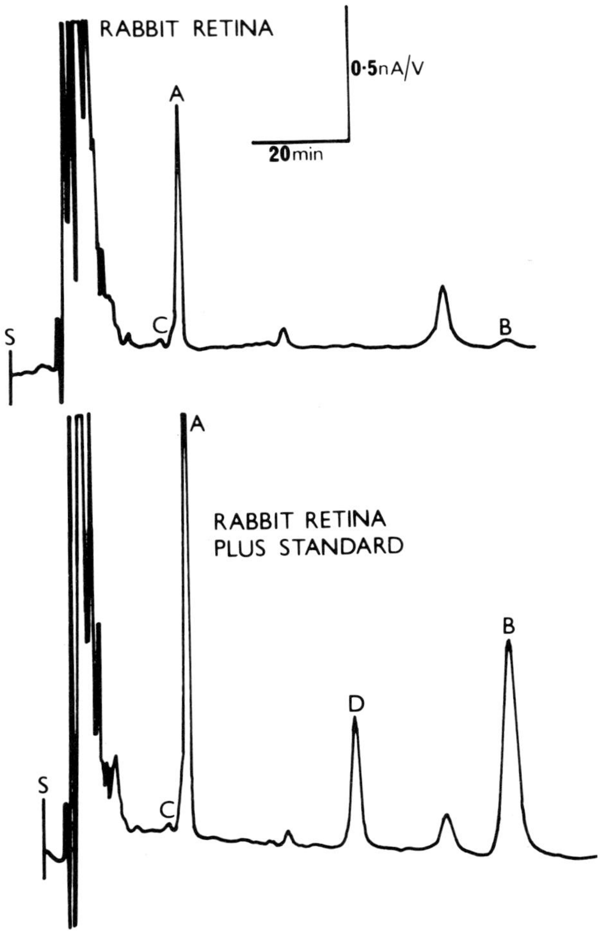

Fig.2.

HPLC and electrochemical detection of various substances in a perchloric acid extract of rabbit retina. Peaks corresponding to dopamine (A) and serotonin (B) were identified in the rabbit retina extract. The lower trace is an extract where internal standards of dopamine (A), serotonin (B) and 5-hydroxyindoleacetic acid (D) were added. It can be seen that serotonin exists in minute amounts in the rabbit retina (from reference 7).

amacrine neurones can be seen to have taken up the exogenous amine (Fig.3). Identical results are obtained if the retinas are incubated with ^{3}H-serotonin and the radioactivity localised by autoradiography (see Fig.3). These results thus show that there are specific neurones which take up exogenous serotonin in the rabbit retina and because autoradiography and immunohistochemistry give identical results it may be concluded that little of the amine is metabolised. This is consistent with results already reported on the metabolism of ^{3}H-tryptophan.

In other experiments rabbit retina pieces were incubated for 60 mins. at $37^{o}C$ in culture conditions with $10^{-7}M$ of either serotonin, tryptamine, tryptophan or 5,6-dihydroxytryptamine before analysis by serotonin immunohistochemistry. As shown in Fig.4, the only positive results were for retinas incubated with exogenous serotonin. In the case of the 5,6-dihydroxytryptamine experiments, there is a hint of some positive staining, but the background non-specific fluorescence is too intense for a definite conclusion to be reached. These results are open to a variety of interpretations. It could be that only serotonin is taken up specifically by a population of amacrine cells. It is also possible that all the substances are taken up, but because of the specificity of the monoclonal antibody used, only serotonin-accumulating neurones are observed. Should the latter be the case, it would also mean that any accumulated tryptophan, tryptamine or 5,6-dihydroxytryptamine is not metabolised to form appreciable amounts of serotonin. In references to tryptophan, it is worth noting that in the metabolism studies where ^{3}H-serotonin is clearly formed from ^{3}H-tryptophan (Table 1), it was necessary to use chopped up pieces of retina to facilitate the uptake and metabolism of the amino acid. In the immunohistochemistry experiments fairly large pieces of retina were used, so it is possible that entry of the amino acid into the tissue decreased.

Cell culture studies

Cell culture is a useful way to study the different properties of CNS cells and their development. In cell culture studies with rat retina, three classes of cell have been recognised: (1) glial cells, the

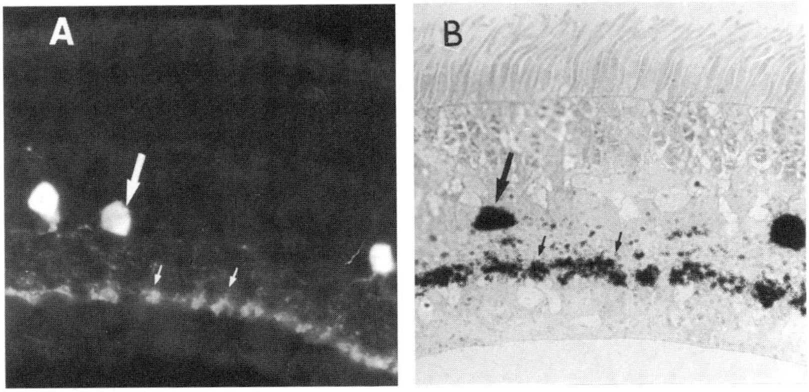

Fig.3. Uptake of exogenous serotonin by rabbit retina as revealed by immunofluorescence studies (A) and autoradiography (B). Similar results are obtained where it can be seen that amacrine cell bodies (large arrows) and processes of these cells (small arrows) take-up the amine

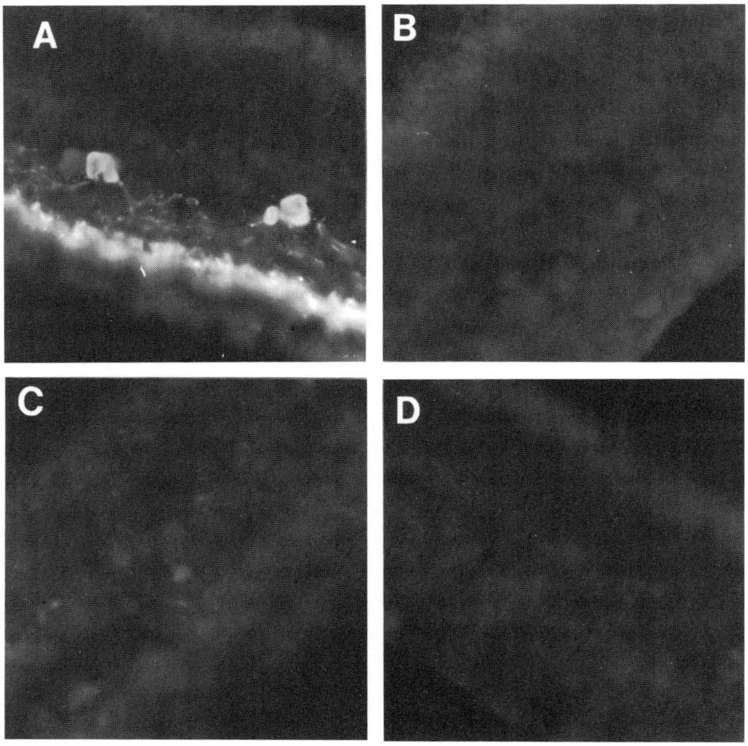

Fig.4. Sections of retinas incubated with serotonin (A), tryptophan (B), 5,6-dihydroxytryptamine (C) or tryptamine (D) and then processed by immunhistochemistry to localise serotonin (see text for details).

surfaces of which bind the monoclonal Ran-2, (2) immature photoreceptors which are spheroid and accumulate ^{3}H-D-aspartate and (3) retinal neurones which are bipolar or multipolar spheroid cells and bind tetanus toxin to their surfaces (18,19). Some of the rat retinal neurones (Class 3) have the capacity to take up exogenous ^{3}H-GABA or ^{3}H-glycine but no evidence could be found for any uptake of ^{3}H-serotonin. In initial studies on rabbit retinal cell cultures, specific cells take up ^{3}H-GABA or ^{3}H-glycine or ^{3}H-serotonin (Fig.5). In the case of the rabbit retinal cell cultures, the different classes of cell have not been completely characterised, but it would appear that all the cells which take up ^{3}H-GABA, ^{3}H-glycine or ^{3}H-serotonin are neurones rather than glial cells which are flattened to the substrate and fairly easily recognisable. The finding that a specific class of rabbit retinal cell in culture takes up exogenous serotonin allows the characterisation of these cells by studying their chemistry, physiology and pharmacology with the main aim of establishing whether the serotonin-accumulating cells are, in fact, serotonergic neurones.

Pharmacological Experiments

In a series of experiments, reserpine, p-chlorophenylalanine and α-methyl-p-tyrosine were injected intravitreally into the right posterior chamber of anaesthetised animals. The volume injected was 10 μℓ and amounted to 5 μg of each substance. Controls received 10 μℓ physiological saline. After 24 hours the animals were killed and the amount of serotonin in the retina of the right eye (injected) was compared with the left (not injected) using HPLC and electrochemical detection. The results showed that (a) injection of physiological saline into the eye did not decrease the retinal serotonin levels, (b) reserpine diminished the serotonin level by 60 ± 10% (n = 3 ± S.E.M.), (c) p-chlorophenylalanine reduced the serotonin level by 40 ± 10% (n = 3 ± S.E.M.) and (d) α-methyl-p-tyrosine decreased the serotonin levels by 25 ± 10% (n = 3 ± S.E.M.).

The results of the injection of reserpine or p-chlorophenylalanine were not unexpected, as the former is known to deplete monoamine stores while

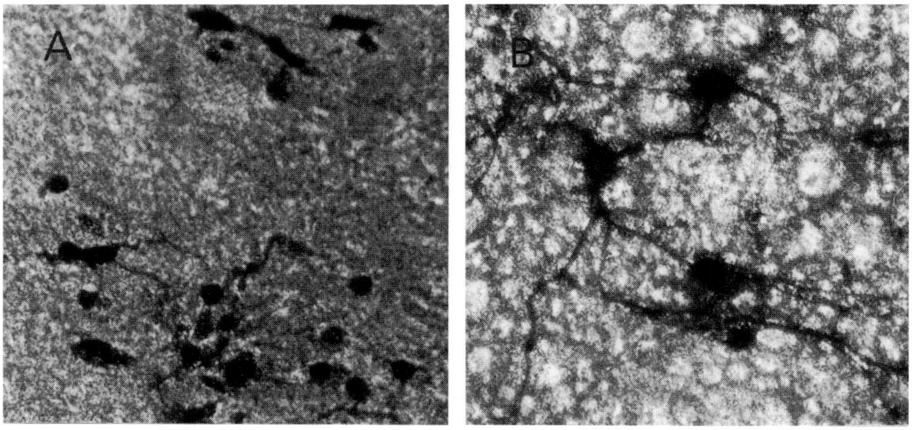

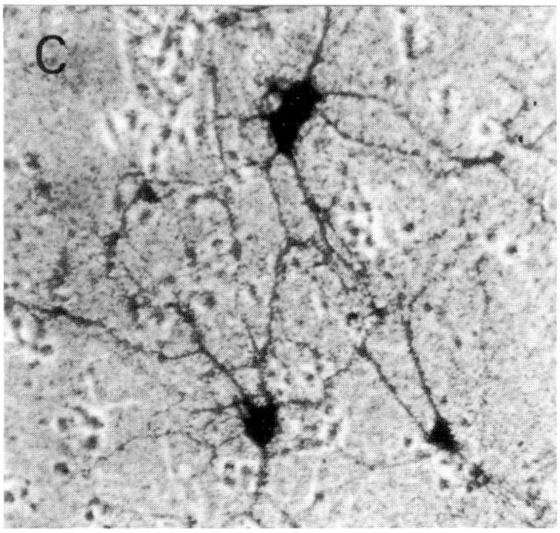

Fig.5.

Autoradiography of ^{3}H-glycine (A), ^{3}H-GABA (B) and ^{3}H-serotonin (C) uptake by six day old cultures of rabbit retinal cells. Cultures were produced from two day old rabbits. It can be seen that serotonin accumulating cells exist in rabbit retinal cultures and these cells appear to be different from those cells which take up exogenous GABA or glycine.

the latter is an inhibitor of the enzyme tryptophan-hydroxylase. The results with α-methyl-p-tyrosine were surprising, because this chemical is an inhibitor of tyrosine-hydroxylase and supposedly does not influence the serotonergic system directly. The findings can be interpreted as suggesting that at least some of the rabbit retinal dopamine and serotonergic neurones are directly linked so that when the dopaminergic neurones are depleted, the serotonergic cells are affected. In the brain, stimulation of dopamine receptors by apomorphine, for example, increases the rate of turn-over of central serotonin in rats (20). There is also evidence that some chick dopamine and serotonin retinal neurones are directly linked (21).

Conclusions

The results discussed suggest that serotonin is a retinal transmitter in the rabbit retina, even though it has not been possible to show directly the localisation of endogenous serotonin. This conclusion is based on the following findings:

(a) Serotonin is synthesised in the retina from tryptophan.

(b) Serotonin is stored in the retina as documented by the reserpine results.

(c) Serotonin is taken up by specific neurones and evidence is provided that the amine is not readily metabolised.

(d) Serotonin is the only identifiable indole derivative in the retina.

(e) The amount of serotonin in the retina is related to the development of the tissues. Serotonin cannot be detected before the eye opens. It is known that a number of transmitter substances are virtually absent from the retina before eye opening has occurred (22,23).

(f) There seems to be a link between some serotonin and dopamine neurones in the retina and this is already established for brain tissue.

Acknowledgement

We are most grateful for financial support from the Wellcome Trust.

References

1. Neal, M.J.: In: Transmitters of the visual system (edt. S.L. Bonting), Pergamon Press, Oxford, pp. 127-143 (1976).

2. Kramer, S.: In: Transmitters of the visual system (edt. S.L. Bonting), Pergamon Press, Oxford, pp. 165-198 (1976).

3. Voaden, M.: In: Transmitters of the visual system (edt. S.L. Bonting), Pergamon Press, Oxford, pp. 107-125 (1976).

4. Iversen, L.L.: Trend in Neuroscience 1, 15-16 (1978).

5. Osborne, N.N.: J.Physiol. (Lond.) 331, 469-479 (1982).

6. Osborne, N.N.: In: Biology of serotonergic transmission (edt. N.N. Osborne), John Wiley and Sons, Chichester, pp.401-430 (1982).

7. Osborne, N.N., Nesselhut, T., Nicholas, D.A., Patel, S. and Cuello, A.C.: J.Neurochem. 39, 1519-1528 (1982).

8. Osborne, N.N., Nesselhut, T., Nicholas, D.A. and Cuello, A.C.: Neurochem.Int. 3, 171-176 (1981).

9. Ehinger, B.: Retina 2, 305-321 (1982).

10. Besharse, J.C.: In: Progress in Retinal Research Vol.1. (edts. N.N. Osborne and G.J. Chader), Pergamon Press, Oxford, pp.81-124 (1982).

11. Gern, W.A. and Ralph, C.L.: Science 204, 183-184 (1979).

12. Koe, B.K. and Weissman, A.J.: Pharmacol.exp.Therap. 154, 499-516 (1966).

13. Bubenik, G.A., Purtill, R.A., Brown, G.M. and Grota, L.J.: Expt. Eye Res. 27, 323-333 (1978).

14. Hamm, H.E. and Menaker, M.: Proc.Natl.Acad.Science 77, 4998-5002 (1980).

15. Binkley, S., Hryshchyshyn, M. and Reilly, K.: Nature 281, 479-481 (1979).

16. Ehinger, B. and Florén, I.: Expt.Eye Research 26, 1-11 (1978).

17. Florén, I. and Hansson, H.C.: Invest.Ophthal. 19, 117-125 (1980).

18. Beale, R., Neuhoff, V. and Osborne, N.N.: In: Biology of Normal and abnormal retinas (edts. R.M. Clayton, J. Haywood and A. Wright), Academic Press, London, pp.9-14 (1983).

19. Beale, R. and Osborne, N.N.: Developmental Brain Research (in press).

20. Scheel-Kruger, J., and Hasselager, E.: Psychopharmacology (Berlin) 36, 189-202 (1974).

21. Parkinson, D. and Rando, R.R.: J.Neuroscience 1, 1211-1217 (1981).

22. Lam, D.M.K., Fung, S-C., and Kong, Y.C.: J.Comp.Neurol. 193, 89-102 (1980).

23. Lam, D.M.K., Fung, S-C., and Kong, Y.C.: J.Neuroscience 1, 1117-1132 (1981).

EFFECT OF 5-HYDROXYANTHRANILIC ACID ON ARACHIDONATE METABOLITES FUNCTION IN PLATELET AGGREGATION SYSTEM

Kiyoshi Goda, Terutoshi Mori, Takashi Ueda and Yahito Kotake
Department of Nutrition, Kobe-Gakuin University
Kobe 673 Japan

Introduction

5-Hydroxyanthranilic acid (5-HAT) was first isolated from rabbit urine by Kotake, and identified as a tryptophan metabolite formed via anthranilic acid by Kotake and Shirai in 1953 (1). 5-HAT has been reported to have various physiological roles in biological systems, but details are still unknown. 5HAT was confirmed to inhibit random microsomal lipid peroxidation, which acts as a radical scavenger and a singlet oxygen acceptor. In an intravascular coagulating system, the well regulated peroxidation of arachidonate in microsomes plays an important role, implying the formation of the cyclic endoperoxide intermediates (PGG_2 and PGH_2). Endoperoxides are converted to thromboxane A_2 (TXA_2) by an enzyme in platelet microsomes, and also to prostacyclin (PGI_2), by an enzyme located in microsomes of blood vessel walls (2,3). TXA_2 induces platelet aggregation, and PGI_2 is the most potent inhibitor of aggregation. TXA_2 and PGI_2 are both unstable substances with opposite effects on platelet aggregation, and may contribute to the maintenance of integrity of vascular endothelium. We have attempted to elucidate the effect of 5-HAT for TXA_2 and PGI_2 function in platelet aggregation.

Progress in Tryptophan and Serotonin Research

Results and Discussion

Platelet aggregation was induced by the addition of arachidonic acid (AA) or ADP <u>in vitro</u> . Aggregation was measured as a change in light transmitted through the platelet suspension. Washed platelet or platelet rich plasma (PRP) were used for aggregation. In Fig.1, AA was added after the pretreatment of washed platelet suspension with indomethacin or 5-HAT at 37°. The platelet aggregation was prevented by the addition of indomethacin, a well known inhibitor of fatty acid cyclooxygenase, but not affected by 5-HAT. The addition of swine aortic microsomes to washed platelets completely inhibited the platelet aggregation induced by AA (Fig. 2), indicating the formation of an anti-aggregatory factor, PGI_2, in this platelet suspension. 5-HAT prevented the inhibition of platelet aggregation by PGI_2 (Fig. 2, 2-4). The extent of prevention is proportional to the concentration of 5-HAT, e.g., the complete prevention at 244 μM of 5-HAT.

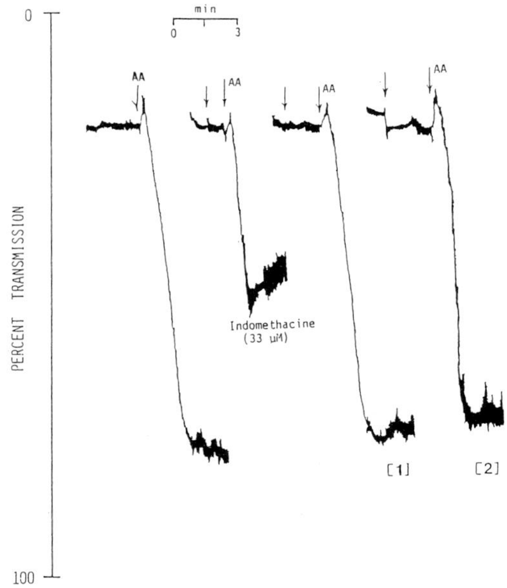

Fig.1. The effect of 5-HAT on human platelet aggregation induced by arachidonic acid(AA). Aggregation was started by the addition of AA, 67μM. 5-HAT : [1] ,150 μM :[2], 200 μM

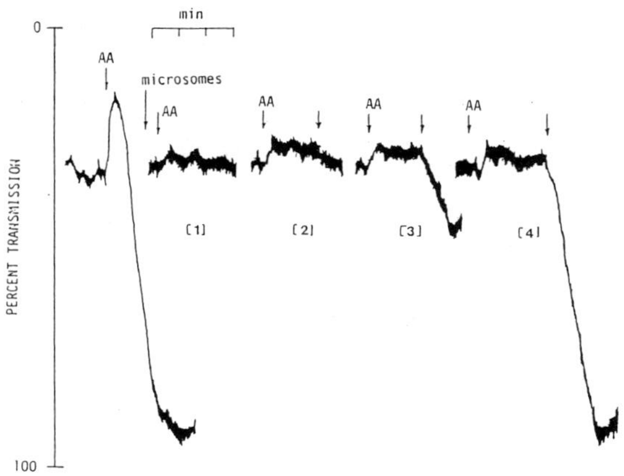

Fig.2. The effect of 5-HAT on on the anti-aggregatory factor, PGI_2, formed by swine aortic microsomes for platelet aggregation. The reaction mixture contained 0.6 mg swine aortic microsomal proteins per ml of the mixture. AA, 137 μM. 5-HAT : (1) none, (2) 31 μM, (3) 121 μM, (4) 244 μM.

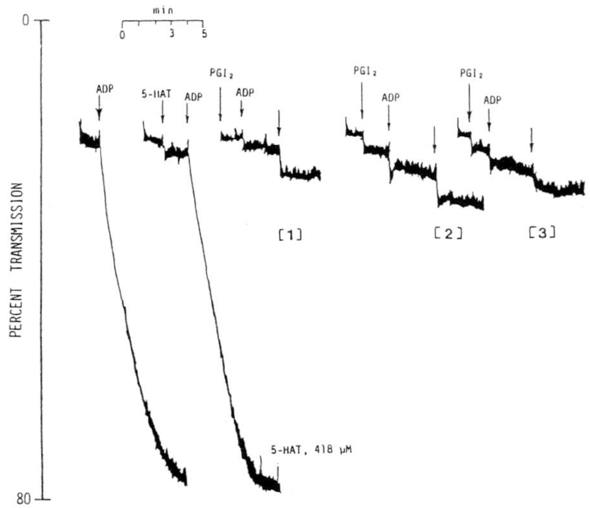

Fig.3. The direct interaction of 5-HAT with PGI_2 in the platelet aggregation system. Human platelet rich plasma was used and the aggregation was started by the addition of ADP. ADP, 67μM. 5-HAT: [1], 418 μM. [2], 835 μM. [3] 167 μM.

5-HAT has a remarkable inhibitory effect on anti-aggregatory behaviour of PGI_2, which we assume to be an effect on PGI_2 function itself or PGI_2 formation. The direct interaction of 5-HAT with PGI_2 was examined (Fig.3). As indicated in Fig.3, 5-HAT shows no interaction with PGI_2, suggesting no effect on PGI_2 function itself in platelet aggregation. According to the above results, 5-HAT greatly influences PGI_2 rather than TXA_2, and has an inhibitory effect on PGI_2 formation in microsomes of blood vessel walls. It is interesting that in vivo the tryptophan metabolite 5-HAT may possibly contribute to the homeostatic function of TXA_2 and PGI_2 in the intravascular system, which have opposite effects to platelet aggregation.

References

1. Kotake, Y., and Shirai, Y.: Hoppe-Seyler's Z.Physiol. Chem. 295, 160-163 (1953).
2. Needleman, P., Moncada, S., Bunting, S., Vane, J.R., Hamberg, M., and Samuelsson, B. : Nature 261, 558-560 (1976).
3. Moncada, S., Gryglewski, R.J., Bunting, S., and Vane, J.R.: Nature 263, 663-665 (1976).

Workshop: Clinical and Behavioural Effects of Tryptophan

INTRODUCTION TO WORKSHOP

Ernest Hartmann

Tufts University School of Medicine
Lemuel Shattuck Hospital, 170 Morton Street, Boston, Massachusetts 02130

In the following section, we shall consider the question of the behavioral and clinical effects of l-tryptophan: In other words, what results does one see when one actually administers l-tryptophan to humans, and, can these effects be clinically useful? (It turns out that no papers about effects on animals are included; we shall be dealing entirely with human beings.)

In many ways, this is a problematic area, lacking the precision and elegance of measurement of many of the more basic tryptophan studies in this volume. In the studies discussed here, the independent variable is simple and consists of the administration of the amino acid l-tryptophan. The dependent variables--the variables through which we examine the effects of tryptophan administration--do not always lend themselves to precise quantitative measurements since we are dealing with such variables as sleep, subjective sleepiness, mood, and so on, in the most difficult of research subjects, the human being.

Before we examine the human data, it would be useful if we knew

Progress in Tryptophan and Serotonin Research

exactly the effects of l-tryptophan administration in animals; unfortunately we do not. There has been a great deal of interest in serotonin and behavior in animals; however, for many reasons other ways of altering brain serotonin have been used in most studies rather than administration of the serotonin precursor, l-tryptophan. What studies there are appear to me fairly consistent: There are data indicating that l-tryptophan administration reduces sleep latency (time to sleep) in animals; that it reduces pain sensitivity; and that it reduces some types of aggressive behavior. There are also related non-behavioral studies in animals showing that tryptophan reduces cardiac arrhythmias under stress and reduces various EEG measures of cortical arousal. These animal studies are roughly comparable with much of the human data to be discussed. To my mind, the findings are all compatible with the view that brain serotonin has a dampening or homeostatic or negative feedback-inducing effect in the brain.

The human studies have almost all been based on various hypotheses about the role of brain serotonin. However, in examining the studies, we should keep in mind that administering tryptophan produces a number of effects aside from increasing brain serotonin. Administering tryptophan can under certain circumstances increase protein synthesis, and can increase the synthesis of tryptamine, niacin, kynurenine metabolites, etc. In addition, we should keep in mind that tryptophan can affect the metabolism and transport of other substances. For instance, animal studies demonstrate that there is competition for transport to brain among at least five neutral amino acids which

Table I

Possibly Clinically Useful Effects of Tryptophan
(All Need Further Investigation)

Antidepressant effects

Adjunct antidepressant effects (combination with standard antidepressants

Reduces sleep latency

Reduces waking time in some insomniacs

Improves sleep patterns the week after administration

Increases sleepiness or drowsiness

Reduces acute sensitivity to pain

Reduces chronic pain

Alleviates migraine headaches

Reverses and prevents certain stupors or comas

Reduces DOPA-induced CNS side effects

Helps some patients with Lesch-Nyhan syndrome (5HTP)

Helps some patients with other forms of mental retardation (5HTP)

Helps some obsessive-compulsive patients

Reduces stress-induced cardiac arrhythmias

Reduces cardiac arrhythmias in patients with apnea

Has some anticonvulsant effects

Reduces "aggressive behavior" in prisoners

Prevents and reverses niacin deficiency

Prevents and reverses some protein deficiency states

Helps identify metabolic abnormalities ("Loading Test")

include tryptophan. In this case, large doses of exogenous tryptophan will not only increase tryptophan transport to brain, but will reduce transport of the other amino acids to brain. One possibly important effect of this would be reduction of phenylalanine and tyrosine available for synthesis of brain dopamine and norepinephrine. I have suggested, for instance, that part of the effect of tryptophan on sleep could well be due to this mechanism.

Table I lists effects of tryptophan noted in humans which are of at least potential clinical use. Looking at the wide array of possible clinical uses, I must admit to being a bit disconcerted. Tryptophan cannot possibly do all this. Some of the "uses" are based on clinical impressions and surveys rather than hard data in terms of placebo controlled studies and may involve simply an active placebo effect. I do not actually believe all of these possible effects will be confirmed.

However, several of the effects I believe are established or at least have been studied enough so that we can discuss them in detail, and those will be the subject of the following papers. In addition, I would like to call attention to two interesting effects of tryptophan about which we have no papers in this volume.

First, Dr. Schmidt et al. have reported the use of tryptophan in patients with sleep apnea and other sleep disorders. The finding was that tryptophan reduced apneic episodes and reduced nocturnal cardiac arrhythmias in these patients (1,2). This effect is compatible with the finding that tryptophan can decrease stress-induced cardiac arrhythmias in dogs. Although the studies are not yet definitive, I believe this is a very important clinical use for the future.

Second, Dr. Lehmann of Sweden has studied the use of tryptophan in reversing coma in several rare conditions. He has found that l-tryptophan will cure or will reverse the comatose condition sometimes found in patients suffering from carcinoid syndrome and also in patients who are comatose after large amounts of l-dopa (3). Carcinoid is the better understood of the conditions. It appears likely that in patients with carcinoid tumors, the serotonin-producing tumors--usually in the gut--are drawing off almost all available tryptophan and leaving little for various uses such as protein synthesis in the brain and elsewhere. Therefore, in this unusual situation adding l-tryptophan restores the possibility of protein synthesis, serotonin synthesis, etc. This interesting and important use of tryptophan occurs in a very specific situation. I do not believe this coma-relieving under such drastic circumstances is in any way incompatible with the more usual finding of the sleep- or sleepiness-inducing effect of l-tryptophan under normal conditions.

Now let us return to the more widely studied effects of tryptophan. These form the basis of the papers immediately following: the effects of l-tryptophan in inducing sleep, sleepiness, and certain mood changes; the effects of l-tryptophan in treating depressive illness, manic illness, and certain types of aggressive behavior; and the effects of tryptophan on acute and chronic pain.

I hope the reader will agree, after examining these papers, that tryptophan has definite effects on human behavior and has several potential uses in alleviating human suffering and illness.

References

1. Schmidt HS, Drugovich ML, Jackson EI: Effects of l-tryptophan in patients with sleep-related respiratory impairment episodes (RIE). Sleep Research 11: 166, 1982.

2. Schmidt HS, Jackson EI: Use of l-tryptophan in disorders of excessive somnolence. Sleep Research 11: 167, 1982.

3. Lehmann J: Tryptophan deficiency stupor--A new psychiatric syndrome. Acta Psychiatrica Scandinavica 65 (Suppl. 300): 1-57, 1982.

TRYPTOPHAN AND HUMAN SLEEP: AN ANALYSIS OF 43 STUDIES

Ernest Hartmann, Daniel Greenwald

Tufts University School of Medicine;
Lemuel Shattuck Hospital, 170 Morton Street, Boston, Massachusetts 02130

My laboratory and others have been interested in the effect of l-tryptophan on sleep and related measures such as sleepiness or tiredness for over 20 years. Our interest first was in the basic chemistry of sleep and tryptophan administration was used as one way of perhaps altering brain serotonin. Our early studies on tryptophan (1,2) were based on faith and guesswork stemming from a variety of animal studies of tryptophan and 5-hydroxy-tryptophan (5-HTP) which suggested that brain serotonin was altered. It was a number of years before studies by Wurtman's group demonstrated that brain serotonin actually was increased by exogenous tryptophan (3).

We have now reviewed and analyzed a total of 43 studies investigating the effects of l-tryptophan on human sleep and sleepiness. These include a few very recent unpublished studies from our laboratory and others (4,5,6, 7,8). For reason of space we cannot list all 43 studies here, but the references can be found by the interested reader in several reviews (9,10). We have examined the results from all these studies in order to answer some basic questions: Does tryptophan have a clear effect on sleep? If so, on which of the many sleep parameters can the results be considered established? How can we account for negative results? What variables will determine the difference between positive and negative results?

These 43 studies have involved a total of 578 subjects receiving l-tryptophan alone, and a number of additional subjects who received l-tryptophan in combination with other medication. The studies are not easy to compare since they were done with differing methodologies in a number of different laboratory settings and for a number of very different reasons. Subjects studied ranged from entirely normal subjects investigated for possible changes in sleep stages after tryptophan to very severe insomniacs and in a few cases subjects who had sleep problems in addition to other

Progress in Tryptophan and Serotonin Research

psychiatric and medical problems. The techniques for study varied from subjective reports by the subject on the effects of tryptophan and placebo (six studies) to objective observational studies in which nurses or investigators made observations every few minutes during the night (five studies), to sleep laboratory studies in which second-by-second recordings could be examined (32 studies). In the latter studies, subjective reports were often obtained in addition to the objective polygraphic measures.

Thus, trying to compare these studies is somewhat like comparing apples and oranges. We shall try to make some sense of the studies first by looking at the different sleep variables that have been studied. In examining a given variable, we shall to some extent necessarily be choosing somewhat similar studies. When discussing effects on sleep stages, we can of course only consider sleep laboratory studies; when discussing sleepiness, we will discuss only studies examining quantified sleepiness scales.

Beginning with the more basic laboratory variables, we can examine first the stages and states of sleep after tryptophan. Here, there is agreement among many studies that tryptophan administration over a wide range of doses--up to perhaps 5 g--produces no dramatic change in sleep stages. Specifically, tryptophan does not reduce REM time or reduce stage 3 and 4 sleep (these are effects produced by most medication that is considered to have hypnotic effects). There is a slight increase, in fact, of stage 4 sleep (deep slow-wave sleep) found generally at all doses studied, though not always reaching statistical significance. At doses of over 5 g, tryptophan is usually reported to reduce REM sleep. There is disagreement about certain more detailed sleep variables such as REM latency.

Turning to the more clinical or insomnia-related variables, Table 1 presents an overview of these results. Here, sleep laboratory as well as non-sleep laboratory studies are included. For a given variable, all studies are included which involve a controlled tryptophan vs. placebo investigation of that variable. Thus, sleep latency (time to sleep) is the variable that has been most frequently studied. We found results on sleep latency in 26 studies. Seventeen of these studies found a significant effect favoring tryptophan over placebo--in other words, reduced

Table 1

Effects of L-Tryptophan Compared to Placebo

Measure	Significant Effect	Trend	No Change	Opposite Trend	Opposite Significant Effect
Reduced sleep latency	17	2	7	0	0
Reduced waking time or number of awakenings	9	3	6	0	0
Increased total sleep	8	4	7	0	0
Improvement in insomnia measures for the days after administration	3	0	0	0	0
Sleepiness scale (increased sleepiness)	4	1	1	0	0

latency with tryptophan. Waking time and number of awakenings were studied in 18 studied and nine of these found a significant reduction of waking time and/or number of awakenings. The other variables were total sleep, improvement in insomnia measures for the period after tryptophan administration--an interesting recently described effect--and increases in sleepiness on quantified scales such as the Stanford Sleepiness Scale. One interesting overall finding (Table 1) is that there were no results in the opposite direction, not even any trends in the opposite direction, despite the many variety of studies included. It is clear that the overall results favor l-tryptophan over placebo, at least in terms of reduced sleep latency, and increased quantified sleepiness. However, even in the category of reduced sleep latency, although there are no opposite results there are several studies that showed no difference between tryptophan and placebo. The same is true with the other variables. Thus, it is worthwhile considering in more detail differences between studies which demonstrate positive tryptophan vs. placebo differences and those which do not.

One parameter that is of obvious importance is dose. Doses of tryptophan from 250 mg to 15 g or from approximately 4 mg/kg to 200 mg/kg have been used in these human studies. My laboratory performed two dose-response studies, the first investigating doses from one to 15 g (11), the second from ¼ g to 1 g (12). Both studies involved very mild insomniacs, basically persons who reported that they took one-half hour to fall asleep at home. These were both sleep laboratory studies involving a total of 25 subjects, 10 in the first and 15 in the second; each subject slept many nights in the sleep laboratory first for adaptation and then taking tryptophan at various doses and placebo in a crossover design. Our results indicated a significant reduction of sleep latency with 1 g or more of l-tryptophan; there was no clear increase in effect when doses were raised higher than 1 g. Doses of ¼ and ½ g produced a trend in the same direction. However, several studies by others have found negative results at doses of 1 g, 2 g, and 4 g of tryptophan (13,14).

In attempting to identify what might account for the difference in results, one factor clearly related to sleep latency results is the type of subject or, more specifically, the original sleep latency of the subject. In a number of studies, all or some of the subjects were normals with a very low sleep latency. Summarizing results from all laboratory studies available, I have been able to obtain individual subject data from five studies in which such data were available on all known subjects who have received placebo as well as 1 g of tryptophan in a controlled study. Results showed that in short latency subjects--basically those with latencies less than one-half hour--there was no clear effect of tryptophan. But in subjects where latency was long, either on placebo or on tryptophan, or both, it was far more often lengthened on placebo (10). Thus, much of the negative data can be explained on the basis of normal subjects, or at least subjects where very short sleep latencies were used. This is hardly surprising; it can be attributed simply to the law of initial values or a "basement effect"--it is impossible to reduce sleep latency using any medication in subjects whose sleep latency is five or ten minutes on placebo because sleep laboratory conditions make it almost impossible to find sleep latencies of less than five minutes.

However, another important point involves time of administration. Tryptophan in the studies reviewed was usually always administered 20 minutes prior to bedtime simply because this has been the usual custom in studying sleeping medications, both in the sleep laboratory and in the clinical world. However, the data from our laboratory and other laboratories on subjective effects and behavioral effects suggest that 20 minutes is not long enough. I will not have time to discuss the sleepiness studies in detail but there is one from our laboratory in which 12 subjects each took in this case 4 g of l-tryptophan, 4 g of l-leucine, or 4 g of placebo and then simply filled out the Stanford Sleepiness Scale (15) every 15 minutes. The results show that tryptophan had a definite effect in producing sleepiness but it was significantly different from placebo only beginning 45 minutes after administration. Fifteen and 30 minutes after administration, subjects did not feel clearly sleepier after tryptophan. Thus, time of administration accentuates the original sleep latency problem. Subjects ideally should be long-latency subjects and tryptophan should be administered at least 30--and preferably 45--minutes before bedtime.

We have looked for other factors that could make a difference in positive and negative results in clinical studies such as age and sex of the subjects. These appear to make only a minor difference: There appear to be less clear or less positive results in older subjects than in young adults and less definite results in women than in men. However, the variables that most clearly influenced results are plotted in Figure 1. The most important variable was type of subject which can also be interpreted as presence or absence, and severity, of insomnia. In this figure, a solid dot represents a significant positive result, whereas an open dot represents a study that either showed no effect or only a trend. When the subjects were mild insomniacs (usually latency-only insomniacs) or normal subjects who report long sleep latencies without complaining of insomnia, the results have been invariably positive. Even the single "negative or unclear" study in this group showed a strong trend favoring tryptophan over placebo. However, when totally normal subjects have been studied, the results have been more mixed, as we have discussed previously. Also, when severe, chronic insomniacs or insomniacs suffering from other illnesses have been studied, the results have likewise been less clear. The interaction of dose of tryptophan and type of subject may be important. In the mild insomniacs or latency-only group, dose makes little difference,

Figure 1

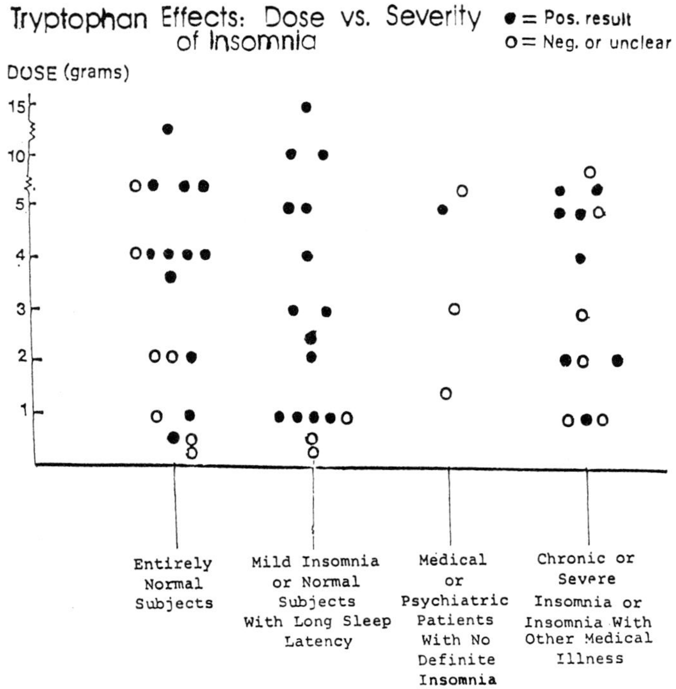

but in the normal subjects, and in the severe insomniacs, there is some indication that higher doses are more likely to produce positive effects than lower doses. Concerning these factors, it appears clearly established that l-tryptophan can have an effect on insomnia, especially reduction of sleep latency and increase of sleepiness, in persons whose sleep is not entirely normal, and it has a clearer effect on those who have mild insomnia or sleep latency problems only.

The results discussed so far have involved a number of different effects, but have mainly involved immediate effects on sleep latency and other sleep measures. These results can most likely be attributed to increases in brain serotonin, though perhaps the reduction of dopamine synthesis from tyrosine and phenylalanine may also be involved. One category of result, which must still be considered preliminary but very exciting, cannot be explained in this way. These are the results involving long-lasting changes in insomnia measures after administration for a few days or a week. These results, called "interval effect" by Schneider-Helmert et al. (16), are very intriguing from a clinical point of view. We have noted that a study by Wyatt et al. in 1970 (17) found similar continuing results in the week after tryptophan administration, though this was not noted by the authors. We have found the same results in a recent study in our laboratory. The result was especially striking to us because it was somewhat unexpected. We were comparing tryptophan 1 g with placebo, with flurazepam, and with secobarbital in separate groups of serious insomniacs. Only flurazepam produced an effect significantly better than placebo during one week of drug administration. Tryptophan produced only a trend towards improved sleep, but reduced latency during the week of administration, but latency continued to improve and there were significant differences between tryptophan and baseline in the second week (the week <u>after</u> drug administration) (4,18). At first, I believed this might be a chance effect, but now in looking over these three studies--as well as some additional studies from Germany which I have not mentioned because they involve tryptophan in combination with other medication--I believe the effect does exist. This is exciting because the effect obviously cannot be explained by the necessarily short-term changes in brain serotonin, or levels of serotonin released, produced by a single dose of l-tryptophan. Rather, they must imply some long-term adaptive mechanism such as probably changes in the sensitivity of receptors.

In summary, I have not presented new data in detail here; I have summarized results from all studies, including a few unpublished ones, involving effects of tryptophan on human sleep. I conclude that tryptophan has effects on sleep, most clearly an immediate reduction in sleep latency and induction of sleepiness found especially in mild insomniacs or situational insomniacs, and also a probable effect in long-term improvement of sleep after administration for a few days or a week. At doses up to 5 g, tryptophan does not produce clearcut changes in stages and states of sleep.

References

1. Hartmann, E.: Psychon. Sci. 8, 295-296 (1967).
2. Hartmann, E.: Psychon. Sci. 8, 479-480 (1967).
3. Fernstrom, J.D., Wurtman, R.J.: Science 173, 149-152 (1971).
4. Lindsley, J.G., Hartmann, E., Spinweber, C., Mitchell, W.: Sleep Research 10, 90 (1981).
5. Lieberman, H.R., Corkin, S., Spring, B.J. et al.: Mood, performance, and pain sensitivity changes induced by food constituents. In, Proceedings of a Conference Held at Massachusetts Institute of Technology on November 9, 1982 entitled, "Research Strategies for Assessing the Behavioral Effects of Foods and Nutrients," 1982.
6. Yogman, M.W., Zeisel, S.H., Roberts, C.: Dietary precursors of serotonin and newborn behavior. In, Proceedings of a Conference Held at Massachusetts Institute of Technology on November 9, 1982 entitled, "Research Strategies for Assessing the Behavioral Effects of Foods and Nutrients," 1982.
7. Nedopil, N., Einhaupl, K., Steinberg, R. et al.: L-tryptophan in chronic insomnia: Results of a 3 month double blind study. In, Sixth European Congress of Sleep Research, Zurich 1982. Abs: p. 115.
8. Leathwood, P.D., Pollet, P.: Diet-induced mood changes in normal populations. In, Proceedings of a Conference Held at Massachusetts Institute of Technology on November 9, 1982 entitled, "Research Strategies for Assessing the Behavioral Effects of Foods and Nutrients."
9. Cole, J., Hartmann, E., Brigham, P.: McLean Hospital Journal 5, 37-71 (1981).
10. Hartmann, E.: Tryptophan and sleep: Who responds to l-tryptophan? In, Perris, C., Struwe, G., Janssen, B. (eds.): Biological Psychiatry 1981, Elsevier Biomedical Press, Amsterdam 1981, pp 613-621.
11. Hartmann, E., Cravens, J., List, S.: Arch. Gen. Psychiatry 31, 394-397 (1974).
12. Hartmann, E., Spinweber, C.: J. Nerv. Ment. Dis. 167, 497-499 (1979).
13. Adam, K., Oswald, I.: Neuropharmacology 18, 1025-1027 (1979).
14. Nicholson, A.N., Stone, B.M.: EEG and Clinical Neurophysiology 47, 539-545 (1979).
15. Hartmann, E., Spinweber, C.L., Ware, J.C.: Sleep Research 5, 57 (1976).
16. Schneider-Helmert, D., Gnirss, F., Schenker, J.: Successful treatment of insomnia by interval therapy with l-tryptophan. In, Sleep 1980, Basel, 5th European Congress of Sleep Research, 1981, pp 421-423.
17. Wyatt, R.J., Engelman, K., Kupfer, D.J. et al.: Lancet 2, 842-846 (1970).
18. Hartmann, E., Lindsley, J.G., Spinweber, C.: Psychopharmacology, in press.

L - TRYPTOPHAN IN CHRONIC INSOMNIA

Norbert Nedopil, Karl Einhäupl, Eckart Rüther, Reinhard Steinberg

Psychiatrische Klinik der Universität München
Direktor: Prof. Dr. H. Hippius, - D-8000 München

Introduction

Serotonergic deficiency has been hypothesized to be one of the major causes of insomnia in men(1). In animals it could be shown that destruction of serotonergic neurons in the n.raphe or pharmacological depletion of serotonin in nerve cells leads to insomnia. Intrathekal application of 5-HT re-instituted sleep in these animals (2,3).

In men, the precursor of 5-HT, l-tryptophan has been reported both to increase the total sleep time(TST) and to decrease the sleep latency (review by 4). A number of studies have also reported beneficial effects of l-tryptophan with and without ß-blockers (5,6,7,8). However most of these studies have not been controlled by placebo and none of them has been carried out for a longer time.

Own double blind cross over studies examining a combination of l-tryptophan and oxprenolol in patients with idiopathic insomnia have not shown any advantages over placebo during a three day period of active medication. But most sleep parameters improved after this period to values significantly better than before treatment. This led to the hypothesis of a long-term effect of the drug, a view which was supported by the findings of other authors(9) and by an own open study on 20 patients(10). Improvement continued in these patients

Progress in Tryptophan and Serotonin Research

during the first two months, although it had not been observed during the first days of treatment.

Method

In order to examine the long-term effects of l-tryptophan in insomnia, a double blind study was carried out on 26 patients over a period of 90 days. Patients received either 1g of l-tryptophan or identical placebo capsules. Patients were examined before treatment to exclude physical or psychiatric disorders. They were free of other psychotropic drugs including hypnotics at least two weeks prior to the drug trial. During the trial they reported on daily sleep protocols their estimated duration of sleep, the subjective quality of sleep (on a 5 point scale ranging from very recreational to very bad), the estimated number of awakenings and the hours awake. 12 patients were selected for sleep polygraphs which were carried out on two consecutive nights 2 weeks prior to the study, during the first nights of medication and after 2,4,8 and 12 weeks. Polygraphs were evaluated according to the guidelines of RECHTSCHAFFEN and KALES. (For the characteristics of the patients see table I)

table I Characteristics of patients

	Verum	Placebo
number	12	14
men	5	6
women	7	8
age (years)	46.9+/- 8.8	42.8+/-7.3
duration of insomnia (years)	12.2+/-10.3	14.8+/-8.2
duration of sleep(hours)	3.1+/- 1.5	4.1+/-1.3
daily sleep protocols(n)	8	8
polygraphs (n)	6	6

Results

All patients grossly underestimated their total sleep time (TST), the difference between the reported duration of sleep and the polygraphically recorded TST varied between 20 and 300 minutes at the beginning of the study. Most patients reported some improvement over time, however, patients receiving l-tryptophan had significantly larger increases of the duration of sleep than patients on placebo(fig1). On the other hand no single parameter of the sleep recordings showed any significant difference between the two treatment groups. There was also no significant change of any parameter during the course of the treatment(fig2): TST, the absolute and relative portions of sleep stages, sleep latency, REM latency, REM-density. The increase of slow wave sleep(stages 3 and 4) occured prior to treatment and is probably due to the withdrawal of other hypnotics, mostly benzodiazepines, during that time.

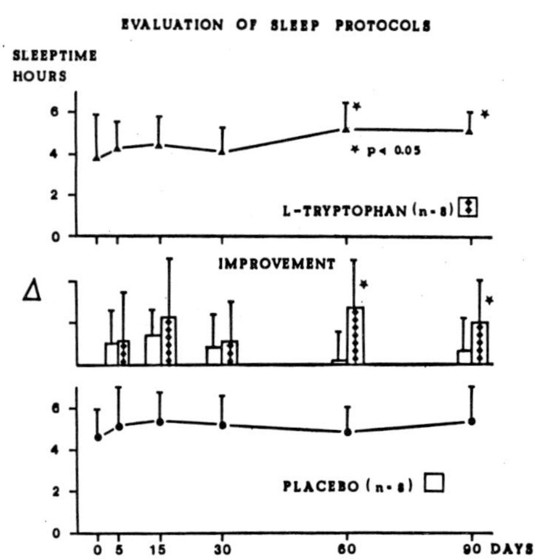

Fig. 1: Duration of sleep according to sleep protocols

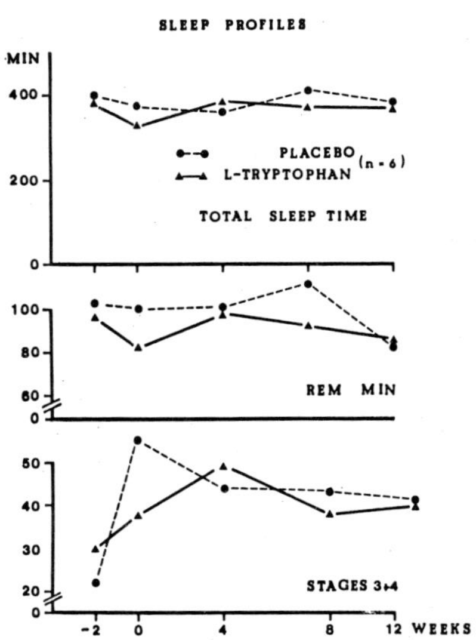

Fig. 2: Parameters of the polygraphs during study

Discussion

Although no difference could be recorded in the sleep profiles of patients with insomnia treated double blindly with either 1g of l-tryptophan or placebo, the subjective ratings of the patients showed an improvement only in the l-tryptophan treated group.
The discrepancy of results between objective parameters and subjective ratings can be explained by the fact that only in patients receiving l-tryptophan the difference between objective and subjective ratings was significantly diminished during the course of the treatment. This difference increases when sleep cycles are disrupted, although the patients may not wake or wake only shortly during such a disruption.
The number of patients studied here is too small to examine the differential effects of the drug and placebo on this sleep architecture. The improvement of subjective ratings of the sleep and the decrease of the difference between objective and subjective parameters, both occuring only in patients treated with l-tryptophan implies a more differentiated view into sleep disturbances than the listing of hours asleep, of portions of sleep stages and of sleep latencies.

References

1. Freeman F.R., Salinas Garcia R.F., Ward J.F.: Eletroencep. Cli. Neurophysiol. 36,657-660 (1974)
2. Jouvet M.: Erg. Physiol 64, 166-307 (1972)
3. Koella W.P.: Physiologie des Schlafes, (1973)
4. Hartmann E.: Am. J. Psychiatry 134, 366-370 (1977)
5. Hartmann E., Cunny R., Chien C.P.:Pharmacopsychologia 19, 114-127 (1971)
6. Schneider-Helmert D.: Intern. Pharmacopsychiat. 16,161-173 (1981)

7. Wyatt R.J.: Biological Psychiatry 5, 33-63 (1972)

8. Nedopil N., Rüther E., Steinberg R.: in Beckmann H. (ed) Biologische Psychiatrie, 346-349 (1982)

9. Gnirrs F., Schneider-Helmert D., Schenker J.: Pharmacopsychiat. 11, 180-185 (1978)

10. Steinberg R., Nedopil N., Rüther E.: in Koella W.P. Sleep 1980, 426-428 (1981)

TRYPTOPHAN (500 mg) DECREASES SUBJECTIVELY PERCEIVED SLEEP LATENCY AND INCREASES SLEEP DEPTH IN MAN

Peter Leathwood and Patricia Pollet

Nestlé Products Technical Assistance Co. Ltd.,
Research Department, CH-1814 La Tour-de-Peilz, Switzerland

Introduction

There have been repeated suggestions that L-tryptophan (TRP) may be a sedative in man, but tests of the effects of TRP on sleep have produced inconsistant results (Adam and Oswald, 1979, Hartman et al., 1974; Wyatt et al., 1970). As there is considerable variability in individual responses to TRP (Leathwood and Pollet, 1982), it is important to use a large group of people for a test of its influence on sleep. Similarly, a variety of factors can effect sleep on any one night, so each treatment should be repeated on several nights. Again, large doses (5 g or more) of TRP produce nausea (Leathwood and Pollet, 1982). This could disrupt sleep, so dose levels should be kept low. Lastly, poor and irregular sleepers are more sensitive than good sleepers to the effects of mild sedatives (Leathwood et al., 1982), so the type of sleeper in the test may influence the results.

We have recently shown that 400 mg of TRP taken in the evening with a small carbohydrate load will triple the plasma TRP to large neutral amino acid (TRP/LNAA) ratio (Ashley et al., 1982). This should be sufficient to produce a significant increase in brain serotonin synthesis (Fernstrom and Wurtman, 1972) which may in turn influence sleep.

Progress in Tryptophan and Serotonin Research

The aim of this study was to test the effects on sleep, in a large group of people claiming to suffer from mild insomnia, of 500 mg of TRP combined with a small carbohydrate load.

Methods

Test samples : In each test sample 500 mg of finely ground L-tryptophan (Hakko Kogyo, Tokyo, Japan) was included in a 10 gm bar of bitter chocolate. The placebo was a similar chocolate bar without TRP.

Experimental design : 76 volunteers (mostly poor or irregular sleepers who were not regularly taking sedatives) were recruited. Each volunteer filled in a short questionnaire covering his or her habitual sleep characteristics and was then given 10 chocolates, each identified only by a code number, and 10 post-sleep questionnaires. The questionnaires ranked sleep latency, sleep quality, sleep depth and sleepiness the next morning on 9-point scales and effect of the treatment (stimulant, no effect, calming) on a 3-point scale. For half the participants the order of administration was : PPTTPPPTTT; for the others TTPPTTTPPP (P = placebo, T = tryptophan).

Analysis of results : Results were first treated by canonical analysis of variance to check within groups homogeneity of responses on the different nights, and then by analysis of variance.

Results and Discussion

72 people completed the study. Mean responses were essentially similar across the 5 nights of each treatment, so for each volunteer the difference between the mean score for each question for each treatment was calculated and used as the basis for the analysis of variance. Mean difference scores (± SEM) for all subjects are shown in Fig. 1. After TRP, perceived sleep latency was significantly shorter, sleep depth increased and the treatment was perceived as calming. On separating the sexes (Fig. 2) it was clear that women were more sensitive to TRP than men.

FIG. 1
EFFECT OF TRYPTOPHAN
ON ALL SUBJECTS

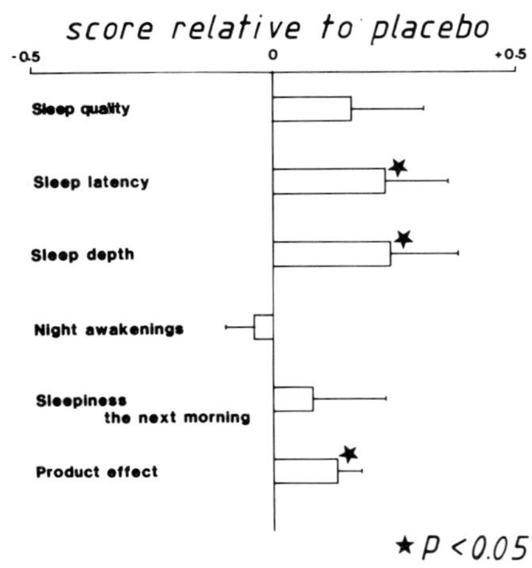

The results were also analysed in terms of the proportions of the population noting an improvement in each sleep parameter. Of all the participants, 59% had shorter perceived sleep latencies after TRP. Women (67% noting improved sleep latency) were more sensitive than men (49%).

FIG. 2
EFFECT OF TRYPTOPHAN
ON MEN AND WOMEN

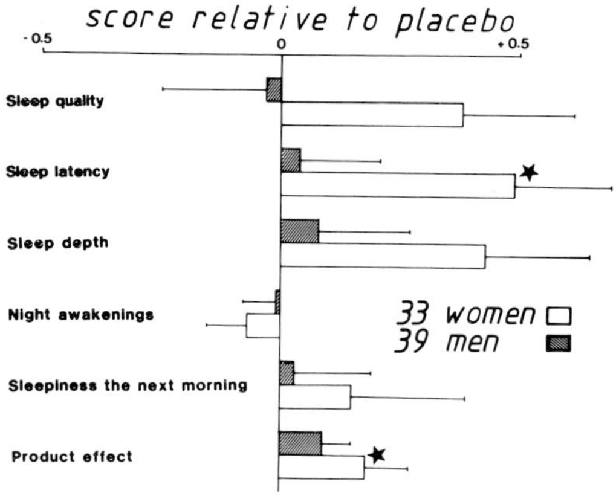

This experiment demonstrates that when TRP is tested under appropriate conditions - i.e. : taken 1 hour before retiring; combined with a small carbohydrate-rich load; using a repeated measures design, and with a large population of poor or irregular sleepers - even 500 mg can significantly effect subjective sleep ratings. Sleep latency was the measure most affected, suggesting that TRP is quick-acting and short lasting.

References

1. Adam, K., Oswald, I.: Neuropharmacol. 18, 1025 (1979).
2. Ashley, D.V.M., Barclay, D.V., Chauffard, F.A. et al.: Am. J. Clin. Nutr. 36, 143 (1982).
3. Fernstrom, J.D., Wurtman, R.J.: Science, 178, 414 (1972).
4. Hartman, E., Cravens, J., List, S.: Arch. Gen. Psychiatry 31, 394 (1974).
5. Leathwood, P.D., Chauffard, F., Heck, E., Munoz-Box, R.: Pharmacol. Biochem. and Behav. 17, 65 (1982).
6. Leathwood, P.D., Pollet, P.: In : Research strategies for assessing the behavioral effects of foods and nutrients. Eds Lieberthan, H.R., Wurtman, R.J., p. 94 (1982).
7. Wyatt, R.J., Engelman, K., Kupfer, D.J., Fram, D.H., Sjoerdsma, A., Snyder, F.: Lancet 2, 842 (1970).

FURTHER EVIDENCE OF THE HYPNOTIC PROPERTIES OF TRYPTOPHAN

Harris R. Lieberman, Suzanne Corkin

Department of Psychology, Massachusetts Institute of Technology
Cambridge, Massachusetts 02139 USA

Richard J. Wurtman

Department of Nutrition and Food Science, Massachusetts Institute of Technology
Cambridge, Massachusetts 02139 USA

Introduction

Evidence continues to accumulate that tryptophan has hypnotic and analgesic properties. Electroencephalographic (EEG) studies have demonstrated that tryptophan, when administered at night, reduces sleep latency and number of awakenings in humans (1). Tryptophan has also recently been shown to alter human mood in a manner consistent with its effects on sleep (2-4). Tryptophan may also reduce pain sensitivity in both healthy and clinical populations (4-6). Tryptophan's hypnotic and analgesic properties have been attributed to its well-documented potentiation of serotonergic neurotransmission (7). Although conclusive evidence confirming this hypothesis has not been presented, little data supporting alternative hypotheses have been advanced.

We recently conducted a study to evaluate the effects of tryptophan on human mood and performance. We also administered tyrosine, which, like tryptophan, is a large neutral amino acid. Tyrosine could have behavioral

Progress in Tryptophan and Serotonin Research

effects that are opposite those of tryptophan, due to its potentiation of catecholaminergic neurotransmission, or because it competes with tryptophan for transport across the blood-brain barrier (7). When systemically administered to mice, tyrosine has also been reported to have properties similar to those of stimulants such as amphetamine and caffeine (8).

Methods

Tryptophan and tyrosine were administered in a double-blind placebo-controlled crossover trial to 20 healthy males on four different days. The subjects ingested either tryptophan (50 mg/kg), tyrosine (100 mg/kg), or one of 2 placebos. In order to evaluate the behavioral effects of these two large neutral amino acids, we administered two self-report mood questionnaires, the Profile of Mood States (POMS) (9), and the Visual Analog Mood Scale (VAMS) (10). Both scales yield specific factor analytically derived subscales. They have been used in a number of previous psychopharmacological investigations and are sensitive to the effects of both stimulants and hypnotics (11-13). We also examined certain cognitive and motor components of performance with the following tests: a simple auditory reaction time (RT) task, a two-choice visual RT task, a tapping task, and a grooved pegboard test. Testing began at 9:00 am, two hours after ingestion of the test substance in pill form. The data were analyzed using analyses of variance and, when appropriate, two-tailed Newman-Keuls tests.

Results

Tryptophan was found to have significant hypnotic properties, as compared to placebo and tyrosine on certain subscales of the POMS and VAMS (Fig. 1). The score from the Alert scale of the VAMS was significantly decreased by tryptophan as compared to placebo ($p < 0.01$), as were the Vigor scale of the POMS ($p < 0.01$) and the Fatigue scale of the POMS ($p < 0.05$). Notably, tryptophan did not significantly alter any other mood subscale, including the Calm subscale of the VAMS and the Anxiety subscale of the POMS. Tyrosine did not alter any of the mood scales.

Neither tryptophan nor tyrosine significantly affected any aspect of performance that we examined. There was a significant difference, however, between tryptophan and tyrosine on the simple RT task in the direction expected, i.e., tryptophan slowed RT relative to tyrosine.

Discussion

Based upon these results and those of other laboratories, it is apparent that tryptophan has hypnotic properties not only at night, as measured by EEG techniques, but also during the day, as measured by self-report mood scales. As noted above, tryptophan did not alter either the Calm scale of the VAMS nor the corresponding Anxiety subscale of the POMS. In contrast, other hypnotics, such as the benzodiazepines, can alter such subscales in normal subjects (11,12), suggesting that tryptophan may have a different mechanism of action than these drugs.

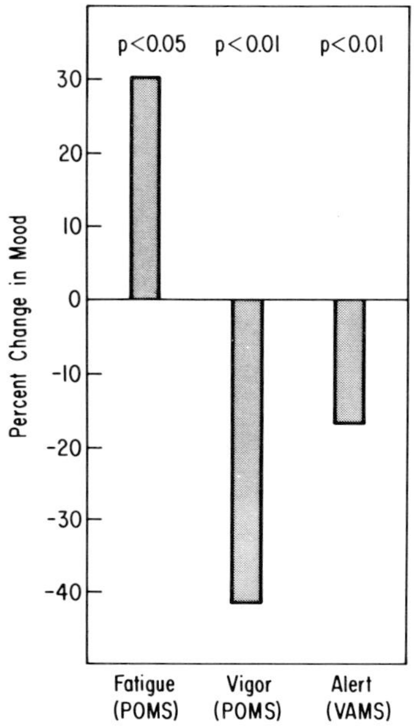

Figure 1. Percentage change in mood state subscales of the POMS and VAMS after ingestion of tryptophan versus a placebo.

Hypnotics also have detrimental effects on performance. It is well known that administration of benzodiazepines or barbiturates impair reaction time and tapping ability, not only immediately after administration, but the next day as well (14). Tryptophan, in contrast, did not impair any of the various aspects of performance that we examined, including reaction time and tapping, although, as noted above, it had significant effects on mood, of the same magnitude as those induced by clinically effective doses of other hypnotics. Furthermore, in laboratory sleep studies, tryptophan has been shown to be an effective hypnotic in doses equal to and even considerably less than the dose we adminstered (1). Further studies are necessary to compare tryptophan with other hypnotics, both with regard to clinical efficacy and side effects, such as impaired performance and next-day drowsiness.

References

1. Hartmann, E.: J. Psychiatr. Res. (in press)

2. Hartmann, E. and Elion, R.: Psychopharmacology, 53, 131 (1977)

3. Hartmann, E. and Spinweber, C. L.: J. of Ner. and Ment. Dis., 167, 497 (1979)

4. Lieberman, H. R., Corkin, S., Spring, B., Growdon, J. H., and Wurtman, R. J. : J. Psychiatr. Res. (in press)

5. Seltzer, S., Stoch, R., Marcus, R., and Jackson, E.: Pain, 13, 385 (1982)

6. Seltzer, S., Dewart, D., Pollack, R. L., and Jackson, E.: J. Psychiatr. Res. (in press)

7. Wurtman, R. J., Hefti, F., and Melamed, E.: Pharmacol. Rev. 32, 315 (1981)

8. Gibson, C. J., Deikel, S. M., Young, S. N., and Binik, Y. M.: Psychopharmacology, 76, 118 (1982)

9. McNair, D. M., Lorr, M., and Droppleman, L. F.: Profile of Mood States Manual (Educational and Industrial Testing Service, San Diego, 1971)

10. Spring, B., Maller, O., Wurtman, R. J., Digman, L., and Cozolino, L.: J. Psychiatr. Res. (in press)

11. File, S. E. and Bond, A. J.: Psychopharmacology, 66, 309 (1979)

12. Spinweber, C. L., and Johnson, L. C.: Psychopharmacology, 76, 5 (1982)

13. Cole, J. O., Pope, H. G., LaBrie, R., and Ionescu-Pioggia, M.: Clin. Pharmacol. Ther., 24, 243 (1978)

14. Johnson, L. C., and Chernik, D. A.: Psychopharmacology, 76, 101 (1982)

Acknowledgements

We thank G. Garfield, R. Beecher, M. Duffy, and C. Wogan for technical assistance.

THE THERAPEUTIC ACTION OF TRYPTOPHAN IN DEPRESSION, MANIA AND AGGRESSION

S.N. Young[1], G. Chouinard[1,2], L. Annable[1], C. Morand[3] and F.R. Ervin[1]

[1]The Department of Psychiatry, McGill University, 1033 Pine Avenue West, Montreal, Quebec, Canada, H3A 1A1

[2]Research Department, Hôpital Louis-H. Lafontaine, Montreal, Quebec

[3]Research Department, Institut Philippe Pinel de Montréal, Montreal, Quebec

Introduction

The use of tryptophan in the affective disorders has elicited much controversy. Even though more than 20 studies have been published on this subject, there is still no consensus on whether tryptophan is an effective antidepressant. The main reason for this is that many studies are difficult to interpret because of lack of placebo-treated control groups. We have now attempted to rectify this situation by performing double-blind placebo-controlled studies on the action of L-tryptophan in depression and acute mania. We also looked at its action in pathological aggression because this is another condition in which brain 5-hydroxytryptamine may be low (1).

The Studies

1) Unipolar depression. Twenty-four newly admitted depressed patients satisfying the Research Diagnostic Criteria for unipolar depression were randomly assigned to treatment with tryptophan (4 g per day, given twice daily) or placebo, for a four-week period. Clinical assessment by the Hamilton Rating Scale for Depression was performed at weekly intervals. There were no significant differences between the mean Hamilton scores for the groups on tryptophan and placebo at any time, although there was a trend for the patients on placebo to respond better

Progress in Tryptophan and Serotonin Research

than those on tryptophan. Thus, five out of the twelve patients on tryptophan were drop-outs due to inadequate response, while only one patient on placebo was a drop-out.

2) Bipolar depression. Eighteen newly admitted depressed patients satisfying the Research Diagnostic Criteria for bipolar depression were randomly assigned to treatment with tryptophan (8 g per day, given twice daily) or placebo, for a four-week period. There was no significant difference between the mean Hamilton scores for the two groups at any time. Two patients were drop-outs in each group and the overall response was similar in each group.

3) Acute mania. Twenty-four patients, newly admitted through the emergency department, satisfied the Research Diagnostic Criteria for acute mania. They were treated with tryptophan (12 g per day given in four doses) for one week. For the second week half the patients were substituted onto placebo under double-blind conditions, while the other half continued to receive tryptophan. All patients were allowed haloperidol on a PRN basis whenever their behaviour was uncontrollable. Assessment was made weekly using a nine-point Clinical Global Impression (CGI).

During the first week of treatment, when all the patients were on tryptophan the mean CGI declined significantly from 7.0 to 4.5. In spite of the highly disturbed nature of the patients when they were admitted to the study, only 14 out of 24 required any haloperidol and only one patient required an appreciable amount. During the second week, 8 of 12 patients on tryptophan continued to improve, while only 1 of 12 on placebo showed any improvement on the CGI. This difference is statistically significant ($p=0.005$) using the Fisher exact probability test.

4) Pathological aggression. The study was carried out on 12 male schizophrenics, who had averaged 13 months at a hospital for mentally ill offenders. All had been convicted of murder or some other person-related crimes and they continued to show episodes of threatening aggression. A double-blind placebo crossover design was used. Tryptophan and placebo were each given for four weeks with a one-week washout period between treatments. One group of 6 patients received 4 g tryptophan per day, while another group of 6 patients received 8 g. The Brief Psych-

iatric Rating Scale (BPRS) was filled in weekly, while a 17-item ward checklist was filled in daily on each patient, to record various types of physical assault, verbal abuse and uncontrolled behaviour.

Pretreatment free plasma tryptophan levels for the aggressive schizophrenics (mean ± SE in μg/ml was 2.52 ± 0.44) was significantly ($p < 0.05$) less than the values we have found previously (2, 3) for non-aggressive schizophrenics (3.83 ± 0.75) or normal controls (3.38 ± 0.74).

Close inspection of the results enabled us to distinguish two groups of patients. One group, characterized by low scores on the Buss-Durkee Hostility Inventory, low lifetime frequency of aggressive events and subnormally reaction GSRs, became worse on tryptophan as indicated by the BPRS factors, depressed mood, conceptual disorganization, and anergia. The second group, who had high Buss-Durkee scores, high lifetime aggression frequency and normal GSRs (i.e. patients who have been characterized as having poor impulse control) improved on tryptophan as indicated by the BPRS factors depression and hostility. For the whole group of 12 patients the ward checklist showed a 30% decrease in incidents while the patients were on tryptophan, relative to placebo, and this decline was statistically significant (x^2=4.72, $p<0.05$). This effect was almost entirely due to the second subgroup of patients.

Discussion

Our studies demonstrate that tryptophan has a therapeutic effect in acute mania and pathological aggression, but they failed to show the superiority of tryptophan to placebo in newly admitted unipolar or bipolar depressed patients. This suggests a similarity between the therapeutic profile of tryptophan and that of lithium. Case reports of the prophylactic action of tryptophan in manic depressive patients (4) are also consistent with such a similarity.

As tryptophan acts presumably through an enhancement of 5-hydroxytryptamine synthesis our data suggest that lithium's known effect on 5-hydroxytryptamine could be an important factor in its therapeutic action. If further studies confirm our findings, tryptophan may become a useful supplement to lithium in patients who do not respond well to that drug, and, because of its lack of toxicity, tryptophan may replace lithium in some circumstances.

References

1. Brown, C.L., Goodwin, F.K., Ballenger, J.C., Goyer, P.F., Major, L.F.: Psychiat. Res. 1, 131-139 (1979).
2. Chouinard, G., Annable, L., Young, S.N., Sourkes, T.L.: Commun. Psychopharmacol. 2, 21-31 (1978).
3. Chouinard, G., Young, S.N., Annable, L., Sourkes, T.L.: Acta Psychiat. Scand. 59, 395-414 (1979).
4. Chouinard, G., Jones, B.D., Young, S.N., Annable, L.: Amer. J. Psychiat 136, 719-720 (1979).

THE EFFECTS OF DIETARY TRYPTOPHAN ON CHRONIC MAXILLOFACIAL PAIN TOLERANCE

Samuel Seltzer, Robert L. Pollack, Dorothy Dewart, Eric Jackson

Temple University School of Dentistry, Philadelphia, PA., USA

Introduction

In the United States, chronic pain represents a cost of $60 billion due to a combination of lost earnings and the cost incurred for diagnoses and attempted treatments (1).

The studies by Fernstrom, Wurtman, and associates (3-7) involving the dietary component tryptophan, and its conversion into the neurotransmitter, serotonin, represents an important stage in the development of treatment modalities solely by nutrition intervention.

Basically, this nutritional regimen rests upon the recognition that elevations in pain thresholds have been correlated with increases in brain serotonin levels which, in turn, were brought about by increases in plasma tryptophan levels (5,7). Conversely, tryptophan-poor corn diets were found to decrease brain serotonin levels, (4, 10).

Seltzer et al (11) confirmed these findings in humans by accompanying the administration of a specific diet with the ingestion of 3 grams daily of tryptophan and significantly raising the pain tolerance threshold in normal, pain-free subjects. These results served as the basis for the present studies that were conducted in the use of nutritional manipulation and tryptophan supplementation in the treatment of subjects with chronic, intractable, maxillofacial pain.

Because of its polar nature tryptophan requires a carrier protein to transport it across the blood-brain barrier. Not only

Progress in Tryptophan and Serotonin Research

does tryptophan have to compete with other large neutral amino acids (LNAA) for access to this same transport protein, but the transport protein, itself, has a lower affinity for tryptophan than for its major competitors. Thus, the amount of tryptophan in the blood entering the brain is a direct function of the ratio of the concentration of tryptophan to the competing LNAA. To compound the situation, tryptophan in food is present in lower amounts (particularly in animal proteins) than are LNAA. Vegetables and some grains do have a more favorable tryptophan/LNAA ratio. The reduced protein content of a diet, therefore, was shown by Fernstrom (8) to be significant because of the decrease in the overall amount of LNAA, thereby increasing the ratio of plasma tryptophan levels to the sum of plasma LNAA.

Fernstrom (7), Colmenares et al (2), and Lipsett et al (9) reported that carbohydrate caused an increase in the relative concentration of blood tryptophan levels. Of all the blood amino acids, tryptophan is the only amino acid that is carried as an albumin-bound complex. While the insulin effect serves to drive the other amino acids into body tissues, the tryptophan remains available to the brain. A high-carbohydrate, reduced-protein diet therefore further improves the tryptophan/LNAA ratio (in favor of the tryptophan) and increases the amount of tryptophan crossing the blood-brain barrier into the brain.

The third major dietary component, fat, also has been shown to exert an effect in the tryptophan-serotonin relationship (9,12), in this case, an inverse relationship. Circulating fatty acids (which are directly proportional to dietary fat intake) reduce the tryptophan-retaining effect of insulin by displacing tryptophan from albumin. In this free form, the insulin effectively removes the tryptophan from the blood stream along with the other LNAA.

Consequently, a reduced protein, reduced fat, and increased carbohydrate intake has been postulated as the most appropriate diet for maximizing brain tryptophan with a resultant increase in the formation of brain serotonin. The diet selected was protein 10%, fat 20%, and carbohydrate 70%, respectively, of the total daily caloric intake.

Methodology

Of forty-three initial patients, 30 patients completed the study: 15 subjects each in the tryptophan and placebo groups. The diagnosis included: temporomandibular joint pain dysfunction syndrome (TMJ), atypical facial neuralgia, migrainous neuralgia, trigeminal neuralgia, phantom tooth pain, ciliary neuralgia, and cervical osteoarthritis. Eleven patients carried dual diagnoses.

At the first appointment, five types of psychological tests were administered to each subject. At each visit subjective pain ratings were obtained by having each subject respond to the question, "On the average (since your last visit) rate your pain on a scale of 0 to 100, where 0 = no pain, and 100 = maximal pain."

Pain perception and pain tolerance levels were assessed with a standard, digital read-out, electrical battery-operated pulp-tester. This was accomplished by applying the probe of the instrument to the labial surface of an upper anterior tooth. Initially, as the voltage was slowly increased, the pain perception threshold was that point when sensation was first perceived; the pain tolerance threshold was the point when the induced pain became unbearable. The scale on the pulp-tester ranges from 0-80.

The nature and purpose of the reduced protein, reduced fat, and high carbohydrate diet was explained to each subject. An individualized diet was prescribed and the subject was counseled on how to implement the diet.

In the double-blind study that was conducted, each subject was provided with capsules containing either 250 milligrams of tryptophan or placebo (cellulose), and instructed to take 2 capsules six times daily. All subjects returned weekly for a clinical examination and a nutrition consultation. On the fourth week, the psychological tests, pain rating evaluation, and pain thresholds were determined as in the initial appointment.

Results

TABLE 1

Mean clinical pain ratings at initial appointment and at week four for Tryptophan and Placebo groups.

	Tryptophan (n=15)		Placebo (n=15)	
	mean	S.D.	mean	S.D.
Initial ratings	57.66	5.71	49.66	6.31
Four-week ratings	25.86	6.50	38.53	8.30
Reduction from week 1 to week 4	31.80		11.13	

The mean reduction in clinical pain rating over the four-week period was 31.80 for the tryptophan subjects, and 11.13 for the placebo subjects (on the 0-100 pain rating scale). The difference is statistically significant at $P<.01$ level.

Of the 15 tryptophan subjects, 13 showed an increase in pain tolerance and 2 showed a decrease. For the control subjects, seven showed an increase and eight a decrease. The mean increase in pain tolerance from Week 1 to Week 4 was 12.53 for the tryptophan subjects compared to 2.60 for the placebo subjects. The difference in the mean change for the tryptophan subjects is statistically significant at the $P<.05$ level, while the difference for the control group was statistically not significant.

Conclusions

The major finding of this study was that the dietary supplementation of tryptophan combined with a high carbohydrate-reduced protein-reduced fat diet resulted in a significant reduction in the pain intensity experienced by chronic pain patients.

TABLE 2

Mean increase in pulpmeter tolerance threshold, and number of subjects in each group that showed higher and lower pulpmeter thresholds from week 1 to week 4, for Tryptophan and Placebo groups.

	Tryptophan (n=15)	Placebo (n=15)
Mean increase in tolerance threshold	12.53	2.60
Subject with Higher tolerance threshold	13	7
Subject with Lower tolerance threshold	2	8

The chronic pain patient is of considerable concern to all health professionals. It is hoped that this study corroborates the feasibility of a treatment modality that is non-invasive and relatively easy to implement.
The results here, however, are the effects achieved during a short-term period of four weeks; we are now initiating another study in which a numer of biochemical parameters will be monitored over a three-year period.

References

1. Bonica, J.J.: Pain Research and Therapy: Past and Current Status and Future Needs. In Pain Discomfort and Humanitarian Care. Elsevier/North Holland, New York, 1980.
2. Colmenares, J.L., Wurtman, R.J., Fernstrom, J.D.: J. Neurochem. 25, 825-829 (1975).
3. Fernstrom, J.D., Wurtman, R.J.: Science 173, 149-152 (1971).
4. Fernstrom, J.D., Wurtman, R.J.: Nature New Biol. 234, 62-64 (1971).
5. Fernstrom, J.D., Larin, F., Wurtman, R.J.: Life Scie. 13, 517-524 (1973).
6. Fernstrom, J.D., Faller, D.V., Shabshelowitz, H.: J. Neural Transm. 36, 113-121 (1975).
7. Fernstrom, J.D.: Metabolism 26, 207-233 (1977).
8. Fernstrom, J.D., Wurtman, R.J., Hammerstrom-Wiklund, B., Rand, W.H., Munro, H.N., Davidson, C.S.: Amer. J. Clin. Nutr. 32, 1912-1922 (1979).
9. Lipsett, D., Madras, B.K., Wurtman, R.J., Munro, H.N.: Life Sci. 12, 57-64 (1973).
10. Lytle, L.D., Messing, R.B., Fisher, L., Phebus, L.: Science 190, 692-694 (1975).
11. Seltzer, S., Stock, R., Marcus, R., Jackson, E.: Pain 13, 385-393 (1982).
12. Wurtman, R.J., Fernstrom, J.D.: Nutr. Revs. 32, 193-200 (1974).

Pathology, Clinical Aspects

MODULATION OF SEROTONIN RECEPTORS BY L-VALINE IN HEPATIC ENCEPHALOPATHY

Elisabeth Kienzl, Peter Riederer, Kurt Jellinger
Ludwig Boltzmann Institute of Clinical Neurobiology
Lainz-Hospital
A-1130 Vienna, Austria

Erich Roth
Dept. of Exp. Surgery I, University School of Medicine
A-1090 Vienna, Austria

Peter Ferenci
Dept. of Gastroenterology I, University School of Medicine
A-1090 Vienna, Austria

Introduction

Hepatic encephalopathy (HE) is characterized by an increased availability of peripheral aromatic amino acids (AAA) to the brain due to decreased metabolism of AAA and loss of branched chain amino acid (BCAA) supply. In particular, L-valine (VAL) therapy normalizes the increased brain tryptophan, serotonin (5-HT) and 5-hydroxyindole acetic acid concentrations and counteracts pathophysiological disturbances with awakening of the patients. The corresponding changes in 5-HT receptor activity also may be influenced by administration of VAL. Using radioreceptor binding techniques with [^{3}H]-5-HT it was shown that 5-HT-1 receptors lose binding density with an increase in affinity while treatment with VAL normalizes both 5-HT-1 receptor binding sites and affinity. Various animal models confirmed that neurotoxic ammonia might cause, at least in part, these changes in membrane activity.
The direct effect of VAL on postsynaptic 5-HT-receptor-units was characterized by setting up a specific assay for VAL-binding to human post mortem neuronal membrane homogenates.

Progress in Tryptophan and Serotonin Research

Method and Results

The in vitro [^{3}H]-5-HT and [^{14}C]-L-VAL binding to brain homogenates were performed as previously described (1,2). Two models of HE have been studied: First, to assess the dependence of [^{3}H]-5-HT density on arterial ammonia concentration, acute hepatic ischemia was induced in portocaval shunted young pigs (Deutsches Landschwein, 18,26 kg) according the model of ABOUNA (3): NH_3 concentration increased from 95,6±45,6 (preoperative) to 1432-748 µg/100 ml (n=7),(8 hours, postoperative). The results confirm that reduction of serotoninergic binding capacity is to a large degree correlated to the slow developing coma (Fig.1). In the second model the effect

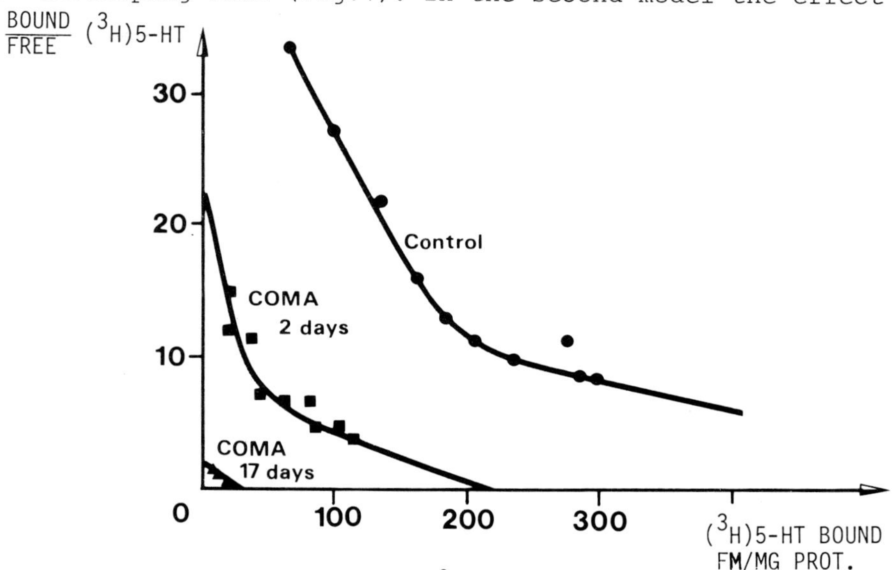

Fig. 1: Scatchard analysis of ^{3}H-5-HT-binding to homogenates of pig cortex;

● control; ■ portocaval shunted pig, duration of coma two days; ▲ portocaval shunted pig, duration of coma 17 days;

method according to (6) using cold 5-HT as unlabeled ligand to define specific binding of hyperammonemia was studied in a hybrid strain of New Zealand white rabbits (4). Intravenous injection of the selective hepatotoxin galactosamine (4,25 mM/kg body weight) induced liver necrosis. The animals developed HE about 24^h after injection and remained in coma for 2 to 6^h before death. The predominant loss of [^{3}H]-5-HT high affinity binding sites indicates that liver cell necrosis modify in a different way as bypass-circulations (Tab. 1).

Tab. 1 (^{3}H)5-HT BINDING TO HOMOGENATES OF RABBIT BRAIN

	B_{max} [fm/mg prot] total spec.bind.	K_D [nM]	B_{max} [fm/mg prot] estimated high affinity spec. binding	K_D [nM]
CONTROLS (n=2)	289 ± 104*	50,60 ± 4,2*	54 ± 2*	2,79 ± 0,92*
UREMIC COMA (Urethra lig) (n=3)	88 ± 59*	24,02 ± 1,5*	13 ± 4*	1,34 ± 0,36*
HEPATIC COMA (Galactosamine)	210	41,92	25	2,0
	360	35,45	0	0
	265	44,12	0	0
	216	41,13	0	0

*) $\bar{x} \pm$ SD

Also 5-HT-1 receptor binding studies performed in brain tissue obtained after autopsy from patients with hepatic coma have shown to react most sensitive to in vitro incubation with L-VAL as compared to control preparation (Fig.2). Measurement of VAL-binding to human brain tissue from controls and HE revealed a marked increase of binding sites with no change in affinity in untreated hepatic coma, while in patients with hepatic coma treated with parenteral nutrition plus VAL no such changes were seen (Tab. 2).

The general VAL-binding characteristics were described by Riederer et al. (5).

Moreover, trazodone, an antidepressant drug, mainly 5-HT antagonist and weak inhibitor of the in vitro uptake of [^{3}H]-5-HT exhibits also a VAL-binding inhibitory activity

Tab. 2 MEASUREMENT OF SEROTONIN RECEPTOR SITES IN HEPATIC COMA (FR. CORTEX)

	N	K_D HIGH AFF. (nM)	K_D LOW AFF. (nM)	B_{MAX} (F MOLES/MG PROT) TOTAL SPEC. BIND.
CONTROLS	8	4,0 ± 1,1	45,0 ± 18,7	1148 ± 271
HEPATIC COMA + L-VALINE	3	2,2 ± 0,4	45,4 ± 13,7	1104 ± 142
HEPATIC COMA	3	2,5 ± 0,8	19,9 ± 13,5	274 ± 196

$\bar{X} \pm$ SD FROM: AROLD N., WESEMANN W. (1982) PERS.COMM.

^{14}C-L-VALINE BINDING TO HUMAN POST MORTEM FRONTAL CORTEX

	N	COMA STAGE	B_{MAX} (PMOL/MG PROT)	B_{MAX} (% CONTROL)	K_D (µM)
CONTROLS	8	-	229 ± 12	100	7,6 ± 0,75
LIVER CIRRHOSIS WITHOUT COMA	2	I/I	534/518	233/226	12,0/6,4
HEPATIC COMA	5	IV	396 ± 49,8*	173	9,4 ± 5,9
HEPATIC COMA (WERNICKE S.)	1	IV	452	197	7,5
HEPATIC COMA + VAL	2	IV→I	220/47	96/21	4,9/7,8

*) $p < 0,001$ compared to controls N = number of brains means ± sem

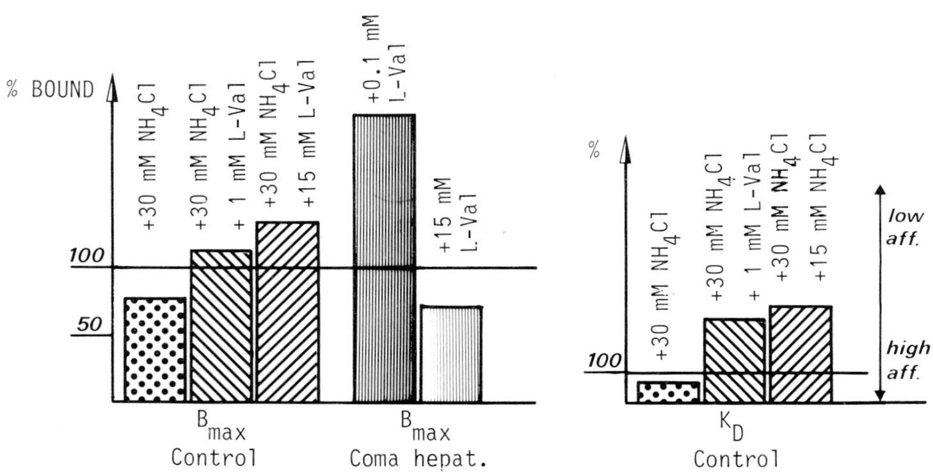

Fig. 2: In-vitro effect of ammonia and L-valine on [^{3}H]-5-HT-binding in human cortex.
In all experiments the concentration of [^{3}H]-5-HT was 10 nM. The endogenous NH_4^+-concentration in the frontal cortex of post mortem coma hepaticum ranged from 4,3 - 5,2 µmol/g tissue.
Baseline values (controls) are given as 100 % line.

(IC_{50}: 3,2.10^{-4}M). Also the distribution of VAL-binding sites in various human brain areas reveals similiarity to 5-HT receptor density (5,6).

Conclusion

Modulation of 5-HT-1 receptors by VAL seems to be of importance as a regulatory principle which is disturbed in hepatic coma. The antagonistic action of VAL on NH_3 induced changes in the synaptic membrane fluidity and function might be explained by its inhibiting action on the passive diffusion of ammonia ions along the electrochemical gradient although the nature of this binding type remains to be further studied. Also binding to an ion dependent specific transport carrier system that will contribute to the modulation of receptor kinetics is suggested (7).

References

1. Kienzl,E.,Riederer,P. Jellinger,K., Wesemann W.: J.Neural Transm. 51, 113-122 (1981).
2. Kienzl,E., Riederer,P., Jellinger,K., Kleinberger, G.: J.Neural Transm. 55, 309-315 (1982).
3. Rauhs R.: Acta Chir. Austr., Suppl. 39 (1981).
4. Ferenci, P., Schafer, D.F., Pappas, S.C., Jones, E.A.: New Aspects of Clinical Nutrition, 485-504,Karger Basel,1983
5. Riederer, P., Kienzl, E., Jellinger, K., Noller, H., Kleinberger, G.: J.Neural Transm., Suppl. 18 (1983), 295-306.
6. Bennett,J.P., Snyder, S.: Mol.Pharm.12, 373-389 (1976).
7. Bruinvels,J., Moleman P.: Essays in Neurochemistry and Neuropharmacology, Vol.5, J.Wiley and Sons,LTD, 1981.

EFFECT OF A B-VITAMIN ON TRYPTOPHAN METABOLISM IN SOUTH AFRICAN BANTU WITH PELLAGRA

L.V. Hankes, C.R. Jansen,* E.P. DeBruin,* and M. Schmaeler

Medical Research Center, Brookhaven National Laboratory, Upton, NY 11973, and *Life Sciences, Atomic Energy Board, Pretoria, South Africa

Introduction

Casal first described pellagra in 1735 and related it to a corn diet. The response of the disease to niacin (1) and the relationships between tryptophan and niacin were uncovered by the studies of Elvehjem and his co-workers (2). Many studies relating the amino acid and vitamin composition of the diet to the etiology of the disease have been published (2,3). In recent studies a 2-g load of L-tryptophan or a 400-mg load of L-kynurenine was used, and analyses for a wide spectrum of urinary metabolites showed that a number of clinical conditions and diseases such as pellagra produce a spectrum of abnormal tryptophan metabolites (4,5). The dependence of the metabolism of kynurenine, a metabolite in the tryptophan-niacin pathway, on vitamin B_6, which had been shown in previous studies in humans (5,6), made it interesting to study in detail the metabolism of kynurenine in pellagra patients before and after vitamin B_6 therapy.

Progress in Tryptophan and Serotonin Research

Patients

The 12 patients selected for this study had nutritionally based pellagra, which had not been treated. They came from an area within a 250-mile radius of Pretoria, South Africa. The patients were adult Africans with what was generally classed as clear cases of pellagra, characterized by well demarcated Casal's collar and lesions on the arms and legs. A small group of physicians well versed in the disease concurred in the diagnosis of each case. The patients were fed their customary maize diet with plain tea for the short duration of the study.

Compounds

The DL-kynurenine-keto-^{14}C used in the study was synthesized by the method of Hankes (7). DL-Tryptophan-7a-^{14}C was converted into N^{α}-acetyl-DL-tryptophan-7a-^{14}C, which was ozonolyzed into DL-kynurenine-ring-2-^{14}C (8). The DL mixtures were resolved into the D and L isomers on paper powder columns (9). The purity of the compounds was checked by paper chromatography. The specific activities were checked by combustion to $^{14}CO_2$ and counting in Ballentine gas tubes (10). The patients were given per os either a ^{14}C-labeled tracer dose (20 to 25 µCi) of L-kynurenine, a ^{14}C-labeled dose plus a 400-mg load of L-kynurenine, or a ^{14}C-labeled dose plus a 400-mg load of L-kynurenine plus 100 mg of vitamin B_6. Samples of patient breath and urine were collected for ^{14}C measurement, and metabolites were isolated from the urines by known methods (11).

Results

After a short time delay the pellagra patients rapidly metabolized the labeled compounds into $^{14}CO_2$. With one exception, the patients given vitamin B_6 with the 400-mg load of L-kynurenine showed a higher metabolism of L-kynurenine-keto-^{14}C than the patients given the load without vitamin B_6. All three of the patients studied with L-kynurenine-ring-2-^{14}C and given vitamin B_6 with the 400-mg load of L-kynurenine showed a higher metabolism of the labeled compound than the patients given the load without vitamin B_6.

Table 1 shows that the average total levels of ^{14}C in the urine and breath were similar whether the patient was given tracer alone, tracer plus load of L-kynurenine, or tracer plus load of L-kynurenine plus vitamin B_6. With a load of L-kynurenine, the $^{14}CO_2$ level was lower and the urine ^{14}C level was correspondingly higher. The administration of vitamin B_6

Table 1
Breath and Urine Carbon-14 Levels of Patients Given ^{14}C L-Kynurenine

No. Patients	L-Kynurenine, 400-mg Load	Vitamin B_6	Percent of ^{14}C in Dose Excreted in 24h: CO_2	Urine	Total
L-Kynurenine-keto-^{14}C					
2			65.50	19.49	84.99
5	X		32.74	38.41	71.15
5	X	X	48.12	16.43	64.55
L-Kynurenine-ring-2-^{14}C					
2			43.50	17.80	61.30
3	X		30.28	39.52	69.80
3	X	X	41.53	29.96	71.49

with the load of L-kynurenine caused an increase in the $^{14}CO_2$ level with a concurrent decrease in the urine ^{14}C level.

Calculation of the labeled metabolites isolated from the urines of the patients given L-kynurenine-keto-^{14}C as percent of the administered ^{14}C dose showed that the addition of vitamin B_6 to the dose of L-kynurenine-keto-^{14}C plus 400-mg load of L-kynurenine caused a significant decrease in the levels of ^{14}C-labeled kynurenine and ^{14}C-labeled hydroxykynurenine in the urine, and a slight decrease in the levels of quinolinic acid. Recalculation of these levels with the ^{14}C level in the urine as a baseline again showed a decrease in the levels of labeled kynurenine and hydroxykynurenine when vitamin B_6 was given, but an increase in the levels of quinolinic acid, N^1-methylnicotinamide, nicotinic acid, and pyridone.

Calculation of the labeled metabolites isolated from the urines of the patients given L-kynurenine-ring-2-^{14}C as percent of the administered ^{14}C dose showed that the addition of vitamin B_6 to the dose of L-kynurenine-ring-2-^{14}C plus 400-mg load of L-kynurenine caused a decrease in the urinary levels of labeled kynurenine, hydroxykynurenine, and xanthurenic acid and an increase in the activity levels of quinolinic acid, picolinic acid, N^1-methylnicotinamide, nicotinic acid, and pyridone. Recalculation of these values with the urinary ^{14}C level as the baseline showed the same relationships among the urinary metabolites, which suggested that the results obtained with the ring-labeled L-kynurenine were more consistent.

Table 2 shows the average urinary values for quinolinic acid, picolinic acid, and nicotinic acid for all the patients with pellagra studied with ^{14}C-labeled L-kynurenines. Very little activity appeared in nicotinic acid even when only the tracer level of the ring-labeled kynurenine was given. In all the

Table 2
24-Hour Excretion of Metabolites by
Patients Given L-Kynurenine

Patients:	2	2	5	3	5	3
400-mg load:	None	None	X	X	X	X
Keto Form:	X		X		X	
Ring Form:		X		X		X
Vitamin B_6:					X	X
Expressed as Percent of Dose						
Quinolinic Acid	2.92 ±0.05	4.30 ±1.12	2.51 ±0.45	2.37 ±0.43	2.56 ±1.35	3.02 ±0.39
Picolinic Acid	0.003 ±0.000	0.73 ±1.00	0.010 ±0.009	0.48 ±0.21	0.004 ±0.002	0.73 ±0.29
Nicotinic Acid	0.01 ±0.00	0.01 ±0.00	0.009 ±0.003	0.006 ±0.003	0.016 ±0.011	0.008 ±0.003
Expressed as Percent of Urinary ^{14}C						
Quinolinic Acid	19.9 ±13.9	24.27 ±1.62	6.78 ±1.51	7.97 ±6.13	27.23 ±18.03	15.90 ±14.68
Picolinic Acid	0.02 ±0.01	5.36 ±7.74	0.028 ±0.018	1.78 ±1.77	0.076 ±0.107	4.25 ±4.46
Nicotinic Acid	0.05 ±0.04	0.07 ±0.01	0.023 ±0.007	0.020 ±0.026	0.21 ±0.22	0.047 ±0.054

groups of patients given ring-labeled L-kynurenine, ^{14}C was significantly incorporated into the urinary picolinic acid, though not in sufficient quantity to account for the increased level of activity seen in the $^{14}CO_2$ after vitamin B_6 administration.

Discussion

Studies of the effects of vitamin B_6 on tryptophan metabolism have shown a mixed response (5,13,14). Normal Caucasians and South African Bantu given vitamin B_6 showed no significant changes in quinolinic acid excretion in their urine (5,13). However, in Caucasians vitamin B_6 deficiency causes elevations in quinolinic acid excretion (15), and the administration of vitamin B_6 to pellagra patients given a 400-mg load of kynurenine causes an additional increase in the quinolinic acid levels in the urine (5). Natives of India given leucine supplements showed elevated excretion of quinolinic acid, which was corrected with vitamin B_6 supplements (14). On the other hand, South African Bantu given L-leucine (3 g 3 times a day for 4 days) showed no changes in urinary quinolinic acid excretion (6). These differences could all be due to basic differences in the customary diets consumed by the subjects.

Previous studies (5,6) which concerned the responses of normal Bantu and Bantu with pellagra to loads of L-kynurenine or of L-kynurenine plus a vitamin B_6 supplement prompted this work. In those studies the urinary metabolites measured were kynurenic acid, xanthurenic acid, o-aminohippuric acid, kynurenine, hydroxykynurenine, quinolinic acid, nicotinic acid, N^1-methylnicotinamide, and pyridone. The basal urinary levels of tryptophan metabolites were all lower in the pellagra patients than in normals. In the normals the 400-mg load of L-kynurenine increased the levels of all the components measured except N^1-methylnicotinamide, nicotinic acid, and pyridone. In the normals the administration of vitamin B_6 with the 400-mg load of L-kynurenine caused a decrease in the urinary levels of all the components except those of nicotinic acid and pyridone, which increased. In similarly treated Bantu with pellagra the urinary levels of kynurenic acid, xanthurenic acid, N^1-methylnicotinamide, quinolinic acid, and pyridone were significantly increased. The large increase in quinolinic acid coupled with significant decreases in kynurenine and hydroxykynurenine suggested that a study with ^{14}C-labeled kynurenines would provide critical information about the effects of vitamin B_6 on the steps in the pathway below the pyrrolase enzyme step.

In the studies with L-kynurenine-ring-2-^{14}C plus a 400-mg load of L-kynurenine plus vitamin B_6, the increased production of $^{14}CO_2$ and the decrease in urinary levels of xanthurenic acid, kynurenine, and hydroxykynurenine were consistent with the idea that vitamin B_6 stimulates the action of the kynureninases and transaminases in the kynurenine-niacin-glutarate pathway.

When only a tracer level of either of the ^{14}C-labeled kynurenines was given, very little activity appeared in the nicotinic acid in the urine. In all the subjects studied with ring-labeled L-kynurenine, the ^{14}C was significantly incorporated into the urinary picolinic acid, though not to a degree sufficient to account for the increased $^{14}CO_2$ activity seen after vitamin B_6 was given. The mechanism of formation for picolinic acid is similar to that for quinolinic acid. Therefore, when vitamin B_6 stimulates the carboxylases in the glutarate pathway, the major portion of the α-amino-β-carboxy-ε-muconic semialdehyde goes on into the glutarate pathway instead of forming picolinic acid. Since the ^{14}C label is in a position not available to the picolinic acid carboxylase, the increased quantity of $^{14}CO_2$ after vitamin B_6 administration must have been due to vitamin B_6 stimulating a coenzyme for the carboxylases in the glutarate pathway or acting as such a coenzyme (12).

References

1. Sydenstricker, V.P.: Am. J. Clin. Nutr. 6 409 (1958).
2. Krehl, W.A. et al.: Science 101 489 (1945).
3. Goldsmith, G.A.: Am. J. Clin. Nutr. 6 479 (1958).
4. Price, J.M. et al.: Adv. Metabol. Disord. 2 159 (1965).
5. Hankes, L.V. et al.: Am. J. Clin. Nutr. 24 730 (1971).
6. Hankes, L.V. et al: Acta Vitaminol. Enzymol. 32 176 (1978).
7. Hankes, L.V.: Biochem. Prep. 11 63 (1966).
8. Warnell, J.L. and Berg, C.P.: J. Am. Chem. Soc. 76 1708 (1954).
9. Hankes, L.V. et al.: Prep. Biochem. 4 47 (1974).
10. Van Slyke, D.D. et al.: J. Biol. Chem. 192 769 (1951).
11. Hankes, L.V. et al.: J. Invest. Dermatol. 58 85 (1972).
12. Smith, S.M. et al.: Biochem. J. 192 673 (1980).
13. Patterson, J.I. et al.: Am. J. Clin. Nutr. 33 2157 (1980).
14. Gopalan, C., and Krishnaswamy, K.: Nutr. Rev. 34 318 (1976).
15. Brown R.R. et al.: J. Nutr. 87 419 (1965).

STUDIES ON THE URINARY EXCRETION OF XANTHURENIC ACID IN DIABETICS

Masayasu Hattori

Osaka National Railways Hospital
Osaka, Japan

Yazo Kotake, Yahito Kotake

Kobe Gakuin University
Kobe, Japan

Hidetsugu Otsuka and Yukio Shibata

Aichi Medical University
Aichi-gun, Japan

Introduction

Yahito Kotake and Y. Matsumara demonstrated in experimental animals that xanthurenic acid (XA) has a diabetogenic action (1, 2). In these studies they supposed that XA may also be closely associated with occurence of diabetes in human. Possible mechanism is as follows: A diet which is high in fat and protein causes vitamin B6 to decrease and such deficiency of vitamin B6 impairs metabolism of tryptophan resulting in producing a large amount of XA (3). XA combines with zinc and insulin (4,5) and the XA-insulin complex has less physiological effect compared with native insulin (6). Thus, diabetic conditions occur. The previous studies, however, suggested another possible mechanism: XA itself directly affects proinsulin synthesis DNA in ß-cells of the pancreas and inhibits proinsulin production then inducing diabetic conditions (7). Tryptophan loading test is a common method to confirm the latent deficiency of vitamin B6 (8). However, it is considered that the ratio of XA to kynurenic acid (KA) or to creatinine

Progress in Tryptophan and Serotonin Research

in urine will alter in diabetic patients. If tryptophan is loaded, XA will be exessively excreted that will be undesirable to the patients, because XA has a diabetogenic action which possibly aggravates the symptom.
The following two studies were carried out in order to substantiate the above assumption.
Study I: In this study 10 diabetic outpatients and 10 healthy normal subjects (the National Railway's employees) were studied, and their urinary levels of XA, KA and creatinine were determined by high pressure liquid chromatography under the conditions described in Tables 1 and 2.

Table 1. Condition of HPLC of XA and KA Determination

PCH column	4.0mmøX25cm
Mobile phase	KH_2PO_4 2g/l solution
Flow rate	1.0ml/min.
Pressure	80-90kg/cm^2
Range	4X10^{-2} AU
Wave length	342 nm
Chart speed	5mm/min.

Table 2. Condition of HPLC of Creatinine Determination

PCH column	4.0mmøX25cm
Mobile phase	KH_2PO_4 0.2/l solution
Flow rate	1.0ml/min.
Pressure	80-90kg/cm^2
Range	8X10^{-2} AU
Wave length	254 nm
Chart speed	5mm/min.

Table 3 and 4 show the urinary quantity of XA and KA (μM/l) and the ratio of XA to KA. The mean value and the standard deviation (S.D.) of the ratio of XA to KA in the normal healthy subjects was 0.17 ± 0.064, but that in the diabetic patients was 0.35 ± 0.099. The difference between the two groups was significant at the level of 0.001.
Tables 5 and 6 present values of XA and creatinine in urine, and the ratio of XA to creatinine. The mean ratio and its S.D. of XA to creatinine in the normal healthy subjects was 0.55 ±

0.22 which was comparatively low, compared with the ratio in the diabetic patients of 0.99 ± 0.32. The difference of the two values was significant at the level of 0.01.

Table 3. XA and KA in Urine of Normal Subjects

NAME	XAμM/L	KAμM/L	XAμM/KAμM
Y.K.	2.1	25.1	0.08
M.M.	0.9	9.5	0.09
I.K.	2.4	14.3	0.17
K.K.	4.9	21.1	0.23
S.A.	3.6	18.5	0.19
K.T.	2.4	13.2	0.18
Y.M.	7.3	50.2	0.15
T.K.	7.3	23.8	0.31
I.F.	4.9	25.1	0.19
I.N.	7.3	58.2	0.12
MEAN	4.3	25.9	0.17

Table 4. XA and KA in Urine of Diabetic Patients

NAME	XAμM/L	KAμM/L	XAμM/KAμM
T.T.	5.7	14.8	0.39
S.M.	9.7	19.0	0.51
N.T.	5.6	21.9	0.26
K.D.	6.1	27.4	0.22
Y.N.	11.5	23.0	0.50
Y.T.	8.5	35.7	0.24
S.T.	9.7	26.4	0.37
H.T.	4.0	9.3	0.43
S.N.	2.4	8.3	0.29
G.H.	1.2	3.6	0.33
MEAN	6.4	18.9	0.35

Table 5. XA and Creatinine in Urine of Normal Subjects

NAME	XAμg/ml	CREAT. mg/ml	XAμg /CREATmg
Y.K.	0.4	1.1	0.36
M.M.	0.2	1.5	0.13
I.K.	0.5	1.5	0.33
K.K.	1.0	1.5	0.67
S.A.	0.7	0.9	0.77
K.T.	0.5	1.1	0.45
Y.M.	1.5	1.8	0.83
T.K.	1.5	2.4	0.63
I.F.	1.0	1.8	0.56
I.N.	1.5	2.0	0.75
MEAN	0.08	1.56	0.55

Table 6. XA and Creatinine in Urine of Diabetic Patients

NAME	XAμg/ml	CREATmg/ml	XAμg /CREATmg
T.T.	1.2	1.1	1.09
S.M.	2.0	2.0	1.00
N.T.	1.2	1.8	0.67
K.D.	1.3	2.4	0.54
Y.N.	2.4	2.2	1.09
Y.T.	1.8	1.3	1.38
S.I.	2.0	2.3	0.87
H.T.	0.8	0.9	0.89
S.N.	0.5	0.3	1.67
G.H.	0.3	0.4	0.75
MEAN	1.4	1.5	0.99

Study II: Admitted into the study were 18 normal healthy subjects (the National Railway's employees) who had never received vitamin B6 before, and 28 diabetic outpatients, half of whom receiving the drug therapy with vitamin B6 and the other half without vitamin B6. Blood samples were taken in the early morning after overnight fast in order to determine serum pyridoxal (PAL) and peridoxal-phosphate (PAL-P) levels. The ratio of the mean level of PAL-P to that of PAL was calculated only with patients in whom both substances were detected. PAL and

PAL-P were separated into two ion exchange columns and were then made to react with KCN, followed by fluorescence analysis (9) as shown in Figure 1.

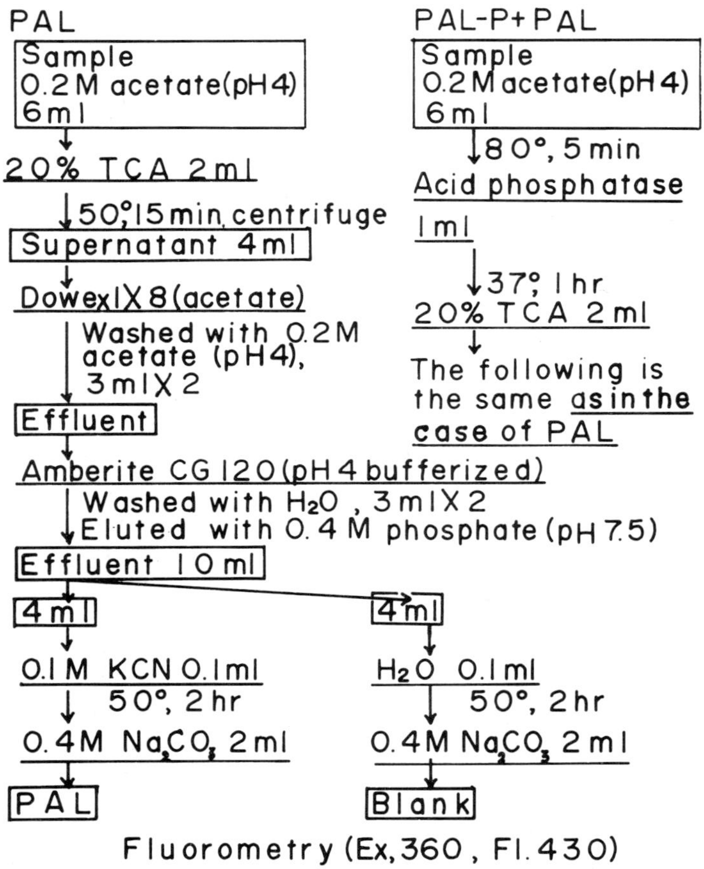

Figure 1. Microdetermination Procedure of PAL and PAL-P
(S. Takanashi, 1970)

Tables 7, 8 and 9 show the quantity of PAL and PAL-P and the ratio PAL-P/PAL.

Table 7. Concentration of Vitamin B6 in Sera of Normal Subjects

NAME	PAL-P mμg/ml	PAL mμg/ml	PAL-P/PAL
T.A.	8.5	5.8	1.45
K.H.	0.8	7.2	0.11
Y.K.	7.6	2.8	2.71
Y.I.	8.0	3.2	2.50
T.G.	9.4	1.0	9.40
T.K.	13.1	3.7	3.54
A.N.	25.1	5.0	5.02
R.Y.	1.6	2.4	0.42

Table 8. Concentration of Vitamin B6 in Sera of Diabetic Patients

NAME	PAL-P mμg/ml	PAL mμg/ml	PAL-P/PL
M.H.	59.9	86.9	0.69
K.H.	37.7	62.6	0.60
Z.T.	2.3	10.0	0.23
T.S.	10.5	10.0	1.05
M.O.	8.7	11.4	0.76
T.M.	20.1	50.1	0.40
Y.O.	10.4	23.4	0.44
T.W.	3.3	6.7	0.44
T.O.	53.3	194.0	0.27
T.S.	6.7	66.8	0.10
U.N.	30.0	60.2	0.50

Table 9. Concentration of Vitamin B6 in Sera of Diabetic Patients 2 hrs after Administration of 90 mg Pyridoxine-HCL

NAME	PAL-P mμg/ml	PAL mμg/ml	PAL-P/PAL
Y.S.	18.3	73	0.25 *
T.I.	8.0	102	0.08 *
J.M.	12.0	106	0.04 *
N.S.	12.2	52.9	0.23 *
S.N.	12.0	274	0.04
M.T.	10.2	239	0.04
K.O.	12.2	256	0.05
F.M.	22.1	234	0.09
M.T.	16.2	309	0.06
M.U.	34.0	205	0.16

*administration of 30 mg pyridoxine hydrochloride

PAL was detected in all of the 14 diabetic patients who had received vitamin B6 and the mean level was 44.45 ± 54.68 mμg/ml, while PAL-P was seen in 11 of them at the mean level of 22.08 ± 20.36 mμg/ml. Eleven of the patients had detectable levels of both substances and the ratio was 0.50 ± 0.27 which was statistically significant.

In the remaining 14 patients without vitamin B6, PAL was detected in seven of them and PAL-P in four. However, detection of both substances was possible only in two of the patients at the ratio of 0.20. No statistical analysis was therefore made. This patient group was further treated with 30 or 90 mg vitamin B6 (pyridoxine hydrochloride), blood samples were taken after two hours. Both PAL and PAL-P were detected in 10 patients. The mean levels of PAL and PAL-P of the 4 patients after administration of 30 mg vitamin B6 were 83.5 ± 25.1 and 12.6 ± 4.2 mμg/ml, respectively. The corresponding figures in the remaining 6 patients after the administration of 90 mg vitamin B6 were 252.8 ± 35.9 and 17.8 ± 9.0 mμg/ml, respectively. The ratios were 0.15 ± 0.11 after 30 mg and 0.07 ± 0.05 after 90 mg vitamin B6, which were both lower than that of the normal healthy subjects.

References

1. Kotake, Yahito: Clin. Chem. 3, 432-446 (1957).
2. Matsumura, Y.: Wakayama Igaku 4, 47-50 (1953).
3. Kotake, Y., Hattori, M.: Jap. J. of Hygiene 34, 474-483 (1979).
4. Murakami, E., Kotake, Y.: J. Biochem. 72, 251-259 (1972).
5. Ueda, T., Goda, K., Mori, T., Murakami, E., Kotake, Y.: J. Biochem. 82, 67-72 (1977).
6. Kotake, Y., Sotokawa, T., Murakami, E., Hisatake, A., Abe, M., Ikeda, Y.: J. Biochem. 63, 578-581 (1968).
7. Noto, Y., Okamoto, H.: Acta Diabet. Lat. 15, 273-282 (1978).
8. Coursin, P.B.: Am. J. Clin. Nutrition 14, 56-61 (1964).
9. Takanashi, S., Matsunaga, I., Tamura, Z.: J. Vitaminol. 16, 132-136 (1970).

URINARY EXCRETION OF XANTHURENIC ACID AND ZINC IN DIABETES

Sayoko Ikeda and Yahito Kotake
Faculty of Nutrition, Kobe-Gakuin University,
Kobe 673, Japan

Introduction

Xanthurenic acid (XA) is a product of tryptophan metabolism(1). The elevation of urinary XA in animals is found on vitamin B-6 deficiency, and urinary XA is known as a valid indicator of vitamin B-6 deficiency (2). Yahito Kotake and the co-workers have presented evidence that XA has a diabetogenic action against animals (3); they have shown that vitamin B-6 deficiency in animals results in experimental diabetes. A significant increase in the urinary excretion of XA has been found in diabetic patients (4) and animals (5).

In recent years, Y. Kotake *et al.* have also demonstrated that XA readily combines with insulin under slightly alkaline conditions and reduces the hormonal activity through its binding towards insulin (6,7). This probably is associated with the diabetogenic activity of XA. On the other hand, the XA-bound insulin with lowered hormonal activity has been shown to be *in vitro* restored by the addition of Zn^{2+} (8, 9); the addition of Zn^{2+} to XA-insulin complex results in the release of XA from its complex by the binding capacity of Zn^{2+} towards XA. In view of this development, the emphasis has been made on the physiological importance of Zn^{2+} on diabetes.

It is, however, uncertain whether XA, as being bound towards Zn^{2+}, is excreted into the urines of diabetic patients and animals. Elevated urinary excretion of XA-Zn^{2+} complex, if any, reveals the physiological status of diabetes. Thus, information on the urinary excretion of XA-Zn^{2+} complex, XA and

Progress in Tryptophan and Serotonin Research

Zn^{2+} on diabetes is urgently needed. Moreover, the precise method for fractionation of these above materials present in urines must be established for this purpose.

The present study was conducted to examine the urinary excretion of XA and zinc in diabetes, and to clarify the chemical properties, especially the affinity against Zn^{2+}, of XA.

Results and Discussion

1) Separation of the Zn^{2+}-Complex and Zn^{2+}-free Form of XA

Chromatography on DEAE-Sepharose CL-6B was shown to be able to resolve into the Zn^{2+}-complex of XA and its metal ion-free form (Fig. 1). In addition, this chromatographic procedure afforded the clear separation of XA from the mixture of XA together with kynurenic acid. Kynurenic acid, a tryptophan metabolite, is known to interfere with the exact determination of the content of XA present in urines (10). Therefore, this chromatographic procedure using DEAE-Sepharose CL-6B will also be a good technique for the exact determination of the content of XA present in urines.

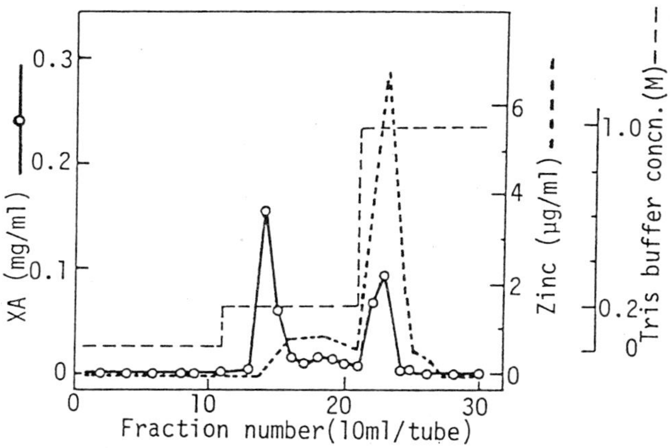

Fig. 1 Separation of the Zn^{2+}-complex and Zn^{2+}-free form of xanthurenic acid by DEAE-Sepharose CL-6B chromatography.

XA was incubated together with $ZnSO_4 \cdot 7H_2O$ in 0.02M Tris-HCl buffer (pH 7.2) for 30 min at room temperature and with a 4:1 XA-to-Zn^{2+} molar ratio. The incubate was subsequently applied on a DEAE-Sepharose CL-6B column (4.0 x 1.5ϕ cm), pre-equilibrated against 0.02M Tris-HCl buffer (pH 7.2).

2) Chemical Properties of XA

XA was found to have a considerably-high affinity against zinc ion. XA substantially did not form the complex together with Ca^{2+}, Mg^{2+}, and mono-valent ions examined. The stability constant of the XA-Zn^{2+} complex, as determined by the gel filtration procedure using Sephadex G-10, was estimated to be 5.12. The binding molar ratio of XA-to-Zn^{2+} in its complex, as determined by the ultraviolet absorption difference spectra resulting from the complex formation, was estimated to be 1:2 or 1:3 at pH 7.2. The complex formation of XA with Zn^{2+} was shown to be highly dependent upon pH (Fig. 2). As seen in this figure, the similar results were found in alloxan and oxine, both which had potent diabetogenic action.

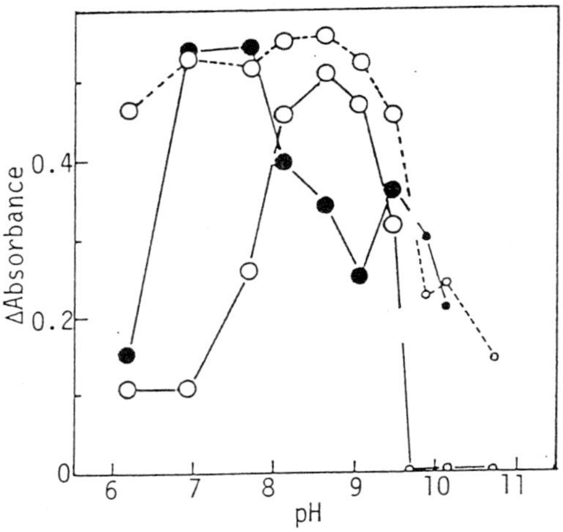

Fig. 2 pH-Dependence of the chelating activities of diabetogenic compounds towards Zn^{2+}.

The reaction was carried out in the following buffers: —○—, 0.33M Tris-acetic acid; and —o—, 0.33M NH_4OH-NH_4Cl. The reaction mixture contained 1.6 x 10^{-7} moles of the indicated compounds, 1.6 x 10^{-6} moles of Zn^{2+} and 1.2 x 10^{-3} moles of the buffer in a total volume of 3.5 ml. The control mixture was prepared by replacing the chelating compound solutions by an appropriate volume of the buffer solution. The difference absorption of the reaction mixture against its control mixture was monitored at the following wave length: XA, 257 nm; oxine, 257 nm; and alloxan 224 nm. —●—, XA; ---○---, oxine; and —○—, alloxan.

3) Urinary Excretion of XA and Zinc

Chromatography of the urines from diabetic patients and animals on DEAE-Sepharose CL-6B revealed that XA was excreted as the Zn^{2+}-complex under neutral or slightly alkaline conditions. The XA-Zn^{2+} complex fraction, obtained by this chromatographic procedure, had detectable amounts of proteinaceous substances, suggesting that XA, as being bound towards both Zn^{2+} and peptide-like substances, was excreted into the urines on diabetes.

Surprisingly enough, a significant increase in urinary zinc was observed on diabetes (Fig. 3). This phenomenon suggests that diabetes may accompany with zinc deficiency. Research is underway to clarify this phenomenon.

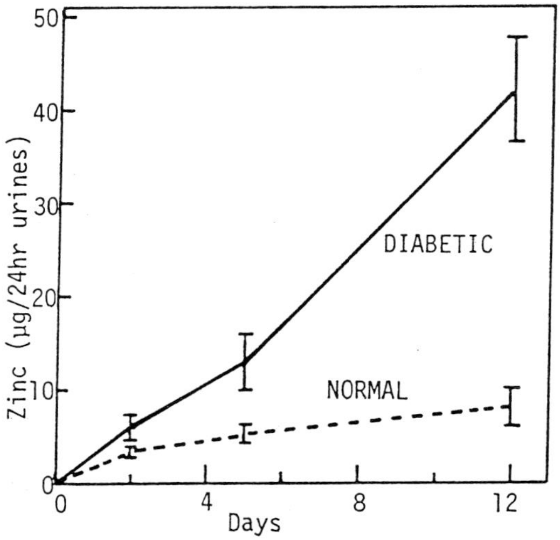

Fig. 3 Changes in the urinary zinc content on streptozotocin-induced diabetic rats.

Diabetes was induced with Wistar strain by an intraperitoneal injection of streptozotocin. The urines from the diabetic rats were collected at appropriate intervals, and assayed for zinc.

References

1. Musajo, L.: Atti Accad. Naz. Lincei 21, 368-371 (1935).
2. Lepkovsky, S., Roboz, E., Haagen-Smit, A.J.: J. Biol. Chem. 149, 195-201 (1943).
3. Kotake, Y., Inada, T.: J. Biochem. 40, 291-294 (1953).
4. Kotake, Y., Tani, S.: J. Biochem. 40, 295-298 (1953).
5. Kotake, Y., Inada, T., Matsumura, Y.: J. Biochem. 41, 255-261 (1954).
6. Kotake, Y., Sotokawa, T., Murakami, E., Hisatake, A., Abe, M., Ikeda, Y.: J. Biochem. 63, 578-581 (1968).
7. Murakami, E., Kotake, Y.: J. Biochem. 72, 251-259 (1972).
8. Kotake, Y., Ueda, T., Mori, T., Murakami, E., Hattori, M.: J. Biochem. 77, 685-687 (1975).
9. Ueda, T., Goda, K., Mori, T., Murakami, E., Kotake, Y.: J. Biochem. 82, 67-72 (1977).
10. Satoh, K., Price, J.M.: J. Biol. Chem. 230, 781-789 (1958).

STUDIES OF THE CLINICAL IMPORTANCE AND MOLECULAR DEFECT OF HEREDITARY VITAMIN B_6 DEPENDENT XANTHURENIC ACIDURIA

Alwin Knapp, Ulrich Grimm

Institut für Medizinische Genetik der Ernst-Moritz-Arndt-Universität, DDR - 2200 Greifswald

Introduction

We carried out tryptophan loading tests in many dermatological patients and in numerous diseases such as chronic urticaria (1), dermatitis seborrhoides (2), pellagra (3), asthma bronchiale, contact eczema and allergic rhinitis (4). We found increased means of kynurenine and xanthurenic acid excretion. On these conditions increased excretion is not only due to secondary factors. Apparently genetic factors play an important role, because often blood-relations show the same excretory pattern. In 1958 we found in these investigations the hereditary vitamin B_6-dependent xanthurenic aciduria (5). Clinically healthy persons were investigated in order to obtain standard values for the excretion of kynurenine and xanthurenic acid after tryptophan loading. Surprisingly a female co-worker had such a high excretion of kynurenine and xanthurenic acid as can often be observed in vitamin B_6 deficiency. An alimentary vitamin B_6-deficiency, which might have led to the increased excretion of xanthurenic acid, could be excluded. The examination of other members of this family revealed that this symptom was inherited and the tryptophan metabolism could be normalized with pyridoxine doses 10 to 20 times as high as the normal daily dose (about 2 mg). Meanwhile already 30 families with xanthurenic aciduria have been found. We investigated the aetiology and tried to assess the incidence and the pathological significance of this inborn error of tryptophan metabolism.

Progress in Tryptophan and Serotonin Research

Biochemical Findings

In xanthurenic aciduria the excretion of xanthurenic acid, kynurenic acid, kynurenine, 3-hydroxy-kynurenine, N-acetyl-kynurenine, N-acetyl-3-hydroxy-kynurenine and xanthurenic acid glucuronide is greatly increased after loading with 200 mg DL- or 100 mg L-tryptophan/kg body weight (for adults 10 g DL- or 5 g L-tryptophan). With both substances we found practically the same results (Tab. 1).
We used colorimetric and fluorometric methods (6).

Table 1. Excretion of xanthurenic acid (μmol/day) after load with 5 g L- and 10 g DL-tryptophan.

Proband	5 g L-tryptophan	10 g DL-tryptophan
1. Ch.M.	488	488
2. H.M.	609	527
3. W.Z.	151	156
4. D.M.	1585	1283
$\bar{x}$	708	609

On an average a healthy person excretes 98 μmol (20 mg) xanthurenic acid and 394 μmol (82 mg) kynurenine per day. Hoes also loaded with 5 g L-tryptophan and found in healthy probands the same results (7). According to our analyses of pedigrees levels of more than 500 μmol xanthurenic acid are indicative of the homozygous form of hereditary xanthurenic aciduria. Levels between 195 μmol and 500 μmol xanthurenic acid are typical of heterozygotes. Tada (8) and we showed that the kynureninase activity in xanthurenic aciduria is distinctly reduced (9). After administration of the coenzyme vitamin B_6 the activity of hepatic kynureninase multiplied and reached nearly normal levels. The pyridoxal-kinase activity and the plasma concentration of pyridoxal-5-phosphate were normal. Also B_6-dependent reactions in metabolism are unaffected, e.g. the erythrocyte aspartataminotransferase activity. Therefore a mutation in the apoenzyme is to be assumed especially at these sites which are responsible for the binding of the coenzyme pyridoxal-5-phosphate. The various excretions of xanthurenic acid and other metabolites after tryptophan load can be referred to the

different severity of the enzyme defect of kynureninase (enzyme polymorphism). We find for example in patients with calcium oxalate stones the mild form of xanthurenic aciduria and in patients with erythema exsudativum multiforme the severe form (10). Every patient always shows the same excretory pattern (Tab. 2.).
These results indicate that the xanthurenic aciduria has been present and will be present for the whole life.

Table 2. Excretion of xanthurenic acid and kynurenine after the first tryptophan loading (1) and 19^x or 6^xx years later (2).

Proband	xanthurenic acid µmol/day 1	2	kynurenine µmol/day 1	2
1	1698	1499^x	-	2005^x
2	1312	1105^x	1009	1825^x
3	1283	1585^xx	1686	1009^xx

Clinical Findings

The hereditary xanthurenic aciduria has no direct correlation to a specific disease. In this respect it is different from other genetic enzymopathies. It is associated with several diseases, however, the disturbance in tryptophan metabolism is not caused by these diseases but is congenital. The xanthurenic aciduria gene causes a disease only in connection with other endogenous and exogenous factors whose nature is still unknown. Accordingly xanthurenic aciduria is also found in individuals who are apparently healthy. Table 3 shows the distribution of diseases in 48 patients with xanthurenic aciduria.

The analysis of the pedigree of 30 families has shown that the heredity of xanthurenic aciduria is autosomal recessive. Because xanthurenic aciduria occurs also in healthy persons, we investigated 382 healthy individuals. Four persons (1 %) had the xanthurenic aciduria gene in homozygous and 26 persons (6,8 %) in heterozygous form (Tab. 4). However, if the families of individuals with positive response are investigated, the incidence of xanthurenic aciduria is much higher, especially in families with a high rate of inbreeding (11).

Table 3. Distribution of diseases in 48 patients with xanthurenic aciduria.

Disease	cases	%
Chronic obstipation	6	13
Calcium oxalate urolithiasis	5	10
Oligophrenia	5	10
Chronic urticaria	4	8
Psychic disorders	3	6
Hypertonia	3	6
Drug eczema	3	6
Ulcus cruris	3	6
Asthma bronchiale	2	4
Cancer	2	4
Erythema exsudativum multiforme	2	4
Healthy persons	10	21

Table 4. Xanthurenic acid excretion (μmol/day) and frequency of xanthurenic aciduria gene in homozygous and heterozygous form.

	N	xanthurenic acid	xanthurenic aciduria gene in homozygous		heterozygous form	
		$\bar{x} \pm s$	N	%	N	%
Healthy persons	382	109 ± 136	4	1.05	26	6.8
Families of individuals with positive response	72	236 ± 287	7	9.7	20	27.8
Families with high consanguinity	36	304 ± 230	7	19.4	14	38.9

Figure 1 shows the distribution of xanthurenic acid excretion under these conditions. These results confirm impressively the heredity of xanthurenic aciduria.

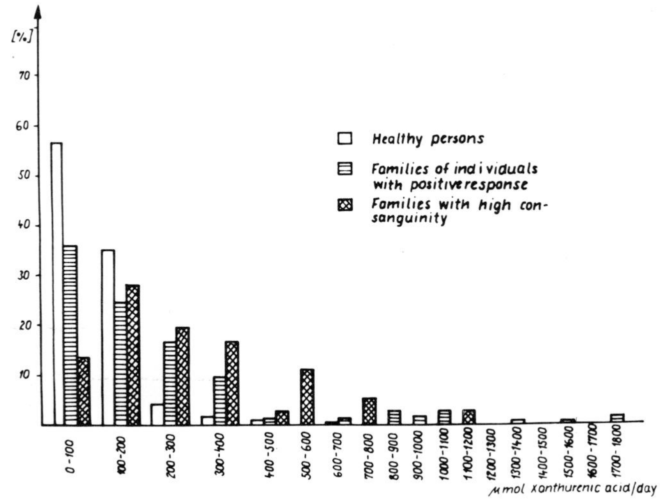

Fig. 1. Distribution of xanthurenic acid excretion in healthy persons, families of individuals with positive response and families with high consanguinity.

Treatment

If a hereditary xanthurenic aciduria has been diagnosed, vitamin B_6 therapy is instituted. Administration of pyridoxine with 20 to 40 mg per day has proved sufficient. This treatment should be continued for life time. In few cases we found a surprising improvement of the clinical symptoms (erythema exsudativum multiforme), in other cases at least an improvement of the clinical symptoms (chronic obstipation) or of the general state of health (psychic disorders, chronic urticaria). If the metabolic disturbance has existed for a longer time, the clinical symptomes can be irreversible. The most effective prophylaxis is given by a B_6 therapy immediately post partum.

Discussion

The discovery of hereditary xanthurenic aciduria in 1958 was the first example of a vitamin dependent inborn error of metabolism. Its possible association with oligophrenia is not surprising since most of the hitherto detected inborn errors of metabolism are associated with oligophrenia (12). The different clinical symptoms and the fact that xanthurenic aciduria occurs in healthy persons and is not always accompanied by oligophrenia support our view that the gene of xanthurenic aciduria causes diseases only in connection with other endogenous and exogenous factors. Of these factors the different severity of the enzyme defect (enzyme polymorphism) and different modes of nutrition in early childhood could be important. We find in xanthurenic aciduria a mild form (calcium oxalate urolithiasis and a severe form (erythema exsudativum multiforme). Other factors too, for example transient infections with fever or severe infection diseases may play a role. If the mother is homozygous, a damage of the fetus in utero is possible. The results so far obtained call for vitamin B_6 prophylaxis of the newborns in all families in which xanthurenic aciduria has been diagnosed.

References

1. Knapp, A., Wolfram, G., Posch, W.: Derm. Mschr. 156, 175-183 (1970).
2. Knapp, A., Meyer, D., Wolfram, G., Vater, M.: Derm. Mschr. 151, 168-173 (1965).
3. Schlenzka, K., Grimm, U.: Derm. Mschr. 160, 848-864 (1974).
4. Knapp, A.: Genetische Stoffwechselstörungen, VEB Gustav Fischer Verlag, Jena 1977.
5. Knapp, A.: Nahrung 2, 568-582 (1958).
6. Grimm, U., Knapp, A., Schmitz, W., Smetan, M., Schmitz, K.-W., Reddemann, H., Schulz, M., Schlenzka, K.: Fortschr. Neurol. Psychiat. 46, 178-201 (1978).
7. Hoes, M.M.A.J.: J. Clin. Chem. Clin. Biochem. 19, 259-264 (1981).
8. Tada, K., Yokoyama, Y., Nakagawa, H., Arakawa, T.: Tohoku J. Exp. Med. 95, 107-114 (1968).
9. Grimm, U.,Wulff, K.: Dt. Gesundh.-Wesen 38, (1983), in press.
10. Knapp, A., Wolfram, G., Heilmann, H.H.: Dt. Gesundh.-Wesen 22, 2449-2455 (1967).
11. Weltischtschew, J.E., Knapp, A., Baraschnew, J.I., Grimm, U., Kasanzewa, L.C., Rawinskaja, I.B.: Dt. Gesundh.-Wesen 32, 2100-2105 (1977).
12. Knapp, A.: Ernährungsforschung 22, 65-67 (1977).

TRYPTOPHAN-NICOTINIC ACID PATHWAY IN CARDIOVASCULAR DISEASES

Vera Rudzíte, Gunta Sileniece, Jánis Jirgensons
Laboratory of Biochemistry Latvian Research Institute of Cardiology, Riga USSR

Introduction

Nicotinic acid has been stated to dilate blood-vessels and to improve coronary blood flow. The ethylester of nicotinic acid - nicetan - is known to have a persistent hypotensive effect (1,2). This work aims at investigating the endogenic formation of nicotinic acid in patients with essential hypertension and ischaemic heart disease.

Material and methods

20 healthy individuals, 69 patients with ischaemic heart disease and 129 patients with essential hypertension were examined.
The endogenic formation of nicotinic acid has been investigated by quantitative analysis of nicotinic acid in the whole blood of fasting patients both before and after L-tryptophan loading with 30 mg/kg body weight. The amount of kynurenine and total (free+albumin bound) tryptophan in the blood serum has also been observed before and after the tryptophan load. Although the cyanogenic bromide method, developed by Koenig, is the most widely used colorimetric methos for estimating nicotinic acid in blood samples, we have used the colorimetric method described by Diab (3). The colour produced by the cyanogen bromide method reaches its maximum intensity within 2 min., after which it starts to fade rapidly. In the method of

Progress in Tryptophan and Serotonin Research

Diab, however, the colour remains stable for more than 4 h: Nicotinic acid is reduced to piperidine-3-carboxylic acid by Zn/HCl. The latter forms a yellow Cu-dithiocarbamate complex with CS_2 and ammoniacal $CuSO_4$. The colour is measured at 445 nm, follows Beer's law and shows good recovery (99,66±0,29%).
For quantitative determination of kynurenine in blood serum we have used the colorimetric method of Joseph and Risby (4). After alkaline cleavage of kynurenine the formed aminoacetophenon is extracted, diazotized and coupled. The intensity of the colour is measured at 560 nm. Recovery of kynurenine added to plasma is 97±5%.
The absorption of tryptophan from the gastro-intestinal tract has been controlled by the quantitative determination of total tryptophan in blood serum before and after the tryptophan load. We used the method of Spies and Chambers (5) for these investigations. This method is based on a colour-forming reaction between tryptophan, p-dimethylaminobenzaldehyde and sodium nitrite in sulfuric acid solution. The colour is measured at 590 nm and remains stable for 2h. The deterioration of the colour-producing capacity amounted no more than 5% in 5 days.
The main trend in every group is indicated by the arithmetic means of the values obtained±standard deviation (SD) of the means. By dividing the difference between arithmetical means obtained in cases of illness by those of healthy individuals a t-value is obtained. Then the statistical significance of the difference was evaluated by consulting statistical tables in which the t distribution was tabulated (6). If p is small, the difference becomes plausible. In cases of $p<0,05$ the difference is statistically significant; in cases of $p>0,05$ the difference is statistically not significant.
We have investigated the influence of kynurenine and its metabolites on isolated frog heart by the method of Straub (7) using L-kynurenine sulfate (Serva), 3-hydroxy-DL-kynurenine (Koch-Light Laboratories Ltd), kynurenic acid (Dr. Theodor Schuchart, München), xanthurenic acid (Dr. Theodor Schuchart,

München), and 3-hydroxyanthranilic acid (Serva) for these experiments.

Results and discussion

Healthy individuals

In healthy and fasting persons the mean amount of nicotinic acid in whole blood was 30,1±3,0 μmol/ml, the mean amount of kynurenine in serum - 3,4±0,3 μmol/ml, the mean amount of to-

Table 1

Hours	Total tryptophan in μmol/ml	Kynurenine in μmol/ml	Nicotinic acid in μmol/ml
0	69,6± 4,9	3,4±0,3	30,1±3,0
2	295,9±26,8 $p<0,05$	17,2±2,1 $p<0,05$	41,0±4,7 $p<0,05$
3	212,2±19,6 $p<0,05$	22,2±3,6 $p<0,05$	47,7±5,7 $p<0,05$
4	167,6±15,6 $p<0,05$	16,8±3,2 $p<0,05$	48,7±3,0 $p<0,05$
5	-	-	43,3±5,6 $p<0,05$
6	-	-	36,3±4,3 $p>0,05$
24	55,8± 2,4 $p>0,1$	4,0±0,2 $p>0,25$	35,7±5,7 $p>0,25$

The mean concentration±SD of total tryptophan and kynurenine in blood serum and nicotinic acid in whole blood in 20 healthy persons before and after an oral tryptophan load of 30mg/kg body weight.

tal tryptophan 69,6±4,9 μmol/ml. These values are confirmed by the data of literature (1, 2, 3, 4, 8). The mean values were obtained from 10 men and 10 women. Differences between sexes have not been observed.
After an oral tryptophan load we saw an increase in the amount of total tryptophan, kynurenine and nicotinic acid (see table). Corresponding data have been reported on total tryptophan and

kynurenine in plasma after an oral tryptophan load of 25mg/kg body weight (8), whereas data of nicotinic acid have not been published, although the formation of nicotinic acid from tryptophan had been demonstrated in classical experiments of Goldberger et al. (1, 9).

Ischaemic heart disease

In our investigations we have examined 17 patients suffering from myocardial infarct (1st group), 39 patients complaining of heaviness and pain in the heart (2nd group), and 13 patients with weakness of sinus node (3rd group). The electrocardiograms of these patients with feeling of heaviness and pain in the heart were either normal or had a myocardial hypoxia and impaired driving of the impulses through one of the branch of the bundle of His. The weakness of sinus node has been identified by the method of endocardial electrostimulation (10, 11). In cases of myocardial infarct the mean concentrations of nicotinic acid ($30,5 \pm 2,2$ µmol/ml) and kynurenine ($3,2 \pm 0,2$ µmol/ml) were normal under fasting conditions. A slight decrease in the concentration of total tryptophan ($56,8 \pm 4,3$ µmol/ml) was not statistically significant ($p>0,1$). After an oral tryptophan load we observed decreased values of nicotinic acid ($p<0,05$) and total tryptophan ($p<0,05$) in blood. The curve of kynurenine did not differ from that of controls ($p>0,5$) (see fig. 1).
The 2nd group with feeling of heaviness and pain in the heart indicated normal concentrations of nicotinic acid ($28,7 \pm 1,9$µmol/ml; $p>0,5$) in whole blood and normal levels of kynurenine ($4,8 \pm 0,4$ µmol/ml; $p>0,1$) in blood serum under fasting conditions. The amount of total tryptophan was decreased ($50,0 \pm 2,4$ µmol/ml; $p<0,05$). After an oral tryptophan load nicotinic acid ($p<0,05$) and total tryptophan ($p<0.05$) were reduced in contrast to kynurenine ($p<0,05$) which was excessively increased. In these cases vasodilatators as well as hypotensive and cardiac drugs were ineffective.
It is known, that the tryptophan load test with measurement of kynurenine, 3-hydroxykynurenine and xanthurenic acid provides

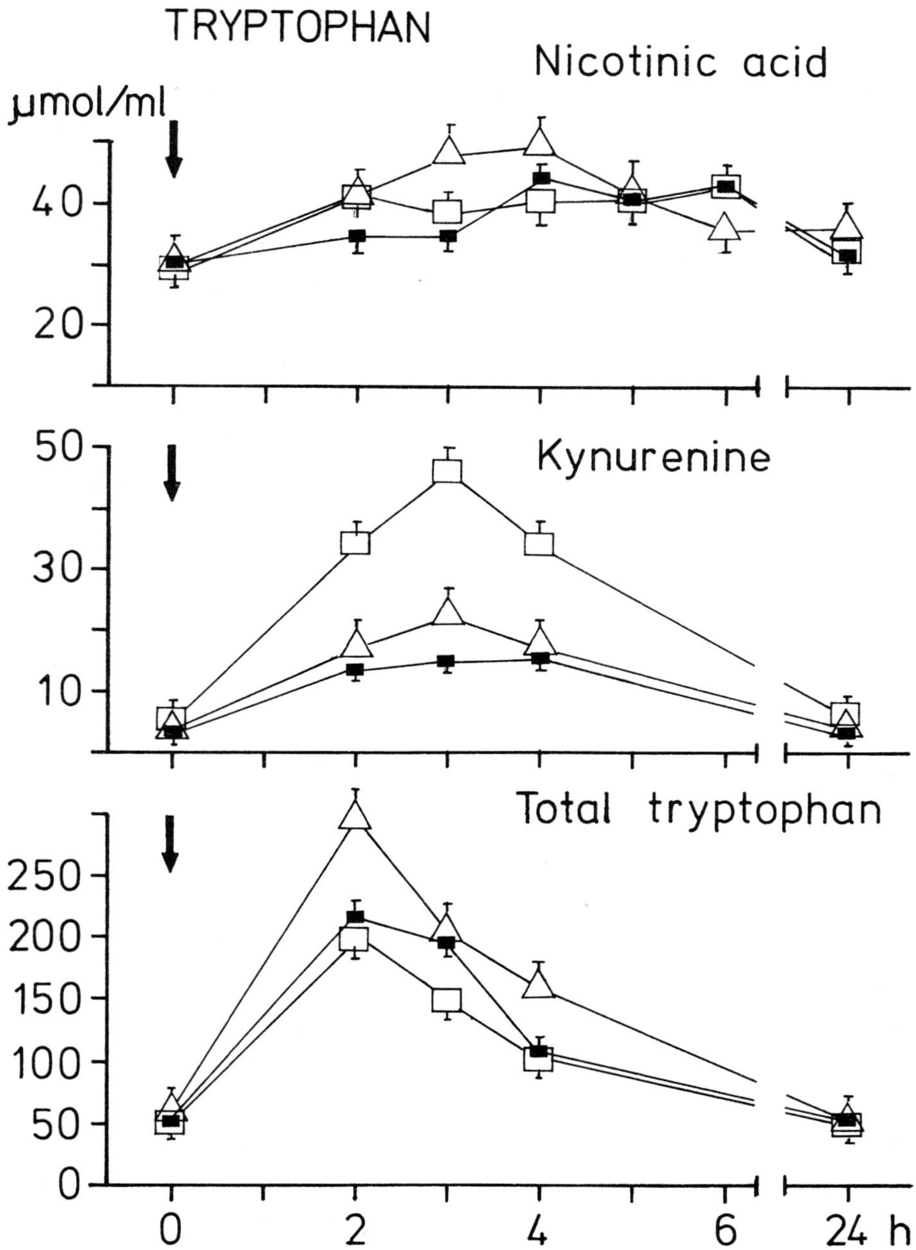

Fig. 1. Blood concentration±SD of nicotinic acid, kynurenine and total tryptophan in healthy individuals(-Δ-) and in patients with ischaemic heart disease (1st group:-■-; 2nd group: -□-) before and after an oral L-tryptophan load of 30 mg/kg body weight.

the most sensitive test for deficiency of pyridoxal-5-phosphate, that is an activated form of vitamin B_6 (12). It is also mentioned, that vitamin B_2 and B_6 as well as adenosintriphosphoric acid (ATP) take part in the activation of vitamin B_6 (13). Moreover, the rise of kynurenine level is related to a failure also of kynurenine-3-hydroxylase - nicotinamide dependent enzyme (1, 9).

We have stated the normalization of impaired tryptophan metabolism in the 2nd group of patients with ischaemic heart disease after 3 - 4 weeks of oral administration of vitamin B_6 and nicotinic acid in doses of 100 mg and vitamin B_2 in a dose of 10 mg 3 times a day and additionally treated with injections of adenosintriphosphoric acid in a dose of 10 - 20 mg once a day. An example of the normalization is given in fig. 2. Reportedly, administration of vitamin B_6 (14) and nicotinic acid (15) is therapeutically very effective in pellagra as well as in pyridoxal-5-phosphate deficiency. It has also been shown that the formation of pyridoxal-5-phosphate is dependent from the activity of the pyridoxal-kinase, an ATP dependent enzyme, and of the pyridoxine-phosphate-oxidase, a flavinmononucleotide (FMN)-dependent enzyme (see fig. 3). The FMN is derived from vitamin B_2 in the presence of ATP (16). Therefore, all components used in our treatment are necessary for the elimination of pyridoxal-5-phosphate deficiency. The normalization of tryptophan metabolism is followed by fading of the feeling of heaviness and pain in the heart.

Our clinical observations encouraged us to investigate the action of kynurenine, 3-hydroxykynurenine, kynurenic-, xanthurenic- and 3-hydroxyanthranilic acid on an isolated frog heart. The kymograms (see fig. 4) showed that L-kynurenine sulfate in the concentration of 3 μmol, 3-hydroxy- DL -kynurenine in the concentration of 3 μmol, and xanthurenic acid in the concentration of 10 μmol initiate attacks of bradycardia that were observed before and after the washing off of the heart. In some cases we could also record cardiac arrests at the beginning of diastole for one minute. 3-Hydroxyanthranilic acid produced extrasystoles in the concentration of 1 μmol (S.10).

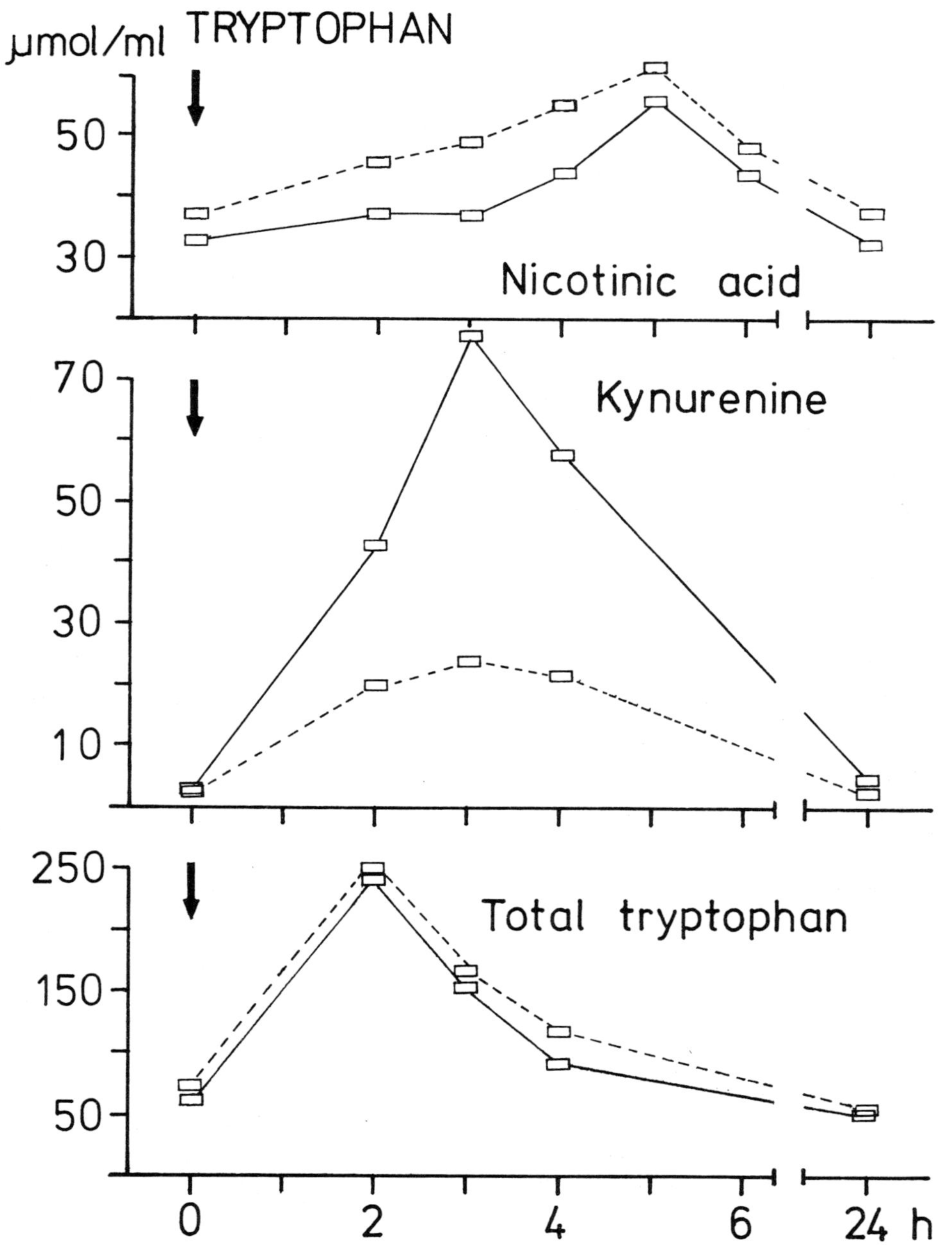

Fig. 2. Values of blood nicotinic acid, kynurenine, and total tryptophan before and after an oral tryptophan load of 30 mg/kg body weight in a patient of the 2nd group with ischaemic heart disease before (-□-) and after (--□--) administration of vitamins B_6 and B_2, nicotinic acid, and ATP.

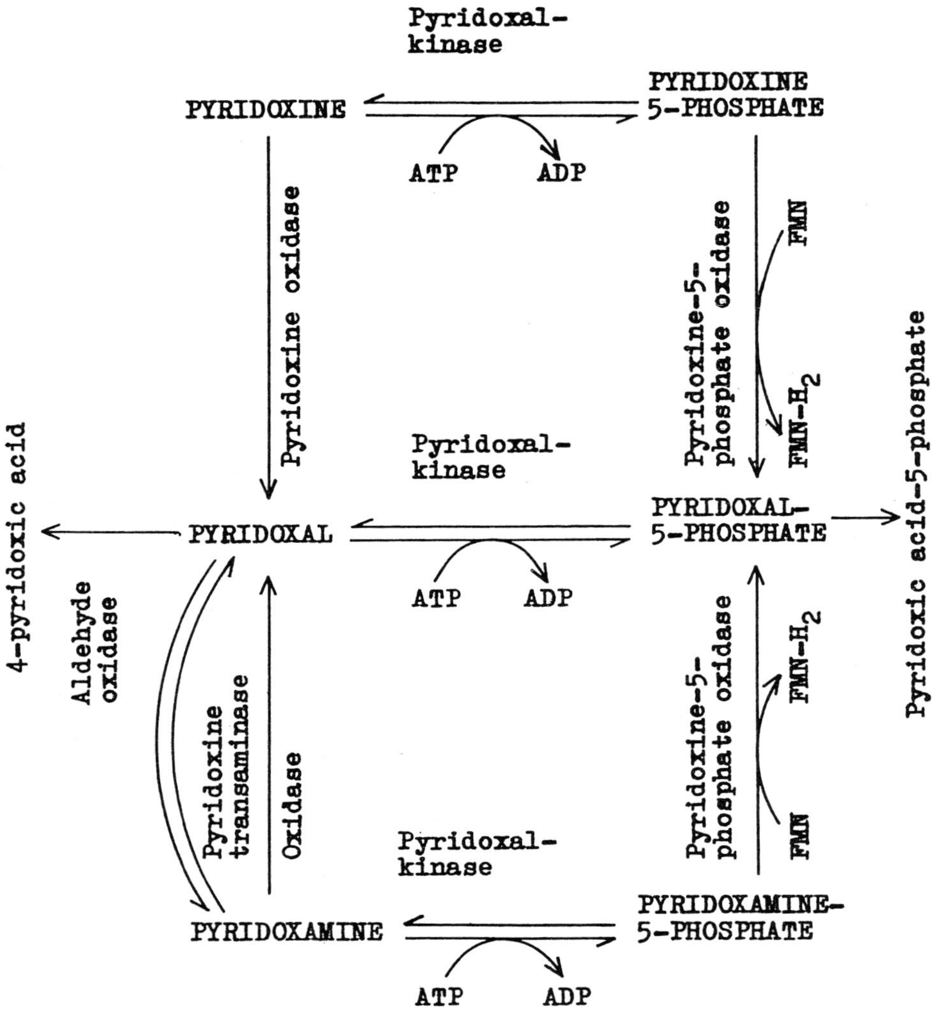

Fig. 3. The conversion of vitamin B_6 in organism

During the second examination of the patients with kynurenine accumulation in the blood serum we found out that the feeling of heaviness and pain in the heart was followed by such an impairment in the heart rhythm as observed experimentally on isolated frog heart after applying kynurenine and its metabo-

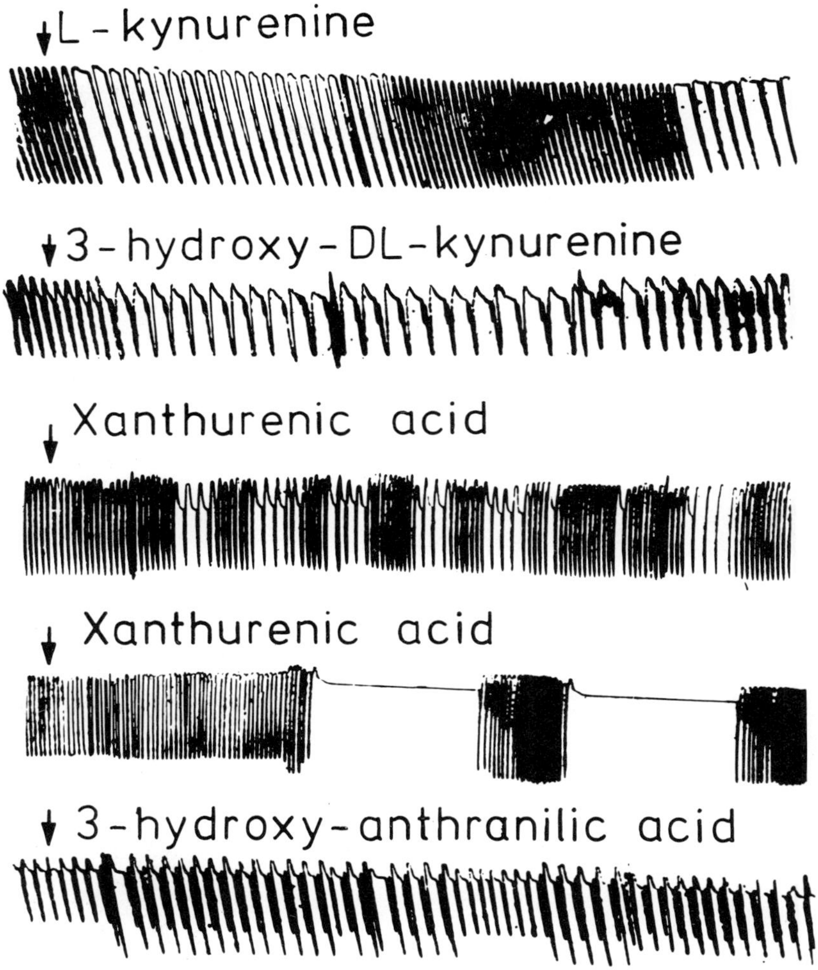

Fig. 4. Kymograms obtained in experiments on isolated frog heart. The impairment of cardiac rhythm caused by kynurenine and its metabolites are shown.

lites. So we we suggested that these impairments could not be identified in cardiograms. Accordingly a group of patients with weakness of sinus node had been investigated. In these fasting patients we observed a normal nicotinic acid concentration (31,4±2,6 μmol/ml), but increased kynurenine (5,4± 0,4 μmol/ml;p<0,05) and decreased total tryptophan levels (52,9±6,4 μmol/ml;p<0,05). After an oral tryptophan load the endogenic formation of nicotinic acid, however, was missing, whereas kynurenine (p<0,05) was increased and total tryptophan(p<0,05) decreased (see fig. 5). In these cases normalization of tryptophan metabolism was not possible by administration of vitamin B_6, nicotinic acid, vitamin B_2, and ATP.

Essential hypertension

In all observed patients with essential hypertension the systolic and diastolic blood pressure were increased. Hypertension occurring as a manifestation of a known disease - disease of kidneys and urinary tract, coarctation of the aorta, phaeochromocytoma, Cushing's syndrome, polyarteritis nodosa - was excluded by clinical observations. The plasma activity of renin and the amount of aldosterone were normal both in lying and walking states. All patients had been suffering from arterial hypertension for 5 - 10 years. The pathogenic cause of this syndrome is unknown. In many of these cases hypotensive drugs were of little effect, especially in regard to feeling of heaviness and pain in the heart associated with hypertension. We observed these patients in order to investigate the endogenic formation of nicotinic acid.

Out of 129 investigated patients we could find a decreased level of nicotinic acid (14,4±1,1 μmol/ml; p<0,05) only in 28 cases. It has been reported that decreased concentration of total serum tryptophan is an early symptom for nicotinic acid deficiency (2). In the fasting patients we observed decreased total tryptophan concentrations in blood serum (48,5±3,4 μmol/ml; p<0,05) in 12 out of 28 cases (1st group). In the other 16 cases (2nd group) the slight decrease of total tryptophan in serum (55,8±4,4 μmol/ml) was not statistically significant (p>0,1). Under fasting the amount of serum kynurenine was normal (3,98± 0,5 μmol/ml; p>0,25) both in patients of 1st and 2nd

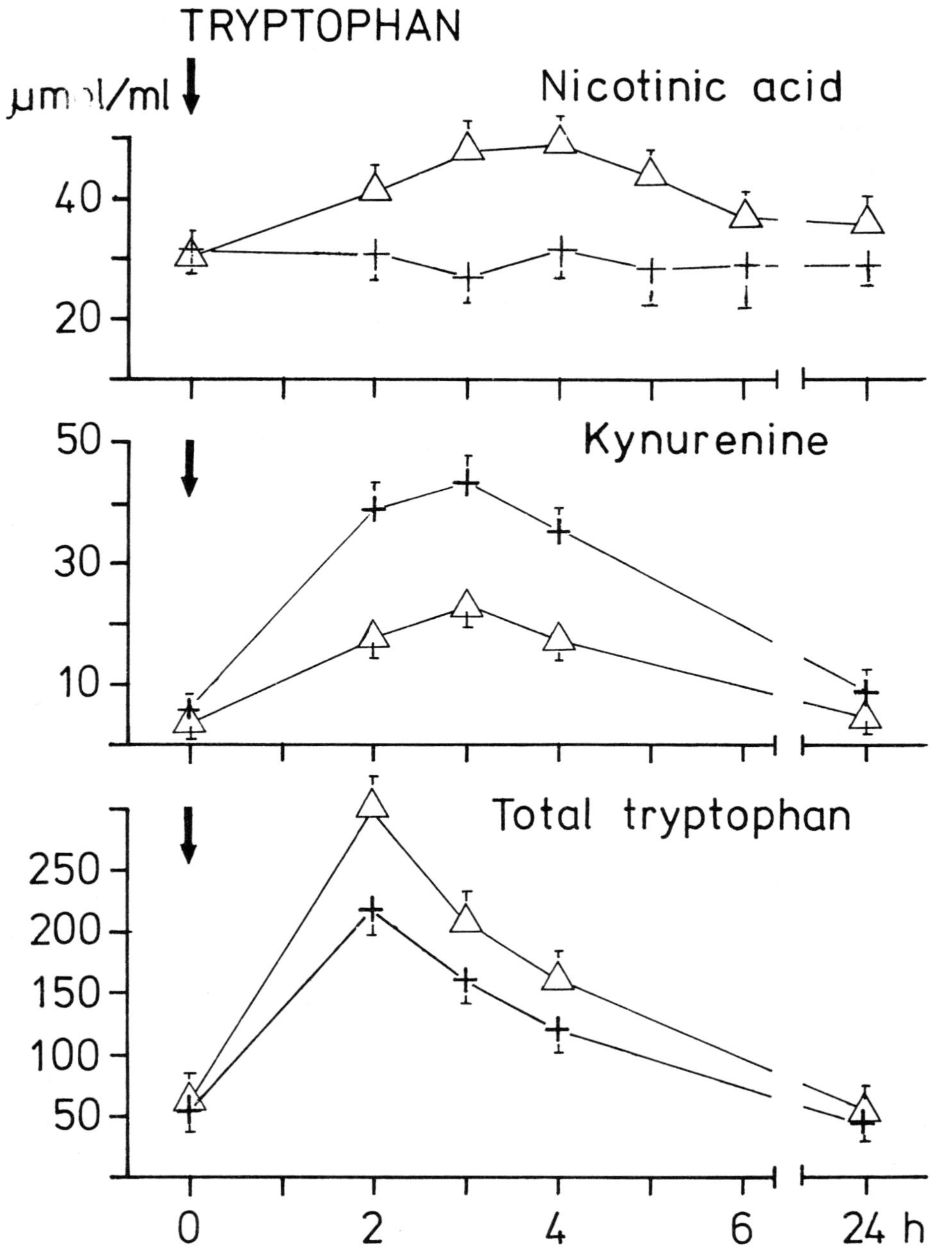

Fig. 5. Blood concentration±SD of nicotinic acid, kynurenine and total tryptophan in healthy individuals(-△-) and in patients with weakness of sinus node(-+-) before and after an oral tryptophan load of 30 mg/kg body weight.

group. After an oral tryptophan load in the 1st group of patients with essential hypertension (see fig. 6) we observed increased serum kynurenine values ($p<0,05$), whereas nicotinic acid ($p<0,05$) and total tryptophan ($p<0,05$) were decreased. In the 2nd group we also noticed a decrease of endogenic formation of nicotinic acid ($p<0,05$). Kynurenine and total tryptophan were not significantly different from controls ($p>0,5$; $p>0,25$). In the 1st group of patients with essential hypertension the impairment in tryptophan-nicotinic acid pathway could be explained by pyridoxal-5-phosphate deficiency. It is difficult to ascertain in what way the endogenic formation of nicotinic acid is impaired in patients of the 2nd group with essential hypertension and also in patients with myocardial infarct (see fig. 1). This question requires much attention, because in further 50 patients with essential hypertension (3rd group) we have observed a decreased endogenic formation of nicotinic acid, associated with a normal tryptophan absorption and a normal kynurenine increase after an oral tryptophan load. These patients (see fig. 6) differed from the patients of the 2nd group only by normal nicotinic acid values in whole blood under fasting conditions ($38,3\pm1,7$ µmol/ml; $p>0,25$). The electrocardiograms of these patients with essential hypertension (2nd and 3rd group) were either with predominance of the left ventricle of the heart or with focal lesions in the myocardium.

In the remaining 51 patients (out of 129) with essential hypertension we observed normal nicotinic acid values ($40,7 \pm3,2$ µmol/ml; $p>0,25$), increased kynurenine ($6,0\pm0,58$ µmol/ml; $p<0,05$) and decreased total tryptophan values ($44,9\pm2,1$ µmol/ml; $p<0,05$). After an oral tryptophan load the formation of nicotinic acid was decreased in 36 cases (4th group) and even missing in 15 cases (5th group). In both groups kynurenine was enhanced ($p<0,05$; $p<0,05$), whereas absorption of tryptophan ($p<0,05$; $p<0,05$) from the gastro-intestinal tract was diminished (see fig. 7). These changes in tryptophan metabolism

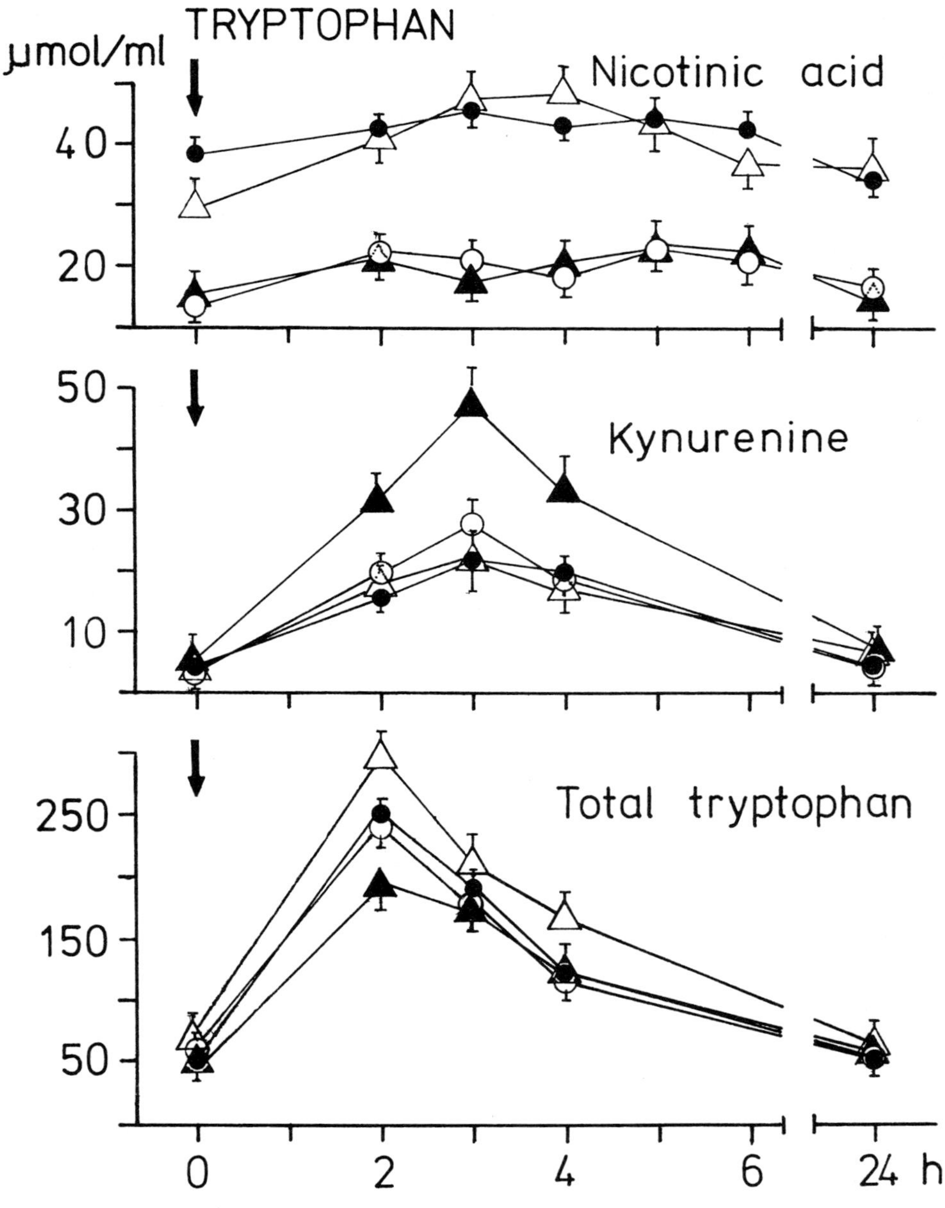

Fig. 6. Blood concentration±SD of nicotinic acid, kynurenine and total tryptophan in healthy individuals (-△-) and in patients with essential hypertension (1st group: -▲-; 2nd group: -○-; 3rd group: -●-) before and after an oral tryptophan load of 30 mg/kg body weight.

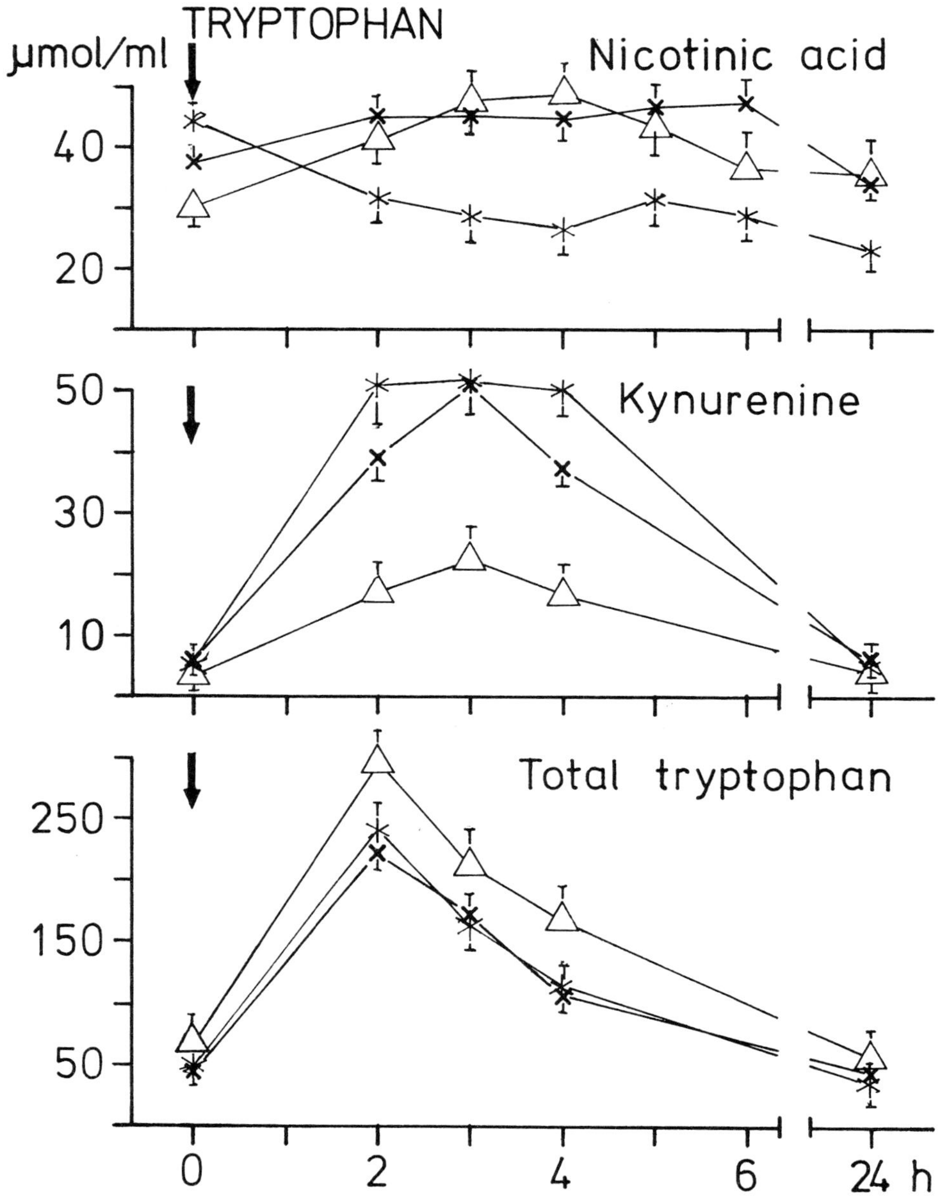

Fig. 7. Blood concentration±SD of nicotinic acid, kynurenine and total tryptophan in healthy individuals (-Δ-) and in patients with essential hypertension (4th group:-x-; 5th group: -*-) before and after an oral tryptophan load of 30 mg/kg body weight.

can be interpreted as a result of pyridoxal-5-phosphate deficiency in the organism. It has also been mentioned that deficiency of pyridoxal-5-phosphate diminishes the absorption of amino acids from the gastro-intestinal tract (17). Our results reaffirmed these data of literature: in cases with increased kynurenine after tryptophan loading the total tryptophan content is decreased in blood serum.

The results obtained by investigations of the tryptophan-nicotinic acid pathway in essential hypertension resemble the changes stated by us in ischaemic heart disease. It is known that arterial hypertension is one of the most frequent causes of ischaemic heart disease (18). Moreover, all patients of the 1st, 4th, and 5th group complained of heaviness and pain in the heart. The electrocardiograms were either with predominance of left ventricle of the heart, or with myocardial hypoxia and disturbed driving of impulses through one of the branch of the bundle of His. Many of these patients complained of attacks of bradycardia and extrasystoles, especially in those patients with lack of endogenic formation of nicotinic acid. At first clinically these complaints were interpreted as the result of hypertension. The use of hypotensive drugs, however, were ineffective in these cases.

The administration of vitamin B_6, nicotinic acid (each 100 mg), and vitamin B_2 (10 mg 3 times a day) and injections of ATP (10 - 20 mg once a day) normalized the impaired tryptophan metabolism after 3 - 4 weeks in the same way as in ischaemic heart disease (see fig. 2) and faded the feeling of heaviness and pain in the heart. The hypotensive drugs could be either diminished in doses or taken off.

It is not clear up today in what way the impaired tryptophan-nicotinic acid pathway influences the arterial blood pressure: Is there any effect of the metabolites on cardiac rhythm, or do they directly influence the blood vessels? To elucidate these questions further investigations are necessary.

Conclusion

The main topic of this work is the discovery that the impairment in tryptophan-nicotinic acid pathway caused by pyridoxal-5-phosphate deficiency in the organism forms the pathogenic basis for a special unknown form of the ischaemic heart disease which can result in the weakness of sinus node in future. These impairments can be associated either with normal arterial blood pressure or elevated systolic and diastolic blood pressure. We have also come to the conclusion that every kind of disease as well as drug therapy and wrong nutrition stimulate the development of this peculiar unknown form of ischaemic heart disease if associated with permanent impairment of the tryptophan-nicotinic acid pathway.

References

1. Рудзит, В., К.,: Триптофан, Медицина, Ленинград I973.
2. Ефремов, В., В.,: Витамин PP(никотиновая кислота). В книге Витамины, под ред. М., И., Смирнова, Медицина, Москва, 336-370 (I974).
3. Diab, A., M.: Arzneim.-Forsch. 27, 2134-2136 (1977).
4. Joseph, M., H., Risby, D.: Clin.Chim.Acta 63, 197-204 (1975).
5. Spies, J., R., Chambers, D., C.,: Anal.Chem. 20, 30-39 (1948).
6. Weber, E.: Grundriss der biologischen Statistik für Naturwissenschaftler, Landwirte und Mediziner, Jena 1957.
7. Васильев, Л., Л., Ветюков, И.А.: Большой практикум по физиологии человека и животных, Высшая школа, Москва I96I.
8. Green, A., R., Aronson, J., K., Curzon, G., Woods, H., F.: Br.J.Pharmacol. 10, 603-610 (1980).
9. Meister, A.: Biochemistry of amino acids, Acad.Press, London, part 2, 841-884 (1965).
10. Бредикис, Ю., Ю., Думечус, А., С.: Эндокардиальная электростимуляция сердца, Мокслас, Вильнюс I979.
11. Volkmann, H.,Paliege, R., Müller, S., Kühnert, H.: Z. gess.inn. Med. 37, 1-7 (1982).
12. Brown, R.R.: Biochemical and Medical Aspects of Tryptophan Metabolism, Elsvier/North-Holland Biochemical Press, Amsterdam, New York, Oxford, 227-235 (1980).
13. Горяченкова, Е., В.: Витамин B_6(пиридоксин). В книге Витамины, под ред. М. И., Смирнова, Медицина, Москва, 236-263 (I974).
14. Hankes, L.V.,Leklem, J.,E., Brown, R., R., Mekel, R., C., P., M.: Am.J.Clin.Nutr. 24, 730-739 (1971).
15. Crapaldi, G., Allegri, G., De Antoni, A., Costa, C., Mugeo, M.: Acta Vitamin.Enzymol. 29, 140-144 (1975).
16. Ленинджер, А.: Биохимия, Мир, Москва I974.
17. Кушак, Р., И.: В книге Руководство по физиологии, под ред. А., М., Уголева, том - Физиология всасывания, Наука, Ленинград, 285-345 (I977).
18. Чазов, Е., И.: Руководство по кардиологии, Медицина, Москва, том-3, I982.

THE DEPENDENCE OF STRESS INDUCED INCREASES IN FUNCTIONAL 5HT ACTIVITIY ON INCREASED TRYPTOPHAN AVAILABILITY: EVIDENCE THAT IN VIVO ELECTROCHEMISTRY MONITORS 5-HYDROXYINDOLE RELEASE

Michael H. Joseph and Guy A. Kennett*
Division of Psychiatry, MRC Clinical Research Centre,
Harrow, Middlesex, U.K

Introduction

Stress, in particular immobilisation stress, in rats is accompanied by increased metabolism of the neurotransmitter 5HT to 5HIAA (1, 2, and references in 3). Since 5HT is not depleted, its rate of synthesis must be increased; this may be accounted for by the increase in brain tryptophan following stress (2). This however raises the question of whether the increased <u>turnover</u> of 5HT is accompanied by increased <u>release</u> from serotonergic neurones. We have previously argued (4) that preventing the tryptophan increase and hence the increase in 5HT synthesis in brain in stress should result in 5HT depletion if release does increase in stress.

We have found that administration of valine, 200mg/kg, a large neutral amino acid that competes with tryptophan for brain uptake, 1 hour beforehand, prevented the increase in brain tryptophan in stress. This did not result in a depletion of 5HT however, rather the increase in 5HIAA was blocked(4). This might suggest that no increase in 5HT release occurs on immobilisation and that the increased 5HT synthesised is metabolised to 5HIAA without release occurring as proposed by Grahame-Smith (5). However, the 5HT mediated functional response to stress, elevated plasma corticosterone was also found to be significantly attenuated by valine (4), suggesting that the serotonergic responses to stress were at least partly <u>dependent upon</u> increased brain tryptophan. We have subsequently shown that the attenuation by valine of the corticosterone response to stress was substantially reversed by co-administration of tryptophan, but

* Present address: Department of Neurochemistry, Institute of Neurology, London WC1, U.K.

Progress in Tryptophan and Serotonin Research

not tyrosine with the valine (3). This demonstrates that the valine effect was mediated via tryptophan and 5HT rather than via tyrosine and the catecholamines.

We aimed to follow this up in two ways. Firstly we wished to investigate more directly whether 5HT release was increased in stress, in a brain region which may be more typical of brain tryptophan metabolism than the hypothalamus (6). For these purposes we validated the technique of *in vivo* voltammetry in the hippocampus as a measure of 5HT release. Secondly we wished to examine whether valine could inhibit 5-hydroxyindole release in response to stress, and whether this could be reversed by tryptophan, or by tyrosine.

Method

Carbon paste/epoxy resin glass microelectrodes were chronically implanted in the rat right dorsal hippocampus; rostral caudal + 3.43m; medial lateral + 2.5mm; dorsal ventral + 1.7mm (7). Ag/AgCl reference and stainless steel auxiliary electrodes were also placed in contact with the rat brain surface. Additional screws and dental acrylic form a mounting, fixing the electrodes to the skull surface, and connections are made to a pedestal socket embedded in the mounting. Linear sweep voltammetry, 0 - 1V in 10 seconds was applied automatically by a control box every 5 minutes. A current response was observed and measured giving a trace with 2 peaks. The 0.35V peak corresponds to that produced by 5HT in vitro (8). Recording is carried out on an unrestrained unanaesthetised rat. Results are expressed as percentage of average signal, from 10 to 65 min inclusive, after initiation of recordings. Repeated scans at 5 minute intervals showed a gradual decline of the 0.35V signal which is unaffected by saline.

Results (8,9)

Validation study

Drugs which release 5HT (PCA, fluoxetine) increase the signal. A drug which inhibits 5HT synthesis (PCPA) reduces the original signal and prevents the behavioural effects of PCA, and the increase of the signal above control levels. Tryptophan which specifically stimulates 5HT synthesis causes a small dose dependant increase in signal, but this increase is delayed with respect to the time course expected from biochemical studies. Monoamine oxidase inhibitors (MAOI) had no effect on the signal. This suggests that 5HIAA is also being measured and the changes after MAOI are equal and opposite. Probenecid confirms that 5HIAA can be detected. This and the tryptophan results lead to the conclusion that extraneuronal 5HT and 5HIAA are being measured (9).

Stress and valine study

Immobilisation stress causes a biphasic increase in the signal, which is unaffected by AMPT pretreatment (10, 11), suggesting that the biphasic increase represents the response of extraneuronal 5-hydroxyindoles in the hippocampus to immobilisation stress. Pretreatment with valine, 200 mg/kg i.p. 1 hr previously, significantly attenuates both parts of this biphasic increase, though stress plus valine still results in a signal significantly increased from control (non-stressed) rats. Administration of tryptophan 100 mg/kg with the valine completely reverses the valine mediated inhibition of the serotonergic stress response. However, administration of tyrosine with the valine does not reverse the effects of valine and in fact significantly potentiates the inhibition of the stress response. Prior treatment of rats with para-chlorophenylalanine (PCPA) substantially attenuates the stress induced increase in signal, while prior treatment with AMPT, an inhibitor of catecholamine synthesis has little effect. This confirms that the increased signal is due to 5-hydroxyindole release, and that it is dependent on increased 5HT synthesis.

Conclusions

The results of the _in vivo_ electrochemical experiments agree with those using corticosterone release in response to stress as an index of functional 5HT activity. Preventing the increase in brain 5HT synthesis on stress using valine attenuates both the increase in 5-hydroxyindole release in the hippocampus and the increase in CRF release in the hypothalamus. The effects of valine are not caused by reduced catecholamine synthesis as whilst tryptophan can reverse the effects of valine, tyrosine cannot, and in fact potentiates them. Preventing the increase in brain 5HT synthesis on stress using PCPA to inhibit tryptophan hydroxylase also attenuates the increased release of 5-hydroxyindoles, confirming that the latter is dependent upon increased 5HT synthesis. If this is true also in humans then factors preventing increased tryptophan availability on stress, e.g. abnormal plasma binding or increased levels of competing amino acids, may compromise 5HT-mediated stress responses. This may lead to a resultant vulnerability to stress which may be relevant to stress precipitated psychotic breakdown (3,11).

References

1. Bliss, E.L., Ailion, J., Zwanziger, J.: J. Pharm. Exp. Ther. 164, 122-134 (1968).
2. Curzon, G., Joseph, M.H., Knott, P.J.: J. Neurochem. 19, 1967-1974 (1972).
3. Joseph, M.H., Kennett, G.A.: Psychopharmacol. 79, 79-81 (1983).
4. Kennett, G.A., Joseph, M.H.: Neuropharmacol. 20, 39-43 (1981).
5. Grahame-Smith, D.G.: J. Neurochem. 18, 1053-1066 (1971).
6. Curzon, G., Marsden, C.A.: J. Neurochem. 25, 251-256 (1975).
7. Konig, J.F.R., Klippel, R.A.: The rat brain, a stereotactic atlas. Kreiger, R.E. New York (1967).
8. Joseph, M.H., Kennett, G.A.: Neuropharmacol. 20, 1361-1364 (1981).
9. Kennett, G.A., Joseph., M.H.: Brain Res. 236, 305-316 (1982).
10. Kennett, G.A., Joseph, M.H.: Neurosci. Lett. Suppl. 7; S56 (1981).
11. Joseph, M.H., Kennett, G.A.: Brain Res. (in press) (1983).

INCREASED AVAILABILITY OF TRYPTOPHAN TO THE BRAIN IN STRESS IS NOT MEDIATED VIA CHANGES IN COMPETING AMINO ACIDS

M.H. Joseph, J.A. Johnson and G.A. Kennett*

Division of Psychiatry, MRC Clinical Research Centre, Watford Road, Harrow, U.K.

* Present address: Dept. of Neurochemistry, Inst. of Neurology, London.

Introduction

The rate of synthesis of the neurotransmitter 5HT in the brain varies with the availability of its precursor, tryptophan, from the plasma. This is altered by the binding of tryptophan to plasma albumin, and by the plasma levels of large neutral amino acids (LNAAs) which compete with tryptophan for brain uptake (Fig.1). The relative importance of these factors under physiological conditions has been the subject of recent controversy. A physiological situation in which brain synthesis of 5HT and tryptophan availability are both increased is immobilisation stress in the rat. Since binding of tryptophan to plasma albumin is reduced in stress (Fig.1),

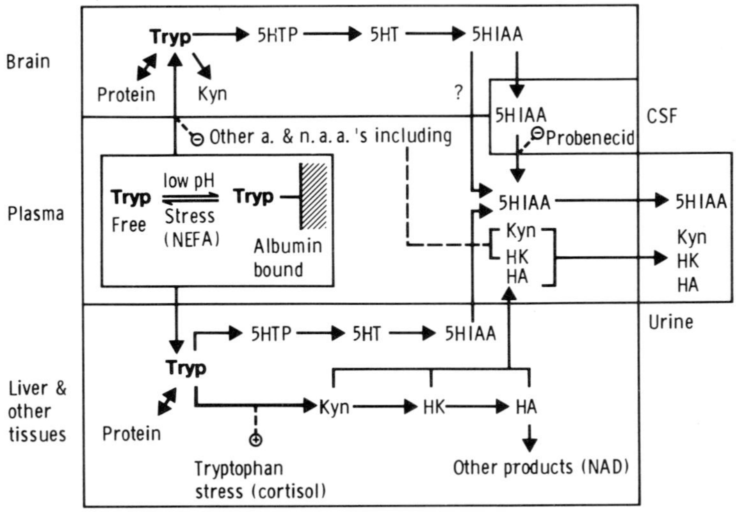

Fig.1. Tryptophan metabolism

Progress in Tryptophan and Serotonin Research

secondary to sympathetic activation, it has been argued that this is the primary cause of the increased brain tryptophan. We have demonstrated that prior administration of valine, one of the competing LNAAs will reverse the stress induced increase in brain tryptophan, and also in 5HT synthesis (1). Increased 5HT release and functional activity are also substantially attenuated (2-4). This effect of valine raises the question of whether competing amino acids are reduced in stress, and whether this accounts for some, or indeed all, of the increased tryptophan availability. Accordingly we have used HPLC to measure competing LNAAs in plasma samples from rats used in our study of valine on immobilisation (1), in which brain tryptophan levels had already been measured. The tryptophan availability ratio has been calculated for each animal, and plotted against brain tryptophan.

Method

Amino acids were determined by isocratic HPLC on a 15cm column of Hypersil ODS5 with fluorescence detection following deproteinisation of plasma with methanol. Tryp, val, phe, ile and leu were determined as their OPT derivatives, using 0.1M phosphate buffer, pH 7.0/methanol (50/50) as eluant, and detection at 365/450nm, by a modification (5) of a published method (6). A typical trace from a plasma sample is shown in Fig 2a. Tyr and tryp were determined by direct HPLC, using 0.1M phosphate buffer, pH 4.5/methanol (90/10) as eluant, and their native fluorescence (254/330nm). A typical trace from a plasma sample is shown in Fig.2b.

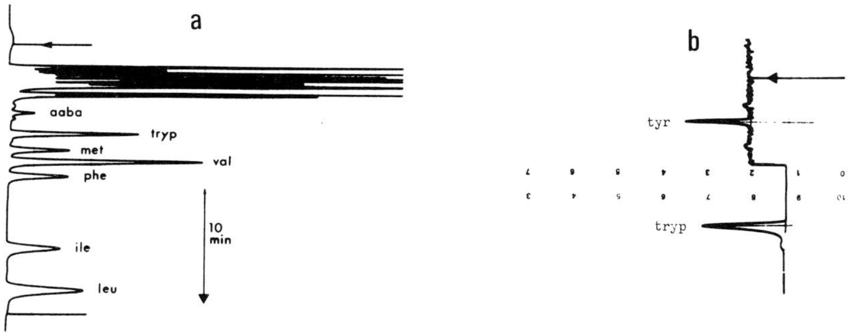

Fig.2. HPLC analysis of rat plasma amino acids
(a) OPT amino acid derivatives
(b) Native fluorescence of aromatic amino acids.

Results

Stress alone tended to reduce the levels of all the LNAAs - changes in tryp and tyr reached significance. Valine alone had little effect on competing LNAAs (other than valine) - tyr was marginally reduced, and phe significantly so. Valine potentiated the effects of stress on competing LNAAs so that all fell significantly. The increse in valine was not significantly affected (Fig.3). The tryptophan availability ratio (Tryp/ tyr + phe + val + ile + leu) was calculated for each animal, and plotted against brain tryptophan. These variables were significantly correlated, but distinct regression lines were obtained for stressed (+ valine) and unstressed (+ valine) animals (Fig.4).

Conclusion

These results indicate that tryptophan availability ratio is a good predictor of brain tryptophan within the stressed or unstressed group, but that changes in it do not account for the difference between the groups.

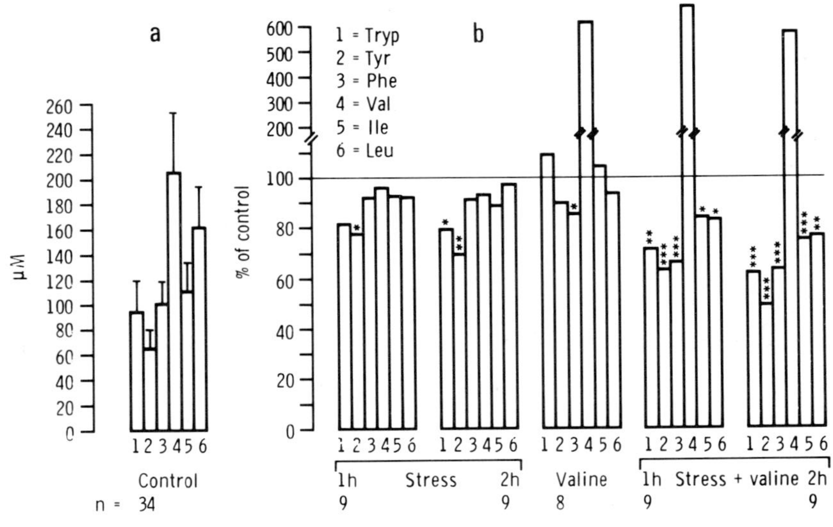

Fig.3. Efforts of stress and/or valine on plasma amino acids
(a) Control levels of amino acids
(b) Effects of stress and/or valine as percentage of control levels * $p < 0.05$ ** $p < 0.01$ *** $p < 0.001$

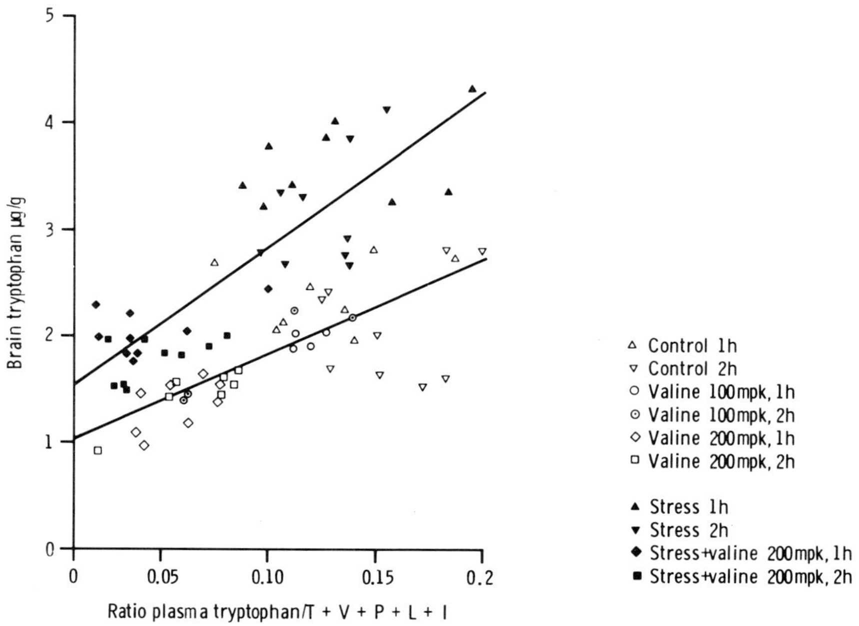

Fig.4. Plasma tryptophan availability and brain tryptophan
Open symbols : unstressed animals, r = 0.79
Closed symbols : stressed animals, r = 0.84

We conclude that reductions in competing amino acids are not responsible for the increased brain tryptophan seen in stress. The increase may be due to reduced binding of tryptophan to plasma albumin, but changes in the permeability of the blood-brain barrier, or in the kinetic parameters of the brain LNAA uptake system cannot be excluded.

Acknowledgements

We thank Philip Davies and the Medical Illustration Department for skillful technical assistance.

References

1. Kennett, G.A., Joseph, M.H.: Neuropharmacol. 20, 39-43 (1981).
2. Joseph, M.H., Kennett, G.A.: Psychopharmacol 79, 79-81 (1983).
3. Joseph, M.H., Kennett, G.A.: This volume.
4. Joseph, M.H., Kennett, G.A.: Brain Research (in press) (1983).
5. Joseph, M.H., Davies, P.: J. Chromatog. (in press) (1983).
6. Lindroth, P., Mopper, K.: Analytical Chem. 51 1667-1674 (1979).

FREE AND BOUND PLASMA TRYPTOPHAN LEVELS IN PSYCHIATRIC DISORDERS

D.P. Agarwal, B. Ziemsen, H.W. Goedde
Institute of Human Genetics, University of Hamburg, Hamburg, FRG

G. Philippu, U. Milech, O. Schrappe
Psychiatrische Klinik in der Universitäts-Nervenklinik, Würzburg, FRG

Introduction

Changes in the free tryptophan levels in plasma of patients suffering from affective disorders have been reported by various authors. However, the available data are quite contradictory (1-9) and a need is felt for more investigations to understand the role of tryptophan metabolism in the pathogenesis of affective illness. In the present communication we report on the free and albumin bound plasma tryptophan levels in patients with various psychiatric disorders and matched control subjects.

Subjects

94 male and 128 female patients with endogenous psychoses and 57 male and 60 female control subjects were studied (age range: 20 to 60 years). The control subjects were patients hospitalized at the Surgical Clinic of the University of Würzburg who showed no signs of psychiatric illness in an exploratory interview conducted a day before blood sampling by Dr. U. Brickwedde of the Psychiatric Clinic of the University of Würzburg. The psychotic patients were hospitalized at the Psychiatric Clinic of the University of Würzburg. All were being treated with

Progress in Tryptophan and Serotonin Research

psychotropic medications (neuroleptics, antidepressants, lithium, tranquilizers, or combination of these drugs). The criteria of Feighner et al. (1972) were used to assign the patients to the following diagnostic groups: schizophrenia (n=52), schizoaffective (n=41), bipolar illness (n=65), and unipolar depression (n=64). The mean duration of treatment for the schizophrenic patients was 10,62 years (range: 2 to 37 years). None of the schizophrenic patients had returned to premorbid levels of psychosocial adjustment.

In the patients with schizoaffective psychosis, the duration of illness ranged from 2 to 16 years. Twenty of these patients had a remitting illness marked by delusions, hallucinations, and affective symptoms; between psychotic episodes, which ranged from two to five in number, all had returned to premorbid levels of psychosocial adjustment. The remaining 21 patients in the schizoaffective psychosis group had a chronic illness (mean duration = 8 years), characterized by delusions and prominent affective features. The bipolar patients had at least one attack of both depression and mania. When blood samples were taken, these patients were in either a manic or depressive state. The unipolar patients had three to eight recurrent episodes.

Subjects with liver disease, alcoholism, or other severe medical diseases were excluded from the present study. Patients were also eliminated who had signs of an organic brain disease ascertained by psychological, neurological, and electroencephalographical examinations. Before blood sampling, all subjects fasted for 10 hours and were asked to rest during this period. Blood was obtained at 8 a.m. by venipuncture. Samples from each diagnostic category, including controls, were assayed together on different days without the knowledge of diagnosis.

Methods

Ten milliliter of blood was drawn into plastic syringes containing 0,1 % sodium EDTA. Platelet-rich plasma (PRP) was prepared by centrifugation at 100 g for 10 min. The PRP samples were transported on ice to Hamburg and were analysed within 24 hours. Platelets were separted by centrifugation at 1.085 g for 15 min. The platelet poor plasma (PPP) was used for the tryptophan assay. Free and albumin bound tryptophan were analysed according to the method of Dencla & Dewey (1967) as modified by Bloxam and Warren (1974). Free tryptophan was obtained by centrifuging PPP through a membrane cone (amicon) at 270 g for 90 min. Aliquots of 100 μl ultrafiltrate were used for the tryptophan assay.

Results

As shown in table 1, significant increase in the free tryptophan was observed in schizophrenics, schizoaffectives and manic-depressives when compared with controls. When various diagnostic groups were compared with each other, following differences in total and free tryptophan were noted: Schizophrenics vs unipolar depressives: $P < 0.01$-0.05; schizoaffectives vs all depressives: $P < 0.005$. Statistically significant higher total and free tryptophan values were noted for patients receiving phenothiazine derivatives and lithium.

When patients diagnosed as manic-depressives were split into either manics or depressives based on the actual clinical state at the time of analysis and the depressed patients were added to the unipolar depressives and compared as a group with manics, higher tryptophan values were noted for the manic subjects ($P < 0.01$).

When free and total tryptophan were correlated to psychiatric symptoms, slightly increased total tryptophan was found in patients with hyperactivity, restlessness, euphoria and delusions ($P < 0.007$-0.001). Increased free tryptophan was more pronounced in patients with hyperactivity, retardation, euphoria, agitations, delusions and hallucinations.

Table 1: Total and free plasma tryptophan levels in psychiatric patients grouped according to disease symptoms or medication

Group	N (male + female)	Plasma tryptophan, μg/ml (mean ± SD)			
a) Disease symptoms		Total	P	Free	P
Controls	117	12.17-3.75		0.61-0.40	
Schizophrenics	52	13.38-3.08	0.002	0.78-0.35	0.004
Schizoaffective-psychotics	41	13,27-2.99	0.005	0.98-0.83	0.00001
Unipolar depressives	65	12.18-1.78		0.65-0.29	
Manic depressives	64	13.01-3.58	0.04	0.84-0.40	0.00001
Manics	49	13.31-3.41		0.85-0.39	
Depressives	79	12.13-2.39	0.02*	0.68-0.32	0.01*
b) Medication					
Butyrophenon	35	13.13-2.87		0.73-0.36	
Phenothiazine Derivatives	44	13.31-3.35	0.005	0.84-0.45	0.001
Tri- and Tetrecyclic Antidepressants	26	13.52-3.04		0.72-0.43	
Lithium	36	13.57-3.86		0.88-0.49	
Tranquilizers	23	12.83-2.41		0.80-0.48	

* Manic vs depressive

Discussion

It is generally believed that the availability of circulating free tryptophan may primarily influence the synthesis of serotonin in the brain. However, the level of free plasma tryptophan is subject to many factors like degree of binding to albumin, liver pyrrolase activity, emotional stress, diet and sex. This may also explain the controversial reports in the literature showing lower, higher or even unchanged free tryptophan levels in endogenous and non-endogenous disorders.

In the present study, our results showing higher free tryptophan values in manic-depressives and schizoaffective psychotics confirm the earlier observations of Niskanen et al. (1976). However, in the present study, higher free tryptophan values were observed in manics than in depressed. No significant difference in free and total tryptophan was observed in depressed which support the findings of Möller et al. (1979) and Riley and Shaw (1976) and contradict the report of Coppen and Wood (1978).

In patients with hyperactivity, significantly increased total and free tryptophan levels were observed. Increased motor activity associated with stress and increased consumption of food might explain such an increase (Curzon, 1979). However, not net increase in the total tryptophan pool was noted.

References

1. Coppen, A., Eccleston, E.G., Peet, M.: Lancet ii, 60-63 (1973).
2. Aylward, M., Maddock, J.: Lancet ii, 936 (1973).
3. Peet, M., Moody, D.P., Worrall, E.P., Walker, P., Naylor, G.J.: Brit. J. Psychiat. 128, 255-258 (1976).
4. Riley, G.J., Shaw, D.M.: Lancet ii, 1249 (1976).
5. Garfinkel, P.E., Warsh, J.J., Stancer, H.C., Siboney, D.: Arch. Gen. Psychiat. 33, 1462-1466 (1976).

6. Möller, S.E., Kirk, L., Honoré, P.: J. Affect. Dis. 1, 69-76 (1979).

7. Niskanen, P., Huttunen, M., Tamminen, T., Jääskelärnen, J.: Brit. J. Psychiat. 128, 67-73 (1976).

8. Möller, S.E., Kirk, L., Honoré, P.: J. Affect. Dis. 2, 47-59 (1980).

9. DeMyer, M.K., Shea, P.K., Hendric, H.C., Yoshimura, N.N.: Arch. Gen. Psychiat. 38, 642-646 (1981).

10. Feighner, J.P., Robnis, E., Guze, S.B., Woodruff, R.A., Winkur, G., Munoz, R.: Arch. Gen. Psychiat. 26, 57-63 (1972).

11. Denckla, W.D., Dewey, H.K.: J. Lab. Clin. Med. 69, 160-169 (1967).

12. Bloxam, D.L., Warren, W.H.: Anal. Biochem. 60, 621-625 (1974).

13. Mann, H.B., Whitney, D.R.: Ann. Math. Statist. 18, 50-60 (1947

14. Cöppen, A., Wood, K.: Psycholog. Med. 8; 49-57 (1978).

15. Curzon, G.: Ann. Biol. 37, 27-33 (1979).

CEREBROSPINAL FLUID 5HIAA AND TRYPTOPHAN IN DEPRESSION AND SUICIDE

Csaba M. Banki

Regional Neuropsychiatric Institute
H-4321 Nagykallo, Hungary

Mihály Arató and Zsuzsa Papp

National Institute for Nervous and Mental Disease
H-1281 Budapest, Hungary

Mihály Kurcz

Biochemical Laboratory, Chinoin Chemical Works
H-1134 Budapest, Hungary

Introduction

There is strong evidence that central serotonin metabolism is involved in human behavioural phenomena. For example, depression has been repeatedly found to be associated with signs of a deficient central serotonin turnover in some patients (1) who may also benefit from specific chemotherapy increasing 5HT metabolism (2). Postmortem studies also suggested that brain 5HT, and its metabolite 5-hydroxyindoleacetic acid (5HIAA) might be decreased in suicide victims who were supposed to have suffered from depression (3). More recent studies indicate that 5HT deficiency may be more closely related to suicidal behaviour itself than to depression (4,5,6). Aggression and violent suicide may be facilitated by central 5HT deficit (7,8,9), a hypothesis supported by experimental and pharmacological data (10).

Although brain biochemistry in man cannot be studied directly, the measurement of 5HIAA concentration in the lumbar cerebrospinal fluid (CSF), if performed in a standardized procedure (11,12), can yield relevant information about the global brain 5HT turnover. In addition, CSF tryptophan (TRY) concentration is a measure of brain TRY availability, one factor determining 5HT synthesis rate (14).

Progress in Tryptophan and Serotonin Research

Methods

We studied only female psychiatric inpatients aged between 21 and 69 years, free from major psychotropic medication for at least 2 weeks before admission. We used DSM-III (Diagnostic and Statistical Manual of Mental Disorders, Third Edition) diagnoses: 32 patients had major depression, and 53 other subjects suffered from non-depressive disorders: schizophrenia, alcohol abuse, somatization disorder, etc. For control we used biochemical data from 32 age-matched neurological inpatients who were free from significant psychopathology (15). Among our 85 patients 36 attempted suicide immediately before admission: 23 took tranquilizer overdoses and 13 used violent methods as neck cutting, firesetting, attempted drowning. Intoxicated patients were investigated after detoxication and at least 6 days drugfree period. Following informed consent 7 ml lumbar CSF was taken at 0.8.00-09.00 a.m., under strictly controlled circumstances: bed rest, fasting, diet restrictions, etc., CSF samples were kept in dark and at -60°C until assayed. Both 5HIAA and TRY were measured by fluorometric procedures described earlier (15), with a reproduction error below 10% and a detection limit about 20-30 nmol/L for 5HIAA and 0.1-0.2 μmol/L for TRY. Results are given as mean ± standard deviation.

Results

1. There was no significant difference among depressed and non-depressed patients and controls for either CSF variable:

	DEPR (n=32)	NDEPR (n=53)	CONTR (n=32)
5HIAA	142 ± 62	151 ± 67	141 ± 54
TRY	4.8 ±2.3	4.9 ± 2.7	4.1 ± 1.4

2. There was significant difference among violent (VI), drug (D) and non-attempter (N) patients, regarding suicide:

	VI (n=13)	D (n=23)	N (n=49)
5HIAA	93 ± 23	141 ± 67	153 ± 69
TRY	5.7 ± 2.2	5.1 ± 3.2	4.1 ± 2.1

Statistical significance has been tested by analysis of variance: 5HIAA differences were significant at 2% while TRY differences at 7% level.

3. To determine the influence of both diagnosis and suicide on the CSF metabolites a 2-way analysis of variance was computed using two variables:

			VI		D		N
5HIAA,	DEPR	(n=5)	98 ± 26	(n=9)	156 ± 69	(n=18)	148 ± 67
	NDEPR	(n=8)	90 ± 22	(n=14)	131 ± 61	(n=31)	156 ± 72
TRY,	DEPR		5.8 ± 2.0		4.9 ± 2.7		4.4 ± 1.9
	NDEPR		5.6 ± 2.2		5.2 ± 3.2		4.0 ± 2.1

As it is obvious from the above data the difference was significant mainly for the suicidal groups at 1% level, while only marginally different for diagnosis at 6% level. Most of the difference was accounted for by 5HIAA, which proved to be markedly lower in violent suicidal patients; the slightly higher CSF TRY in the same group only approached statistical significance.

4. In controls there was a significant correlation between CSF 5HIAA and TRY concentration (r=0.36), which was absent in both depressed and non-depressed patients.

5. A separate discriminat function analysis showed that VI and D patients differed mainly in their CSF 5HIAA concentration, while D and N patients had more divergence in their CSF TRY.

Discussion

The above results support other findings (4,6,8) indicating a central serotonin hypofunction in suicidal patients. It is remarkable that CSF 5HIAA was found lower only in those subjects who used violent, ie. active, aggressive means in their attempt; since similar relationsship has been reported between manifest aggression and CSF 5HIAA (8) it may be conceived that violent suicide may represent one type of aggressive behaviour as strongly proposed by psychoanalists. The facilitation of aggression by serotonin antagonist or depleting drugs has been consistently proved in animals (10), and quite recently there is growing evidence that the situation may be similar in man (16,17). On the other hand, the slight elevation of

CSF TRY would be hard to interpret: TRY may be increased in some physical (13) and psychiatric (18) conditions, but the pathological significance of this small biochemical alteration remains to be elucidated.

References

1. Van Praag, H.M.: Depression and Schizophrenia, a Contribution on their Chemical Pathologies. Spectrum, New York 1977.
2. Van Praag, H.M., De Haan, A.: Acta psychiat.scand.suppl. 280, 89 (1980).
3. Gottfries, C.G.: Acta psychiat.scand.suppl. 280, 49 (1980).
4. Åsberg, M., Träskman, L., Thorén, P.: Arch. Gen. Psychiat. 33, 1193 (1976).
5. Träskman, L., Åsberg, M., Bertilsson, L.: Arch. Gen. Psychiat. 38, 631 (1981).
6. Oreland, L., Wiberg, A., Åsberg, M.: Psychiat. Res. 4, 21 (1981).
7. Ågren, H.: Psychiat. Res. 3, 225 (1980).
8. Brown, G.L., Ebert, M.H., Goyer, P.F., Jimerson, D.C.: Amer. J. Psychiat. 139, 741 (1982).
9. Åsberg, M.: Paper read at the 4th Nobel Conference, Skokloster, June 18-19, (1982).
10. Eichelman, B. In: Psychopharmacology of Aggression ed.: M. Sandler Raven, New York 1979, pp 61-93.
11. Banki, C.M., Molnar, G.: Psychiat. Res. 5, 23 (1981).
12. Bertilsson, L., Åsberg, M., Lantto, I.: Psychiat. Res. 6, 77 (1982).
13. Young, S.N., Lal, S., Sourkes, S.: J.Neurol.Neurosurg.Psychiat. 38, 322 (1975).
14. Curzon, G., Marsden, C.A.: J.Neurochem. 25, 251 (1975).
15. Banki, C.M., Vojnik, M., Molnar, G.: J.Affect Disord. 3, 81 (1981).
16. Åsberg, M.: Paper read at the 7th World Congress of Psychiatry, Vienna, July 11-16, (1983).
17. Goodwin, F.: Paper read at the same Congress.
18. Banki, C.M., Molnar, G.: Psychiat. Clin. 14, 167 (1981).

THE INFLUENCE OF L5-HYDROXYTRYPTOPHAN (L5-HTP) ON CLINICAL AND BIOCHEMICAL PARAMETERS IN DEPRESSIVE PATIENTS

M. Raba*, P. Riederer, W. Danielczyk, and D. Seemann**
* Postfach 1242, D-8011 Kirchheim / ** Ludwig-Boltzmann-Institut, Neurolog.Abteilung, Wolkersbergenstraße 1, A-1130 Wien

1. INTRODUCTION

In the pathogenesis of depression biogenic amines play an important role. Especially in elderly people a reduction of neurotransmitters up to 52 % could be shown (2, 8). Serotonin potentiating agents were shown to be effective in depressive patients initially exhibiting decreased 5-HIAA-levels in the liquor whereas patients with high 5-HIAA-levels were successfully treated with Noradrenalin potentiating drugs (6, 7). These findings correspond to Ashcroft (1), Coppen (3, 4), and Schildkraut (9) regarding the involvement of Serotonin- and Noradrenalin-metabolism in depressive disorders. Thus 30 depressive patients were treated with L5-HTP over a period of 6 weeks based on the idea to substitute a potential deficit in the Serotonin line.

2. MATERIALS AND METHODS

In this open, non-comparative trial the Hamilton Depression Scale (HAMD) was used to evaluate the clinical status at baseline, after 3, and 6 weeks with 100 mg of L5-HTP three times daily (t.i.d.). At weekly intervals the sleeping behaviour was checked using ordinal scales. Besides, blood pressure, heart rate, and body weight were monitored. At baseline and at the end of treatment lab tests were performed to study a possible influence of L5-HTP. Adverse reactions had to be documented. Parallely 6 patients received single doses of 100 mg of L5-HTP to study the kinetic profile.

3. PATIENT POPULATION

After completion data of 30 depressive patients were available. Mean age was 71,8 years (2 male/28 female). As expected elderly patients were predominant. On the average the duration of disease was 63,8 months

In a second approach three levels were generated. It could be shown that the lowest level increased from n = 8 (26,7 %) to reach n = 20 (74,1 %) after 6 weeks:

Table 2: STRATIFIED HAMD - DATA

score levels	baseline		3 weeks		6 weeks	
≤ 60	8	26,7 %	17	58,6 %	20	74,1 %
61 - 120	15	50,0 %	10	34,5 %	6	22,2 %
> 120	7	23,3 %	2	6,9 %	1	3,7 %
total	30		29		27	
no data	-		1		3	

In a third approach, in accordance with CIPS, matched subscales - depressive inhibition/depressive agitation/anxiety/somatogenic disturbances - all showed highly significant ($p < 0.01$) success rates.

Summarizing the results on quality of sleep there was an increase in very good/good from initially n = 4 (13,3 %) to finally n = 18 (66,7 %).

In a parallel experiment the influence of single doses of 100 mg of L5-HTP was studied. L5-HTP levels increased from baseline (0,395 µg/ml) to maximum values after 2 hours (1,59 µg/ml) to reach nearly baseline after 4 hours (0,41 µg/ml).

5.2 Safety

In 93,3 % of the cases the investigators graded the tolerability to be very good/good, the patients reference values was 90,0 %.

Monitoring of blood pressure, heart rate, and body weight did not show any clinically relevant changes.

In 4 patients (13,3 %) the treatment had to be stopped prematurely due to adverse events. In 8 patients (28,7 %) side effects were registered, most of them being gastrointestinal and transient.

A relevant change in lab tests could not be seen during the course of the (min. 5/max. 408 months). 93 % of the patients had been pre-treated, and in 96 % this pre-treatment was not sufficient. 93 % of the patients exhibited concomitant diseases all receiving suitable concomitant medications.

4. DOSAGE OF L5-HTP

The dosage was stepwise increased over 1 week (from 100 mg to 300 mg daily). As from the 7th day the dosage was kept constant (100 mg t.i.d.).

5. RESULTS

5.1 Efficacy

Data processing regarding HAMD was carried out in accordance with the Collegium Internationale Psychiatriae Scalarum (CIPS) (5). Baseline mean was calculated to be 90,9 and decreased during the course of treatment to 49,7 ($p < 0.01$). Item success rates are listed below:

Table 1: SUCCESS RATES OF INDIVIDUAL ITEMS OF HAMD

rank	item	success rate
01	hypnagogic disturbances	76,9 %
01	hypnotic disturbances	76,9 %
03	degree of depression	74,1 %
04	suicidality	72,7 %
05	hypochondria	72,0 %
06	psychic anxiety	70,4 %
07	guilty conscience	66,7 %
08	inhibition	65,4 %
09	disturbances of sleep (morning)	64,0 %
10	agitation	57,1 %
11	variation of day-night rhythm (severity)	56,0 %
12	somatogenic anxiety	54,2 %
13	somatogenic symptoms (gastrointestinal)	52,9 %
14	genital symptoms *	50,0 %
15	paranoid symptoms	47,6 %
16	somatogenic symptoms (in general)	46,2 %
17	hallucination/obsession	45,5 %
18	work and activities	44,4 %
19	depersonalisation/derealisation	41,2 %
20	variation of day-night rhythm (global)	40,0 %
21	body weight (global)	36,4 %
22	body weight (quantitative)	20,0 %
23	discernment of disease *	11,1 %

* very low cell frequency

study. As expected 5-HIAA-levels showed a highly significant increase ($p < 0.01$) from 4,5 µg/ml to 79,2 µg/ml during treatment.

6. SUMMARY

In this open, non-comparative clinical trial 30 depressive patients were treated with 100 mg L5-HTP t.i.d. over a period of 6 weeks.

Global HAMD data as well as subscales showed a significant improvement ($p < 0.01$) regarding time. Safety data showed no clinically relevant changes in blood pressure, heart rate, body weight, and lab tests. Tolerability of L5-HTP was excellent. Determination of 5-HIAA-levels showed good compliance. Blood levels of L5-HTP showed a peak after 2 hours.

In conclusion, a therapy with L5-HTP obviously is a valuable tool in depression. Further controlled trials have to confirm these findings.

REFERENCES

1. Ashcroft G.W., Crawford T.B.B., Eccleston D., Sharman D.F., Mac Cougall D.J., Stanton J.B., Binns J.K.: 5-Hydroxyindole compounds in the cerebrospinal fluid of patients with psychiatric or neurological disease. Lancet II, 1049 (1966).

2. Carlsson A.: The impact of catecholamine research on medical science and practice. In: Usdin J.E. (ed), catecholamines: Basic and clinical frontiers. Pergamon Press New York, 4-19 (1979).

3. Coppen A., Shaw D.M., Mallerson A., Eccleston E., Gundy G.: Tryptamine metabolism in depression. Brit. J. Psychiat. 111, 993-998 (1965).

4. Coppen A.: Serotonin in the affective disorders. Factors in depression (Kline N.S. ed), Raven Press New York, 33-44 (1974).

5. Internationale Skalen für Psychiatrie, Beltz Test GmbH, Weinheim (1981).

6. Praag H.M. van: Neuroendocrine disorders in depressions and their significance for the monoamine hypothesis of depression. Acta Psychiat. scand. 57, 389-404 (1978).

7. Praag H.M. van: Indoleamines in depression. In: Burrows G.D.: Handbook of of studies on depression. Excerpta Medica Amsterdam/London/New York, 303-324 (1977).

8. Riederer P., Wuketich St.: Time course of nigrostriatal degeneration in Parkinsons's disease. J. Neural Transm. 38, 277-301 (1976).

9. Schildkraut J.J.: The catecholamine hypothesis of affective disorders. Amer. J. Psychiat. 122, 509-522 (1965).

PLASMA TRYPTOPHAN LEVELS AND 5-HYDROXYTRYPTAMINE SYNTHESIS IN THE BRAIN AND THE SPINAL CORD IN ARTHRITIC RATS.

Jeanne Weil-Fugazza, Françoise Godefroy, Magali Bineau-Thurotte and Jean-Marie Besson.

Unité de Recherches de Neurophysiologie Pharmacologique de l'INSERM (U 161), 2 rue d'Alésia, 75014 Paris, France.

Introduction

We have previously reported that in rats suffering from experimentally polyarthritis induced by intradermal injection of killed Mycobacterium butyricum suspended in mineral oil (Freund's adjuvant), there is a marked increase in 5-hydroxytryptamine (5-HT) metabolism in the CNS 15 days after the induction of the arthritism. This increase appeared to be related to an increase in free tryptophan (TRP) level in plasma. However we have observed that in spite of the fact that the free TRP level in plasma was still highly enhanced 21 days after the induction of polyarthritis the TRP, 5-HT and 5-hydroxyindole acetic acid (5-HIAA) had almost returned to basal values in the forebrain. In the spinal cord the TRP level had also returned to normal values but the 5-HT and the 5-HIAA levels remained markedly increased (1). These latter observations suggest that in the spinal cord of such chronic suffering animals, there is an activation of the 5-HT synthesis which may be unrelated to changes in TRP availability. Therefore, in order to test this hypothesis, we measured in the present work, by mean of an isotopic method, the rate of 5-HT synthesis, in the forebrain, brainstem, ventral and dorsal parts of the spinal cord, in rats sacrificed at different times after the induction of the polyarthritis. Determinations of plasma TRP and Tyrosine (TYR) levels and TRP levels in the CNS were also performed in these rats. Details of the methods used have been previously described (1,2).

Progress in Tryptophan and Serotonin Research

Results

As previously reported (1) polyarthritis is associated with a marked increase in plasma free TRP level and a marked decrease in total plasma TRP level and the present study indicates that these modifications are still highly significant 42 days after the induction of the polyarthritis (Fig. 1). No significant changes in TYR level occur 15 days after the inoculation, but this level is significantly higher than in normal rats at 42 days (Fig. 1). As shown in Fig. 2, 15 days after the induction of polyarthritis the endogenous levels of TRP are significantly increased in all the parts of the CNS examined and this increase is associated with an increase in 5-HT synthesis which is significant in the dorsal part of the spinal cord and the brainstem. Thereafter endogenous TRP levels tend to return to normal values but a slight increase in these levels is still observed at 42 days. 5-HT synthesis also tends to return to normal values in the forebrain and the brainstem. By contrast this synthesis remains highly increased in the dorsal cord and a significant increase is observed in the ventral cord at 28 days and 42 days.

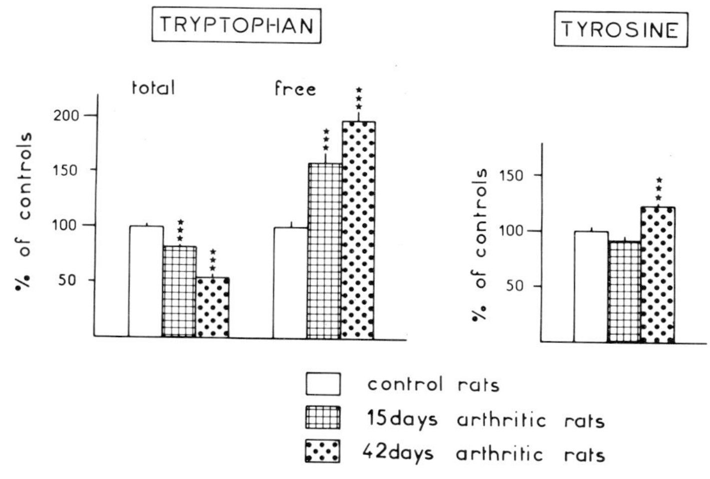

Fig. 1 : Plasma Tryptophan and Tyrosine levels in "arthritic rats" sacrificed 15 and 42 days after the induction of polyarthritis.

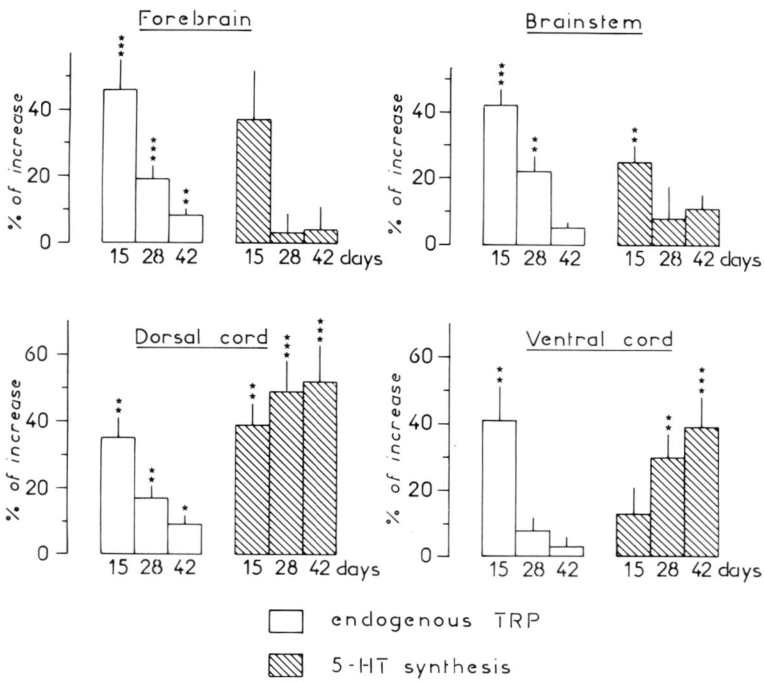

Fig. 2 : Endogenous Tryptophan levels and 5-HT synthesis in the CNS in "arthritics" rats 15, 21 and 42 days after the induction of polyarthritis.

Conclusion

This study confirmed that 15 days after the induction of polyarthritis in rat there is an increase in 5-HT synthesis in the CNS related to an increase in TRP availability. Thereafter despite the marked increase in the free TRP levels observed in the plasma, the TRP levels in the CNS tend to return to basal values. This turn is probably related to the competition of tyrosine (and perhaps other amino-acids) with the TRP transport to the brain (3). Indeed an increase in the plasma TYR level was observed at 42 days. The observation that at this time the 5-HT synthesis is still markedly increased at the spinal level and seemed to occur , at least partly, independently of the variations of TRP levels in the CNS, is in good agreement with previous reports showing that the bulbospinal serotonergic system is activated in anaesthetized animals submitted to pain stimuli. It has been reported an increase in the release of 5-HT in the lumbar CSF following intense stimulation of the sciatic nerves in the cat (4) and we observed an increase in 5-HT synthesis, not associated with an increase in TRP levels, in the spinal cord of rats submitted to an intense noxious stimuli (5). In the same manner it has been reported increases in 5-HIAA levels in the CSF of suffering patients (6, see also Costa et al., this meeting). However it must be determined whether the increase in 5-HT synthesis observed in "arthritic" rats is associated with an increase in the amount of 5-HT release from nerves terminals. Indeed, it could be postulated that an increase in 5-HT synthesis related to increases in TRP levels would enhance the amount of 5-HT released from nerve terminals (3) but such an hypothesis has been disputed (7,8). Experiments based on the evaluation of the potassium evoked release of 5-HT from dorsal cord slices seem to indicate that the increase in 5-HT synthesis observed at 15 days is not associated with such an increase. Further experiments will be made 42 days after the induction of the polyarthritis, since at this time the increase in 5-HT synthesis in the spinal cord appeared to be, at least partly, independent of the variations of TRP.

References

1. Weil-Fugazza, J., Godefroy, F., Coudert, D., Besson, J.M.: Pain 9, 319-325 (1980).
2. Godefroy, F., Weil-Fugazza, J., Coudert, D., Besson, J.M.: Brain Res. 199, 415-424 (1980).
3. Fernstrom, J.D.: Ann. Res. Med. 32, 413-425 (1981).
4. Yaksh, T.L., Tyce, G.M.: J. Physiol., Paris 77, 483-487 (1981).
5. Weil-Fugazza, J., Godefroy, F., Chitour, D., Le Bars, D.: C.R. Acad. Sc., Paris 293, 89-92 (1981).
6. Ghia, J.N., Mueller, R.A., Ducan, G.H., Scott, D.S., Mao, W.: Anesth. Analg. 60, 854-861 (1981).
7. Green, A.R., Grahame-Smith, D.G.: Nature, London 260, 487-491 (1976).
8. Elks, M.L., Youngblood, W.W., Kizer, J.S.: Brain Res. 172, 471-486. (1979).

EFFECTS OF ANTIRHEUMATIC DRUGS AND TRICYCLIC ANTIDEPRESSANTS ON TOTAL AND FREE SERUM TRYPTOPHAN LEVELS IN ARTHRITIC RATS.

Françoise Godefroy, Jeanne Weil-Fugazza, Jean-Marie Besson.

Unité de Recherches de Neurophysiologie Pharmacologique de l'INSERM (U 161), 2 rue d'Alésia, 75014 Paris, France.

Stephen H. Butler.

School of Medicine, Department of Anesthesiology, RN 10, University of Washington, Seattle, Washington 98195, USA.

Introduction

It has been reported that in patients with rheumatoid arthritis, there is an increase in the protein-bound plasma tryptophan (TRP) and a decrease in the free TRP levels. Moreover it was shown in experiments performed on these patients, that several common antirheumatic compounds, in spite of their diverse structures, were all capable of reducing this excessive binding of TRP. So it has been theorized that it might be at this point that one mechanism of action of the antirheumatic agents should be located (1). Surprisingly, and contrary to the observation made in humans, we have shown that in rats suffering from experimentally induced polyarthritis, there is a marked increase in plasma free TRP levels (2). The discrepancy between this observation and that made in humans led us to investigate the effect of chronic administration of aspirin and phenylbutazone on the plasma TRP levels in these arthritic rats. Indeed such a treatment significantly inhibits the development of the "arthritic" syndrome in these animals (3). Since we had observed that chronic treatment with amitriptyline reduced pain sensitivity and the arthritic process in this animal model (4), and since it has been reported that tricyclic antidepressants may decrease pain sensitivity in arthritic patients (5,6), the effects of amitriptyline and imipramine were also investigated.

Progress in Tryptophan and Serotonin Research

Methods

Polyarthritis was induced in Sprague-Dawley rats, weighing 150-175g, by intradermal injection of killed Mycobacterium butyricum suspended in mineral oil (Freund's adjuvant) (3). The treatments were started 21 days after the induction of polyarthritis. The drugs were administered daily for 4 weeks. Normal and arthritic rats receiving saline were run in parallel as controls. Weekly assessments of clinical signs of arthritis were made. Clinical observations include the circumference of the tibio-tarsal or "ankle" joints and a numerical arthritis score based on the evaluation of the swelling and the deformity of the hind paws, the swelling and the deformity of the tail, inflammation of the penis, conjunctivitis and bleeding from the nose. At the end of the four weeks period treatment the rats were sacrificed and blood was collected from trunk vessels and TRP levels estimated as previously described (2).

Results

The present study confirmed the decrease in total and the increase in free TRP levels induced by polyarthritis in rats (Fig. 1, 2).

a- Chronic treatment with aspirin or phenylbutazone in arthritic rats.
As previously reported (3) it was observed that chronic administration of aspirin and phenylbutazone significantly inhibited the development of the arthritis. As shown in Fig. 1, phenylbutazone (100 mg/kg, orally) induced a large decrease in free TRP level ($P < 0.001$) and an increase in total TRP level ($P < 0.001$), whereas aspirin (200 mg/kg, orally) did not modify the free TRP level but enhanced the decrease in total TRP level ($P < 0.01$) observed in control arthritic rats.

b- Chronic treatment with amitriptyline or imipramine in arthritic rats.
The chronic administration of amitriptyline induced a slight but significant inhibition of the arthritis. Although the chronic treatment with imipramine tends to induce a reduction of the arthritis, this effect was not significant. The chronic administration of both amitriptyline (10 mg/kg, IP) and imipramine (10 mg/kg, IP) decreased the free TRP level ($P < 0.05$) but did not modify the total TRP level (Fig. 2).

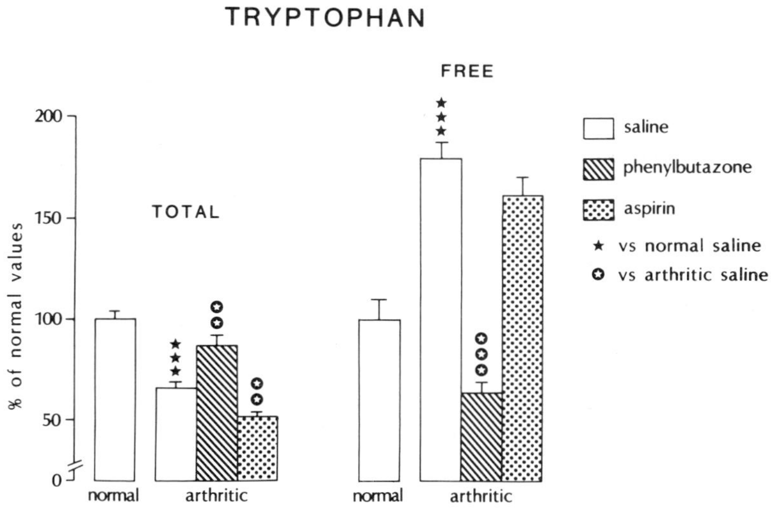

Fig. 1 -Effect of chronic administration of phenylbutazone (100 mg/kg, P.O.) or aspirin (200 mg/kg, P.O.) on the total and free TRP levels in arthritic rats.

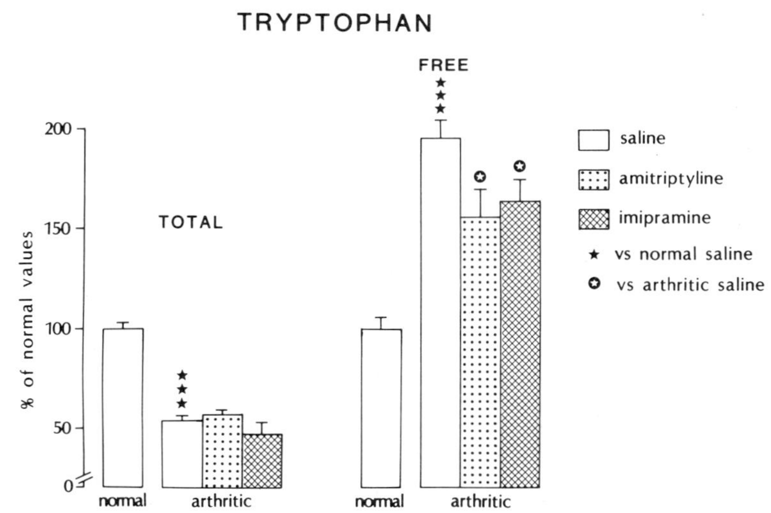

Fig. 2 -Effect of chronic administration of amitriptyline (10 mg/kg, I.P.) or imipramine (10 mg/kg, I.P.) on the total and free TRP levels in arthritic rats.

Conclusion

These results show that in arthritic rats phenylbutazone and aspirin modify the equilibrium between bound and free TRP in opposite manner whereas these two drugs markedly reduce the arthritic syndrome in these rats. Although there are obviously important differences between human rheumatoid diseases and induced polyarthritis in rats, these opposite results question the hypothesis of a relationship between the ability of antirheumatic drugs to reduce the binding of TRP on plasma protein and their antirheumatic properties. Moreover experiments now in progress show that chronic treatment with aspirin and phenylbutazone tends to counteract the increase in plasma tyrosine levels observed in these arthritic rats (see Weil-Fugazza et al., this meeting). This indicates that antirheumatic drugs may modify not only plasma TRP levels but also other free amino acids levels. On the other hand the results observed in arthritic rats receiving amitriptyline and imipramine indicate that these drugs may modify the levels of free TRP in plasma in some pathological states. However, the observations made in arthritic rats treated with aspirin and phenylbutazone question the hypothesis of a relationship between the reduction of arthritis induced by amitriptyline and the effect of this drug on plasma TRP.

References

1. Aylward, M., Maddock, J., Wheeldon, N., Parker, R.J. : Rhumatology and Rehabilitation 14, 101-114 (1975).

2. Weil-Fugazza, J., Godefroy, F., Coudert, D., Besson, J.M. : Pain 9, 319-325 (1980).

3. Newbould, B.B. : Brit. J. Pharmacol. Chemother. 21, 127-130 (1963).

4. Butler, S., Weil-Fugazza, J., Godefroy, F., Coudert, D., Besson, J.M. : Neuroscience Letters, suppl. 10, S97 (1982).

5. Scott, W.A. : Practitioner 202, 802-807 (1969).

6. Gringas, M.A. : J. intern. Med. Res. 4, 41-49 (1976).

TRYPTOPHAN, SEROTONIN AND 5-HYDROXYINDOLEACETIC ACID LEVELS IN HUMAN CSF IN RELATION TO PAIN.

Carlo Costa, Francesco Ceccherelli*, Antonio Bettero, Gino Marin*, Ludovico Mancusi* and Graziella Allegri

Istituto di Chimica Farmaceutica - Centro di Studio sulla Chimica del Farmaco e dei Prodotti Biologicamente Attivi del CNR and Istituto di Anestesia e Rianimazione - Università - Padova - Italy

Introduction

In several studies the involvement of the serotoninergic system in pain modulation is reported (1,2). The more confirmed hypothesis is that the decrease of serotoninergic function into the brain or spinal cord causes an increase of sensitivity and reactivity to noxious stimuli, whereas an increase of serotoninergic neurotransmission is correlated to analgesia. Some Authors studying the serotoninergic pathway in chronic patients and acute pain have measured the 5-hydroxyindoleacetic acid in CSF and found an increase in the patients in respect to the controls (3). To obtain a more complete picture on the function of the serotoninergic system in chronic pain patients we have determined simultaneously tryptophan (TRP), serotonin (5-HT) and 5-hydroxyindoleacetic acid (5-OHIAA) in the CSF. In the same patients we have also determined the serum tryptophan (total and free) levels to verify a relation between the serum and CSF values.

Materials and methods

Patients. We have considered 30 subjects, males and females, divided into two groups: 14 patients without pain

Progress in Tryptophan and Serotonin Research

(ranging from 30 to 66 years) and 16 chronic patients (ranging from 30 to 69 years) affected by lumbar discopathy or prolapsed disc, suffering in pain at least 4 months. All the subjects examined were on free diet and had not touched food in the previous 12 hours.

Measurement of pain. The chronic pain patients were submitted, before blood and CSF sampling, to the Huskisson test (4) as a subjective measure of pain intensity (mean values 47 mm) and were also questioned as to the lenght of sleep.

Blood and CSF sampling measurements. After adequate information of the process and obtainment of the permission of the patients, 8 ml of venous blood are sampled and kept at 4°C. Immediately afterwards, 2 ml of CSF are collected with needle n. 22 positioned at lumbar level (L2-L3; L3-L4) by dripping into a test tube, and frozen in dry ice. The serum TRP (total and free) and CSF TRP, 5-HT and 5-OHIAA determinations were carried out with a new fully automated HPLC method arranged by us,based on the direct measure of the native fluorescence with previous retention of the protein fraction in a precolumn system and elution with isocratic gradient of acetonitrile-phosphate buffer. The fluorimetric detector was set at 295 nm and 336 nm for excitation and emission, respectively. The chromatograms of standard solutions and CSF sample with the retention time, are reported in Fig. 1.

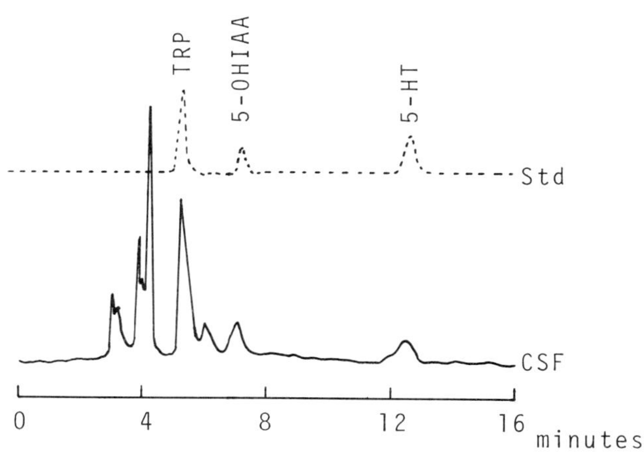

Fig. 1 - Chromatograms of standard and CSF sample solutions obtained at 295/336 nm.

Results and discussion

The results found in serum and CSF samples of healthy volunteers and chronic patients are reported in Table 1.

Table 1. Mean values (± S.E.M.) of serum TRP and CSF TRP, 5-HT and 5-OHIAA in healthy volunteers and chronic pain patients.

	SERUM mg/l		CSF ng/ml		
Patients	TRP total	TRP free	TRP	5-HT	5-OHIAA
without pain (n. 14)	9.67 ±0.50	1.54 ±0.17	332.7 ±15.5	3.13 ±0.65	2.98 ±0.39
chronic pain (n. 16)	10.52 ±0.43	1.44 ±0.09	349.0 ±17.2	*	5.60° ±2.70

* at levels lower than detection limit by injecting 20 μl of sample. - ° Values at $P < 0.001$.

The values are expressed as means of mg/l for the serum and as mean of ng/ml for the CSF samples.No significant differences are found between the mean values of serum TRP (total and free) in the two groups of patients considered, even if the free serum TRP fraction seems to decrease in the chronic pain patients in respect to the patients without pain. Instead the data obtained in CSF are very interesting. In the chronic pain patients, a low increase of TRP appears as if the uptake were favoured through the hematoencephalic barrier, whereas there is a very high decrease of 5-HT and an increase of 5-OHIAA statistically significant at P 0.001 in agreement with the results by Ghia et al. (3). In 80% of these suffering patients a decrease of sleep lenght caused by a chronic insomnia has also been found. These preliminary results seem to show an increased catabolism of 5HT in chronic pain patients, supporting an involvement of serotonin in antinociception and insomniac situation and the important role of tryptophan and serotoninergic system in modulation pain.

References

1. Messing, R.B., Lytle, L.D.: Pain 4, 1 (1977).

2. Brase, D.A.: "Neurochemical Mechanism of Opiates and Endorphins", in: Adv. Biochem. Psycopharmacol. Vol. 20, eds. H.H. Loh and D.H. Ross, Raven Press, New York 1979.

3. Ghia, J.N., Mueller, R.A., Duncan, G.H., Scott, D.S., Mao, W.: Anesth. Analg., 60, 854 (1981).

4. Huskisson, E.C.: Lancet, 9, 1127 (1974).

DYNAMICS OF PLASMA TRYPTOPHAN FOLLOWING ANTIMIGRAINE DRUGS

Silvia Salmon, Paola Del Re, Daniela Caleri, Marcello Fanciullacci and Federigo Sicuteri

Department of Internal Medicine and Clinical Pharmacology, University of Florence, Viale Morgagni 85, Florence

Introduction

In vitro experimental (1) and clinical (2) data indicate that plasma free tryptophan (TP), the fraction not bound to albumin, plays a role in regulating the action of a variety of therapeutic agents. TP availability, i.e. the concentration of free plasma TP, is still considered the main mechanism for controlling both brain TP levels and 5HT synthesis within the brain (3, 4), in normal and pathological conditions (5).

Serotonin, together with endorphins, is involved in the control of endogenous pain modulation (6). A neurochemical impairment of these neuromodulators of the antinociceptive system has been hypothesized in the pathogenesis of migraine and related headaches (7).

Altered TP levels have been found in migraneous patients (8) and in stress conditions (9).

The aim of the present research is:

1) to observe the displacing action of drugs such as: a) cinnarizine (a calcium antagonist); b) captopril (inhibitor of both angiotensin converting enzyme and enkephalinase) and c) zomepirac (inhibitor of prostaglandin synthesis) on tryptophan plasma albumin binding.
   The effects in the prophylactic and/or symptomatic therapy of the above mentioned drugs in migraine and headache were previously reported (10, 11, 12).
2) to correlate plasma free TP changes following drugs administration with the drug pharmacokinetics as previously determined by other authors (13, 14, 15).

Progress in Tryptophan and Serotonin Research

Materials and methods

Thirty-four (34) volunteer subjects (mean age = 33.19 years), hospitalized in the Division of Internal Medicine and Clinical Pharmacology of the University of Florence for clinical investigations, participated in the study. The subjects were drugs-free for seven days and fasting overnight prior to the initiation of the study.
Subjects were then divided into three groups:
Group I (N=10) were orally given a placebo tablet on the first day of the study followed by 150 mg of Cinnarizine (Janssen-Belgium) on the second day;
Group II (N=16) were given 100 mg of Captopril (Squibb) p.os;
Group III (N=8) received 100 mg of Zomepirac sodium (Cilag).
Blood sample were taken prior (baseline value) and at pre-set intervals following drug or placebo administration.
Free (F), total (T) and bound (B) TP (B = T - F) levels were determined with a spectrofotofluorimetric method and ultrafiltration technique according to Eccleston's technique (16).
Mean TP levels were obtained and statistical evaluation by the Student **t** (paired test) was performed.
Mean TP values were compared for each drug with the respective pharmacokinetics curves previously determined by other authors: Cinnarizine (13), Captopril (14), Zomepirac (15).

Results

Mean plasma free and bound TP of the three groups acutely orally administered Cinnarizine, Captopril, Zomepirac and placebo are summarized in Table 1, and represented grafically in Fig. 1, 2, 3.
Placebo showed no significant differences when compared to baseline value.

TABLE 1 FREE AND BOUND TP (Mean levels ± S.E.)

	BASE		30'		60'		120'		180'		240'		24h	
	F	B	F	B	F	B	F	B	F	B	F	B	F	B
CINNARIZINE (10)	1.73	10.24	2.57	9.06	2.33	9.00	2.32	8.50	2.00	9.00	-	-	1.49	10.36
	±0.2	±0.8	±0.2	±0.8	±0.2	±0.7	±0.2	±0.6	±0.2	±0.9			±0.2	±0.7
CAPTOPRIL (17)	1.00	9.32	1.49	8.38	1.46	8.43	1.30	8.30	1.00	9.71	-	-	1.16	10.89
	±0.13	±0.8	±0.15	±0.7	±0.10	±0.8	±0.14	±0.8	±0.17	±0.96			±0.26	±1,2
ZOMEPIRAC (8)	1.5	9.10	2.38	7.37	1.70	7.82	1.67	7.89	2.01	7.48	1.57	8.06	-	-
	±0.13	±1.1	±0.6	±1.4	±0.2	±1.2	±0.1	±1.2	±0.2	±1.1	±0.01	±1.3		
PLACEBO (10)	1.88	9.13	2.21	8.32	1.84	8.11	2.14	7.76	2.08	7.61	-	-	1.94	9.29
	±0.2	±0.8	±0.2	±0.5	±0.2	±0.7	±0.2	±0.7	±0.2	±0.7			±0.2	±0.7

( ) in brackets number of subjects

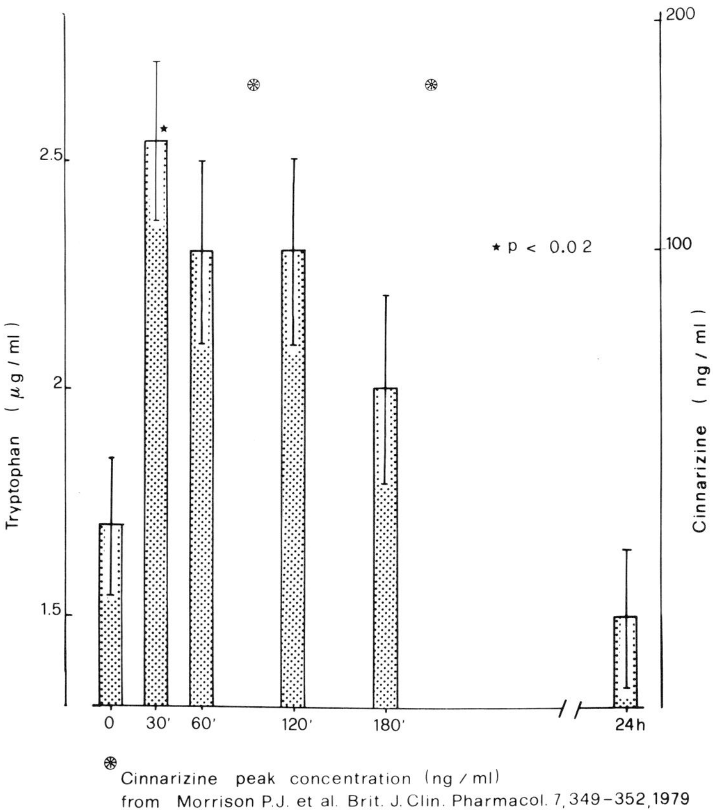

Fig. 1:

a) Free plasma TP (Mean levels ± S.E.) in 10 subjects following Cinnarizine oral acute treatment (150 mg)
The increase is statistically significant ($p<0.02$) 30' following the drug administration.

b) Cinnarizine plasma levels (peak concentrations) following 75 mg oral administration of the drug in 12 subjects.

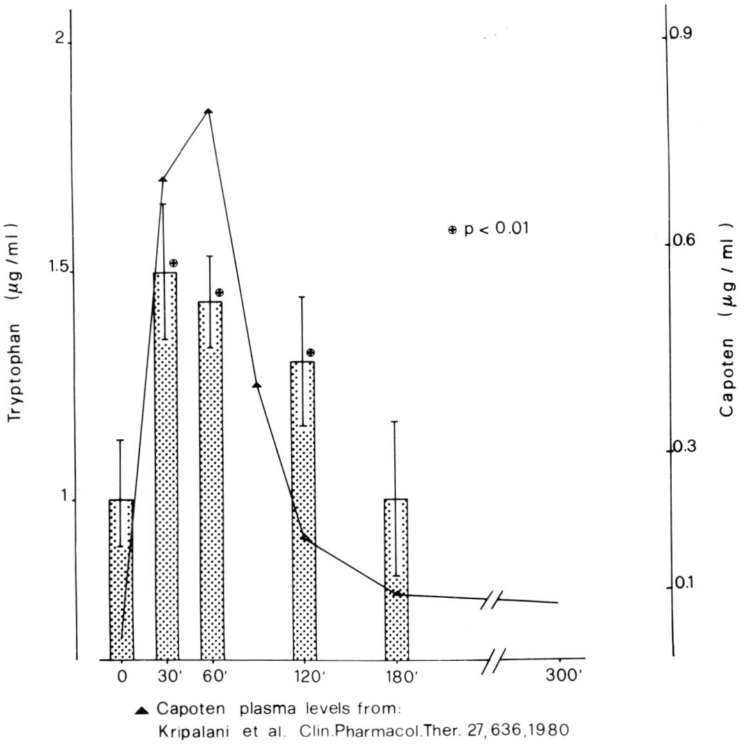

Fig. 2:

a) Free plasma TP (Mean levels ± S.E.) in 16 subjects following Captopril (Capoten) oral treatment (75 mg).
   Mean increases are statistically significant at 30', 60' and 120' ($p<0.05$) following drug administration.

b) Capoten plasma levels (pharmacokinetics curve) following a single oral dose (100 mg) of Capoten labelled 36 S in 10 subjects.
   Capoten peak plasma concentrations were achieved within 180'.

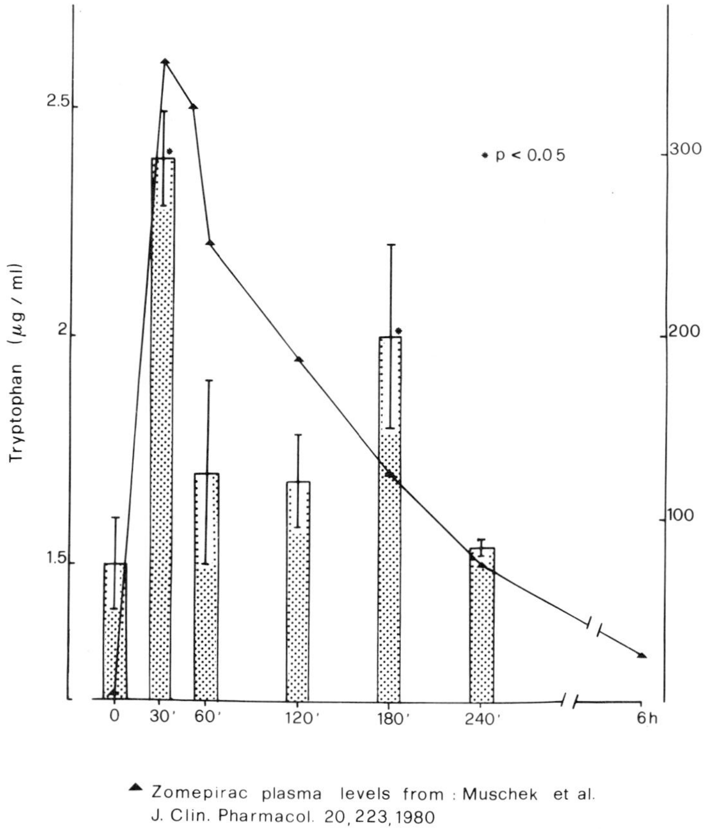

Fig. 3:

a) Free plasma TP (Mean levels ± S.E.) in 8 subjects following Zomepirac oral treatment (100 mg).
   Free plasma TP significantly increases ($p<0.05$) 30' and 180' following the drug administration.

b) Zomepirac plasma levels (pharmacokinetics curve) following two 50 mg Zomepirac tablets oral administration.
   Peak plasma concentrations of Zomepirac were achieved within 1 hour.

Conclusions

Current literature shows that the three drugs studied, Cinnarizine, Capoten and Zomepirac, are rapidly and completely absorbed after oral administration (13, 14, 15). Zomepirac is highly bound to plasma proteins, however this binding may be decreased by salycilates as they are well known to displace TP from its stereospecific binding site on albumin (15). Several drugs, showing very different molecular structures as well as pharmacological properties, which are bound to a higher extent to serum albumin, are capable of displacing tryptophan from its single binding site on the human serum albumine molecule (HSA). According to Muller's in-vitro observations, the displacing activities are not dependent upon the affinity of the drugs for HSA, but rather due to an allosteric or a competitive mechanism (1).
In this research, we have compared pre-existing kinetic data, of those drugs effective in the treatment of migraine, with plasma free TP, following acute oral administration.
The results of the plasma TP levels show a correspondence between the significant increases of free TP and the maximum peak concentrations of all the drug studied. Only Zomepirac shows a significant increase of free TP at 180' following drug administration, to which a high concentration of the drug in plasma does not correspond.
Placebo treatment did not induce any significant increase.
The mechanism of transport through the blood brain barrier and the correlation between free and bound TP with respect to brain 5HT levels have not yet been completely defined (17,5): however, in view of the ability of certain neurons to synthetize their particular neurotransmitters dependent upon plasma concentration levels of their precursors (17), the displacing activity on plasma bound TP might be relevant in the analgesizing mechanism of the drugs studied and confirm the already suggested hypothesis of a central mechanism of action.

References

1) Muller W.F. and Wollert U.: Res. Commun. in Chem. Pathology and Pharmacology, 10 (3), 565-568, (1975)
2) Paoletti R., Sirtori C. and Spano P.F.: Ann. Rev. 15, 73-81, (52 ref) (1975)
3) Fenstrom J.D., Wurtman R.J.: Science 173, 149-151, (1971)
4) Tagliamonte A., Biggio G., Gessa G.L.: Farmacol. Ter, 11, 251-255, (1971)
5) Wurtman R.J., Hefti F. and Mchamed E.: Amer. Soc. Pharmaco. Exp. Therapeutics, 32, 4, 315-335, (1981)
6) Basbaum A. and Fields H.L.: Ann. Neurol. 4, 451-462, (1978)
7) Sicuteri F.: Advances in Pain Research and Therapy, ed Bonica, 1, 871-880, (1976)
8) Salmon S., Fanciullacci M., Bonciani M. and Sicuteri F.: Headache, 17, 238-241, (1978)
9) Kennett G.A. and Joseph M.H.: Neuropharmacology 20, 39-43, (1981)
10) Salmon S., Michelacci S., Sicuteri F.: Intnl. Congress "Headache 80" Abstract Book, 114, (1980)
11) Sicuteri F.: Cephalalgia 1, 77-82, (1981)
12) Diamond S.: J. Clinical Pharmacology, 298-302, (april 1980)
13) Morrison P.J., Bradbrook I.D., Rogers H.S.: British J. Clin. Pharmacol. 7, 349-352, (1979)
14) Kripalani K.J., Mc Kinstry D.N. et al.: Clin. Pharmacol. Ther. 27, 636-641, (1980)
15) Muschek L.D. and Grindel J.M.: J. Clin. Pharmacol. 20, 223-229, (1980)
16) Eccleston E.G.: Clin. Chim. Acta, 48, 269-282, (1973)
17) Curzon G.: J. Neural Transm. (suppl) 15, 81-92, (1979).

COMPARISON OF MORPHINE AND MEPTAZINOL EFFECTS ON RAT TRYPTOPHAN DISPOSITION AND METABOLISM

Myrddin Evans, Abdulla A-B Badawy, Anthony Dacey

South Glamorgan Health Authority, Addiction Unit Research Laboratory,
Whitchurch Hospital, Cardiff CF4 7XB, Wales, U.K.

Acute administration of the opioid analgesic meptazinol enhances rat brain 5-hydroxytryptamine (5-HT) synthesis and liver tryptophan pyrrolase activity by increasing the availability of circulating tryptophan to the brain and liver respectively, whereas chronic treatment with the drug fails to influence tryptophan metabolism and disposition.

Introduction

There is strong evidence implicating 5-HT in the production of analgesia by, and the development of tolerance to and physical dependence on, morphine and certain other opioids (1,2). The opioid drug meptazinol [m(3-ethyl-1-methyl hexahydro-1H azepin-3-yl)phenol] hydrochloride is a potent analgesic in both experimental animals and man, but, presumably because of its opiate antagonistic activity, it is unlikely to be used by opiate addicts or to cause dependence in man [see (3) and references cited therein]. However, whereas the effects of morphine on cerebral 5-HT metabolism are well documented (4,5), those likely to be caused by meptazinol have not been examined, as far as we could ascertain. We have therefore performed a comparative study of the acute and chronic effects of administration of meptazinol and morphine on rat liver and brain tryptophan metabolism and related metabolic systems in an attempt to examine further the relationship between 5-HT and analgesia and dependence. The results of such a study are reported here.

Materials and Methods

All details concerning animals (male Wistar rats), chemicals, drug administration and chemical, enzymic and other determinations have been described or referred to previously in detail elsewhere (5,6).

Progress in Tryptophan and Serotonin Research

Results and Discussion

The acute effects of morphine and meptazinol on rat tryptophan metabolism and disposition and related metabolic systems are shown in Table 1, which also includes, for comparative purposes, those of the non-opiate analgesic paracetamol (acetaminophen) and the opiate antagonist naloxone. As shown, all analgesics significantly increased the concentrations of brain 5-HT and 5-HIAA, most likely by elevating that of brain tryptophan. Naloxone was ineffective in these respects. These results strongly suggest that analgesics enhance brain 5-HT synthesis (and possibly also turnover) by increasing brain tryptophan concentration, and provide support to the role of 5-HT in analgesia. This increase in brain tryptophan concentration is most likely the result of an increased availability of circulating tryptophan to the brain, as is suggested by the observed increases in both concentration of, and percentage, free serum tryptophan. These latter increases are most likely caused by displacement of bound tryptophan, either directly (paracetamol and morphine) or via NEFA (meptazinol). Increased tryptophan availability by morphine and meptazinol appears also to influence the liver, by enhancing tryptophan pyrrolase activity by a substrate- (or cofactor-) type mechanism (that involving increased haem saturation). Pyrrolase was not influenced by naloxone [see(5)] and was only moderately inhibited by paracetamol.

Chronic morphine administration has previously been shown (5) to enhance rat brain 5-HT synthesis by increasing tryptophan availability to the brain after inhibition of liver tryptophan pyrrolase activity. All these effects are confirmed by the results in Table 2, which also shows that, by contrast, chronic meptazinol administration in daily doses comparable to those of morphine (50mg/kg each) failed to influence tryptophan disposition or its metabolism in liver or brain. It could be argued that the failure of meptazinol to act under these conditions is due to its rapid hepatic metabolism after oral intake (7). This is, however, unlikely, because of the finding (results not shown) that chronic administration of daily doses of up to 500mg/kg each failed to influence the above parameters. In other experiments also not shown here, it was found that chronic administration of naloxone (1mg/kg/day for 21 days) also failed to influence tryptophan

Table 1. Acute Effects of Opiates and Paracetamol on Rat Tryptophan Metabolism and Disposition

Treatment	0.9% NaCl	Morphine	Meptazinol	Paracetamol	Naloxone
Test Dose ..	(2ml/kg)	(50mg/kg)	(50mg/kg)	(50mg/kg)	(5mg/kg)
Brain 5-HT	0.46 ± 0.014	$0.64 \pm 0.033^{***}$	$0.56 \pm 0.029^{+++}$	$0.58 \pm 0.022^{***}$	0.43 ± 0.014
Brain 5-HIAA	0.32 ± 0.013	$0.50 \pm 0.012^{***}$	$0.39 \pm 0.013^{**}$	$0.39 \pm 0.028^{+}$	0.33 ± 0.017
Brain Trp	2.40 ± 0.10	$3.19 \pm 0.20^{*}$	$3.03 \pm 0.11^{**}$	$2.68 \pm 0.06^{+}$	2.50 ± 0.18
Free serum Trp	1.02 ± 0.02	$1.67 \pm 0.06^{***}$	$1.43 \pm 0.07^{***}$	$1.25 \pm 0.06^{**}$	1.09 ± 0.04
Total serum Trp	33.62 ± 0.89	$31.17 \pm 0.79^{+}$	$26.68 \pm 0.52^{***}$	31.33 ± 1.23	$31.42 \pm 0.59^{+}$
Free serum Trp (%)	3.03 ± 0.09	$5.36 \pm 0.30^{***}$	$5.36 \pm 0.23^{***}$	$3.99 \pm 0.20^{**}$	$3.47 \pm 0.11^{+++}$
Liver Trp	3.02 ± 0.05	$4.02 \pm 0.35^{++}$	$3.35 \pm 0.10^{+++}$	2.65 ± 0.22	3.33 ± 0.14
Liver Trp pyrrolase:					
Holoenzyme	1.40 ± 0.16	$3.00 \pm 0.18^{***}$	$3.50 \pm 0.46^{***}$	1.50 ± 0.08	1.50 ± 0.09
Total enzyme	2.90 ± 0.10	$3.80 \pm 0.22^{*}$	$5.70 \pm 0.50^{**}$	2.60 ± 0.11	3.20 ± 0.20
Apoenzyme	1.50 ± 0.04	-	$2.20 \pm 0.11^{**}$	$1.10 \pm 0.06^{**}$	1.70 ± 0.27
Haem saturation (%)	48 ± 1	$79 \pm 1^{***}$	$61 \pm 3^{*}$	-	47 ± 5
Serum NEFA	0.39 ± 0.04	$0.14 \pm 0.02^{***}$	$0.57 \pm 0.06^{+}$	0.38 ± 0.03	0.40 ± 0.04
Serum albumin	32.7 ± 0.7	33.4 ± 1.6	$35.4 \pm 0.4^{*}$	33.0 ± 0.4	34.0 ± 0.5
Serum corticosterone	35.2 ± 1.7	$63.6 \pm 3.6^{***}$	$30.3 \pm 1.2^{+}$	$30.6 \pm 1.0^{+}$	35.3 ± 1.8
Serum glucose	137 ± 4	$223 \pm 22^{**}$	$151 \pm 6^{+}$	137 ± 4	141 ± 4

Rats received an intraperitoneal injection of drugs or an equal volume of 0.9% (w/v) NaCl and were killed 2h later. Parameters given were determined as described under Materials and Methods and are means $\pm$ S.E.M. for each group of 4 (pyrrolase parameters) or 6 rats (all other determinations). The significance of the differences between test and control results is indicated as follows: $^{+}P < 0.05$; $^{++}P < 0.025$; $^{+++}P < 0.02$; $^{*}P < 0.01$; $^{**}P < 0.005$; $^{***}P < 0.001$. Liver tryptophan pyrrolase activities are in µmol of kynurenine formed/h per g wet wt., whereas the percentage haem saturation of the enzyme is 100 X holoenzyme activity/total enzyme activity. Other expressions are as follows: serum and tissue tryptophan and hydroxyindoles (µg/ml or per g wet wt.); serum non-esterified fatty acids or NEFA (mM); serum albumin (g/l); serum corticosterone (µg/l); serum glucose (mg/dl).

Table 2. Chronic Effects on Morphine and Meptazinol on Tryptophan Metabolism

Test Treatment	Control	Morphine	Meptazinol
Brain Trp	1.88 ± 0.08	2.33 ± 0.07**	1.80 ± 0.03
Brain 5-HT	0.56 ± 0.006	0.65 ± 0.008***	0.55 ± 0.015
Brain 5-HIAA	0.36 ± 0.009	0.44 ± 0.008***	0.35 ± 0.007
Free serum Trp	1.46 ± 0.05	1.83 ± 0.07**	1.36 ± 0.02
Total serum Trp	19.34 ± 0.50	23.86 ± 1.30*	17.72 ± 0.45
Free serum Trp (%)	7.55 ± 0.26	7.67 ± 0.30	7.67 ± 0.22
Pyrrolase Holoenzyme	2.10 ± 0.09	2.20 ± 0.07	2.40 ± 0.12
Pyrrolase Total enzyme	5.20 ± 0.29	2.60 ± 0.06***	5.40 ± 0.12
Pyrrolase Apoenzyme	3.10 ± 0.22	0.40 ± 0.05***	3.00 ± 0.12

All details, methods and expression and comparisons of results are as in Table 1, except that drugs were administered chronically for 3 weeks in drinking water as described previously for morphine (11). The significance of the differences is indicated as follows: *$P < 0.01$; **$P < 0.005$; ***$P < 0.001$.

metabolism and disposition, whereas that of paracetamol (50mg/kg/day each) only caused a very slight increase in brain indoles. These results therefore demonstrate that meptazinol, which is thought to have a very low abuse liability, fails to influence chronically tryptophan metabolism. Whether this distinction between meptazinol and morphine could form the basis of an opioid-abuse-liability test requires further comparative studies with drugs.

We thank Wyeth Laboratories and staff at Taplow, U.K. for support, Endo Laboratories and the DuPont Organization for a gift of naloxone and Messrs C. J. Morgan and N. R. Davis for skilful assistance

References

1. Way,E.L.,Ho,I.K.,Loh,H.H.:Adv.Biochem.Psychopharmacol.10,219-231 (1974)
2. Bryant,R.M.,Olley,J.E.,Tyers,M.B.:Br.J.Pharmacol. 77, 615-624 (1982).
3. Evans, M., Robson, P.J., Chadd, M.A., Evans, C.M., Fry, D.M.: Postgrad. Med. J. 59 (Suppl. 1), 78-84 (1983).
4. Messing, R. B., Flinchbaugh, C., Waymire, J.C.: Psychopharmacology 17, 391-396 (1978).
5. Badawy,A.A.-B.,Punjani,N.F.,Evans,M.: Biochem. J. 196, 161-170 (1981).
6. Badawy,A.A.-B.,Evans,M.,Punjani,N.F.:Br.J.Pharmacol.74, 489-494 (1981).
7. Franklin, R. A.: Br. J. Pharmacol. 59, 565-569 (1977).

DECREASED CONCENTRATIONS OF TRYPTOPHAN AND SEROTONIN IN THE BRAIN OF DEVELOPING HYPERPHENYLALANINEMIC RATS

A CONTRIBUTION TO THE UNDERSTANDING OF THE METABOLIC DEFECTS IN PHENYLKETONURIA

G. Huether, K. Schott and V. Neuhoff

Max-Planck-Institut für experimentelle Medizin
D-3400 Göttingen

Introduction

The most frequent inborn error of amino acid metabolism, phenylketonuria, is characterized by the postnatal accumulation of phenylalanine (Phe) and a simultaneous depletion of other essential amino acids in the blood and in the brain tissue. A similar situation is experimentally induced in suckling rats by the daily injection of Phe plus alpha-methyl-Phe (an in vivo inhibitor of Phe-hydroxylase). In both conditions, human PKU and chronic experimental hyperphenylalaninemia (hyper-Phe), the low concentrations of in particular the large neutral amino acids in the brain is thought to result from the competitive inhibition of their uptake at the blood-brain barrier by the more than twentyfold elevated Phe-concentration in the blood. The decreased availability of large neutral amino acids in the presence of high Phe-concentrations hampers the protein synthesis in the developing brain. As a consequence, normal development is disturbed, the brain becomes microcephalic, and among others, its myelin content is reduced (1,2).

The mental retardation is the most significant feature of untreated PKU-children. Therefore, research in this field was almost exclusively concerned with the brain. The decreased supply to the developing brain by the reduced concentrations

Progress in Tryptophan and Serotonin Research

of certain amino acids in the blood of phenylketonuric children (3) has not found much consideration. It has recently been shown that the decrease in the concentrations of essential amino acids in the serum of hyper-Phe rats is even greater than that in the brain tissue (4). This suggests that the depletion of amino acids in the brain is not caused by the competitive inhibition of their uptake, but results from the a priori decreased supply to the developing brain by the respective amino acids.

Tryptophan (Trp) is the most depleted amino acid in the hyper-Phe brain (4). It may therefore serve as an example to study the metabolic defects that cause the insufficient supply of certain amino acids to the developing brain of PKU-children. For this purpose, rats were rendered hyperphenylalaninaemic by the daily injection of alpha-methyl-Phe and Phe as described elsewhere (1). The amino acid concentrations of various body tissues were measured by thin layer micro-chromatography of the dansylated amino acids (5). Aromatic amines and their metabolites in the brain were determined by HPLC and electrochemical

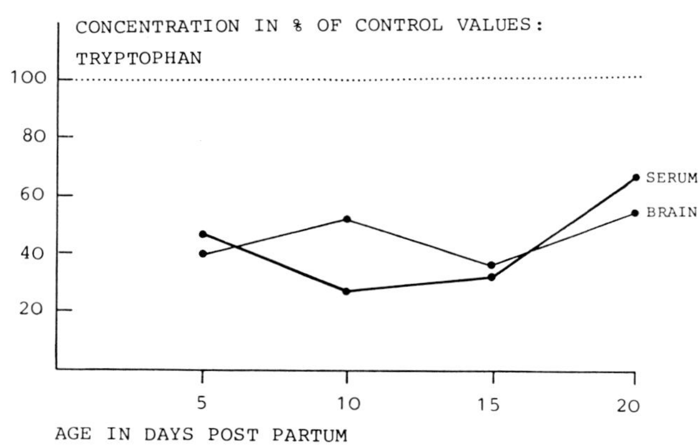

Fig. 1: Alterations caused by hyper-Phe in the concentration of Trp in the blood and in the brain of developing rats. N = 6, the differences are significant at all ages (t-test, $p < 0.01$). For absolute values ± S.D. see (4).

detection (6).

Results and Discussion

The concentrations of several amino acids are decreased in the brain of hyper-Phe rats. Trp is the most depleted amino acid in the developing hyper-Phe brain (Fig. 1). The concentrations of serotonin and its major degradation product, 5-hydroxyindole-acetic acid, in the forebrain are also decreased in the hyper-Phe state (Fig. 2). This decrease is in the same order of magnitude as the depletion of Trp. This observation is in accordance with the concept that the synthesis of serotonin in the mammalian brain depends on the availability of Trp (7). Obviously, the decreased availability of the precursor amino acid in the hyper-Phe brain limits the synthesis and the degradation of serotonin.

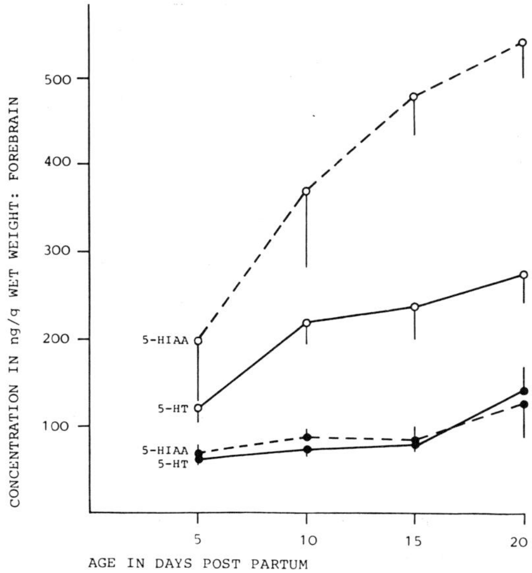

Fig. 2: The concentrations of serotonin (5-HT) and 5-hydroxy-indoleacetic acid (5-HIAA) in the brain of developing rats. Open circles: controls, full dots: hyper-Phe rats. Values are means ± S.D. The decreases are statistically significant at all ages (t-test, $p < 0.01$). The measurements were performed by HPLC with electrochemical detection (6).

It has been shown by others (8) that protein synthesis in the developing brain is hampered by the decreased supply with certain amino acids in hyper-Phe. Therefore, several processes of cell differentiation are severely disturbed, the brain becomes microcephalic, and its myelin content is reduced (1,2).

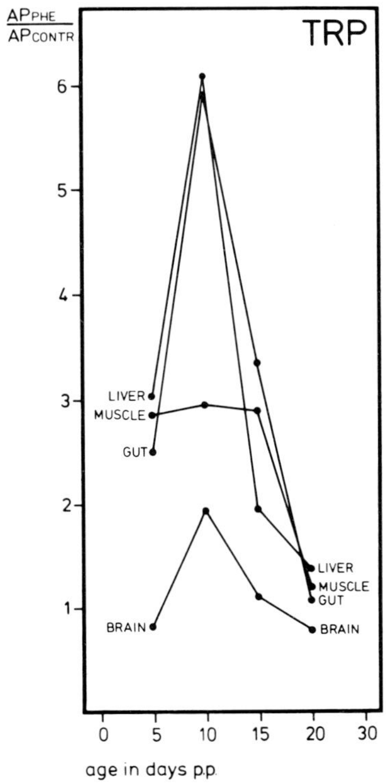

The amino acid depletion in the hyper-Phe brain is thought to result from the competitive inhibition of the uptake of other large neutral amino acids at the blood-brain barrier in the presence of high circulating Phe-concentrations (for references see 4). However, the depletion of amino acids in the hyper-Phe brain - even that of Trp - is less than the depletion of these amino acids in the blood (see Fig. 1 and ref. 4). Therefore, under the steady state conditions of chronic hyper-Phe, the decreased amino acid pool of the developing brain is primarily a reflection of the decreased amino acid supply to the developing brain.

Fig. 3: Alterations caused by hyper-Phe in the accumulative potency for Trp of several developing rat tissues. The concentration of Trp in each tissue (µmol/kg) per that in the serum (µmol/l) of hyper-Phe rats (AP_{Phe}) was divided by the respective tissue to serum ratio of controls (AP_{Contr}). $AP_{Phe}/AP_{Contr} = 1$: no alteration of accumulative potency, $AP_{Phe}/AP_{Contr} > 1$: increased accumulative potency in hyper-Phe.
Note: After day 10 p.p., the amino acid imbalance imposed to the developing rats gets more and more effectively abolished (metabolic adaptions for counterregulation, see 4).

The depletion of other amino acids in the blood is a well known feature of untreated PKU-children (3). Their disturbed brain development may be a consequence of the insufficient amino acid supply from the periphery. It is remarkable that the most depleted amino acid in the blood of untreated PKU-children is not - as one would expect - tyrosine, but are arginine and threonine of the essential amino acids, and proline (only 25% of normal levels!) of the nonessential amino acids (3). This phenomenon has not found much consideration in the past. It is hardly compatible with the theory of a defect hydroxylase locus being the solely and primary cause of this inborn error of amino acid metabolism, and it indicates a more generally affected state of amino acid utilization and protein metabolism in PKU.
In hyper-Phe rats, the depletion of Trp and other amino acids in the blood and in the brain is paralleled by an increased accumulation of these amino acids in the peripheral tissues (for Trp see Fig. 3). This observation again indicates an altered steady state of protein synthesis and protein degradation in the hyper-Phe state and/or a general influence of hyper-Phe on the amino acid metabolism and compartmentation in the living organism.
Lysine is a rate limiting amino acid for protein anabolism during the phase of rapid growth in several species (9). The additional injection of lysine to hyper-Phe rats counteracts the depletion of all amino acids in the blood and in the brain (and the growth retardation of the developing brain) in suckling hyper-Phe rats (for the effects on Trp-levels see Fig. 4). This effect happens in spite of the fact that the concentration of Phe in the blood of the hyper-Phe rats rises even somewhat by the lysine-injections.

In summary, the depletion of Trp and other amino acids in the brain of hyper-Phe rats - under in vivo steady state conditions - is only unsatisfyingly explained by competitive effects in the amino acid uptake at the blood-brain barrier. The depletion of Trp (and of other amino acids) in the blood, and the increased accumulation of these amino acids in peripheral tissues of the hyper-Phe organism indicate an shifted balance between protein synthesis and protein degradation. We would suggest that both, Phe and lysine, are rate limiting substrates for protein synthesis during the hyperplastic and hypertrophic phase of growth. An increased availability of Phe would rise the requirement of lysine, and vice versa. Under this aspect, the preferential decrease in the concentration of lysine in the blood of PKU-heterozygote women during the first trimester of pregnancy deserves special interest (10). One can not exclude that the process of metabolic maturation of the developing fetus in a heterozygote PKU-mother becomes more or less reversibly adapted to an altered availability of Phe and lysine during early ontogenesis. Under this aspect, the features of postnatal amino acid metabolism in the untreated PKU-baby, e.g., the elevating Phe-concentration, the decreasing concentrations of Trp and other amino acids in the blood, the alterations of amino acid utilization and compartmentation, and maybe even the insufficient expression of liver-Phe hydroxylase may be considered as epiphenomena of an altered metabolic maturation during intrauterine life, i.e. responses rather than failures.

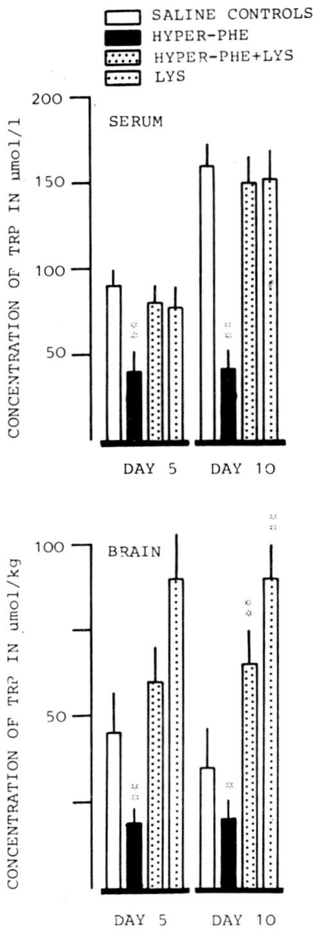

Fig.4: The concentration of Trp in the blood and in the brain of amino acid imbalanced rats at day 5 and 10 p.p. Values are means ± S.D. (N = 9). Asteriks indicate significant differences from respective control values (*p 0.05, **p 0.01, t-test). 40 µl of the following solutions per g body weight were injected in 12 hrs. intervals starting at day 3 p.p. to the developing rats: 0.9% NaCl (CONTROLS), 1.08 g alpha-methyl-Phe plus 2.14 g Phe in 100 ml 0.5% NaCl (HYPER-PHE), this solution plus 1.0 g L-lysine/100 ml (HYPER-PHE+LYS), 1.0 g L-lysine/100 ml 0.5% NaCl (LYS). All solutions were adjusted to pH 7.2 - 7.4.

References

1. Huether, G., Kaus, R., Neuhoff, V.: Neuropediatrics 13, 177-182 (1982).
2. Huether, G., Kaus, R., Neuhoff, V.: Neuropediatrics 14, 12-19 (1983).
3. Linneweh, F., Ehrlich, M.: Klin. Wochenschr. 40, 225-226 (1962).
4. Huether, G., Schott, K. Sprotte, U., Thoemke, F., Neuhoff, V.: Int. J. Devtl. Neurosci. (in press, 1983).
5. Schott, K., Huether, G., Neuhoff, V.: Biochem. Med. 29, 285-292 (1983).
6. Huether, G., Schott, K., Thoemke, F., Sprotte, U., Steinhaus, U., Neuhoff, V.: Neurochem. Int. (in press, 1983).
7. Fernstrom, J.D.: Physiological Reviews 63, 484-545 (1983).
8. Hughes, J.V., Johnson, T.C.: J. Neurochem. 26, 1105-1113 (1976).
9. Zimmerman, R.A., Scott, H.M.: J. Nutr. 87, 13-18 (1965).
10. Cabalska, B., Nowaczewska, J., Duczynska, N.: Biochem. Med. 27, 296-277 (1982).

This work was supported by a grant of the Deutsche Forschungsgemeinschaft (SFB 33).

COFACTOR DEPENDENCY OF SEROTONIN AND CATECHOLAMINE BIOSYNTHESIS IN PATIENTS WITH ATYPICAL PHENYLKETONURIA AND ENDOGENOUS DEPRESSION

A. Niederwieser[1], H.-Ch. Curtius[1], R.A. Levine[2], W. Lovenberg[2], B. Woggon[3], J. Angst[3], H. Müldner[4] and D. Leupold[5]

1 Universitäts-Kinderklinik, CH-8032 Zürich, Steinwiesstr. 75, (Switzerland)
2 National Institutes of Health, Bethesda, MD (USA)
3 Department of Psychiatry, University of Zurich (Switzerland)
4 Landeskrankenhaus Wiesloch (FRG)
5 Department of Pediatrics, University of Ulm (FRG)

The two key enzymes in the serotonin and dopamin biosynthesis, tryptophan 5-hydroxylase and tyrosine 3-hydroxylase, as well as liver phenylalanine 4-hydroxylase, need tetrahydrobiopterin (BH_4) as a cofactor. There are a few neurological and even affective diseases, where a deficiency in the cofactor BH_4 has been proven (atypical phenylketonuria) or may be assumed (dystonia, endogenous depression, Parkinson's disease). Theoretically, such diseases could be treated simply by cofactor substitution. However, peripherally administered BH_4 enters the brain only to a small extent. Pharmacologically high doses will be necessary to produce a central effect in the patients. Preliminary experiments have shown that high-dose BH_4 treatment can be effective in some cases of Parkinson's disease (1,2), dystonia (3), and, as described below, in atypical phenylketonuria and endogenous depression.

Progress in Tryptophan and Serotonin Research

BH_4-deficient hyperphenylalaninemia

The group of various forms of hyperphenylalaninemia includes rare diseases, also named "malignant hyperphenylalaninemia"(4), caused by inborn defects in biopterin metabolism, namely, dihydropteridine reductase deficiency (5), "dihydrobiopterin synthetase" deficiency (6) and GTP cyclohydrolase I deficiency (7). Typical clinical symptoms (4) are muscular hypotonia of the trunc, hypertonia of extremities, progressive retardation in development leading to early death despite dietary restriction of the phenylalanine intake. The neurological pathology seems to be due mainly to a secondary defect of serotonin and catecholamine neurotransmitter biosynthesis.

A trial with high-dose BH_4 (8) was effective in a 5 year old patient with "dihydrobiopterin synthetase" deficiency (9). The girl had been treated since 1978 with a neurotransmitter replacement therapy consisting of L-Dopa, 5-hydroxytryptophan and Carbidopa and BH_4 dihydrochloride, 1.2 mg/kg/day under a normal, phenylalanine unrestricted diet. 36 Hours after withdrawal of all drugs, the child fell into a crisis with muscular hypotonia, ataxia, and symptoms resembling severe drug intoxication. Within 3-4 hours after administration of BH_4 dihydrochloride (22 mg/kg, orally), all symptoms disappeared and the girl was fully active. Only 4 days later, the clinical symptoms reappeared and were again removed by BH_4. A sharp rise in urinary serotonin and dopamine (measured by mass fragmentography using deuterated internal standards) to values above the normal range was observed (9), indicating effective stimulation of peripheral serotonin and dopamine production by BH_4. From the disappearance of all clinical symptoms, it was concluded that sufficient BH_4 entered the brain. Some other patients with the same disease did not respond as good to BH_4 (10) or not at all.

Endogenous depression

There are indications that subgroups of patients with endogenous depression suffer from a decreased activity of central serotonin and/or noradrenalin systems. The stimulation of serotonin production by BH_4 observed in the above patient with atypical phenylketonuria led us to perform therapeutic trials with BH_4 also in patients with endogenous depression. After a single oral dose of BH_4, improvements in mood were observed in two out of 3 patients (1). Out of 3 other patients, one was responding with elevation in mood after BH_4 was given orally at a dose of 1 g early in the morning for 3 days.

This patient (11) unresponding to a variety of antidepressive drugs during 10 years of hospital care, had responded to 5-hydroxytryptophan, but long term treatment was impossible because uncontrollable manic symptoms appeared in addition to nausea and vomiting. Tryptophan had no effect. For two months after the first trial with BH_4, this patient received no antidepressive drugs. Then, BH_4 was administered at a dose of 1 g daily. After 5 days with no response (intermittent vomiting was noted), tryptophan and tyrosine were given in addition to BH_4, in case nausea and vomiting had been restricting her intake of these neurotransmitter precursor amino acids. The next day a dramatic clinical improvement occured that lasted over a week. The dose of BH_4 was decreased in the search for an effective maintenance dose. The patient's mood worsened gradually when under 100 mg. Whereas no correlation was found between mood and urinary free serotonin or dopamine, the elevation in mood coincided with a significant increase of subnormal values of 5-hydroxyindole acetic acid (5HIAA), homovanillic acid (HVA) and biopterin concentrations in cerebrospinal fluid (11). After BH_4 and tryptophan were discontinued on day 49, the metabolites in CSF returned to the low levels found before treatment. The ratios of metabolite:biopterin in CSF were constant, that means that the turnover of biogenic amine was directly proportional to brain BH_4 levels.

Conclusion

These results indicate that some patients with endogenous depression may suffer from cofactor deficiency in brain. The use of BH_4, alone or in combination with tryptophan or traditional therapy, should be considered a novel approach to antidepressive therapy.

Acknowledgement

This work was supported by the Swiss National Science Foundation, project no. 3.266-0.82.

References

1. Curtius, H.-Ch., Müldner, H., Niederwieser, A.: J, Neural Transmission 55, 301-308 (1982).
2. Narabayashi, H., Kondo, T., Nagatsu, T., Sugimoto, T., Matsuura, S.: Proc. Japan Acad. 58, 283-287 (1982).
3. LeWitt, P.A., Newman, R.P., Miller, L.P., Lovenberg, W., Eldridge, R.: New Engl. J. Med. 308, 157-158 (1983).
4. Danks, D.M., Bartholomé, K., Clayton, B.E., Curtius, H.-Ch., Gröbe, H., Kaufman, S., Leeming, R., Pfleiderer, W., Rembold, H., Rey, F.: J. Inher. Metab. Dis. 1, 49-53 (1978).
5. Kaufman, S., Holtzman, N.A., Milstien, S., Butler, I.J., Krumholz, A.: New Engl. J. Med. 293, 785-790 (1975).
6. Niederwieser, A., Curtius, H.-Ch., Bettoni, O., Bieri, J., Schircks, B., Viscontini, M., Schaub, J.: Lancet I 131-133 (1979).
7. Niederwieser, A., Staudenmann, W., Wang, M., Curtius, H.-Ch., Atares, M., Cardesa-Gracia, J.: Eur. J. Pediatr. 138, 97 (1982).

8. Kaufman, S., Kapatos, G., McInnes, R.R., Schulman, J.D., Rizzo, W.B.: Pediatrics 70, 376-380 (1982).

9. Niederwieser, A., Curtius, H.-Ch., Wang, M., Leupold, D.: Eur. J. Pediatr. 138, 110-112 (1982).

10. Endres, W., Niederwieser, A., Curtius, H.-Ch., Wang, M., Ohrt, B., Schaub, J.: Helv. paediat. Acta 37, 489-498 (1982).

11. Curtius, H.-Ch., Niederwieser, A., Levine, R.A., Lovenberg, W., Woggon, B., Angst, J.: Lancet I, 657-658 (1983).

TRYPTOPHAN, 5-HYDROXYTRYPTOPHAN AND DOPA METABOLISM IN THE RAT WITH EXPERIMENTAL PHENYLKETONURIA

D.J. Byrd, M. Arslan, H. Meuter, G. Bodammer
Kinderklinik, Medizinische Hochschule Hannover,
Postfach 610180, 3000 Hannover 61, FRG.

D. Hunneman
Physiologisches Institut der Universität Göttingen,
Humboldt-Allee 7, 3400 Göttingen, FRG.

Introduction

The metabolism of tryptophan (Trp), 5-hydroxytryptophan (5HTrp) and Dopa is impaired in untreated phenylketonuria (PKU). Plasma Trp and the plasma response to an oral load of Trp are subnormal(1). Serotonin (5HT) in blood and 5-hydroxyindoleacetic acid (5HIAA) in urine are below normal (2). Dopamine (DA) and its metabolites are found at low concentrations (3). The exact relationship between these observations and the elevated concentrations of phenylalanine (Phe) and its metabolites in blood and tissues is unknown. The subnormal biosynthesis of 5HT and Da is thought to play a role in the pathogenesis of mental retardation in untreat--ed PKU. It may also be responsible at least in part for the adverse reaction to diet termination in some treated patients (4). It is for these reasons that the underlaying mechanism of these metabolic defects continues to remain a subject of investigative interest.

For resons of ethics, many of the experiments aimed at elucidation of the mechanism can not be carried out on humans and recourse has been taken to animal or cellular models of PKU (5). 4-Chlorophenylalanine (PCP) and alpha-methylphenylalanine are used to inhibit phenylalanine hydroxylase in vi-

Progress in Tryptophan and Serotonin Research

vo and thus induce hyperphenylalaninemia in animals (6,7). With both compounds the residual enzyme activity of 20 to 25% is sufficient to metabolise Phe at such a rate that through treatment with them alone only moderate hyperphenylalaninemia can be induced. Excess Phe given orally or by injection to pretreated animals increases plasma Phe to high levels similar to those seen in untreated PKU and also causes excretion of Phe metabolites in the urine.

Little attention has been paid to the metabolism of Trp, 5HTrp and Dopa in the animal model. The purpose of this paper is to report on the metabolism of these amino acids in a PCP rat model and compare it to that in untreated PKU.

Material and Methods

Chronic experimental PKU was induced in adult, male black-hooded rats (220 - 250 g) by feeding chow containing 0.1 % (w/w) PCP and 3 % (w/w) L-Phe (Karl Roth, Karlsruhe,FRG). Water was given ad lib. Heparinised blood was obtained by puncture of the bubular veniplexus (ether narcosis) or after decapitation. Urine was collected over ascorbic acid and under paraffin oil while animals were kept in metabolic cages. Brains were removed rapidly after decapitation, weighed and placed in liquid N_2 until work up. Whole brains were homogenised in lithium citrate buffer (see below) and the supernatant used for determination of free amino acids. Amino acids in saline solution were given either by stomach tube or by i.p. injection.

All free amino acids were determined with the help of the Biotronik (Puchheim,FRG) amino acid analysers LC 2000 and LC 6000 (DC 4A resin, lithium-citrate-5-buffer system, ninhydrin). Buffers and conditions of analysis were according to the Biotronik instruction manual. In this system Dopa elutes between cystathionine and isoleucine and 5HTrp between ammonia and ornithine. Samples were deproteinated with 3 % sulfosalicylic acid.

Urinary organic acids from ether extracts were determined as trimethylsilyl derivatives by gas chromatography/mass fragmentography with the help of a Varian MAT model CH 7A mass spectrometer (70 ev) coupled via Bieman-Watson separator to a Varian Aerograph model 2740 gas chromatograph (Bremen and Darmstadt, FRG). The column was 5 % SE 52 on Chromsorb WHP, 100/200 mesh; the carrier He and temperature: 100° - 6°/min. up to 240°C.

Results and Discussion

After 14 days of treatment, plasma Phe in the experimental animals rose to a mean value of 1040 ± 349 uMol/l (N = 15), controls: 161 ± 14 uMol/l (N = 5). Tyrosine (Tyr) was near normal at 120 ± 55 uMol/l (controls 117 ± 21 uMol/l). The concentrations of Phe and Tyr were elevated in the brains of PKU animals (tab. 1) and the ratio Phe/Tyr rose from 0.26 to 3.0.

	PKU	PKU + Trp	Controls
Trp	48.3 ± 4.7	59.3 ± 3.3	58.2 ± 4.2
Phe	2460 ± 370	2710 ± 510	78.1 ± 11.0
Tyr	910.3 ± 40.6	890.1 ± 90.0	302.1 ± 29.6

Table 1 : Aromatic amino acids in brain of rats with experimental PKU and controls. Mean ± SD in nMol/g wet wt.. PKU+Trp = exptl. PKU with supplement of 1g/kg L-Trp in chow. N = 5/group.

The gas chromatographic analysis of urine revealed the presence of typical Phe metabolites: phenyllactic acid (PLA) (95 ± 7 uMol/24 h); phenylacetic acid (72 ± 25 uMol/24 h) and phenylpyruvic acid (PPyA) (24 ± 9 uMol/24 h). Controls excreted phenylacetic acid (9 ± 3 uMol/24 h) and hippuric acid (11 ± 4 uMol/24 h).

Basal plasma Trp and the response of plasma Trp to an oral or i.p. Trp load was subnormal in the PKU animals (tab. 2).

min.	0	30	60	90	120
Con	73.1±13.7	129.8±24.3	118.3±1.6	99.0±19.7	81.7±22.1
PKU	20.7± 6.4	65.5±11.9	57.7±7.8	50.7± 9.8	41.3±13.3

Table 2: Plasma Trp in rats with experimental PKU and controls (Con) before and after an i.p. L-Trp load (30 mg/kg b.wt). Mean values ± SD in uMol/l. N = 5/group.

In PKU animals, the maximum in plasma Trp at 30 min. post loading was only ca. 50 % of control value, but the rate of elimination of Trp from blood was slower. The concentration of Trp in brains of PKU rats was decreased but became normal under Trp supplementation (tab. 1). The PKU rats excreted more IAA and ILA than controls (tab. 3) (9).

The plasma responses to 5HTrp (tab. 4) and Dopa (tab. 5) were similar to that following Trp. The excretion of 5HIAA after 5HTrp was two to three times higher in PKU animals (tab. 3). A similar response of Dopac to Dopa was observed: PKU (40.73 ± 2.40 uMol/24 h), controls (9.40 ± 2.78 uMol/24h.

Conclusions: 1) Indolic Trp metabolism in the rat with experimental PKU is similar to that in human PKU; 2) There may exist in PKU alternate pathways for the metabolism of Trp, 5HTrp and Dopa which bypass the corresponding amines

	IAA 1)	ILA 1)	5HIAA 1)	5HIAA 2)	5HIAA 3)
A	0.37±0.08	0.30±0.21	0.025±0.011	5.83±2.90	11.61±1.93
B	1.93±0.55	0.87±0.21	0.013±0.008	16.9±3.7	21.7±2.9

Table 3: Urinary indolic acids in rats with experimental PKU (B) and controls (A). 1) Basal excretion and excretion of 5HIAA: 2) without and 3) with pretreatment with carbidopa (see tab. 4). Mean ± SD in uMol/24h. N = 5/group.

min.	0	30	60	90	120
Con	0.00	673.0±39.6	351.0±57.0	166.3±92.7	103.7±61.7
PKU	0.00	305.1±55.8	189.0±25.4	81.7±46.4	48.1±31.9

Table 4: Plasma 5HTrp in rats with experimental PKU and controls (Con) after an oral load of 5HTrp (30 mg/kg b.wt.) and pretreatment with carbidopa (60 mg/kg chow). Mean values ± SD in uMol/l. Without carbidopa 5HTrp could not be detected after load of same size(see tab. 3 for urinary 5HIAA).

(10,11) and 3) The chemical animal model of PKU appears to offer a tool to study factors affecting Trp and Dopa metabolism in hyperphenylalaninemia.

min.	0	30	60	120
Con	0.00	7.26±1.55	5.10±1.50	2.70±1.06
PKU	0.00	3.29±0.70	1.79±0.66	1.20 (N = 1)+

Table 5: Profile of Dopa in plasma of rats with experimental PKU and controls (Con) after an oral load of L-Dopa (60 mg/kg b.wt.). Mean values ± SD in uMol/l. N = 3/group. +) Detected in only one animal at t = 120 min. See text for urinary excretion of Dopac after load.

References:

1. Yarbro, M.T., Anderson, J.A.: J. Pediatr. 68: 895 - 904 (1966).
2. Pare, C.M., Sandler, M., Stacey, R.S.: Lancet I: 551 (1957).
3. Nadler, H.L., Hsia, D.Y.Y.: Proc. Soc. Exp. Biol. Med. 107: 721 (1961).
4. Olek, K., Wadenbach, P., Byrd, D.J.: Klin. Wschr. 58: 135 - 140 (1980).
5. Hommes, F.A., ed. (1979): Models for the study on inborn errors of metabolism. Elsevier, North Holland Biomedical Press, Amsterdam.
6. Lipton, M.A., Gordon, R., Guroff, G., Udenfriend, S.: Science 156: 248 - 250 (1967).
7. Greengard, O., Yoss, M.S., Del Valle, J.A. Science 192, 1007-1008 (1976).
8. Lane, J.D., Schöne, B., Langenbeck, K., Neuhoff, V.: Cata. Biochim. 627: 144 - 156 (1980).
9. Armstrong, M.D., Robinson, K.S.: Arch. Biochem. Biophys. 52: 287 - 288 (1954).
10. Kido, R., Noguchi, T., Minatogawa, Y.: Acta vitamin. enzymol (Milano) 29: 344 - 345 (1975).
11. Lees, G.J., Weiner, N.: J. Neurochem. 20: 389 - 403 (1973).

THE ROLE OF TRYPTOPHAN OXIDATION IN OCULAR TISSUE DAMAGE.

Seymour Zigman, Department of Ophthalmology, University of Rochester School of Medicine and Dentistry, Rochester, NY USA.

Introduction

Maximal tryptophan photochemical interaction with near-ultraviolet (UV) radiation occurs within the eye. In particular, the ocular lens absorbs nearly all near-UV energy entering the eye by virtue of its high content of protein and yellow pigments. The pigments improve vision in diurnal animals and recently have been found to protect the posterior portion of the eye from near-UV damage. Indeed, even the filtering pigments present in the lenses of diurnal animals are tryptophan oxidation products.

Tryptophan photochemistry has an importance in the functioning of the eye that is secondary only to that of the visual photopigments. While the photopigments are essential in the process of vision, tryptophan oxidation products are central factors in its functional demise. The destruction of tryptophan by photooxidation is due to its high extinction coefficient for UV light, and the tail of absorption into the longer wavelength range. These properties of tryptophan lead to the formation of both stable and unstable photoproducts due to its exposure to environmental near-ultraviolet radiation. Such photoproducts have abiotic properties. This presentation will compare the tryptophan oxidation products that are present in the ocular tissues with those that are formed by exposing tryptophan to ultraviolet light. It will then indicate how basic biological and biochemical processes are adversely influenced by these photoproducts so as to lead to

Progress in Tryptophan and Serotonin Research

ocular tissue damage. Finally, the expression of certain destructive changes in animal and human eyes leading to impaired vision will be related to the presence and action of tryptophan oxidation products.

Tryptophan and its oxidation products have been found in all ocular tissues and fluids. The distribution of these substances in the ocular tissues is shown in Figure 1. Our attention has been focused on the lens, which contains and is bathed by fluids containing tryptophan and numerous oxidation products thereof. The conservative state of the lens due to its encapsulation and poor proteolytic breakdown of proteins makes this tissue a likely site for the buildup of tryptophan oxidation products and photobiological consequences. Figure 1 also lists the precursors, effectors and products that are known to be present in the lens; also listed are protectors aginst the oxidation of tryptophan. Note that near-UV light is a major effector in the lens, which absorbs it strongly. Lower concentrations of oxidizable substances are present in the aqueous humor and in the vitreous humor because both of the fluid compartments return oxidized products to the circulation. The retina contains an abundance of both tryptophan and oxygen, but due to a rapid circulation, little buildup of oxidized tryptophan products seems to occur in this tissue. Neither the vitreous humor nor the retina are normally exposed to near-ultraviolet light, because the lens totally absorbs it anterior to them. Thus, in normal eyes photooxidation would appear not to alter tryptophan in the posterior tissues. However, a large number of humans have had their lens removed due to cataract surgery, and so without an intact lens, these posterior ocular tissues become susceptible to the products of tryptophan produced by near-UV light.

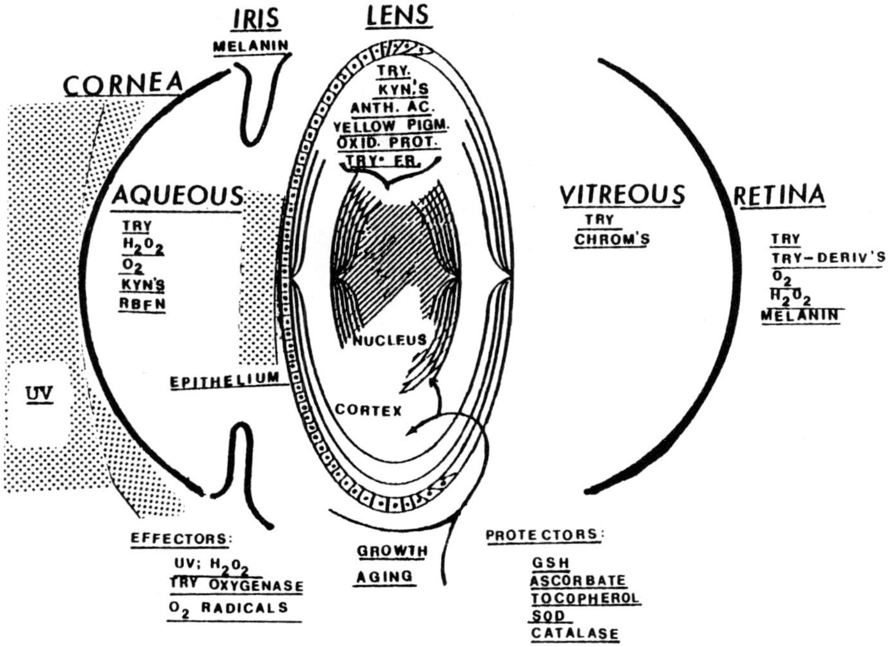

Figure 1.

Distribution of tryptophan and its oxidation products among the ocular tissues. Also listed are the oxidants and antioxidants present. Abbreviations:
TRY, tryptophan; KYN, kynurenine; ANTH.AC, anthranilic acid; OXID. PROT., oxidized protein; TRY.FR, tryptophan free radicals; GSH, reduced glutathione; SOD, superoxide dismutase; RBFN, riboflavin; CHROM'S, chromophores.

Free Radicals

Our particular interest is in the photooxidative processes induced by near-ultraviolet light in the range of 300-400 nm. UV-A (320-400 nm) penetrates the cornea and is nearly totally absorbed by the lens. The most easily detected products of the photooxidation of tryptophan are free radicals, and a series of studies has shown that the human lens contains both stable and unstable free radicals, which resemble or are identified as excited states of tryptophan. Several laboratories have been able to detect these free radicals in human lenses only at low temperatures (1, 2), while other free radicals are detectable at room temperature, and have a very slow decay rate ($t_{1/2}$ of 100 hrs; 3). Figure 2 presents ESR spectra of free radicals that are produced by exposing lens protein plus tryptophan solutions to UV-A. A summary of the free radicals studied thus far is found in Table I. More recent studies of the stable free radicals of human lenses have indicated that they are present at higher levels in the outer cortical region is the younger part of the lens since, with growth, the newer fibers are layed down superficially to the older fibers. The anterior cortex is exposed to the most intense UV-light. The free radicals of tryptophan that are produced in the presence of lens proteins also bind to the proteins which develop a free radical signal similar to that of the free radicals. As this occurs, the signal size per unit protein decreases so that the tryptophan free radicals are partially quenched. Thus paralleling lens aging, the tryptophan free radical intensity **decreases.** It is postulated that this process leads to the formation of covalent cross-links between protein chains which contributes to protein aggregation, light scattering and cataracts.

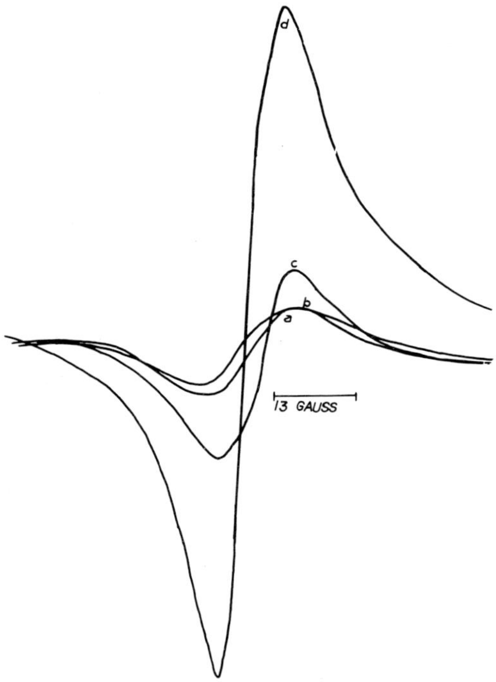

Figure 2.

ESR spectra of calf lens γI-crystallin; a) untreated; b) plus 0.1% tryptophan; c) near-UV irradiated as in Figure 1. ESR spectrometer settings gain, 100; modulation, 5; temperature, $10^{o}C$.; sweep 50 mm.

Table I. Excited states in human lenses.

Sample	Temperature	Signal peak Gauss	G value	Line width Gauss	Microwave Frequency GHz	Power mW	Lifetime (UV-Induced)
(tryptophan)	77 K	(1430)		(25)	—	—	—
Human lens nucleus		3240 radical	2·005	25	9·1	10	4 seconds
Human lens nucleus	77 K	3240	2·007	25	9·1	75	2 minutes
(tryptophan)	295 K	(3251)	(2·0057)	(10)	(9·1)	(2)	100 hours
whole human lens: cortex and nucleus	295 K	3251	2·0052	10	9·1	2	

Near-UV Photobiology of Tryptophan

Photoproducts of tryptophan are produced by exposing neutral aqueous aerobic solutions of tryptophan (10^{-3}M) to near-UV radiation (365 nm; with a spread from 300-400 nm). The irradiance at 365 was maintained at 5-10 mW/cm^2, an irradiance not greater than that found in sunlight. Figure 3 illustrates the separation of such photoproducts on a Sephadex G-10 gel filtration column, as described previously by Sun and Zigman (4). There are six distinct peaks, the last and largest of which is unaltered tryptophan itself. While all of the peaks are not identified, three of them have been: HPI (see legend of Figure 3); NFK (see legend of Figure 3); PPI (see legend of Figure 3). The other peaks are labelled PT-1 (photoproduct of tryptophan 1) and PT-2 (photoproduct of tryptophan 2). Figure 3 also indicates the properties of these molecular species and also summarizes their differential biological actions. PT-1 seems inactive biologically; PT-2 is an an antagonist of cell division, whether in the presence of additional near-UV light or in its absence; HPI has no influence on cell division, whether in the presence or absence of UV-radiation; NFK is known to be an enzyme photosensitizer (5); PPI acts both as a cell division antagonist an an enzyme and protein inhibitor, probaly due to the fact that it is an organic peroxide and mimics the action of hydrogen peroxide in biological systems. PT-2 inhibits both cell division and growth of fertilized sea urchin eggs and mouse embryonic fibroblasts. Table II indicates that the action of this growth inhibiting photoproduct interferes with the synthesis of macromolecules in these cells. The inhibition of macromolecule synthesis has also been shown to occur in the lens. PT-2 appears to be a better macromolecule synthesis inhibitor than are the other tryptophan products under investigation. The interaction of tryptophan oxidation products with proteins is nicely demonstrated by

SEPHADEX G-10 SEPARATION —
NEAR UV PHOTOPRODUCTS OF TRY

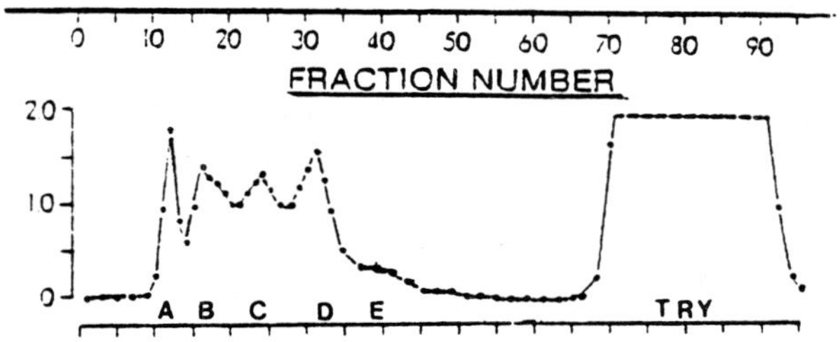

PROPERTIES AND ACTIONS

	MOL. WT.	MAX. ABS.	MAX. FLUOR.	R_F.	ACTION
PT I (A)	384	230;290	310/365	.340	NONE
PT II (B)	290	225;255; 315	310/390	.344	STOPS CELL DIVN. +OR− N-UV
HPI (C)	225	240;290	290/334	.330	INHIBIT ENZYMES-N-UV PHOTSENSITIZER
NFK (D)	200	230;265; 320	320/430	.220	ENZYME PHOTOSENS. FOR N-UV
PPI (E)	225	240;290	290/334	.385	OXID. PROTEINS; KILL CELLS
	(D)	(NM)			

Figure 3.

Separation of tryptophan photoproducts using Sephadex G-10 column gel filtration. Biological actions and physicochemical properties are also listed.

Abbreviations:

MOL. WT, approximate molecular weight; ABS, absorption; FLUOR, fluorescence; R_f, partition coefficient on TLC (silica gel) plates using methanol, water (4:1) solvent; HPI, 3a-hexahydropyrroloindole; PPI, 3a-hydroperoxy-hexahydropyrroloindole.

(a) *Mouse embryonic fibroblasts*	Precursor incorporated		
Additive	*^{3}H-phenylalanine*	*^{3}H-uridine*	*^{3}H-thymidine*
	(counts per minute per dish acid insoluble)		
Dark control tryptophan	16.000 ± 750	2.500 ± 250	12,000 ± 750
Six hour Near-UV exposed tryptophan	4.500 ± 335	1.600 ± 126	3,500 ± 333
Percent inhibition of:			
(a) incorporation	72%	36%	71%
(b) uptake into soluble	37%	22%	14%

(b) *Fertilized sea urchin eggs*			
Additive	(counts per minute per O.D. unit acid insoluble)		
Dark control tryptophan	1,150 ± 180	413 ± 49	520 ± 37
Twenty-four hour Near-UV exposed tryptophan	400 ± 56	210 ± 34	180 ± 39
Percent inhibition of:			
(a) incorporation	65%	51%	66%
(b) uptake into acid soluble	15%	23%	17%

Table II. Inhibition of macromolecule synthesis by near-UV light- exposed tryptophan.

observing the changes in the ocular lens with age and in certain types of cataracts. An important factor to be considered is the exposure of the individuals to sunlight. A number of investigators have examined the protein constituents of the human ocular lens, which for the most part have been altered by interactions with tryptophan oxidation products, either in free-form or within the protein polypeptide chain. With increasing age and with near-UV exposure, the slightly yellow human lens pigment increases its density to become dark yellow and even brown. Accompanying this change in pigmentation is the development of a longer wavelength fluorescence than that of free or protein-contained tryptophan. This characteristic fluorescence has an excitation in the near-ultraviolet (UV-A; 360 nm excitation) and an emission over 400 nm (ie: 440 nm). This non-tryptophan fluorescence is measurable in the whole lens, and is associated mainly with the insoluble fraction (or the highly aggregated proteins plus fiber cell membrane fragments). In some cases, the highly aggregated formerly soluble protein has become associated with fiber cell membranes, so as to cause a high degree of non-tryptophan fluorescence that is due to crosslinking reactions involving tryptophan oxidation products. Figure 4 shows the distribution of the abnormal fluorescence among the protein fractions of differing size that are extracted from lenses of humans active either in high sunlight or low sunlight environments. This fluorescence is present in abundance in the highly aggregated fraction, mainly in the sunlight exposed individuals. Several of the fluorescent crosslinks have been identified by Dillon _et al_ (6), Garcia-Castineiras _et al_ (7), and Augusteyn (8) to be beta-carbolines, N-formylkynurenine, bityrosine, and anthranilic acid (associated with the highly aggregated lens proteins). The fluorescence associated with the highly aggregated human lens proteins is thus due to tryptophan oxidation products (other than bityrosine).

Not only are structural proteins adversely influenced by tryptophan photoproducts, but enzymes with important functions in maintaining the transparency of the lens have also been found to be UV-sensitized by tryptophan oxidation products. Catalase loses its activity when exposed to near-UV light in

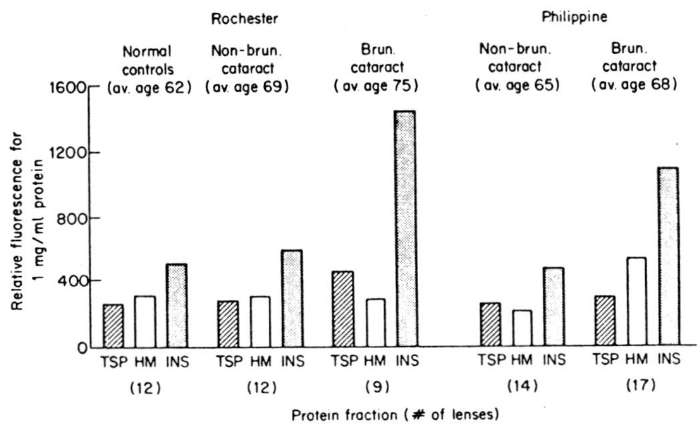

Figure 4.

Non-tryptophan fluorescence of the protein fractions of cataractous human lenses (number of lenses in parentheses).

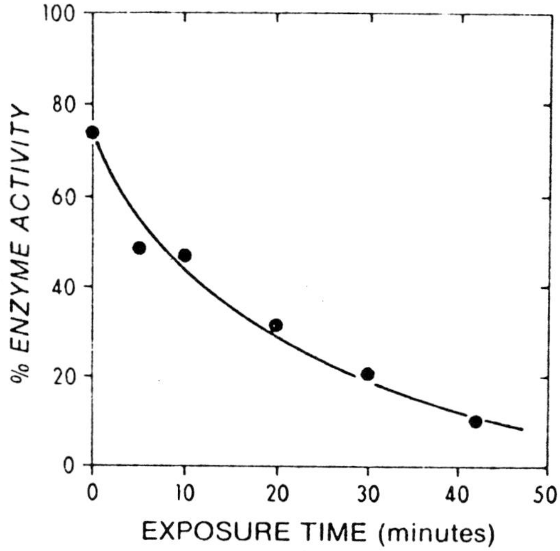

Figure 5.

Time course of near-UV inhibition of 15-PGDH activity in the presence of 6 μM PT II. 100% activity is the activity of the enzyme in dark with no PT II added.

the presence of tryptophan or its products (9). Chemical findings indicate changes in the protein of enzymes as well as their prosthetic groups. Since hydrogen peroxide is continually being generated in the lens by metabolism and can also enter it from the aqueous humor (Spector et al (10), catalase is needed to prevent damage to the lens by H_2O_2 oxidation. If catalase is inactivated by near-UV light, an enhancement of H_2O_2 concentration will result causing protein oxidation and complete metabolic breakdown leading to cataract.

Another enzyme prostaglandin dehydrogenase, is quite important to the retina in the control of inflammation since it destroys an excess of prostaglandins that enhance the inflammatory process. Figure 5 shows the near-UV photosensitivity of prostaglandin dehydrogenase. Glutathione reductase is another enzyme whose inactivation would be detrimental to cells and tissues. This enzyme aids in the oxidation-reduction balance, since it serves as a reducing agent for glutathione (GSH) which is very much involved in in the protection of lens protein sulfhydril groups. The oxidation of GSH to its disulfide would be most damaging to the lens by virtue of its influence both on structural and enzymatic proteins. For this enzyme it was also found that purified HPI was quite effective in reducing the activity of the enzyme by photosensitization (11).

Several other enzymes have been found to be photosensitized to near-UV light by tryptophan photoproducts, which include: cytochrome oxidase, xanthine oxidase and Na/K ATPase.

Certain types of cataract are known to be induced in animals due to salt and osmotic imbalance resulting from a decreased activity of Na/K ATPase in the lens. When this enzyme is deficient, the lens swells and structural anomalies such as liquid lakes and irregularities of refractive index lead to its opacification. If the tryptophan present was converted to near-UV photosensitizers, the inhibition of Na/K ATPase in the lens by exposure to near-UV light would enhance the formation of cortical cataracts (12). Na/K ATPase is also essential for the transduction of light absorption into the visual impulse in the photoreceptor outer segments of the retina. The adverse influence of near-UV and tryptophan photoproducts on the retina would only apply when the retina is exposed to ultraviolet light, as in the eyes with the lens removed (13). Figure 6 indicates the loss of ATPase activity in the bovine retina due to near-UV light exposure with tryptophan added to the incubation medium.

Altered Properties of Ocular Fluids

The posterior portion of the eye would be much more susceptible to damage from near-UV radiation due to tryptophan product photosensitization after the lens has been removed, since the lens is the major anterior eye UV filter. There is reason to believe that photochemical alterations of tryptophan can result from the exposure of the vitreous humor to near-UV radiation, as would occur in a lensless eye. This is based on findings that the aqueous humor, the anterior fluid of the eye, readily develops tryptophan photoproducts due to near-UV exposure. These photoproducts were found to bind to the serum proteins in the aqueous humor and markedly altered their chemical and physical properties (see 14). Similarly to the aqueous humor, the vitreous humor is 98% water with low concentrations of salts, amino acids, nucleotides, and a collagenous type of protein (that serves

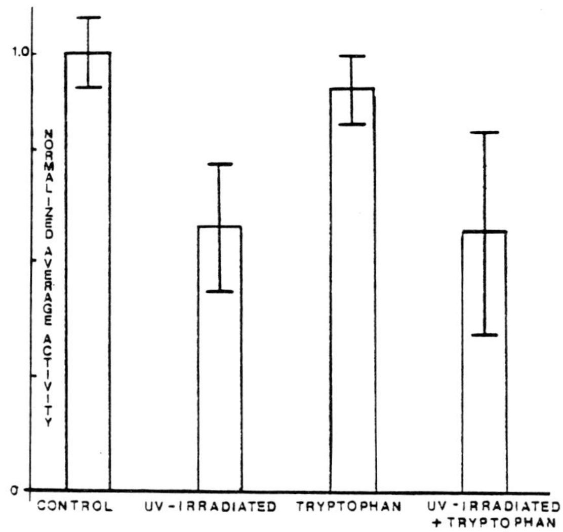

Figure 6.

Inhibition of ATPase activity in bovine retina by UV-A (320-400 nm) radiation, plus and minus tryptophan as a photosensitizer (*in vitro*). Assayed by luciferin-luciferase phosphorescence as measured using a luminometer.

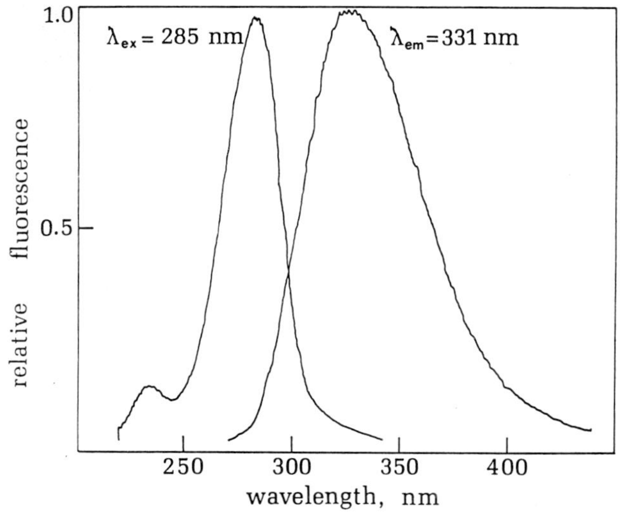

Figure 7.

Fluorescence spectrum of fresh bovine vitreous humor as measured in a recording spectrofluorometer (Perkin-Elmer 512) using 1 cm cells.

to maintain the viscous gel state).

The vitreous humors of animals and man emit a fluorescence characteristic of tryptophan, with excitation maxima at 285 nm and emission maxima at 331 nm (see Figure 7). When the vitreous humor is exposed in vitro to near-UV radiation, a loss of tryptophan fluorescence results within 5 hrs of exposure, as shown in Figure 8. The decay of this fluorescence is also shown. To assess whether this process takes place in the vitreous humor in vivo, guinea pigs with one lensless eye and one normal eye were exposed to ambient near-UV radiation. The vitreous humors were examined for tryptophan fluorescence. Compared with the normal eye, lensless eye vitreous humors exhibited markedly diminished tryptophan fluorescence (see Figure 9). This shows that ambient near-UV can readily photooxidize vitreous humor tryptophan in an eye without a lens.

Retinal Effects of Near-UV Radiation Plus Tryptophan Products

No major study has as yet been reported that assesses the adverse influence of tryptophan products plus near-UV light on the physical integrity or normal electrophysiological function of the retina. There is no doubt that the additional tryptophan products plus greater near-UV irradiances reaching the retina in lensless eyes would lead to photochemical, structural, and functional defects. Indeed, the retinas of monkeys and of squirrels have been found to be damaged by near-UV radiation only when the lens was absent (see Ham, et al 15), and Collier *et al* 16). While the mechanism of this damage is not yet known, a major working hypothesis should include damage relative to near-UV photosensitization by tryptophan products.

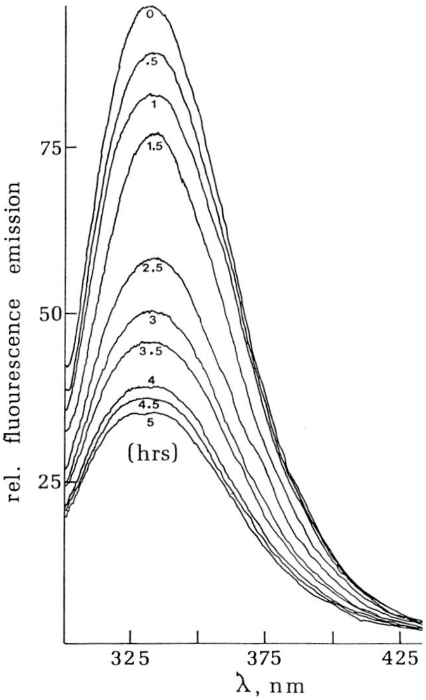

Fig.8

Loss of bovine vitreous humor TRY fluorescence due to exposure to UV-A (320-400 nm) radiation (in vitro).

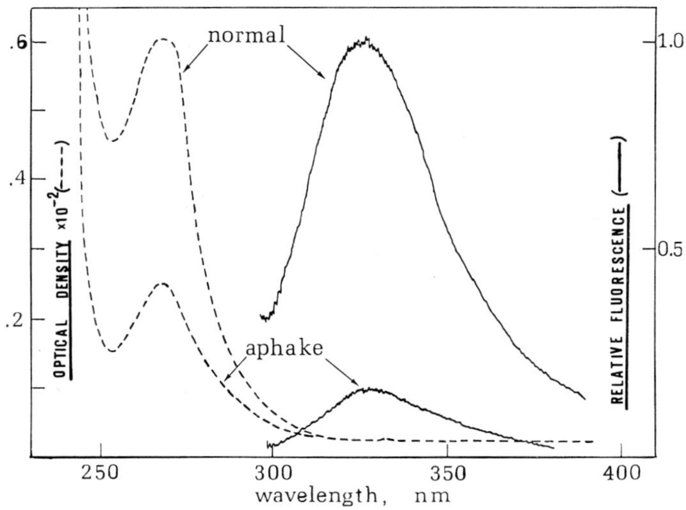

Figure 9.

Loss of guinea pig vitreous humor TRY absorbance and fluorescence due to 3 months of *in vivo* exposure to UV-A (320-400 nm) radiation at 1 mW/cm^2. One eye was normal (*ie*: lens intact) while the other eye had no lens present (*ie*: Aphakic) due to experimental lensectomy).

The types of damage observed in lensless squirrel eye retinas relative to near-UV exposure could be related to the types of photochemical change induced by tryptophan products. This damage will now be described in some detail. Figure 10 presents morphologically observed damage to squirrel retinas due to the ambient exposure to 40W BLB lamps of animals with one lensless and one normal eye. Figure 11 shows that the lensless eye is deficient in its electrical activity, as estimated by electroretinography. These figures show that filtration of near-UV light by the lens pigments prevents gross destruction of the photoreceptors and other retina cell types and loss of electrical activity (16).

While these studies do not show a direct relationship between retinal damage due to near-UV radiation and tryptophan photochemistry, the damage observed resembles that expected for tissues photosensitized by tryptophan. The previous discussion has shown that tryptophan products plus near-UV radiation interfere with macromolecular synthesis in various cells. Lack of macromolecular synthesis is also a cause of certain retinal degenerations. Loss of retinal electrical activity has been associated with transductional defects which would certainly occur if the Na/K ATPase activity were diminished. Decreases of this enzyme's activity have been described above for retinal tissue exposed in vitro to near-UV energy with added tryptophan present. Decreased activity of enzymes relative to the oxidation-reduction balance in cells could result in oxidative damage relative to peroxide generation (catalase) and generalized -SH enzyme inhibition (glutathione reductase). Light damage to the retina has been associated with the inflammatory process, which is stimulated by inhibition of prostaglandin breakdown enzymes (prostaglandin dehydrogenase). Other mechanisms for near-UV damage that must also be considered are retinoic acid photosensitization and toxic melanin-derived free radical reactions.

Conclusion

The most pronounced and well-known influence of try photooxidation seems to occur in the lens, which absorbs this radiation and contains a multitude of products derived from its oxidation and photooxidation. Products formed in the aqueous humor can diffuse into the lens to enhance the changes leading to certain types of cataract. Brown cataracts do appear frequently in humans living near the equator (where near-UV light is intense). Retinal damage has been observed in animals exposed to near-UV light after the lens has been removed. Future work on try near-UV photochemistry will provide clinicians with information useful in the prevention of cataract and retinal degeneration.

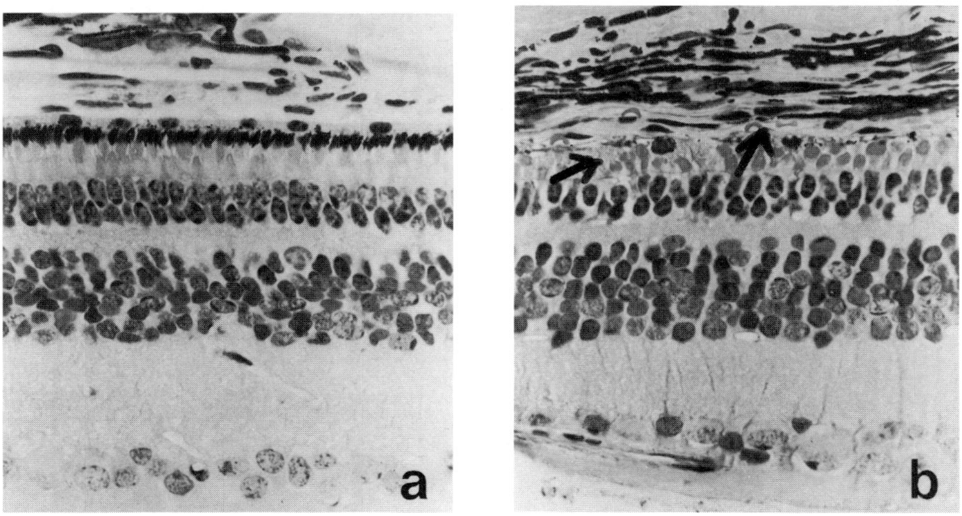

Figure 10.

Morphologically observed damage to the retinas of a grey squirrel (Sciurid) exposed for 4 months in vivo to UV-A radiation (320-400 nm) at 1 mW/cm²; a) normal eye; b) aphakic eye. Fixed in glutaraldehyde; stained with hematoxylin plus eosin; magnification 400 X. Note photoreceptor and pigment epithelium damaged.↗

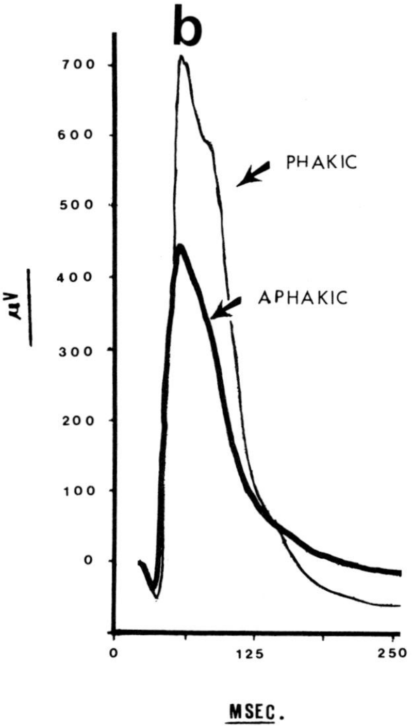

Figure 11
Electroretinographic representation of electrical activity of the eyes of a grey squirrel (Sciurid) exposed to UV-A (320-400 nm) radiation (1 mW/cm^2) for 3 months.
Note loss of intensity of the b-wave.

References

1. Weiter, J., and Finch, E,: Nature (London) 254, 536-537 (1975)
2. Borkman,R.F., Lerman, S.:Exp. Eye Res. 25, 303-309 (1977).
3. Zigman, S.: Documenta Ophth. 8, 267-274 (1976).
4. Sun, M., Zigman, S.:Photochem. and Photobiol. 29, 893-897 (1978).
5. Walrant, P., Santus, R.: Photochem. and Photobiol. 19, 411-417 (1974).
6. Dillon, J., Spector, A., Nakanishi, K.:Nature (London) 259, 422-423 (1976).
7. Garcia-Castineiras, S., Dillon, J., Spector, A.:Exp. Eye Res. 26, 461-476 (1978).
8. Augusteyn, R.: Ophthal. Res. 7, 217-224 (1975).
9. Zigman, S., Yulo, T., Greiss, G.: Mol. Cell. Biochem. 11, 149-154 (1976).

10. Spector, A., Garner, W.H.: Exp. Eye Res. 33, 673-681 (1981).

11. Kalustian, A., Zigman, S., Sun, M.:Proc. Amer. Soc. Photobiol. (Burlington,VT) pp 112-113 (1978).

12. Kinoshita, J.H.: (Protor Lecture) Invest. Ophthal. 13, 713 (1974).

13. Zigman, S.:Documenta Ophthalmological (in press, 1983).

14. Zigman, S., Schultz, J.B., Yulo, T. and Grover, D.: Israel J. of Med. Sci. 8, 1590-1595 (1972).

15. Ham, W.T., Mueller, H.A., Ruffolo, J.J., Guerry,D., Guerry, R.K.: Am. J. Ophthal. 93, 299-306 (1982).

16. Collier, R., Zigman, S., Waldron, W.: (Suppl.) Invest. Ophthalmology and Vis. Sci. 23, 250 (1983).

TRYPTOPHAN AND KYNURENINE IN SENILE CATARACT

Graziella Allegri, Mario R. Angi*, Carlo Costa, Antonio Bettero

Istituto di Chimica Farmaceutica - Centro di Studio sulla Chimica del Farmaco e dei Prodotti Biologicamente Attivi del CNR e *Clinica Oculistica - Università - Padova - Italy

Introduction

In previous articles (1-3) the authors have shown that there was an increase in the urinary excretion of tryptophan (Trp) metabolites, both as total per cent and as kynurenines, in patients affected by senile cataract after an oral load of 50 mg/kg b.w. L-Trp. In addition, no significant differences were found in the free and total serum Trp levels between fasting normal subjects and cataract patients, whereas a significant decrease of the content of free Trp appeared in the lens following the evolution of the cataract and the progressive formation of the pigment. On the other hand, Cotlier et al. (4,5) found the total and free Trp were raised in the plasma of senile cataract and diabetic patients. In this paper, in order to confirm our previous findings, we have investigated the concentration of serum Trp in another group of cataract patients and in diabetics, in comparison with age and sex matched controls by using a new highly sensitive technique involving HPLC. Trp and kynurenine (Kyn) content has also been determined in aqueous humour of cataract patients and clear lens controls.

Materials and methods

14 Male Patients (mean age 55) affected by senile cataract,

Progress in Tryptophan and Serotonin Research

16 diabetic patients (mean age 54), 4 diabetics with cataract, and 15 controls (mean age 56) were studied for the analysis of the serum Trp after admission to the Department of Ophthalmology and Internal Medicine at the University of Padua. Blood samples were collected between 7.50 and 8.20 a.m. in fasting subjects, and treated as previously reported (6). Total and free Trp and Kyn were measured by HPLC with fluorescence and UV detection, respectively (Costa C. et al., this volume). Aqueous humour samples for determination of Trp and Kyn were collected by paracentesis under local anaesthesia from fasting subjects: 3 healthy volunteers (mean age 64), and 5 cataract patients (mean age 67). Aqueous humour samples were also obtained from 3 patients in coma vigil (averaging in age 62 y) before and after 3 hours following an oral load (50 mg/kg b.w.) of L-Trp, administered through a nasogastric tube; the coma was caused by a subarachnoidal hemorrage in healthy subjects.

Results and discussion

Table 1 reports the mean ± S.E.M. of the total and free Trp serum levels in controls, senile cataract patients and diabetics with cataracts and in diabetics with clear lenses. It can be seen that the Trp values show no significant differences among the groups examined. However, Cotlier et al. (4,5) found higher levels in plasma of cataractous patients and diabetics with cataract suggesting that this increase may reflect a partial vitamin B6 deficiency. But, we did not observe high urinary excretion of xanthurenic acid (XA), an index of the vitamin B6 deficiency, after load of L-Trp in cataract patients (2). The mean concentrations (mg/l±S.E.M.) of Trp and Kyn in the aqueous humour of the controls were 8.20±1.81 and 0.21±0.08, respectively; in cataractous aqueous humour they were 8.19±0.95 and 0.93±0.26. The comparison of these data between the two groups of subjects shows that the aqueous humour Trp content is similar, whereas the Kyn concentration is significantly higher in cataract patients ($P > 0.001$). Preliminary results showed that after an oral dose of L-Trp (50 mg/kg b.w.) both in controls and in cataract patients, Trp and Kyn levels increased in aqueous

Table 1. Mean (±S.E.M.) total and free serum tryptophan levels.

Subjects	Serum Trp mg/l total	Serum Trp mg/l free
Controls (15)	13.86±0.66	2.31±0.18
Cataract patients (14)	12.87±0.76	2.46±0.24
Diabetics (16)	12.77±0.64	1.87±0.26
Cataract diabetics (4)	13.05±0.84	1.83±0.20

humour. Therefore the concentration of these two amino acids in the aqueous humour seems to be related to their plasma levels. To confirm this, the levels of Trp and Kyn in the serum and in aqueous humour of 3 comatose patients have been determined before and 3 hours after an oral load of

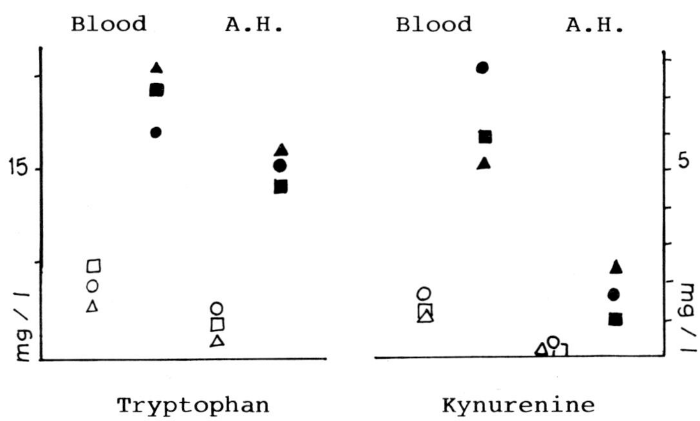

Fig. 1. Trp and Kyn blood and aqueous humour (A.H.) levels before (□○△) and after (■●▲) an oral load of L-Trp (50 mg/kg b.w.). Symbols represent the same patient both in the blood and in the aqueous humour.

L-Trp. Fig. 1 shows the evident rise of these two amino acids after loading. Therefore we conclude that high serum levels of Kyn could lead to higher levels of this compound in the aqueous humour and then in the lens; once inside the lens it can be transformed into other compounds contributing to the formation of the yellow-brown pigments, that are one of the major features of senile cataracts.

References

1. Allegri, G., Angi, M.R.: Lancet ii, 158 (1981).
2. Costa, C., Angi, M.R., De Carli, M., Vanzan, S., Allegri, G.: Boll. Soc. It. Biol. Sper. LVIII, 333 (1982).
3. Allegri, G., Angi, M.R., Costa, C., Vanzan, S.: Boll. Soc. It. Biol. Sper. LVIII, 340 (1982).
4. Cotlier, E., Sharma, Y.R., Lancet i, 607 (1980).
5. Cotlier, E., Sharma, Y.R., Zuckerman, J., Puklin, J., Teasley, B., Irvine, J.: Exp. Eye Res. 33, 247 (1981).
6. De Antoni, A., Allegri, G., Costa, C., Vanzan, S., Bertolin, A., Carretti, N., Zanardo, V.: Acta Vitaminol. Enzymol. 2, 17 (1980).

URINARY 5-HYDROXYINDOLEACETIC ACID AND SERUM PHENOBARBITAL OF CHILDREN SUFFERING FROM EPILEPSY

Dragoslav Milovanović

Institute of Analytical Chemistry, Faculty of Pharmacy, Belgrade, Yugoslavia

Branka Stanković, Ljiljana Milovanović, Zagorka Koričanac, Dragan Trpković

Institute of Analytical Chemistry, Faculty of Pharmacy, Gelgrade, Yugoslavia

Introduction

Epilepsy is a common disease of childhood. Seizures, which characterize this long known and sometimes unpredictable illness, are only a symptom of biochemical disturbances of organic damage to the central nervous system (CNS).

Some investigations associate epilepsy with brain amines (1). Tryptophan, 5-hydroxytryptamine (5-HT) and 5-hydroxyindoleacetic acid (5-HIAA) have been subjects of investigation in patients with epilepsy (1, 2, 3). Determination of 5-HIAA in the urine serves as a readily measurable index of the metabolized tryptophan via the 5-hydroxyindole route (4, 5).

Investigating 5-HIAA in mental retardation we have observed smaller concentrations of this acid in a group of children who were receiving anticonvulsant medication (data communicated at the Congress in Kyoto, 1980). Wishing to test this observation, we undertook to investigate urinary 5-HIAA and serum phenobarbital in children with epilepsy.

Progress in Tryptophan and Serotonin Research

Materials and Methods

For the examination of urinary 5-HIAA a group of children with epilepsy was selected. Indications for treatment were made after investigating each child, including EEG and examination of the fundus oculi. Anticonvulsant therapy (only phenobarbital) was given every day in the evening. Children who were receiving any other therapy were not included in this investigation.

The somatic, neurological and psychiatric status of the children studied was normal. At least three days before urine and blood sampling for analysis were taken, patients had not been allowed to eat bananas, nuts etc.

Seventy nine children of both sexes (28 female, 51 male) and different age (49.37% up to 10 years of age, range 1.5 to 18) were included in this investigation. In each patient there were investigated: onset ($\overline{X}$=6.52 ± 4.82 years of age) and duration of the disease ($\overline{X}$=3.37 ± 3.25 years of age). Daily doses of phenobarbital were 15 to 200 mg, the mean values and SD being 73 ± 9.40. These doses calculated on the basis of body weight were $\overline{X}$=2.34 ± 1.19, range 0.89 - 6.23 mg/kg.

The control group included thirty nine healthy children of similar age - up to ten years of age - 29; from ten to eighteen - 10; girls - 21, boys - 18.

5-HIAA was determined in the morning urine, using the spectrophotometric method (5). At the same time, blood for analysis was also taken, on an empty stomach from the cubital vein in an interictal period. Phenobarbital levels were determined with an automatic analyser (Gilford 3500) by the enzyme multiplied immunoassay technique (EMIT) (6).

Results and Discussion

The values of urine 5-HIAA and serum phenobarbital are presented in Figure 1.

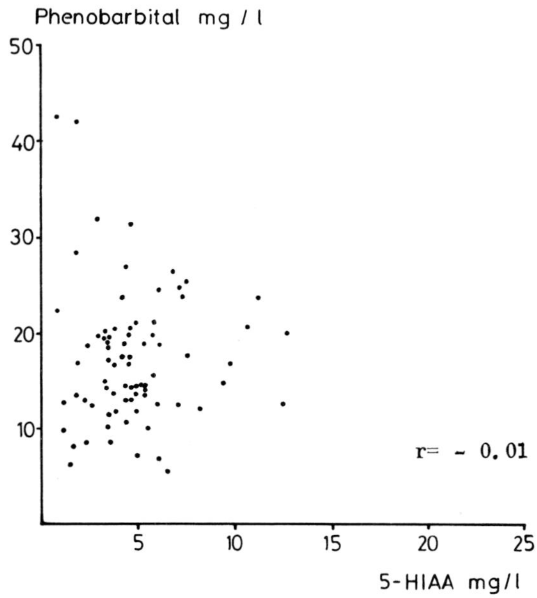

Figure 1. Relationship between the values of urinary 5-HIAA and serum phenobarbital in patients with epilepsy

Although no correlation between 5-HIAA and phenobarbital was found (r= - 0.01), a certain regularity in the grouping of data can be noticed: (1). The highest values of phenobarbital correspond to the lowest concentrations of 5-HIAA. (2) If the levels of serum phenobarbital are somewhat lower, the concentrations of urine 5-HIAA will be the highest. One of the possible explanations is that higher values of phenobarbital have a repressing effect on tryptophan metabolism, which at lower concentrations of the drug in the blood is not observed.
The levels of serum phenobarbital are: 17.31 ± 6.92 ($\overline{X}$ ± SD), from 5.4 - 42.5 mg/l. In most children they were in the optimal therapeutic range.
The levels of 5-HIAA in relation to the clinical characteri-

stics are summarized in Table 1.

Significantly higher concentrations of urine 5-HIAA were found in children with longer duration of the disease, with relatively rare seizures and in the group of patients with partial epilepsy. In 34.18 % of the seventy nine patients examined 5-HIAA was raised. This rise is statistically significant ($P<0.01$).

Table 1. Clinical Characteristics and 5-Hydroxyindoleacetic Acid in the Urine in Children with Epilepsy

		5-HIAA (mg/l) N	X	SD	T-test		P
Duration of epilepsy (years)	up to 1	28	4.29	2.24	1.19	>	0.05
	1 - 3	19	4.82	2.33	2.47	<	0.05
	> 3	32	5.70	4.46	2.42	>	0.05
Type of epilepsy	GE	40	4.91	3.84	1.85	>	0.05
	PE	33	5.16	2.83	2.74	<	0.01
	FC	6	4.58	3.07	1.33	>	0.05
Seizure frequency	weekly or monthly	37	4.50	2.43	1.75	>	0.05
	yearly	15	5.94	2.94	4.18	<	0.001
No seizures 1 - 3 or more years		27	5.14	4.49	1.58	>	0.05
Total number		79	4.99	3.36	3.06	<	0.01
Control group		39	3.76	0.85			

GE: Generalised epilepsy; PE: Partial epilepsy with or without loss of consciousness; FC: Febrile convulsions

The reduced values of 5-HIAA in CSF ought to be caused by the action of anticonvulsant drugs (2). Lower levels of 5-HT in the blood of patients with epilepsy, estimated at 8 and 12 a.m., were found in the second determination. Such a finding is explained by the inhibitory action of anticonvulsant drugs or by the increased values of tryptophan in the CNS (7). A negative

correlation between CSF amine metabolites and serum anticonvulsant concentrations would be expected if either phenobarbital or diphenylhydantoin affected 5-HIAA (1).
A reduced excretion of 5-HIAA in the urine was found in patients with epilepsy, but the values were somewhat higher in the patients with more frequent seizure (3).
The presented results of investigation of urine 5-HIAA point indirectly to a possible disturbance of tryptophan metabolism in children with epilepsy. Such findings can only partly be explained by the suggested repressing action of phenobarbital on tryptophan metabolism. A more likely explanation may be given by the following factors: duration of the disease, type of epilepsy, number of seizures.

Conclusion

Urinary 5-hydroxyinoleacetic acid and serum phenobarbital were determined in seventy nine children with epilepsy. The obtained data suggest that higher concentrations of phenobarbital in the blood have a repressing effect on tryptophan metabolism. Concentrations of urinary 5-HIAA are dependent upon some clinical characteristics of the disease. Out of 79 patients investigated, 5-HIAA was raised in 34.18 %.

References

1. Shaywitz, B.A., Cohen, D.J., Bowers, M.B.: Neurology 25, 72-79 (1975).
2. Young, S.N., Gauthier, S., Anderson, G.M., Purdy, W.C.: Journal of Neurology, Neurosurgery, and Psychiatry 43, 438-445 (1980).
3. Avanesova, T.S., Sviridova, E.I.: Zurn, Nevropatologii i psihiatrii im. Korsakova 80, 857-863 (1980).
4. Haverback, B.J., Sjoerdsma, A., Terry, L.L.: The New England Journal of Medicine 255, 270-272 (1956).

5. Udenfriend, S., Titus, E., Weissbach, H.: J. Biol. Chem. 216, 499-505 (1955).

6. Dietzler, D.N., Hoelting, C.R., Leckie, M.P., Smith, C.H., Tieber, V.L.: Amer. J. of Clin. Pathol. 74, 41-50 (1980).

7. Chiodi, F., Paris, L., Casacchia, M.: Boll. Soc. It. Biol. Sper. 57, 2372-2376 (1981).

ANTICONVULSANT DRUGS PRODUCE DIFFERENT EFFECTS ON PLASMA TRYPTOPHAN CONCENTRATIONS IN EPILEPTIC PATIENTS

Judith Pratt, Peter Jenner, Edward H. Reynolds, Simon D. Shorvon
University Department of Neurology, Institute of Psychiatry and King's College Hospital Medical School, Denmark Hill, London. SE5, UK.

Introduction

Some anticonvulsant drugs elevate cerebral 5-hydroxytryptamine (5-HT) and 5-hydroxyindoleacetic acid (5-HIAA) concentrations in rodents (1) probably as a result of reduced 5-HT turnover(2). However, anticonvulsants produce inconsistent changes in 5-HIAA and tryptophan concentrations in the cerebrospinal fluid of epileptics (3,4). In both animals and man altered 5-HT metabolism generally occurs at dosage levels greater than those required for anticonvulsant activity (2,3). Since free plasma tryptophan concentrations influence brain tryptophan concentrations and 5-HT function (5) we have examined the effects of some anticonvulsant drugs on plasma tryptophan concentrations in epileptic patients in comparison to untreated epileptics and normal volunteers.

Patients and Methods

Patients consisted of 108 drug treated epileptics (mean age 30.1 years, range 6-75) and 24 untreated epileptics (mean age 29.9 years, range 12-61). Seventy patients were receiving carbamazepine (CBZ), diphenylhydantoin (DPH), or phenobarbitone (PHB). The remainder received primidone alone or multiple drug therapy. In the drug treated group 47 had grand mal epilepsy, 32 partial epilepsy and 27 grand mal accompanied by partial epilepsy. The control group consisted of 33 normal volunteers (mean age 29.0 years range 16-62). Venous blood (20 ml) was taken for analysis of total and free plasma tryptophan concentrations (6). No dietary or fluid restrictions were imposed on the subjects.

Results

Untreated epileptics had slightly lower total and bound, but not free, plasma tryptophan concentrations than normal volunteers (Table 1). In patients receiving anticonvulsant medication the most prominent

Progress in Tryptophan and Serotonin Research

effect was for CBZ therapy to enhance total, bound and free plasma tryptophan concentrations in comparison with untreated epileptics and normal volunteers (Table 1). In contrast, DPH and PHB reduced free plasma tryptophan concentration in comparison to values found in both control groups. DPH, but not PHB, caused a decline in total and bound tryptophan concentrations compared to normal volunteers (Table 1).

Table 1 Plasma tryptophan concentrations (μg/ml ± 1 SEM)

Group	n	Total	Bound	Free
Untreated epileptics	24	9.15±0.35^x	7.63±0.37*	1.52±0.12
Normal volunteers	33	10.13±0.25	8.78±0.23	1.34±0.05
Diphenylhydantoin	39	8.66±0.27**	7.45±0.22**	1.21±0.08*†
Carbamazepine	24	12.72±0.71**‡	9.94±0.52^x‡	2.75±0.19**‡
Phenobarbitone	7	8.89±0.52	7.89±0.51	1.00±0.09*†

xp < 0.01, **p < 0.001 compared with normal volunteers
†p < 0.05, ‡ p < 0.001 compared with untreated epileptics

The effect of CBZ on total, bound and free plasma tryptophan concentrations appeared to be related to serum drug concentrations. Values for free plasma tryptophan were 2.55 ± 0.20 μg/ml (N = 9), 2.90 ± 0.39 μg/ml (N = 9) and 3.51 μg/ml (N = 2) respectively in patients with subtherapeutic, therapeutic and toxic serum concentrations of CBZ in comparison to concentrations of 1.34 ± 0.05 μg/ml in normal volunteers. In contrast, DPH produced a slight decline in free tryptophan as drug serum concentrations increased although total and bound tryptophan was unchanged.

Support for a differential effect of CBZ and DPH on plasma tryptophan concentrations was indicated in a study of 10 untreated epileptics who were re-examined on drug therapy. Carbamazepine (N = 7) increased total and free plasma tryptophan concentrations from 8.36 ± 0.71 μg/ml and 1.38 ± 0.19 μg/ml respectively to 10.62 ± 0.34 and 2.19 ± 0.23 μg/ml ($p < 0.05$). DPH (N = 3) therapy slightly reduced total and free plasma tryptophan concentrations (10.32 ± 0.78 and 1.53 ± 0.05 μg/ml to 7.99 ± 0.55 and 0.99 ± 0.16 μg/ml).

Type of seizure or the degree of seizure control had little effect on tryptophan parameters. There were no differences in plasma tryptophan concentrations in patients experiencing seizures the preceeding month (poorly controlled) and those who had not (well controlled). Patients with grand mal or grand mal accompanied by partial epilepsy had similar plasma tryptophan concentrations to those in control groups. Epileptics suffering from partial epilepsy, however, had slightly reduced bound plasma tryptophan concentrations compared to normal volunteers although total and free tryptophan were unchanged.

Conclusions

Overall, epilepsy did not influence plasma tryptophan concentrations. CBZ and DPH produced opposite effects on plasma tryptophan concentrations by an unknown mechanism. CBZ is structurally related to tricyclic antidepressant drugs which may elevate plasma tryptophan concentrations by inhibition of liver tryptophan pyrrolase (7). CBZ may have similar effects. DPH may decrease bound plasma tryptophan through competition with the amino acid for its albumin binding site and both DPH and PHB may enhance the activity of tryptophan pyrrolase by enzyme induction.

There is little difference in the antiepileptic efficacy of DPH and CBZ and so their differential effects on plasma tryptophan concentrations are unlikely to be related to anticonvulsant activity. However, the two drugs may have different chronic effects on mental function. In comparative studies of CBZ and DPH in normal volunteers and epileptic patients,tests of cognitive function have consistently shown less impairment in CBZ treated patients (8,9). The opposite effects of CBZ and DPH on tryptophan concentrations may more closely reflect different actions of the drugs on mental function.

References

1. Bonnycastle, D.D., Giarman, N.J., Paasonen, M.K.: Br. J. Pharmac. 12, 228-231 (1957).

2. Pratt, J.A., Jenner, P., Marsden, C.D. In: Neurotransmitters Seizures and Epilepsy (Morselli, P.L. et al, eds), pp. 227-238. Raven Press, New York (1981).

3. Chadwick, D., Jenner, P., Reynolds, E.H.: Lancet i, 473-476 (1975).

4. Young, S.N., Gauthier, S., Anderson, G.M., Purdy, W.C.: J. Neurol.

Neurosurg. Psychiat. 43, 438-445 (1980).

5. Curzon, G.: J. Neural. Trans. Suppl. 15, 81-92 (1979).

6. Eccleston, E.G.: Clin. Chim. Acta 48, 269-272 (1973).

7. Badawy, A.A-B., Evans, M : Biochem. Pharmac. 30, 1211-1216 (1981).

8. Dodrill, C.B., Troupin, A.S.: Neurology 27, 1023-1028 (1977).

9. Thompson, P., Huppert, F., Trimble, M.: Acta Neurologica Scand. 62, 75-81 (1980).

THE ROLE OF METABOLISM IN 3-METHYLINDOLE-INDUCED ACUTE LUNG INJURY[1]

James R. Carlson, Mark R. Nocerini, Roger G. Breeze
Washington State University
Pullman, Washington 99164, USA

I. Introduction

Acute bovine pulmonary edema and emphysema (ABPE) is a naturally-occurring disease of cattle that occurs in many countries, especially the United States, Canada, Great Britain, and other parts of Europe. The disease usually occurs in mature cows within 2 to 4 days after abrupt pasture change from sparse dry forage to lush green pasture. We have shown that tryptophan (TRP) is fermented by anaerobic bacteria in the rumen to indoleacetic acid (IAA) with subsequent decarboxylation to 3-methylindole (3MI) by a newly identified <u>Lactobacillus</u> sp. (1,2). 3-Methylindole is absorbed from the rumen into the blood and it is responsible for the development of ABPE. In addition to the rumen, 3MI is a constituent of cigarette smoke (3), and it is present and absorbed from the large intestine of man and other animals (4). Acute lung injury after exposure to 3MI occurs in cattle, sheep, goats, and horses, but not in rats, mice, hamsters, rabbits, pigs and some other species. The risk of exposure to 3MI in man is unknown. Detailed reviews of the pathogenesis of ABPE and mechanism of action of 3MI have recently been published (5,6,7).

Injury to the lung occurs within minutes after exposure to 3MI and in ruminants, the alveolar Type I epithelial cells and non-ciliated bronchiolar epithelial (Clara) cells are most severely affected (8,9). Necrotic changes in the alveolar lining cells lead to pulmonary edema and interstitial emphysema. Both cell types slough off the basement membrane

[1]Supported in part by Washington State University Agricultural Research Center, project 1893 and NIH grant HL-13645.

Progress in Tryptophan and Serotonin Research

leaving denuded areas. Within hours after exposure, there is a proliferation of smooth endoplasmic reticulum (SER) in surviving cells. Other cell types in the lung appear to be less sensitive to the effects of 3MI. In horses, only the non-ciliated bronchiolar cells of the small airways are affected resulting in small airway obstruction. Alveolar epithelial cell damage and pulmonary edema are not observed in horses which demonstrates both cellular and species differences in sensitivity to 3MI exposure (7).

After oral or intravenous administration of 3MI, the plasma concentration remains relatively low; and 3MI is rapidly cleared from the blood with a half-life of approximately 20-25 minutes. Tissue concentrations are also low and do not exceed the concentration in the plasma indicating that 3MI does not concentrate in other body organs and tissues (10). Administration of ^{14}C-3MI revealed that radioactivity is rapidly excreted in the urine in the form of at least 10 metabolites. The primary metabolic pathway results in the excretion of 60-80% of the radioactivity as 3-methyloxindole (3MOI) and related metabolites. A secondary pathway leads to the excretion of smaller quantities of indole-3-carboxylic acid (11).

II. Role of Metabolism in Acute Lung Injury

A. Metabolic pathways and toxicity of metabolites

Evidence indicates that 3MI itself is not directly responsible for acute lung injury. Pretreatment of goats with phenobarbital (PB), an inducer of mixed function oxidase (MFO), results in more rapid metabolism of 3MI and enhanced lung injury after 3MI administration. Pretreatment with piperonyl butoxide (BT), an inhibitor of MFO, slows 3MI metabolism and prevents acute lung injury after 3MI administration. These findings demonstrate that metabolism of 3MI is associated with acute lung injury (12). The metabolic step responsible for lung injury is apparently an early step in the metabolism of 3MI. [^{3}H]-3-Methyloxindole and [^{3}H]-indole-3-carbinol, early metabolites in the respective metabolic pathways of 3MI, were administered to goats with or without ^{14}C-3MI (13). Neither 3MOI or indole-3-carbinol caused lung injury when given at a dose of 0.04 g/kg body weight (3MI causes acute lung injury at this dose). Chromatography of the urine of goats given 3MOI resulted in radioactive elution patterns that were similar to those from goats infused with 3MI. The results indicate that the major urinary metabolites of 3MOI and indole-3-carbinol are not toxic, and they implicate a metabolic step prior to the formation of these compounds in the mechanism of lung injury.

B. Formation of reactive intermediate(s)

The formation of a reactive intermediate of 3MI was studied in covalent binding experiments (14,15,16). Lungs of cattle and goats given ^{3}H-3MI or ^{14}C-3MI contained more bound radioactivity than any other organs studied, including liver. Incubation of crude microsomal preparations from goat lungs with ^{14}C-3MI resulted in higher covalent binding than in liver or kidney preparations and a non-toxic metabolite, ^{3}H-3MOI, did not result in covalent binding. Recently, additional evidence for formation of a reactive intermediate has been reported with ESR spin-trapping techniques using phenyl-t-butyl nitrone as the trapping agent (17). In short term incubations with lung microsomes, a nitrogen centered free radical was detected after addition of 3MI. The results of both in-vitro and in-vivo studies clearly indicate that a reactive intermediate is formed during the metabolism of 3MI in lung tissue.

C. Evidence for the involvement of MFO in the metabolism of 3MI.

Several lines of evidence implicate the MFO system in the metabolism and activation of 3MI. The proliferation of SER in affected cells and the chemical nature of the urinary metabolites are compatible with this concept. More direct evidence includes the fact that MFO inhibitors and inducers such as BT,PB, and 3-methylcholanthrene (3MC) alter the severity of lung injury (12,18). Similarly, in-vitro covalent binding is dependent on the presence of oxygen and NADPH, and it is inhibited by carbon monoxide, cytochrome C, and BT.

We conducted competitive substrate experiments to determine the effect of a variety of xenobiotics on in-vitro covalent binding of 3MI. In goat lung microsomes, PB, biphenyl, dimethylnitrosamine, lasiocarpine, and 3MC all competitively inhibited covalent binding of 3MI. The K_I for biphenyl and PB are lower than the K_I for 3MC and other inhibitors suggesting a greater affinity of biphenyl and PB for the forms of cytochrome P-450 responsible for activation of 3MI. Apparently, more than one type of substrate can interact competitively with 3MI during activation. Aniline, benzopyrene, and benzphetamine were uncompetitive or non-competitive inhibitors.

D. Role of reactive intermediates in acute lung injury

Reactive intermediates that are formed by MFO metabolism could cause lung injury by covalent binding to tissue macromolecules with disruption of cellular function or through peroxidation reactions. Preliminary studies investigating possible oxidation or peroxidation mechanisms have been negative. There is no indication that 3MI infusion alters the evolution of hydrocarbon gases (ethane, pentane, etc.) from goats suggesting that peroxidation of fatty acids is not increased (unpublished data).

The role of reactive intermediates in the mechanism of toxicity can be established by: (1) correlation of dose and time of covalent binding with toxicity, (2) correlation of organ and cell specificity of toxicity with covalent binding, (3) correlation of covalent binding with toxicity after treatment with inducers and inhibitors, (4) correlation of binding and toxicity with concentrations of sulfhydryl compounds such as glutathione, (5) correlation of covalent binding and toxicity with structurally related compounds (19).

We investigated several of these relationships to determine whether there was evidence for activated intermediates in the mechanism of 3MI-induced acute lung injury. Groups of goats were given doses of ^{14}C-3MI from 0.01 to 0.04 g/kg BW, and they were killed after 4 hours to determine tissue covalent binding. Covalent binding was closely related to dose in the lung but not in the liver, blood, or other tissues. This experiment demonstrated that covalent binding is dose dependent in the target organ.

Another study was conducted to investigate the relationship between in-vivo covalent binding, tissue glutathione status, and 3MI-induced lung injury (15). Groups of 6 goats were pretreated with cysteine to sustain tissue glutathione levels or with diethylmaleate to deplete tissue glutathione prior to ^{14}C-3MI infusion. As in previous experiments, covalent binding was the highest in the lung compared to liver and kidney. Cysteine pretreatment significantly decreased covalent binding in the lung, prolonged survival time, and reduced the severity of lung injury. Diethylmaleate pretreatment significantly increased covalent binding, shortened survival times and enhanced lung injury. 3-Methylindole alone depleted glutathione concentrations in the lung, liver and kidney to 36, 66, and 75% of control levels, respectively.

These experiments indicate that covalent binding of ^{14}C-3MI is dose dependent, and highest in the target organ. Binding is also related to the severity of lung injury after metabolic intervention to alter glutathione status. These observations support the hypothesis that 3MI-induced pneumotoxicity results from the formation of an activated intermediate and that glutathione plays a role in its detoxification.

E. Role of glutathione in detoxification of the reactive intermediate.

Glutathione reduces covalent binding of ^{14}C-3MI both in-vitro and in-vivo and influences the severity of 3MI-induced acute lung injury. It is possible that glutathione conjugates with the reactive intermediate resulting in lower binding to cellular macromolecules and detoxification. We investigated this hypothesis by measuring covalent binding in microsomal preparations and by analysis of the supernatant for conjugation products. Washed goat lung microsomes were incubated for 15 minutes with a NADPH generating system, ^{3}H- or ^{14}C-3MI and different concentrations of glutathione. The reaction was stopped with methanol precipitation, and the pellet was washed exhaustively with organic solvents to determine covalent binding. The supernatant was subjected to reverse phase high pressure liquid chromatography (HPLC). Elution patterns of radioactive compounds demonstrated the presence of a new metabolite that was present only when glutathione was included in the incubation mixture. The same results were obtained when either ^3H- or ^{14}C-3MI was used. The quantity of this metabolite increased as the concentration of glutathione increased. Also, covalent binding was inversely related to the content of the new metabolite. Further studies are in progress using ^{35}S-glutathione and ^3H-3MI in double isotope experiments along with gas chromatography linked mass spectroscopy for identification of this compound. The results are consistent with the hypothesis that glutathione conjugates with the reactive intermediate of 3MI resulting in detoxification by decreasing covalent binding to cellular macromolecules.

III. Summary

Recent results from our laboratory indicate that metabolism of 3MI is responsible for acute lung injury in ruminants and horses. Metabolism is mediated by the MFO system in the lung and a reactive intermediate is formed during an early step in the metabolic process. The reactive intermediate covalently binds to tissue macromolecules resulting in lung injury. This binding is dose dependent, organ specific; and it can be altered by metabolic intervention that also influences the severity of lung injury after administration of 3MI. Glutathione can decrease the toxicity of 3MI presumably by conjugating with the reactive intermediate in a detoxification process. Preliminary indications suggest the formation of a glutathione adduct and identification of this compound may yield information on the chemical nature of the reactive intermediate.

References

1. Yokoyama, M. T., Carlson, J. R.: Appl. Microbiol. 27, 540-549 (1974).
2. Yokoyama, M. T., Carlson, J. R., Holdeman, L. V.: Appl. Environ. Microbiol. 34, 837-842 (1977).
3. Wynder, E. L., Hoffman, D.: Tobacco and Tobacco Smoke. Academic Press, New York 377-379 (1967).
4. Fordtran, J. S., Scroggie W. B., Polter, D. E.: J. Lab. Clin. Med. 64, 125-132 (1964).

5. Carlson, J. R., Bray, T. M.: Advances in Nutritional Research. Academic Press, New York 5, 31-55 (1983).

6. Carlson, J. R., Breeze, R. G. Handbook of Natural Toxins, Marcel Dekker, Inc., New York 1, 85-115 (1983).

7. Breeze, R. G., Carlson, J. R.: Adv. Vet. Sci. & Comp. Med., Academic Press, New York 26, 201-231 (1982).

8. Bradley, B. J., Carlson, J. R.: Am. J. Pathol. 99, 551-560 (1980).

9. Huang, T. W., Carlson, J. R., Bray, T. M., Bradley, B. J.: Am. J. Pathol. 87, 647-666 (1977).

10. Bradley, B. J., Carlson, J. R.: Life Sciences 30, 455-463 (1981).

11. Hammond, A. C., Carlson, J. R., Willett, J. D.: Life Sci. 25, 1301-1306 (1979).

12. Bray, T. M., Carlson, J. R.: Am. J. Vet. Res. 40, 1268-1272 (1979).

13. Potchoiba, M. J., Carlson, J. R., Breeze, R. G.: Am. J. Vet. Res. 43, 1418-1423 (1982).

14. Hanafy, M. S. M., Bogan, J. A.: Life Sci. 27, 1225-1231 (1980).

15. Nocerini, M. R., Carlson, J. R., Breeze, R. G.: Life Sci. 32, 449-458 (1983).

16. Bray, T. M., Carlson, J. R.: Fed. Proc. 38, 1329 (1979).

17. Kubow, S., Bray, T. M., Janzen, E. G.: Fed. Proc. 42, 816 (1983).

18. Leung, C. T., Carlson, J. R., Breeze, R. G.: Can. J. Physiol. Pharm. 61, (In press) (1983).

19. Boyd, M. R.: C.R.C. Critical Reviews in Toxicology 7, 103-176 (1980).

BETA-CARBOLINE LEVELS IN BLOOD OF SMOKERS AND NON-SMOKERS

Pierre Baumann, Liliane Koeb

Clinique Psychiatrique Universitaire de Lausanne
(Dir. Prof. Chr. Müller)
CH-1008 Prilly-Lausanne / Switzerland

Laurent Rivier

Institut de Biologie et de Physiologie végétales de l'Université
CH-1015 Lausanne / Switzerland

Introduction

Betacarbolines (BC) have been since a long time a field of investigation for ethnopharmacologists (1). Indeed, simple BC have been identified in numerous South American hallucinogenic preparations, like Ayahuasca (2). Soon, speculations have developed about the mechanism of action of these psychoactive compounds and about their natural occurrence in the body. Recently, the state of research in the field of BC was summarized (3). Some of the most prominent biochemical properties of BC are their activity as inhibitors of MAO (4), as inhibitors of the uptake of neurotransmitters (5) and as ligands to one subclass of benzodiazepine receptors (6).

Attempts to demonstrate the presence of BC in the human organism by various analytical methods have been seriously challenged, as careful investigations have shown multiple sources of artefactual formation of BC. Indeed, some tetra-hydro-BC (THBC) are readily formed *in vitro* by the Pictet-Spengler reaction, at room temperature, in presence of indoles and aldehydes deriving either from the general metabolism or from - especially for aldehydes - impurities in the reagents (7).

Progress in Tryptophan and Serotonin Research

Nevertheless, the presence of THBC in the human or animal organism, either in urine, in plasma, in platelets, in CSF or in brain has been claimed (8 - 15).

For the present clinical biochemical approach, the attention was focused on two BC, namely harman and norharman, known for their potent activities at the benzodiazepine-GABA receptor complex. They differ from the THBC in that the ring formed by condensation of the indole with the aldehyde is aromatic. Spontaneous artefactual formation of these unsaturated compounds during the analytical procedure is much unlikely, as drastic conditions are needed to form harman-like compounds from THBC by oxidative mechanism (16). Tobacco smoke is an often cited exogenous source of harman and norharman, formed by pyrolysis from aldehydes and tryptophan present in tobacco (17, 18). It has been hypothetised that the occurrence of these two BC might contribute to the psychoactivity of the tobacco smoking (19). Both compounds have been demonstrated qualitatively in rat brain, in human blood platelets (20, 21) and urine (22). However, there is no quantitative data available for their occurrence in man. The aim of the present study was to determine the presence of harman and norharman and quantify them in blood plasma and thrombocytes of both smoking and non-smoking volunteers.

Material and Methods

Gas chromatography - mass spectrometry : The analyses were performed on a GC-MS computer system (Hewlett-Packard), model 5985; GC was carried out in the splitless mode, on a 25 m x 0.33 mm i.d., OV-1701 fused silica WCOT column (No 7425, Chrompack). The carrier gas pressure was 0.65 kg/cm^2 corresponding to a flow of He in the column of 1,5 ml/min. Typical oven conditions were : initial temperature : 100°C followed by heating rate : 25°C/min; final temperature : 220°C for 10 min. Injection and transfer line temperature : 260°C. The conditions for obtaining mass spectra and for Selected Ion Monitoring (SIM) were published elsewhere (16).

Standard solutions. Harman (Fluka), norharman (EGA-Chemie) and 3-N-methyl- 1,2,4 trihydro- γ-carboline (THGC) (Dr G. Schöllnhammer, Troponwerke / Cologne) are dissolved in 0.1 N HCl and frozen as stock solutions of 1 ng/μl.

Generally, 10 ml blood were collected in heparinized plastic tubes. (All subsequent extraction steps were carried out in glass tubes). After centrifugation for 15 min at 190 g, at room temperature, the thrombocytes were counted in the supernatant, of which 3 ml were centrifuged again at 12 000 g at 4°C. The resultant pellet with the thrombocytes was frozen until analysis. The lower phase originating from the first centrifugation (containing erythrocytes and some remaining plasma) was centrifuged again for ten min at 3000 g, at room temperature. The plasma was pooled together with the plasma obtained after elimination of the thrombocytes. When plasma only was analysed, whole blood was centrifuged for ten min at 3000 g (room temperature).

Analysis in thrombocytes : 50 ng THGC in solution and 1 ml 0.1 N HCl were added to the pellet. The sonified mixture was then analysed like plasma, after alcalinisation with 1.5 ml 0.25 N NaOH.

Analysis in plasma : 1 ml plasma was alcalinized with 1 ml 0.25 N NaOH and proceeded as described earlier (23).

Results

Methodological Aspects. An internal standard, similar in structure to harman and norharman, but whose presence was very unlikely in the human organism, was chosen : 1,2,4-trihydro-γ-carboline (THGC), and monitored at m/z = 143.2 (100 %) (unpublished) during SIM. Harman was monitored at 182.2 (100 %) and 181.2 (35 %), norharman at 168.2 (100 %) (16).

The following columns were found to be unsuitable : glass capillary columns : SP 2250, OV-101, OV-1; fused silica columns: SE-54, CP Sil 5, SP 2100.
The best peak shapes were obtained with a OV-1701 fused silica WCOT column. Typical retention times were : THGC : 4.9min; harman : 5.35 min; norharman : 5.62 min. Due to the relatively low chromatographic performance of norharman, the investigations were focused on harman.

As supposedly harman and norharman may be natural constituents of human blood, calibration curves could not be established with normal plasma. Thus, 50 ng THGC and increasing amounts of harman and norharman were added to a phosphate buffer (pH 7.4) and extracted like plasma. The lower limit for quantitation of harman was determined as 0.2 ng/ml, and 2 ng/ml for norharman. It was observed that harman-like products were even found in buffer solutions (fig. 1). For this reason, levels of harman measured below 0.2 ng/ml were considered to be irrelevant.
On the other hand, the calibration curve appeared to be linear for the range 0-2 ng/ml for harman (fig. 1a), and for 0-10 ng/ml for norharman (fig. 1b).

Adding 10 ng harman and norharman to either 1 ml buffer or plasma, the ratio of the recoveries of these compounds from plasma relatively to buffer was 84.9 $\pm$5.6 % (n = 5) for harman and 92.2 $\pm$ 8.0 % (n = 4) for norharman (cf also fig. 1b).

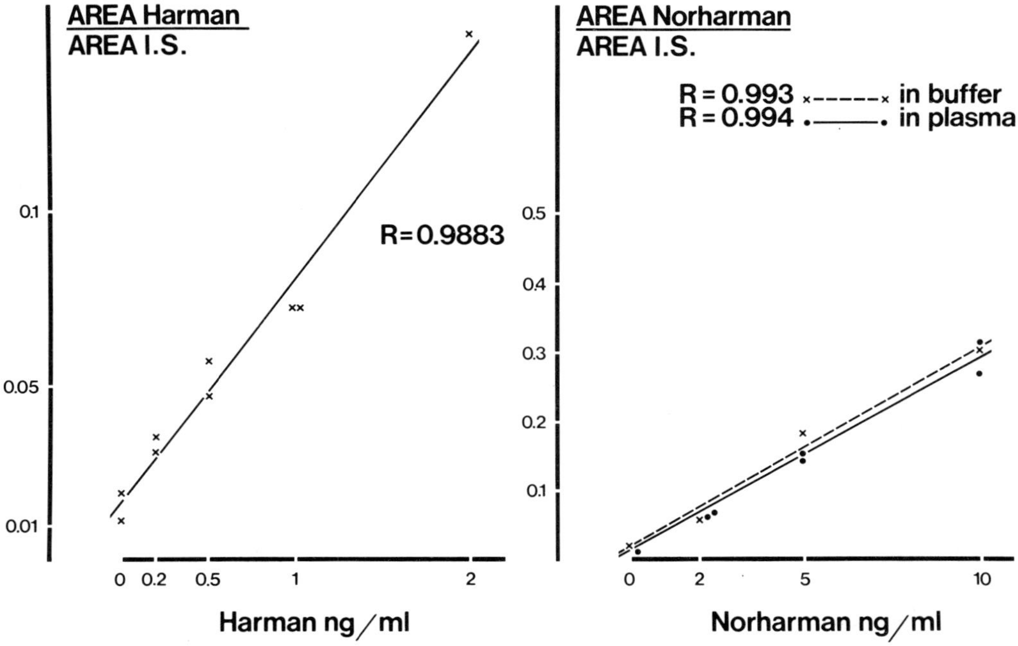

Fig. 1 Calibration curves for the quantification of a) harman (left) and b) norharman (right).

Investigations with human subjects. A pharmacokinetic investigation was carried out with two volunteers, in "baseline" conditions and after smoking, one (M.) smoking about 40 cigarettes a day, the other (L.) being a pipe smoker. No smoking was allowed the night before the experiment. The blood samplings were carried out by a non-smoker in a carefully aerated room. The ten min smoking session was held in another room. For baseline values, heparinized blood was sampled 10,5 and 0 min before smoking. Then, the subjects smoked for ten min, either two cigarettes (M.) or a pipe (about 3 g tobacco; subject L.). Blood samplings were taken 10, 15, 20 and 25 min after the beginning of smoking. The results (fig. 2) clearly show that levels of harman never exceed 0.2 ng/ml plasma. Consequently, it is considered that the levels measured before and after smoking do not exceed background levels.

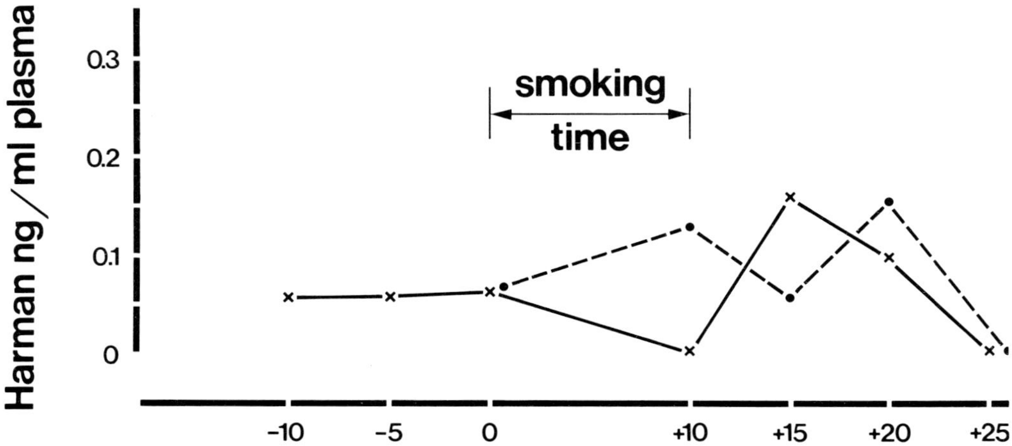

Fig. 2 Levels of harman in blood plasma of healthy subjects before and after smoking. Pipe smoker L. : x——x, cigarette smoker M. : ● ---- ●

Harman was also analysed in a group composed of eight male and four female volunteers. Nine were smokers. One female subject was a psychiatric patient under chronic treatment with chloralhydrate (4 g/day). Blood was drawn from all subjects in the morning and from six of the smokers at 4 p.m. too, i.e. after

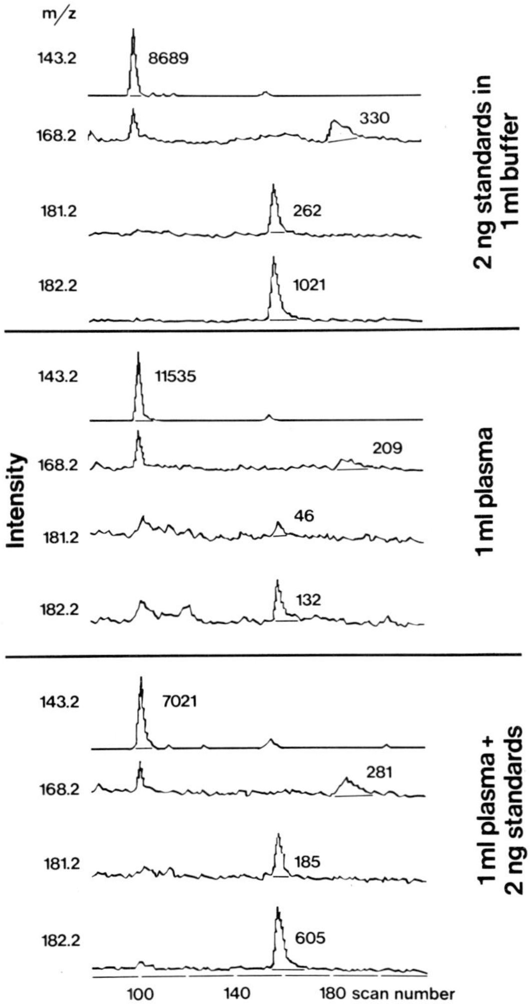

Fig. 3 Reconstructed SIM trace of standards (harman, norharman, THGC) in buffer (upper graph), of an extract of 1 ml blood plasma of a smoker (middle graph), and of an extract of 1 ml of the same plasma to which standards were added. For explanations, see text.

a day of smoking. In no subject the level of harman exceeded 0.2 ng/ml plasma or 0.2 ng/10^8 thrombocytes. A typical example of a plasma analysis of smoker L., who participated also in the first experiment, is shown in fig. 3. The tracing in the middle of the graph shows a small peak for harman (area 132) at m/z = 182.2, at 181.2, but also for norharman (area 209) at m/z = 168.2 in an extract of 1 ml plasma sampled at 4 p.m., with the internal standard THGC, (m/z = 143.2). Again, the presence of harman in plasma could be possibly excluded, as it does not exceed the level generally found in plasma-free samples as a result of artefact. For comparison, the upper graph in fig. 3 represents the peaks of the BC when 2 ng harman and norharman, and 50 ng of THGC are added to 1 ml buffer, while the lower part of the fig. 3 indicates the presence of these compounds when added to 1 ml of the same subject's plasma.

Whatsoever, smoking does not increase the level of BC in blood.

Discussion

The method described allows quantitative analysis of endogenous level down to 0.2 ng harman and 2 ng norharman in 1 ml plasma or 10^8 thrombocytes. The levels of harman detected and quantified in smokers and non-smokers, but also in a patient treated with chloral do not exceed the blanks. Moreover, in two pharmacokinetic experiments, the levels of harman were not increased meaningfully by smoking. Earlier, the presence of harman in platelets, in the arcuate nucleus and in urine of the rat, and in urine of man respectively, was demonstrated qualitatively only (20, 22). Due to their lipophilic properties, harman and norharman are probably unevenly distributed in the body. Thus, low plasma levels would not necessarily reflect the absence of BC in brain. The origin of low levels of BC in blanks is supposedly from the ubiquitous presence of tobacco smoke constituents. E.g., extreme precautions have to be taken for the analysis of nicotine in plasma (24).

In the pharmacokinetic experiment, this time schedule has been chosen, as harman and norharman exert their maximal effect on the brain within 10 min after an i.v. injection in the rat and peak levels of nicotine are already observed 10 min after smoking (26).

Other possible sources for harman and norharman leading to significant levels of these compounds in human blood may be eliminated :

- In one patient treated with chloral hydrate, no significant presence of harman and norharman was demonstrated. This hypnotic aldehyde was thought to enhance the formation of BC with indoles. However, chloral hydrate is readily metabolized in human to trichloroacetic acid and trichloroethanol, but chloraldehyde itself is not present in plasma in measurable amounts (27). This rules out its possible role as a precursor for BC.
- Formaldehyde is formed in the liver from secondary and tertiary amines not only from a reaction mediated by cytochrome P-450, but also from drugs like methenamine (28). Whether the fact that tryptophan also binds to oxidized liver microsomal cytochrome P-450 (29) is of any importance for a facilitated formation of BC is, until now, pure speculation.

No data are available about the metabolism of harman and norharman in the organism. Metabolic pathways have only been outlined for THBC (30). Oxidative degradation implying microsomal mixed function oxidase seems probable. At least, part of this enzyme system is inhibited non-competitively by norharman (and harman ?), which shows a type II spectrum of binding with microsomes (31).

These considerations together with the results of this investigation lead to the conclusion that albeit BC like harman and norharman exert manifold effects on neurotransmitter systems of the brain but also on peripheral enzyme complexes, it is doubtful whether they occur in the human organism in amounts of physiological significance.

Acknowledgements : We are grateful to Dr. J. Schöpf for the medical supervision of the pharmacokinetic experiments and to Mrs. Chr. Bertschi for secretarial assistance.

References

1. Allen, J.R.F., Holmstedt, B.R. : Phytochemistry 19, 1573-1582 (1980).
2. Rivier, L., Lindgren, J.E. : Economic Botany 26, 101-129 (1972).
3. Bloom, F., Barchas, J., Sandler, M., Usdin, E. (eds) : Beta-Carbolines and Tetrahydroisoquinolines; Alan R. Liss, Inc., New York (1982).
4. Sandler, M., Glover, V., Armando, I., Clow, A. : in ref. 3, 215-226.
5. Buckholtz, N.S. : in ref. 3, 183-190.
6. Braestrup, C., Nielsen, M., Olsen, C.E. : Proc. Natl. Acad. Sci. 77, 2288-2292 (1980).
7. Bosin, T.R., Holmstedt, B., Lundman, A., Beck, O. : Analytical Biochemistry 128, 287-293 (1983).
8. Bosin, T.R., Holmstedt, B., Lundman, A., Beck, O. : in ref. 3, 15-27.
9. Beck, O., Lundman, A. : submitted.
10. Allen, J.R.F., Beck, O., Borg, St., Schröder, R. : Eur. J. Mass Spectrom. in Biochem., Med. and Environm. Res. 1 171-177 (1980).
11. Peura, P., Kari, I., Airaksinen, M.M. : Biomed. Mass Spectrom. 7 / 11-12, 553-555 (1980).
12. Kari, I., Peura, P., Airaksinen, M.M. : Biomed. Mass Spectrom. 7 / 11-12, 549-552 (1980).
13. Barker, S.A., Harrison, R.E., Brown, G.B., Christian, S.T.: Biochem. and Biophys. Res. Commun. 87, 146-154 (1979).
14. Beck, O., Bosin, T.R., Lundman, A., Borg, St. : Biochem. Pharmacology, 31, 2517-2521 (1982).
15. Honecker, H., Coper, H., Fähndrich, C., Rommelspacher, H. : J. Clin. Chem. Clin. Biochem. 18, 133-135 (1980).

16. Baumann, P., Rivier, L., Perey, M. : in A. Frigerio (ed), Recent Developments in Mass Spectrometry in Biochemistry, Medicine and Environmental Research 8, 7-18 (1983).

17. Poindexter, E.H. Jr., Carpenter, R.D. : Phytochemistry 1, 215-221 (1962).

18. Snook, M.E., Chortyk, O.T. : J. of Chromatogr. 245, 331-338 (1982).

19. Janiger, O., Dobkins de Rios, M. : Economic Botany 30, 149-151 (1976).

20. Shoemaker, D.W., Cummins, J.T., Bidder, T.G., Boettger, H.G., Evans, M. : Naunyn-Schmiedeberg's Arch. Pharmacol. 310, 227-230 (1980).

21. Bidder, T.G., Shoemaker, D.W., Boettger, H.G., Evans, M., Cummins, J.T. : Life Sciences 25, 157-164 (1979).

22. Rommelspacher, H., Strauss, S., Lindemann, J. : Febs Letters 109, 209-212 (1980).

23. Baumann, P., Koeb, L., Tinguely, D., Rivier, L. : Eur. J. Mass Spectrom. in Biochem. Med. and Environm. Res. 2 19-26 (1982).

24. Curvall, M., Kazemi-Vala, E., Enzell, C.R. : J. Chromatogr. 232, 283-293 (1982).

25. Fehske, K.J., Müller, W.E. : Brain Research 238, 286-291 (1982).

26. Hengen, N., Hengen, M. : Clin. Chem. 24, 50-53 (1978).

27. Breimer, D.D. : Clinical Pharmacokinetics 2, 93-109 (1977).

28. Ariëns, E.J., Hanselaar, A.G.J.M., Henderson, P. Th., Simonis, A.M. : Eur. J. Clin. Pharmacol. 23, 373-375 (1982).

29. Orme-Johnson, N.R., Orme-Johnson, W.H. : Methods in Enzymology, vol. LII, Biomembranes; S. Fleischer, L. Packer (eds), Academic Press, New York, 252-279 (1978).

30. Greiner, B., Rommelspacher, H. : in ref 3, 201-208.

31. Fujino, T., Matsuyama, A., Nagao, M., Sugimura, T. : Chem.-Biol. Interactions 32, 1-12 (1980).

THE RELATIONSHIP BETWEEN TRYPTOPHAN, TRYPTOPHAN METABOLITES AND BLADDER CANCER

Hans Wolf

Department of Urology, KK Hvidovre University Hospital,
University of Copenhagen, DK-2650 Hvidovre, Denmark.

Early Suggestions of a Relationship between Tryptophan and Bladder Cancer

In 1947 Ekman and Strömbeck demonstrated that patients with bladder cancer excreted increased amounts of unidentified diazotisable aromatic amines in urine (1). It was later recognized that most of the diazotisable aromatic amines normally found in human urine are metabolites of tryptophan (2), so that a possible link between tryptophan and bladder cancer was theoretically established.

Such a link was also experimentally established, when in 1950 Dunning and her coworkers (3) published their findings that rats of the Fischer strain in addition to liver and ear tumors developed nearly 100 per cent bladder tumors when fed 2-acetylaminofluorene and D,L-tryptophan, whereas no bladder tumors were detected in rats receiving 2-acetylaminofluorene alone. They suggested that the combined elimination products of 2-acetylaminofluorene and tryptophan might be responsible for the induction of bladder tumors.

In addition, the chemical composition of many of the tryptophan metabolites of the kynurenine pathway is very similar to known human bladder carcinogens, such as 2-naphtylamine, benzidine etc., so that also from analogy a relationship between tryptophan metabolites and bladder cancer could be expected

Progress in Tryptophan and Serotonin Research

to exist (Fig.1).

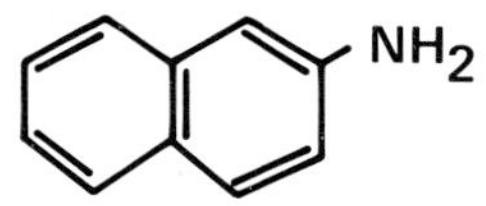

2–NAPHTYLAMINE

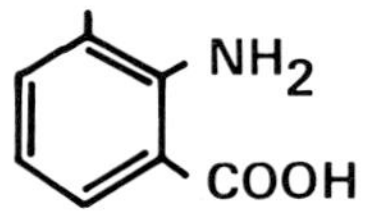

3–HYDROXY–ANTHRANILIC ACID

Fig.1. The chemical composition of an established bladder carcinogen (2-naphtylamine)and a tryptophan metabolite (3-hydroxyanthranilic acid).

Studies on the Excretion of Tryptophan Metabolites in Bladder Tumor Patients

If tryptophan metabolites of the kynurenine pathway were etiologically related to a proportion of bladder tumor cases, it would be expected that at least some bladder tumor patients were excreting increased amounts of one or several tryptophan metabolites in the urine. Therefore, studies on the excretion of tryptophan metabolites among bladder tumor patients and control subjects were initiated. Price and his coworkers (4,5) in Wisconsin and Boyland and his coworkers in England (6) found that several bladder tumor patients excreted increased amounts of several of the aromatic degradation products of tryptophan, especially when a tryptophan loading dose was given (Fig.2).

The percentage of bladder tumor patients with abnormal tryptophan metabolism was in some studies inversely related to the percentage of bladder tumors that could be attributed industrial exposure (7), but recent studies have shown that also in non-industrial areas only few patients with bladder cancer showed abnormal tryptophan metabolism (8,9). From these studies will appear, that the results of these excretion studies are rather divergent.

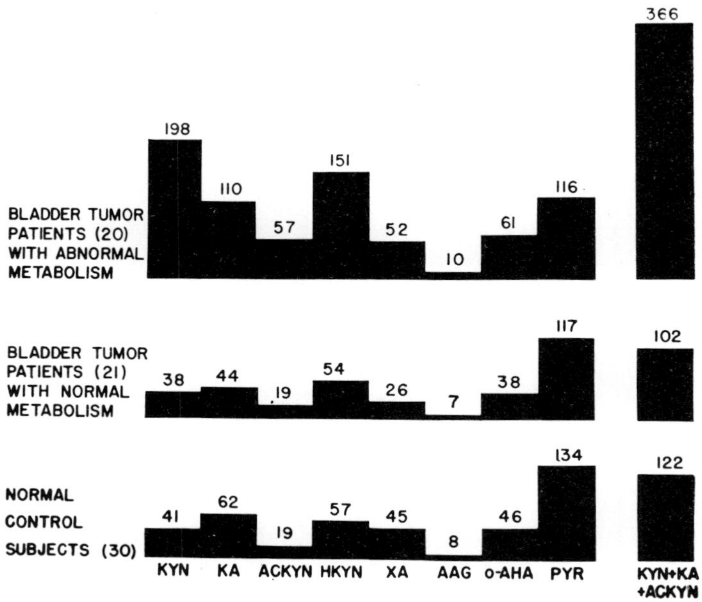

Fig.2. Average urinary excretion of tryptophan metabolites (μmoles excreted) after a loading dose of 2 g L-tryptophan. Abbreviation:KYN,kynurenine; KA, kynurenic acid; ACKYN, acetylkynurenine; HKYN, 3-hydroxykynurenine; XA,xanthurenic acid; AAG, anthranilic acid glucuronide; O-AHA, o - aminohippuric acid; PYR, N-methyl-2-pyridone-5-carboxamide.

Vitamin B_6 supplementation could in most of these studies partly correct the abnormal tryptophan metabolism found in bladder tumor patients (5). It is well-known that vitamin B_6 is an important cofactor for some of the enzymatic reactions (kynureninase and kynurenine transaminase) of the kynurenine pathway, and that vitamin B_6 deficiency results in a similar abnormality of tryptophan metabolism as seen in some bladder tumor patients (10). Therefore, a slight vitamin B_6 deficiency has been suspected to be present in some bladder tumor patients.

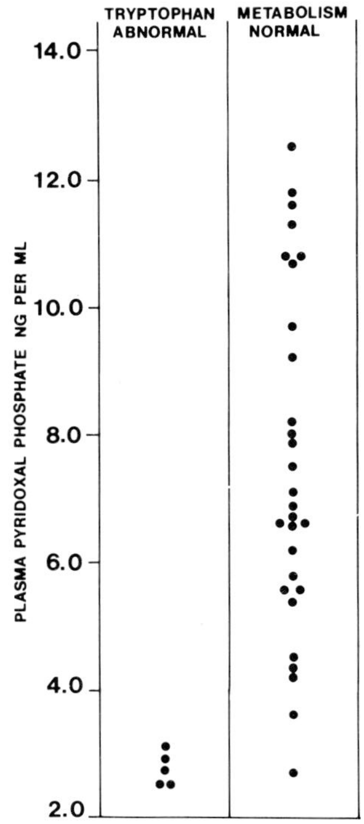

Fig.3. Plasma pyridoxal phosphate concentration in bladder tumor patients in relation to abnormal or normal tryptophan metabolism in the same patients.

From Fig.3 will appear that in fact all patients with abnormal tryptophan metabolism had at the same time very low plasma pyridoxal phosphate concentration, indicating vitamin B_6 deficiency. The mechanism, however, may be more complex, since often it was not possible to correct tryptophan metabolism completely with vitamin B_6 (5,11). In addition, the latter group found depressed hepatic kynureninase activity in bladder tumor patients with abnormal tryptophan metabolism despite added pyridoxal phosphate to the assay system (11).

On background of such studies and the finding that especially recurrent bladder tumors often were associated with abnormal tryptophan metabolism (12), clinical trials using vitamin B_6 as a tumor prophylactic agent were initiated. However, no convincing prophylactic effect of vitamin B_6 has so far been demonstrated (13).

However, it is doubtful whether it is possible at all from such excretion studies to judge whether abnormal tryptophan metabolism is involved in bladder carcinogenesis, since 1) we do not know whether tryptophan metabolism also was abnormal during the induction period of bladder cancer, which may well be 15 to 30 years, and 2) the abnormal tryptophan metabolism and/or vitamin B_6 deficiency may be a secondary phenomenon due to the disease itself. Thus, it is well-known that many diverse diseases and clinical states are associated with abnormal tryptophan metabolism (14), and none of these have so far been associated with an increased incidence of bladder cancer. Therefore, it probably has to be concluded that these excretion studies do not uniformly support the idea that tryptophan metabolites of the kynurenine pathway may induce bladder cancer in man.

Experimental Studies in Animals

Direct application of tryptophan metabolites to the bladder wall

Along with the above excretion studies in bladder tumor patients and control subjects experimental studies in animals were also conducted. Boyland and his coworkers (15) surgically implanted pellets of different chemicals (such as paraffin wax or cholesterol) as a vehicle for various test substances into mouse bladders. With known bladder carcinogens as test substances they obtained a significant number of bladder tumors in this animal model. With 3-hydroxykynurenine and 3-hydroxyanthranilic acid as test substances they also obtained a bladder tumor incidence of 20 per cent compared to 8 per cent in animals implanted with the vehicle alone.

Similar results were obtained by Bryan and his coworkers in Wisconsin (16). In systematic studies performed by this group, it was shown that with the above technique, when cholesterol was used as a vehicle, a continuous exposure for a long period of the bladder wall to a high concentration of the test substance was obtained. In addition, it was shown that the liberated aromatic amine was absorbed and accumulated to a great extend in the bladder wall (17), so that carcinogenesis theoretically could progress there. Such accumulation could also be demonstrated in autoradiographic studies of the bladder wall after exposure to ^{3}H-labelled-3-hydroxyanthranilic acid (18).

The incidence of bladder carcinomas in mice exposed to various tryptophan metabolites with the pellet implantation technique using cholesterol as vehicle will appear from Table I, which summarizes the results obtained by Bryan and his coworkers. The potent bladder carcinogen, 2-naphtylhydroxylamine, gave an incidence of 29 per cent carcinomas and served as a positive control group. It will be seen that 8 of the tryptophan metabolites tested gave a significant number of bladder carcinomas with this technique.

However, many reservations have been expressed about the meaning of tumors produced by the bladder implantation technique. The relative role of the vehicle and the test substance in carcinogenesis is uncertain. Therefore, this animal model has not gained universal recognition as a test system for carcinogenesis.

	$T_{\frac{1}{2}}$ days	Carcinomas			Per cent obtained at other institutions
		Number/ no.mice	per cent	p value	
Sham operation	—	0/26	0	-	
Cholesterol alone	—	7/87	8	-	6.5 - 9.0
2-Naphtylhydroxylamine	54	15/52	29	0.002	21 - 33
3-Hydroxykynurenine	5.2	12/40	30	0.002	24
3-Hydroxyanthranilic acid	5.3	17/78	22	0.01	20
Xanthurenic acid (XA)	47	11/55	20	0.04	
8-Methyl ether of XA	0.7	18/54	33	0.0002	
8-Hydroxyquinaldic acid	6.1	10/52	19	0.05	
L-Kynurenine	3.2	14/42	33	0.002	
Acetyl-L-kynurenine	28	21/94	22	0.01	
Quinaldic acid	4.7	12/37	32	0.003	
Indican	—	13/78	17	0.09	
O-aminohippuric acid	7.8	7/45	16	0.1	
Anthranilic acid	36	10/75	13	0.2	
Kynurenic acid	6.3	9/68	13	0.2	8

Table I. 50 per cent elution time ($T_{\frac{1}{2}}$days) and incidence of bladder carcinomas in mice living more than 175 days following implantation of 2-naphtylhydroxylamine (an ultimate bladder carcinogen) or various metabolites of tryptophan suspended in cholesterol pellets.

Feeding experiments

In feeding experiments with tryptophan or tryptophan metabolites alone, it has **not** been possible to produce bladder tumors in animals. However, the exposure of the bladder mucosa in such experiments is minimal, since both tryptophan and tryptophan metabolites are mostly metabolized to non-aromatic products that reach the bladder in only small concentrations (17,19). Nevertheless, Radomski and his coworkers (20) by feeding dogs large doses of D,L-tryptophan observed marked focal hyperplasia of the transitional epithelium which they considered suggesting a cocarcinogenic effect of tryptophan.

Promotion by Dietary D,L-Tryptophan (2%) of Dietary N- (4- (5-nitro-2-furyl)-2-thiazolyl) formamide (0.2%) (FANFT) Induced Bladder Carcinogenesis in Rats

Treatment	Bladder tumours/ number of animals examined histologically	Tumour incidence %
FANFT for 6 weeks	4/20	20
Tryptophan	0/19	0
FANFT for 6 weeks, then tryptophan	10/19	53

(Cohen, Arai, Jacobs, Friedell, 1979)

Table II. Promotion by dietary D,L-tryptophan (2%) of N-(4-(5-nitro-2-furyl)-2-thiazolyl) formamide (FANFT) -induced bladder carcinogenesis in male rats.

In recent years, the initiation-promotion model of chemical carcinogenesis which originally was applied by Berenblum (21) to experimental skin carcinogenesis has also been applied to experimental bladder carcinogenesis (22,23). It is perhaps in such a light that the early studies of Dunning and her colleques (3) should be viewed. A promotor type of action of D,L-tryptophan in dogs was suggested by Radomski and his co-workers (24). However, although between 25 and 50 per cent of the dogs developed bladder tumors when fed large doses of D,L-tryptophan for more than 3 years after a single dose of 4-aminobiphenyl or 2-naphtylamine, the number of animals used in the experimental group and the control group was too small to reach a statistically valid conclusion.

From Table II will be seen that in rat experiments a promoting effect of dietary D,L-tryptophan could be demonstrated. However, a less pronounced promoting effect was found when L-tryptophan, the naturally occurring isomer, was used (25).

Conclusions

Although the role of tryptophan or tryptophan metabolites in bladder carcinogenesis has been studied intensely for more than 30 years, no definite answer to the question can yet be given. However, on basis of these studies we still suspect this naturally occurring essential amino acid or its metabolites to be etiologically to human bladder cancer, probably as promoting agents.

References

1. Ekman, B., Strömbeck, J.P.: Acta physiol.scand. 14, 43 (1947).
2. Brown, R.R., Price, J.M.: J. biol. Chem. 219, 985 (1956).
3. Dunning, W.F., Curtis, M.R., Maun, M.E.: Cancer Res. 10, 454 (1950).
4. Brown, R.R., Price, J.M., Wear, J.B.: Proc. Am. Assoc. Cancer Res. 2, 7 (1955).
5. Brown, R.R., Price, J.M., Satter, E.J., Wear, J.B.: Acta Un. Int. Contr. Cancer 16, 299 (1960).
6. Boyland, E., Williams, D.C.: Biochem. J. 64, 587 (1956).
7. Brown, R.R., Price, J.M., Friedell, G.H., Burney, S.W.: J. Natl. Cancer Inst. 43, 295 (1969).
8. Wolf, H., Brown, R.R.: Scand. J. Urol. Nephrol. 13, 143 (1979).
9. Friedlander, E., Morrison, A.S.: J. Natl. Cancer Inst. 67, 347 (1981).
10. Yess, N., Price, J.M., Brown, R.R., Swan, P.B., Linkswiler, H.: J. Nutr. 84, 229 (1964).
11. Gailani, S., Murphy, G., Kenny, G., Nussbaum, A., Silvernail, P.: Cancer Res. 33, 107 (1973).
12. Yoshida, O., Brown, R.R., Bryan, G.T.: Cancer 22, 773 (1970).
13. Byar, D., Blackard, C.: Urology 10, 556 (1977).

14. Wolf, H.: Scand. J. clin. Lab. Invest. suppl. 136 (1974).

15. Allen, M.J., Boyland, E., Dukes, C.E., Horning, E.S., Watson, J.G.: Brit. J. Cancer 11, 212 (1957).

16. Bryan, G.T., Brown, R.R., Price, J.M.: Cancer Res. 24, 596 (1964).

17. Bryan, G.T., Morris, C.R., Brown, R.R.: Cancer Res. 28, 186 (1968).

18. Röhl, L., Hochberg, K., Kochen, W.: Scand. J. Urol. Nephrol. 3, 214 (1969).

19. Wolf, H., Brown, R.R.: Clin. Sci. 41, 237 (1971).

20. Radomski, J.L., Glass, E.M., Deichmann, W.B.: Cancer Res. 31, 1690 (1971).

21. Berenblum, J.: Cancer Res. 1, 807 (1941).

22. Chiowaniec, J., Hicks, R.M.: Brit. J. Cancer 39, 355 (1979).

23. Cohen, S.M., Arai, M., Jacobs, J.B., Friedell, G.H.: Cancer Res. 39, 1207 (1979).

24. Radomski, J.L., Radomski, T., Mac Donald, W.E.: J. Natl. Cancer Inst. 58, 1831 (1977).

25. Fukushima, S., Friedell, G.H., Jacobs, J.B., Cohen, S.M.: Cancer Res. 41, 3100 (1981).

ROLE OF INDOLEAMINE 2,3-DIOXYGENASE IN THE DEFENCE MECHANISM AGAINST TUMOR GROWTH

Ryotaro Yoshida, Osamu Takikawa, Hiroaki Yasui, Osamu Hayaishi

Department of Medical Chemistry, Kyoto University Faculty of Medicine, Kyoto 606, Japan

and Kimio Yasuhira

Department of Pathology, Chest Disease Research Institute, Kyoto University, Kyoto 606, Japan

Introduction

Indoleamine 2,3-dioxygenase [indoleamine:oxygen 2,3-oxido-reductase (decyclizing] (IDO) is ubiquitously distributed in various organs of mammals except in the liver (1), and is remarkably (approximately 100-fold) induced in the mouse lung after a single intraperitoneal injection of bacterial lipopolysaccharide (LPS) (2), during influenza virus infection (3) or by the addition of mouse interferon to the lung slices (4). In the case of tumor transplantation, tumor cells are able to grow in a syngeneic strain of mice but are rejected from the allogeneic strains of animals. In an attempt to find out the biological function of IDO, we treated animals with an intraperitoneal injection of Meth-A or Lewis lung cancer (3LL) cells, and determined the IDO activity in the intraperitoneal cells from various strains of mice.

Results

1) IDO induction and growth of Meth-A in intraperitoneal cavity. When Meth-A cells were given intraperitoneally to

Progress in Tryptophan and Serotonin Research

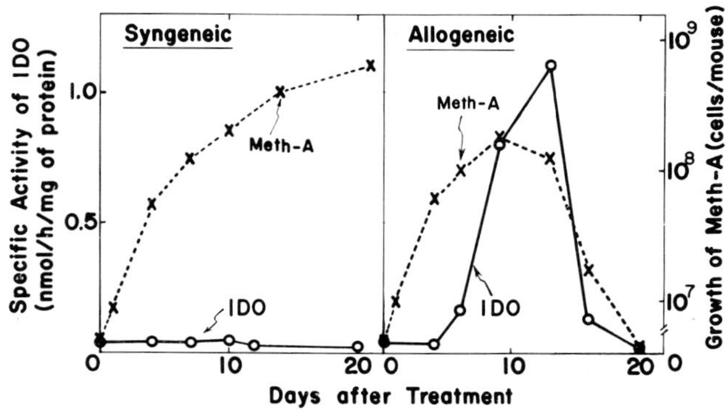

Fig. 1. IDO induction and growth of Meth-A in an intraperitoneal cavity.

the syngeneic (BALB/c) strain of mice (Fig. 1), the tumor cells continued to grow in the intraperitoneal cavity for 2 to 3 weeks and the mice did not recuperate. The IDO activity in the intraperitoneal cells was almost undetectable and did not change significantly by the treatment. By contrast, when Meth-A cells were injected intraperitoneally to the allogeneic (e.g., C57BL strain) animals, the growth of Meth-A began within 24 h after transplantation but the tumor cells ceased to grow on the 9th day. Then, rapid elimination of the tumor cells from the intraperitoneal cavity commenced around the 14th day and on the 20th day none was found. The IDO activity in the intraperitoneal cells began to increase from the 4th day, reached a peak on the 10th day and persisted until 14th day. Thereafter, the enzyme activity rapidly decreased and was undetectable around 20th day. Similar enzyme induction was observed by an intraperitoneal transplantation of 3LL cells to allogeneic (e.g., BALB/c) strain of mice but not to syngeneic animals (C57BL strain of mice).

2) Enrichment of cells containing induced IDO. In the case of syngeneic strain of mice, there was no infiltration of lymphocytes after the Meth-A transplantation. By contrast, the time course of enzyme induction in the allogeneic

animals coincided well with that of the intraperitoneal infiltration of small lymphocytes. When the intraperitoneal cell suspension obtained on the 8th day after the treatment was applied to a meter glass (1 X g, for 60 min), the cells were distributed with two major fractions. The lower fraction was centrifuged at 25 X g for 20 min. The pellet was resuspended in the original volume of medium 199, and then centrifuged again. The upper fraction in the meter glass was centrifuged in the same way. Thus, we obtained four cell fractions; two pellets and two supernatant fractions. In these fractions the IDO activity and the cell population were determined (Fig. 2). The IDO activity exclusively localized in the tumor cell-rich fraction, but the fraction still contained a small amount of lymphocytes.

3) Identification of cells containing induced IDO. In order to determine whether the IDO was induced in Meth-A cells or in a certain type of lymphocytes, we examined the effects of antisera for tumor cells or lymphocytes in the presence of rabbit complement on the IDO activity in the peritoneal exudate cells (Fig. 3). Although the incubation

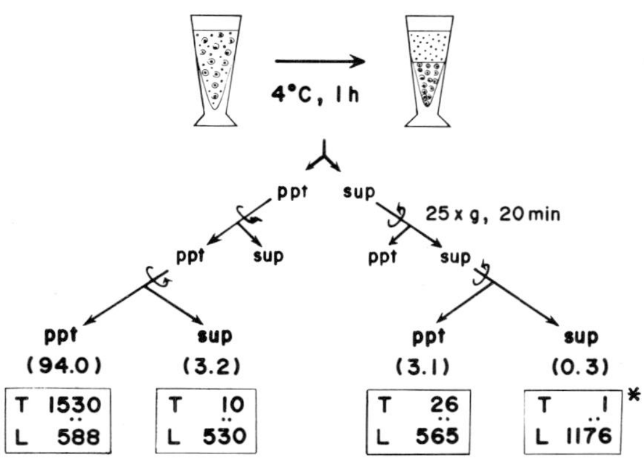

Fig. 2. Enrichment of cells containing induced IDO. Data in parentheses represent the recovery (%) of the total activity. T, Meth-A cells; L, small lymphocytes. * 10^4 cells are expressed as 1.

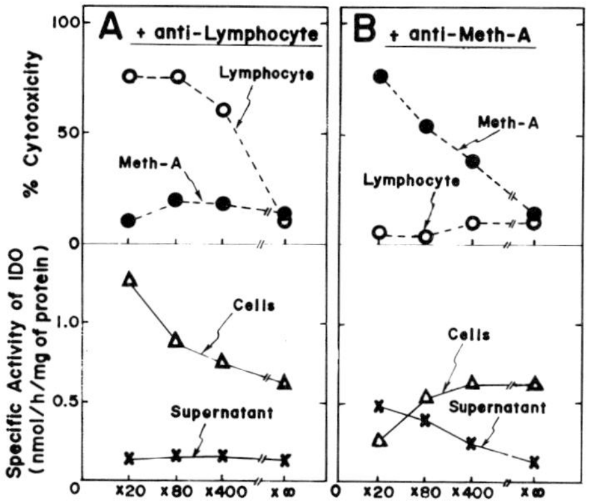

Fig. 3. Identification of cells containing induced IDO.

of lymphocytes with the antiserum for lymphocytes (Fig.3-A) did not affect on the induced IDO, the IDO activity in the intraperitoneal cells was dose-dependently reduced by the treatment with the antiserum for tumor cells (Fig. 3-B). Fig. 3-B also showed that significant enzyme activity was detected in the supernatant fractions after the treatment. Recently, Takikawa et al. in our laboratory indicated that the IDO induced by LPS treatment caused a significant decrease in the plasma tryptophan concentration and about 5-fold increase in the plasma kynurenine level. These results indicated that indoleamine 2,3-dioxygenase was markedly induced in the tumor cells transplanted to the allogeneic strains of mice but not in tumor cells transplanted to the syngeneic animals, and implied that the decrease in the intracellular concentration of tryptophan (substrate of the enzyme) caused the inhibition of tumor growth.

References

1. Hayaishi, O., Hirata, F., Fujiwara, M., Ohnishi, T., Nukiwa, T.: Proceedings of the Tenth Federation of European Biochemical Societies Meeting, pp. 131-144, Federation of European Biochemical Societies, Amsterdam (1975).
2. Yoshida, R., Hayaishi, O.: Proc. Natl. Acad. Sci. U.S.A. 75, 3998-4000 (1978).
3. Yoshida, R., Urade, Y., Tokuda, M., Hayaishi, O.: Proc. Natl. Acad. Sci. U.S.A. 76, 4084-4086 (1979).
4. Yoshida, R., Imanishi, J., Oku, T., Kishida, T., Hayaishi, O.: Proc. Natl. Acad. Sci. U.S.A. 78, 129-132 (1981).

THE RELATIONSHIP BETWEEN PLASMA KYNURENINE AND INDOLEAMINE 2,3-DIOXYGENASE ACTIVITY IN THE EXTRAHEPATIC TISSUES

Osamu Takikawa#, Ryotaro Yoshida, Hiroaki Yasui and Osamu Hayaishi

Department of Medical Chemistry, Kyoto University Faculty of Medicine, Kyoto 606, Japan

Introduction

Indoleamine 2,3-dioxygenase (IDO) is a heme-containing enzyme with MW=41,000 (1) and is distributed in various organs of mammals except in the liver (2). Purified IDO catalyzes an oxygenation reaction essentially identical with that of L-tryptophan 2,3-dioxygenase (tryptophan pyrrolase) in the liver, but exhibits a broad substrate specificity for not only tryptophan but also 5-OH-tryptophan, serotonin and tryptamine (1). Recently, Yoshida et al. (3,4) have demonstrated that IDO in the lung and other various tissues of mouse was induced 10 to 120-fold under the pathological conditions, i.e. an intraperitoneal administration of gram negative bacterial cell wall (lipopolysaccharide (LPS) and the infection of influenza virus. To clarify the physiological substrate of the induced IDO, we determined the corresponding metabolites in the mouse body and the urine during the induction of IDO using high performance liquid chromatography (HPLC). In this paper, we report that tryptophan is the main substrate of the induced IDO, and that against the general concept of the tryptophan metabolism (5), the key enzyme which determines how much tryptophan enters the kynurenine pathway is extrahepatic IDO rather than hepatic tryptophan pyrrolase.

#Present address: Department of Biochemistry, Wakayama Medical College, Wakayama 640, Japan

Progress in Tryptophan and Serotonin Research

Results

1) Induction of IDO: As reported previously (3), the IDO activity of the lung of ICR mouse increased to a maximum (about 10 times that of the control) around 1 day after a single intraperitoneal injection of LPS from *S. abortus equi* (5 μg/mouse), and thereafter gradually decreased (Fig. 1A). Essentially similar time course of the induction was observed with IDO distributed in the alimentary tract (not shown).
2) Plasma Kynurenine: Under the same conditions, the plasma kynurenine level also increased 5 to 6-fold and gradually decreased as shown in Fig. 1B. The time course of the plasma kynurenine level was almost identical to that of the IDO activity in the lung. Other corresponding metabolites derived from 5-OH-tryptophan, serotonin and tryptamine were undetectable in the plasma even at the point of the maximal IDO induction.
3) Half-life of Blood Kynurenine: The blood kynurenine was pulse-labelled by a trace amount of ^{14}C-kynurenine and the decline of the specific radioactivity of kynurenine was chased. The half-life (about 10 min), thus determined, of blood kynurenine at the maximal IDO induction was almost comparable to that of the control, indicating that the elevation of plasma kynurenine level was not due to an inhibition of the further metabolism of kynurenine but was resulted from the net increase in the kynurenine production.
4) Urinary Metabolites of Kynurenine: Analysis by HPLC showed that in the case of the control ICR mouse, the urinary metabolites of kynurenine were kynurenic acid and xanthurenic acid in the ratio of 1 : 7. As shown in Fig. 1C, the urinary xanthurenic acid level increased about 3-fold by the same LPS treatment, while the amount of kynurenic acid was almost constant. The time course of the increment of xanthurenic acid level correlated well with that of the plasma kynurenine level, and also with that of the pulmonary IDO activity. Kynurenine itself and the further metabolites of kynurenine such as 3-OH-

kynurenine, N(α)-acetyl-kynurenine, 3-OH-anthranilic acid and anthranilic acid were not detectable in the urine even at the highest level of plasma kynurenine. These facts indicated that the increased kynurenine was converted into xanthurenic acid via 3-OH-kynurenine, and then was excreted in the urine.

5) Tryptophan Pyrrolase in the Liver: In contrast to the induction of the extraphepatic IDO, the LPS injection rather suppressed the hepatic tryptophan pyrrolase activity to less than 50% of the control activity (Fig. 1D), which did not correlate at all with the levels of plasma kynurenine and

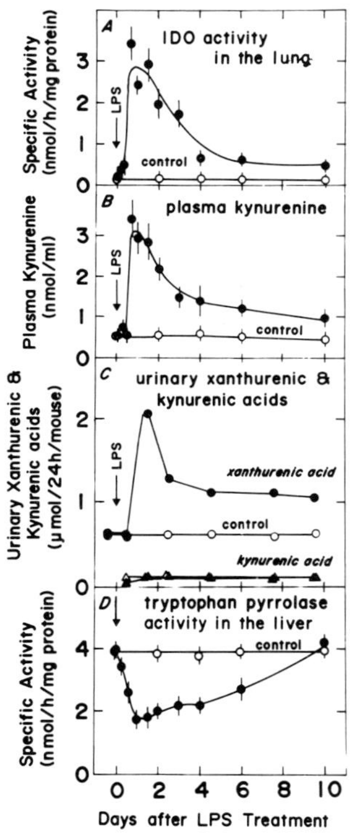

Fig. 1 Effects of the IDO inducer (LPS) on the IDO activity in the lung (A), the plasma kynurenine level (B), the levels of urinary xanthurenic and kynurenic acids (C) and the tryptophan pyrrolase activity in the liver (D). Each points in (A), (B) and (D) represents mean ± SEM for five mice. For (C), a group of ten mice was treated by LPS and the urine was collected every 24h in a metabolic cage.

urinary xanthurenic acid. Furthermore, we found that the induction of tryptophan pyrrolase by hydrocortisone increased neither the plasma kynurenine nor the urinary metabolites of kynurenine (not shown). Thus, the contribution of the hepatic tryptophan pyrrolase on the kynurenine production was minor, and the levels of plasma kynurenine and urinary xanthurenic acid reflected well the IDO activity in extrahepatic tissues.

Conclusions

On the basis of the above-mentioned results, we concluded that tryptophan is the physiological substrate of the induced IDO, and that the key enzyme on the kynurenine pathway is IDO rather than tryptophan pyrrolase.

References

1. Shimuzu, T., Nomiyama, S., Hirata, F., Hayaishi, O.: J. Biol. Chem. 253, 4700-4706 (1978).
2. Hayaishi, O.: J. Biochem. (Tokyo) 79, 13-21 (1976).
3. Yoshida, R., Hayaishi, O.: Proc. Natl. Acad. Sci. USA. 75, 3998-4000 (1978).
4. Yoshida, R., Urade, Y., Tokuda, M., Hayaishi, O.: Proc. Natl. Acad. Sci. USA. 75, 4084-4086 (1979).
5. Wolf, H.: Studies on Tryptophan Metabolism in Man, The Scandinavian Journal of Clinical and Laboratory Investigation, Supplementum 136. Copenhagen 1974.

L-TRYPTOPHAN METABOLISM ALONG THE KYNURENINE PATHWAY IN HUMAN MALIGNANT MELANOMA

Carlo Costa, Mario Lise*, Donato Nitti*, Antonio De Antoni, Alessandro Zogno, Edda Cardin dè Stefani**, Graziella Allegri

Istituto di Chimica Farmaceutica - Centro di Studio sulla Chimica del Farmaco e dei Prodotti Biologicamente Attivi del CNR, *Istituto di Patologia Chirurgica I e **Istituto di Dermatologia - Università - Padova - Italy

Introduction

In previous researches into the role of tryptophan in the biogenesis of melanins (1-3) we demonstrated that tryptophan, labelled either in the benzene ring-^{14}C(U) or in methylene-^{14}C, was incorporated into the melanin of mouse Harding-Passey melanoma. The incorporated radioactivity was almost equal to that obtained with tyrosine-2-^{14}C, administered as a comparison. How tryptophan is incorporated into melanins is not yet known. According to our data (4,5) and those reported by other authors (6), the possible pathways appear to be two: the "via 5-hydroxytryptophan" (4,6) and the "via kynurenine" through the 3-hydroxykynurenine derivative (4). Therefore, in an effort to observe whether tryptophan is involved in melanogenesis also in man we have considered it interesting to study tryptophan metabolism along the kynurenine pathway in patients with cutaneous malignant melanoma.

Materials and methods

54 patients (24 males and 30 females, mean age 50 years), with cutaneous malignant melanoma at various clinical

Progress in Tryptophan and Serotonin Research

stages, admitted to the Institute of Surgical Pathology of Padova University, were studied before surgical treatment and 37 patients (18 males and 19 females, mean age 48 years) after surgical treatment. No drug administration was effected during the investigations. The patients were divided in relation to the local invasion of primary tumor according to Clark's classification (7): 3 women were at 2nd level, 5 men and 10 women at 3rd level, 12 men and 10 women at 4th level, 6 men and 12 women at 5th level. In 2 men and in 3 women with lymph node metastasis the localization of the primary tumor was not known. A group of 21 healthy laboratory workers (11 males and 10 females) were used as controls. They were allowed free diet, without excesses or irregularities. After collection of the basal urine that showed negligible levels of tryptophan metabolites with our chromatographic method, the subjects received 50 mg/kg b.w. of L-tryptophan (Trp) administered orally in a glass of orange juice between 7 a.m. and 8 a.m. The urine was collected for a 10-h period after the load and kept at 0-4° C during collection. The time collection was based on experimental findings such as the blood concentration curve of Trp after loading (8). The analysis of the Trp metabolites was carried out immediately, as previously reported (8,9). The following 10 Trp metabolites were determined: kynurenine (Kyn), Nα-acetylkynurenine (ACKyn), kynurenic acid (KA), anthranilic acid (AA), o-aminohyppuric acid (o-AHA), 3-hydroxykynurenine (3-OHKyn), xanthurenic acid (XA), 8-methyl ether of xanthurenic acid (OCH3XA), 3-hydroxy acetylkynurenine (3-OHACKyn) and 3-hydroxyanthranilic acid (3-OHAA).

Results and discussion

As with the load of 50 mg/kg b.w. of L-Trp the values of the urinary metabolites are homogeneous between the sexes, they have been put together. The mean total urinary excretion of the metabolites, expressed as μmoles appeared to be: 917±62 and 911±55 μmoles respectively for patients before and after surgical treatment, In the control group it was 615±54 μmoles. There were no significant differences in the values between the groups of patients before and

Table 1. Mean urinary (±S.E.) tryptophan metabolites by healthy subjects and patients with malignant melanoma after a loading dose of 50 mg/kg b.w. L-tryptophan (µmoles excreted/10 hours).

Metabolite	Controls (21)*	Patients with melanoma before surgery (54)*	after (37)*
	µmoles	µmoles	µmoles
Kyn	189±20.0	338±33.2	321±44.5
ACKyn	95±13.5	123±11.2	125±11.2
KA	148±26.2	178±11.0	184±17.9
o-AHA	41± 5.8	23± 2.2	22± 2.5
3-OHKyn	40± 4.3	76± 9.2	87±18.7
XA	47± 4.9	108±11.3	89±10.3
$O-CH_3XA$	-	22± 5.1	18± 6.0
3-OHAA	52± 6.0	53± 5.6	61± 7.1
	P < 0.005		
Total	615±54.2	917±62.1	911±55.0

* Number of subjects is indicated in parenthesis.

after surgical treatment, whereas such values were significantly higher (at $P < 0.005$) in respect to the controls. In addition, larger statistically significant differences between patients and controls in the excretion of Kyn, 3-OHKyn, XA and OCH3XA have been found. Considering the excretion of these metabolites according to Clark levels of the tumor it appeared that before surgery higher amounts were excreted in patients with melanoma at 3rd and 4th Clark's level, whereas after surgery the values of the metabolites were almost similar at all Clark levels. The high excretion of XA and OCH3XA should evidence a deficiency of vitamin B6 in the patients with cutaneous malignant melanoma. In conclusion, an abnormal Trp metabolism is present in patients with cutaneous malignant melanoma, increasing with the progress of the disease.

In addition, surgical treatment does not seem to modify the abnormality of the metabolism of this amino acid.

References

1. Allegri, G., De Antoni, A., Baccichetti, F., Costa, C.: Gazz. Chim. Ital. 102, 426 (1972).
2. De Antoni, A., Allegri, G., Costa, C., Bordin, F.: Experientia 30, 600 (1974).
3. Costa, C., Allegri, G., De Antoni, A.: Acta Vitamin. Enzymol. 29, 223 (1975).
4. Musajo, L., Benassi, C.A., Allegri, G., Levorato, E., De Antoni, A.: Chim. e Ind. 48, 1221 (1966).
5. Benassi, C.A., Veronese, F.M., De Antoni, A., Pajetta, P.: Gazz. Chim. Ital. 97, 3 (1967).
6. Viscontini, M., Mattern, G.: Helv. Chim. Acta, 53, 832 (1970).
7. Clark W.H.:Cancer Res. 29, 705 (1969).
8. Costa, C., De Antoni, A, Allegri, G., Vanzan, S.: La Ricerca Clin. Lab. 9, 165 (1979).
9. Coppini, D., Benassi, C.A., Montorsi, M.: Clin. Chem. 5, 391 (1959).

Nutritional Aspects

CONTENT AND STABILITY OF TRYPTOPHAN IN FOODS

Henrik K. Nielsen and Richard F. Hurrell
Nestlé Products Technical Assistance Co. Ltd.
CH-1814 La Tour-de-Peilz, Switzerland

Content

Tryptophan is an essential amino acid for man. It cannot be synthesized in the body and therefore must be supplied by the diet. Food proteins in general contain 0.7 - 1.8 g/16g N of tryptophan. FAO/WHO (1) recommend that a good quality protein should contain at least 1.0 g/16g N. Although some 30% of plant proteins are deficient in tryptophan it is very rarely the first limiting amino acid. This is normally either lysine or the sulphur amino acids, which also tend to vary more widely in their content in foods than does tryptophan.

Table 1 shows the tryptophan content of different food categories and was compiled from data provided by the FAO (2). The different foodstuffs have been ranked according to their mean tryptophan content. Only in one instance, in a small number of seemingly unimportant nuts and seeds, was tryptophan found to be the first limiting amino acid. Of more importance is the relatively large proportion of fruits, cereals and legumes which are deficient in tryptophan. This is particularly significant for cereals which are the staple foods in many parts of the world and legumes which are eaten as protein supplements. Considering the legumes, broad bean, mung bean, lentil, chick pea and the common garden pea contain arount 90% of the recommended tryptophan level, whereas the pigeon pea contains only 60%.

Progress in Tryptophan and Serotonin Research

Table 1. Tryptophan in different food categories

Category (number of samples)	Mean trp content (g/16g N)	% with trp as first limiting amino acid	% deficient in in trp (< 1.0g/16gN)
Milk and eggs (5)	1.45	0	0
Nuts and seeds (40)	1.39	8	20
Roots (8)	1.32	0	13
Vegetables (39)	1.25	0	17
Cereals (15)	1.21	0	27
Meat, poultry and fish (25)	1.14	0	0
Legumes (22)	1.09	0	41
Fruits (9)	0.92	0	67

Among the cereals, the low level of tryptophan in maize (around 70% of its recommended level) is by far the most important both in terms of protein quality and in relation to pellegra, the nicotinic acid deficiency desease associated with maize eating populations (3). The human body can synthesize nicotinic acid from tryptophan and although milk and eggs for instance are low in nicotinic acid they have long been recognized as beneficial to patients with pellegra because of their high tryptophan content. The low level of tryptophan in maize protein does little to overcome the non-availability of its nicotinic acid.

Stability

The levels of tryptophan in Table 1 are in raw foodstuffs whereas most foods are cooked or industrially processed before consumption. During food processing and storage many chemical reactions can occur between the different food components (4). Proteins can react with reducing sugars, polyphenols, fats and their oxidation products as well as with various chemical additives. These reactions lead to colour and flavour formation, reduction in nutritional value and occasionally to products with potentially toxic effects (5). In this short review we will concentrate on the nutritional consequences. Lysine, methionine, cyst(e)ine and tryptophan are often reported to be the most reactive amino acids in the protein chain. Our results however indicate that under normal conditions of processing and storage tryptophan is relatively stable.

Protein - oxidizing lipid reactions. These reactions can occur for instance during the storage of cereals and fish products. When unsaturated fatty acids oxidize there is a formation of free radicals and hydroperoxides followed by a range of secondary breakdown products such as aldehydes and ketones. Tryptophan and sulphur amino acids could be oxidized by the free radicals and hydroperoxides. The secondary breakdown products are more likely to react with lysine but could also react with other amino acids.

In order to investigate these reactions, we have stored a model system of whey protein, methyl linolenate (C 18:3) and water in sealed tins under different conditions. Table 2 shows the percentage tryptophan, lysine and methionine remaining in samples stored for 4 weeks at 37°C under conditions of excess oxygen or where oxygen was restricted to one mole per mole of lipid. The tryptophan content of these two samples fell by 4% and 14% compared to the much greater losses of lysine (26% and 60% respectively) and the almost complete oxidation of methionine to methionine sulphoxide. The latter amino acid is only partly bioavailable.

Table 2 shows the available tryptophan content of the same two samples as measured in the rat assay. Much larger falls in tryptophan (11% and 44% respectively) were found than in the chemical assay. The difference between the two assays could be accounted for by a large decrease in the overall

Table 2. Remaining tryptophan*, lysine and methionine in whey protein-methyl linolenate samples stored at 37°C for 4 weeks with and without excess oxygen.

	oxygen limitation	excess oxygen
Tryptophan after alk. hydr. (A)	96% ± 1%	86% ± 1%
A x protein digestibility	88% ± 1%	60% ± 2%
Available tryptophan with rats	89% ± 6%	56% ± 6%
Lysine	74% ± 1%	40% ± 1%
Methionine	18% ± 2%	6% ± 2%

*Tryptophan was determined by HPLC after alkaline hydrolysis (6), methionine by ion exchange chromatography of the same alkaline hydrolysates and lysine by the FDNB-technique (7). Available tryptophan was a slope ratio assay (8) based on (9).

protein digestibility presumably due to crosslinkages being formed between the protein chains via the fat degradation products. It can be seen that the chemically determined tryptophan value multiplied by protein digestibility is in good agreement with the rat assay value.

It is difficult of course to relate these relatively extreme model systems to what could actually happen in real foods. These experiments stress however the importance of oxygen level in fat oxidation and even a small limitation of oxygen to 400 ml of air per gram of oil significantly reduced the amino acid degradation. It would appear therefore that during storage of food products, where oxygen uptake is far smaller, significant tryptophan degradation is unlikely to occur.

Polyphenol - protein reactions. Phenolic acids such as chlorogenic acid are widely distributed in plants. Sunflower seeds for instance which contain about 16% protein contain up to 2% chlorogenic acid and smaller quantities of caffeic acid and other phenolic acids. During protein extraction oxidative browning reactions take place and coloured products result. In the presence of alkali or polyphenol oxidase phenolic acids are transformed into the corresponding quinone. This can polymerize to give brown pigments but it can also react with lysine, methionine and tryptophan (10).

We have reacted an aqueous solution of casein (5%) with caffeic acid (0.5%) for 3 h at 20°C and then fed the freeze-dried casein-caffeic acid complexes to rats. We found losses of available lysine, methionine and tryptophan of 44%, 26% and 13% respectively at pH 10 and 19%, 13% and 8% at pH 7 with polyphenol oxidase.

Again the losses of lysine and methionine were greater than the losses of tryptophan which appeared to be mainly due to a general reduction in protein digestibility. The loss of available methionine was probably due also to its oxidation to methionine sulphoxide which under certain conditions was almost complete. Lysine however appeared to form covalent lysine - caffeoquinone compounds.

Reducing sugar - protein reactions. Maillard reactions. The reactions of proteins with reducing sugar (the Maillard reactions) are the major source of nutritional damage to food proteins during processing and storage. The Maillard reactions occur during the baking of bread, the production of breakfast cereals, the heating of meats, but most importantly during the processing and storage of milk products since milk is the only naturally-occurring protein food that has a high content of reducing sugar (4).

Maillard reactions can be conveniently divided into early and advanced Maillard reactions. The first step involves the condensation reaction between the carbonyl group of the reducing sugar and the amino group of lysine. The reaction rapidly proceeds via Schiff's bases to the deoxyketosyl compound or Amadori product. This is the major form of blocked lysine after early Maillard reactions and is believed to be unavailable as a lysine source for the organism. At this stage there is no colour formation. The deoxyketosyl compounds are degraded during the advanced Maillard reactions which follow. These reactions are not well-defined but are responsible for the numerous flavours and odours as well as browning.

Some of our earlier results (11) using α-N-acetyl-tryptophan in solution had indicated that the advanced Maillard reaction products could react with tryptophan. However this does not appear to happen readily in food products. We have stored milk powders for several weeks at 70°C and investigated the fate of lysine and tryptophan (Fig 1). After only 2 weeks storage 54% of lysine was blocked as lactulosyl-lysine although the product still retained its natural colour.

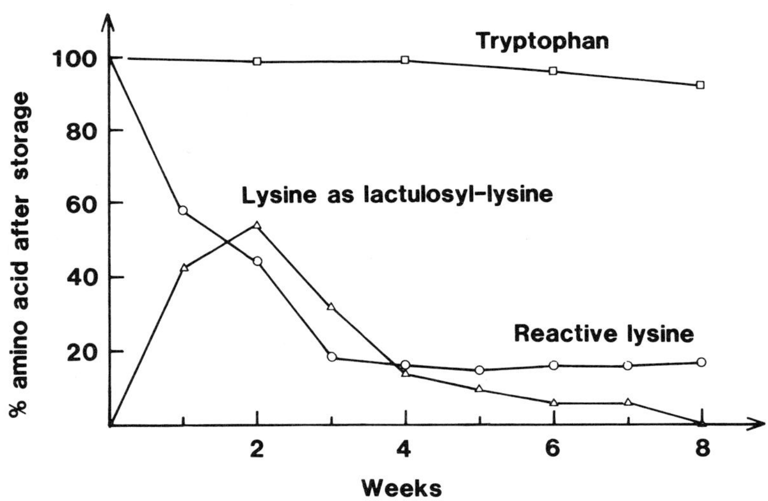

Fig. 1. Tryptophan (6), reactive lysine (12) and lysine as lactulosyl-lysine (12) of whole milk powder stored at 70°C.

This is a classic example of early Maillard damage. From 3 weeks onwards the product became brown as lactulosyl-lysine degraded and advanced Maillard reactions took place. Lysine was reduced to 16% of its original value although tryptophan measured chemically remained unchanged.

In a subsequent experiment, available tryptophan was measured with rats in two milk powders which had undergone either early Maillard or advanced Maillard damage. As shown in Table 3, the available tryptophan, as well as the chemical values were virtually unchanged although lysine losses were 21% and 80% respectively. The loss of tryptophan due to Maillard reactions would therefore appears to have little or no nutritional significance.

Table 3. Tryptophan and lysine remaining* in whole milk powder after 'early' and 'advanced' Maillard reactions.

	'early' Maillard (9 weeks, 50°C)	'advanced' Maillard (4 weeks, 60°C)
Tryptophan after alk. hydr.	101% ± 1%	101% ± 1%
Available tryptophan with rats	109% ± 7%	92% ± 7%
Reactive lysine	79% ± 2%	20% ± 2%

* Tryptophan determined as described in Table 2 and reactive lysine by the furosine technique (12).

Conclusion

In conclusion, it appears that tryptophan is rarely the first limiting amino acid in food proteins. It is deficient however in some legumes and cereals; its low level in maize being probably the most important example. As far as food processing and storage are concerned the losses of tryptophan appear to be small and of only minor nutritional importance, when compared to the much larger losses of lysine and methionine.

References

1. FAO/WHO Ad Hoc Expert Committee: Energy and Protein Requirements. FAO Nutrition Meetings Report Series No 52. Rome (1973).

2. FAO: Amino-acid content of foods and biological data on proteins. Rome (1970).

3. Carpenter, K.J.: Pellegra. Academic Press. New York (1981)

4. Hurrell, R.F.: p. 369-88 in Food and Health: Science and Technology. Eds Birch, G.G,; Parker, K.J. Applied Science Publishers Ltd, London (1980).

5. Aeschbacher, H.U.: p. 349-62 in Mutagens in Our Environment. Alan R. Liss Inc. New York (1982).

6. Nielsen, H.K., Hurrell, R.F.: This volume, 111.

7. Booth, V.H. : J. Sci. Fd. Agric. 22, 658-64 (1971).

8. Finney, D.J.: Statistical Methods in Biological Assay, 2nd. ed. Hafner Publishing Co. New York (1964).

9. Gupta, J.D., Elvehjem, C.A.: J. Nutr. 62, 313-324 (1957).

10. Hurrell, R.F., Finot, P.A. Cuq, J.L.: Br. J. Nutr. 47, 191-211 (1982).

11. Finot, P.A., Magnenat, E., Guignard, G., Hurrell, R.F.: Internat. J. Vit. Nutr. Res. 52, 226 (1982).

12. Finot, P.A., Deutsch, R., Bujard, E.: p. 345-55 in Progress in Food and Nutrition Science, Vol. 5, Maillard reactions in foods. Ed. Eriksson, C. Pergamon Press. Oxford (1981).

INCREASED AROMATIC AMINE (KYNURENINE) IN DRIED FISH MEAT UNDER SUNLIGHT

Hitoshi Takahashi

Department of Pharmacology, Institute for Medical Immunology, Kumamoto University Medical School, Kumamoto, Japan 860

Introduction and Methods

Dried fish under sunlight has been a very common food among the people in oriental countries. In such preparation of food photolytic degradation of tryptophan residue to kynurenine in proteins could be suspected, because photooxidative degradation of some amino acid residues in protein is known to occur under irradiation of ultra-violet light in the presence of photosensitizing substances such as methylene blue, porphyrins, etc. To study this problem the homogenate of raw or dried fish meat (around 5 g) was treated with 100 and 20 ml of ether-ethanol mixture (1:1) to remove lipids. Precipitated protein portion was then subjected to acid hydrolysis in 150 ml of 6 N HCl for 6 hours under refluxing. The hydrolysate was filtered through a glass filter and evaporated almost to dryness to remove HCl, and finally dissolved in about 100 ml of water. Kynurenine in the hydrolysate was determinend by the same method used for the aromatic amines in urine by Brown and Price (1).

Results and Discussion

In this assay addition of 2 mg of L-tryptophan to the protein precipitate before hydrolysis did not alter kynurenine value at all and over all recovery of 50 µg of L-kynurenine sulfate (34 µg of L-kynurenine) added before hydrolysis was between

Progress in Tryptophan and Serotonin Research

Table 1, Kynurenine contents in fish meat dried under sunlight

Squid (4)	26.7 --- 35.7 µg/g
Horse mackerel (4)	18.1 --- 22.9 µg/g
Dolphin fish (3)	5.5 --- 9.8 µg/g
Flying fish (3)	2.1 --- 5.1 µg/g
Shirasuboshi (4)	6.1 --- 7.7 µg/g
Sardine (2)	20.1 --- 24.2 µg/g
Sardine steamed (3)	23.2 --- 28.5 µg/g

78 to 95 % at the end of assay. This indicates that tryptophan is not converted to kynurenine during hydrolysis and kynurenine is resistant to acid hydrolysis.

With this procedure kynurenine contents in several kinds of dried fish most commonly eaten in Japan were determined. The results are shown in Table 1.

Exposing fresh meat of dolphin fish to direct sunlight for 8 hours increased the average kynurenine content to 3.5 µg/g from 0.9 µg/g (wet weight).

The possibility of lipid peroxide involvement in this degradation is under investigation. At this time it is hard to say that increase of aromatic amine in dried fish is hazardous to our health, however, it is definitely a reduction of nutritional value of fish protein by destructing an essential amino acid, tryptophan.

Reference

1. Brown, R. R., Price, J. M.: J. Biol. Chem. 218, 985--997 (1956).

TRYPTOPHAN REQUIREMENT OF FARM ANIMALS

Hans Steinhart

Institut für Ernährungsphysiologie der Technischen
Universität München
D-8050 Freising-Weihenstephan

Estimation of Amino Acid Requirements

The requirements of amino acids for farm animals cannot be regarded as fixed as they depend on many parameters which have to be defined in advance. For example there is a difference between the requirement for a growing pig and a pregnant sow. In the growing pig about 80 - 90 % of the amino acids are used to synthesize body proteins whereas only 10 - 20 % are used for maintenance (1). In the pregnant sow during the first two weeks of pregnancy most of the amino acids are used for maintenance and in the later weeks used for the synthesis of foetuses. The requirements also depend on the level of energy in the diet. In energy deficient rations, part of the protein is used to meet the energy requirement of the animal. Another important factor is the supply of balanced proteins in the feeds. Imbalanced proteins alter the requirements of the amino acids because there is competition for absorption from the intestine and cell walls. Competition between tryptophan (Trp) and the large neutral amino acids (LNAA) Ile, Leu, Val, Phe and Tyr is known to occur not only during absorption (2) but also at the blood-brain-barrier (3). In addition, Trp impedes the absorption of the most important amino acids in animal nutrition, Lys and Met (4, 5). Menke and Gruber (6) also showed that the second and third most limiting amino acids in a diet have an influence on the utilization of the first limiting amino acid.

Progress in Tryptophan and Serotonin Research

Another problem concerning Trp is the protein turnover. The different essential amino acids resulting from protein catabolism are not all reused in the same proportions for the synthesis of new proteins. Waterlow et al. (7) found that under normal nutritional conditions about 80 % of the amino acids from protein catabolism can be re-utilized. Lys (81 %) and Met (91 %) are however reused to a greater extent than the average of the amino acids (8). In contrast, the reutilization of Trp is very low (9). This means that the metabolic consequences of a deficiency of Trp in the diet is manifested earlier than with a deficiency of other amino acids.

Another important point is that Trp, together with Lys and Met, is one of the most labile amino acids. It is especially sensitive to irradiation and oxidation (10). This lability is not so important in human nutrition (11) because humans have a varied diet but it can become a problem in animal nutrition because farm animals normally get the same feedstuffs for a long period. It seems probable that the oxidation products of Trp are somewhat toxic if experienced by an animal for any length of time.

The animal nutritionist has difficulties in estimating the correct amount of Trp in single feedstuffs. There are many analytical procedures available to determine the content of Trp in single feedstuffs. However, the high standard deviations in amino acid tables show that until now it has been difficult to accurately determine the content of Trp.

A final point regarding Trp is that corn and corn products, the most commonly used feedstuffs, are very low in Trp.

An important question in animal nutrition is how to experimentally determine the amino acid requirement of animals. There are two different approaches. First, the determination of metabolic parameters, such as changes in the concentration of the limiting amino acid in the blood or in the muscle, determination of urea in the blood,excretion of N-metabolites in the urine and N-deposit. Second the determination of production parameters like growth, feed:gain ratio, protein content in the carcass or N-retention. It is well known that metabolic parameters are more sensitive than production parameters to the amino acid supply. That means that the tolerance width of production parameters is larger than that of metabolic parameters for amino acid imbalanced diets. Because production parameters are more important in animal production than metabolic parameters, more experiments employ production parameters to determine the amino acid requirement.

Following these reflections some experiments with broiler chickens are described, dealing with the effects of oxidized Trp, and imbalanced diets.

Influence of Tryptophan Oxidation Products on Growth And Physiological Parameters of Chickens.

This experiment was carried out to investigate growth, feed efficiency and blood Trp content of chickens fed diets containing oxidized isolated soy bean protein as the only protein source (12). During the oxidation by H_2O_2, 87 % of the Trp of the soy bean protein was destroyed. The feeds were supplemented with different amounts of L-Trp to determine whether any effects were caused by Trp deficiency or by the Trp oxidation products. One-day old male broiler chickens were housed in cages, and 12 chickens used per treatment. The animals were fed ad libitum and water was supplied automatically. The chickens were killed after 26 days by decapitation. The metabolisable energy (ME) was 14.1 MJ/kg feed. Eighteen grams of crude protein were used per MJ ME. With the exception of Trp the diets contained a balanced amino acid composition. The control group (group 0) received non-oxidized soy bean protein. The diets contained the following amounts of Trp: control group, 0.22 %; group 1, 0.03 %; group 2, 0.10 %; group 3, 0.22 %, and group 4, 0.47 %. The growth curves of the chickens are shown in Figure 1. The regression for group 1 shows only a slow growth. Although the animals received only 0.03 % Trp, all animals survived. In

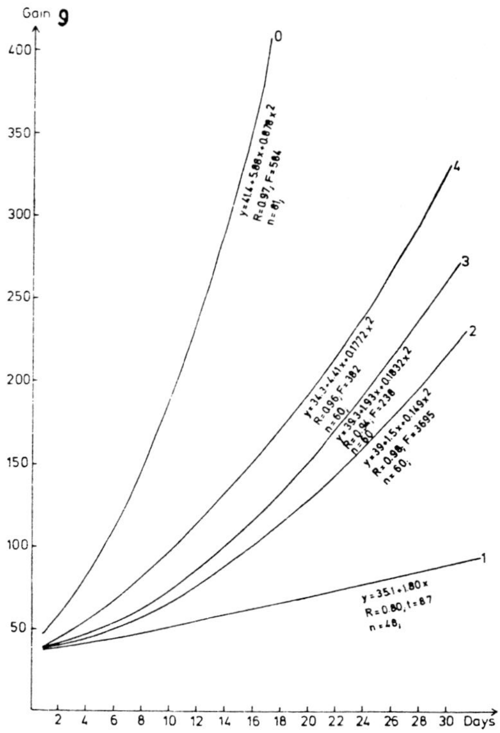

Figure 1: Growth curves of the chickens

contrast to these results are those from experiments that were conducted with a corn-gelatine diet. This diet also contained only 0.03 % Trp but the chickens died within 10 days. It is possible that in the soy bean experiment some of the products resulting from the oxidation of soy bean protein by H_2O_2 may be metabolized in a similar way to Trp. These products which have not yet been identified may have substituted for Trp.

The growth curves already changed with the low Trp supplementation in the feed. Although the Trp supply was identical in the control group and in group 3, the animals of the control group grew significantly faster than in group 3. The higher the Trp content in the oxidized diets, the better the chickens grew. However, the animals in group 4 that received twice the Trp of the control animals grew significantly slower than the control group. From these results it may be concluded that there are toxic decomposition products of Trp. These products could easily be antagonists to Trp as suggested by Meister (13).

The feed efficiency (g feed/g gain) is presented in Table 1. Over the experimental period (1st to 20th day) the feed efficiency was highest in the control group (1:1.6) and also in group 4 (1:1.8). It decreased very strongly in groups 2 and 3, and was lowest in group 1 (1:4.1). These results correspond

Table 1: Feed efficiency (g feed/g gain)

group	1st-5th	6th-12th	13th-17th	18th-20th	1st-20th
0	1.26	1.99	1.49	1.65	1.60
1	4.85	3.12	4.83	3.55	4.09
2	2.64	3.86	1.32	1.51	2.33
3	3.27	1.66	1.89	1.76	2.15
4	2.45	1.82	1.62	1.32	1.80

closely with the growth curves. They cannot be explained with imbalances of Trp in the diets because in this case, group 0 and group 3 must be similar.

The physiological parameter, total plasma Trp content of slaughtered chickens is summarized in Table 2.

Table 2: Trp content in the plasma of slaughtered chickens (nmol Trp/ml plasma)

group	$\bar{x} \pm s$	group	$\bar{x} \pm s$
0	40.93 ± 2.64^{a}	3	33.90 ± 3.15^{de}
1	9.88 ± 2.73^{b}	4	41.53 ± 3.49^{a}
2	29.43 ± 2.27^{ce}		

Mean values that have the same superscript are not significantly different (test of Tukey, $P < 0.05$)

These data also correspond very well with the production parameters. Total plasma Trp was very low in group 1, indicating a severe deficiency of Trp. There were no significant differences between the animals in groups 2 and 3, the values were, however, lower than those in the control group. The plasma Trp content in group 4 was similar to that in group 0, although the Trp content of the feed was different by a factor two.

As a consequence of these results it can be concluded that Trp oxidation products have a strong influence on the utilization of Trp in the diet. This influence cannot be explained by imbalances. It is therefore very important to avoid Trp oxidation during treatment or storage of feeds especially if these feeds are to be used over long periods.

Tryptophan and Large Neutral Amino Acids

The aim of this experiment was to investigate whether different ratios of Trp and large neutral amino acids (LNAA) (Ile, Leu, Val, Phe, Tyr) in diets have an effect on growth, daily gain, daily feed intake and feed efficiency in broiler chickens (14). The experiment was conducted with male broiler chickens. The animals, 90 in each group, were fed six diets from the first day. The experiment lasted 28 days. Feed and water were not limited. Bodyweight and feed consumption were measured weekly. The experiment was conducted bifactorially. The content of Trp in the diet was either 0.15 % (group A) or 0.21 % (group B), and the content of LNAA was either 3.37 % (1), 4.03 % (2) or 4.83 % (3). The recommended requirement of Trp was met in group B, and that of LNAA in group 2. The Trp content in group A was 30 % below, the LNAA content in group 1 16 % below, and in group 3 20 % above requirement. The crude protein content was 21.5 %, and the ME 12 MJ/kg diet. The production parameters are given in Table 3.

The weights after 28 days and the daily gains of the animals with sufficient intake of all amino acids (group B/2) showed normal values. A lower dietary concentration of Trp caused lower weights and lower daily gains (group A/2). An excessive concentration of LNAA resulted in lower weights and reduced daily gains in both Trp groups and this effect was most striking in the group with the highest Trp content. In both Trp groups the effects seen with lower concentration of LNAA were not as strong as the surplus of LNAA.

The results of the daily feed intake show the same trends as the data already mentioned. The differences between the six

Table 3: Growth parameters of the chickens from the 1st to the 28th day of the experiment

		LNAA 1	LNAA 2	LNAA 3
weight (g)				
Trp	A	451.7 ± 29.7^{c}	471.7 ± 34.7^{c}	418.7 ± 24.7^{b}
	B	472.0 ± 18.3^{c}	670.2 ± 26.5^{d}	272.3 ± 26.0^{a}
average daily gains (g)				
Trp	A	15.4 ± 1.1^{c}	16.2 ± 1.3^{c}	14.2 ± 0.9^{b}
	B	16.2 ± 0.7^{c}	23.5 ± 1.0^{d}	8.8 ± 1.0^{a}
average daily feed intake (g)				
Trp	A	32.0 ± 2.2^{c}	32.4 ± 3.0^{c}	29.3 ± 2.1^{b}
	B	34.3 ± 2.1^{c}	43.4 ± 1.1^{d}	20.9 ± 1.9^{a}
feed efficiency (g feed/g gain)				
Trp	A	2.07 ± 0.06^{bc}	2.01 ± 0.15^{b}	2.06 ± 0.05^{bc}
	B	2.12 ± 0.08^{c}	1.85 ± 0.09^{a}	2.38 ± 0.08^{d}

groups are, however, not so great when the feed efficiency is considered. There is no effect of LNAA in the groups with low Trp supply (A) but the differences between the other groups are significant (B), although not as great as seen in daily gains, feed intake or final body weight. It may therefore be assumed that the effect of an excess of LNAA on weight gain is primarily produced by a depression of feed intake. A simple explanation of these results based on imbalance effects is not possible because the most imbalanced diets, i.e. groups A/2 and A/3 did not show the greatest effects. The imbalances in amino acid supply were not great enough to explain the reported differences in the growth parameters. Antagonistic effects between Trp and LNAA have also to be taken into account.

References

1. Mitchell, H. H.: Some species and age differences in amino acid requirements. In: Albanese, A. A.: Protein and Amino acid Requirements of Mammals. Academic Press, New York 1950.
2. Wiseman, G.: Absorption of protein digestion products. In: Smith, D. H.: Intestinal Digestion, Biomembranes, Vol 4 A, Plenum Press, New York 1974.
3. Wurtman, R. J.: Nutritional control of brain tryptophan and serotonin. In: Hayaishi, O., Ishimura, Y., and Kido, R.: Biochemical and Medical Aspects of Tryptophan Metabolism, Elsevier North-Holland, Amsterdam 1980.
4. Munck, B. G.: Biochim. Biophys. Acta 126, 299-307 (1966).
5. Munck, B. G., Rasmussen, S.N.: Biochim. Biophys. Acta 389, 261-280 (1975).
6. Menke, K. H., Gruber, G.: Effect of tryptophan on protein utilization and energy intake of pigs. This volume, 571.
7. Waterlow, J. C., Garlick, P.J., Millward, D.: Protein turnover in mammalian tissues and in the whole body. North-Holland Publ. Comp., Amsterdam 1978.
8. Bock, H.-D.: Rolle und Umfang des Aminosäure-Pools bei Schwein und Geflügel im Wachstum. Adademie der Landwirtschaftswissenschaften der DDR 1980.
9. Munro, H. N.: Free amino acid pools and their role in regulation. In: Munro, H. N.: Mammalian Protein Metabolism. Academic Press, London, New York 1970.
10. Steinhart, H.: Landwirtsch. Forsch. 32, 63-73 (1979).
11. Nielsen, H. K., Hurrell, R. F.: Content and stability of tryptophan in foods. This volume, 527.
12. Steinhart, H., Kirchgeßner, M.: Z. Tierphysiol., Tierernährg. u. Futtermittelkde. 40, 332 - 343 (1978).
13. Meister, M.: Biochemistry of the Amino Acids, Vol. II, Second edition, Academic Press, New York 1965.
14. Steinhart, H., Kirchgeßner, M.: Z. Tierphysiol., Tierernährg. u. Futtermittelkde., in press (1983).

PATTERN OF RADIOACTIVITY FROM ENDOGENOUS ^{3}H-TRYPTOPHAN EXCRETED IN THE URINE OF FASTED OR RATS FED DIETS CONTAINING DIFFERENT LEVELS OF CASEIN

Paul-André Finot and Edith Magnenat

Research Department, Nestlé Products Technical Assistance Co. Lt.
CH-1814 La Tour de Peilz, Switzerland

Introduction

The free pool (circulating and intracellular) of an amino acids is essentially the sum of the dietary contribution (exogenous) and that arising from protein degradation (endogenous). The contribution of exogenous TRP to circulating TRP level in humans has been studied by Ashley et al. (1) using ^{15}N-TRP. Endogenous TRP comes mainly from degradation of tissue protein and hence the endogenous TRP concentration can be influenced by nutrition (e.g. level of dietary protein), behaviour (e.g. feeding pattern) and hormonal factors. In this experiment we investigated the effect of nutrition on liberation of endogenous TRP, from protein-bound sources by measuring urinary excretion and retention in the liver, kidney and brain of radioactivity in rats previously injected with ^{3}H-TRP, and fed different diets.

Experiment

Twelve Sprague-Dawley rats (128 ± 3 g) were fed a diet containing 18 % casein for 24 h prior to injecting with 15 μci of L-[5-^{3}H]tryptophan (20 μci/mmole) at 15.00 h on day 1. Afterwards, they were placed in metabolic cages in a room lighted for 12 h daily (06.00-18.00 h) and continued to receive the same diet for three additional days to allow excretion of radioactivity not incorporated into tissue proteins. On the evening of day 4, the rats were separated in 4 groups of three rats each. One group was fasted, and the other groups received one of three diets containing 0, 18 and 70 % casein respectively. Urine was collected each

Progress in Tryptophan and Serotonin Research

day at 7.00, 12.00, 18.00 and 24.00 h. After killing, at 18.00 h on day 8, the livers, kidneys and brains were removed for measurement of retained radioactivity.

Results and discussion.

Weight gain (g) (from day 1 to day 8) of the fasted, 0 %, 18 % and 70 % casein groups were - 15.2 ± 2.1, 4.5 ± 4.2, 51.5 ± 2.0 and 49.2 ± 1.8 respectively. The weight (g) of the livers was 3.92 ±0.05, 5.26 ± 0.64, 8.68 ± 0.52 and 8.88 ± 0.48 and that of the kidneys was 1.15 ± 0.07, 1.20 ± 0.04, 1.66 ± 0.06 and 1.96 ± 0.12 respectively.

	fasted animals	0 % casein	18 % casein	70 % casein
radioactivity excreted in the urine from day 4 to day 8 (% of injected radioactivity)				
	9.88 ± 1.24	6.34 ± 0.52	8.04 ± 0.46	14.6 ± 0.7
radioactivity remaining in the organs (% of injected radioactivity)				
Liver	5.09 ± 0.32	3.82 ± 0.26	4.34 ± 0.25	4.25 ± 0.05
Kidney	1.36 ± 0.17	1.07 ± 0.05	0.98 ± 0.03	0.98 ± 0.05
Brain	0.72 ± 0.10	0.57 ± 0.06	0.40 ± 0.05	0.39 ± 0.05
specific radioactivity in the organs (10^{-3} x DPM/g)				
Liver	418 ± 22	238 ± 41	161 ± 7	154 ± 7
Kidney	379 ± 38	289 ± 24	190 ± 2	162 ± 3
Brain	158 ± 11	124 ± 5	98.0 ± 2,5	90.2 ± 3.2

Table 1. Radioactivity excreted in the urine and retained in the liver, kidneys and brain of rats injected ^{3}H-TRP.

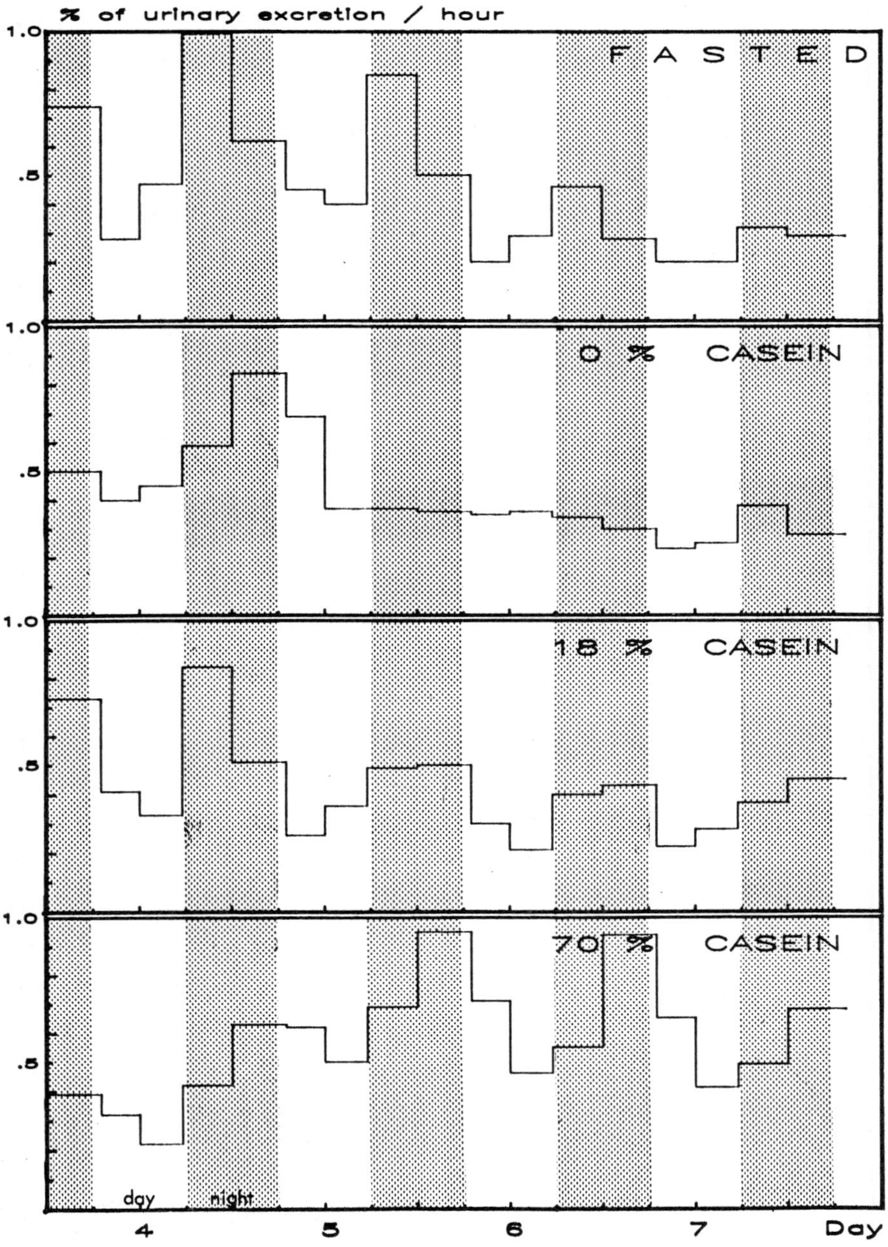

Fig. 1 : Diurnal pattern of radioactive urinary excretion after injection on day 1 of ^{3}H-TRP

In the fasted and 0 % casein groups therefore protein was being catabolized and in the other groups synthesized. Urinary radioactivity excretion decreased with decreasing protein level, suggesting a "sparing" effect of endogenous TRP by dietary carbohydrate. Fasting rats appeared to utilize endogenous proteins as an energy source, thus increasing the loss of endogenous TRP in urine. The radioactivity remaining in the organs, was the same in rats fed diets with or without protein, despite great differences in organ weight suggesting that their endogenous TRP pools were conserved. This is confirmed by the specific radioactivity data (Table 1). Although protein-catabolic rats had higher tissue specific activity than protein-anabolic rats, this simply reflects dilution of the endogenous pool by exogenous TRP. Thus, endogenous TRP of liver, kidney and brain appears to be relatively stable and little renewed even in severe catabolic states.

The pattern of urinary excretion (fig. 1) shows that during fasting, urinary TRP burning was highest in the early part of the night, when rat's energy needs are highest. In contrast, the 0 % casein group, endogenous TRP was constantly excreted over day and night. Rats eating 18 % casein had higher endogenous TRP excretion during the night than during the day while those eating 70 % casein has highest urinary TRP excretion during the latter part of the night. These diurnal patterns of endogenous TRP excretion suggest that the loss of endogenous TRP is controlled by dietary protein level and the state of metabolism of the animals.

Wannemacher and Dinterman (2) have concluded in a similar study that diurnal rhythms of phenylalanine oxidation is also controlled by dietary proteins. Carbohydrates tend to mute the responses.

1. Ashley, D., Finot, P.A., Liardon, R.: This volume, 587.
2. Wannemacher, W., Dinterman, E.: Biochem. J. 190, 663-671 (1980).

PROTEIN/CARBOHYDRATE SELECTION IN RATS FOLLOWING ADMINISTRATION OF TRYPTOPHAN, GLUCOSE OR A MIXTURE OF AMINO ACIDS

Léon Arimanana, D. Ashley, D. Furniss and Peter Leathwood

Nestlé Products Technical Assistance Co. Ltd.,
Research Department, CH-1814 La Tour-de-Peilz, Switzerland

Introduction

Pharmacological studies suggest that, in rats offered a choice of high and low protein diets, changing brain serotonin (5-hydroxytryptamine; 5-HT) can influence protein/carbohydrate selection. When brain 5-HT is depleted by treatment with p-chlorophenylalanine or 5,7-dihydroxytryptamine, energy intake is unaffected but protein intake falls (1). If synaptic levels of 5-HT are increased by injecting small doses of fenfluramine (6), there is a transient rise in the proportion of calories eaten as protein (% P-E).

The objective of this study was to test the effects on protein selection of (a) tryptophan (TRP) the precursor of 5-HT; (b) glucose (GLU) which in these particular experimental conditions should accelerate TRP entry into the brain (2), and (c) a mixture of amino acids (MAA), as used by Gessa et al. (3), which should compete with TRP for entry into the brain. Experiment 1 shows that these treatments do produce reliable changes in brain 5-HT metabolism; experiment 2 tests their effects on food intake and protein selection.

Progress in Tryptophan and Serotonin Research

Methods

Experiment 1 : 4 groups of 6 male Sprague-Dawley rats (250 g) were starved overnight, then at 14h.45 given by gavage either NaCl (0.9%); TRP (0.52 mmole/kg); MAA (4.1 mmole/kg) as used by Gessa et al. (3), or GLU (26 mmole/kg). At 16h.30 the rats were killed, brains were frozen at -20°C and, within 2 weeks, brain TRP, 5-HT and 5-hydroxyindoleacetic acid (5-HIAA) were measured using HPLC with electrochemical detection (4).

Experiment 2 : 8 male Sprague-Dawley rats (250 g) were individually housed in cages each equipped with 2 food pellet delivery systems, one stocked with protein-free pellets weighing 75 mg each, the other with similar sized pellets containing 55% casein. Removal of a pellet from a food cup was detected by a photocell, recorded by a Hewlett-Packard 9845A desk top computer, and another pellet immediately delivered. After 6 days in these cages, when daily intake and selection had stabilised, the experiment began. The rats were removed from the cages and starved overnight. At 14h.45 groups of 2 rats received NaCl, TRP, MAA or GLU (for quantities see above). At 16h.30 rats were returned to the feeding cages and intake recorded for 24 h. Using a latin square design this procedure was repeated at 3-day intervals so that each rat received each treatment. Results were submitted to analysis of variance followed by one-tail "t" tests to check the significance of changes in the predicted directions.

Results and Discussion

Experiment 1 : Analysis of variance showed signifiant ($p < 0.01$) inter-group differences for each of the

3 brain parameters measured. The results are summarised in Table 1.

Table 1 : Effects of Tryptophan, Mixed Amino Acids or Glucose on Brain Tryptophan, 5-Hydroxytryptamine and 5-Hydroxyindole acetic acid in the rat.

Treatment	Brain TRP (μg/gm ± SEM)	Brain 5-HT (μg/gm ± SEM)	Brain 5-HIAA (μg/gm ± SEM)
NaCl	3.7 ± 0.20	0.40 ± 0.015	0.37 ± 0.023
TRP	5.2 ± 0.28**	0.45 ± 0.031	0.51 ± 0.059
MAA	1.7 ± 0.13**	0.28 ± 0.022**	0.24 ± 0.024*
GLU	4.5 ± 0.16*	0.47 ± 0.015*	0.38 ± 0.022

different from NaCl control : * $p < 0.05$; ** $p < 0.01$.

Pretreatment with tryptophan significantly increased brain TRP but not 5-HT or 5-HIAA. MAA significantly lowered all 3 and glucose increased brain TRP and 5-HT ($p < 0.05$).

Experiment 2 : Analysis of variance showed significant ($p < 0.05$) intertreatment differences in food intake and %P-E selected one hour after offering a choice of 0% and 55% casein pellets. The results are summarised in Fig. 1.

After 1 h food intake was lowest following glucose administration and highest after tryptophan. % P-E selection was higher after pretreatment with tryptophan or glucose than after MAA ($p < 0.05$). The changes were transient and after 24 h the differences in food intake and protein selection had disappeared.

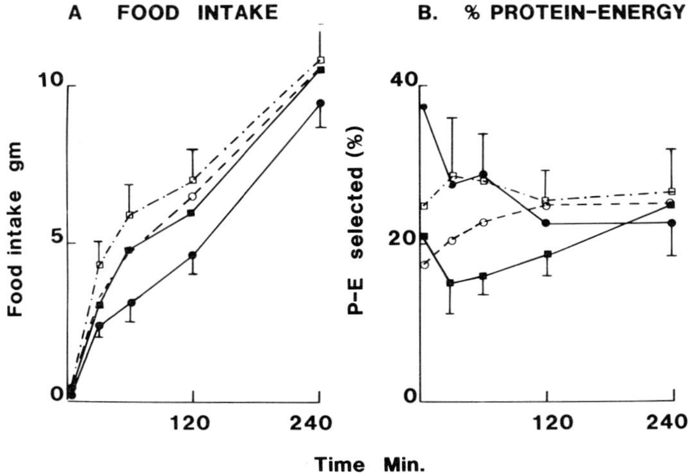

Fig. 1 : Effects of NaCl --o--; TRP --□--; GLU —●—; and MAA —■— on : food intake and % P-E selection.

These results, obtained under carefully chosen experimental conditions to ensure reliable changes in brain 5-HT metabolism during the period that food selection was measured, are consistent with the prediction (5) that accelerating brain 5-HT synthesis should lead to a preference for protein and lowering brain 5-HT should diminish protein preference.

References

1. Ashley, D.V.M., Coscina, D., Anderson, G.H. Life Sci. 24, 973 (1979).
2. Fernstrom, J.D., Wurtman, R.J. Science 174, 1023 (1971).
3. Gessa, G.L., Biggio, G., Fadda, F., Corsini, G.U., Tagliamonte, A. J. Neurochem. 22, 869 (1974).
4. Leathwood, P.D., Ashley, D.V.M. In : The Body Weight Regulatory System. Eds L.A. Cioffi, W.P.T. James and T.B. van Itallie. Raven Press, New York, p. 263 (1981).
5. Leathwood, P.D., Ashley, D.V.M. Appetite, in press (1983).
6. Wurtman, J.J., Wurtman, R.J. Science 198, 1178 (1977).

INFLUENCE OF TRYPTOPHAN ON GASTRIC MUCOSAL PROTEIN SYNTHESIS

A.P.N. Majumdar
Institute of Medical Biochemistry, University of Aarhus, DK-8000 Aarhus C, Denmark

Protein synthesis in mammalian cells is known to be influenced by a variety of dietary and hormonal factors. Among the dietary factors amino acids appear to have a regulatory effect on protein synthesis. Whereas fasting causes disaggregation of hepatic polysomes resulting in a lowered protein synthesis (1-3), refeeding of a complete amino acid mixture or only L-tryptophan reverses the situation (1, 3). In recent years, we have further demonstrated that in both well-fed adrenalectomized and adrenalectomized-diabetic rats force-feeding of L-tryptophan stimulates amino acid incorporation into albumin, fibrinogen, transferrin and ferritin (4-6). Such an observation further suggests that the tryptophan-mediated stimulation of hepatic protein synthesis is independent of adrenal hormones and insulin.

The structural and functional properties of the gastric mucosa are affected by a number of dietary and hormonal factors. Total deprivation of food diminishes gastric mucosal protein synthesis (7, 8), which has, in part, been attributed to disaggregation of polysomes (8). Adrenalectomy results in diminution of size and content of the zymogen granules of the chief cells (9) and lowers protein and RNA content in the gastric mucosa (10). In view of the earlier observation that L-tryptophan stimulates hepatic protein synthesis, we decided to investigate the role of the amino acid in regulating gastric mucosal protein synthesis in intact and adrenalectomized rats.

Materials and Methods

Adult Wistar rats of either sex were used. In the first series of experiments the animals were bilaterally adrenalectomized or sham-operated 1 month before use and were maintained on ad. libitum rat Chow pellets and water throughout. The drinking water of adrenalectomized rats contained 0.9% NaCl, whereas sham-operated animals received tap water for drinking. The rats received by stomach-tube either L-tryptophan (30 mg/100 g body wt.) or an equivalent volume of water alone and were killed 24 or 48 h later. L-[4,5-^{3}H]leucine (10 µCi/100 g body wt.; 46 Ci/mmol, Radiochemical Center, Amersham, England) was injected intraperitoneally 30 min before sacrifice. For determination of protein specific radioactivity (cpm/mg protein) of the stomach about 100 mg of tissues were homogenized in 5 ml 10% trichloroacetic acid. After centrifugation the precipitates were treated the same way as described earlier (11). An aliquot of the supernatant was counted for trichloroacetic acid-soluble radioactivity.

In the next series of experiments, intact rats were used. Groups of rats were either maintained on ad. libitum Chow pellets (fed control group) or fasted for 4 days with access to water. After 4 days of fasting they were supplied with either a complete liquid diet, Vamin^{R} or a tryptophan-free liquid diet, which was the same as that of the complete-diet with the exception that tryptophan (1 mg/liter) was replaced by an equivalent amount of glucose. Each animal received 60 ml of the nutrient solution (40 kcal). The animals were maintained on the diet for 4 days.

After being killed the stomach was removed, opened and washed thoroughly in 0.9% NaCl. It was frozen immediately on solid CO_2 and stored at -20°C. For isolation of polyribosomes gastric mucosal scrapings from each rat were homogenized in 6 ml homogenizing buffer containing 0.05 M Tris-HCl, pH 7.4, 0.005 M $MgCl_2$, 0.150 M KCl, 0.25 M sucrose and 25 µg/ml polyvinyl sulfate. Polyribosomes were isolated as described earlier (8). Conditions for amino acid incorporation in vitro and the preparation of cell sap from normal rat liver were the same as reported earlier (8). Rat liver cell sap was used as a source for tRNA and activating enzymes. Experiments were statistically evaluated with Student's t test for non-paired values, taking $p < 0.05$ as the level of significance.

Results

The results of the first series of experiments are given in Table 1. In this study, overall protein synthesis in the stomach was assessed by measuring the rate of [^{3}H]leucine incorporation into total protein in vivo after a 30 min pulse.

Table 1. Effect of a single tube-feeding of L-tryptophan on protein synthesis in the stomach of well-fed adrenalectomized rats.

	Specific radioactivity of protein	
Groups and treatment	cpm/mg protein	Corrected value (% incorporation)
Sham-operated (water-fed)	1163 ± 93 (207; p < 0.005)	88.1 ± 2.8 (113; p < 0.05)
Adrenalectomized (water-fed)	560 ± 120 (100)	77.7 ± 6.1 (100)
Adrenalectomized:		
24 h after tryp.	1339 ± 64 (239; p < 0.001)	89.4 ± 0.81 (115; p < 0.05)
48 h after tryp.	928 ± 115 (165; NS)	84.8 ± 2.9 (109; NS)

The values represent the mean ± SEM of 8-10 experiments. Figures in parenthesis represent percentage of water-fed adrenalectomized control and p value. NS = non significant. Data are reproduced from Scand. J. Gastroent. 14, 949-954 (1979).

One month after adrenalectomy the incorporation of [^{3}H] leucine into total stomach protein was decreased by 52% when compared with the sham-operated control (Table 1). In adrenalectomized rats, 24 and 48 h after tryptophan force-feeding [^{3}H]leucine incorporation into total stomach protein was increased by 139% and 65%, respectively, compared to the water-fed adrenalectomized group. Since the rate of amino acid incorporation into protein _in vivo_ can be affected by changes in precursor pool size, the data was corrected for acid-soluble radioactivity (12). Although the magnitude of difference between the groups decreased greatly, the pattern of difference between the groups remained the same as observed for uncorrected data (Table 1).

In the next series of experiments the effects of fasting and subsequent refeeding of a complete or tryptophan-free diet on the ability of gastric mucosal ribosomes to synthesize protein in a cell-free system were investigated. It should be mentioned that the gastric mucosal ribosomes from all groups were incubated with normal rat liver cell sap as a source for tRNA and activating enzymes, along with other com-

Table 2. Effects of 4 days of fasting and subsequent feeding for 4 days of complete or tryptophan-free diet on endogenous mRNA-directed protein synthesis by gastric mucosal ribosomes in a cell-free system.

Treatment	Incorporation $\times 10^{-3}$ cpm/mg rRNA
Fed (control)	55.5 ± 6.7 (100)
Fasted	31.3 ± 2.3 (56; $p < 0.025$)
Complete diet	41.8 ± 5.9 (75; NS)*
Tryptophan-free diet	29.1 ± 2.5 (52; $p < 0.025$)

Protein synthesis was measured by using [^{14}C] protein hydrolysate as described previously (8). Values represent the mean ± SEM of three preparations of ribosomes from each group. Figures in parenthesis represent percentage of the fed control and p value. *$p < 0.05$ when compared with the tryptophan-free diet group. The data are reproduced from Ann. Nutr. Metab. 26, 264-272 (1982).

ponents of protein synthesis. Since the present in vitro system differs only with respect to ribosomes, the difference in amino acid incorporation between the groups could be interpreted as due to changes in protein synthesizing capacity of the ribosomes.

After 4 days of fasting protein synthesis by gastric mucosal ribosomes was significantly decreased by 45%, compared to the fed control (Table 2). Refeeding of fasted rats for 4 days with the complete liquid diet produced a 34% enhancement in ribosomal protein synthesis, when compared with the fasted group. Tryptophan-free diet was found to be ineffective in this matter (Table 2).

Protein to polyphenylalanine ratio, which represents a ratio of polysomes to monosomes (13) was determined for the gastric mucosal ribosomes, prepared from different groups.

Table 3. Effects of 4 days of fasting and subsequent feeding of a complete or tryptophan-free diet on endogenous mRNA and poly-U-directed protein synthesis by gastric mucosal ribosomes in a cell-free system.

Treatment	Incorporation, × 10^{-3} cpm/mg rRNA			protein polyphenylalanine ratio
	minus poly-U	plus poly-U	polyphenylalanine	
Fed (control)	14.5 (100)	389.7 (100)	375.2 (100)	0.039 (100)
Fasted	7.1 (49)	322.4 (83)	315.3 (84)	0.023 (58)
Complete diet	12.5 (86)	355.9 (91)	343.5 (92)	0.036 (93)
Tryptophan-free diet	7.6 (52)	307.2 (79)	299.6 (80)	0.025 (65)

Incubations at 37°C were performed with ^{14}C phenylalanine in the presence and absence of poly-U (400 µg/ml) as described earlier (8). Results are averages of two experiments. Figures in parenthesis represent percentage of fed control. Data are reproduced from Ann. Nutr. Metab. 26, 264-272 (1982)

Protein to polyphenylalanine ratios were measured by assaying gastric mucosal ribosomes for protein synthesis in the absence and presence of poly-U. [^{14}C]phenylalanine was used as a tracer.

Four days of fasting caused a marked reduction in protein to polyphenylalanine ratio, but could be significantly increased by refeeding of the 4-day fasted rats with the complete liquid diet (Table 3). Again, the tryptophan-free liquid was found to be ineffective in rectifying the situation (Table 3).

Discussion

A variety of pathophysiological conditions can affect the structural and functional properties of the stomach. The results of the present investigation demonstrate that a prolonged lack of adrenal hormones or a total deprivation of food markedly lowers the rate of protein synthesis in the stomach. Adrenalectomy or hypophysectomy has also been shown to lower gastric mucosal pepsin activity (9), which could be normalized by long-term administration of glucocorticoids (14). In adrenalectomized rats, force-feeding of a large dose of tryptophan has also been found to stimulate gastric mucosal pepsin activity (15). Prior administration of actinomycin-D, an inhibitor of RNA synthesis, abolishes the tryptophan-mediated stimulation of gastric mucosal pepsin activity in adrenalectomized rats (15). This, together with the observation of a higher protein synthesis in the stomach of adrenalectomized rats after tryptophan administration suggests that the stimulation of gastric mucosal pepsin activity by tryptophan could, in part, be due to an increased synthesis of the enzyme.

Of essential amino acids, L-tryptophan can be considered unique in the diversity of functions it provides (16-19). One of the extensively studied properties of tryptophan is its ability to stimulate hepatic protein synthesis (20). The present results demonstrate that the gastric mucosal protein synthesis is also responsive to L-tryptophan. Furthermore, the observation of the inability of the tryptophan-free liquid diet to promote protein synthesis in the gastric mucosa of fasted rats indicates the physiological importance of dietary tryptophan for protein synthesis in the stomach. The non-stimulatory effect of the tryptophan-free diet could, in part, be attributed to its inability to induce polysome formation in the gastric mucosa. This assumption is made on the basis of the observation that protein to polyphenylalanine ratio, which represents a ratio of polysomes to monosomes (13), decreases after fasting and remains at that level after refeeding of the 4-day fasted rats with the tryptophan-free diet for 4 days. On the other hand, the complete liquid diet is found to be effective in stimulating protein to polyphenylalanine ratio. The maintenance of protein synthesis can be considered essential for normal metabolic activity of a tissue. Thus the present observation that the tryptophan-free liquid diet is ineffective in normalizing the rate of protein synthesis in the gastric mucosa suggests that dietary tryptophan may play a significant role in maintaining the structural and functional properties of the gastric mucosa.

Summary

Bilateral adrenalectomy or a total deprivation of food in intact rats caused a marked reduction in the rate of protein synthesis in the stomach. Force-feeding of a large dose of tryptophan to well-fed adrenalectomized rats produced a significant increment in protein synthesis in the stomach. In intact fasted rats refeeding of a nutritionally complete liquid diet, but not a tryptophan-free diet, was found to be effective in promoting the rate of protein synthesis in the stomach.

Acknowledgements

The work was supported by grants from the Danish Medical Research Council and Fonden til Lægevidenskabens Fremme, Denmark.

References

1. Fleck, A., Shepherd, J., Munro, H.N.: Science 150, 628-629 (1965).
2. Staehlein, T., Verney, E., Sidransky, H.: Biochim. Biophys. Acta 145, 105-109 (1967).
3. Sidransky, H., Sarma, D.S.R., Bongiorno, M., Verney, E.: J. Biol. Chem. 243, 1123-1132 (1968).
4. Jørgensen, A.J.F., Majumdar, A.P.N.: Biochem. Med. 13, 231-240 (1975).
5. Jørgensen, A.J.F., Majumdar, A.P.N.: Biochem. Med. 16, 37-46 (1976).
6. Majumdar, A.P.N., Jørgensen, A.P.N.: Biochem. Med. 17, 116-119 (1977).
7. Majumdar, A.P.N., Goltermann, N.: Gastroenterology 73, 1060-1064 (1977).
8. Majumdar, A.P.N.: Biochem. Med. 22, 288-298 (1979).
9. Abrams, G.D., Baker, B.L.: Gastroenterology 27, 462-468 (1954).
10. Majumdar, A.P.N.: Scand. J. Gastroent. 16, 151-156 (1981).
11. Majumdar, A.P.N., Greenfield, S., Røigaard-Petersen, H.: Scand. J. Clin. Lab. Invest. 31, 219-224 (1973).
12. Majumdar, A.P.N., Nakhla, A.M.: Brit. J. Pharmacol. 66, 211-215 (1979).
13. Murthy, M.R.V.: Biochim. Biophys. Acta 119, 599-613 (1966).
14. Hirschowitz, B.L., Streenten, D.H.P., Pollard, H.M.: Endocrinology 59, 419-424 (1956).

15. Majumdar, A.P.N.: Scand. J. Gastroent. 14, 949-954 (1979).

16. Floyd, J.C.Jr., Fajans, S.S., Conn, J.W., Knoff, R.E., Rull, J.: J. Clin. Invest. 45, 1487-1502 (1966).

17. Fernstrom, J.D., Wurtman, R.J.: Scient. Am. 230, 84-91 (1974).

18. Krehl, W.A., Teply, L.J., Sarma, P.S., Elvehjem, G.A.: Science 101, 489-490 (1945).

19. Wittman, J.S.III.: J. Nutr. 106, 631-635 (1976).

20. Majumdar, A.P.N.: Nutr. Rep. Int. 26, 509-522 (1982).

TRYPTOPHAN AND INTESTINAL SECRETION

Helen J. Cooke

Department of Physiology, University of Nevada
Reno, Nevada, U.S.A.

Introduction

Actively transported amino acids and sugars increase transmural electrical potential difference (PD), short-circuit current (Isc) and active Na absorption across the small intestine from a variety of species _in vitro_ (1). Studies from this laboratory have shown that 2 mM tryptophan increased the PD, Isc, and active Na absorption in isolated segments of newborn rabbit jejunum; whereas, 10 mM tryptophan, after causing an initial increase in these parameters, rapidly reduced PD and Isc to a plateau at 50-60 min (2,3). Tissue conductance was reduced significantly after 10 mM tryptophan (2). The aim of the present investigation was to elucidate the nature of the changes in PD, Isc, and tissue conductance evoked by 10 mM tryptophan.

Methods

Flat sheets of jejunum from 3-5 day-old rabbits were set up in flux chambers as described previously (2,3). Ten mM tryptophan was present in the solutions bathing two segments of jejunum from each rabbit; two other segments from the same rabbit served as controls, and the solutions contained 10 mM mannitol to make the osmolarity of control solutions identical to that of experimental solutions. Unidirectional mucosal-to-serosal (Jms) and serosal-to-mucosal (Jsm) fluxes of mannitol or Na and Cl were measured simultaneously on paired, short-circuited segments of jejunum. Tracer quantities of ^{14}C-mannitol or ^{22}NaCl and Na^{36}Cl were added to one side of the chamber and three 1.0 ml samples taken from the opposite side at 10 min

Progress in Tryptophan and Serotonin Research

intervals 60 min after adding the isotope. At the termination of the experiment, 1.0 ml was taken from the opposite side.

Results

Figure 1 shows the effects of 10 mM tryptophan on unidirectional and net Na and Cl fluxes. In control tissues, a small net Na flux was present. In the presence of tryptophan, unidirectional Na fluxes were reduced and net Na flux was abolished (Fig. 1A). Chloride appeared to be distributed passively across the intestine in controls (Fig. 1B). In the presence of tryptophan, unidirectional Cl fluxes increased, and active Cl secretion was evoked. Tryptophan reduced the tissue conductance from 56.5 ± 1.3 to 39.3 ± 0.7 $mS \cdot cm^{-2}$.

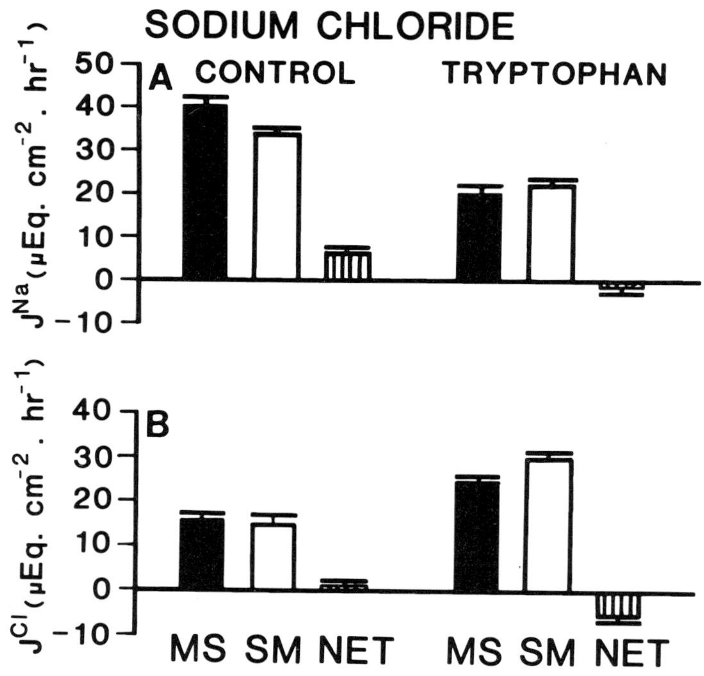

Fig. 1. Na and Cl fluxes in the presence and absence of 10 mM tryptophan in newborn rabbit jejunum. Means ± SE are given for 14 tissues in 7 rabbits. Ms, mucosal-to-serosal; SM, serosal-to-mucosal.

The effect of ion substitution on unidirectional and net Na and Cl fluxes is shown in Fig. 2. In Cl-free solutions (Fig. 2), tryptophan reduced the unidirectional Na fluxes and increased slightly net Na flux. Tissue conductance was 37.1 ± 1.4 $mS \cdot cm^{-2}$ in controls compared to 27.8 ± 1.3 $mS \cdot cm^{-2}$ with tryptophan. In Na-free solutions (Fig. 2), unidirectional Cl fluxes did not differ, and no net Cl flux was evident. When tryptophan was added, unidirectional Cl fluxes increased slightly, and no net Cl flux was seen. Tissue conductance averaged 19.3 ± 1.2 $mS \cdot cm^{-2}$ after tryptophan and was not different from controls in Na-free solutions.

In control tissues, mannitol Jms and Jsm averaged 0.15 ± 0.02 and 0.17 ± 0.02 $\mu M \cdot cm^{-2} \cdot hr^{-1}$, respectively. In the presence of tryptophan, mannitol fluxes averaged 0.26 ± 0.03 and 0.27 ± 0.02 $\mu M \cdot cm^{-2} \cdot hr^{-1}$, respectively.

Analysis of the tissue conductance and transmural mannitol flow showed that two pathways for ion flow were present in control tissues. The conductance of the transcellular pathway was approximately 73% of the total tissue conductance of newborn rabbit jejunum; whereas, the conductance of the paracellular pathway was 27% of the total. Tryptophan reduced the

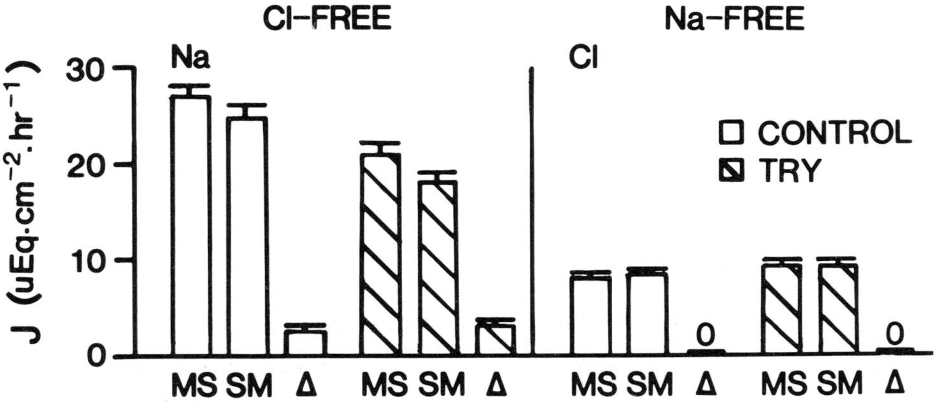

Fig. 2. Effect of 10 mM tryptophan on Na and Cl fluxes when the bathing solutions were Cl-free or Na-free, respectively. Means ± SE are given for 8-12 rabbits. Ms, mucosal-to-serosal; SM, serosal-to-mucosal; Δ, net flux.

total tissue conductance by decreasing the conductance of the cellular pathway to 26% of the total and increasing the conductance of the paracellular pathway.

Discussion

Tryptophan increases electrogenic Na absorption (2) followed by inhibition of active Na absorption and stimulation of active Cl secretion in newborn rabbit jejunum. Chloride ion appears to be necessary for the latter effect on net Na flux, and Na ion appears to be a requirement for evoking Cl secretion. The results are consistent with a previous hypothesis that tryptophan stimulates electrogenic Na absorption simultaneously with stimulating a neutral NaCl secretory mechanism (2).

The effects of tryptophan on Na and Cl fluxes in the newborn rabbit jejunum appear to result from a decrease in ionic conductance of the cellular pathway, coupled with an increase in the permeability of the paracellular pathway. The effects of tryptophan on ion transport are reminiscent of the effects of theophylline on rabbit ileum (4).

References

1. Schultz, G., Curran, P. F.: Physiol. Rev. 50, 637-718 (1970).
2. Cooke, H. J., Cooke, A. R.: Am. J. Physiol. 242, G308-G312 (1982).
3. Cooke, H. J., Pfankuche, L., Cooke, A. R.: Am. J. Physiol. 239, G306-G310 (1980).
4. Powell, D. W.: Am. J. Physiol. 227, 1436-1443 (1974).

TRYPTOPHAN METABOLISM IN UNDERNOURISHED DEVELOPING RAT BRAIN

S. Kalyanasundaram and P.S.V. Ramanamurthy
National Institute of Nutrition, Hyderabad-500007, India.

Introduction

Undernutrition during the rapid phase of brain growth reduces the cell number, protein and myelin in experimental animals (1,2). Alterations in the activity of these animals in many psychological tests have been reported (3). Also the concentrations of neurotransmitters like dopamine, norepinephrine, serotonin and histamine were reported to be altered in these animals (4,5). As the concentration of tryptophan is limiting in brain it has been suggested that tryptophan may regulate synthesis of brain proteins (6) and also serotonin (7). Whether the altered serotonin levels in undernourished animals is due to an alteration in the availability of tryptophan or to an alteration in its metabolism was investigated in the present study.

Methods

Female rats (Wistar/NIN strain) after conception were fed either (a) 20 % protein ad libitum (control) (b) 7.5 % protein diet ad libitum (experimental) or (c) were pair-fed with animals receiving 7.5 % protein diet (pair-fed). This feeding schedule was maintained throughout gestational and lactational period and the young ones were weaned on to the respective diets. The young ones were killed at weekly intervals from birth till 35 days and the whole brains were removed and processed for the estimation of tryptophan (8),

Progress in Tryptophan and Serotonin Research

tryptophan-5-hydroxylase (9), aromatic aminotransferase (10) and monoamine oxidase (11). For studies on utilisation of tryptophan for protein synthesis, the animals were injected ^{3}H-tryptophan (10 μCi/100 g body weight) intraperitoneally and were killed at 30, 60, 120 and 240 min after injection. The whole (total) brain homogenate radioactivity and the radioactivity of the protein precipitate were determined.

Results and Discussion

Tryptophan levels: The levels observed in the control group are in good agreement with the normal pattern of development reported (12,13). In the experimental and pair-fed animals, the tryptophan levels at birth are higher when compared to controls. This may suggest that during gestational period the experimental groups may be mobilising sufficient amounts of this amino acid for the fetus. However, from day 14 onwards the malnourished animals showed a significant decrease in brain tryptophan levels, which coincides with the decrease in serotonin levels (5).

MAO activity: The enzyme activity increases from birth to 28 days in all the groups. But the increase in the enzyme activity is significantly higher in both the undernourished groups. These findings may suggest that in undernourished animals, the decreased brain serotonin concentration may be not only due to decreased tryptophan levels but also to an increased metabolism.

Tryptophan-5-hydroxylase activity: This showed an increase in malnourished animals from day 14 onwards. The time at which this increase was observed coincides with the decrease in tryptophan and serotonin levels (5). This increase in the enzyme activity may signify an adaptive mechanism for the decreased serotonin levels, since similar increase in the enzyme activity was observed when brain serotonin levels were lowered by treatment with drugs (14).

Aromatic aminotransferase: The activity was assayed using tyrosine as a substrate since it has been suggested that the end product of transamination of tyrosine, p-hydroxy-phenyl-pyruvic acid, may be a substrate for tryptophan aminotransferase (15). Aminotransferase activity as observed by us indicated that the activity is reduced in malnourished animals. It is therefore, possible that utilisation of the amino acids for energy purpose is decreased in undernourished animals. It is likely that the tryptophan aminotransferase activity might also be decreased in undernourished brain, since the concentration of tryptophan and p-hydroxy-phenyl-pyruvic acid are low in undernourished animals.

Incorporation of tryptophan into protein: (i) The amount of radioactivity in the whole brain homogenate is increased in undernourished animals which may be due to an increased influx of the amino acid or reduced efflux of tryptophan and its metabolites, and (ii) at all time points, the amount of available label incorporated into protein is higher in the undernourished animals which may either be due to the decreased pool size of tryptophan in these animals or due to increased recycling of this amino acid.

In summary our findings suggest that there are alterations in tryptophan metabolism in undernourished developing rat brain. The reduced brain tryptophan levels and the increased monoamine oxidase activity may be the factors responsible for the reduced serotonin levels observed in these animals. Alterations in other metabolic pathways like transamination and protein synthesis may be an adaptive mechanism for the conservation of this amino acid which is in short supply in undernourished animals (16,17).

References

1. Winick, M., Noble, A.: J. Nutr. 89, 300-306 (1966).
2. Wiggins, R.C., Miller, S.L., Benjamins, J.A., Krigman, M.R., Morrel, P.: Brain Res. 107, 257-273 (1976).
3. Frankova, S., Barnes, R.H.: J. Nutr. 96, 477-484 (1968).
4. Sobotka, T.J., Cook, H.P., Brodie, R.E.: Brain Res. 65, 443-457 (1974).
5. Ramanamurthy, P.S.V.: J. Neurochem. 28, 253-254 (1977).
6. Allen, R.E., Raines, P.L., Regan, D.M.: Biochem. Biophys. Acta. 190, 323-336 (1969).
7. Fernstrom, J.D., Wurtman, R.J.: Science 173, 149-152 (1971).
8. Denckla, W.D., Dewey, H.K.: J. Lab. Clin. Med. 69, 160-169 (1967).
9. Gal, E.M., Patterson, K.: Anal. Biochem. 52, 625-629 (1973).
10. Diamondstone, T.I.: Anal. Biochem. 16, 395-401 (1966).
11. Nagatsu, T.: Biochem. J. 59, 606-612 (1966).
12. Bourgoin, S., Fairre-Bauman, A., Benol, P., Glowinski, J. Hamon, M.: J. Neurochem. 23, 319-328 (1974).
13. Tyce, G.M., Flock, E.V., Owen, C.A. Jr.: Progress in Brain Research, Vol. 9. The Developing Brain (Hinwich, W.A., Himwich, H.E. eds.), pp. 198-203, Elsvier, Amsterdam, 1964.
14. Ross, S.B., Oie, F.: Neurosci. Lett. 5, 215-220 (1977).
15. Lees, G.J.: Life Sci. 20, 1749-1762 (1977).
16. Freedman, L.S., Samuels, S., Fish, I., Schwartz, S.A., Lange, B., Katz, M., Morgano, L.: Science 207, 902-904 (1980).
17. McFarlane, I.G., von Holt, C.: Biochem. J. 111, 557-563 (1969).

EFFECT OF TRYPTOPHAN ON PROTEIN UTILIZATION AND ENERGY INTAKE OF PIGS

Karl-Heinz Menke, Friedrich Gruber
Institut für Tierernährung der Universität Hohenheim
D-7000 Stuttgart 70

Introduction

Proteins deficient in one or several essential amino acids (AA) may not only lead to a higher excretion of urinary nitrogen and thereby to a decrease in efficiency of utilization, but also to reduced feed intake. This is especially true for diets high in maize proteins which are known to be deficient in tryptophan (1). The following experiments have been set up to elucidate the effect of tryptophan on voluntary feed intake and protein utilization at different levels of other essential AA in the diet.

Methods

N-balance experiments have been conducted with young growing pigs of 14-25 kg body weight, held in metabolism cages (2) and fed rations consisting mainly of maize and maize gluten, supplemented with L-tryptophan (98%), L-isoleucine (99% purity, both Fa.Merck), L-lysine and L-methionine (both 99% purity, Fa.Animedica), alone or in combinations (for details of ration composition see 3). In total 40 groups of 6 animals each were used in 8 experiments. Periods of total collection of feces and urine were two-times 7 days. Means of N-retention (RN,g), and intake of digestible N (IDN,g) were used for calculation of efficiency of protein utilization (k_N), using a basal value of N-metabolism of 0.23 g N/kg$^{0.75}$:

Progress in Tryptophan and Serotonin Research

$$k_N = (RN + 0.23\ W^{0.75})/IDN$$

Feed intake is expressed as intake of metabolizable energy per metabolic body weight ($M = MJ\ ME/kg^{0.75}$), in order to allow comparisons with data from literature.

Results

The effects of changes in AA-pattern on protein utilization (k_N) and energy intake (M) are shown in figures 1-8. Levels of AA intake are expressed as ratios:

$$Q = \frac{\text{AA in the feed protein (g/16 gN)}}{\text{AA in a reference protein (g/16 gN)}}$$

The AA-pattern of this reference protein has been derived from N-balance experiments with 64 groups of pigs using a curvilinear model of amino acid functions and an iterative regression procedure(4).The maximum efficiencies were found to be at (g/16gN) 10.5 lysine, 2.0 tryptophan, 3.2 methionine + cystine, and 4.1 isoleucine. AA-levels recommended by ARC (5) reach 90 (lys), 83 (trp), 100 (met+cys) and 99% (ile) of the maximum efficiencies of AA-functions in this model.
In experiment 1 (figure 1) tryptophan (trp) had no effect on efficiency of protein utilization (k_N, left column) but increased significantly energy intake (M, right column, groups 4:5), whereas lysine (lys) improved efficiency but not energy intake (groups 1:2 and 3:4). The effect of trp appears again when comparing groups 3 and 1 in this experiment. In experiment 2 (fig.2) trp had a negative effect on efficiency and a not significant positive effect on energy intake (groups 2:3). The higher trp-levels in groups 3 and 4 seem to be the condition, however, for the positive effect of lys on energy intake, since lys did not affect energy intake at a lower level of trp (groups 1:2). Isoleucine (ile) improved efficiency and decreased energy intake in this experiment (4:5).

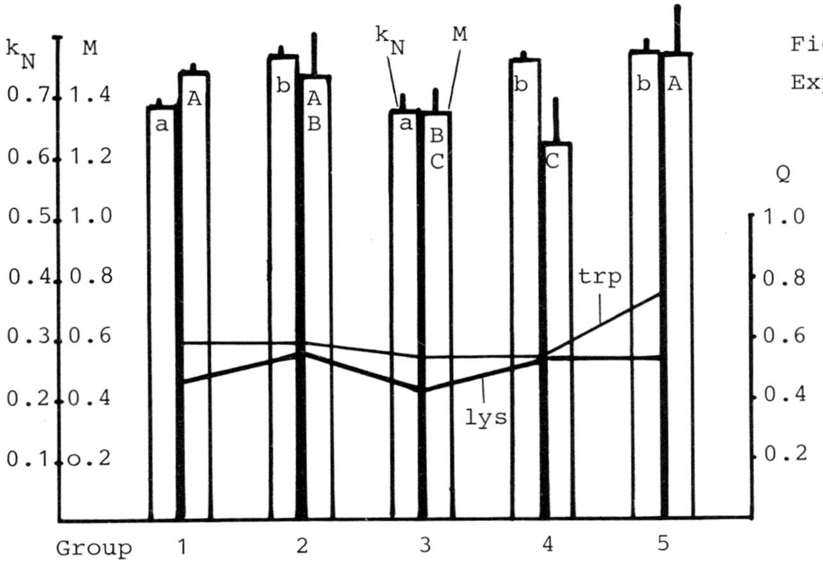

Fig. 1
Expt 1

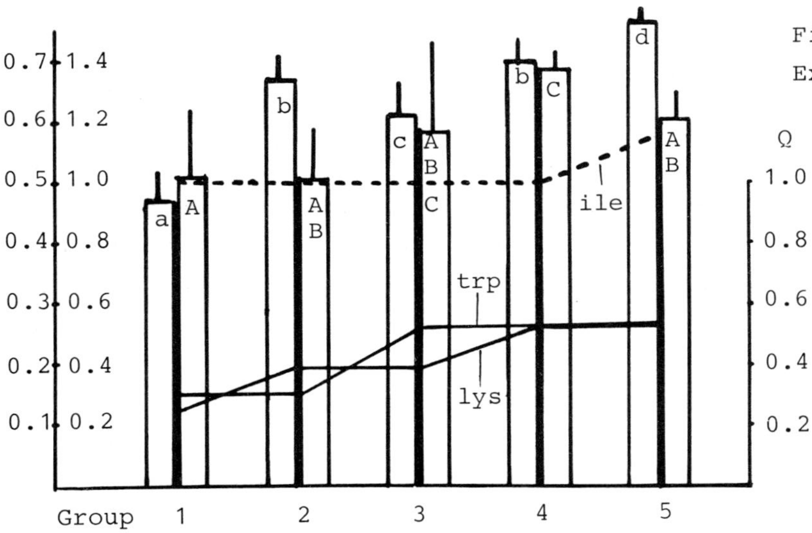

Fig. 2
Expt 2

Fig.1-8. Effect of level of tryptophan, lysine, isoleucine, methionine and cystine on protein utilization (k_N, left column) and energy intake (M, right column). Differences are significant $p<0.05$ when letters are different (a:b; A:B; etc). Q = Ratio of amino acid content of protein of the diet, to content of the same amino acid in a reference protein (further explanations see text).

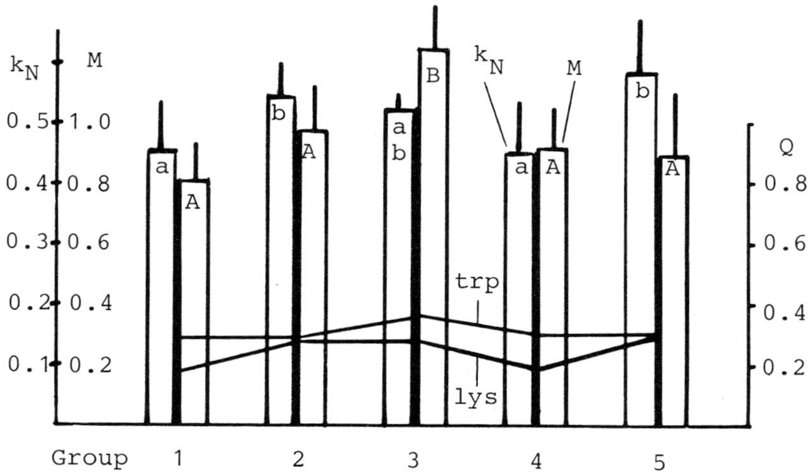

Fig.3
Expt 3

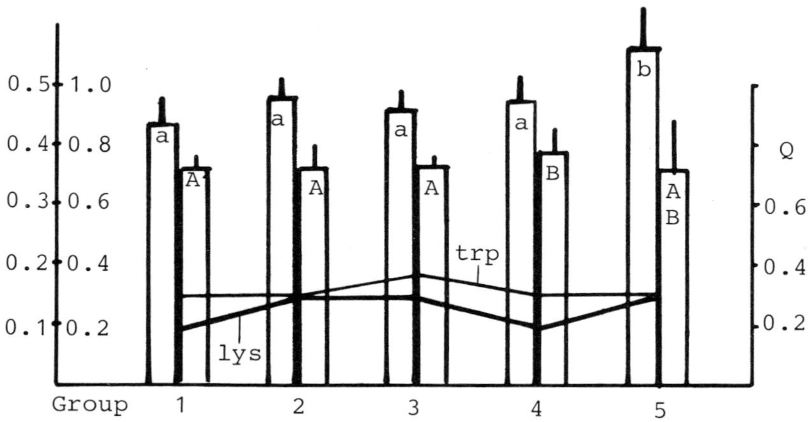

Fig.4
Expt 4

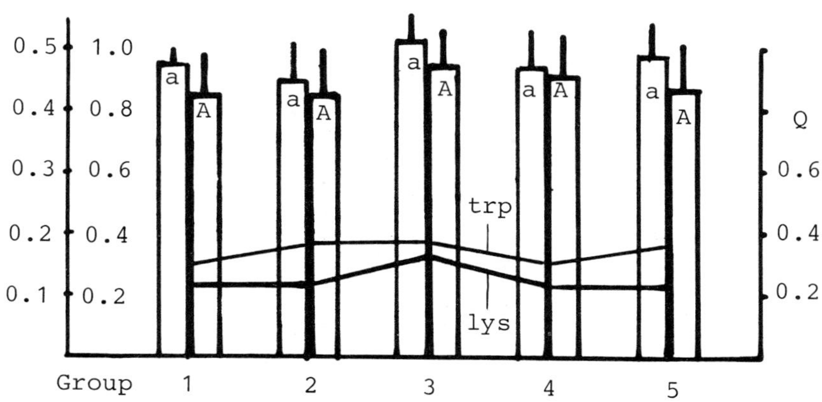

Fig.5
Expt 5

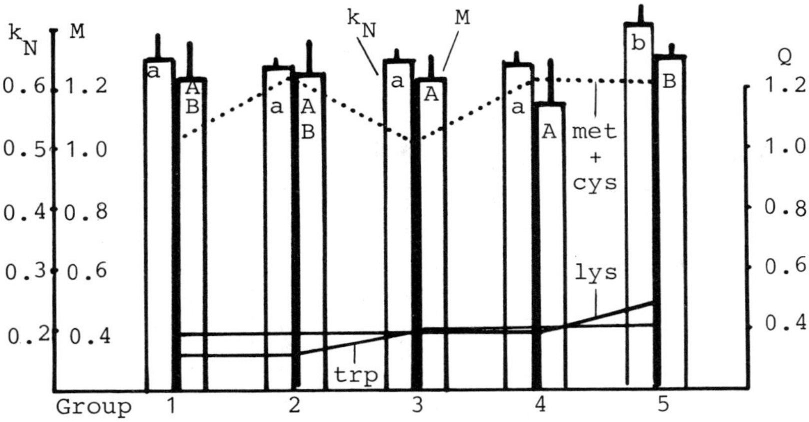

Fig. 6
Expt 6

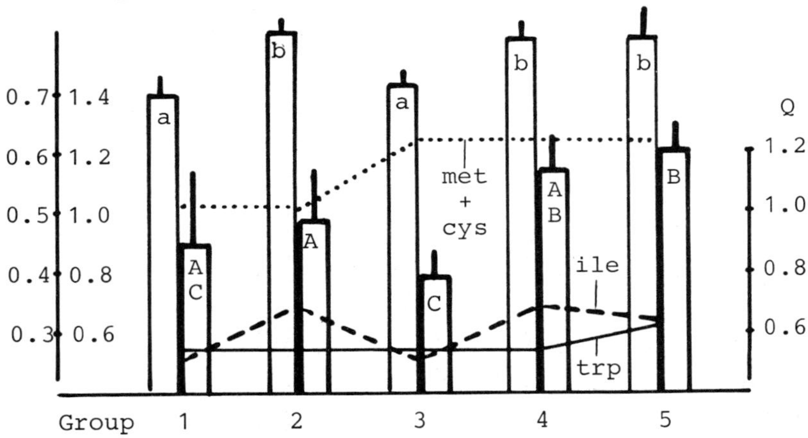

Fig. 7
Expt 7

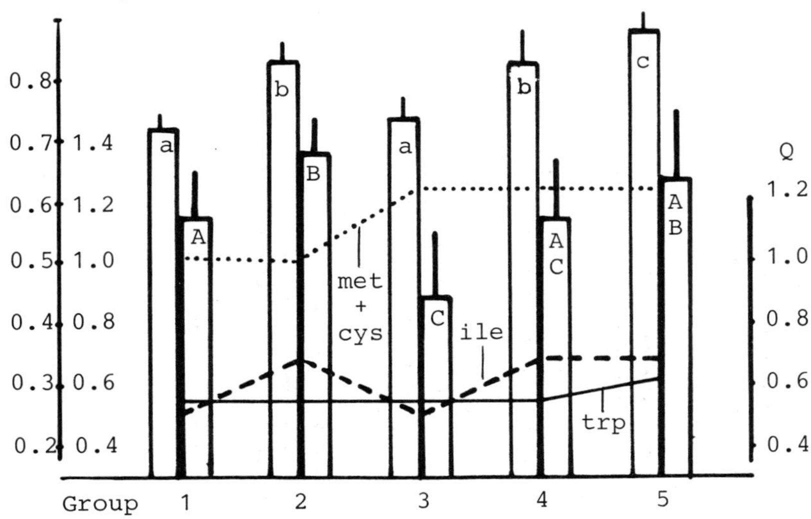

Fig. 8
Expt 8

The positive effect of trp on energy intake is confirmed in experiment 3 (2:3) but not in experiments 4 (2:3) and 5 (1:2 and 4:5), presumably because of the very low level of lys in these rations. In the experiments 3-5 the groups 4 and 5 were fed lower levels of protein (ca 9g digestible protein/MJ ME) than groups 1-3 (ca 12 g/MJ). It appears from comparison of groups 1:2 and 4:5 that there is no interaction between the effects of supplementation of these AA and protein level in the diet, within the range of 9 to 12 g digestible protein/MJ ME. In experiments 6-8 the effects of supplementation of trp, methionine (met) and isoleucine (ile) have been studied at higher levels of lys (4-7 g/16 gN). There is a tendency to increase energy intake after trp-supplementation in experiments 7 and 8 (groups 4:5), but again no effect on efficiency occurs. Ile improved efficiency significantly when increased from 2.1 to 2.8 g/16 gN in experiments 7 and 8 (groups 1:2 and 3:4). Effects on energy intake were less pronounced and not always significant. Supplementation of met did not affect efficiency in experiments 6-8, but tends to impair energy intake, especially in experiment 8 (groups 1:3). Results are compiled in table 1. It can be seen that lys had positive effects on efficiency of protein utilization in all 10 comparisons, only two of them did not reach the level of significance. The effect of lys on energy intake was significant only if trp was present in levels higher than 0.8 g/16 gN. Trp alone had no positive effect on utilization of protein, but improved energy intake significantly in 3 of 9 cases. Combined supplementation of lys and trp enhanced efficiency and energy intake in 6 of 7 cases. Lys and trp seem to be dependent on each other in their effects on energy intake. Iterative regression analysis with k_N=y as dependend variable resulted in the amino acid function: $AF_{trp}=0.3 + 1.4\ Q^z - 0.7\ Q^{2z}$, in which the linear regression coefficient b_1 is low compared to that for lys (1.82), ile (2.6) or met+cys (4.8), indicating a rather low effect of trp on efficiency of utilization when being the first limiting AA (z=1). However, when not being the first limiting AA, trp may still enhance the positive effect of the first

Table 1. Effect of amino acid supplementation on efficiency of protein utilization (k_N) and energy intake (M)

Expt., Groups	Change in amino acid content (g/16 gN)						Effect on	
	Lysin	Trypt.	Met+C	Ile	Thr	Leu	k_N	M[1)]
	Effect of lysine:							
3, 1:2	1.9→3.0	0.6	4.0	3.6	3.2	18.8	+	(+)
3, 4:5	1.9→3.1	0.6	4.0	3.6	3.2	18.8	+	o
4, 1:2	1.9→3.0	0.6	4.0	3.6	3.2	18.8	(+)	o
4, 4:5	1.9→3.1	0.6	4.0	3.6	3.2	18.8	+	(-)
5, 2:3	2.4→3.4	0.7	4.1	3.9	3.4	14.5	(+)	(+)
2, 1:2	2.5→4.0	0.6	4.6	4.2	3.6	14.4	+	o
2, 3:4	4.0→5.5	1.0	4.5	4.1	3.5	14.1	+	+
6, 4:5	4.0→5.1	0.8	3.8	4.1	3.2	10.4	+	+
1, 1:2	4.8→5.9	1.2	3.4	4.0	3.4	8.3	+	o
1, 3:4	4.6→5.6	1.1	3.7	3.7	3.3	9.1	+	(-)
	Effect of tryptophan:							
5, 1:2	2.4	0.6→0.7	4.2	3.9	3.4	14.6	(-)	o
5, 4:5	2.4	0.6→0.7	4.2	3.9	3.4	14.6	o	o
3, 2:3	3.0	0.6→0.7	4.0	3.6	3.2	18.7	o	+
4, 2:3	3.0	0.6→0.7	4.0	3.6	3.2	18.7	o	o
2, 2:3	4.0	0.6→1.0	4.5	4.1	3.6	14.3	-	(+)
6, 1:3	4.0	0.6→0.8	3.3	4.1	3.2	10.5	0	0
1, 3:1	4.7	1.1→1.2	3.5	3.8	3.3	8.7	o	o
1, 4:5	5.6	1.1→1.6	3.7	3.7	3.3	9.1	o	+
7, 4:5	6.8	1.1→1.2	4.0	2.7	3.9	9.6	o	o
	Effect of lysine + tryptophan							
3, 1:3	1.9→3.0	0.6→0.7	4.0	3.6	3.2	18.8	+	+
4, 1:3	1.9→3.0	0.6→0.7	4.0	3.6	3.2	18.8	o	o
5, 1:3	2.4→3.4	0.6→0.7	4.1	3.9	3.4	14.5	(+)	(+)
6, 2:5	4.0→5.1	0.6→0.8	3.9	4.1	3.2	10.4	+	(+)
2, 2:4	4.0→5.5	0.6→1.0	4.5	4.1	3.5	14.1	+	+
1, 3:5	4.6→5.6	1.1→1.6	3.7	3.7	3.3	9.1	+	+
1, 3:2	4.6→5.9	1.1→1.2	3.5	3.8	3.3	8.7	+	(+)

1) M=MJ ME/$kg^{0.75}$; + = significant positive; - = sign. negative; (+),(-)= difference >2%,but not sign; o = no effect

Table 1. (continued)

Expt., Groups	Change in amino acid content (g/16 gN) Lysin	Trypt.	Met+C	Ile	Thr	Leu	Effect on k_N	M
	Effect of isoleucine and methionine							
2, 4:5	5.5	1.0	4.4	4.0→4.8	3.5	14.0	+	-
7, 1:2	6.9	1.1	3.3	2.1→2.8	3.9	9.7	+	(+)
7, 3:4	6.9	1.1	4.0	2.1→2.8	3.9	9.7	+	+
8, 1:2	6.9	1.1	3.3	2.1→2.8	3.9	9.7	+	+
8, 3:4	6.9	1.1	4.0	2.1→2.8	3.9	9.6	+	(+)
6, 1:2	4.0	0.6	3.3→4.0	4.1	3.2	10.5	o	o
6, 3:4	4.0	0.8	3.3→3.9	4.1	3.2	10.5	o	(-)
7, 1:3	6.9	1.1	3.3→4.0	2.1	3.9	9.7	o	o
8, 1:3	6.9	1.1	3.3→4.0	2.1	3.9	9.7	o	-
8, 2:4	6.9	1.1	3.2→4.0	2.8	3.9	9.6	o	-
7, 1:4	6.9	1.1	3.3→4.0	2.1→2.8	3.9	9.7	+	(+)
7, 3:5	6.9	1.1→1.2	4.0	2.1→2.6	3.9	9.6	+	+
2, 3:5	4.0→5.5	1.0	4.5	4.1→4.8	3.5	14.1	+	o
6, 3:5	4.0→5.1	0.8	3.3→3.9	4.1	3.2	10.4	+	+
6, 1:4	4.0	0.6→0.8	3.3→3.9	4.1	3.2	10.5	o	o

limiting AA lys, as is indicated by the high value for the exponent z_{trp}=0.44 (compared to 0.27, 0.18 and 0.28 for the other AA). This exponent describes the reduction in the effect of an AA when changing from the 1st into the 2nd limiting position(4).

References

1. Cornelius,H.:Untersuchungen zur Bewertung von Maisproteinen in Rationen für wachsende Schweine. Diss.,Stgt-Hohenheim,1981.

2. Berschauer,F.,Oster,A.:Z.Tierphys.Tierern.Fmk.45,122-132(1981).

3. Gruber,F.:Der Einfluss von Aminosäuren-Zulagen auf die Futteraufnahme und die Proteinverwertung von Ferkeln. Diss., Stuttgart-Hohenheim, 1981.

4. Menke,K.H.,Gruber,F.,Gaus,G.:4th Int.Symp.Prot.Metabol.a.Nutr. Europ.Ass.Anim.Prod.;Clermont-Ferrand, 5-9 September 1983(i.pre

5. ARC:The Nutrient Requirements of Pigs. 1981;Commonwealth Agric. Bureaux, Farnham Royal, Slough SL2 3BN, UK

A HIGH CARBOHYDRATE, PROTEIN-FREE MEAL, EITHER AT BREAKFAST OR IN THE EVENING DOES NOT ALTER PLASMA TRYPTOPHAN TO LARGE NEUTRAL AMINO ACID RATIOS IN HEALTHY MEN.

D.V. Ashley and P.D. Leathwood
Research Department, Nestlé Products Technical Assistance Co. Ltd.
CH-1814 La Tour de Peilz, Switzerland

Introduction

A carbohydrate-rich, protein-free meal, when fed to fasted rats, increases brain tryptophan (TRP) levels and can accelerate brain 5-hydroxytryptamine (5-HT) synthesis (1). These effects are probably mediated by insulin-dependent decreases in plasma concentrations of the large neutral amino acids (LNAA) which compete with TRP for the same transport mechanism into the brain. Other animal studies show that 2-3 fold increases in the plasma TRP/LNAA ratio are necessary to cause significant changes in brain 5-HT metabolism (2). These findings have led to the speculation that any "high-carbohydrate, protein-poor meal elevates brain tryptophan, accelerating serotonin synthesis" and that "a high protein meal depresses serotonin synthesis" (3). The aim of the present experiments was to ascertain the effects of a low protein (1.6%), carbohydrate-rich; a high protein (20%), and a TRP-containing (0.3-0.4%) carbohydrate-rich meal, eaten in the morning or in the evening on the plasma TRP/LNAA ratio in man.

Experimental Protocol

Six healthy young men (22-42 yrs old) participated in the breakfast study and 8 in the evening study. The experimental meals were prepared either as a strawberry-flavoured beverage (breakfast) or as an asparagus-flavoured soup (evening) pre-packaged in 100 g lots, supplying

500 kcal and 25% fat (4). Experiments were carried out at weekly intervals using a Latin Square design so that each participant received all treatments. In the breakfast study, on the morning prior to the test day, participants reported to the clinical research facility before eating and were fed standard meals for breakfast (0800h), lunch (1230h) and supper (1830h). At 0800h the following morning, after placement of an indwelling catheter, a fasting blood sample was taken and subjects ate one of the test meals. Venous blood samples were taken at 30 minute intervals for 2 h and after 3 h. In the evening study, subjects reported at 0730h on the morning of the test day for breakfast, and at 1230h for lunch. At 2100h, pre-test meal blood samples (5 ml) were taken, and then 6 of the 8 subjects were fed one of the 3 soups. The other 2 subjects continued to fast. Blood was again taken 1 and 2 h after drinking the soups. The plasma removed after centrifugation of blood samples, was immediately frozen at -20°C until analysed for plasma amino acids (4).

Results and Discussion

Tryptophan uptake into the brain and hence brain 5-HT synthesis, can be altered by changing either plasma TRP levels, or concentrations of one or more of its brain-transport competitive LNAA, or both (1). The limits to which this may apply, however, has often been overlooked. In experiments on rats fed various diets (2) small increases in the TRP/LNAA ratio did not reliably predict brain TRP or 5-HT levels, whereas 2-3 fold increases did. As the human brain responds to increases in dietary TRP by synthesizing more 5-HT, as indicated by increased cerebral spinal fluid 5-HIAA concentrations (5), it seems that a similar mechanism could operate in man and animals. We therefore determined the changes in plasma TRP/LNAA ratio ensuing after eating meals of different composition either in the morning or evening.

These results strongly suggest that brain 5-HT concentrations could be increased only by the TRP-containing carbohydrate diets. Although TRP/ LNAA ratios increased from 0.13 to 0.15 ($p < 0.04$) after a carbohydrate breakfast (Fig. 1a) and decreased to 0.10 ($p < 0.04$)

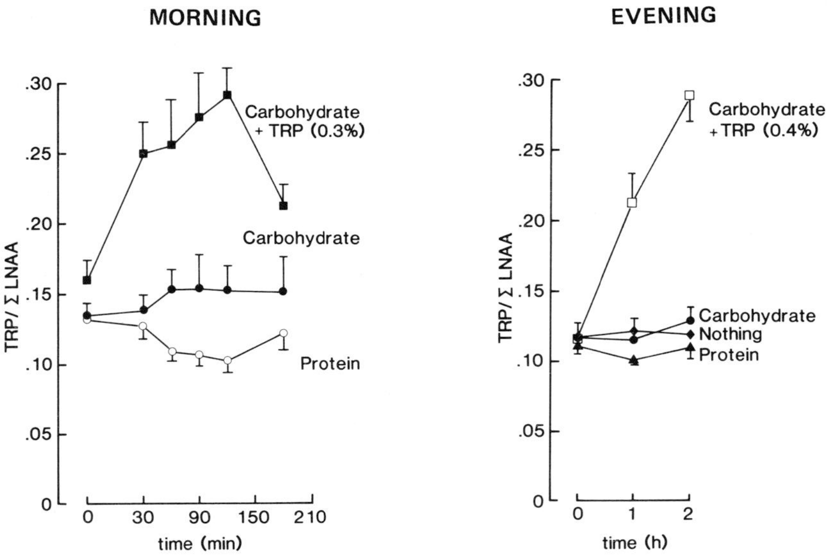

Figure 1 : Plasma TRP/LNAA changes in humans after consumption of (a) a breakfast meal; (b) an evening meal.

after the protein breakfast, these changes were almost certainly insufficient to alter brain TRP uptake and brain 5-HT synthesis. In the evening, (Fig. 1b) neither of the carbohydrate-rich, nor the protein-rich meal altered plasma TRP/LNAA ratios. However, the TRP-containing breakfast and TRP-containing evening meal increased plasma TRP/LNAA ratios 2- to 3-fold, the minimum necessary for reliable increases in brain 5-HT synthesis to occur (2).

In an earlier study on humans (6) who had fed for 5 days on high or low protein diets (Fig. 2), the increase in plasma TRP/LNAA ratios, after a carbohydrate meal eaten at 0700h, was similar to that found in our study, but again, well below the minimum necessary to predict a reliable increase in brain TRP uptake. Plasma TRP/LNAA ratios were inversely proportional to long-term levels of protein intake (see Fig. 3) perhaps because prolonged protein feeding caused an expansion of the plasma LNAA pool. Taken together, the results show that only the carbohydrate-TRP containing meals, which caused 2-3 fold increases in plasma TRP/LNAA ratios, can be expected reliably to alter brain 5-HT metabolism in man.

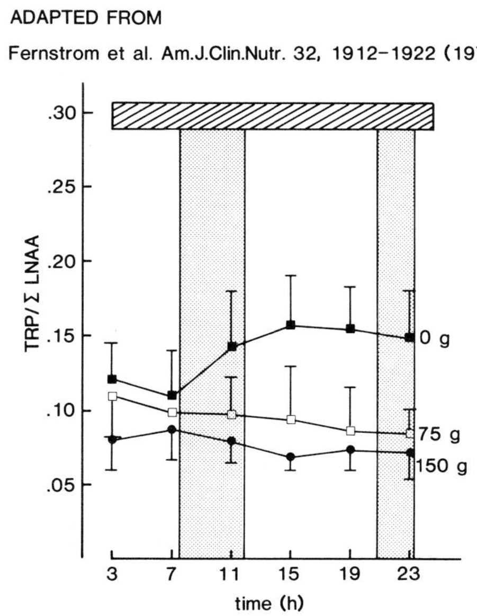

Figure 2 : Diurnal variations in plasma TRP/LNAA ratios in normal human subjects ingesting different levels of protein. Data are taken from Fernstrom and Wurtman 1979 (6); and transposed to the same scales as in Fig. 1. The hatched area represents the times at which measurements were made in the present study.

References

1. Fernstrom, J.D., Wurtman, R.J. : Science, 174, 1023 (1971).
2. Fernstrom, J.D., Wurtman, R.J. : Science, 178, 414 (1972).
3. Wurtman, R.J.: Lancet, i (8334), 7145 (1983).
4. Ashley, D.V., Barclay, D.V., Chauffard, F., Moennoz, D., Leathwood P.D. : Am. J. Clin. Nutr. 36, 143 (1982).
5. Dunner, D.L., Goodwin, F.K. : Arch. Gen. Psychiat. 26, 364 (1972)
6. Fernstrom, J.D., Wurtman, R.J., Hammarstrom-Wikland, B., Rand, W.M., Munro, H., Davison, C.S.: Am. J. Clin. Nutr. 32, 1912 (1979)

EFFECTS OF LARGE NEUTRAL AMINO ACIDS ON TRYPTOPHAN TRANSPORT INTO THE BRAIN DURING DEVELOPMENT

D.V. Ashley, M. Fleury, S. Hardwick, P.D. Leathwood, D. Moennoz
Research Department, Nestlé Products Technical Assistance Co. Ltd.
CH-1814 La Tour-de-Peilz, Switzerland

Introduction

In adult rats, plasma tryptophan (TRP) competes with the other plasma large neutral amino acids, valine, leucine, isoleucine, phenylalanine, tyrosine and methionine (LNAA), for transport into the brain. Thus changes in plasma TRP/LNAA concentration ratio can influence brain TRP levels (1). The degree to which LNAA compete with TRP for entry into the brain during development is not known. In this experiment, we examined the influence of oral doses of TRP and LNAA on brain TRP levels in rat pups aged 5, 11 and 21 days.

Materials and Methods

In a preliminary study (Table 1) we showed that in 8-day old rats, as in adult rats (1), brain levels of TRP and its principal metabolites, 5-hydroxytryptamine (5-HT) and 5-hydroxyindole acetic acid (5-HIAA), had all increased 2 h after oral administration of TRP (100 mg/kg). This time interval was used in the subsequent full-scale study.

In the full-scale study, a total of 92 rat pups were used. At 0700h on the appropriate day (i.e. either 5, 11 or 21 days after birth) litters were separated from their mothers and randomly assigned to one of 4 groups of at least 7 pups each. At 0830h pups were given a solution of either saline, TRP or a mixture of LNAA (without TRP) or LNAA + TRP, by means of a Gilson pipette directed to the back of the throat. The composition of these mixtures are shown in Table 2. Two hours later,

Progress in Tryptophan and Serotonin Research

Table 1 : Effect of oral TRP on brain TRP, 5-HT and 5-HIAA of 8 day old rats

Treatment		Brain TRP µg/g	Brain 5-HT µg/g	Brain 5-HIAA µg/g
	0h (n = 4)	14.3 ± 1.5	0.119 ± 0.017	0.415 ± 0.056
Saline	1h (n = 4)	15.8 ± 1.5	0.128 ± 0.019	0.827 ± 0.079
TRP	1h (n = 4)	26.2 ± 1.7 ($p < 0.001$)	0.129 ± 0.017	0.691 ± 0.004
Saline	2h (n = 4)	16.0 ± 0.5	0.124 ± 0.012	0.751 ± 0.123
TRP	2h (n = 3)	32.4 ± 2.5 ($p < 0.001$)	0.159 ± 0.008	1.08 ± 0.017 ($p < 0.02$)

Table 2 : Composition of the experimental solutions

Solution*	Composition g/l						
	L-TRP	L-VAL	L-MET	L-ILE	L-LEU	L-PHE	L-TYR
0.9 % NaCl	-	-	-	-	-	-	-
TRP	12.5	-	-	-	-	-	-
TRP + LNAA	12.5	17.53	9.05	14.82	12.89	10.35	0.23
LNAA	-	17.53	9.05	14.32	12.89	10.25	0.23

* All solutions were made up in saline. All rats received 8.6 ml/kg body weight of the solutions.

pups were killed, blood plasma and brains were collected and frozen at -40°C until analysed. Plasma amino acids were measured by HPLC of their OPT derivatives with fluorescence detection (2). Brain TRP, 5-HT and 5-HIAA were measured by HPLC with electrochemical detection (3).

Results and Discussion

Plasma TRP concentrations increased significantly ($p < 0.01$) after administration of TRP or TRP + LNAA solutions. Differences between these group plasma TRP levels were significant only on day 5 ($p < 0.01$). LNAA administration alone did not alter plasma TRP levels (fig. 1).

Plasma TRP/LNAA ratios (fig. 1) the index suggested (4) to best reflect brain TRP concentrations (4), decreased ($p < 0.01$) after giving LNAA. TRP administration increased 3-4 fold plasma TRP/LNAA ratios. Mean plasma TRP/LNAA ratios in the TRP/LNAA group fell intermediate between those of the control and TRP treated groups though the increase on day 21 was not statistically significant.

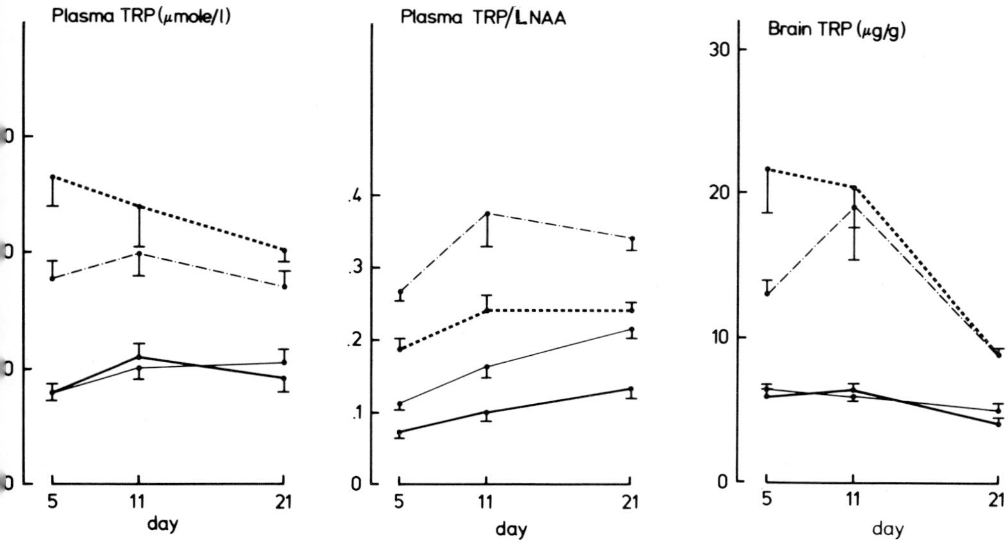

Fig. 1 Profiles of plasma TRP, plasma TRP/LNAA ratios and brain TRP in suckling rats 2 h after receiving either saline (——•——), TRP (—·—•—·—··), TRP+LNAA (- - -•- - -) or LNAA (▬▬●▬▬)

Administration of LNAA to suckling rats did not lower brain TRP levels up to 21 days (fig. 1). Furthermore, when TRP + LNAA were given, brain TRP concentrations were similar to those in rats given TRP alone, despite large differences in their TRP/LNAA ratios. Examination of the graphic profiles over all groups suggested closer correlation of the brain TRP changes with plasma TRP than with plasma TRP/LNAA ratios.

Correlation analysis confirmed these observations (Table 3). Correlations based on group means (r_β) showed a closer correlation with plasma TRP than with plasma TRP/LNAA ratios.

Within groups, on day 5, both plasma TRP and plasma TRP/LNAA ratios had equal predictive value for brain TRP concentrations. By day 21, however, only the plasma TRP/LNAA ratio correlated significantly with brain TRP levels. Thus, with increasing age, LNAA competition with TRP in plasma becomes a more important predictor of brain TRP levels.

Table 3 : Summary of correlation between brain TRP, plasma TRP and plasma TRP/LNAA ratios

	Brain TRP vs Plasma TRP	Brain TRP vs Plasma TRP/LNAA
Between Groups (r_β)		
Day 5 (n = 4)	0.99	0.58
Day 11 (n = 4)	0.99	0.81
Day 21 (n = 4)	0.98	0.80
Within groups (r_ω)		
Day 5 (n = 31)	0.62	0.57
Day 11 (n = 33)	0.63	0.87
Day 21 (n = 38)	0.05	0.49

r_β significant at 5 % if $r < 0.95$
r_ω significant at 5 % if $r < 0.37$

References

1. Arimanana, L., Ashley, D., Furniss, D., Leathwood, P.D.: This volume, 549.
2. Fleury, M., Ashley, D.V. : Analy. Biochem. in press (1983).
3. Moennoz, D., Ashley, D.V. : Experienta 38, 757 (1982).
4. Wurtman, R.J., Hefti, F., Melamed, E.: Pharmacol. Rev. 32, 15 (1980).

CONTRIBUTION OF EXOGENOUS N^{15}-TRYPTOPHAN TO PLASMA AND RED BLOOD CELL TRYPTOPHAN AND KYNURENINE IN HEALTHY HUMANS.

D.V. Ashley, P.A. Finot, R. Liardon

Research Department, Nestlé Products Technical Assistance Co. Ltd.
CH-1814 La Tour-de-Peilz, Switzerland

Introduction

Tryptophan (TRP), the precursor of the neurotransmitter serotonin (5-HT) has found increasing usage as a dietary supplement for inducing sleep (1) and may be of potential value in control of appetite and mood (2). Individual responses differ (1) however, and clear identifiers of TRP responders have yet to be developed. Little information is available on the relative contribution of an exogenous TRP load to its different metabolic routes in humans. As metabolic induction rates in responders and non-responders may differ, we attempted to estimate the relative contribution of N^{15} labelled TRP contained in a carbohydrate breakfast meal (3) to plasma and red blood cell TRP and kynurenine (KYN) in healthy humans.

Experimental Protocol

On the day prior to the experiment, the six participating males were fed standard meals for breakfast, lunch and supper. At 0800h the following morning, blood was taken from the fasting volunteers who were then fed for breakfast a 500 kcal carbohydrate meal containing 0.3% TRP with N^{15}/N^{14} ratio = 5. After the meal, blood was taken at 30 min intervals for 2 h and after 3 h. Plasma was removed after centrifugation and the erythrocytes frozen at -20°C until analysed.

Progress in Tryptophan and Serotonin Research

Analytical Methods

TRP in plasma was analysed fluorimetrically as previously described (4). Erythrocytes (3 ml) were washed twice with 5 ml 0.9% NaCl and centrifuged at 3000 rpm for 10 min. The clean erythrocytes were then suspended in 1 ml 0.9% NaCl. 1 M acetic acid (2 ml) was mixed with 1 ml of the erythrocyte suspension, left to stand for 10 min and then remixed and centrifuged at 12000 rpm for 10 min. Amino acids from plasma or erythrocyte extract were eluted from Dowex $50/4H^+$ resin column, and prepared for isotopic enrichment determination according to a previously described method (5).

Determination of N^{15}-isotope enrichment in plasma or erythrocytes was determined by selected ion monitoring (SIM) GC-MS. After being converted into N-trifluoroacetyl esters, TRP and KYN, along with other free amino acids, were separated on a fused silica capillary column. Isotopic enrichment was determined by measuring the intensity ratio of the molecular ions of labelled and unlabelled TRP and KYN derivatives. Alternatively, the same determination was done by monitoring the main fragment ions (indole or phenyl moiety). In either case, the measured ratio was corrected for naturally occurring C^{13} isotope. The accuracy of the method was checked by analysing a series of standard mixtures of labelled and unlabelled TRP. Perfect linearity was observed when plotting molecular ion or main fragment intensity ratio against N^{15}/N^{14}-TRP concentration ratios. Correlation coefficients varied between 0.9991 and 0.9999. An average standard deviation of 0.2% was found for the isotopic enrichment determination.

Results and Discussion

N^{15}/N^{14}-TRP and KYN enrichment in plasma and of TRP in erythrocytes is shown in Fig. 1. Enrichment in plasma TRP was similar for all subjects. Maximum enrichment, occurring 60-90 min after meal consumption was only 30%, implying that the free TRP pool was 3-4 times larger than the ingested dose.

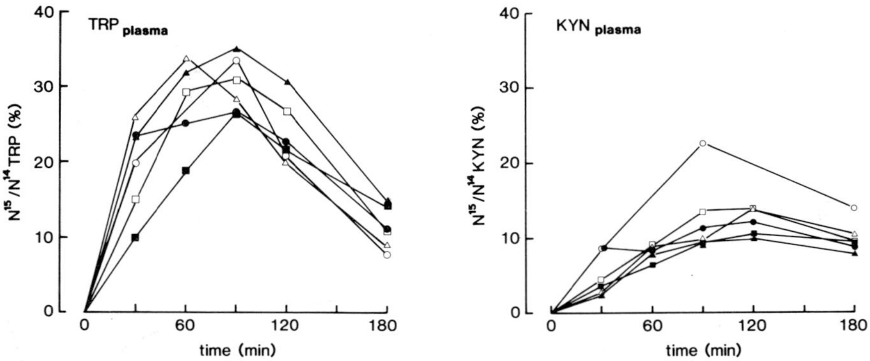

Fig. 1 : N^{15}/N^{14} enrichment of plasma TRP and plasma KYN after a 500 kcal carbohydrate breakfast meal (individual subjects).

About 50% of the exogenous TRP load was metabolised via the KYN route, as indicated by KYN enrichment (Fig. 1). One subject had markedly higher enrichment than in the others.

Erythrocyte TRP was little enriched suggesting that these cells exchange TRP only slowly with plasma. KYN enrichment could not be detected in erythrocytes.

Total plasma TRP concentrations were 40% increased after 30 min (Fig. 2) returning to baseline at 180 min. Calculating $N^{14}/N^{14}+N^{15}$ from the isotopic enrichment data, exogenous TRP accounted for 16-18% of the total plasma TRP concentration 30 min after the meal (Fig. 2). This implies that after 30 min, endogenous TRP (probably from tissue sources) contributed to the increased total plasma TRP concentrations and accounted for some 22% of the total TRP in plasma. After 3 h, exogenous TRP still accounted for 10% of the total plasma TRP concentration and 10% of the total plasma KYN concentration, implying that the exogenous pool was continually being diluted by the endogenous pool.

This experiment establishes a methodology which will be useful for demonstrating metabolic differences between TRP responders and non-responders. Yuwiler et al., (6) suggested that subjects with the most sluggish increases in plasma KYN were the most behaviourally affected by TRP, implying that slow rates of induction of KYN synthesis make

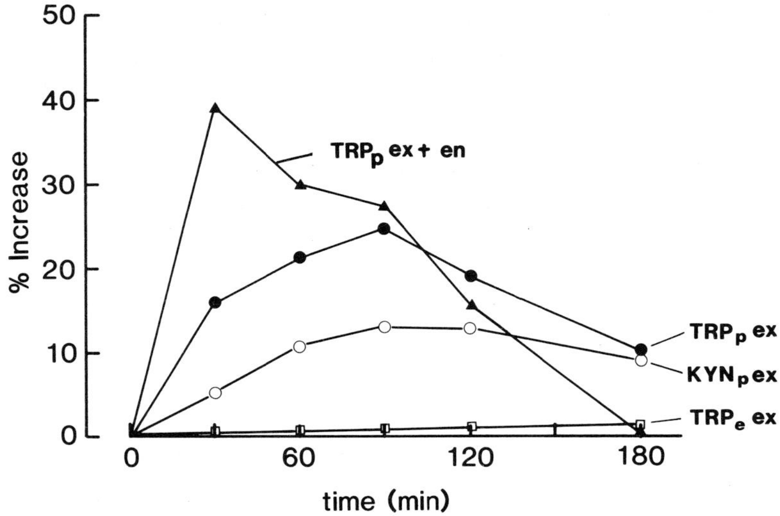

Fig. 2 : Relative increases in total plasma TRP (▲—▲) as compared to increases in plasma TRP (●—●) and KYN (o—o) due to exogenous sources. Exogenous TRP contributes little to erythrocyte TRP (□—□).

available more TRP for uptake by the brain. Substrate induction of KYN synthesis occurs rapidly, almost concomitantly with the appearance of exogenous TRP in the plasma (Fig. 2). An earlier report (5), using larger TRP loads showed maximum increases in total plasma KYN to occur after 3 h. Although total plasma KYN was not measured in the present experiment, one can speculate that induction of tryptophan pyrrolase, the enzyme catalysing conversion of TRP to KYN in the liver, occurs in 2 phases; an early phase due to exogenous substrate and a later phase due to other factors (probably hormones). In this later phase, the substrate for KYN synthesis is most likely of endogenous origin.

References

1. Pollet, P.,Leathwood, P:D.: This volume, 311.
2. Wurtman, R.J., Hefti, F., Melamed, E.: Pharmacol. Rev. 32, 15 (1980).
3. Ashley, D.V., Leathwood, P.D.: This volume, 579.
4. Bloxam, D., Warren, W.H.: Anal. Biochem. 60, 621 (1974).
5. Bier, D.M., Christopherson, H.L.: Anal. Biochem. 94, 242 (1979).
6. Yuwiler, A., Brammer, G.L., Morley, J.E., Raleigh, M.J., Flammery, J.W., Geller, E.: Arch. Gen. Psych. 38, 619 (1981).

CARBOHYDRATE MEAL INCREASES BRAIN 5-HYDROXYTRYPTAMINE SYNTHESIS IN THE ADULT RAT ONLY AFTER PROLONGED FASTING

D.V. Ashley, P.D. Leathwood, D. Moennoz

Research Department, Nestlé Products Technical Assistance Co. Ltd.
CH-1814 La Tour-de-Peilz, Switzerland

Introduction

It has been proposed that in rats offered a choice of diets containing different quantities of protein, brain 5-hydroxytryptamine (5-HT) acts as a variable ratio sensor monitoring the selection of carbohydrate and protein intake (1). This idea developed from the observation that brain 5-HT synthesis increased significantly within 2 h after 15 h fasted rats began to consume a carbohydrate-rich, protein-free meal, but failed to increase (or decreased) when the rats ate meals containing substantial quantities of protein (2). But, prolonged fasting itself increases turnover of 5-HT in the brain (3). The present experiments were carried out to determine if dietary effects on brain 5-HT synthesis also occurred when the period of fast was reduced to 8 h or 3 h.

Experimental protocol

Experiment 1; 8 hour fast : 30 male Sprague-Dawley rats (∿ 200 g) were randomly divided into 5 groups and housed individually in a room lighted for 12 h daily (1000h-2200h). All rats were given stock diet (NAFAG SA, Switzerland). At 2300h the following evening, food was removed and the rats fasted until 0700h the next morning, when they were given a semi-synthetic diet containing 0, 5, 10 or 20% protein or the stock diet. The procedure was repeated on three consecutive nights to habituate the rats to the taste of the diets. On the third evening, however the group "stock diet" continued to fast. On the morning of the fourth day at

Progress in Tryptophan and Serotonin Research

1000h, the rats were killed, their brains removed, frozen on solid CO_2, and stored at -40°C until analysed for TRP, 5-HT and 5-hydroxyindole acetic acid (5-HIAA) by HPLC with electrochemical detection.

Experiment 2; 3 hour fast : 30 male Sprague-Dawley rats (∿ 200 g), again randomly divided into 5 groups, were housed individually in a room lighted for 1200h (0600-1800h). All rats received stock (NAFAG SA, Switzerland). On the following evening at 2400h, food was removed and the rats left to fast until 0300h. At that time they were given either the 0, 5, 10 or 20% protein diets, or stock. At 0600h the stock diet was refed to all groups. The procedure was repeated on the two following evenings. On the third evening the "stock" group continued to fast. On the morning of the fourth day, at 0600h, the rats were killed. Blood plasma was analysed for amino acids (4) and brains for TRP, 5-HT and 5-HIAA.

Results and Discussion

Experiment 1 : After an 8 h fast, brain TRP levels were higher in rats feeding from the 0% protein diet. Rats consuming 10 or 20% protein had similar brain TRP levels to fasting controls (Fig 1). Brain 5-HT levels were the same in all groups. Brain 5-HIAA levels however were inversely proportional to dietary protein content and were significantly higher in the 0% protein, 5% protein and fasted groups, than in the 10 or 20% protein groups ($p < 0.05$).

Experiment 2 : In this study, in addition to brain TRP, 5-HT and 5-HIAA, we also measured plasma amino acids. After the 3 h fast, the 0% casein diet had no influence (compared to continued fasting) on brain TRP. The 10 and 20% protein diets significantly lowered brain TRP. None of the diets had any effect on brain 5-HT, but brain 5-HIAA levels were significantly lowered by the 5, 10 and 20% protein diets.

Plasma tryptophan to large neutral amino acid ratios (TRP/LNAA) were

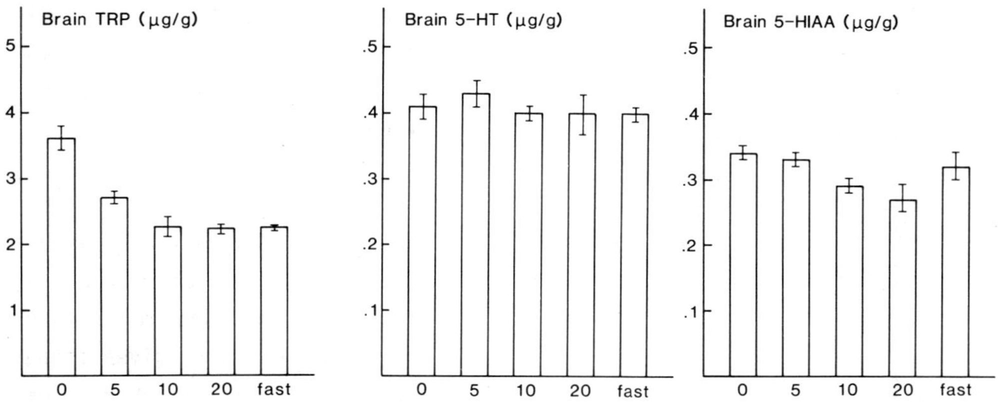

Fig. 1 : Brain TRP, 5-HT and 5-HIAA levels in rats having consumed either 0, 5, 10 or 20% protein diets or nothing during 3 h after 8 h fast.

Table 1 : Brain TRP, 5-HT and 5-HIAA levels in rats having consumed 0, 5, 10 or 20% protein diets or nothing (fast) during 3 h after a 3 h fast

Group	TRP (μg/g)	5-HT (μg/g)	5-HIAA (μg/g
0% protein	4.0 ± 0.3[a]	0.42 ± 0.02	0.29 ± 0.03[a]
5% protein	3.2 ± 0.2[b]	0.39 ± 0.02	0.24 ± 0.02[b]
10% protein	2.8 ± 0.1[b]	0.40 ± 0.01	0.25 ± 0.01[b]
20% protein	2.8 ± 0.2[b]	0.40 ± 0.01	0.25 ± 0.01[b]
Fast	3.6 ± 0.2[a]	0.41 ± 0.01	0.31 ± 0.02[a]

Mean ± SEM (n = 6). Means with different superscripts are significantly different from each other ($p < 0.05$).

inversely proportional to the dietary protein content (Table 2). Fasting rats and rats consuming the 20% protein diet had identical plasma TRP/LNAA ratios. Consequently, plasma TRP/LNAA ratios correlated weakly with brain TRP levels ($r = 0.50$).

Table 2 : [1]Plasma TRP and TRP/LNAA ratios in rats having consumed 0, 5, 10 or 20% protein diets or nothing (fast) during 3 h after a 3 h fast

Group	TRP (μmole/l)	LNAA (μmole/l)[2]	TRP/LNAA
0% protein	97.0 ± 3.9[a]	379.8 ± 30.1[a]	0.26 ± 0.02[a]
5% protein	101.8 ± 3.2[a]	440.8 ± 33.7[a]	0.24 ± 0.01[ab]
10% protein	100.4 ± 5.3[a]	491.9 ± 21.1[a]	0.20 ± 0.01[b]
20% protein	115.9 ± 2.8[b]	742.3 ± 41.2[b]	0.16 ± 0.01[c]
Fast	94.9 ± 4.0[a]	622.0 ± 13.9[b]	0.15 ± 0.01[c]

[1] Mean ± SEM (n = 6). Means with different superscripts are significantly different from each other ($p < 0.05$).

[2] LNAA: Large neutral amino acids in plasma which compete with plasma TRP for the same carrier for entry into the brain ([valine] + [leucine] + [isoleucine] + [phenylalanine] + [tyrosine] + [methionine])

It was concluded that : (1) Carbohydrate consumption after a short fast increases plasma TRP/LNAA ratios but does not affect brain TRP or 5-HT levels (2). The change in 5-HT metabolism after high protein diets was independent of plasma TRP/LNAA ratios.

References

1. Leathwood, P.D., Ashley, D.V. In: The Body weight Regulatory system. Cioffi, G., James, W.P.T., Van Ittalie, T.B. (eds) Raven Press p.263 (1981).

2. Fernstrom, J.D., Wurtman, R.J.: Science, 178, 414 (1972).

3. Curzon, G., Joseph, M.H., Knott, P.J.: J. Neurochem. 19, 1967 (1972).

4. Fleury, M., Ashley D.V.: Anal. Biochem. (in press).

EXOGENOUS TRYPTOPHAN: A STUDY OF ITS EFFECTS ON THE LEVELS OF AROMATIC AMINE(S) IN CEREBRAL AND HEPATIC TISSUES IN MICE

D. Ghosh, Imeh U. Affiah, James Guilford and Lavaniel L. Henderson
Biology Department, Texas Southern University, Houston, Texas, 77004

Introduction

Since the discovery of tryptophan in 1901 by Hopkins and Cole (1), studies on this amino acid and its metabolites have been numerous. A few years after its isolation, it was shown to be essential for mammalian nutrition (2) and has since been demonstrated to be a precursor for a number of physiologically active compounds in plants and animals; indole-3-acetic acid, 5-hydroxytryptamine (5-HT, serotonin) and melatonin are only a few examples from an extensive list of tryptophan derivatives. Most of the tryptophan in the mammalian liver is metabolized through kynurenine; this pathway does also lead to the formation of nicotineamide and its subsequent transformation to nicotineamide nucleotides (NAD and NADP).

Studies on cerebral metabolism of L-tryptophan have mostly centered on the biosynthesis of 5-HT and its degradation by classic monoamine oxidases (MAO). In recent years, however, a novel enzyme, indoleamine-2,3-dioxygenase (IDO) has been discovered which cleaves the pyrrole ring in tryptophan, tryptamine as well as a number of other D and L indole isomers (3,4). Three aromatic amines which could be derived from tryptophan are kynurenine, kynuramine and anthranilic acid. L-kynurenine and kynuramine are known to be formed in discrete areas of the rat brain apparently through IDO catalysis (5). Kynuramine has further been implicated as an inhibitor of α-adrenoreceptors (6). Data reported in this paper document the levels of aromatic amine(s) in cerebral and hepatic tissues derived from exogenous tryptophan.

Progress in Tryptophan and Serotonin Research

Results and Discussion

Experiments were performed to compare the contents of total aromatic amine(s) in cerebral and hepatic tissues of mice employing the Bratton-

Table I

Effects of Three Different Doses of L-Tryptophan on Aromatic Amine(s) Reactivities in Cerebral and Hepatic Tissues of Young and Old BALB/cAn. Tex. Male Mice (Values Are Expressed in Nanomoles of Kynurenine Equivalence/Gram Tissue)

Dose of L-tryptophan	3-Month Old		15-Month Old	
	Brain	Liver	Brain	Liver
(1) 450 mg/kg	287.6±0.6(5)	938.00±2.5(5)	185±1.3(5)	1188±3.1(5)
(2) 900 mg/kg	340.0±3.1(5)	1398.30±0.3(5)	241±1.5(5)	1151±2.5(5)
(3) 1800 mg/kg	544.3±1.5(5)	7586.05±3.0(5)	482±0.0(5)	7128±2.5(5)

Marshall reagents (7,8). The animals (BALB/cAn. Tex. male mice) were injected intraperitoneally with several doses of L-tryptophan, for example, 450 mg/kg, 900 mg/kg and 1800 mg/kg. Initially, two time-intervals, $1\frac{1}{2}$ hours and 4 hours were utilized for sacrificing animals from the time of administration of the amino acid; the $1\frac{1}{2}$-effect of aromatic amine(s) reactivities in cerebral samples was negligible and this time-span was then abandoned in further experimentation. In a dose-response study involving the young (3-month) and the old (15-month) mice, both groups produced a graduated but non-linear increase in aromatic amine(s) reactivities in cerebral and hepatic tissues (Table I).

The 4-hour effect resulted, on the average, in a 7-fold increase in hepatic tissues compared to cerebral samples. Cerebral aromatic amine(s) contents were persistently higher in the young mice--being 25% more, on the average, than the older ones. This difference could be due to either more transport of aromatic amine(s) from systemic circulation into cerebrospinal fluid (CSF) or higher cerebral IDO activity in the young mice *per se* or both. Also, the capacity of the brain to accumulate aromatic amine(s) seems to be confined within a narrow range.

A comparison of the cerebral aromatic amine(s) contents from the fasted and fed animals showed the former yielding approximately 60% more of the reactivity than the latter in two doses of L-tryptophan (900 mg/kg and 1800 mg/kg) administered.

(Supported in part by USPHS NIH Grant HL07434-01 to Dr. Joseph Buckley, Dean, College of Pharmacy, University of Houston Central Campus.)

References

(1) Hopkins, F. G., Cole, S. W.: J. Physiol. (London) 27, 418 (1901).

(2) Wilcock, E. G., Hopkins, F. G.: J. Physiol. (London) 35, 88 (1906).

(3) Hirata, F., Hayaishi, O.: Acta. Vitamin. Enzymol. (Milano) 29, 288 (1975).

(4) Hayaishi, O., Hirata, F.: Ibid. 29, 291 (1975).

(5) Gal, E. M., Sherman, A.: Neurochem. Res. 5, 223 (1980).

(6) Johnson, T. D., Clarke, D. E.: European J. Pharmacol. 72, 351 (1981).

(7) Bratton, A. C., Marshall, E. K.: J. Biol. Chem. 128, 537 (1939).

(8) Nishikimi, M.: Biochem. Biophys. Res. Commun. 63, 92 (1975).

INFLUENCE OF EXCESS LEUCINE INTAKE ON THE CONVERSION OF TRYPTOPHAN TO NAD IN RATS FED LOW PROTEIN DIET

Takeshi Suzuki, Shunsuke Yuyama, Atsuko Sasaki, Mieko Yamada and Rikako Kumagai

Department of Nutrition and Biochemistry, The Institute of Public Health, Minato-ku, Tokyo 108, Japan

Introduction

In several previous studies, we have shown that the administration of excess leucine to rats, fed low protein diets, increased pool size of quinolinic acid (Q) and decreased that of N^1-methylnicotinamide (MNA). These results suggested that the influx of tryptophan (Trp) via Q into the NAD cycle, at least in part, was lowered by the excess leucine intake (1). In this experiment, to confirm the depression of NAD biosynthesis, the incorporation of ^{14}C-Trp into nicotinic acid (NiA)- and nicoinamide (Nam)-nucleotides was examined. In the rats fed low protein diet, excess leucine intake decreased both biosynthesis and metabolism of NAD.

Experimental Method

Treatment of animals -- All rats were housed individually and prefed on 18% casein diet for 1 week; subsequently the basal diet was fed for 1 week. The rats were then divided into two groups (control and leucine excess groups) and fed the experimental diets for 2 weeks. The rats of the leucine excess group were administered 500mg of L-leucine and 500 μg of nicotinic acid added daily to the basal diet, and control group rats were given 400mg of nonessential amino acids mixture (isonitrogenous to added leucine) and 500μg of nicotinic acid daily.

Progress in Tryptophan and Serotonin Research

At the end of the feeding period the rats were injected 2μCi of ^{14}C-ring uniformly labeled-tryptophan intraperitoneally. At 10, 20 and 40min after injection, each of 5 rats were decapitated, their liver excised and immediately homogenized in 9 vol of 1N perchloric acid at 0°.
The incorporation of ^{14}C-Trp into nucleotides was studied using thin-layer chromatography of nucleotides (NiA- and Nam-nucleotides) in the neutralized supernatant (separation by HPTLC on cellulose plate) (to be published). Pre-coated HPTLC cellulose plates (20x10cm, without fluorescent indicator, E. Merck 5786) were used. Sample application on the plate into narrow bands was performed using Linomat III (Camag), and the locations of separated nucleotides on the plates were identified and photocopied using Reprostar (Camag) under UV light (254nm). After the locations of the nucleotides were identified, the sections of the thin layer containing each nucleotide were scraped from the plate, and redioactivity was determined with a gel scintillator (Lumagel SB) by the liquid scintillation spectrophotometer. Direct quantitative assessment of nucleotides on the HPTLC plate was attempted. Before the thin layer was scraped, the UV quenching bands were scanned using a HPTLC Scanner, and Chromatopac C-R1A (Shimadzu) was used to integrate peak areas.

Results and Discussion

Thin-layer chromatography of NiA- and Nam-nucleotides. --
The sample solution (100-300μl) was applied evenly in a 90mm strip to the plate using a Linomat III. The plate was developed with a solvent system devised by us (1M ammonium acetate buffer, pH 6.0, 15ml; iso-butyric acid, 20ml; n-propanol, 5ml; acetone, 40ml; methanol, 10ml; 9 % aqueous ammonia, 15ml). With this method 5 nucleotides were separated well enough to be scraped off. Rf values of authentic standards were as follows: NMN, 0.50; NaMN, 0.39; NAD, 0.18 and NADP, 0.12.

Incorporation of labeled Trp into NAD and other nucleotides -- Total amount of labeled nucleotides in the liver of the excess leucine fed rats was higher than that of control rats at each time after injection. It was assumed that this result showed a decreased output of nucleotides from the NAD cycle rather than increased synthesis. Incorporation of labeled Trp into NAD increased with time in the excess leucine diet group, and its increment was higher in the later phase (from 10 to 20 min). On the other hand, in the control group the increment of labeled Trp incorporation was high in the initial phase. The incorporation of labeled Trp into other nucleotides showed no marked differences with either diet group (Fig. 1).

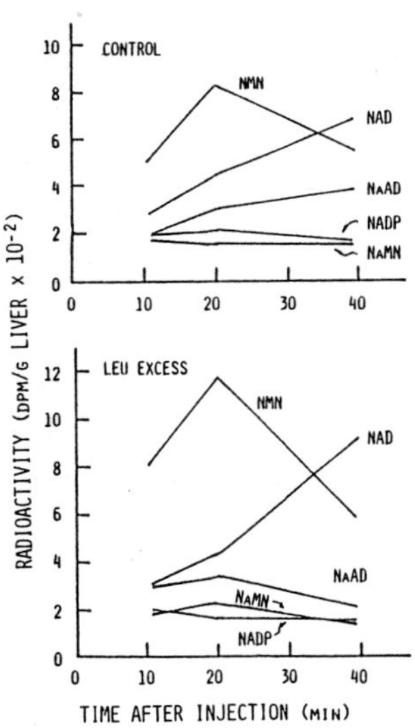

Fig. 1

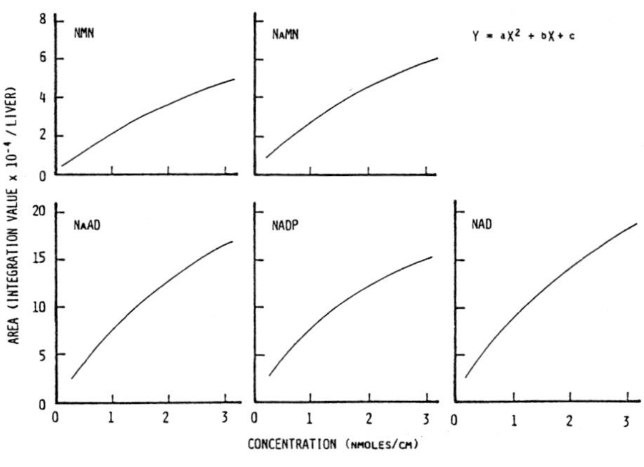

Fig. 2

Direct quantitative determination of nucleotides on HPTLCplate -- For the quantitative determination, the intensity of the quenching band under the UV light was able to be acquired as an integrated peak area by the TLC-Scanner equipped with an analog recorder and integrator, and these areas were transferred into concentration values. When the concentration is plotted vs. peak area obtained at different concentrations of the authentic standard, slightly curved calibration graphs are obtained

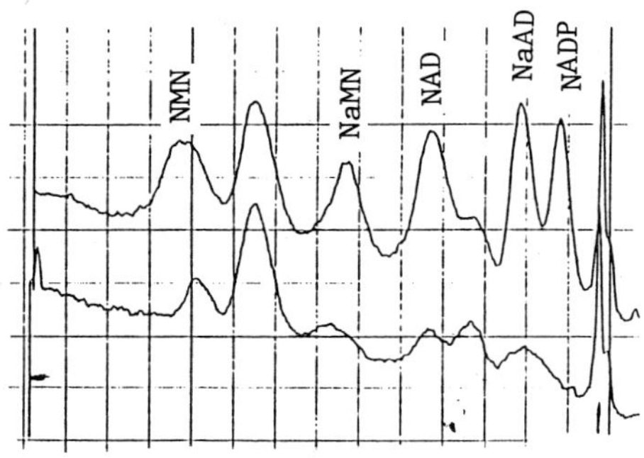

Fig.3

(Fig. 2). The nonlinear relationship between concentration and integrated peak area can be adapted to a square function. Preliminary determination of nucleotides from liver extracts was attempted. Five µl of liver extract applied at 10mm lenghts on the plate at two tracks, and on one track 2 µl of authentic standard of appropriate concentration was added. Fig. 3 shows typical chromatograms of authentic standard (upper curve) and of liver extract (lower curve). Besides the UV-absorption bands of the 5 nucleotides there were some more, not yet identified bands. Chromatographic separation was not sufficient for quantitative assessment. So further investigations have been performed.

Summary

The excess leucine supply to the rats fed a low protein diet resulted in a delay of NAD biosynthesis from Trp and a suppression of metabolism of NAD and other nucleotides.

References

1. T. Suzuki, S. Yuyama, A. Sasaki, M. Yamada, C. Haba, H. Yoshida in: O. Hayaishi, Y. Ishimura, R. Kido, eds.: Biochemical and Medical Aspects of Tryptophan Metabolism, p 337, Elsevier, Amsterdam. New York 1980

THE PELLAGRAGENIC ACTION OF A DIETARY EXCESS OF LEUCINE: EFFECTS OF A HIGH-LEUCINE DIET ON TRYPTOPHAN METABOLISM.

David A. Bender
Courtauld Institute of Biochemistry, The Middlesex Hospital Medical School, London W1P 7PN, U.K.

Bahieldin I. Magboul,
Food Research Laboratories, Khartoum, Democratic Republic of the Sudan.

Introduction

Pellagra continues to be a major problem of public health nutrition in many developing countries, not only among people whose dietary staple is maize, the cereal traditionally associated with pellagra, but also in India among those whose diet is based on jowar (Sorghum vulgare). Gopalan & Srikantia (1960) noted that the tryptophan content of jowar is higher than that of maize, and nearly as high as that of rice, a cereal that is not associated with pellagra. They suggested that an important factor in the pellagragenicity of jowar might be the relative excess of leucine in the protein of this cereal. Further studies from the Indian Institute for Medical Research have apparently confirmed the pellagragenic action of a dietary excess of leucine: Gopalan et al (1969) induced black tongue disease (the canine equivalent of pellagra) in dogs by manipulating the leucine content of the diet, and Belavady & Udayasekhara Rao (1979) showed a correlation between the leucine content of different cultivars of jowar and the induction of black tongue disease in dogs.

Other workers (Nakagawa et al, 1975, Manson & Carpenter, 1978a,b) have been unable to demonstrate any effect of a dietary excess of leucine on niacin nutritional status in man, rats, chicks or dogs.

These studies were undertaken in order to attempt to resolve this conflict.

Progress in Tryptophan and Serotonin Research

Methods

Rats were fed for 7 weeks from weaning on maize-gelatine-sucrose based diets providing: (a) minimally adequate amounts of tryptophan and nicotinamide together, but neither alone; (b) adequate nicotinamide with a minimal amount of tryptophan; (c) adequate tryptophan with no added nicotinamide. In the experimental diets 15g L-leucine was added / kg diet, at the expense of carbohydrate. (Magboul & Bender, 1983).

The concentration of nicotinamide nucleotides in whole blood was determined fluorimetrically (Bender et al, 1982) and the activities of the enzymes of the tryptophan oxidative pathway were determined by the methods that have been described previously (Bender, 1980), both with and without the addition of 7.6 mM L-leucine to the incubation medium.

The metabolism of tryptophan was assessed in vivo by measuring the release of $^{14}CO_2$ after the administration of (methylene-^{14}C)tryptophan as an index of kynureninase activity, and (benzene ring U-^{14}C)tryptophan as an index of picolinate carboxylase activity.

Results

The addition of 15g leucine/kg diet led to reduced blood concentrations of nicotinamide nucleotides (NAD + NADP) when the rats were wholly or partially dependent on synthesis from tryptophan:

Diet providing (mg/kg):		Blood NAD + NADP (µmol/l) control		high-leucine diet	
tryptophan	nicotinamide	mean	sem	mean	sem
a) 980	1.5	59	2.4	30	0.7 ***
b) 900	4.0	45	1.7	40	1.8
c) 1030	0	75	1.6	57	1.9 ***

(*** significantly different from control by t-test, $P < 0.001$)

Addition of leucine (7.6 mM) to the incubation buffer led to an increase in the activity of tryptophan oxygenase and significant inhibition of kynureninase, which was competitive with respect to the substrate:

	Enzyme activity (nmol/min/g liver)			
	Control		+ leucine	
	mean	sem	mean	sem
Tryptophan oxygenase	9.9	1.1	17.5	1.1 **
Kynurenine hydroxylase	20.3	0.3	20.6	0.5
Kynureninase	24.5	0.9	19.9	0.7 **
3-Hydroxyanthranilate oxidase	113	1.2	114	1.4
Picolinate carboxylase	13.8	0.2	14.0	0.2
Quinolinate phosphoribosyltransferase	0.52	0.02	0.51	0.03

(** significantly different from control by t-test, $0.01 > P > 0.001$)

The addition of leucine to diet (c) led to a significant increase in the urinary excretion of kynurenine:

control	14.1 ± 1.8 mmol/mol creatinine
high-leucine diet	18.9 ± 2.1 mmol/mol creatinine

In rats fed on diet (c) the addition of leucine led to impaired and delayed production of $^{14}CO_2$ from(methylene-^{14}C)tryptophan, indicating competitive inhibition of kynureninase:

	Control		high-leucine diet	
	mean	sem	mean	sem
total $^{14}CO_2$ in 4h (10^3 dpm)	159	18	91	12 *
time to peak of $^{14}CO_2$ production (min)	70		80	
peak height (10^3 dpm in 10 min interval)	16	1.7	8.5	0.8 **
rate constant for decay (min^{-1}, $x10^{-3}$)	14.25	0.08	10.92	0.11 ***

(significance of differences from control, by t-test:
* $0.05 > P > 0.01$, ** $0.01 > P > 0.001$, *** $P < 0.001$)

In rats fed on diet (c) the addition of leucine led to increased production of $^{14}CO_2$ from (benzene ring U-^{14}C)tryptophan, indicating increased activity in vivo of picolinate carboxylase, despite the reduction in kynureninase activity:

	Control mean	Control sem	high-leucine diet mean	high-leucine diet sem
total $^{14}CO_2$ in 4h (10^3 dpm)	56.9	3.3	82.8	8.6 *
time to peak of $^{14}CO_2$ production (min)	60		60	
peak height (10^3 dpm in 10 min interval)	5.2	0.3	7.9	0.7 **
rate constant for decay(min^{-1}, $x10^{-3}$)	12.2	0.05	12.87	0.07 ***

(significance of differences from control, by t-test:
* $0.05 > P > 0.01$, ** $0.01 > P > 0.001$, *** $P < 0.001$)

These results confirm that a dietary excess of leucine depletes tissue nicotinamide nucleotides by inhibiting their synthesis from tryptophan, and therefore might be pellagragenic. Not only is kynureninase inhibited, but there is increased total oxidation of tryptophan, at the expense of NAD synthesis, as a result of increased activity of picolinate carboxylase.

References

Belavady, B. & Udayasekhara Rao, P. (1979) Ind. J. Exp. Biol. 17 659
Bender, D.A. (1980) Biochem. Pharmacol. 29 707
Bender, D.A., Magboul, B.I. & Wynick, D. (1980) Brit. J. Nutr. 48 119
Gopalan, C. & Srikantia, S.G. (1960) Lancet 1 954
Magboul, B.I. & Bender, D.A. (1983) Brit. J. Nutr. - in press
Manson, J.A. & Carpenter, K.J. (1978a) J. Nutr. 108 1883
Manson, J.A. & Carpenter, K.J. (1978b) J. Nutr. 108 1889
Nakagawa, I., Ohguri, S., Sasaki, A., Kajimoto, M., Sasaki, M. & Takahashi, T. (1975) J. Nutr. 105 1241

EFFECT OF CADMIUM ON SOME KEY ENZYMES OF TRYPTOPHAN - NAD PATHWAY IN RATS *IN VIVO*

Shunsuke Yuyama, Atuko Sasaki and Takeshi Suzuki
Department of Nutrition and Biochemistry
The Institute of Public Health, Tokyo, Japan

Introduction

It is known that dietary tryptophan (Trp) can be converted to NAD in the body. The efficiency of convertion of Trp to NAD was reported to be influenced by nutritional factors and to be regulated by the activities of enzymes concerned in the Trp-NAD pathway (1.2.3). In cadmium (Cd) administered rats, the efficiency of conversion of Trp to NAD is expected to be regulated by activities of enzymes concerning the Trp-NAD pathway. Concerning with the tryptophan oxygenase (TO), Kench et al (4) studied the TO activity in liver cytosol of chicks and rats which is stimulated by Cd *in vivo* and *in vitro*. In the present paper, concerning with the Trp-NAD pathway, the alteration of activities of liver TO, 3-hydroxyanthranilate oxygenase (3-HAO), quinolinate phosphoribosyltransferase (QPRT), nicotinamide methyltransferase (NMT) and kidney picolinate carboxylase (PC) have been observed. And also contents of Cd in liver and kidney, and in liver 38,700 x g homogenate supernatant have been measured. In addition, the effects of Cd ion applied to incubation mixture for the above mentioned enzymes have been studied.

Results and Discussion

1) *Effects of Cd intake on the body weights, liver and kidney weights* : 20 male Wistar strain rats (5 rats litter mate, 3

Progress in Tryptophan and Serotonin Research

week-old) were fed a basal diet (18 % casein) for 2 weeks before starting the experiment. Drinking water was provided *ad libitum*. $CdCl_2$ was added to the basal diet at levels of 0.5, 2.0 and 5.0 mg Cd / day for 2 weeks. Those animals were pair-fed for 5.0 mg Cd group. Body weights, liver and kidney weights are shown in Table 1. The weight gain for 2 weeks depressed in association with increasing levels of dietary Cd by 100% for basal diet group, 88% for 0.5mg Cd group, 64% for 2.0mg Cd group and 28% for 5.0mg Cd group, respectively. Thus the growth appears not to be related to the food intake but to the dietary Cd levels. There was significant and gradual reduction in relative body weights, liver and kidney weights of rats with increasing dietary Cd levels.

2) Effects of Cd intake on some key enzymes of Trp-NAD pathway, of the Cd contents of tissue and the Cd concentrations of incubation mixture for enzyme assay : In Table 2 are given liver and kidney enzymes activities. TO, 3-HAO and QPRT activities of liver 38,700 x g homogenate supernatant, NMT activity of liver $(NH_4)_2SO_4$ fraction (30-60% saturation) and PC activity of kidney 105,000 x g homogenate supernatant were assayed by the methods of Piras (5), Ichiyama (6), Iwai (7), Cantoni (8) and Ikeda (9), respectively. The enzyme activities are expressed as specific activities. TO activity tended to increase with 0.5mg Cd group, significantly decrease with 2.0mg Cd group, and the Cd concentrations of incubation mixture for enzyme assay ranged from 3.95 x 10^{-8}- 2.79 x 10^{-6}M (Table 4). 3-HAO, QPRT, NMT and PC activities significantly elevated with increasing levels of dietary Cd. In Table 3,4 are shown the Cd contents of tissues and the Cd concentrations of incubation mixture. From these results, the above mentioned enzyme activities were found to be altered by Cd administration. The alteration of these enzyme activities *in vivo* are closely related to the chemical form of Cd in tissue when the Cd was absorbed into body. According to Andoh et al.(10), (10), at 2 hr after the administration, the Cd-metallothionein occupied about 80% of absorbed Cd, but 20% of remainder did not exist as combined form and was soluble in the cytoplasm frac-

tion. Therefore, the ratio of Cd ions in the liver and kidney homogenate supernatant fraction may supposed to be very low.

3) Effects of Cd ions applied to incubation mixture on the above mentioned enzymes *in vitro*: The activities of enzymes concerned in the Trp-NAD pathway were found to be changed by Cd ions added to incubation mixture as shown in Fig.1-5. Therefore, the enzyme activities were essentially different with changing Cd concentrations of incubation mixture *in vivo* and *in vitro* as can be seen in Table 2,4 and Fig. 1-5. Accordingly, the alterations of these enzyme activities *in vivo* did not seem to be caused by Cd transferred in the body, but to be probably caused by the metabolic alteration, the transformation of enzyme protein which was caused by Cd administration. The possible mechanism for this transformation, however, is not yet known. From these data [2),3)], the elevations of 3-HAO, QPRT,NMT and PC activities indicate that enzymatic efficiency of conversion of Trp to NAD and enzymatic ability of oxidation of Trp *via* glutarate pathway may be supposed to accelerate with the increasing levels of dietary Cd. However, it is impossible to draw definite conclusions concerning the inflow of Trp to NAD and to glutarate pathway in the whole body from the increased alteration of enzyme activities.

Table 1. Effects of the cadmium administration on weights[a)] of the body, liver and kidney of rats.

		Initial body weight (g)	Middle body weight (g)	Final body weight (g)	Liver weight (g)	Kidney weight (g)
Basal diet	(5)	116.4 ± 9.4	155.2 ± 11.7	198.6 ± 11.0	9.50 ± 1.14	1.65 ± 0.11
0.5 mg cadmium administration	(5)	116.6 ± 7.3	142.6 ± 6.3	189.0 ± 8.3	8.45 ± 0.72	1.58 ± 0.09
2.0 mg cadmium administration	(5)	116.6 ± 1.3	144.0 ± 3.2	170.0 ± 7.8**	7.42 ± 0.79**	1.43 ± 0.15*
5.0 mg cadmium administration	(5)	117.4 ± 3.3	124.4 ± 7.2**	140.0 ± 7.3**	6.72 ± 0.53**	1.25 ± 0.09**

a) The body weight was measured weekly after starting the experiment.
Figures were expressed as mean ± SD; number of rats in parentheses. Asterisk mark, * and ** means the significant difference at $P < 0.05$ and $P < 0.01$, respectively.

Table 2. Effects of cadmium administration on the activities of tryptophan oxygenase, 3-hydroxyanthranilate oxygenase, quinolinate phosphoribosyltransferase, nicotinamide methyltransferase in liver and picolinate carboxylase activity in kidney of rats.

Daily Dose of Cadmium		Tryptophan oxygenase (nmoles/mg protein/hr)	3-Hydroxy-anthranilate-oxygenase (nmoles/mg protein/min)	Quinolinate phosphoribosyl-transferase (nmoles/mg protein/min)	Nicotinamide methyltransferase (nmoles/mg protein/hr)	Picolinate carboxylase (nmoles/mg protein/min)
0 mg	(5)	3.72 ± 0.697	29.00 ± 8.038	0.73 ± 0.159	19.04 ± 5.770	2.75 ± 0.600
0.5 mg	(5)	4.11 ± 1.064	32.14 ± 10.150	0.80 ± 0.211	23.39 ± 5.139	3.36 ± 1.182
2.0 mg	(5)	2.65 ± 0.631*	79.74 ± 18.126**	0.85 ± 0.232	27.74 ± 6.565	4.68 ± 1.692*
5.0 mg	(5)	3.06 ± 1.118	91.65 ± 32.217**	1.08 ± 0.253*	28.43 ± 3.713*	4.97 ± 1.091**

Figures were expresses as mean ± SD; number of rats in parentheses. Asterisk mark, * and ** means the significant difference at $P < 0.05$ and $P < 0.01$, respectively.

Table 3. Effects of cadmium administration on the cadmium content of liver and kidney. One lobe of liver and one kidney from each of two rats were analyzed for cadmium content with a Hitachi 207 atomic spectrometer.

Daily Dose of Cadmium	Cadmium in liver		Cadmium in kidney	
	μg/whole tissue	μg/g tissue	μg/whole tissue	μg/g tissue
0 mg	0.984	0.104	0.010	0.006
0.5 mg	27.157	3.213	9.350	5.533
2.0 mg	95.639	12.890	28.391	17.127
5.0 mg	182.105	27.091	58.750	39.171

Figures were expressed the average of two rats.

Table 4. Effects of cadmium administration on the cadmium concentration of incubation mixture. In rats administered 0, 0.5, 2.0 and 5.0 mg per day of cadmium, cadmium concentration of liver 38,700 x g supernatant was 1.19×10^{-7} M, 1.26×10^{-6} M, 3.46×10^{-6} M and 8.38×10^{-6} M, respectively. The values of cadmium concentration in the incubation mixture were calculated from these values. Nicotinamide methyltransferase; the cadmium concentration of ammonium sulfate fraction (30–60% saturation) was calculated from the ratio of protein concentration to cadmium concentration of liver 38,700 x g supernatant. Picolinate carboxylase; the cadmium concentration of 105,000 x g supernatant was calculated from the ratio of protein concentration to cadmium content of kidney.

Daily Dose of Cadmium	Tryptophan oxygenase [a)]	3-Hydroxy-anthranilate oxygenase [a)]	Quinolinate phosphoribosyl-transferase [a)]	Nicotinamide methyltrans-ferase [a)]	Picolinate carboxylase [b)]
	M	M	M	M	M
0 mg	3.95×10^{-8}	1.07×10^{-8}	1.19×10^{-7}	0.77×10^{-7}	0.42×10^{-9}
0.5 mg	4.20×10^{-7}	1.15×10^{-7}	1.27×10^{-6}	0.82×10^{-6}	3.70×10^{-7}
2.0 mg	1.15×10^{-6}	3.14×10^{-7}	3.36×10^{-6}	2.25×10^{-6}	1.15×10^{-6}
5.0 mg	2.79×10^{-6}	7.62×10^{-7}	8.38×10^{-6}	5.45×10^{-6}	2.26×10^{-6}

a) Mean values were obtained for liver supernatant from each groups (n = 5).
b) Mean values were obtained for kidney from two rats.

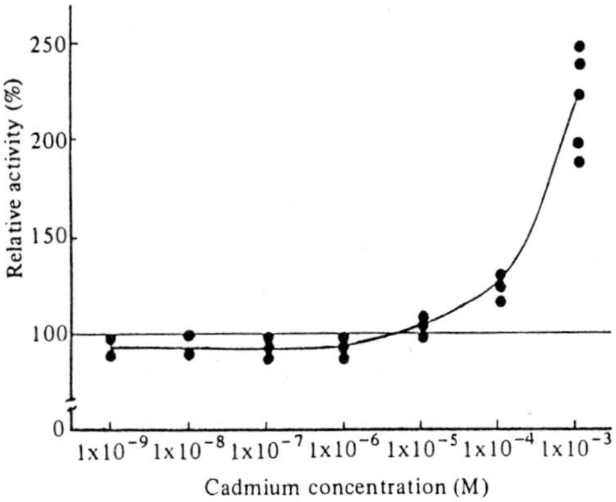

Fig. 1. Effects of varying concentrations of cadmium on tryptophan oxygenase activity. The enzyme activity was determined spectrophotometrically according the method of Prias, et al.

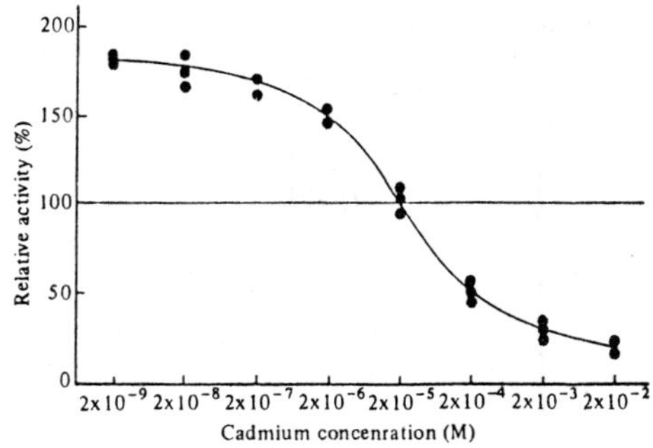

Fig. 3. Effects of varying concentrations of cadmium on quinolinate phosphoribosyltransferase activity. The enzyme activity was determined according to the microradioassay method of Iwai, et al.

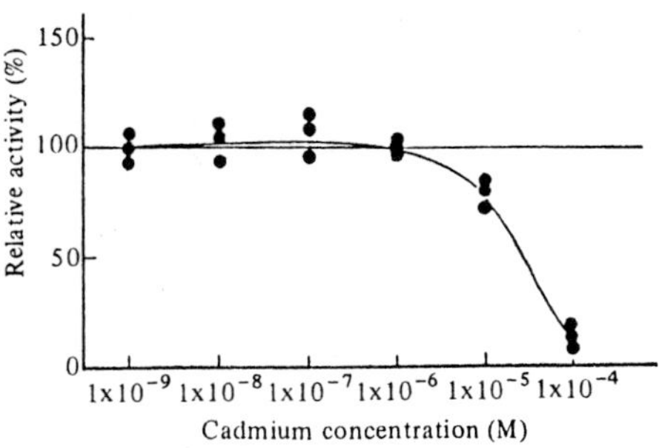

Fig. 2. Effects of varying concentrations of cadmium on 3-hydroxyanthranilate oxygenase activity. The enzyme activity was determined spectrophotometrically according to the method of Ichiyama, et al.

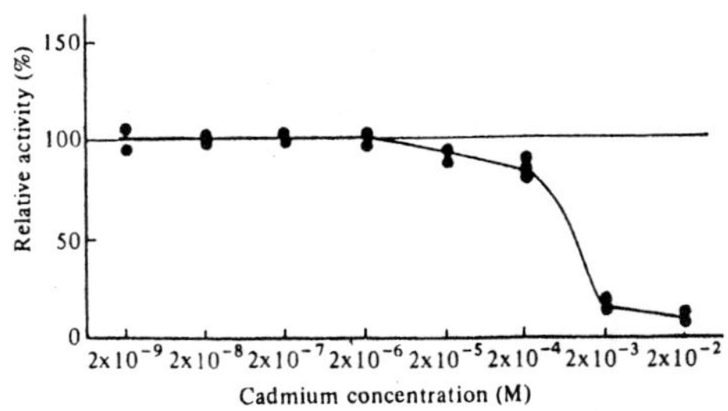

Fig. 4. Effects of varying concentrations of cadmium on nicotinamide methyltransferase activity. The enzyme activity was determined according to modified method of Cantoni.

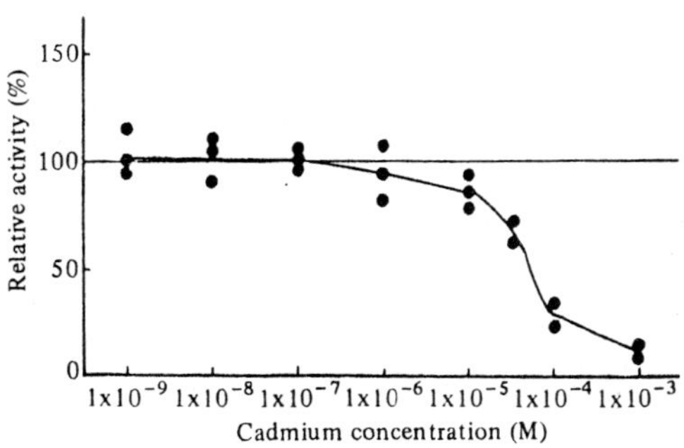

Fig. 5. Effects of varying concentrations of cadmium on picolinate carboxylase activity. The enzyme activity was determined spectrophotometrically according to the method of Ikeda, et al.

References

1. Satyanarayana, U., Narasinga Rao, B.S.: Br. J. Nutr. 38, 39-45 (1977).
2. Satyanarayana, U., Narasinge Rao, B.S.: J. Nutr. 107, 2213-2218 (1977).
3. Krishnaswamy, K., Bapurao,S. : Br. J. Nutr. 39, 61-64 (1978).
4. Kench, J.E., Gubb, P.J.D. : Biochem. J. 120, 270 (1970).
5. Piras, M.M., Knox, W.E. : J. Biol. Chem. 242, 2952-2958 (1967).
6. Ichiyama, A., Nakamura, S., Kawai, H., Honjo, T., Nishizuka, Y., Hayaishi, O. : J. Biol. Chem. 240, 740-749 (1965).
7. Iwai, K., Taguchi, H. : J. Nutr. Sci. Vitaminol. 19, 491-499 (1973).
8. Cantoni, G.L. : J. Biol. Chem. 189, 203-216 (1951).
9. Ikeda, M., Tsuji, H., Nakamura, S., Ichiyama, A., Nishizuka, Y., Hayaishi, O. : J. Biol. Chem. 240, 1395-1401 (1965).
10. Andoh, M., Sayato, Y., Tonomura, M. : Jpn. J. Hyg. Chem. 22, 24-30 (1976).

THE INFLUENCE OF INCREASED AVAILABILITY OF TRYPTOPHAN ON THE FORMATION OF TRYPTAMINE AND SEROTONIN DURING EARLY ONTOGENESIS

Gerald Huether
Max-Planck-Institut für experimentelle Medizin
D-3400 Göttingen, Germany

Tryptophan (Trp) is the precursor of important modulators (tryptamine, serotonin, melatonin) in animals and men. Many metabolic and behavioural functions in mammals are controlled by catecholaminergic and serotoninergic mechanisms (Fig. 1). Most of these functions are thus indirectly dependent on an appropriate supply of Trp to the adult organism. During early developmental stages, these integrative circuits are not yet expressed. It is possible that their elaboration during growth and maturation is dependent on an appropriate supply of Trp during early ontogenesis.

Already the early blastula stage, cells located at the vegetal pole of the sea urchin embryo produce serotonin. The Trp-hydroxylating and decarboxylating pathways are not saturated, and if more Trp is made available to such cells, they will produce more serotonin or tryptamine. It has been suggested (1) that the rapid formation of serotonin at the onset of morphogenesis may be related to the onset of massive collagen synthesis at the expense of amino acids derived from yolk proteins. For collagen synthesis, however, only very little Trp would be used (collagen is poor in Trp) and the free Trp would be available for the formation of serotonin. Increases and fluctuations in the concentrations of indoleamines are characteristic of certain blastula and gastrula stages (2,3). Tryptamine and serotonin play an important role in the process of cell differentiation and morphogenesis during early embryogenesis. Serotonin is involved in the regulation of cell division (cleavage, separation of mitotic centres). Intracellular flow, cell shape, morphogenetic and pulsatory movements, the primary invagination, neurulation, torsion and

Progress in Tryptophan and Serotonin Research

flexure of the embryo have all been shown to be triggered by indoleamines (see for references 1,4,5). Serotonin also causes alterations in the intracellular levels of cyclic nucleotides and in the state of phosphorylation of developing cells and, by doing so, is also involved in the control of transcription and translation of genetic information (6). Even a modulating effect of serotonin on polynucleotide metabolism has been suggested (7), based on the observation that serotonin increases protein synthesis in fibroblasts via stimulation of RNA-synthesis.

In view of these observations, it is not astonishing that the injection of Trp into incubating eggs of the chick has been found to interfere with embryonal brain growth, somitogenesis, the decomposition of yolk granules, or the outgrowth of extraembryonic blastoderm (8,9). If 5-hydroxytryptophan was injected only once into a hen's egg at the third day of incubation, a permanent decrease in brain serotonin concentrations was observed throughout embryogenesis. After hatching, brain serotonin levels were elevated (measured until day 30 (10)). In the developing mammalian embryo, the expression of Trp-hydroxylase can be modified by certain treatments of the mother (decrease by injection of p-chloro-phenylalanine, this decrease is prevented by the additional injection of Trp (11)). An appropriate synthesis of serotonin (with respect to both, the amount and the timing) in the developing brain is a condition of normal brain development. Cell division, neuronal and glial differentiation, cell migration and synaptogenesis in the developing brain are controlled by serotonin ("guidance" function of serotoninergic neurons during brain development (4)).

It is unknown, how the synthesis of indoleamines is regulated during embryogenesis. Alterations in the availability of Trp during early ontogenesis may be generated in a self-regulated process. An generally increased supply of Trp to the developing embryo would then lead to the precocious and/or in -

increased formation of tryptamine and serotonin by the developing cells, and may therefore modify normal development.

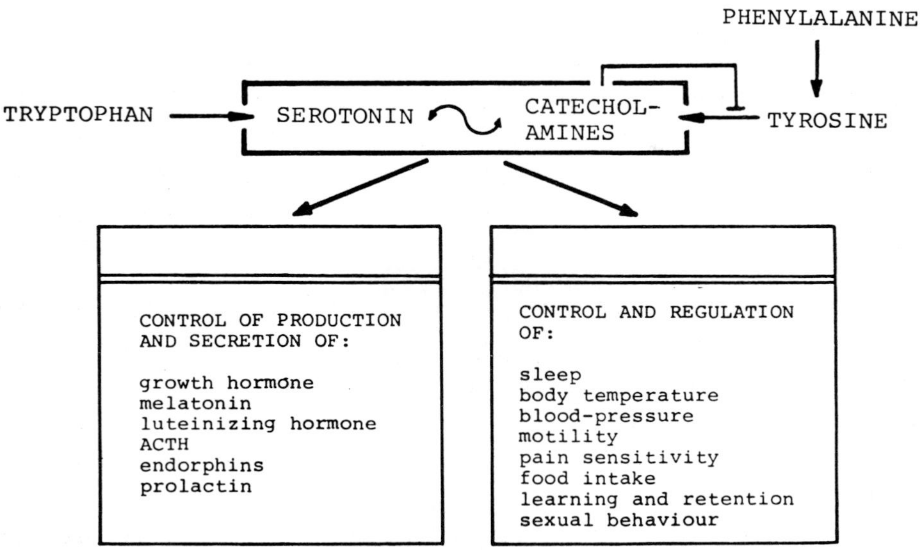

Fig. 1: Many functions in the adult mammal are regulated by the balance of serotoninergic and catecholaminergic mechanisms. The synthesis of catecholamines is controlled by product inhibition. The formation of serotonin depends on the availability of Trp. All functions listed in this figures have been shown to be influenced by alterations in the supply of Trp to the adult mammal (for references see (14)).

To test this assumption, the influence of an increased availability of Trp on the formation of its decarboxylated metabolites was investigated during chick and rat development.

The hen's egg is a virtually closed system. Trp was injected into the subembryonal fluid of eggs incubated at 37.8°C for

24 hrs. For this purpose L-tryptophan was dissolved by sonication in the thin albumen of a parallel incubated egg (10 mg/ml). 200 µl (containing 2 mg Trp, controls: no Trp added) were injected per egg. The accumulation of Trp and of tryptamine was measured at day 3 and day 5 of incubation (Hamburger-Hamilton stages 17/18 and 26, respectively (12)).

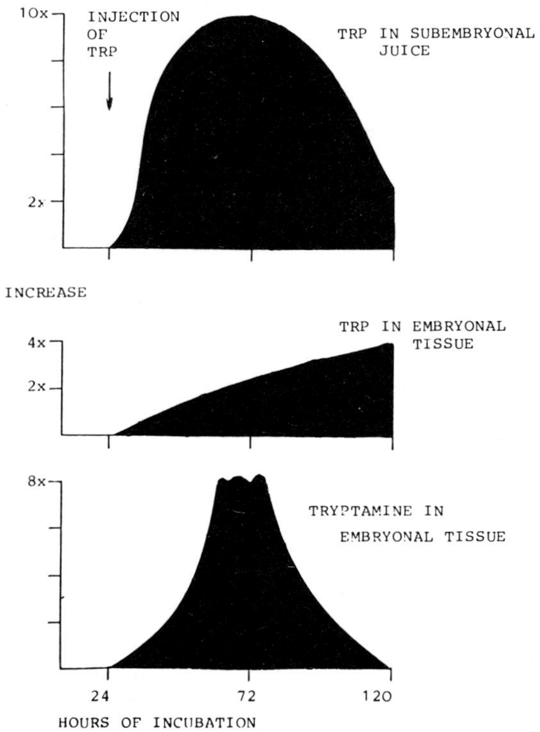

Fig. 2: Accumulation of Trp in the subembryonal fluid and in the embryonal tissue of eggs injected 2 mg Trp in 200 µl thin albumen into the subembryonal juice at 24 hrs. of incubation (Trp was measured by thin layer micro-chromatography of the dansylated amino acids (15)). Initially (day 3) an excess of tryptamine is formed by the developing embryo (measured by HPLC and electrochemical detection, detailed description of the method in (16)). Peak values varied between 10 and 50 times the control levels. N = 6.

At day 3 of incubation, the embryo accumulates Trp not in direct relation to the concentration in the subembryonal juice, however, it converts a certain amount of Trp to tryptamine (see Fig. 2).

At day 5 of incubation, the Trp excess in the subembryonal fluid has almost disappeared (from about 10-fold elevation at day 3 to a 3-fold elevation at day 5). The concentration of tryptamine does not increase any more. Trp is accumulated in the embryonal tissue to a greater extent than it is elevated in the fluid (body 5-fold, head 4-fold).

The altered concentration of Trp in the embryonal tissue reflects the alterations in the steady state of uptake and utilization of this amino acid. Likewise, the alterations in the concentration of tryptamine in the Trp-injected chick embryo reflect the altered steady state of synthesis and degradation of this indoleamine. Evidently, the Trp-decarboxylating pathway is not saturated in the early developing chick embryo. An increased availability of Trp automatically leads to an increased formation of tryptamine. Tryptamine is an phylogenetically old modulator of intercellular communication, affecting the metabolic state and the function of developing cells. The exogenously imposed amino acid imbalance causes not only Trp metabolism to accelerate, protein synthesis and the concentration of free amino acids in the embryonal tissue are also altered by the increased supply of Trp (data to be published elsewhere). It is reasonable to expect that these metabolic responses will lead to the formation of "endogenous imbalances", which influence subsequent processes in the metabolic ontogeny.

The experimental rise in the availability of Trp to the developing chick embryo is only a model to illustrate the metabolic effects on developing cells following an altered supply of essential substrates during early embryogenesis. The question arises, how constant the composition of the egg is normally, e.g. with respect to the absolute amount of Trp, the

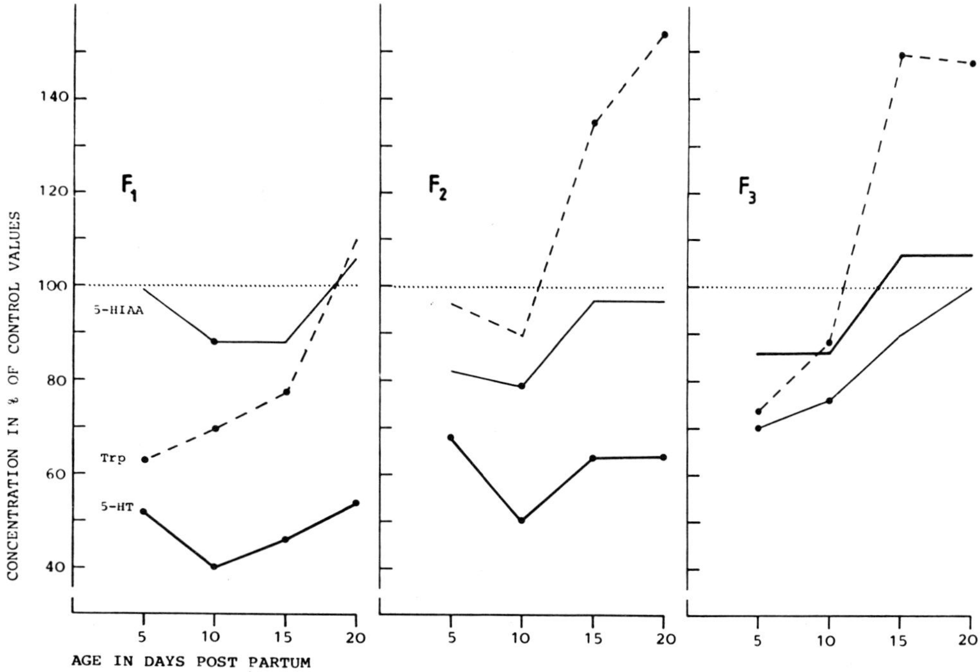

Fig. 3: Alterations in the concentrations of Trp, Serotonin (5-HT) and 5-hydroxyindoleacetic acid (5-HIAA) in the brain of developing rats bred for three consecutive generations on a Trp-enriched diet. The Trp-content in the diet was raised about 4-fold. No alterations of the control values were seen through three generations bred on the normal diet. F_1 = first generation (parents placed on the Trp-enriched diet a fortnight before mating). F_2 = second generation (offspring of F_1, kept on the Trp-enriched diet), F_3 = third generation (offspring of F_2, kept on the Trp-enriched diet). Dots indicate that the difference from the respective control value is statistically significant (t-test, $p < 0.05$). Trp was measured by thin layer micro-chromatography of the dansylated amino acids (15). Indoles were measured by HPLC, electrochemical detection (for detailed method see (16)). The absolute values ± S.D. of the control rats are published elsewhere (16).

ratio between Trp and other amino acids present in the yolk and in the white, and the rate of utilization during embryonal protein synthesis. This question has not found much consideration in the past. According to the best of our knowledge, the amino acid composition has only been investigated in the hen's egg by Rupe and Farmer (13). This study shows that during the first 250 hrs. of incubation, the yolk is the primary source of amino acids for the embryonal protein anabolism. During this period, individual amino acids of the yolk proteins are preferentially utilized by the embryo, until they become exhausted in the yolk-derived subembryonal fluid. After 250 hrs. of incubation, the amino acids of the white are preferentially utilized. Yolk proteins are particularly rich in Lys, Arg, Thr, His and Tyr. Met, Phe, and Trp are available in greater amounts from the proteins of the white.

Interestingly enough, the amino acid composition of the hatched chicken is nearly identic with the amino acid composition of the white (13).

One cannot exclude that shifts in the nutritional supply in a given population will cause shifts in the amount and the proportions of essential substrates (e.g. Trp) made available to the descendants by their parents. It is possible that such shifts during early ontogenesis will influence and modify all processes of metabolic and phenotypic maturation which are dependent on a particular amino acid.

To test this assumption, rats were bred for three consecutive generations on a Trp-enriched diet. For this purpose, Wistar rats were housed and bred under standardized conditions. The powdered, complete standard diet (ssniff R, Alleindiät für Ratten, Versuchstierdiäten GmbH, Soest, Germany) was thoroughly mixed with L-tryptophan, moistened, pelleted, and dried at room temperature. Control diet: no amino acids added, Trp-diet: 10 g Trp added per Kg diet (normal content o.35 g% raised to 1.34 g%, 3.8-fold increase in Trp-content). The

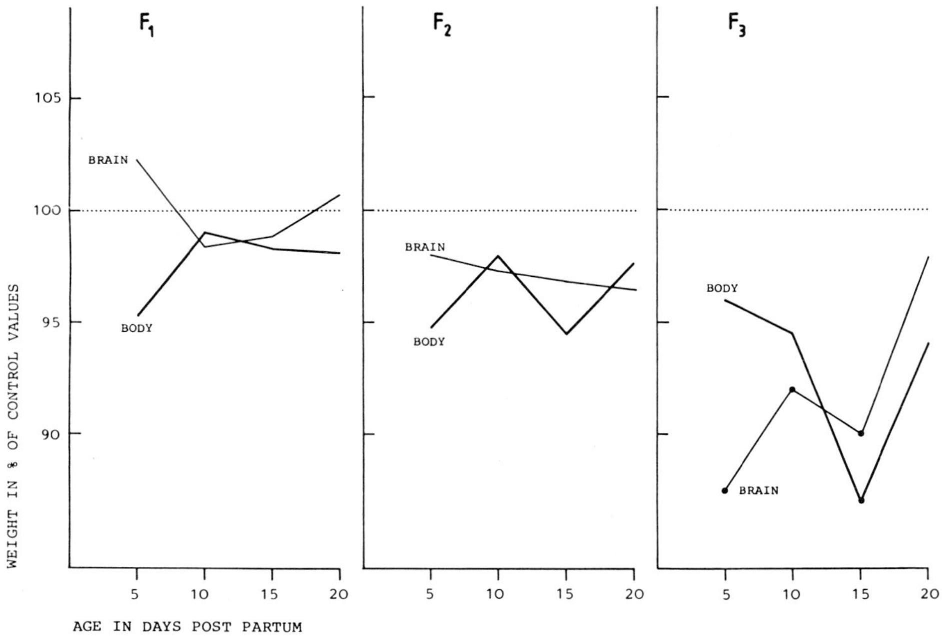

Fig. 4: Alterations in the postnatal weights of the body and of the brain of developing rats bred for three consecutive generations on Trp-enriched diets. For additional details see legend of Fig. 3.

concentrations of Trp, serotonin and 5-hydroxyindoleacetic acid in the brain of each generation's offspring were measured at days 5, 10, 15 and 20 post partum (Fig. 3).

In the first generation (offspring of normal rats placed on the Trp-enriched diet two weeks before mating) the concentration of Trp in the brain was significantly decreased until 15 days of age (i.e. the whole suckling period), and so was the concentration of serotonin in the forebrain. The level of the major serotonin metabolite, 5-hydroxyindoleacetic acid, was significantly decreased only at day 10 p.p. In the offspring of the F_2 generation a recovery from the depletion of Trp and

Table 1: Weight of body and daily food intake of the third generation of rats grown and bred on a Trp-enriched (10 g/kg) diet. The measurements were performed at the age of 4 months for 2 weeks. No alterations of body weights were observed during this interval. Values are means ± S.D., N = 6 (6 males, ♂, 6 females, ♀). Significant differences from the normal male or female values (t-test): *p 0.01

DIET		BODY WEIGHT (g)	FOOD INTAKE (g/day·rat)	(g/day·kg body weight)
Normal	♀	243 ± 9	17.7 ± 0.9	73 ± 4
	♂	433 ± 27	27.0 ± 1.1	62 ± 4
Trp-enriched	♀	208 ± 19*	13.6 ± 2.1*	66 ± 11
	♂	331 ± 12*	22.9 ± 2.8*	70 ± 9

of serotonin in the brain was observed. In the third generation, the serotonin concentration in the forebrain was no longer different from controls. However, in this generation, the initial brain weight and the postnatal weight gain of the body and the brain was significantly decreased (Fig. 4). The adult rats of the third generation were significantly smaller than the control rats bred on normal diets, and so was their daily food intake (Table 1). The statistically significant sex differences in daily food intake per kg body weight in the normal rat population was no longer seen after three generations on the Trp-enriched diet.

In summary, the results indicate that an altered nutritional supply of tryptophan during the development of an individual may cause various metabolic alterations. Such responses seem to last and to become continuously modified through several generations - until a new steady state is finally reached.

In the course of this adjustment, the expression of certain phenotypic parameters may be either activated or suppressed. In nature, an increased availability of Trp to the individuals of a certain population would probably never occur. Here, any nutritional changes would be much more complex, involving simultaneously several substrates and cofactors, generally superimposed by other environmental alterations. Therefore, the present study can only be taken as a model to illustrate that the development of a certain phenotype can be shifted through generations in response to an altered nutritional supply during ontogenesis.

References

1. Gustavson, T., Toneby, M.: Exptl. Cell Res. 62, 102-117 (1970).
2. Baker, P.C.: Acta Embryol Morph. Exp. 8, 197-204 (1965).
3. Buznikov, G.A., Chudakova, J.V., Znezdia, N.D.: Embryol. Exp. Morph. 12, 563-573 (1964).
4. Lauder, J.M., Wallace, J.A., Krebs, H., Petrusz, P., McCarthy, K.: Brain Res. Bull. 9, 605-625 (1982).
5. Buznikov, G.A., Shmukler, Y.B.: Neurochem. Res. 6, 55-68 (1981).
6. Blum, J.J.: Biogenic Amines as Physiologic Regulators, ed. by J.J. Blum. Englewood Cliffs, Prentice-Hall 1970, pp. 95-118.
7. Boncek, R.J., Speropoulos, A.J., Noble, N.L.: Proc. Soc. Exp. Biol. (New York) 140, 599-603 (1972).
8. Ahmad, G., Zamenhof, S.: Life Sci. 22, 963-970 (1978).
9 Emanuelsson, H., Palen, K.: Wilh. Roux Arch. Entw. Mech. Org. 177, 1-17 (1975).
10. Eiduson, S.: J. Neurochem. 13, 923-932 (1966).
11. Lauder, J.M., Sze, P.Y., Krebs, H.: Developmental Neurosci. 4, 291-295 (1981).
12. Hamburger, V., Hamilton H.: J. Morphol. 88, 49-92 (1951).
13. Rupe, C.O., Farmer, C.J.: J. Biol. Chem. 213, 899-906 (1951).
14. Huether, G., Sprotte, U., Thoemke, F., Neuhoff, V.: Funkt. Biol. Med. 2, 18-30 (1983).
15. Schott, K., Huether, G., Neuhoff, V.: Biochem. Med. (in press, 1983).
16. Huether, G., Schott, K., Thoemke, F., Sprotte, U., Steinhaus, U., Neuhoff, V.: Neurochem. Int. (submitted, 1983).

TRYPTOPHAN METABOLISM IN MAMMALS

TRYPTOPHAN METABOLISM AND ITS CONTROL IN MAMMALIAN LIVER

Christopher I. Pogson, Rosario A. Munoz-Clares, Jonathan S. Cook and Stephen A. Smith.

Biological Laboratory, University of Kent, Canterbury, Kent and Department of Biochemistry, University of Manchester, Oxford Road, Manchester, M13 9PL, U.K.

Tryptophan differs from most other aminoacids in the number of pathways through which it may be metabolised. Qualitatively, the synthesis of 5-hydroxytryptamine, tryptamine and indole 3-acetate are of considerable importance. Quantitatively, however, the flux to these products is very small indeed in comparison with the pathway of tryptophan oxidation in the liver (1). Entry into this pathway, the 'Kynurenine pathway', is through the irreversible reaction catalysed by tryptophan 2,3-dioxygenase. Because tryptophan transport across the liver plasma membrane is apparently rapid, tryptophan dioxygenase is a major factor controlling the plasma concentration of tryptophan which is, in turn, important in the regulation of 5-hydroxytryptamine synthesis in the brain. Flux through the kynurenine pathway is therefore crucial in the maintenance of tryptophan homeostasis, and factors affecting this flux physiologically or pharmacologically are *a priori* of interest.

Previous studies of hepatic tryptophan metabolism have been performed both *in vivo* (2,3) and with isolated, perfused livers (4,5). Such studies are, however, constrained by technical and analytical problems. The use of isolated liver cell preparation overcomes many of these difficulties. Large numbers (up to 150) of similar samples may be incubated

Progress in Tryptophan and Serotonin Research

simultaneously for varying periods and in the presence of a range of substrates and putative effectors(6,7).

Tryptophan transport into the liver cell is non-concentrative with no evidence for active uptake mechanisms (4-6). The process is relatively rapid; it is unlikely that it would ever be a limiting factor in the metabolism of the aminoacid (6). Rates of metabolism of the various radiolabelled tryptophans are linear for at least 90 min. after a brief initial lag period. This linearity is not only convenient but also indicates that dilution of specific activity through intracellular proteolysis is negligible.

Tryptophan in plasma occurs both free and bound to albumin(8); at physiological concentrations less than 20% is free (9). The ratio of free to bound is altered by substances that can compete for albumin binding sites, such as fatty acids and various drugs (9,10).

Addition of increasing concentrations of fatty acid-free albumin to cell incubations reveals a progressive inhibition of tryptophan dioxygenase flux, this being most pronounced at low, near-physiological concentrations of the aminoacid (6). Plots of the decreases in flux and free tryptophan in the medium against albumin concentration are similar, suggesting that liver will normally draw only on the free pool for metabolism. At a fixed albumin concentration, addition of palmitate displaces bound tryptophan so that the free concentration increases. This is again reflected in a proportionate increase in catabolic flux (6); the effect could be of some significance in states where blood fatty acid concentrations are elevated (e.g. starvation, diabetes).

In the absence of albumin, half-maximal rates of dioxygenase activity in cells are seen at 0.30mM L-tryptophan, close to

TABLE 1

Effects of nutritional and hormonal factors on tryptophan metabolism in isolated rat liver cells

Cells were prepared and incubated as described in (7). Tryptophan was added 20 min after pyruvate (10mM); the albumin concentration was 2% (w/v). Rates are nmol/h per mg dry wt. ± S.E.M. with the number of independent observations in parentheses. Metabolic flux: I, Through tryptophan dioxygenase; II, Through kynureninase; III, To acetyl CoA. IV, Quinolinate accumulation (III-II). A. 0.1mM tryptophan; B. 2.5mM tryptophan.

	Rats			
Metabolic flux	Fed (n=4)	48h starved (n=3)	Chronic Diabetic (n=5)	Adrenalectomized (n=3)
A				
I	2.71 ± 0.36	4.10 ± 0.40	6.82 ± 0.85	2.27 ± 0.30
II	1.60 ± 0.13(59%)	2.90 ± 0.16(71%)	5.27 ± 0.73(77%)	2.23 ± 0.23(98%)
III	0.70 ± 0.09(26%)	0.90 ± 0.10(22%)	4.84 ± 0.26(71%)	0.70 ± 0.21(31%)
IV	0.90 ± 0.11(33%)	2.00 ± 0.13(49%)	0.43 ± 0.13(6%)	1.53 ± 0.25(67%)
B				
I	19.00 ± 3.00	28.80 ± 2.80	49.84 ± 5.88	15.78 ± 1.56
II	9.20 ± 1.60(48%)	14.90 ± 2.42(52%)	29.86 ± 4.89(60%)	14.82 ± 1.64(94%)
III	1.41 ± 0.60(7%)	2.40 ± 0.42(8%)	20.54 ± 2.30(41%)	1.91 ± 0.34(12%)
IV	7.79 ± 1.24(41%)	12.50 ± 1.50(44%)	9.32 ± 4.08(19%)	12.94 ± 2.16(82%)

the K_m value reported for the isolated enzyme (11). Rates at 2.5mM-L-tryptophan are near maximal; comparison of these with measurements of total enzyme activity in liver homogenates suggests either that all enzyme in cells must be in the holo-form or that activating factors are present in the cell but not in dilute extracts. If the former, then it remains to be resolved whether conversion of apo- to holo-enzyme occurs during cell preparation or is normally complete *in vivo*.

Tryptophan dioxygenase activity in cells is not acutely affected by glucagon, insulin or glucocorticoids added to cell incubations. The enzyme is, however, well-known to be inducible and flux measurements in cells from appropriately pre-treated animals bear this out (Table 1; 6,7,13). Starvation has little effect, especially if results are expressed on the basis of DNA content. Chronic (10 day) streptozotocin-diabetes increases flux by more than 100% as expected (see 12). Adrenalectomy, perhaps surprisingly, does not significantly decrease tryptophan catabolism - a finding which agrees with studies *in vivo* (14).

Cells from fed, starved and diabetic animals exhibit lower flux rates through kynureninase than through the dioxygenase (Table 1). This is consistent with a loss of kynurenine and its transamination products into the medium (5,15), especially at the higher concentration of tryptophan. Accumulation of a number of metabolites could in theory explain the difference between the kynureninase flux and that through acetylCoA. Direct measurement shows, however, that quinolinate alone can account for these results with rat liver cells (7). The build-up of quinolinate is responsible for the well-known inhibition of gluconeogenesis by tryptophan(16), the locus of inhibition being phosphoenolpyruvate carboxykinase(17). Tryptophan is a less potent agent in cells from chronically-diabetic rats (18); this is explained by the diversion of carbon towards acetylCoA

TABLE 2

The effect of Redox state on Tryptophan Dioxygenase Flux

Cells were incubated with 5mM glucose, 5mM L-glutamate, 0.1mM L-tryptamine and 2% (w/v) bovine albumin in Krebs-Henseleit buffer. Ethanol and xylitol were added at 10mM. Flux is expressed as nmol/h per mg dry wt. Results are means from three experiments. I, Total NAD^+/NADH; II, Total $NADP^+$/NADPH; III, Lactate/Pyruvate; IV, Calculated free NAD^+/NADH; V, Flux.

	Control	Ethanol	Xylitol	Ethanol plus xylitol
I	36	9	19	8
II	0.4	0.3	0.4	0.4
III	3.1	22.4	8.7	42.8
IV	2906	402	1035	212
V	2.99	2.51	2.47	1.58

in this condition (Table 1). The most likely reason lies in the induction by diabetes of the first enzyme of the sequence to acetylCoA, picolinate carboxylase (19,20). The conversion of tryptophan to citric acid cycle intermediates should thus be seen as occurring through a series of separate reaction sequences with tryptophan dioxygenase and picolinate carboxylase as significant regulators of flux.

Tryptophan inhibits gluconeogenesis in mouse as well as rat liver(21); it is considerably less effective, however, in the gerbil and totally ineffective in the guinea pig and sheep (22). To what extent this interaction is important in other species, including man, is not known.

Tryptophan dioxygenase is inhibited allosterically by the reduced pyridine nucleotides; it has been suggested that this inhibition may be a physiological control on catabolic flux (23). If this were the case, then tryptophan metabolism should vary with changes in cytosolic redox state. We have performed experiments to test this possibility. Ethanol and xylitol both decrease the total NAD^+/NADH and the cytosolic free NAD^+/NADH ratios (Table 2). The changes in dioxygenase activity are only significant when both ethanol and xylitol are present, and do not correlate well with changes in cellular redox state. Calculations show that the concentration of free NADH in the cytosol is unlikely to be much more than 10μM; this is too low for it to be effective as an inhibitor(23).

In vivo, pyrazole is converted to 4-hydroxypyrazole which is a potent inhibitor of isolated tryptophan dioxygenase (24). In liver cells, flux through the dioxygenase is decreased by 50% at 0.15mM-4-hydroxypyrazole(25). Dioxygenase flux is calculated as the sum of $^{14}CO_2$ and ^{14}C-labelled aliphatic products (e.g. formate) in the medium. The ratio of these two varies with the absolute rate, the percentage of CO_2 being higher at low flux rates. We noted that 4-hydroxypyrazole not only inhibits the dioxygenase but also greatly decreases the $^{14}CO_2$/[^{14}C] aliphatic products ratio. Experiments with [^{14}C]formate suggested that the inhibitor acts by blocking catalase, another heme-containing enzyme, which can oxidise formate. When methionine is present, the relative amount of $^{14}CO_2$ is greatly increased, and 4-hydroxypyrazole has no effect on the ratio. The stimulation of formate oxidation is explained by the increased availability of tetrahydrofolate, released from the 'methyl trap' by methionine (26). Because methionine is normally present *in vivo*, we conclude that catalase probably plays little or no role in the conversion of tryptophan- derived formate to CO_2 (25).

The isolated cell incubation system is particularly convenient for screening the actions of compounds of pharmacological interest. Carbidopa and benserazide are clinically-useful inhibitors of L-aromatic aminoacid decarboxylase. Because they are effective through their ability to react with pyridoxal phosphate, it was not unexpected that they would inhibit tryptophan oxidation. Carbidopa in particular decreases kynureninase flux by 35% at 50μM, and inhibits the dioxygenase at concentrations about 0.1mM. In addition, oxidation of alanine (derived from the kynurenine side chain) was strongly inhibited. Benserazide produces generally similar patterns but is less effective (27). These results highlight the problems of interpretation of experiments with whole animals solely involving measurement of CO_2 formation from tryptophan.

6-Chlorotryptophan is converted in the liver to 4-chloro-3-hydroxyanthranilate, a potent inhibitor of 3-hydroxyanthranilate oxidase in vitro (28). Exposure of liver cells to 6-chloro-tryptophan has no effect on flux from tryptophan to hydroxy-anthranilate, but decreases the accumulation of quinolinate, as evidenced by the diminished sensitivity of gluconeogenesis to tryptophan. Pyrazinamide is metabolized to a strong inhibitor of picolinate carboxylase (29). Measurements of tryptophan metabolism in cells incubated with pyrazinamide confirm this, showing decreased flux through the lower part of the catabolic sequence. This block at the level of carboxylase is again reflected in accumulation of quinolinate. In this instance, gluconeogenesis becomes more sensitive to tryptophan because of the greater net conversion of tryptophan to quinolinate.

Acknowledgements: We are grateful to the Medical Research Council, the Science & Engineering Research Council, the British Diabetic Association, the Smith Kline & French Foundation, and the European Molecular Biology Organisation.

References

1. Young, S.N., St.Arnaud-McKenzie, D., Sourkes, T.L.: Biochem. Pharmacol. 27, 763-767 (1978).
2. Young, S.N., Oravec, M., Sourkes, T.L.: J. Biol. Chem. 249, 3932-3936 (1974).
3. Joseph, M.H., Young, S.N., Curzon, G.: Biochem. Pharmacol. 25, 2599-2604 (1976).

4. Ng, C.Y., Hagino, Y., Swan, P.B., Henderson, L.M.: J. Nutr. 99, 465-473 (1970).
5. Green, A.R., Woods, H.F., Joseph, M.H.: Br. J. Pharmacol. 57, 103-114 (1976)
6. Smith, S.A., Pogson, C.I.: Biochem. J. 186, 977-986 (1980).
7. Smith, S.A., Carr, F.P.A., Pogson, C.I.: Biochem. J. 192, 673-686 (1980).
8. McMenamy, R.H., Lund, C.C., Van Marike, J., Oncley, J.L.: Arch. Biochem. Biophys. 93, 135-139 (1961).
9. Curzon, G., Knott, P.J.: Br. J. Pharmacol. 50, 197-204 (1974).
10. Curzon, G., Friedel, J., Knott, P.J.: Nature 242, 198-200 (1973).
11. Munoz-Clares, R.A., Cook, J.S., Smith, S.A., Pogson, C.I.: FEBS Lett. 117, 265-268 (1980).
12. Smith, S.A., Marston, F.A.O., Dickson, A.J., Pogson, C.I.: Biochem. Pharmacol. 28, 1645-1651 (1979).
13. Smith, S.A., Pogson, C.I.: Biochem. J. 200, 605-609 (1981).
14. Madras, B.K., Sourkes, T.L.: Arch. Biochem. Biophys. 125, 829-836 (1968).
15. Yeh, J.K., Brown, R.R.: J. Nutr. 107, 261-271 (1977).
16. Ray, P.D., Foster, D.O., Lardy, H.A.: J. Biol. Chem. 241, 3904-3908 (1966).
17. Snoke, R.E., Johnston, J.B., Lardy, H.A.: Eur. J. Biochem. 24, 342-346 (1971).
18. Smith, S.A., Elliott, K.R.F., Pogson, C.I.: Biochem. J. 176, 817-825 (1978).
19. Mehler, A.H., McDaniel, E.G., Hundley, J.M.: J. Biol. Chem. 232, 323-330 (1958).
20. Ikeda, M., Tsuji, H., Nakamura, S., Ichiyama, A., Nishizuka, Y., Hayaishi, O.: J. Biol. Chem. 240, 1395-1401 (1965).
21. Pogson, C.I., Crisp, D.M., Smith, S.A.: Acta. Vitaminol. Enzymol. 29, 323-335 (1975).
22. Munoz-Clares, R.A., Lloyd, P., Lomax, M.A., Smith, S.A., Pogson, C.I.: Arch. Biochem. Biophys. 209, 713-713 (1981).
23. Cho-Chung, Y.S., Pitot, H.C.: J. Biol. Chem. 242, 1192-1198 (1967).
24. Rouach, H., Ribiere, C., Nordmann, J., Nordmann, R.: FEBS Lett. 101, 149-152 (1979).
25. Cook, J.S., Pogson, C.I.: Biochem. J. 204, 307-312 (1982).
26. Billings, R.E., Noker, P.E., Tephly, T.R.: Arch. Biochem. Biophys. 208, 108-120 (1981).
27. Smith, S.A., Pogson, C.I.: Biochem. Pharmacol. 30, 623-628 (1981).
28. Parli, C.J., Krieter, P., Schmidt, B.: Arch. Biochem. Biophys. 203, 161-166 (1980).
29. Nasu, S., Yamaguchi, K., Sakakibara, S., Imai, H., Ueda, I.: Biochim. Biophys. Acta. 677, 109-119 (1981).

TRYPTOPHAN METABOLISM BY THE ISOLATED, PERFUSED LIVER AND KIDNEY OF RAT
(Quantification of the Metabolism by the Compartmental Analysis)

Hidetsugu Otsuka, Sadaki Nomura, Fumio Takeuchi, Ryoko Yokomine
Hitoshi Fujimoto, Miyako Ugata and Yukio Shibata
Department of Biochemistry, Aichi Medical University
Nagakute, Aichi 480-11, Japan

Introduction

Considering the signification of metabolic reactions in organisms, it is important to appreciate *"quantitative assessment of metabolic flux"*, *"distribution of metabolic flux in competitive pathway"* and *"control and regulation of metabolic flux"*. Quantitative assessment of tryptophan metabolism in steady state has not been well documented though various types of experiments have been reported. We tried to study quantitative assessment of tryptophan metabolism *via* kynurenine pathway and distribution of its metabolic flux in normal and vitamin B_6-deficient rat liver and kidney by the method of compartment analysis.

Methods

Perfusion Technique —— Perfusion of the isolated rat liver and kidney were carried out according to the techniques of Miller *et al.*(1) and Weiss *et al.* (2), respectively. The perfusion medium was Krebs-Ringer's bicarbonate buffer containing 10 mM glucose (3). The medium was gassed with a mixture of 95% O_2 and 5% CO_2. A perfusion period of 30-min with the medium containing L-tryptophan(0.1, 0.5, 1.0 mM) or L-kynurenine(25, 50, 500 μM) was allowed at 37° C before the start of the experiment. The perfusate was not circulated for 30-min pre-perfusion. After stabilizing, 100 ml of the medium for liver and 20 ml for kidney were perfused circularly. The flow rate was maintained constantly 2-3 ml/min per g tissue.
Analysis of perfusion Medium —— Quantitative analysis of L-tryptophan and its metabolites in perfusate was carried out with a reversed-phase high-performance liquid chromatography method described by us (4).

Progress in Tryptophan and Serotonin Research

Vitamin B_6-deficient Animals —— Male albino rats of Wistar strain aged 4 weeks, weighing 80-90 g, were fed *ad lib.* on vitamin B_6-free chow for 60 days. Vitamin B_6-deficient rats, weighing 170-230 g, were used for the experiment.

Results

Perfusion of L-Tryptophan into Liver —— Among L-tryptophan metabolites *via* kynurenine pathway, exudation of L-kynurenine, kynurenic acid and anthranilic acid were at a detectable level during a perfusion period of 60-min, but exudation of 3-hydroxy-L-kynurenine, 3-hydroxyanthranilic acid, 5-hydroxyanthranilic acid and xanthurenic acid were at a barely detectable level (data are not given). Disappearant rate of L-tryptophan and cumulative rate of L-kynurenine are shown in Table I.

Table I. Cumulative rate of L-kynurenine in perfusate

Initial Concentration in Perfusate $[Trp_{out}]$ (mM)	$\frac{d[Kyn_{out}]}{dt}$ (nmol / min / g Liver)		$\frac{d[Trp_{out}]}{dt}$ (nmol / min / g Liver)	
	Normal	Vitamin B_6 -deficiency	Normal	Vitamin B_6 -deficiency
1.0	4.52	25.87	121.40	161.48
0.5	1.50	11.74	59.56	98.68
0.1	0.37	2.54	19.21	52.11

Perfusion of L-kynurenine into liver —— Exudation of kynurenic acid and anthranilic acid were at a detectable level, but 3-hydroxy-L-kynurenine, 3-hydroxyanthranilic acid, 5-hydroxyanthranilic acid and xanthurenic acid were at a barely detectable level (data are not given). Disappearant rate of L-kynurenine from the perfusate is shown in Table II.

Table II. Disappearant rate of L-kynurenine in perfusate

Initial Concentration in Perfusate $[Kyn_{out}]$ (μM)	$\frac{d[Kyn_{out}]}{dt}$ (nmol / min / g Liver)	
	Normal	Vitamin B_6 -deficiency
500	42.00	110.37
50	8.30	13.92
25	4.37	7.41

Activities of the enzymes in Liver —— Activities of enzymes were assayed in liver homogenate. Activities of tryptophan pyrrolase and 3-hydroxyanthranilate dioxygenase per g tissue showed two-fold increase in vitamin B_6-deficiency (data are not given).

Perfusion of L-kynurenine into kidney —— Among L-kynurenine metabolites, exudation of anthranilic, kynurenic and xanthurenic acid were at a detectable level, but exudation of 3-hydroxy-L-kynurenine, 3-hydroxyanthranilic acid and 5-hydroxyanthranilic acid were at a barely detectable level in perfusate during a perfusion period of 60-min. Urinary metabolites during perfusion showed a similar tendency as mentioned above (data are not given).

Execution of Compartmental Analysis —— Simplified exchange of substances in two compartments, such as between perfusate and liver cell, is shown in scheme 1. The rate of L-kynurenine appearance in perfusate on perfusion of

Scheme 1.

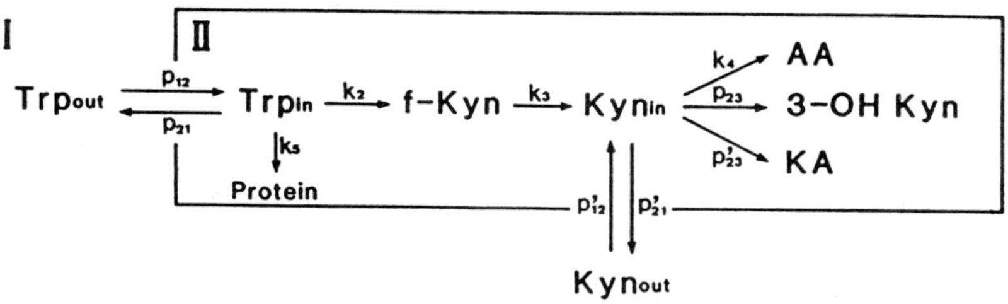

L-tryptophan is expressed as follows (5):

$$\frac{d[kyn_{out}]}{dt} = \frac{P'_{21}}{L_1}[kyn_{in}] \qquad (1)$$

The rate of L-kynurenine disappearance from perfusate on perfusion of L-kynurenine is expressed as follows (5):

$$\frac{d[kyn_{out}]}{dt} = \frac{P'_{12}}{L_1}[kyn_{out}] \qquad (2)$$

About 10 g of liver were perfused with 100 ml of medium. Therefore, L_1 and L_2 were regarded as 10 and 1 cm^3, respectively. And it was supposed that P'_{12} approximated equally to P'_{21}. The simultaneous equations (1) and (2) were solved using the experimental results shown in Table I and II. And then, solutions of P' for normal and vitamin B_6-deficient rat liver were given by 1.7×10^{-2} and 2.9×10^{-2} $cm^3.min^{-1}$, respectively. The estimated concentration of L-kynurenine in liver and calculated flux of enzymes are shown in Table III.

Concentration of L-kynurenine in liver —— Correlation between L-tryptophan concentration in perfusate and L-kynurenine concentration in liver is shown in Fig 1. Polynominal regression analysis was carried out by computer. The experimental results well fitted are as follows:

Normal;

$$f(X) = 11.033 + 106.53X - 12.584X^2 + 301.05X^3 - 104.44X^4$$

Table III . Fluxes of some enzymes related to L-kynurenine metabolism

	[Trp_{out}] (mM)	[Kyn_{in}] (μM)	Activity of Enzymes (nmol / min / g Liver)		
			Kynureninase	Kynurenine 3-hydroxylase	Kynurenine aminotransferase
Normal	0.1	21.8	2.83	23.80	1.04
			(0.12 :	1 :	0.04)
	1.0	265.7	31.00	47.55	11.96
			(0.65 :	1 :	0.25)
Vitamin B_6 -deficiency	0.1	85.5	7.02	36.94	3.68
			(0.19 :	1 :	0.10)
	1.0	900.8	20.49	46.46	35.52
			(0.44 :	1 :	0.76)

Vitamin B_6-deficiency;

$$f(X)= 25.833 + 636.19X - 363.45X^2 + 2297.2X^3 - 2754.7X^4 + 1051.3X^5$$

X; Concentration of L-tryptophan in perfusate (mM),

f(X); Concentration of L-kynurenine in liver (μM).

Fig 1. o;Normal, •;V.B_6-def.

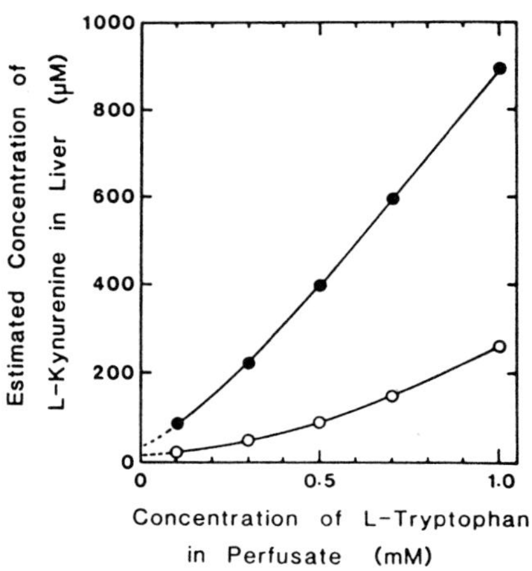

Concentration of free-L-tryptophan in blood was reported to be approximately to 0.01 mM. Substituting 0.01 mM for X, L-kynurenine concentration in liver, f(X), for normal and vitamin B_6-deficiency were obtained by 12.1 and 32.2 μM, respectively. It was derived by compartmental analysis that L-kynurenine concentration in liver of vitamin B_6-deficient animals increased. This result corresponded with the fact of two-fold activation of hepatic tryptophan pyrrolase in vitamin B_6-deficient animals. Xanthurenic acid formation in short-time perfusion was dependent upon kidney.

References

1. Miller, L.L., *et al.*: J. Exp. Med. 94, 431-453 (1951).
2. Weiss, C., *et al.*: Am. J. Physiol. 196, 1115-1118 (1959).
3. Williamson, J.R., *et al.*: J. Biol. Chem. 244, 4607-4616 (1969).
4. Otsuka, H., *et al.*: J. Aichi Med. Ass. 8, 221-224 (1980).
5. Simon, W.: Mathematical Techniques for Physiology and Medicine, PP. 84, Academic Press/New York, London 1972.

EFFECTS OF OESTROGENS ON THE METABOLISM OF TRYPTOPHAN: IMPLICATIONS FOR THE INTERPRETATION OF THE TRYPTOPHAN LOAD TEST FOR VITAMIN B_6 NUTRITIONAL STATUS.

David A. Bender
Courtauld Institute of Biochemistry, The Middlesex Hospital Medical School
London W1P 7PN, U.K.

Introduction

The ability to metabolise a test dose of tryptophan (the tryptophan load test) is widely accepted as the most convenient and sensitive method of assessing vitamin B_6 nutritional and metabolic status. The test depends on the fact that kynureninase is a pyridoxal phosphate dependent enzyme, and therefore in vitamin B_6 deficiency its activity is impaired, so that onward metabolism is inhibited and abnormally large amounts of kynurenine, hydroxykynurenine, kynurenic acid and xanthurenic acid are excreted.

A large number of studies (eg Rose & Braidman, 1971) have suggested that oestrogens may induce vitamin B_6 deficiency, since in women receiving oestrogens as oral contraceptives or as menopausal hormone replacement therapy there is impaired metabolism of a tryptophan load. However, other indices of vitamin B_6 nutritional status, such as the ability to metabolise a test dose of methionine, the urinary excretion of the end-product of vitamin B_6 metabolism (4-pyridoxic acid), the plasma concentration of pyridoxal phosphate and the activation of erythrocyte aspartate aminotransferase by added pyridoxal phosphate are generally unaffected by oestrogen administration.

A previous study from this laboratory (Bender & Wynick, 1981) demonstrated that kynureninase is directly inhibited by oestrogen conjugates in a manner that is competitive with respect to the substrate and uncompetitive

Progress in Tryptophan and Serotonin Research

with respect to the cofactor, pyridoxal phosphate. It was suggested that such inhibition would account for the observations of abnormal tryptophan metabolism in women receiving oestrogens without the need to postulate oestrogen-induced vitamin B_6 deficiency.

Patnaik & Sarangai (1980) claimed to have demonstrated induction of tryptophan oxygenase by oestradiol. However, Bender et al (1983) were unable to reproduce their findings, and indeed showed no increase in the activity of this enzyme, or even a slight fall in activity, after the administration of oestradiol to rats in a variety of conditions, including ovariectomised and adrenalectomised animals, and those which had received a dose of tryptophan before the oestrogen. They concluded that tryptophan oxygenase is not induced by oestrogens, and that the apparent induction reported by Patnaik & Sarangai (1980) was the result of a methodological artefact, due to the inhibition of kynureninase by oestrogen metabolites, and hence increased recovery of the product of tryptophan oxygenase action.

Methods

The metabolism of tryptophan was assessed in vivo in rats by measuring the production of $^{14}CO_2$ after the administration of (^{14}C)tryptophan. The label from (ring-2-^{14}C)tryptophan is released (indirectly) as $^{14}CO_2$ as the result of tryptophan oxygenase action, and that from (methylene-^{14}C)tryptophan as the result of kynureninase action.

Female rats received an intraperitoneal injection of (^{14}C)tryptophan and the production of $^{14}CO_2$ was measured at 10 minute intervals over 4 hours. Each animal received a dose of (^{14}C)tryptophan on two occasions, once after an injection of 5% ethanol:saline and on the following day one hour after the administration of 500 μg oestradiol/kg body weight in 5% ethanol:saline.

Results

The administration of oestradiol led to inhibition of the production of

$^{14}CO_2$ from (ring-2-^{14}C)tryptophan, indicating inhibition of tryptophan oxygenase:

	Control		+ oestradiol		
	mean	sem	mean	sem	
total $^{14}CO_2$ in 4h (10^3 dpm)	28.8	4.1	26.6	3.5	
min to peak of $^{14}CO_2$ production	60		80		
peak height (10^3 dpm in 10 min)	3.36	0.63	2.59	0.26	
rate constant for decay (min^{-1} x10^{-3})	18.54	0.52	15.9	0.71	**

The administration of oestradiol led to a delay in the time of the peak production of $^{14}CO_2$ from (methylene-^{14}C)tryptophan, but no diminution in the total amount of $^{14}CO_2$ produced, an effect that is compatible with competitive inhibition of kynureninase by oestradiol or a metabolite:

	Control		+ oestradiol		
	mean	sem	mean	sem	
total $^{14}CO_2$ in 4h (10^3 dpm)	24.6	4.1	26.6	3.5	
min to peak of $^{14}CO_2$ production	80		110		
peak height (10^3 dpm in 10 min)	2.18	0.18	2.07	0.67	
rate constant for decay (min^{-1} x10^{-3})	13.22	0.12	12.03	0.19	***

(Significance of differences from control, by t-test:

** $0.01 > P > 0.001$, *** $P < 0.001$)

These results confirm that there is inhibition rather than induction of tryptophan oxygenase as a result of oestrogen administration, and that the inhibition of kynureninase by oestrogen metabolites that has been reported previously (Bender & Wynick, 1981) also occurs _in vivo_.

More importantly, these results suggest that in the case of women who are receiving oestrogens the tryptophan load test may not be a reliable indicator of vitamin B_6 nutritional status, since there is a direct effect of oestrogens or their metabolites on kynureninase, the enzyme that is the basis of the tryptophan load test.

References

Bender, D.A. & Wynick, D. (1981) Brit. J. Nutr. 45 269
Bender, D.A., Laing, A.E., Vale, J.A., Papadaki, L. & Pugh, M. (1983) Biochem. Pharmacol. - in press
Patnaik, S.K. & Sarangai, S. (1980) J. Biochem.(Tokyo) 87 1249
Rose, D.P. & Braidman, I. (1971) Amer. J. Clin, Nutr. 24 673

This study was supported by a grant from the Medical Research Council of Great Britain.

THE FUNCTIONS AND REGULATION OF TRYPTOPHAN PYRROLASE

Abdulla A.-B. Badawy

South Glamorgan Health Authority, Addiction Unit Research Laboratory, Whitchurch Hospital, Cardiff CF4 7XB, Wales, U.K.

Tryptophan pyrrolase (TP: tryptophan 2,3-dioxygenase, EC 1.13.11.11) is the haem-dependent liver cytosolic enzyme that catalyses the first and rate-limiting step of the kynurenine-nicotinic acid pathway of tryptophan degradation. TP is regulated by at least three mechanisms: (a) hormonal induction by glucocorticoids; (b) substrate activation and stabilization by tryptophan; (c) cofactor activation by haem. The nature of, and differences between, these three mechanisms will be discussed, as will be evidence that TP activation by its amino acid substrate may involve an action of tryptophan on the haem-biosynthetic pathway. The above regulatory mechanisms appear to be involved in, or to be designed to facilitate, the following functions of TP: (a) detoxication of tryptophan; (b) utilization of haem; (c) control of tryptophan availability for its hepatic and extrahepatic functions. Functions in which the role of TP is established are synthesis of liver NAD^+ and brain 5-hydroxytryptamine. Strong evidence is now emerging that both tryptophan and TP may play important roles in the regulation of hepatic haem biosynthesis through control of availability of the so-called regulatory-haem pool. The possible role of TP in the control of tryptophan availability for other functions or effects, such as protein synthesis or regulation of gluconeogenesis or foetal development, requires investigation. It is suggested that TP is an important regulatory enzyme capable of influencing a variety of physiological processes.

Introduction

Tryptophan pyrrolase (TP: tryptophan 2,3-dioxygenase, EC 1.13.11.11) is the haem-dependent liver cytosolic enzyme catalyzing the oxidative cleavage of the pyrrole moiety of L-tryptophan during the first and rate-limiting step of the kynurenine-nicotinic acid pathway of tryptophan degradation. This enzyme has been the subject of many investigations that have led to our present knowledge of some of the mechanisms involved in enzyme regulation. The functions and regulation of TP have been reviewed by the author in 1977 (1). The present review will include only a short account

Progress in Tryptophan and Serotonin Research

of the differences between the TP regulatory mechanisms and of recent evidence on the nature of the substrate mechanism, but will concentrate more on a discussion of the functions of the enzyme, particularly in relation to synthesis of liver NAD^+ and haem. The need for further and extensive investigations of the possible role of TP in a number of other physiological processes will also be emphasized. Before discussing TP regulation and functions, it may be useful to describe briefly the activity of the enzyme.

Tryptophan pyrrolase activity

In the liver of man, rat and certain, but not all, other animal species [see ref. (2) and references cited therein], TP exists in at least two forms. The already active reduced holoenzyme does not require the addition of haematin for demonstration of its activity _in vitro_ (3), whereas the haem-free predominant form (apoenzyme) does. The activation of the rat liver enzyme _in vitro_ consists of two steps, both of which require the participation of tryptophan. The inactive apoenzyme is first conjugated with haem to form the oxidized holoenzyme. This is then reduced to the active form. Oxidation and reduction here have always been assumed to involve the redox state of the haem iron. Evidence, however, suggests that the catalytically active enzyme exists in two redox states [see (4) and references cited therein]: a cuprous ferri-haem form and a cuprous ferro-haem form. In assaying tryptophan pyrrolase _in vitro_, measurements of both the apoenzyme and holoenzyme activities are important, not only to find out which form(s) is inhibited or activated by a particular agent or treatment, but also to examine the presence or absence of the enzyme [see also (2)]. The significance of the absence of the TP apoenzyme from certain animal species will be discussed later on.

Regulation of tryptophan pyrrolase

By using the catalytic property of the rat liver enzyme, labelled amino acid incorporation into its purified protein and immunological titration techniques, it has been possible clearly and conclusively to distinguish between two mechanisms (5,6,7). The hormonal induction mechanism involves induction of synthesis of new apoenzyme, probably by increasing the amount of TP m-RNA (8), whereas the substrate mechanism consists of decreased degradation of pre-existing apo-(TP) in the presence of the normal rate of its synthesis. It has been suggested (6) that tryptophan stabilizes TP by causing an initial activation of its apoenzyme in vivo [for this effect, see also (9,10)]. This is unlikely, because simple activation (by administration to rats of haematin or 5-aminolaevulinate) does not stabilize the enzyme (10). That stabilization requires the presence of tryptophan itself, rather than changes in TP activity or in saturation of the enzyme with haem is suggested by the finding (10) that stabilization is achieved by a dose of tryptophan (50mg/kg body wt.) that does not enhance or activate TP in the fed rat, irrespective of whether the enzyme is in the basal state or has previously been induced or activated.

Since the demonstration that rat liver TP can be activated by administration of 5-aminolaevulinate (11,12,13) or haematin (10), considerable evidence that the enzyme is regulated by its haem cofactor has been obtained under a variety of experimental conditions involving altered haem metabolism which will be discussed later on in relation to haem utilization by TP. As mentioned above, TP activation by its cofactor does not lead to stabilization of the enzyme (10).

The possible mechanism(s) by which tryptophan causes the initial and the overall activation of TP merits discussion here. Because tryptophan does not enhance TP synthesis (5), it was suggested (10) that the ability of translational inhibitors to block the tryptophan activation of TP must involve prevention of an effect of tryptophan on synthesis of a substance other than apo-(TP), possibly haem itself. A mechanism of action of tryptophan involving enhanced haem synthesis is certainly compatible with the two earlier findings that tryptophan administration to rats increases the concentration of liver microsomal haem (9) and the activity of the rate-limiting enzyme of haem biosynthesis, 5-aminolaevulinate synthase (14). We have, however, failed to confirm this latter finding, but have, on the contrary, demonstrated (15) that tryptophan decreases liver 5-aminolaevulinate synthase activity in both sexes of two different strains of rats. Such a decrease by tryptophan was, however, not unexpected, because of strong evidence (reviewed below) that rat liver 5-aminolaevulinate synthase activity exhibits an inverse relationship with the haem saturation of TP. The tryptophan-induced increase in this saturation, which is sensitive to translational inhibitors (5), must, therefore, involve an action of tryptophan on haem biosynthesis exerted at a step(s) beyond that catalyzed by 5-aminolaevulinate synthase. Evidence supporting this possibility is provided by the findings (15) that the tryptophan-induced increase in the haem saturation of TP and the concomitant decrease in 5-aminolaevulinate synthase activity can be prevented by Pb^{2+}, and that tryptophan itself can potentiate the similar reciprocal effects on these two parameters caused by administration of 5-aminolaevulinate. These results therefore suggest that tryptophan may promote the conversion of 5-aminolaevulinate into haem, possibly by enhancing 5-aminolaevulinate dehydratase activity. The precise step(s) at which tryptophan acts clearly requires investigation.

Some evidence has been obtained in support of possible regulation of TP by other mechanisms, such as feedback control by NAD(P)H (16,17) and coregulation by insulin (18). These mechanisms and also possible regulation by other hormones require detailed investigation. A summary of similarities and differences between the three major mechanisms of TP regulation is given in Table 1.

Table 1. Comparison of the Tryptophan Pyrrolase Regulatory Mechanisms

Aspect	Mechanism .... Hormonal	Substrate	Cofactor
Effector	Glucocorticoids	L-Tryptophan	Haem
Early saturation with haem	–	+	+
General saturation with haem	–	+	+
Haem-saturation ratio	< 1	> 1	> 1
Increased haem concn. or synthesis	–	+	+
Stabilization of TP	–	+	–
Increased synthesis of TP	(++++)	(++)	(+)

(+) to (++++), Arbitrary scale of increase; +, positive effect or presence; –, negative effect or absence. The haem-saturation ratio is the ratio of holoenzyme activity/apoenzyme activity. This Table is modified from (19) by permission.

Functions of tryptophan pyrrolase

Once the TP reaction has occurred, it could be considered that three major events have taken place: (a) degradation of tryptophan, and, hence, (b) limitation (or control) of its availability; (c) utilization of haem. These three functions of TP have important implications to a number of physiological processes, and the following is an account of the role of TP in this context.

Degradation of tryptophan as a detoxication mechanism: The importance of the liver in tryptophan degradation is illustrated by the finding in the rat (20) that the liver can degrade all its tryptophan every 7.5min. The hepatic kynurenine-nicotinic acid is therefore the quantitatively most important pathway of tryptophan degradation in the rat as well as in man, and the same is true for certain, but not all, other animal species (2). Species in which the above pathway is deficient are known to be sensitive to tryptophan toxicity and to possess neither the apoenzyme form of TP nor its hormonal induction mechanism [see (2) and references cited therein]. It is noteworthy that absence of the hormonal induction mechanism from the rat following adrenalectomy renders this species sensitive to tryptophan toxicity (6), and that, in species sensitive to this toxicity, tryptophan administration causes a very rapid activation of TP and is effective in achieving this at much smaller dose levels as compared with the responses of the rat enzyme (2). It may therefore be suggested that degradation of tryptophan by TP could be considered a detoxication mechanism designed to decrease the production of toxic metabolites formed by alternative routes. This concept may have important implications in relation to veterinary care and animal farming. The nature of TP and tryptophan metabolism in many other animal species remains to be examined, as is the possible toxicity of certain tryptophan metabolites in relation to human pregnancy and neonatal care [for a discussion of these latter aspects, see (2)].

Control of tryptophan availability for liver NAD^+ synthesis: The conversion of tryptophan into nicotinic acid by the hepatic kynurenine pathway, of which TP is the first and rate-limiting enzyme, raises the possibility that TP activity could play an important role in NAD^+ synthesis from tryptophan, in addition to the role of tryptophan availability itself. It has been suggested (21,22,23) that only tryptophan availability is the important factor, but evidence in favour of this conclusion is difficult to interpret for several reasons, including the use of multiple-dietary-deficient animals. By using fed or 24h-starved rats, we were able to demonstrate (24) clearly an important role for liver TP in NAD^+ synthesis from tryptophan under two different sets of experimental conditions. Thus, we found (24) that only corticosterone, which induces TP, but not deoxycorticosterone, which does not, increases the concentrations of all four pyridine dinucleotides in livers of rats fed a nutritionally-adequate diet. We were also able to dissociate the role of TP from that of tryptophan availability in the 24h-starved rat, which, because of the physiological changes it exhibits, served as an ideal model for this study. Thus, as shown in Table 2, the 24h-starved rat exhibits the following changes: (a) increased release of corticosterone into the circulation; (b) enhanced lipolysis leading to increased free serum tryptophan concentration; (c) this leads to an increased availability of tryptophan to the liver; (d) liver TP activity is enhanced by a hormonal-type mechanism; (e) liver [NADP(H)] are increased; this preferential increase in the NADP couple may be significant in relation to the need for synthetic reactions in starvation. It is clear therefore that the role of TP activity and tryptophan availability can be examined and possibly dissociated in the 24h-starved rat by using specific and selective blockers or inhibitors of one or more of the above changes. The results summarized in Table 2 show that this is so and, thus, establish clearly an important role for liver TP activity in NAD^+ synthesis from tryptophan in the rat.

Table 2. Starvation, Tryptophan Pyrrolase and NAD^+ Synthesis in the Rat

Effect of starvation	Glucose	Propranolol	Actinomycin D	Allopurinol
Increased [serum corticosterone]	+	–	–	–
Increased [free serum tryptophan]	+	+	–	–
Increased liver [tryptophan]	+	+	–	–
Enhanced liver TP activity	+	–	+	+
Increased liver [NADP(H)]	+	–	+	+

+, Prevention or reversal of effect; –, lack of prevention or reversal of effect. See also (24) for detailed data.

Control of tryptophan availability for cerebral 5-hydroxytryptamine synthesis: The inverse relationship between liver TP activity and brain 5-HT synthesis is well known and documented and has been discussed by the author in 1977 (1). Strong evidence supporting this relationship has since been obtained from the results of experiments involving chronic administration of dieldrin (25), ethanol (26,27,28), other drugs of dependence (29) and glucose or sucrose (30), or subsequent withdrawal of ethanol (31) and other drugs of dependence (29), as well as those involving acute administration of antidepressants (32,33), salicylate (34) or cortisol (35). It is noteworthy that this latter study (35) has also established that the decrease in rat brain 5-HT synthesis observed during TP induction by cortisol is caused by a decrease in tryptophan availability to the brain brought about by an enhanced hepatic catabolism of the amino acid, rather than by inhibition of its cerebral uptake by kynurenine. This and evidence from the above studies therefore strongly suggest that the major, if not the only, mechanism by which liver TP activity influences brain 5-HT synthesis is that of altered tryptophan availability to the brain. The above studies have also shown that liver TP is a major target for drugs of dependence and antidepressant drugs, and, that this hepatic enzyme may therefore play an important, albeit indirect, role in drug dependence and depressive illness. Accordingly, further work with TP in relation to these two areas of mental illness may be fruitful.

Possible control of tryptophan availability for other functions and effects: The results reviewed in the above two sections suggest that liver TP controls intrahepatic and extrahepatic tryptophan availability for two well known functions (synthesis of liver NAD^+ and brain 5-HT). It follows therefore that liver TP may also control tryptophan availability for other functions, such as protein synthesis and degradation [(35) and references cited therein] and possible control of cellular differentiation and development [(36); see also G. Huether, elsewhere in this publication], and effects, such as possible control of gluconeogenesis [(1); see also C. I. Pogson, R. A. Munoz-Clares, J. S. Cook & S. A. Smith, elsewhere in this publication]. Further work with liver TP in relation to these physiological processes may therefore be fruitful, as would be any work with the enzyme in relation to changes in which tryptophan can be shown to be involved.

Table 3. Haem-Saturation Ratio of Rat Liver Tryptophan Pyrrolase

State of saturation with haem	Holoenzyme activity	Total enzyme activity	Haem-saturation ratio
Basal (fed rats)	2	5	0.66
Unaltered (1)	4	10	0.66
(2)	6	15	0.66
Increased (1)	4	5	4.00
(2)	6	10	1.50
Decreased (1)	1	5	0.25
(2)	2	10	0.25
(3)	3	15	0.25

The values given are based on experimental findings[see (19) for these]. The holoenzyme and total enzyme activities are those obtained respectively in the absence and the presence of added haematin and are expressed in µmol of kynurenine formed/h per g wet wt. of liver. The apoenzyme activity, obtained by difference, has been used to calculate the haem-saturation ratio (ratio of holoenzyme activity/apoenzyme activity). This Table is reproduced here from ref. (19) by permission.

Haem utilization and possible control of hepatic haem biosynthesis: Although TP in rat liver utilizes only 1.6% of total hepatic haem, work performed almost exclusively in this Laboratory and reviewed by the author in 1979 (19) has shown clearly and conclusively that, because of the versatility of, and the absence of methodological or metabolic limitations to, this utilization by the enzyme, TP is the most suitable and sensitive hepatic haemoprotein for assessing delicate changes in liver haem concentration. This assessment is best expressed by the so-called haem-saturation ratio (ratio of holoenzyme activity/apoenzyme activity). The sensitivity and versatility of TP in this context are illustrated by the ease with which this ratio can be altered experimentally as shown in Table 3 [see (19) for details]. Because the haem saturation of rat liver TP is influenced by conditions involving changes in the activity of 5-aminolaevulinate synthase (the rate-limiting enzyme of haem biosynthesis), the possibility was raised that TP utilizes the so-called regulatory-haem pool, which is thought to be responsible for the feedback control exerted at the above enzymatic step. Considerable evidence in support of this possibility has been obtained from the results of experiments with 2-allyl-2-isopropylacetamide (37,38,39), haematin (40), tryptophan (15) and direct and indirect displacers of bound tryptophan (41). It is noteworthy that, in addition, the following conclusions were made: (a) the haem saturation of TP is not a simple expression of availability of the regulatory-haem pool, but is a means of controlling it; (b) this pool represses 5-aminolaevulinate synthase in the fed rat because it remains unutilized by TP; (c) tryptophan itself may play a significant role in haem metabolism, and the latter may be subject to control by dietary and pharmacological factors that influence lipid metabolism and tryptophan disposition. Further work in this area may therefore be fruitful.

Reviewed work from this Laboratory has been performed in collaboration with various colleagues and with the support of the Addiction Research Foundation for Wales, Medical Council on Alcoholism, Medical Research Council, Welsh Office and Wellcome Trust.

References

1. Badawy, A. A.-B.: Life Sci. 21, 755-768 (1977).
2. Badawy, A. A.-B., Evans, M.: Biochem. J. 158, 79-88 (1976).
3. Feigelson, P., Greengard, O.: J. Biol. Chem. 236, 153-157 (1961).
4. Brady, F. O., Feigelson, P.: J. Biol. Chem. 250, 5041-5048 (1975).
5. Schimke,R.T., Sweeney,E.W., Berlin,C.M.: J. Biol. Chem. 240,322-331(1965).
6. Knox, W. E.: Adv. Enzyme Regul. 4, 287-297 (1966).
7. Schimke, R. T.: Curr. Top. Cell. Regul. 1, 77-124 (1969).
8. DeLap,L., Feigelson,P.: Biochem.Biophys.Res.Commun. 82,142-149 (1978).
9. Greengard, O., Feigelson, P.: J. Biol. Chem. 236, 158-161 (1961).
10. Badawy, A. A.-B., Evans, M.: Biochem. J. 150, 511-520 (1975).
11. Wetterberg,L.,Yuwiler,A.,Geller,E.: Life Sci. 8, 1047-1049 (1969).
12. Druyan, R., Kelly, A.: Biochem. J. 129, 1095-1099 (1972).
13. Badawy, A. A.-B., Evans, M.: Biochem. J. 136, 885-892 (1973).
14. Marver, H. S., Tschudy, D. P., Perlroth, M. G., Collins, A.: Science 154, 501-503 (1966).
15. Badawy,A.A.-B.,Welch,A.N.,Morgan,C.J.: Biochem. J. 198, 309-314 (1981).
16. Cho-Chung, Y. S., Pitot, H. C.: J. Biol. Chem. 242, 1192-1198 (1967).
17. Badawy, A. A.-B., Evans, M.: Biochem. J. 156, 381-390 (1976).

18. Gebhardt, R., Metzler, H., Mecke, D.: Hoppe-Seyler's Z. Physiol. Chem. 362, 198-199 (1981).

19. Badawy, A. A.-B.: Biochem. Soc. Trans. 7, 575-583 (1979).

20. Young, S. N., Sourkes, T. L.: J. Biol. Chem. 250, 5009-5014 (1975).

21. Powanda,M.C., Wannemacher,R.W., Jr.: J. Nutr. 100, 1471-1478 (1970).

22. Greengard, P., Sigg, E. B., Fratta, I., Zak, S. B.: J. Pharmacol. Exp. Ther. 154, 624-631 (1966).

23. Greengard,P.,Kalinsky,H.,Manning,T.J.,Zak,S.B.:J.Biol.Chem.243, 4216-4221 (1968).

24. Badawy, A. A.-B.: Biochem. J. 196, 217-224 (1981).

25. Kohli,K.K.,Chandrasekaran,V.P.,Venkitasubramanian,T.A.: J.Neurochem. 28, 1397-1398 (1977).

26. Badawy,A.A.-B.,Punjani,N.F.,Evans,M.:Biochem.J.178,575-580 (1979).

27. Punjani,N.F.,Badawy,A.A.-B.,Evans,M.: Biochem.J. 184, 165-168 (1979).

28. Badawy,A.A.-B.,Evans,M.,Punjani,N.F.:Br.J.Pharmacol. 68,22-24 (1980).

29. Badawy,A.A.-B.,Punjani,N.F.,Evans,M.: Biochem.J. 196, 161-170 (1981).

30. Badawy, A. A.-B., Punjani, N. F., Evans, M.: J. Pharmacol. Methods 3, 167-171 (1980).

31. Badawy, A. A.-B., Punjani, N. F., Evans, C. M., Evans, M.: Biochem. J. 192, 449-455 (1980).

32. Badawy, A. A.-B., Evans, M.: Biochem. Pharmacol. 30, 1211-1216 (1981).

33. Badawy, A. A.-B., Evans, M.: Br. J. Pharmacol. 77, 59-67 (1982).

34. Badawy, A. A.-B.: Br. J. Pharmacol. 76, 211-213 (1982).

35. Young, S. N.: Br. J. Pharmacol. 74, 695-700 (1981).

36. Henderson, M., Kitos, P. A.: Teratology 26, 173-181 (1982).

37. Badawy, A. A.-B., Morgan, C. J.: Biochem. J. 186, 763-772 (1980).

38. Morgan, C. J., Badawy, A. A.-B.: Biochem. J. 186, 993-996 (1980).

39. Badawy, A. A.-B., Welch, A. N., Morgan, C. J.: Biochem. J. 206, 441-449 (1982).

40. Welch, A. N., Badawy, A. A.-B.: Biochem. J. 192, 403-410 (1980).

41. Badawy, A. A.-B., Morgan, C. J.: Biochem. J. 206, 451-460 (1982).

KYNURENINE AMINOTRANSFERASE ACTIVITY IN HUMAN AND OTHER MAMMALIAN TISSUES

Ryo Kido
Department of Biochemistry, Wakayama Medical College
Wakayama 640, Japan

It has been believed for many years that in mammals, the conversion of kynurenine to kynurenic acid is catalyzed only by the enzyme kynurenine aminotransferase (EC 2.6.1.7, L-kynurenine:2-oxoglutarate aminotransferase) (Fig. 1).
Recently, this enzyme has been shown to be identical with 2-aminoadipate aminotransferase (EC 2.6.1.39) in rat kidney (1). On the other hand, it has been reported that several apparently different enzymes catalyze the transamination of kynurenine with various 2-oxoacids in mammalian species (2,4,5,6). Under the present circumstances the category of the enzyme which is involved in the formation of kynurenic acid is confused because of considerable interspecies differences. In this contribution, the species difference of kynurenine aminotransferase will be discussed centering on the recent studies in our laboratory.

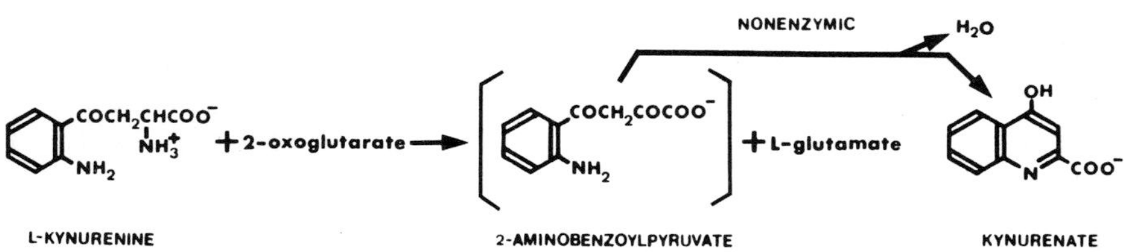

Fig. 1. Kynurenine aminotransferase [EC 2.6.1.7, L-kynurenine : 2-oxoglutarate aminotransferase (cyclizing)]

Progress in Tryptophan and Serotonin Research

Table 1. Rat enzymes with kynurenine aminotransferase activity

Enzyme no.	pI	pH optimum	Amino acceptor	Molecular weight	Identical with
1	5.2	8.0-8.5	Pyruvate	79000	Glu T (EC 2.6.1.15) HPT-2 (EC 2.6.1.-)
2	6.8	6.0-6.5	2-Oxoglutarate	100000	2-Aminoadipate aminotransferase (EC 2.6.1.39)
3	8.0	8.0-8.5	2-Oxo-4-methyl-thiobutyrate Glyoxylate Pyruvate Hydroxybutyrate	80000	AGT-1 (EC 2.6.1.44) SPT (EC 2.6.1.51) HPT-1 (EC 2.6.1.-)
4	9.4	8.0-8.5	2-Oxoglutarate 2-Oxoglutarate	100000	m-GOT (EC 2.6.1.1) m-TGT (EC 2.6.1.5)

Rat enzymes

The available data indicate that in rat tissues, there are at least four enzymes capable of catalyzing the kynurenine-2-oxo-acid transaminase reaction (2,4,7,8,9,10,11); these enzymes were designated 1, 2, 3 and 4.

A summary of some properties of these enzymes is given in Table 1. Enzyme 1 has a pI of 5.2 and a pH optimum between 8.0-8.5. The enzyme is active with pyruvate as amino acceptor but not with 2-oxoglutarate, and utilizes various aromatic amino acids as amino donors. The apparent Km values are about 1.4 mM and 1.3 mM for kynurenine and pyruvate, respectively. The molecular weight is estimated approximately 79000 (3). This type of enzyme is present in liver (8), kidney (11), brain (7) and small intestine (9). This enzyme is identical with histidine-pyruvate aminotransferase isoenzyme 2 (EC 2.6. 1.-) (3) and with glutamine-oxoacid aminotransferase (EC 2.6. 1.15) (12).

Enzyme 2 possesses a pI of 6.8 and a pH optimum between 6.0-6.5. It is specific for 2-oxoglutarate as amino acceptor and localized in liver and kidney. The molecular weight is about 100000. This is historically called kynurenine aminotransferase (EC 2.6.1.7) which is identical with 2-aminoadipate aminotransferase (EC 2.6.1.39) (1). Enzyme 3 is present only in liver and inducible by glucagon (4). It has a dimeric structure formed by identical subunits. The enzyme has a pI of 8.0 and a pH optimum between 8.0-8.5. 2-Oxoacids are effective in the following order of activity: 2-oxo-4-methylthiobutyrate > glyoxylate > pyruvate > hydroxypyruvate > 2-oxoglutarate. This enzyme is identical with alanine-glyoxylate aminotransferase (EC 2.6.1.44), serine-pyruvate aminotransferase (EC 2.6.1.51) and histidine-pyruvate aminotransferase (EC 2.6.1.-) (4,13,14). Enzyme 4 is widely distributed in rat tissues. The enzyme has a molecular weight of 100000, a pI of 9.4 and a pH optimum between 8.0-8.5. The enzyme is specific for 2-oxoglutarate as amino acceptor, and identical with mitochondrial aspartate-2-oxoglutarate aminotransferase (EC 2.6.1.1) and mitochondrial tyrosine-2-oxoglutarate aminotransferase (EC 2.6.1.5) (9,15,16,17)

Rabbit enzyme

When liver or kidney extract of rabbits is subjected to isoelectric focusing on a pH 3.5-10 Ampholine gradient, only a single activity peak of kynurenine aminotransferase appears. This activity peak coincides with that of 2-aminoadipate aminotransferase with a pI of 5.8 and a pH optimum 7.0-7.5. The enzyme has a molecular weight of 90000 and utilizes 2-oxoglutarate, pyruvate, glyoxylate and oxaloacetate as effective amino acceptors. It is interesting to note that these 2-oxoacids can serve as amino acceptors for kynurenine transamination.

Dog enzyme

In dog, the liver is the only tissue with detectable kynurenine aminotransferase activity. The enzyme is identical with 2-aminoadipate aminotransferase (EC 2.6.1.39) and iodotyrosine aminotransferase, having a pH optimum of 7.5 and isoelectric points of 7.8 and 8.0.

Bovine enzyme

Isoelectric focusing of bovine liver extract shows only a single activity peak of kynurenine-pyruvate aminotransferase which coincides with those of kynurenine-glyoxylate aminotransferase and glutamine phenylpyruvate aminotransferase. These data suggest the identity of bovine kynurenine aminotransferase with glutamine aminotransferase (EC 2.6.1.15). No kynurenine 2-oxoglutarate aminotransferase activity is measurable in liver. Studies on the purification, characterization and identification of kynurenine aminotransferase from bovine liver is required.

Monkey enzyme

Isoelectric focusing of the mitochondrial extract from monkey liver does not show detectable kynurenine-2-oxoglutarate aminotransferase activity. However, an enzyme has been purified with kynurenine aminotransferase activity from liver (4). This enzyme is active with glyoxylate, pyruvate, hydroxypyruvate and 2-oxo-4-methylthiobutyrate in order of activity. It has a molecular weight of 80000, an isoelectric point of pH 8.0 and pH optimum of pH 8.0-8.5. This enzyme might be identical with alanine-glyoxylate aminotransferase isoenzyme 1 (EC 2.6.1.44). Further studies are required.

Human enzymes

As shown in Fig. 2 the isoelectric focusing of human liver extract results in the appearance of a single activity peak of alanine-glyoxylate aminotransferase with pI 8.3, coinciding with kynurenine-glyoxylate, oxaloacetate or 2-oxo-glutarate aminotransferase activities. An additional very small peak of kynurenine-2-oxoglutarate aminotransferase activity is also found with pI 5.0. Kynurenine-glyoxylate aminotransferase has a molecular weight of 90000 with two identical subunits and pH optimum at 9.0-9.5. The enzyme catalizes transamination between kynurenine and several 2-oxoacids. The order of effectivemess of 2-oxoacids is glyoxylate oxalo-acetate > hydroxypyruvate > pyruvate > 4-methylthio-2-oxobutyrate > 2-oxobutyrate. Kynurenine-glyoxylate aminotransferase activity is competitively inhibited by the addition of L-alanine or L-serine.

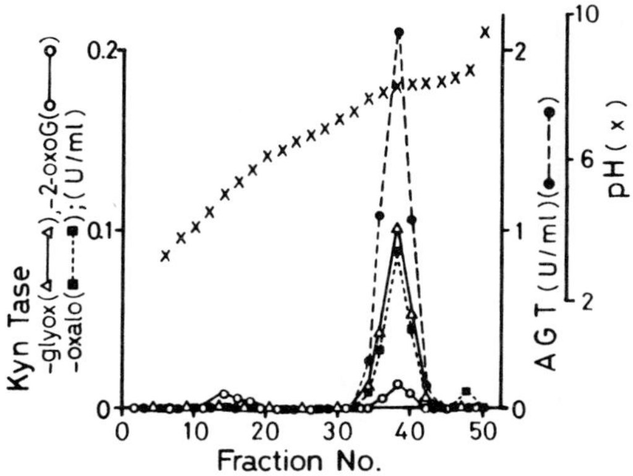

Fig. 2. Isoelectric focusing profile of human liver extract

This enzyme is identical with alanine-glyoxylate aminotransferase (EC 2.6.1.44) and serine-pyruvate aminotransferase (EC 2.6.1.51) (6).

Conclusion and Comments

The kynurenine aminotransferase activity appears very differently among species. In an attempt to elucidate the entity of kynurenine aminotransferase we must consider an overlap in substrate specificity among the enzymes purified. To make matters worse it is probable that kynurenine aminotransferase has been studied by many groups of workers utilizing only a pair of kynurenine and 2-oxoglutarate. The above considerations attest to the difficulty of working with kynurenine aminotransferase that exhibits wide substrate specificity and interspecies differences. It is imperative that kynurenine aminotransferase in mammalian tissues be reinvestigated depending on species and substrate specificity.

References

1. Tobes, M.C., Mason, M.: J. Biol. Chem. 252, 4591-4599 (1977).
2. Noguchi, T., Minatogawa, Y., Okuno, E., Nakatani, M., Morimoto, M., Kido, R.: Biochem. J. 151, 399-406 (1975).
3. Noguchi, T., Kido, R.: Hoppe-Seyler's Z. Physiol. Chem. 357, 649-656 (1976).
4. Harada, I., Noguchi, T., Kido, R.: Hoppe-Seyler's Z. Physiol. Chem. 359, 481-488 (1978).
5. Tobes, M.C., Mason, M.: Life Sci. 22, 793-801 (1978).
6. Okuno, E., Minatogawa, Y., Nakamura, M., Kamoda, N., Nakanishi, J., Makino, M., Kido, R.: Biochem. J. 189, 581-590 (1980).
7. Minatogawa, Y., Noguchi, T., Kido, R.: J. Neurochem. 23, 271-272 (1974).
8. Nakatani, M., Morimoto, M., Noguchi, T., Kido, R.: Biochem. J. 143, 303-310 (1974).
9. Noguchi, T., Nakamura, J., Kido, R.: Life Sci. 13, 1001-1010 (1973).
10. Noguchi, T., Nakatani, M., Minatogawa, Y., Kido, R.: J. Neurochem. 25, 579-582 (1975).
11. Noguchi, T., Nakatani, M., Minatogawa, Y., Morimoto, M., Kido, R.: Hoppe-Seyler's Z. Physiol. Chem. 356, 1245-1250 (1975).
12. Noguchi, T., Okuno, E., Kido, R.: Biochem. J. 161, 177-179 (1977).
13. Noguchi, T., Okuno, E., Takada, Y., Minatogawa, Y., Okai, K., Kido, R.: Biochem. J. 169, 113-122 (1978).
14. Noguchi, T., Okuno, E., Kido, R.: Biochem. J. 159, 607-613 (1976).
15. Nakamura, J., Noguchi, T., Kido, R.: Biochem. J. 135, 815-818 (1973).
16. Miller, J.F., Litwack, G.: J. Biol. Chem. 246, 3234-3240 (1971).
17. Rosenberg, J.C., Sapico, V., Litwack, G.: Physiol. Chem. & Physics. 6, 139-157 (1974).

INTRACELLULAR LOCALIZATION AND CHARACTERIZATION OF 3-HYDROXYKYNURENINASE IN HUMAN LIVER

Joji Inada, Masanori Kimura, Etsuo Okuno, Ryo Kido
Department of Biochemistry, Wakayama Medical College
Wakayama 640, Japan

Introduction

3-Hydroxykynureninase (3-hydroxykynurenine hydrolase, EC 3.7.1.3), a key enzyme in the biosynthetic pathway from tryptophan to the nicotinamide nucleotides has been studied in a number of different species (1,2,3,4,5,6,7,8,9,10,11). The enzyme from human tissues, however, remains unknown. In the present investigation, we studied intracellular localization and characterization of human hepatic 3-hydroxykynureninase.

Results

1) Localization of the enzyme. It was found that at least 75% of 3-hydroxykynureninase and kynureninase were localized in the supernatant fraction and the remainder in the particulate fractions by the method of de Duve et al. (12). On the sucrose-gradient centrifugation the sedimentation profile of the 3-hydroxykynureninase was almost identical with that of mitochondria marked by glutamate dehydrogenase (GDH) (Fig. 1). Furthermore, none of the used marker enzymes of subcellular organelles namely catalase, acid phosphatase and esterase coincided with the 3-hydroxykynureninase on sucrose gradient centrifugation and Percoll-sucrose gradient centrifugation (data are not shown).

Progress in Tryptophan and Serotonin Research

Fig. 1. Sucrose density gradient (64-27%) centrifugation profile of human liver post nuclear fraction

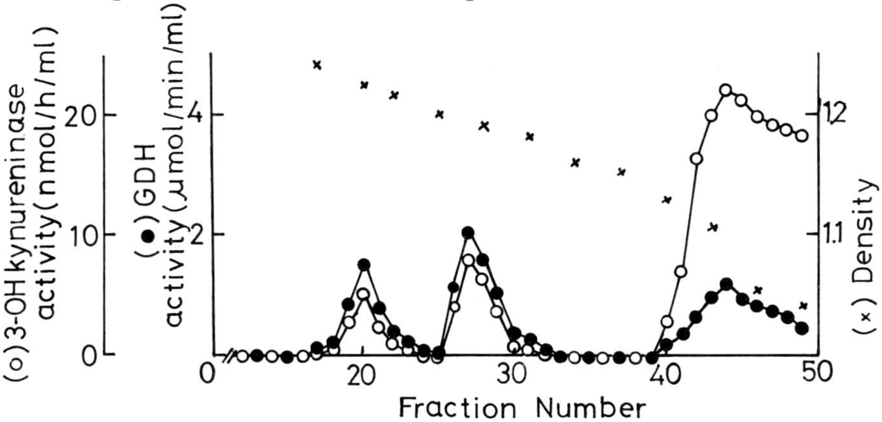

2) Purification of 3-hydroxykynureninase. Human liver was homogenized in 9 vol of 0.25 M-sucrose and the supernatant was obtained by centrifugation at 10,000g for 30 min. The enzyme from supernatant was purified 55.7 fold with an overall recovery of 18% by acid treatment (pH 5.0), heat treatment (60°C for 1 min), ammonium sulfate fractionation (35-70% saturation), Affi-Gel Blue column, DE-52 column, hydroxyapatite column, and Sephadex G-100 gel filtration. The enzyme in mitochondria was prepared from heavy mitochondrial fraction, and isoelectric focusing and Sephadex G-100 gel filtration were applied. The ratio of 3-hydroxykynureninase and kynureninase activities was almost constant at about 15:1.

3) Properties of 3-hydroxykynureninase. The molecular weight of the enzyme purified from supernatant was estimated as approx. 130,000 by Sephadex G-100 gel filtration and 118,000 by sucrose density gradient centrifugation. Those of mitochondrial enzymes were 130,000 and 120,000. Both supernatant and mitochondrial enzymes showed the same ratio of the activity for 3-hydroxykynurenine to that for kynurenine, and possessed the same isoelectric point of pH 5.9. Both 3-hydroxykynureninase and kynureninase activities of the purified enzyme from supernatant were inhibited by 58.7% and 62.5% in the presence of 0.5 mM phenylhydrazine, respectively.

They were also inhibited by 15.3%, 36.3% or 5.7%, 12.8% in the presence of 0.5 mM hydrazine or semicarbazide. The preparation from the heavy mitochondrial extract showed similar inhibition profiles. Effect of pH on 3-hydroxykynureninase were determined at various concentrations of pyridoxal 5'-phosphate (PLP) with 3-hydroxykynurenine and kynurenine as shown in Fig. 2(a,b).

The apparent pH optima for 3-hydroxykynurenine were pH 7.3 at 8 µM PLP and pH 7.8 at 80 µM PLP. With regard to kynurenine those were pH 8.0 at 8 µM PLP and pH 9.0 at 80 µM PLP. The optimum pH shifted to higher pH with the increase of the PLP concentration on both substrates.

The Michaelis constants were determined from double-reciprocal plots of the relationship between the reaction velocity and the substrate or cofactor concentration. The apparent Km values of the supernatant enzymes were as follows: 3-hydroxykynurenine, 7.7×10^{-5} M; kynurenine, 1.0×10^{-3} M; and pyridoxal 5'-phosphate, 2.5×10^{-6} M with 3-hydroxykynurenine. Those of the mitochondrial enzymes were 7.8×10^{-5} M for 3-hydroxykynurenine, 1.3×10^{-3} M for kynurenine and 2.9×10^{-6} M for pyridoxal 5'-phosphate with 3-hydroxykynurenine. On the basis of the above findings, we conclude that identical en-

Fig. 2. Effect of pH on 3-hydroxykynureninase (a) and kynureninase (b) activity

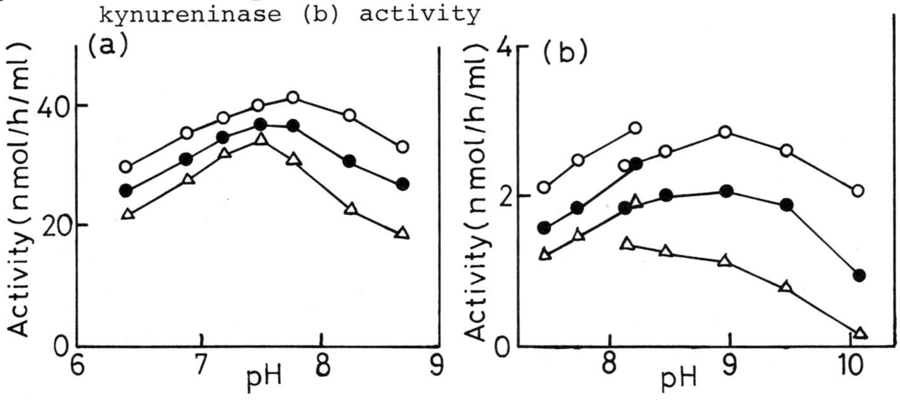

Potassium phosphate buffer was used at pH 6.4-8.7 and Tris/HCl buffer at pH 8.2-10.1. Pyridoxal 5'-phosphate concentration, 8 µM (△), 32 µM (●) and 80 µM (○) are shown.

zymes exist in the cytosol and mitochondria of human liver and these enzymes are specifically active for 3-hydroxykynurenine.

References

1. Wiss, O.: Hoppe-Seyler's Z. Physiol. Chem. 293, 106-111 (1953).
2. Gaertner, F.H., Shetty, A.S.: Biochim. Biophys. Acta 482, 453-460 (1977).
3. Knox, W.G.: Biochem. J. 53, 379-385 (1953).
4. Wiss, O., Weber, F.: Z. Physiol. Chem. 304, 232-240 (1956).
5. Tanizawa, K., Soda, K.: J. Biochem. 85, 901-906 (1979).
6. Hayaishi, O., Stanier, R.Y.: J. Bacteriol. 62, 691-709 (1951).
7. Moriguchi, M., Yamamoto, T., Soda, K.: Biochem. Biophys. Res. Commun. 44, 752-757 (1971).
8. Moriguchi, M., Kanamoto, T., Soda, K.: Biochemistry 12, 2969-2974 (1973).
9. Jakoby, W.B., Bonner, D.M.: J. Biol. Chem. 205, 699-707 (1953).
10. Jakoby, W.B., Bonner, D.M.: J. Biol. Chem. 205, 709-715 (1953).
11. Schott, H-H., Krause, U.: Hoppe-Seyler's Z. Physiol. Chem. 360, 481-488 (1979).
12. De Duve, C., Pressman, B.C., Gianetto, R., Wattiaux, R., Appelmans, F.: Biochem. J. 60, 604-617 (1955).

TIME-COURSE OF 5-HTP METABOLISM IN RABBITS: RELEVANCE OF TRANSAMINATING PATHWAY.

Vincenzo Politi, Giovanna De Luca, Giovanni Di Stazio, Guido Palamara and Mario Materazzi

Polifarma Research Centre - Via Tor Sapienza, 138 - 00155 Roma (Italy)

Introduction

It is well known that aromatic L-aminoacid decarboxylase (EC 4.1.1.28) has a pivotal role in many central nervous functions, allowing the biosynthesis of neurotransmitters like dopamine and serotonin from their precursors, L-dopa and L-5-hydroxytryptophan (5-HTP), respectively. Some years ago, drugs such as benserazide and carbidopa were introduced in therapy for the treatment of Parkinsonism, in addition to L-dopa, because they can rise cerebral dopamine levels, inhibiting dopa decarboxylation outside the blood brain barrier.

More recently, 5-HTP was introduced in therapy as a precursor of serotonin in human cerebral diseases like migraine and depression, but reports are somehow conflicting (1). A few non steroidal antiinflammatory drugs, like fentiazac, flufenamic and mefenamic acid, were shown to be inhibitors of aromatic L-aminoacid decarboxylase "in vitro"(2) as well as "in vivo"(3). Because some of these drugs also showed anti-migrainous properties (4,5), it seemed worthwile to investigate any possible connection between 5-HTP metabolism and analgesic effects.

Progress in Tryptophan and Serotonin Research

Methods

New Zealand rabbits, fasted overnight, were anaesthetized with uretane and randomized in the following groups: a) control; b) 5-HTP 10 mg/Kg; c) Carbidopa 1.5 + 5-HTP 10 mg/Kg; d) Fentiazac 80 + 5-HTP 10 mg/Kg. Drugs were administered by gastric probe after dissolution in water. Blood was collected in 2 ml samples at 0.5, 1, 2, 3, 5, 6, 8 hours after drug administration with heparinized syringes. After addition of EDTA and ascorbic acid, it was centrifuged for 10' at 2000 x g. Proteins in the supernatant were precipitated by TCA 12% and, after centrifugation, 10 microliters were injected in HPLC Lichrosorb RP8 columns. Indole derivatives were detected fluorimetrically: excitation was set at 300 nm; emission at 340 nm.

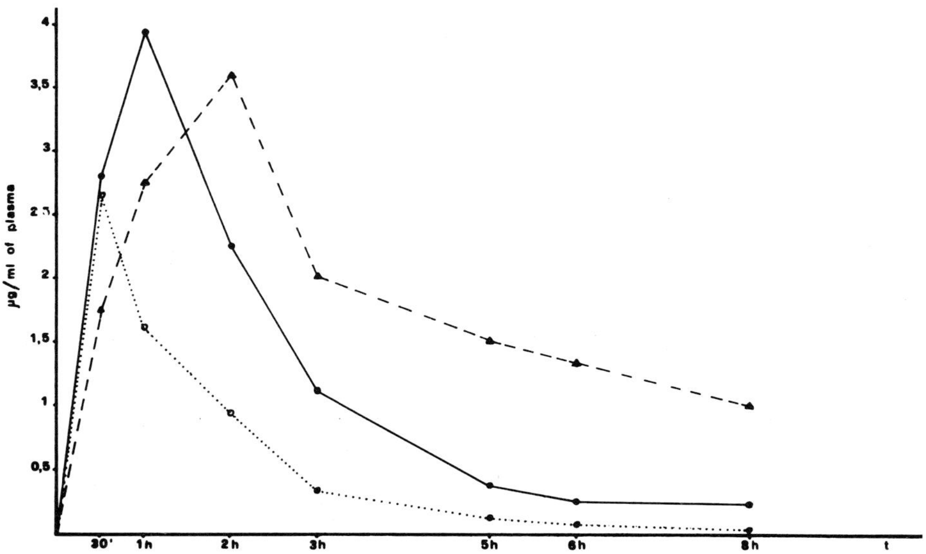

Figure 1. 5-HTP plasmatic levels in rabbits after oral administration of 5-HTP 10 mg/Kg (o·········o), carbidopa 1.5 mg/Kg + 5-HTP 10 mg/Kg (●——●) and fentiazac 80 mg/Kg + 5-HTP 10 mg/Kg (▲----▲)

Results

In preliminary experiments, it was shown that the method used allows the recovery of at least 95% of the 5-HTP present in plasma. Furthermore, this compound appears to be well separated from tryptophan, serotonin and other minor indole metabolites.

Fig. 1 shows the time-course of 5-HTP in plasma of rabbits treated with 5-HTP alone, 5-HTP plus carbidopa and 5-HTP plus fentiazac. While control plasma have undetectable levels of 5-HTP (not shown), 5-HTP treated animals display a peak 30' after drug administration. Carbidopa shifts peak levels to 1 hour and fentiazac to 2 hours.

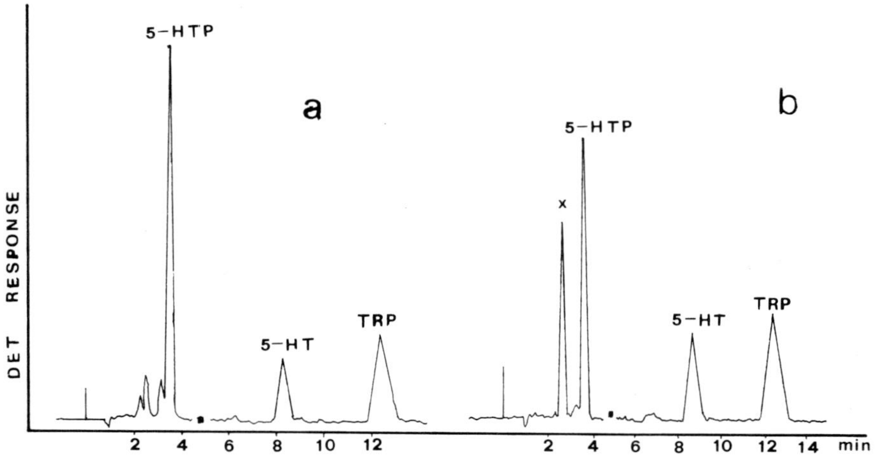

Figure 2. Plasma sample chromatograms related to carbidopa(a) and fentiazac (b) treated rabbits.
x: unknown.

Because fentiazac appears to slow the return to basal 5-HTP levels, HPLC chromatograms were analyzed in more details. As can be seen in fig 2, which shows indole metabolites after 1 hour of drug administration, plasma of carbidopa-treated rabbits lack a substance having a retention time of 2,5 minutes, which is distinct from indole derivatives, available from chemical dealers.

It was therefore supposed that fentiazac, inhibiting aromatic-l-aminoacid decarboxylase, shifts 5-HTP to its transaminated product, i.e. 5-hydroxy-indolepyruvic acid(5-HIPA), while carbidopa also blocks this pathway due to hydrazine binding with pyridoxal group.
To test this hypothesis 5-HTP was incubated with porcine mitochondrial a-spartate aminotransferase(EC 2.6.1.1.),an enzyme which can transaminate most of the aromatic aminoacids. Results showed that 5-HIPA, which is slowly formed by enzyme activity, has the same retention time as the "unknown" indole metabolite, when injected in HPLC. Some unpublished preliminary results also suggest that in humans treated with 5-HTP, one of the main metabolites is 5-HIPA. This raises an interesting question about a possible therapeutic role of indole α-Ketoacids. Of course, 5-HIPA may only be a "parking" metabolite for serotonin biosynthesis; but it cannot be ruled out that it may have some pharmacological(analgesic?) effect by itself, because of the relevance of transaminations in hypoxia and ischemia.

References

1. Van Praag H.M.: In: Aromatic Amino Acid Hydroxylases and Mental Disease, J. Wiley & Sons, 1979.
2. Barboni E., Borri Voltattorni C et al.: IRCS Med. Sci. 5, 199 (1977).
3. Politi V., Materazzi M et al.: International Congress of Inflammation, Bologna 1978, Abstract Book P 79.
4. Feleppa M., Florio C. et al.: La Settimana Medica 68, 367-372 (1980).
5. Vardi Y., Rabey et al.: Neurology 26, 447-450 (1976).

TRYPTOPHAN AMINOTRANSFERASE ACTIVITY IN RAT LIVER

John Stanley, Arthur Nicholas, Ian Thompson, Chris. Pogson.
Department of Biochemistry, University of Manchester,
Oxford Road, Manchester, M13 9PL, U.K.

Introduction

Tryptophan aminotransferase can be assayed by measuring the increase in extinction at 328nm due to the formation of a complex between the enol form of indolepyruvate and borate(1). A glutathione-dependent tautomerase in rat liver catalyses the conversion of the keto form of indolepyruvate into the enol form(2). However, this assay is non-linear over the first few minutes because tautomerisation of indolepyruvate is rate limiting. A new radiochemical assay for tyrosine aminotransferase has been developed in this laboratory based on the release of ^3H from L-[side chain 2,3-^3H]tyrosine (3). In principle this method could be used to assay tryptophan aminotransferase.

The identity of liver tryptophan aminotransferase is controversial. Studies of the substrate specificity of purified tyrosine aminotransferase suggest that tryptophan aminotransferase activity in crude rat liver extracts may represent, at least in part, a side reaction of tyrosine aminotransferase (4). By contrast, the greater sensitivity of tyrosine aminotransferase activity to induction by hydrocortisone and to inhibition in vitamin B_6 deficiency indicates that two separate enzymes may exist (5,6). This question is reexamined in the present work.

Methods

Crude rat liver extracts were prepared by homogenising 1g tissue in 4ml 50mM potassium phosphate pH 7.6, containing 2mM 2-oxoglutarate, 1mM EDTA

Progress in Tryptophan and Serotonin Research

and 0.2mM pyridoxal phosphate. Extracts were centrifuged at 105,000g for 30 min at 4° and supernatants used for assay. Tyrosine aminotransferase was assayed by the method described in (3). Trytpophan aminotransferase was assayed spectrophotometrically in 0.5M sodium borate pH 8.0, containing 40mM L-tryptophan, 6.7mM 2-oxoglutarate and 0.04mM pyridoxal phosphate at 328nm. Assay temperature was 30° and cuvette volume 3ml. 300μl liver supernatant were assayed. Tryptophan aminotransferase was assayed radiochemically in 100mM potassium phosphate pH 7.6, containing 10mM 2-oxoglutarate, 0.8mM pyridoxal phosphate, 40mM L-tryptophan (containing 1μCi L-[side chain 2,3-^{3}H] tryptophan) and 5μl liver supernatant in a total volume of 50μl. Reaction was initiated by addition of prewarmed supernatant and stopped 60 min later with 600μl Norit GSX (100 mg/ml suspension in distilled water). After thorough mixing samples were centrifuged at 12,000g for 2 min. Then 400μl supernatant were counted in 2ml PCS. Assay temperature was 30°C.

Some rats were injected intraperitoneally with either 750mg tryptophan/Kg body weight in 0.9% (w/v) NaCl containing 0.1% (w/v) Tween 80 or 1mg triamcinolone/100g body weight in 0.9% (w/v) NaCl. Injections were administered at 5 p.m. and 9 a.m. the following morning. Rats were sacrificed at 12 a.m.

A tyrosine aminotransferase antiserum was prepared by immunising rabbits with the purified rat liver enzyme. Antiserum was concentrated by ammonium sulphate precipitation. Tyrosine aminotransferase was precipitated by incubating 100μl rat liver supernatant with 3μl antiserum at 25° for 1 hr and then overnight at 4°. Tyrosine aminotransferase was removed by centrifugation at 12,000g for 2 min.

Results

(1) K_m of tryptophan aminotransferase for L-tryptophan. Using the radiochemical assay and a range of L-tryptophan concentrations with a constant radiochemical specific activity, a Lineweaver-Burk plot gives a K_m for L-tryptophan of 92.1mM. Under the same conditions the

spectrophotometric assay gives a value of 25.2mM.

(2) Competition studies. When tryptophan aminotransferase was assayed at 10mM L-tryptophan by the radiochemical method, 10mM kynurenine or hydroxykynurenine inhibited its activity by 79% and 97% respectively.

(3) Immunoprecipitation studies. Under conditions which completely precipitate tyrosine aminotransferase there is no effect of the antiserum on tryptophan aminotransferase assayed by the radiochemical method. However, when tryptophan aminotransferase was assayed by the spectrophotometric method the antiserum inhibited its activity by 66%.

(4) Induction studies. There is no significant difference between the ratios of the activities of tyrosine aminotransferase to tryptophan aminotransferase in fed, 48hr starved, triamcinolone and tryptophan treated rats when the enzyme is assayed by the spectrophotometric method. When tryptophan aminotransferase is assayed by the radiochemical method the ratio is significantly increased in the tryptophan treated rats.

Discussion

The K_m of tryptophan aminotransferase for L-tryptophan is 3 to 4 times greater when the enzyme is assayed by the radiochemical assay. Furthermore tyrosine aminotransferase antiserum inhibits the activity of tryptophan aminotransferase by 66% when the enzyme was assayed by the spectrophotometric method. Finally tryptophan aminotransferase appears to be more sensitive to induction by tryptophan when the enzyme is assayed radiochemically. Taken together these results strongly suggest that the two methods do not measure the same activity.

The radiochemical method measures the loss of ^3H for L-[side chain 2,3-^3H]tryptophan which during transamination will exchange with the solvent. In this method charcoal serves to adsorb aromatic compounds including tryptophan so strictly speaking counts remaining in the supernatant after mixing with charcoal and centrifugation are non-aromatic products. Since tryptophan aminotransferase activity measured by the radiochemical method is inhibited by kynurenine and hydroxykynurenine it is possible that the method measures the ability of kynureninase to catalyse cleavage of the side chain of tryptophan. The side chain would appear as alanine (7) which is a non-aromatic product. This explanation would account for the very high K_m of tryptophan aminotransferase for L-tryptophan when the enzyme is assayed radiochemically.

Experiments with an antiserum to tyrosine aminotransferase have demonstrated that about 66% of tryptophan aminotransferase activity measured spectrophotometrically is tyrosine aminotransferase. This finding is in agreement with the induction studies which demonstrate that the two enzyme activities are equally sensitive to starvation, triamcinolone and tryptophan. However about 33% of the tryptophan aminotransferase activity measured spectrophotometrically remains to be identified.

The authors thank Dr. A.J. Dickson for advice in the preparation and use of antiserum against tyrosine aminotransferase and the Medical Research Council for financial support.

References

1. Lin, E.C.C., Pitt, B.M., Civen, M., Knox, W.E.: J. Biol. Chem. 233, 668-673 (1958).
2. Spencer, R.P., Knox, W.E.: Arch. Biochem. Biophys. 96, 115-124 (1962).
3. Marston, F.A.O., Pogson, C.I.: FEBS Lett. 83, 277-280 (1977).
4. Jacoby, G.A., La Du, B.N. J. Biol. Chem. 239, 419-424.
5. Civen, M., Knox, W.E.: Science 129, 1672-1673 (1959).
6. Lin, E.C.C., Civen, M., Knox, W.E. J. Biol. Chem. 233, 1183-1185 (1958).
7. Haglund, K., Marks, M., Ford, J.: Mol. Cell. Biochem. 18, 71-74 (1977).

THE OXYGENATION OF TRYPTOPHAN AND INDOLES WITH SUPEROXIDE ION AND DIOXYGEN IN THE PRESENCE OF COPPER AND RUTHENIUM CATALYSTS

Éva Balogh-Hergovich and Gábor Speier

Research Group for Petrochemistry of the Hungarian Academy of Sciences and Department of Organic Chemistry, Veszprém University of Chemical Engineering,
8201-Veszprém, Hungary

Introduction

Tryptophan-2,3-dioxygenase catalyses the oxidative C-C scission in the indole ring of tryptophan and N-formylkynurenine is formed (1). It was shown, that the enzyme contains iron and copper as cofactors and the catalytic activity is dependent on the presence of copper(I). The formation of intermediate dioxygen-complex(es) was also assumed (2). The enzyme is believed to require superoxide ion as a substrate (3). Model reactions of tryptophan with superoxide ion and dioxygen using copper catalyst have been investigated (4, 5, 6).

Results

1) Substituted indoles (1) react with superoxide ion generated from potassium superoxide using phase transfer catalysts (18-crown-6 or triethyl(benzyl)ammonium chloride) in tetrahydrofuran to form three types of products. When R^2= H or Ph o-formyl or o-acylaminoketones (6) were formed, where R^2= Me 2-quinolones (7) were isolated. In some cases N-acetylanthranilic acid (8f) was formed. Tryptophan showed inertness towards superoxide ion. All reactions were found to be chemiluminescent for which the singlet → triplet transition of (6) is presumably responsible. The products formed can be rationalized by the mechanism shown in Scheme 1.

Progress in Tryptophan and Serotonin Research

In the case of (1f) the o-acylaminoketone (6f) is further oxidized by O_2^- to N-acetylanthranilic acid (8f) and if R^2= Me (1c, 1e) 2-quinolones (7) are formed through intramolecular condensation. Conducting these reactions under dioxygen instead of argon some dioxygen was consumed. This suggests that a small fraction of the indoles was autoxidized according to the well established mechanism, which is a minor competitive pathway to that outlined in Scheme 1.

2) Since ruthenium stands below iron in the Periodic Table, $RuCl_2(PPh_3)_3$ was tested as a catalyst for the oxygenation of 3-methylindole (1a). It was found, indeed, that using the above ruthenium catalyst 3-methylindole (1a) was cleaved by dioxygen to o-formamidoacetophenone (6a) in dichloromethane or tetrahydrofuran in 35-50% yields.

3) Copper(I) chloride in pyridine and dichloromethane catalyzed the oxygenation of 3-methylindole (1a) to o-formamidoacetophenone (6a) and tryptophan (1h) to N-formylkynurenine (6h) in ~70% and 3% yields respectively (5, 6).

Using copper(II) chloride as oxidant under argon in acetonitrile 3-methylindole (1a) underwent a fast reaction according to Scheme 2 giving copper(I) chloride, hydrogen chloride and the organic products o-formamidoacetophenone (6a), 3-methyloxindole (9), 3-(3-methyl-2-indolyl)-3-methyloxindole (10), 2-chloro-3-methylindole (11), and 3,3'-dimethyl-2,2'-biindole (12) in yields as indicated. The same reaction under dioxygen led to only o-formamidoacetophenone (6a) in 75% yield (7). This suggests that in the reaction of 3-methylindole (1a) with copper(II) chloride, as a consequence of hydrogen abstraction, the radical (2a) was formed, which in the absence of dioxygen dimerized to (12) or was chlorinated by excess copper(II) chloride to (11). In the presence of dioxygen or even of traces of it, however, oxygenation of the radical (2a) occured to yield (6a), (9), and (10). A similar mechanism may be operative in the copper(I) chloride catalyzed oxygenation reaction too.

Superoxide ion generates by hydrogen abstraction indolyl radicals (2) which react further with O_2^- to give the indoxyl hydroperoxides (3). This is followed by an intramolecular nucleophilic addition to give the dioxetanes (4), which break down *via* (5) into the *o*-formyl or *o*-acylaminoketones (6).

Scheme 1

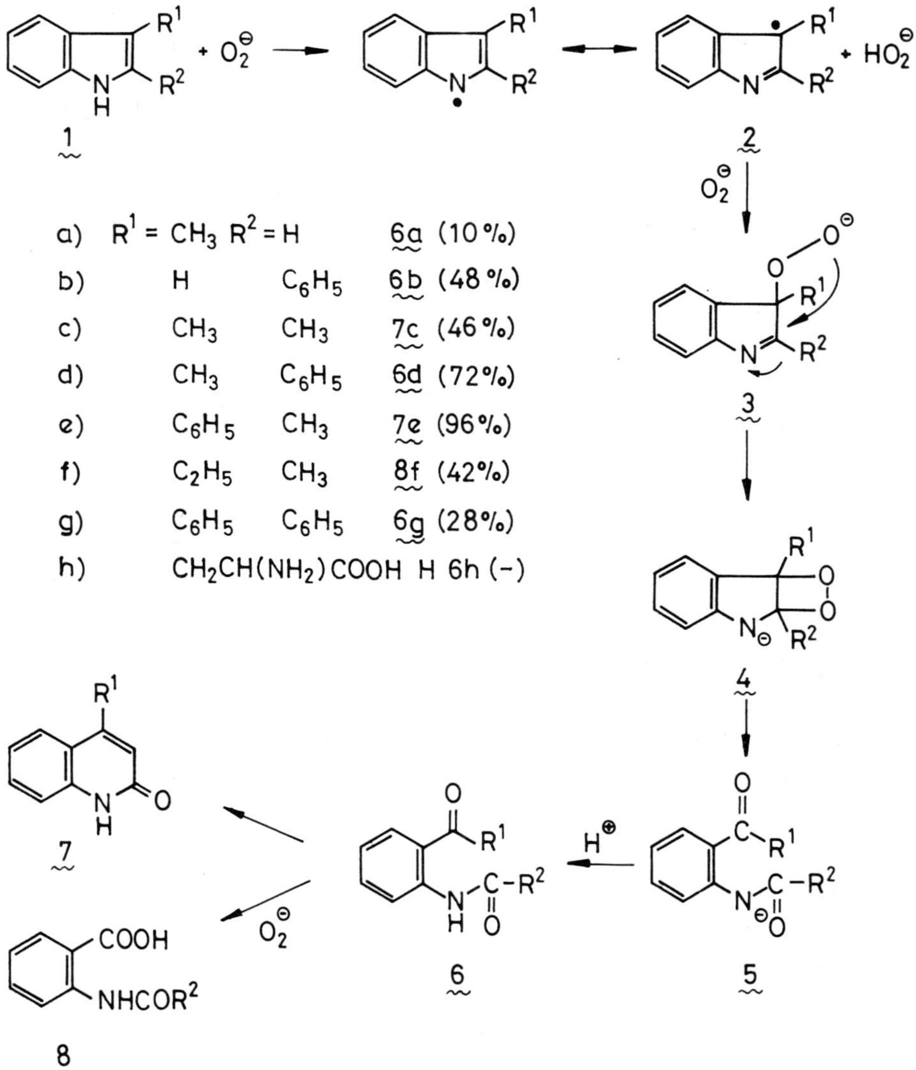

Scheme 2

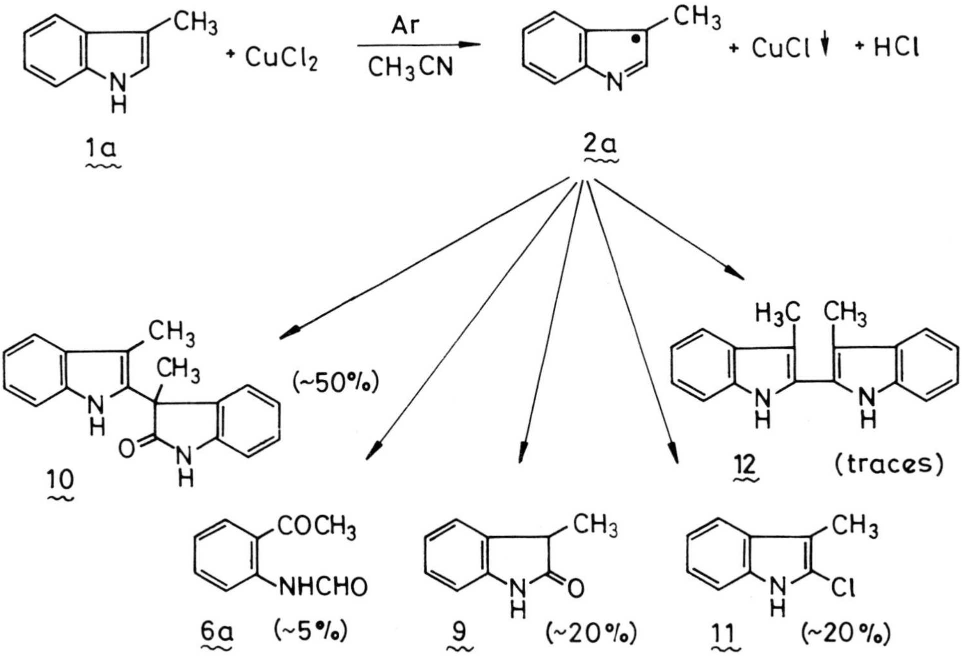

References

1. Hayaishi, O.: Molecular Mechanism of Oxygen Activation, Academic Press, New York 1974.
2. Brady, F.O.: Bioinorg. Chem. 5, 167 (1975).
3. Hirata, F., Hayaishi, O.: Superoxide and Superoxide Dismutases (Eds. Michelson, A. M., McCord, J. M., Fridovich, I.) Academic Press, London 1977.
4. Balogh-Hergovich, É., Speier, G.: Tetrahedron Lett. 4473 (1982).
5. Balogh-Hergovich, É., Speier, G.: J. Inorg. Biochem. 13, 297 (1980).
6. Tsuji, J., Takayanagi, H.: Chem. Lett. 65 (1980).
7. Tsuji, J., Takayanagi, H., Yamamoto, K.: Bull. Chem. Soc. Jpn. 54, 2369 (1981).

SEROTONIN- AND MELATONIN-DEPENDENT LIGHT EMISSION INDUCED BY XANTHINE OXIDASE

TOMIHIKO UEMURA and KEN KADOTA
Division of Neurochemistry, Psychiatric Research Institute of Tokyo, 2-1-8, Kamikitazawa, Setagaya-ku, Tokyo 156, Japan

Introduction

Xanthine-xanthine oxidase (X-XO) system can metabolize many compounds both aerobically and anaerobically. Aerobically it metabolizes methional (1), tryptophan (2), salicylic acid, phenol (3) by hydroxyl radical ($OH^{\bullet}$) derived from superoxide anion (O_2^-), luminol by both $OH^{\bullet}$ and O_2^- (4), ascorbic acid and α-tocopherol (5). O_2^--dependent metabolism of melatonin and 5-HT (5-hydroxytryptamine, serotonin) by indoleamine 2,3-dioxygenase has been well established (6). This enzyme couples well with XO system both in vitro (7) and in vivo (8) so as to support the metabolism of indoleamines. In this communication, we report the activity of bovine milk XO as a catalyst for the oxidative degradation of 5-HT and melatonin in the presence of Fe-EDTA complex in alkaline media probably via a transient chemiluminescent compounds.

Results

1) Effect of various indole compounds and pH on the chemiluminescence (CL) intensity induced by X-XO system.
Among the various indole compounds examined, 5-HT and melatonin were the most active to induce CL monitored by a Packard Tri-Carb model 3330 Liquid Scintillation Spectrometer with the coincidence turn off in [^{3}H]-range. The

Progress in Tryptophan and Serotonin Research

addition of $FeSO_4$ as Fe-EDTA complex decreased the CL intensity of 5-HT but increased strongly that of melatonin. The replacement of H_2O by D_2O did not enhance the light intensity of 5-HT and melatonin. Fig.1 shows the time courses of the light emission from both indole compounds continuously monitored at various pH by an Aloka Luminescence Reader, BLR-101. The maximum emission was seen at pH 10.2 for 5-HT and at 9.5 for melatonin, respectively. (Note that the scale of the light intensity of melatonin at pH 9.5 was one third smaller than the others.) It was seen about 5 min after the initiation of the reaction and disappeared gradually within 10 to 20 min.

2) Effects of various scavengers for the active oxy-species on the CL.

Both O_2^- scavengers (catechol, Tiron, SOD) and H_2O_2 scavenger (catalase) suppressed the CL as do GSH and ascorbate. The $OH^{\bullet}$ scavengers (mannitol, sodium benzoate, dimethylsulfoxide) exerted no effect on the CL induced by 5-HT and

Effect of pH on serotonin- or melatonin-induced chemiluminescence of the xanthine-xanthine oxidase system.

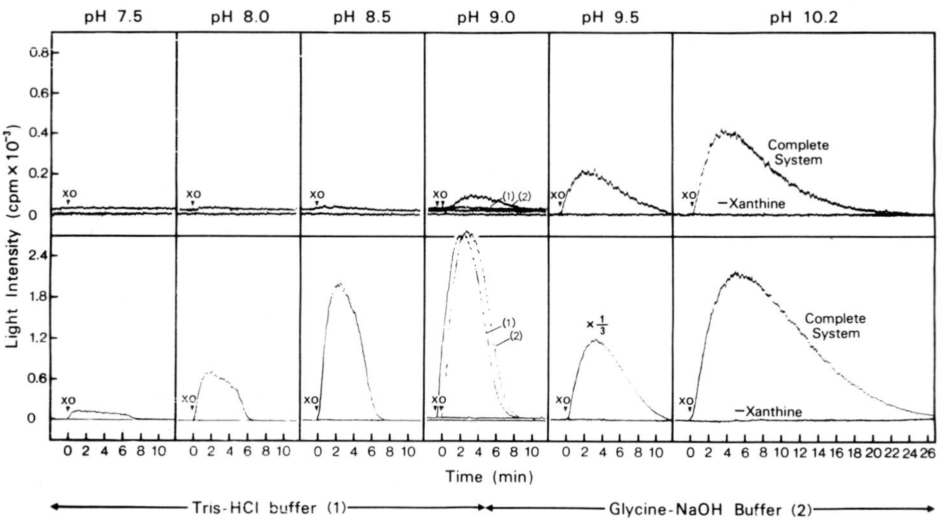

The reaction mixture, in 0.912 ml final vol. contained 0.05 M buffer, 100 µM EDTA-K_2, 66 µM X, 20.75 nM XO and 100 µM 5-HT or 200 µM melatonin. Temp. 30°C.

a very slight inhibitory effect on that induced by melatonin. Singlet oxygen quenchers (NaN_3, 2,5-dimethylfuran) exerted an inhibitory effect on 5-HT-dependent CL without any effect on melatonin-dependent CL. These results indicate that the two reactive oxy-species, i.e. O_2^- and H_2O_2, are necessary to induce CL from 5-HT and melatonin.

3) 5-HT disappearance and a pink coloured product formation catalyzed by XO (Fig.2).

The incubation of 5-HT with X-XO system at pH 10.2 in the presence of Fe-EDTA complex caused a pink coloured product with a concomitant loss of the amount of 5-HT. It could not be formed in the control mixture lacking one of the three components, i.e. X, XO and Fe-EDTA complex, despite

Disappearance of 5-HT and appearance of a pink coloured product in the presence or absence of $FeSO_4$ catalyzed by xanthine oxidase.

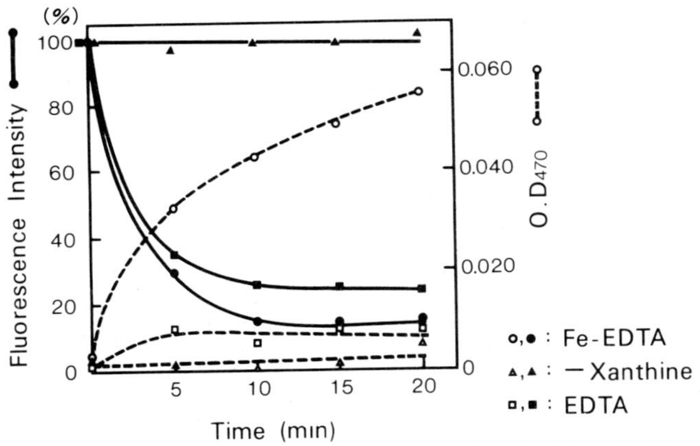

For fluorescence, the reaction mixture, in 0.42 ml final vol. contained 0.42 mM X, 150 µM EDTA-K_2, 0.05 M glycine-NaOH buffer, pH 10.2, 100 µM $FeSO_4$ if necessary, 12 µg of 5-HT and 10 µg of XO. An aliquot of 0.1 ml of the reaction mixture was added to 2.9 ml of 3 N HCl. After centrifugation, the fluorescence was read at 550 nm with activation at 295 nm. For optical density, the reaction mixture, in 1.68 ml final vol. contained the same components as described except for the addition of 240 µg of 5-HT and 40 µg of XO. The reaction was stopped by adding 20 µl of 100 % TCA. After centrifugation, the optical density was read at 470 nm against a reference cuvette where the supernatant from the reaction mixture with no addition of XO was added.

the loss of the 5-HT fluorescence even in the absence of extra-added $FeSO_4$. The same experiment for melatonin showed the loss of its fluorescence regardless of the addition of $FeSO_4$ without a pink colour formation.

Discussion

The CL process is transient and may represent only a small portion of the total chemical change. The time courses of the pink colour appearance, of the fluorescence disappearance and those of CL suggest that CL may be an initial event for 5-HT to be transformed into a pink coloured product which may be an initial oxidation product(s) of 5-HT. We are now attempting to characterize a metabolite(s) of 5-HT and melatonin produced by X-XO system in the presence of Fe-EDTA complex.

References

1. Beauchamp, C. and Fridovich, I. J. Biol. Chem. 245, 4641-4646 (1970).
2. McCord, J. M. and Day, E. D. FEBS Lett. 86, 139-142 (1978).
3. Richmond, R., Halliwell, B., Chauhan, J. and Darbre, A. Anal. Biochem. 118, 328-335 (1981).
4. Hodgson, E. K. and Fridovich, I. Photochem. Photobiol. 18, 451-455 (1973).
5. Nishikimi, M. and Yagi, K. Biochemical and Medical Aspects of Active Oxygen (Hayaishi, O. and Asada, K. eds.) 79-87, Japan Scientific Societies Press (1977).
6. Hayaishi, O. J. Biochem. (Tokyo) 79, 13p-21p (1976).
7. Hirata, F., Ohnishi, T. and Hayaishi, O. Biochemical and Medical Aspects of Active Oxygen, 101-109, Japan Scientific Societies Press (1977).
8. Taniguchi, T., Hirata, F. and Hayaishi, O. Biochemical and Medical Aspects of Active Oxygen, 111-117, Japan Scientific Societies Press (1977).

THE IN VIVO EFFECT OF SODIUM FLUORIDE ON THE KEY ENZYMES OF TRYPTOPHAN METABOLISM

François Geeraerts, Lucien Schimpfessel, Robert Crokaert

Laboratorium voor Biochemie, Vrije Universiteit Brussel, B-1090 Brussels, Belgium.

Introduction

Sodium fluoride is known to interact with peripheral tryptophan (TRP) metabolism by lowering the activity of the serotonin pathway (1) and by inhibiting, at least *in vitro*, the key enzyme of the kynurenine pathway, TRP-pyrrolase (2). These observations and the fact that more and more fluor containing drugs are made available, prompted us to study the *in vivo* effect of NaF on these two pathways by examining the influence on liver TRP-pyrrolase and on the brain TRP-hydroxylating system.

Experimental

1) Sample preparation.

Sodium fluoride (10 mg/kg in 0.9 % NaCl) was administered to a first series of male Wistar rats (250-300 g) by stomach intubation. Control rats received only the vehicle. A second series received together with the NaF, 50 mg L-TRP/kg, also by stomach intubation. A central decarboxylase inhibitor (hydroxy-benzyl-hydrazine) was injected I.P.

Progress in Tryptophan and Serotonin Research

(100 mg/kg) to all animals at t = 0. At t = 120 min, the animals were killed by decapitation. Livers and brain stems were dissected at 4°C, homogenized with a Potter-Elvehjem homogenizer, respectively in 3 volumes phosphate buffer (0.02 M, pH 7.4, containing 0.14 M KCl) and 2.5 volumes TRIS-acetate (0.05 M, pH 7.4, containing 1 mM dithiotreitol) and centrifuged in a Beckman LC-65 ultracentrifuge at 30,000 g for 60 min at 4°C. The supernatant was used as enzyme source after removal of the upper milky layer.

2) Enzyme determinations.

The activity of the two forms of liver TRP-pyrrolase, the inactive apoenzyme (AE) and the active haem-conjugated complete enzyme (CE) were determined by monitoring the total enzyme activity (TE = CE + AE) at 360 nm in the presence of added haematin, and the CE activity in the absence of external haem. The AE activity was calculated by difference (3,4). The *in vivo* hydroxylation of tryptophan was determined by measuring the amount of 5-hydroxy-TRP formed during the 2 h period by a HPLC-fluorometric analysis of the deproteinized 30,000 g brain stem supernatant as published earlier (5). The liver dihydropteridine reductase activity was measured by monitoring the coupled oxidation of NADH at 340 nm with 6-MPH_4 as substrate according to Craine et al. (6).

Results and Discussion

The results obtained for TRP-pyrrolase without and with induction by TRP are shown in Table I.

From this, it appears that, also *in vivo*, NaF lowers the haem saturation of TRP pyrrolase : it inhibits the conjugation of the apoenzyme with the prosthetic haem group, just as was observed *in vitro* (2). Addition (*in vitro*) of an excess of haem restores the activation of the AE, especially when the AE is "protected" by TRP. This suggests that NaF has a direct effect on the conjugation of the AE with the

Table I : Effect of NaF on liver TRP-pyrrolase activity.

1) Without TRP administration

	Controls (n = 10)			NaF (n = 10)					
				with 2 μM haem[2]			with 20 μM haem[2]		
	Activity	%	HSR[1]	Activity	%	HSR[1]	Activity	%	HSR[1]
TE	3.32			2.80			2.80		
CE	0.83	25	.333	0.42	15	.176	0.57	20.2	.255
AE	2.49	75		2.38	85		2.23	79.8	

2) With 50 mg L-TRP/kg

	Controls (n = 10)			NaF (n = 10)					
				with 2 μM haem[2]			with 20 μM haem[2]		
	Activity	%	HSR[1]	Activity	%	HSR[1]	Activity	%	HSR[1]
TE	8.30			3.57			3.60		
CE	4.40	53	1.128	0.68	19	.234	1.83	51	1.04
AE	3.90	47		2.89	81		1.77	49	

Results are expressed in μmol kynurenine formed/h per g wet weight of liver.
[1]HSR = Haem Saturation Ratio = CE/AE.
[2]in reaction mixture.

Table II : Effect of NaF on brain stem TRP hydroxylation.

	nmoles of 5-OH-TRP formed/h per g of protein	
Controls (n = 6)	15.80	$p < 0.05$
NaF (n = 5)	11.12	

prosthetic group. Further, *in vitro*, it could be observed that NaF does not inhibit the enzymatic action of complete TRP-pyrrolase.

As shown in Table II, NaF is interfering with the TRP hydroxylating system too. In a same period of time (2 hours) NaF-treated animals formed only 70.4 % 5-hydroxy-tryptophan of the amount formed by the controls. These results extend the former observations (1) where evidence was presented that NaF is inhibiting the peripheral hydroxylation of tryptophan.

In order to locate the block of this hydroxylation, the *in vitro* effect of NaF on the activity of dihydropteridine reductase, the enzyme regenerating the pterine cofactor of the hydroxylase, was measured. No effect could be detected. This could indicate that the interaction NaF - TRP hydroxylation system occurs at the hydroxylase level itself.

From these data, it can be concluded that *in vivo*, sodium fluoride decreases the velocity of the two most important pathways of tryptophan metabolism, by inhibiting the tryptophan-pyrrolase and tryptophan-hydroxylase activities.

References

1. Geeraerts, F., Schimpfessel, L., Crokaert, R.: Fluoride 14, 155-160 (1981).
2. Geeraerts, F., Schimpfessel, L., Crokaert, R.: Arch. Int. Physiol. Biochim. 90, B118-B119 (1982).
3. Badawy, A.-A.B., Evans, M.: Biochem. J. 138, 445-451 (1974).
4. Piras, M.M., Knox, W.E.: J. Biol. Chem. 242, 2952-2958 (1967).
5. Geeraerts, F., Schimpfessel, L., Crokaert, R.: Chromatographia 15, 449-450 (1982).
6. Craine, J.E., Hall, E.S., Kaufman, S.: J. Biol. Chem. 247, 6082-6091 (1972).

ROLES OF ACTIVE OXYGEN AND SUPEROXIDE DISMUTASE IN ANTHRANILAMIDE HYDROXYLATION AND ITS DERIVED PHENOXAZINONE-LIKE PIGMENT FORMATION BY RAT LIVER MICROSOMAL MONOOXYGENASE SYSTEM

Yoshiji Ohta, Rikio Shinohara, Junko Naito and Isao Ishiguro
Department of Biochemistry, Fujita-Gakuen University School of Medicine, Toyoake, Aichi 470-11, Japan

Introduction

Anthranilamide (An.Amide), one of the metabolites of anthranilic acid, is markedly hydroxylated by rat liver microsomal monooxygenase using NADPH as an electron donor in the presence of superoxide dismutase (SOD) (1,2). The reaction mechanism of this hydroxylation was studied with a reconstituted system of rat liver microsomal monooxygenase (2). The following reaction mechanism was proposed: A trace of 3-hydroxyanthranilamide (3-OH An.Amide) is produced in the initial stage of reaction. This metabolite reacts with O_2^- produced by NADPH-cytochrome c reductase (fp_2) to form a reaction intermediate. The intermediate stimulates O_2^- formation by coupling with fp_2 to lead to a marked NADPH oxidation and H_2O_2 formation. Consequently, An.Amide is hardly hydroxylated by cytochrome P-450 (P-450). However, in the presence of SOD, flow of electrons from NADPH to P-450 is increased since O_2^- produced by fp_2 is removed and thus 3-OH An.Amide cannot react with O_2^-, resulting in the easy hydroxylation of An.Amide by P-450.

Threrefore, the present work has been undertaken to identify the reaction intermediate of 3-OH An.Amide with O_2^- by the reaction of the intermediate with various quinone and semiquinone trappers and by the formation of a pigment derived from the intermediate. The results indicate that 3-OH An.Amide reacts with O_2^- produced by fp_2 to form its quinone which is trans-

Progress in Tryptophan and Serotonin Research

formed into the semiquinone radical by coupling with fp_2.

Results and Discussion

When a trace of 3-OH An.Amide was added to NADPH-fp_2 system, marked increases in oxidation of NADPH and formation of O_2^- and H_2O_2 with lag time were observed. But these increases were intensively inhibited by the prior addition of SOD to this system. These findings suggest that 3-OH An.Amide reacts with O_2^- produced by fp_2 to form a reaction intermediate which stimulates O_2^- formation by coupling with fp_2 to lead to a marked NADPH oxidation and H_2O_2 formation. Hence, the reaction of the intermediate with various quinone and semiquinone trappers was studied.

Reduced glutathione (GSH) and cysteine, which are known to react with electrophiles (3) and semiquinone radicals (4) as well as O_2^- (5), strongly inhibited NADPH oxidation catalyzed by fp_2 with 3-OH An.Amide under conditions that the reaction intermediate of 3-OH An.Amide with O_2^- produced by fp_2 could be coupled with fp_2 and that O_2^- generated by the coupling reaction could be removed by SOD added on the way of reaction. Ascorbic acid, which is known to react with semiquinones and quinones (6) as well as O_2^- (7), also 3-OH An.Amide-stimulated NADPH oxidation in NADPH-fp_2 system under the same conditions as used for GSH and cysteine. Lysine, which is known to react with o-quinones (3), also inhibited 3-OH An.Amide-stimulated NADPH oxidation and H_2O_2 formation in NADPH-fp_2 system. These results suggest that the intermediate of 3-OH An.Amide with O_2^- is its quinone form and this quinone is transformed into the semiquinone radical by coupling with fp_2.

Then, the formation of a pigment derived from the intermediate was further examined since o-aminophenols are known to condense oxidatively into phenoxazinone pigments via their quinone forms as reaction intermediates (8). When 3-OH An.Amide was added in excess to NADPH-fp_2 system, the formation of a yellow pigment was observed. The pigment extracted from the reaction mixture with water-saturated butanol showed an absorption spectrum having a peak at 440 nm. The absorption spectrum was very similar to that of the phenoxazinone pigment which was made by direct oxidation of 3-OH An.Amide with ferricyanide according to the method of Butenandt et al. (9). Both pigments also showed the same R_f value on the thin layer chromatogram of Silica gel developed with the solvent of butanol/acetic acid/water (4:1:1). These findings indicate that the pigment made from 3-OH An.Amide in NADPH-fp_2 system is a phenoxazinone-like one and therefore a quinone form of 3-OH An.Amide is able to be produced as a reaction intermediate in NADPH-fp_2 system. Moreover, the formation of the phenoxazinone-like pigment in

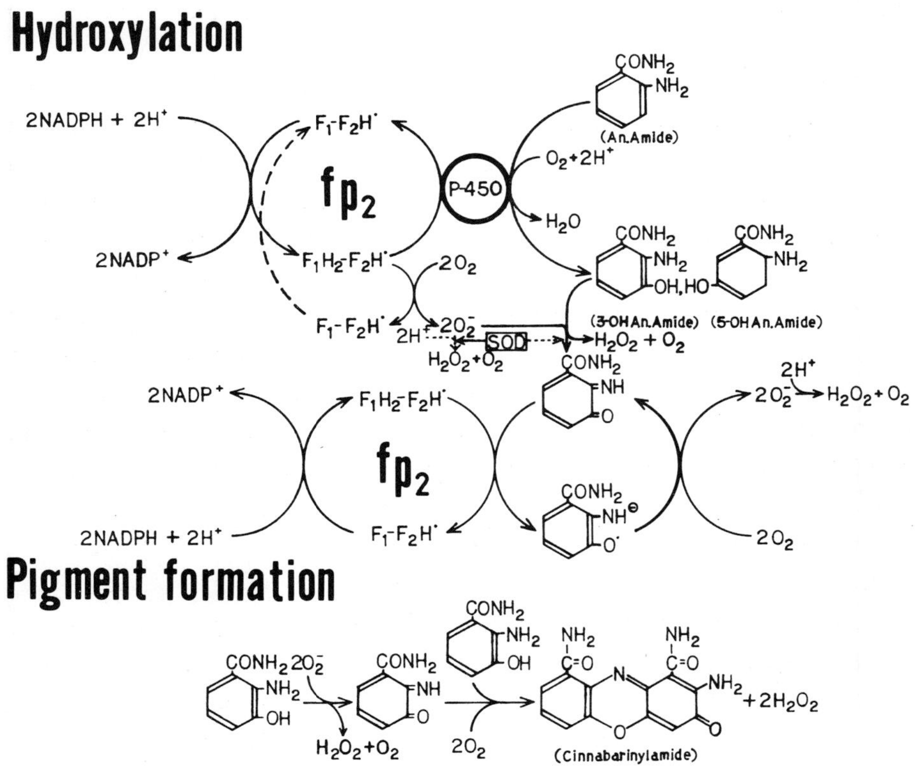

Fig. 1. Proposed schemes for An.Amide hydroxylation and its derived phenoxazinone-like pigment formation by rat liver microsomal monooxygenase system

NADPH-fp_2 system was strongly inhibited by SOD added before or after the start of reaction. On the other hand, the pigment was also formed in acetaldehyde-xanthine oxidase system, O_2^- generating system, and its formation was inhibited by SOD, indicating that the formation of the phenoxazinone-like pigment depends on the reaction of 3-OH An.Amide with O_2^-, by which a quinone form of 3-OH An.Amide can be produced.

From these results, we propose the schemes for An.Amide hydroxylation and its derived phenoxazinone-like pigment formation by rat liver microsomal monooxygenase system as shown in Fig. 1.

References

1. Ishiguro, I., Naito, J., Shinohara, R., Ishikura, A., Terai, T.: Biochim. Biophys. Acta 365, 148-157 (1974).
2. Shinohara, R., Ohta, Y., Naito, J., Ishiguro, I.: in Biochemical and Medical Aspects of Tryptophan Metabolism. (eds. by Hayaishi, o., Ishimura, Y., Kido, R.), Elsevier/North-Holland, Amsterdam, p. 325 (1980).
3. Bolt, H.M., Kappus, H.: J. Steroid Biochem. 5, 179-184 (1974).
4. Scheulen, M., Wollenberg, P., Bolt, H.M., Kappus, H., Remmer, H.: Biochem. Biophys. Res. Commun. 66, 1396-1400 (1975).
5. Asada, K., Kanematsu, S.: Agr. Biol. Chem. 40, 1891-1892 (1976).
6. Demmer, D.W., Heikkila, R.E., Panganamala, R.V., Cohen, G., Cornwell, D.G.: Physiol. Chem. Physics 3, 426-430 (1971).
7. Nishikimi, M.: Biochem. Biophys. Res. Commun. 63, 463-468 (1975).
8. Suba Rao, P.V., Vaidyanathan, C.S.: Biochem. J. 99, 317-322 (1966).
9. Butenandt, A., Schiedt, U., Biekert, E.: Liebigs Ann. Chem. 588, 106-116 (1954).

PROTECTIVE EFFECT OF TRYPTOPHAN UPON THE INFLUENCE OF N-METHYLFORMAMIDE ON THE DEVELOPMENT, THE NAD SYNTHESIS, AND THE ACTIVITY OF THE ADENOSINE DIPHOSPHATE RIBOSYL TRANSFERASE OF RAT EMBRYOS.

Hans Kröger, Regina Grätz, Helga Grahn

Robert Koch-Institut, Nordufer 20
D-1000 Berlin 65

Introduction

Embryotoxicity can be provoked by numerous substances. One of these is N-methylformamide, as several investigations have shown (1 - 5). Only little is known about the action mechanism of these substances.

Results

Applying N-methylformamide (MFA) intraperitoneally to rats on the 11th day of pregnancy causes resorption of the fetuses to about 44 % (Table 1). As can be seen from the comparison of weight to the control animals, there is a growth retardation. L-Methionine, especially in combination with nicotinamide, increases the resorption rate of MFA. Tryptophan on the other hand has a protective effect.

From these results came the hypothesis that MFA could have an influence on the NAD metabolism of the rat embryos. As can be seen in Table 2 nicotinamide - as known - gives a higher rise of NAD than L-tryptophan. In both cases no change in NAD synthesis can be observed under the influence of MFA.

Besides the role as coenzyme in oxidation-reduction processes the overwhelming amount of NAD is involved in adenoribosylation

Progress in Tryptophan and Serotonin Research

Experiment	Total N mothers	Total N embryos	Embryos alive %	Embryos resorbed %	% Weight MFA + Subst. / NaCl-control	% Weight MFA + Subst. / MFA	% Weight MFA + Subst. / Substance	% Weight Substance / NaCl-control	% Weight Substance / MFA
NaCl (0.15 M)	12	141	96.7	3.3	100	-	-	100	119
N-Methylformamide	13	138	55.8	44.2	80.6	100	-	-	-
N-Methylformamide + Nicotinamide	10	107	57.9	42.1	64.4	76.6	66.8	-	-
N-Methylformamide + L-Methionine	7	95	15.9	84.1	81.0	100.1	76.7	-	-
N-Methylformamide + L-Tryptophan	11	120	79.9	20.1	89.3	106.2	82.5	-	-
N-Methylformamide + Nicotinamide + L-Methionine	6	60	1.0	99.0	54.5	64.2	56.4	-	-
N-Methylformamide + L-Tryptophan + L-Methionine	9	95	49.3	50.7	78.8	98.0	78.4	-	-
Methylformamide + N-Methylnicotinamide	7	81	32.1	67.9	80.5	95.8	79.9	-	-
Nicotinamide	7	71	96.3	3.7	-	-	-	96.3	114.6
L-Methionine	6	65	98.4	1.6	-	-	-	107.0	121.2
L-Tryptophan	10	103	96.3	3.7	-	-	-	108.2	128.8
Nicotinamide + L-Methionine	8	82	93.9	6.1	-	-	-	96.4	114.7
L-Tryptophan + L-Methionine	9	90	98.3	1.7	-	-	-	101.8	121.1
N-Methylnicotinamide	7	88	88.4	11.6	-	-	-	101.3	120.5

Table 1: Influence of different substances upon the embryonal development of rats.

The substances were administered i.p. to the animals on the 11. day of pregnancy and the embryos delivered by Ceasarean section on the 21. day. The substances (500 mg/kg) were given to the animals twice in intervals of two hours. In the case of a combination with N-methylformamide (MFA) (1 ml/kg), the substances were administered 1 hour before MFA and 1 hour afterwards MFA.

Table 2: NAD content of rat embryos under influence of different substances.

	Control	+ MFA
NaCl	0.199	0.222*
Nicotinamide	0.365	0.329
L-Methionine	0.245	0.252
L-Tryptophan	0.248	0.234
Nicotinamide + L-methionine	0.329	0.353
L-Tryptophan + L-methionine	0.245	0.257
N-Methylnicotinamide	0.235	0.268

The substances were i.p. administered to the animals on the 11th day of pregnancy and the embryos delivered by Caesarian section for the determination of NAD 8 h later. N-Methylformamide (MFA): 1 ml/kg. All the other substances: 500 mg/kg 1 h before and 1 h after administration of N-methylformamide. For each determination 5 - 8 animals were used.
*Given as µmoles NAD + $NADH_2$/g tissue.

of proteins (S. Shall, personal communication). To test the effect of MFA on this enzyme, nuclei of 11 day old embryos were isolated and the activity of adenosine diphosphate ribosyl transferase (ADPR) determined in vitro. MFA considerably inhibits the activity of this enzyme (Fig. 1).

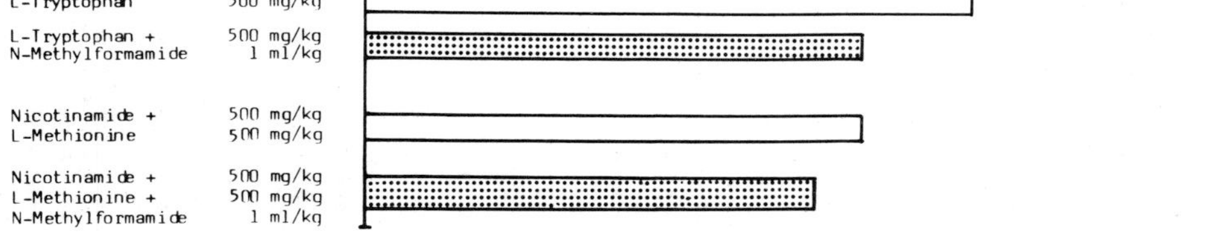

Fig. 1: Activity of the ADPR-transferase in cell nuclei of 11 days old rat embryos under different treatment. The animals were killed two hours after i.p. application of the substances.

L-tryptophan reduces this effect more than nicotinamide + L-methionine. On the other hand nicotinamide + L-methionine have an inhibitory effect by itself.

Discussion

Quite a number of proteins are adenoribosylated. Several reports indicate an involvement of adenoribosylation in regulation and differentiation (6, 7). For the first time, Caplan and Rosenberg (8) concluded that this metabolism is involved in differentiation. There are also several papers about adenoribosylation and repair processes (9, 10).

We have demonstrated that the embryotoxicity of MFA is increased by using nicotinamide + methionine but reduced by tryptophan. This is probably due to the influence on the enzyme ADPR transferase.

References

1. Thiersch, J.B., J. Reprod. Fertil. 4, 219 (1962).
2. Oettel, J., Frohberg, H., Naunyn Schmiedebergs Arch. Pharmacol. 274, 363 (1964).
3. Roll, R., Bär, F., Arzneim. Forsch. 17, 610 (1967).
4. Kreybig, Th. v. et al., Arzneim. Forsch. 18, 645 (1968).
5. Lorke, D., Berichte des Institutes für Arzneimittel des BGA 1, 54 (1978).
6. Hilz, H., Stone, P.R., Rev. Physiol. Biochem. Pharmacol. 76, 1 (1976).
7. Purnell, M.R., Stone, P.R., Whish, W.J.D., Biochem.Soc. Transactions 8, 215 (1980).
8. Caplan, A.J., Rosenberg, M.J., Proc. Natl. Acad. Sci. USA 72, 1852 (1975).
9. Whish, W.J.D., Davis, M.J., Shall, S., Biochem. Biophys. Res. Commun. 45, 722 (1975).
10. Durkacz, B.W., Omidigi, O., Gray, D.A., Shall, S., Nature 282, 593 (1980).

INDUCTION OF TYROSINE AMINOTRANSFERASE BY TRYPTOPHAN IN RAT LIVER

Hans Kröger, Regina Grätz

Robert Koch-Institut, Nordufer 20
D-1000 Berlin 65

Introduction

The enzyme tyrosine aminotransferase (TAT) can be induced by tyrosine and cortisone. It was shown for the first time by Kenney and Flora (1) that the TAT can be induced by tryptophan too. They found that methionine also increases the activity of this enzyme to a small extent.

We have analyzed the tryptophan effect in more detail. We focused our interest especially on the question of whether the tryptophan exerts its effect via the NAD-adenoribosylation metabolism.

Results

A small increase of TAT can be observed 6 hrs after administration of DL-tryptophan (Fig. 1). This increase persists only over a very short period.

L-methionine - on the other hand - already gives maximal increase of enzyme activity after 1 hour (Fig. 2). Normal values are reached 6 hrs after the injection of this amino acid.

A considerable increase of TAT activity can be obtained when the animals receive methionine at different times after (Fig. 3) the administration of tryptophan. The highest values are achieved

Progress in Tryptophan and Serotonin Research

when methionine is administered 2 - 3 hours after the application of tryptophan

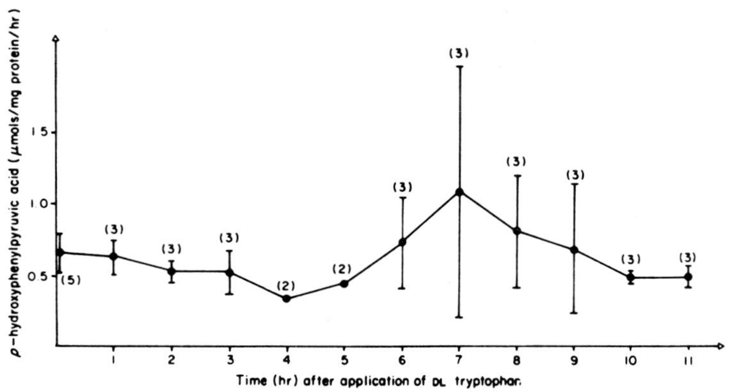

Fig. 1. TAT activity after administration of DL-tryptophan (300 mg/kg).

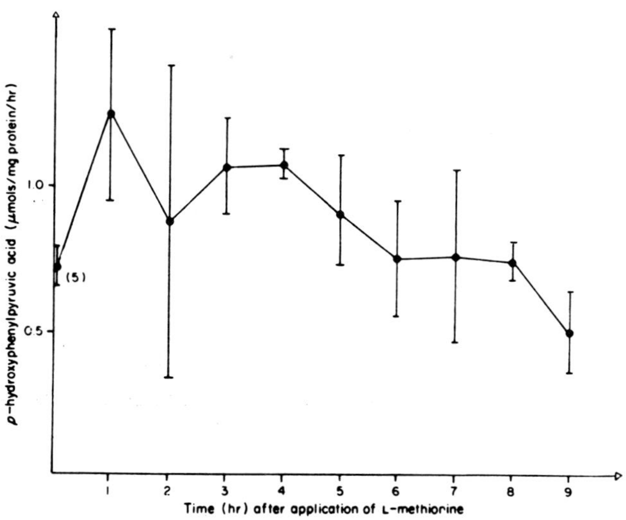

Fig. 2. TAT activity after administration of L-methionine (300 mg/kg). Unless otherwise indicated, data were calculated from values for three animals each.

We know from other experiments that the hepatitis induced by D-galactosamine can be prevented by a combination of tryptophan + methionine (3). Furthermore we showed that adenosine diphosphate ribosyl transferase (ADPR), the enzyme of the adenoribosylation, is inhibited by galactosamine (4). We concluded from these results that NAD adenoribosylation metabolism plays an important role in the development of galactosamine hepatitis.

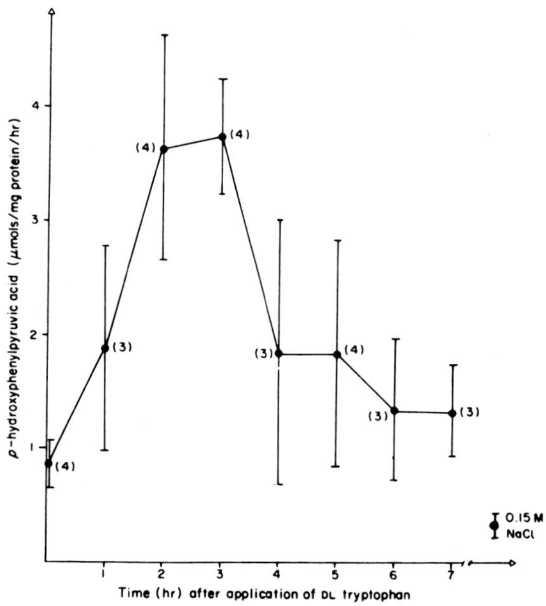

Fig. 3. Induction of tyrosine aminotransferase in the presence of DL-tryptophan and L-methionine. Administration at time zero: 300 mg/kg DL-tryptophan, followed by 500 mg/kg L-methionine at the times indicated. The animals were killed 4 hr later.

It is known from experiments of Reynolds and Reutter (5) that the induction of the TAT by hydrocortisone is inhibited by D-galactosamine. We can confirm these results. Furthermore we found that there is a dose dependency of this effect.

After injection of L-tyrosine in normal animals the activity of the TAT increases about tenfold. This effect is completely inhibited by D-galactosamine.

There is a considerable increase of TAT after injection of tryptophan and methionine. In this case D-galactosamine also blocks the enzyme rise (Table 1).

Treatment of the animals		µmol/mg protein/hr p-hydroxyphenylpyruvic acid
DL-Tryptophan + L-Methionine	(500 mg/kg) (500 mg/kg)	3.54 ± 0.12 (4)
DL-Tryptophan + L-Methionine + D-Galactosamine-HCl	(500 mg/kg) (300 mg/kg) (375 mg/kg)	0.99 ± 0.12 (4)
D-Galactosamine-HCl	(375 mg/kg)	0.88 ± 0.40 (5)
0.15 M NaCl	(10 ml/kg)	0.51 ± 0.03 (4)

Tab. 1. Influence of D-galactosamine-HCl on the induction of the tyrosine aminotransferase by DL-tryptophan and L-methionine in liver of normal rats. Termination 4 hrs after application.

We analyzed the influence of different inhibitors of the ADPR transferase. It can be seen from table 2 that these substances have an inhibitory effect on the induction of TAT by tryptophan and methionine.

Discussion

Tryptophan is known to induce a number of enzymes (for references see (6)). The effect of tryptophan on tyrosine aminotransferase is considerably increased by methionine. This might be due to an influence of methionine on the NAD regulation. It can be seen from the inhibitory effect of antagonists of ADPR transferase that the NAD metabolism might be involved.

Treatment of the animals		µmol/mg protein/hr p-hydroxyphenylpyruvic acid
DL-Tryptophan	(500 mg/kg)	7.07 ± 0.34 (4)
DL-Tryptophan + 3-Aminobenzamide	(500 mg/kg) (25 mg/kg)	3.48 ± 1.15 (4)
DL-Tryptophan + Theophilline	(500 mg/kg) (50 mg/kg)	5.09 ± 0.62 (4)
DL-Tryptophan + Thymidine	(500 mg/kg) (300 mg/kg)	4.16 ± 1.46 (5)
DL-Tryptophan + Nicotinamide	(500 mg/kg) (500 mg/kg)	5.25 ± 0.24 (4)
0.15 M NaCl	(10 ml/kg)	1.30 ± 0.35 (5)

Tab. 2: Influence of antagonists of ADPR transferase on the induction of tyrosine aminotransferase by tryptophan in the rat liver of normal male rats (18 hrs hunger).
The animals were killed 8 hrs after the application of the substances.

We suppose that the tryptophan effect is connected with the ADP ribosylation of certain proteins. It has been shown that the adenoribosylation is involved in the processes of regulation and differentiation (7, 8). Because the activity of TAT is altered by tryptophan one might speculate that this substance is converted to NAD via chinolic acid. The NAD may be a substrate for the ADPR transferase.

References

1. Kenney, F.T., Flora, R.M. J. Biol. Chem. 236, 2699 (1961).
2. Kröger, H., Greuer, B., Biochem. Z. 341, 190 (1965).
3. Kröger, H., Grätz, R., Museteanu, C., Haase, J., Naturwissenschaften 66, 476 (1974).
4. Kröger, H., Grahn, H., Grätz, R., Int. J. Biochem; in press.

5. Reynolds, R.D., Reutter, W.: J. Biochem. Chem. 248, 1562 (1973).
6. Kröger, H., Grätz, R.: Int. J. Biochem. 12, 575 (1980).
7. Caplan, A.J., Rosenberg, M.J.: Proc. Natl. Acad. Sci. USA 72, 1582 (1975).
8. Miwa, M., Sugimura, T., Inui, N., Takayama, S.: Cancer Res. 33, 1306 (1973).

Tryptophan Metabolism in Insects

TRYPTOPHAN METABOLISM IN DROSOPHILA

David T. Sullivan

Department of Biology, Syracuse University
Syracuse, New York 13210

Introduction

Studies on the genes and their respective enzymes which are responsible for the metabolism of tryptophan in Drosophila occupy important positions in the history of biochemical genetics. As has been well documented, the essential relationship between Mendelian hereditary factors, genes, and their biochemical products, enzymes, was firmly established through a series of studies which began with the imaginal disc transplant experiments of Beadle, Ephrussi and their co-workers (1), (2). These studies led to the formal identification of substances which were interpreted to be intermediates along a pathway that led to a specific phenotype. The disc transplant studies were soon complemented by the extensive work of Butenandt and his students (3), (4), (5) which resulted in the chemical identification of the $\underline{v}^{+}$ and $\underline{cn}^{+}$ substances as kynurenine and 3-hydroxykynurenine respectively, and finally to the identification and characterization of xanthommatin as the principal tryptophan derived eye pigment.

Tryptophan metabolism in insects has been viewed as being synonymous with ommochrome eye pigment synthesis. This topic has been the subject of several comprehensive reviews (6), (7). However it is clear that the first several steps in this pathway are functionally important for tryptophan

Progress in Tryptophan and Serotonin Research

catabolism as well as for providing precursors for ommochromes. An interesting evolutionary consideration emerges when the catabolism of tryptophan in insects is compared to tryptophan catabolism in other phylogenetic groups. In many other organisms tryptophan catabolism leads directly to the synthesis of nicotinic acid derivatives which are precursors to several metabolically essential cofactors. Early in the evolution of insects the ability to synthesize nicotinic acid must have been lost. This has resulted in the insects being dependent on their diets for sufficient supplies of nicotinic acid. Since the tryptophan to ommochrome pathway is so ubiquitous in insects, it is interesting to speculate upon whether there is any evolutionary connection between the loss of the ability to synthesize nicotinic acid and gains in the ommochrome synthetic capability. It is clear that there is great diversity among insects in the timing and localization of the enzymes for ommochrome biosynthesis. The excellent and comprehensive review of Linzen (6) should be consulted for a detailed comparison of ommochrome biosynthesis in various groups. One should be hopeful that by comparative studies we may be able to answer the interesting but difficult question of whether the loss of the ability to synthesize nicotinic acid allowed for the increased importance and subsequent evolution of programmed ommochrome biosynthesis or conversely whether the demands made on tryptophan catabolism for ommochrome biosynthesis created a situation whereby it became advantageous to dispense with the ability to synthesize nicotinic acid.

In either case these evolutionary events have provided the opportunity for the evolution of a myriad of developmental programs for the pathway which converts tryptophan to xanthommatin. In this paper I will be particularly concerned with summarizing present information regarding the utilization of genetic information which specifies a developmental program in *Drosophila*.

Enzymology and Genetics of Tryptophan Metabolism

The pathway leading from tryptophan to the principal ommochrome of *Drosophila*, xanthommatin, is shown in Figure 1.

TRYPTOPHAN —(tryptophan oxygenase)→ N-FORMYL KYNURENINE

—(kynurenine formamidase)→ KYNURENINE —(kynurenine hydroxylase)→

3-HYDROXYKYNURENINE —(phenoxazinone synthetase)→ XANTHOMMATIN

Figure 1. Tryptophan to ommochrome pathway.

The pathway consists of at least four steps. Enzyme activities which are able to catalyze each of the four reactions have been identified in *Drosophila*. Noteworthy however, is the fact that not one of these enzymes has yet been structurally or enzymatically characterized.
The first enzyme is tryptophan oxygenase (E.C.1.13.1.12). This is likely the rate limiting step in the pathway. In all respects for which information is available, the *Drosophila* enzyme appears similar to the more well characterized mammalian enzyme. Tryptophan oxygenase is the product of the vermilion, *v*, locus. In all likelihood this locus contains the single structural gene for this enzyme.
Mutants of *v* have reduced or negligible tryptophan oxygenase activity. Several mutant alleles, most notably v^k, have a tryptophan oxygenase activity with altered properties. In addition, tryptophan oxygenase activity is sensitive to dosage of v^+ alleles (8).
Several lines of evidence have indicated that the levels of

tryptophan oxygenase *in vivo* can be experimentally elevated by genetic or nutritional manipulations (9), (10). The mechanism by which this elevation of activity is brought about has not been determined. This is primarily due to the lack of a specific antibody with which to study enzyme turnover.

The second enzyme, kynurenine formamidase, has been studied by Moore and Sullivan (11), (12). In *Drosophila*, but not all insects, two forms of activity have been found. These differ from one another in that form I has a molecular weight of 60,000 daltons while form II has a molecular weight of about 30,000 daltons. The two forms do not appear to be interconvertible and each has its own characteristic P_i, and thermal denaturation properties. In addition, two genetic loci which affect the dosage sensitivity of kynurenine formamidase activity have been found by segmental aneuploidy (13). A locus in cytogenetic region, 25A-27E, affects the level of Form I. A second locus in region 91B-93F specifically affects Form II activity. The discovery of two loci provides a plausible explanation, albeit circumstantial, for why the locus coding for this enzyme has not been discovered as an eye color mutant.

The third enzyme in the pathway is kynurenine hydroxylase (EC.1.14.1.2). This enzyme was initially described in *Drosophila* by Ghosh and Forrest (13) and subsequently in our laboratory (14), (15). Biochemically this enzyme resembles kynurenine hydroxylase from other sources. It requires NADPH as a cofactor but can use NADH about 10% as efficiently. Recent studies in our laboratory have shown that FAD is required by *Drosophila* kynurenine hydroxylase. Partially purified, soluble kynurenine hydroxylase that has been exhaustively dialyzed can be restored to an active state by FAD. In fact full activity of solubilized enzyme is dependent on the presence of FAD in the assay (15). In *Drosophila*, as in other organisms, this enzyme is subcellularly located in mitochondria. Distilled water washes of purified mitochondria release vesicle like particles that contain the kynurenine hydroxylase activity. This is consistent with an outer mitochondrial membrane localization as is the case for this enzyme from Neurospora (17) and mammalian sources.

It is likely that the cinnabar, cn, locus contains the structural gene for kynurenine hydroxylase. Two lines of circumstantial evidence bear on this point. First, kynurenine hydroxylase activity is proportional to gene dosage of cn^+ (14). Second, in a large scale mutagenesis experiment, a number of mutant cn alleles were obtained which define a single complementation group (17). Other eye color mutants which might have been candidates for affecting kynurenine hydroxylase have been shown to have no measurable effects on activity of this enzyme (14).

The final enzyme in the ommochrome biosynthetic pathway has been named phenoxazinone synthetase. This activity was first measured by Phillips and Forrest (19) and has been further studied by Phillips (20), by Howells and co-workers (21), and by Wiley and Forrest (22). The sum of these investigations seem to point to the existence of an enzyme, phenoxazinone synthetase, which can catalyze the oxidative condensation of 3-hydroxykynurenine to xanthommatin. However, this step in the pathway remains the most poorly understood. Early observations had suggested that the oxidative condensation reaction was promoted by small molecules, possibly quinones (23). Some further light on the nature of the reaction has come from the recent work of Wiley and Forrest (22) who have demonstrated that the oxidative condensation of 3-hydroxykynurenine can be promoted by reduced xanthommatin, and in addition they have demonstrated an enzymatic activity which can catalyze this reaction. It is noteworthy that no mutants have yet been found which lack the enzymatic ability to convert 3-hydroxykynurenine to xanthommatin. Consequently, at the present time the final steps of ommochrome biosynthesis remain poorly defined. Two viable hypothesis remain and these may not be mutually exclusive.

Tissue Interaction in Tryptophan Metabolism

In Table 1 is presented a classification of known ommochrome deficient mutants which were discovered through alterations of eye color phenotype. Only two of these, v, and cn, are likely to be structural genes for one of the enzymes of the pathway. The question arises as to what functions are controlled by the wild type alleles of these other genes which cause an altered eye phenotype when mutant. Certain pleiotropic effects appear common to mutants at these loci. All of them appear to affect the concentration of at least some intermediates in the pathway as well as resulting in failure of eye pigment synthesis. Mutations at these loci generally affect Malpighian tubule and/or testes sheath color as well as eye color. Some of them result in deficiencies of the other class of eye pigments, pteridines, as well as ommochromes.

It is evident, through the work of several laboratories, that the enzymes of tryptophan catabolism are not directly affected by mutation of these loci. Consequently, we sought alternative explanations for these mutant eye color phenotypes. Drawing upon suggestions that had been in the eye color mutant literature for some time, we proposed that the reason these mutants failed to make pigments was because they were deficient in the transport of a metabolite(s) which is a pigment precursor (20). The formulation of this

TABLE I

Properties of ommochrome deficient mutants	
Class I	structural gene loci *v*, *cn*
Class II	eye ommochrome effects only *cd*
Class III	pleiotropic effects on tryptophan-metabolites. *ca*, *cho*, *kar* *lt* *ltd* *rc* *st*
Class IV	pleiotropic tissue effects on tryptophan and pteridine metabolism *bri*, *car*, *dor*, *g* *ma*, *p*, *pn*, *z*, *w*.

hypothesis seemed to require that either the relevant metabolite would be tryptophan, or that the ommochrome pathway would not be intact in any one cell type. The principal site of kynurenine production and storage in *Drosophila* is the fat body (24). The developmental profile of tryptophan oxygenase and kynurenine formamidase, Figure 2, are in the form of a bimodal curve which is typical of enzymes of intermediary metabolism. The principal tissue in which these enzymes are found is the fat body (10), (11). Kynurenine hydroxylase has a developmental profile different than tryptophan oxygenase, Figure 3. There is a peak of activity in third instar larvae, but the second peak occurs in mid pupae (14). We have recently determined that the Malpighian tubules are the only larval cells in which kynurenine hyroxylase activity can be measured (15). The enzyme in pupae is principally (greater than 90%) confined to the developing eye. There may be a small amount persisting in the Malpighian tubules at the pupal stage as well.

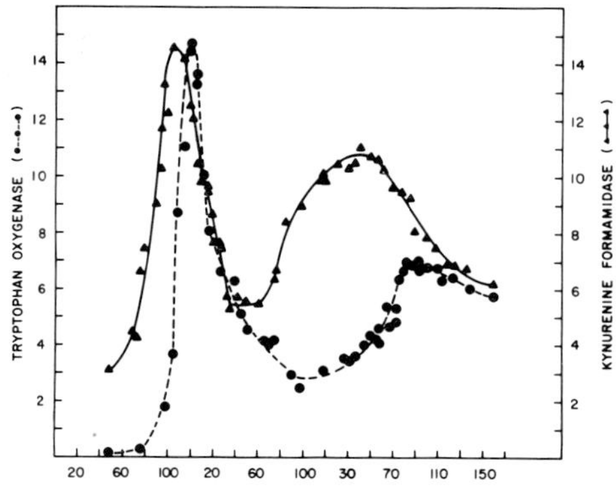

Figure 2. Developmental profile of tryptophan oxygenase and kynurenine formamidase.

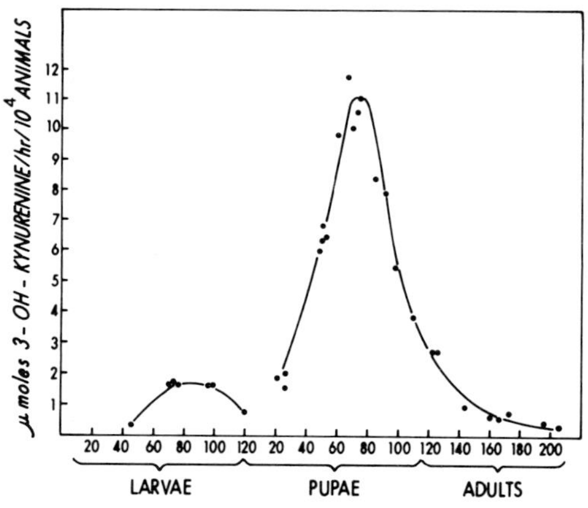

Figure 3. Developmental profile of kynurenine hydroxylase.

We have not been able to detect the conversion of tryptophan to kynurenine in dissected Malpighian tubules, although controversy on this point exists.

As a result of these developmental considerations we envision that the following series of tissue interactions occur to enable the coordinated synthesis of xanthommatin and its precursors. The existence of the large number of eye color mutants will allow us to test part of this model. Our model, Figure 4, proposes that in larvae kynurenine from the fat

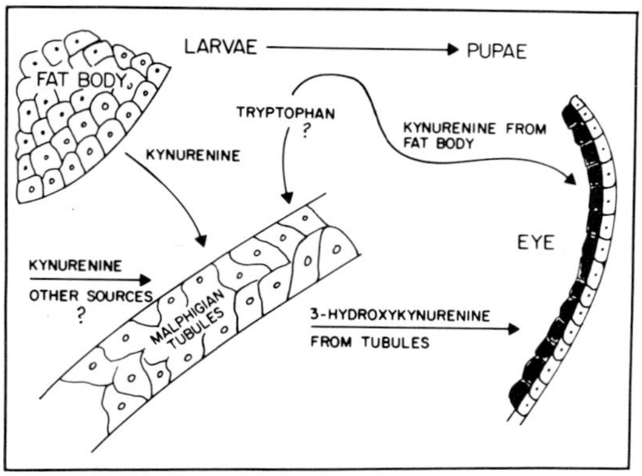

Figure 4. Tissues interactions in tryptophan metabolism.

body (possibly other sources) is released gradually into the hemolymph from which it is transported into cells of the Malpighian tubules where, in $\underline{cn}^{+}$ animals, it can be converted to 3-hydroxykynurenine. This compound accounts, at least in part, for the yellow color of the tubules. In pupae, kynurenine from the fat body (possibly tryptophan and/or 3-hydroxykynurenine) must be transported into the developing eye in order for the production of 3-hydroxykynurenine, and subsequently xanthommation to occur.

Biochemical Genetics of Tryptophan Transport

Our attempts to test this model have involved measuring

the uptake and conversion of the tryptophan metabolites in eyes and Malpighian tubules dissected from wild type and eye color mutant animals (25), Table 2.

From these data we conclude that wild type tubules have the ability to take kynurenine into their cells and convert it to 3-hydroxykynurenine. Dissected tubules from the mutant strains st, ltd, ca, or w do not take up kynurenine nor do they produce 3-hydroxykynurenine. These mutants do have normal amounts of kynurenine hydroxylase in their Malpighian tubules (14) but apparently this enzyme has no access to its substrate. A similar but genetically nonidentical

TABLE 2

Uptake and Conversion of [^{3}H] Kynurenine into Malpighian Tubules and Developing Eyes of Eye Color Mutant Strains

Genotype	Tubules		Eyes	
	Kyn.	3OHK	Kyn.	3OHK
+	9.8	16.3	1.8	6.2
cn	10.9	0.5	3.6	0.3
st	1.3	0.7	1.1	1.8
ltd	1.1	0.3	1.0	1.9
ca	1.2	0.8	0.8	4.2
w	1.5	2.2	0.8	1.1
cd	-	-	1.7	4.8

Data are expressed as d.p.m. x 10^{-3}

situation is found in the developing eyes of pupae. Wild type and cd eyes have the ability to transport kynurenine and convert it to 3-hydroxykynurenine. Eyes dissected from st, ltd or w do not. The mutant ca seems to be an example of a case where the tubules are blocked in kynurenine transport but the eyes are not. This deficiency is consistent with other aspects of the ca phenotype.

The rate of kynurenine uptake by Malpighian tubules dissected from wild type, (closed circles), w (open circles), and the double mutant v, cn larvae (open triangles) has been compared and this data is presented in Figure 5. Kynurenine transport in w tubules is greatly impaired. Kynurenine transport in v, cn tubules is essentially identical to

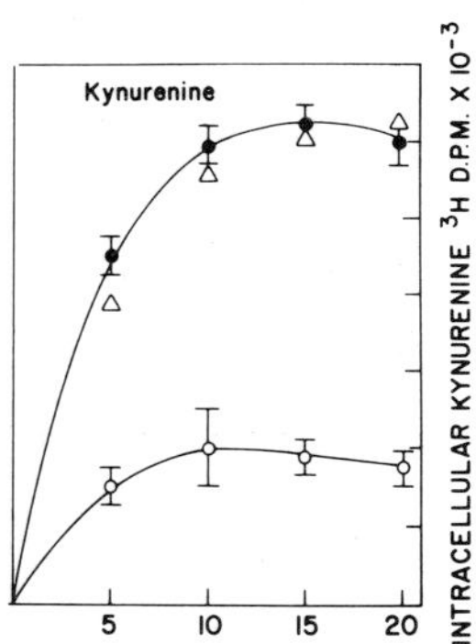

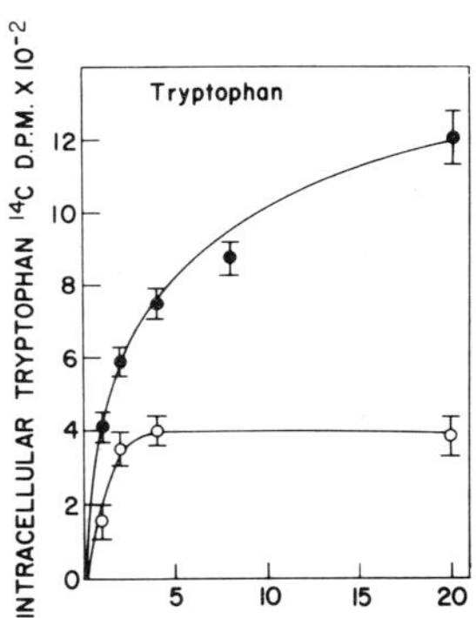

Figure 5. Figure 6.

Kynurenine transport by Malpighian tubules (for details see text).

wild type tubules indicating that kynurenine metabolism is not a driving force in kynurenine transport. What appears to be a small amount of kynurenine transport in w tubules is entry into a slowly equilibrating extracellular compartment (26).

Our comparisons between wild type and white tubules were extended to include tryptophan transport. Somewhat surprisingly, white tubules are unable to transport appreciable tryptophan, Figure 6 (wild type closed circles; white open circles). This result was somewhat puzzling because we reasoned that since tryptophan is an essential nutrient, transport would be required. Since w is such a well studied mutant, we can be confident that there are no appreciable viability affects in w mutants. We hypothesized that there must be at least two tryptophan transport routes. One leads to a large cellular pool and the other leads to a small pool, not measurable in our studies, but which is the pool used for acylation of trp-tRNA. To test this hypothesis, we have measured the incorporation of tryptophan from the medium into cellular protein by cultured tubules from w and wild type. The data shown in Figure 7, in which open figures are wild type and closed figures w reveal no difference in the rate of incorporation of tryptophan into proteins by w and wild type tubules.

We have now become convinced that there exist in *Drosophila* specific transport systems for tryptophan and its catabolites. The normal functioning of these transport systems is required for ommochrome biosynthesis. We find quite striking the amount of genetic information (as

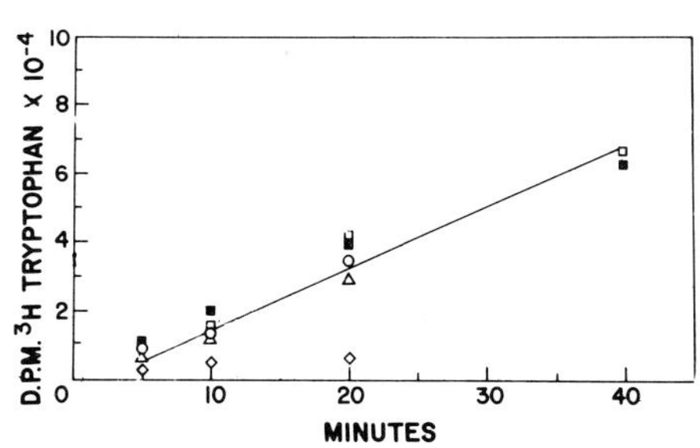

Figure 7. Incorporation of tryptophan into Malpighian tubule protein.

judged by number of independent eye color mutant loci) which is involved in the construction of these transport systems. Our current efforts are aimed at identifying the protein products of these loci. As a first step we have begun to identify those proteins in the membrane of Malpighian tubules which are specific to Malpighian tubule cells (27). By comparing the pattern of proteins in tubules with the patterns of membrane proteins from other cell types we can identify those spots only found in the Malpighian tubules. To date, we have identified approximately 80 such spots. We are now in the process of trying to associate specific spots with the products of specific eye color mutant loci. We hope this strategy will identify the protein products of some of these loci that affect the transport of tryptophan and its metabolites.

References

1. Beadle, G. W., Ephrussi, B.: Genetics. 21, 225-247 (1936).
2. Ephussi, B. Cold Spring Harb. Symp.Quant. Biol. 10, 40-48 (1942).
3. Butenandt, A., Weidel, W., Becker, E. Naturwissenschaften 28, 63-64 (1940).
4. Butenandt, A., Weidel, W., Schlossberger, H. Z. Naturforsch 4b, 242-244 (1949).
5. Butenandt, A., Schafer, W. Recent progress in the chemistry of natural and synthetic coloring matters and related fields. Academic Press, New York, London, San Francisco (1962).
6. Linzen, B. Adv. Insect Physiol. 10, 117-246 (1974).
7. Phillips, J. P., Forrest, H. S. Biology of Drosophila, Vol. 2d, ed. by Ashburner, M. and T. R. F. Wright. Academic Press, New York (1980).
8. Baillie, D. L., Chovnick, A. Molec. Gen. Genet. 112, 341-353 (1971).
9. Rizki, T. M., Rizki, R. M. J. Cell Biology, 17, 87-92 (1963).
10. Sullivan, D. T., Kitos, R. S. Insect Biochem. 6, 649-655 (1976).
11. Moore, G. P., Sullivan, D. T. Biochem. et Biophys. Acta 397, 468-477 (1975).
12. Moore, G. P., Sullivan, D. T., Biochem. Genetics 16, 619-654 (1978).

13. Ghosh, D., Forrest, H. S., Genetics 55, 423-431 (1967).
14. Sullivan, D. T., Kitos, R. J., Sullivan, M. C. Genetics 75, 654-661 (1973).
15. Grillo, S. L. PhD Dissertation, Syracuse University, Syracuse, NY.
16. Cassady, W. E., Wagner, R. P. J. Cell Biol. 49, 536-541.
17. Paton, D. R., Sullivan, D. T. Biochem. Genetics 16, 855-863 (1978).
18. Sullivan, D. T., Grillo, S. L., Kitos, R. J. J. Exp. Zool. 188, 225-233 (1974).
19. Phillips, J. P., Forrest, H. S. Biochem. Genetics 4, 489-498 (1970).
20. Phillips, J. P., Forrest, H. S., Kulkarni, A. D. Genetics 73, 45-56 (1973).
21. Yamamoto, M., Howells, A. J. and Ryall, R. L. Biochem. Genetics, 14, 1077-1090 (1976).
22. Wiley, K., Forrest, H. S. Biochem. Genetics 19, 1211-1221 (1981).
23. Butenandt, A., Biekert, E., Linzen, B. Hoppe-Seylers Z. Physiol. Chem. 305, 284-289 (1956).
24. Rizki, T. M. Biology of Drosophila, Vol. 2c, ed. by Ashburner, M. and T. R. F. Wright. Academic Press, New York (1980).
25. Sullivan, D. T., Sullivan, M. C. Biochem. Genetics 13, 603-613 (1975).
26. Sullivan, D. T., Bell, L. A., Paton, D. R., Sullivan, M. C. Biochem. Genetics 18, 1109-1130 (1980).
27. Sullivan, D. T., Lukacs, B., Mitchell, B.F. (submitted).

ROLES OF ACTIVE OXYGEN AND CATALASE IN THE FORMATION OF PHENOXAZINONE PIGMENTS CATALYZED BY MANGANESE ION

Hisamitsu Ogawa[1], Yoichi Nagamura[2], Isao Ishiguro[3]

1) Department of Biology and 2) Department of Clinical Chemistry, School of Hygiene, and 3) Department of Biochemistry, School of Medicine, Fujita-Gakuen University Toyoake, Aichi 470-11, Japan

Introduction

Cinnabarinic acid (CBA) synthase stimulated by manganese (Mn) ion (1) was purified to homogeneity from Malpighian tubules of the silkworm, Bombyx mori, by using ammonium sulfate precipitation, gel filtration, and CM-cellulose chromatography. The enzyme activity was accompanied by catalase activity in parallel at all steps of purification; the two activities could not be separated from each other. The purified protein was concluded to be catalase. The mechanism of CBA formation in the MN ion-catalase system was studied as a model of the reaction forming phenoxazinone pigments, because catalase is known to participate in the conversion of 3-hydroxyanthranilic acid (3-OH-AA) into CBA in the presence of Mn ion in vitro (2, 3).

Results and Discussion

CBA formation in the reaction mixture containing 60 mM potassium phosphate buffer (pH 7.0), 1.5 mM 3-OH-AA, 0.4 mM $MnSO_4$ and 20 µg catalase in a final volume of 1.0 ml started after a 10 min-lag phase, thereafter progressing linearly. In the reaction mixture without catalase, the pigment formation arrived at a plateau about 20 min after the same lag phase was

Progress in Tryptophan and Serotonin Research

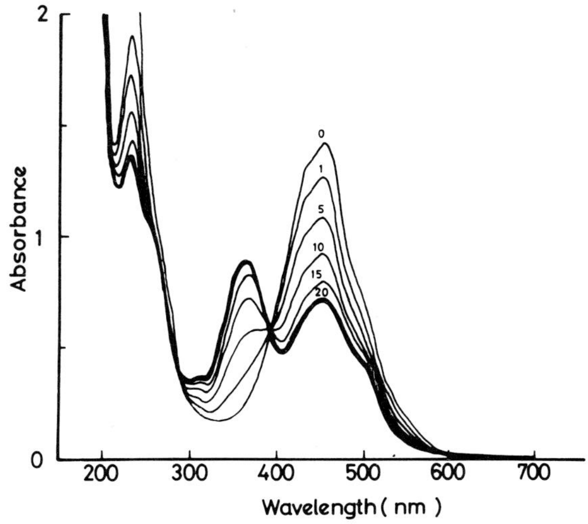

Fig. 1. Changes in absorption spectra during reaction between CBA and H_2O_2 with or without catalase. Reaction mixture contained 60 µM CBA and 0.4 mM H_2O_2 in 60 mM potassium phosphate buffer (pH 7.0) in a final volume of 3.0 ml. Spectral scans were made at 0, 1, 5, 10, 15 and 20 min after addition of H_2O_2 at 20°C. At 20 min, 10 ug catalase was added to the reaction mixture and spectral scans were again made at 10, 20, 30 and 40 min after addition of catalase. The scanning speed was 120 nm/min. CBA was prepared chemically according to Prinz and Savage (4).

observed. When this reaction was followed spectrophotometrically, only the absorption of CBA (at 460 nm) was seen which increased with time in the complete system. In the reaction system without catalase, the appearance of a new absorption peak at 360 nm was observed while absorbance at 460 nm was arriving at a plateau. This new peak appeared when CBA was incubated with a low concentration of hydrogen peroxide (H_2O_2); the spectra taken at different times showed isosbestic points at 290 nm and 390nm. Addition of catalase stopped this absorption change (Fig. 1). H_2O_2 was not detected during incubation of the complete system, but it increased steadily, after a 10 min lag phase, in the nonenzymatic system. CBA was decomposed by H_2O_2 and one of the decomposition products was identified with 3-OH-AA. After preincubation of 3-OH-AA, the addition of Mn ion shortened the initial lag phase of CBA formation. When

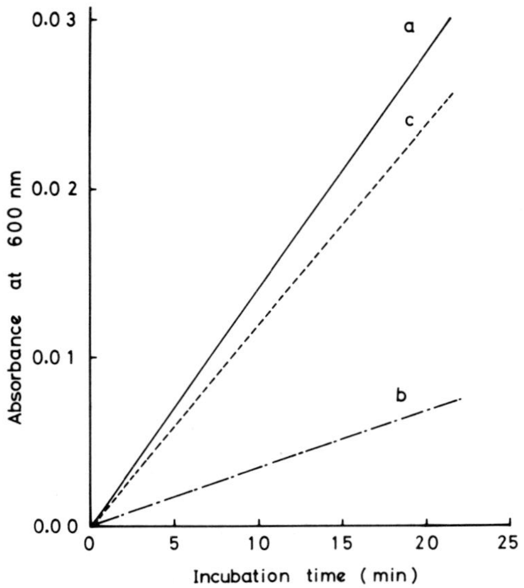

Fig. 2. Reduction of NBT during aerobic incubation of 3-OH-AA. Reaction mixture contained 60 mM potassium phosphate buffer (pH 7.0), 1.5 mM 3-OH-AA, and 0.15 mM NBT in a final volume of 1.0 ml. Assays were carried out with no addition (a), with 3 µg SOD (b), and with boiled SOD (c).

Mn ion and superoxide anion (O_2^-) were both present, the lag phase disappeared. Aerobic incubation of 3-OH-AA under neutral conditions resulted in the generation of O_2^-, as demonstrated by the reduction of nitroblue tetrazolium (NBT) which was inhibited by superoxide dismutase (SOD) (Fig. 2). Presumably, Mn(II) ion participating in CBA formation is oxidized to Mn(III) by O_2^-. Mn(III) was shown to induce the oxidation of 3-OH-AA without a lag phase, and it enhanced the oxidation rate caused by Mn(II), which indicates that a cycle of Mn(II) ion/ Mn(III) ion occurs during oxidation of 3-OH-AA. The pigment formation in the presence of Mn ion is inhibited by reducing agents such as ascorbic acid, glutathione, and cysteine. From these results, the oxidative condensation of 3-OH-AA to CBA in the presence of Mn ion and catalase is presumed to depend on a series of reactions as follows: (i) 3-OH-AA is oxidized to a radical form under aerobic conditions and generates O_2^-. (ii) O_2^- oxidizes Mn(II) ion to Mn(III) ion, yielding H_2O_2.

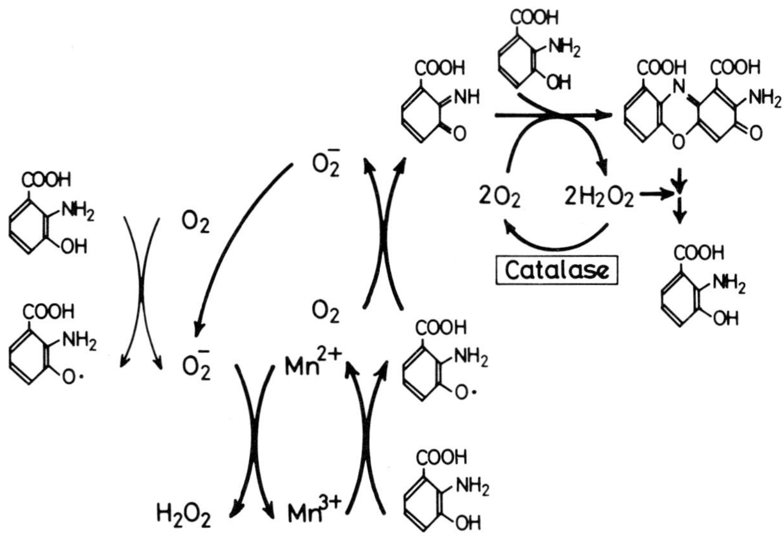

Fig. 3. Proposed reaction mechanism of CBA formation catalyzed by Mn ion with catalase. Fine line, initiation ; thick line, propagation of the pigment formation.

(iii) Mn(III) ion oxidizes 3-OH-AA to a radical form, being reduced to Mn(II). (iv) The radical is further oxidized to the quinoneimine and generates O_2^-. (v) The quinoneimine condenses with 3-OH-AA to CBA, whereby H_2O_2 is formed. (vi) CBA is decomposed by the H_2O_2 formed in the reaction process; one of the decomposition products is 3-OH-AA. (vii) Catalase protects CBA from decomposition by H_2O_2 (Fig. 3).

References

1. Subba Rao, P.V., Vaidyanathan, C.S.: Biochem. J. 99, 317-322 (1966).
2. Savage, N., Prinz, W.: Biochem. J. 161, 551-554 (1977).
3. Ryall, R.L., Ryall, R.G., Howells, A.J.: Insect Biochem. 6, 135-142 (1976).
4. Prinz, W., Savage, N.: Hoppe-Seyler's Z. Physiol. Chem. 358, 1161-1163 (1977).

OMMOCHROME-POTASSIUM URATE COMPLEX AS PIGMENT OF RED FAECES OF *LOCUSTA MIGRATORIA* (INSECTA, ORTHOPTERA, ACRIDIDAE).

Alain E. Bouthier
Laboratoire de Zoologie, Ecole Normale Supérieure, 46 rue d'Ulm, F-75230 Paris Cedex 05

Like numerous other insects, the Acridians yield red or pink faeces in a cyclic rhythm strictly correlated with the moulting cycle. Many authors had noticed this phenomenon : only as observation (1), or by chemical extraction (2), with identification of ommochromes (3) or of xanthommatin and epidermal ommochromes (4).

Methods

Chauvin's extraction procedure (2) is used : selected faeces are extracted with 100% alcohol (x3) to remove chlorophyll, acetone, thereafter diethyl ether (x3). The pigment is extracted with water at 40°C with shaking. The mixture is decanted and filtered through paper. The pigment is then adsorbed on microcrystalline $CaCO_3$ (10% w/v) and the supernatant A is separated by centrifugation (12,000 g, 10 min). The resulting pellet is washed with water, then centrifuged (12,000 g, 10 min). The pigment is desorbed with 3M aqueous pyridine until complete $CaCO_3$ decolouration. Supernatants B are obtained by successive centrifugations (2,500 g, 10 min), pooled and evaporated under reduced pressure ; numerous red-brick crystals precipitate. Supernatant A acidification at pH 2,0 with HCl and ascorbic acid causes an abundant deposit of brown pigment which is separated by centrifuging (12,000g, 10min). The resulting supernatant C is treated with SO_2 producing in the cold an important deposit of crystals. Natural or artificial crystals were observed with a scanning electron microscope CAMECA MEB 07 (Univ. P. & M. Curie, Paris) after gold/palladium metallization.

Progress in Tryptophan and Serotonin Research

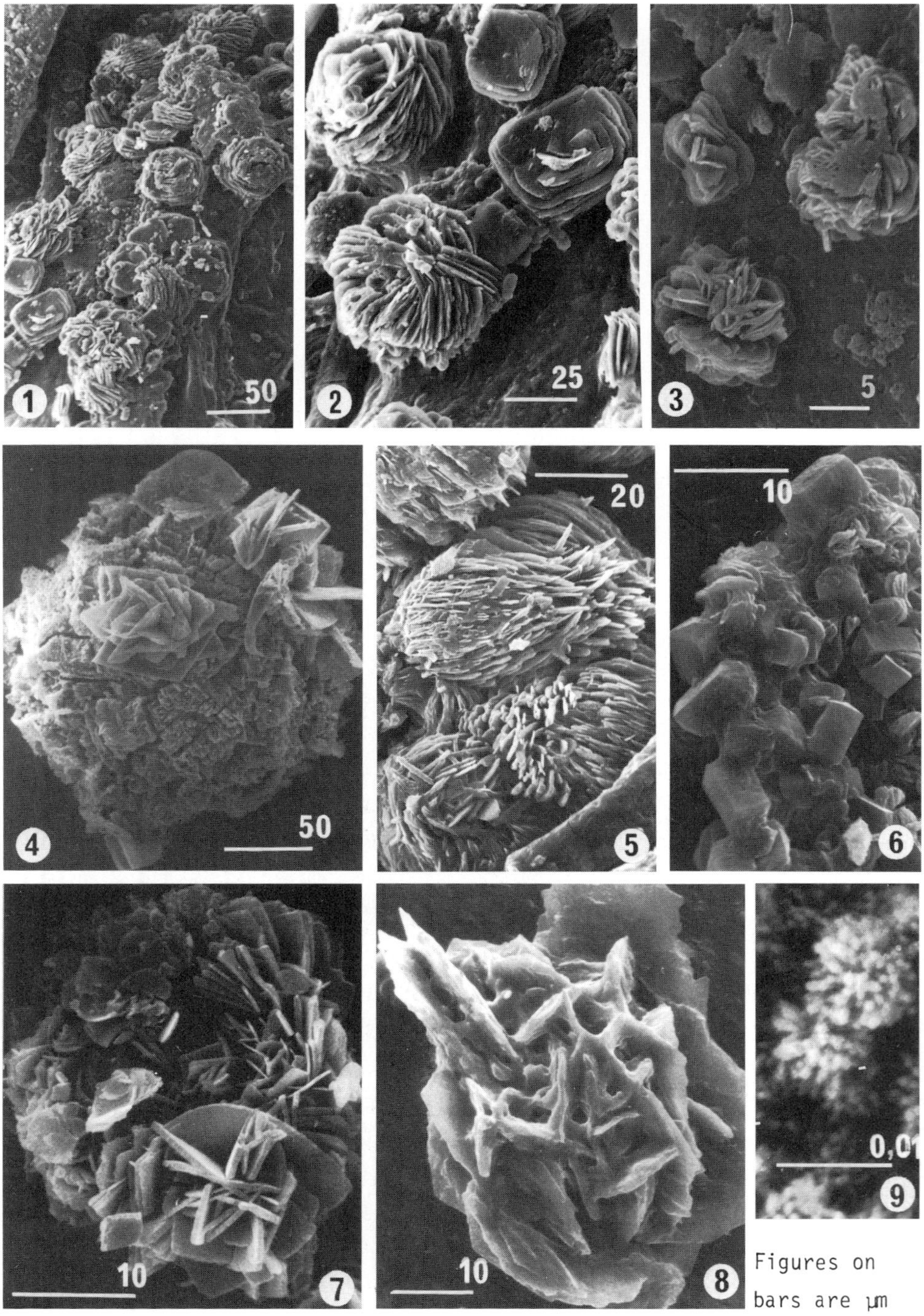

Figures on bars are µm

Results

1) Observation of the crystals *in situ*.

Several crystalline forms coexist on the surface of coloured faeces :

- small roses (diameter 50-60 up to 110 µm) (fig. 1, 2), on red faeces produced at larval or imaginal moults ;
- small roses (diameter 8-12 µm) homothetic of the preceding ones, on pink faeces produced by aged adults (fig. 3); these two forms resemble those observed for acridiommatin 1 (5, fig. 4) and their size is directly related to the intensity of the red colouration of the faeces ;
- stacking of square or rectangular plates somewhat concave and with rounded corners (30x50 or 30x30 µm) (fig. 2);
- "artichokes" (diameter 40-45 µm) formed by aggregation of fine needles or small slabs (length 20-25 µm) (fig. 5);
- another form is present: bipyramidal parallelepipedic rods (short edges 3 µm long, long edges 4,5 to 7,5 µm long) (fig. 6).

All these crystals, directly visible on the surface of faeces, are not covered by the peritrophic membrane. Thus they are probably formed in the hind gut.

2) Observation of the crystals formed *in vitro*.

The more common form (supernatant B) corresponds to the two first natural types (small roses) with a comparable size range (fig. 7). The crystals obtained from supernatant C are slightly different and smaller.

3) Composition of crystals.

The crystals are soluble in 0,1 M Na_2HPO_4 or in 0,3 N NaOH. In this case spectrophotometric recording indicates presence of uric acid (absorption maxima at 234 and 287 nm and rapid degradation by uricase *in vitro*). In a 300 µg sample of pure natural crystals we have found 148 µg urate (uricase method according to 6) and 33 µg K^+ (atomic absorption spectrometry) : the stoichiometry of K-urate is obtained approximately (another experiment gives 390 µg urate for 100 µg K^+). The crystals should contain 60% K-urate and therefore 40% ommochromes. Crystals from supernatants B and C contain 60% urate (75% K-urate) and 70 % urate (85% K-urate), respectively. Elementary analysis of natural crystals reveals lack of Na^+, S or P and a composition very different from that of pure acridiommation 1 or 2 (5) (Table).

Table. Elementary analysis of crystals obtained from locust red faeces, uric acid, and acridiommatins 1 and 2.

	C	H	N	O	residue
Natural crystals	33.60	3.18	29.98	30.76	2.48
Uric acid	35.71	2.38	33.33	28.57	
Acridiommatin 1	45.40	4.83	6.10	37.87	5.80
Acridiommatin 2	51.92	5.17	6.80	28.71	7.40

Only the presence of urate ($C_6N_4O_3H_3$) in the crystals can explain the high nitrogen level. Paper chromatography of the material shows the presence of acridiommatin 1 alone, as the spectrum in 6N HCl reveals the typical maximum at 472 nm. Paper chromatography gives further results : in collidin/0,5 M KH_2PO_4 (1:3, organic phase) on Whatman 3MM or S. & S. 2043bm, acridiommatin 1 presents a Rf of 0,06, acridiommatin 2 0,14, K-urate 0,17, natural crystals 0,29, (B) crystals 0,34 and (C) crystals 0,36. Crystals can be produced from a 0,1 M Na_2HPO_4 solution of 75% K-urate and 25% pure acridiommation 1 into which SO_2 is bubbled: after standing in the cold, red-brick crystals deposit. SEM observation shows them to be analogous to those described above: small roses (size 13-42 µm) (fig. 8) very distinct from pure K-urate crystals obtained in saturated aqueous solution (fig. 9).

Conclusion

The results exposed above allows us to conclude that small crystalline roses present on red faeces of Locusta are composed of a complex between K-urate and acridiommatin 1. Acridiommatin 2 and xanthommatin perhaps form other types of crystals. A higher uric acid level was previously reported in red faeces of silkworm without mention of an ommochrome complex (7).

References

1. Künckel d'Herculais, J.: C. R. Acad. Sc. Paris 114, 240-242 (1892).
2. Chauvin, R.: C. R. Soc. Biol. Paris 130, 1194-1195 (1939).
3. Goodwin, T.W., Srisukh, S.: Biochem. J. 47, 549-554 (1950).
4. Linzen, B.: Z. Naturforsch. 21b, 1038-1047 (1966).
5. Bouthier, A.: C. R. Acad. Sc. Paris D 279, 1265-1268 (1974).
6. Kageyama, N.: Clin. chim. Acta 31, 421-426 (1971).
7. Kawano, T., Matsumura, E.: Yakugaku Zasshi 98, 663-667 (1978).

CHANGES IN FLUORESCENCE AND SIZE OF THE CYTOPLASMIC GRANULES IN THE MALPIGHIAN TUBULES OF DROSOPHILA MELANOGASTER LINKED WITH TRYPTOPHAN METABOLISM

Shizuo Yagi

Department of Biology, College of General Education,
University of Tokushima, Tokushima 770, Japan

Introduction

Tryptophan can be metabolized to ommochrome or brown pigments. Malpighian tubules in *D. melanogaster* are known to be source of two precursors of the brown pigment, kynurenine and 3-hydroxykynurenine (1 - 6). The author found differences in fluorescence and size of the cytoplasmic granules in the Malpighian tubules among various strains of *D. melanogaster*. This communication reports microfluorometric studies on changes of the autofluorescent granules in larval Malpighian tubules of eye color mutants after administration of tryptophan metabolites.

Results

The cytoplasmic granules, stainable with Nile blue sulfate, appear in Malpighian tubules of *D. melanogaster* larvae within 24 h of hatching and the granules emit fluorescence on excitation at 365 nm. Large granules which reach a diameter of 3 - 4 μ are observed in wild type, *Oregon-R*, and two eye color mutant strains, *brown* and *cardinal brown*. The eye color mutant strains, *vermilion brown* and *cinnabar brown* contain small granules of about 1 μ in diameter, and their eye discs develop into brown eyes if the pigment precursors are presented to them. On the other hand, the eye color mutant

Progress in Tryptophan and Serotonin Research

strains, *scarlet brown*, *lightoid brown*, and *white* are devoid of granules; their eye discs are unable to utilize the brown eye color precursors (1-4,6-10). A modified microspectrofluorometer attached to a microspectrophotometer (Zeiss, UMSP-1) recorded the emission spectra of the granules in the tubules with a field of 20 μ in diameter. A Nikon microfluorometer with a comparative photometric device was used to determine the fluorescence intensity of the granules by visual means at 530 nm. The cytoplasmic granules of *Oregon-R*, *brown* and *cardinal brown* emitted weak fluorescence, the emission maximum being around 490 nm. In contrast, *vermilion brown* and *cinnabar brown* emitted strong fluorescence. The peak of the emission spectrum of *vermilion brown* was around 440 nm and that of *cinnabar brown* was around 470 nm (Fig. 1). Then, the effects on the granules of feeding tryptophan metabolites to the larvae were examined by adding 3 mM tryptophan, 3 mM kynurenine or 3 mM 3-hydroxykynurenine to the glucose-agar medium. The fluorescence and size of the granules shifted towards the direction of wild type, *Oregon-R* when *v bw* larvae were given kynurenine or 3-hydroxykynurenine, and *cn bw* larvae were given 3-hydroxykynurenine (Fig. 2 and 3, Table 1). The fluorescence changes occured about 3 h after giving the compounds. Then 2 h later, the granules began to increase in size. Sooner or later, the granules reached steady normal values. Neither fluorescence and size alteration were detected

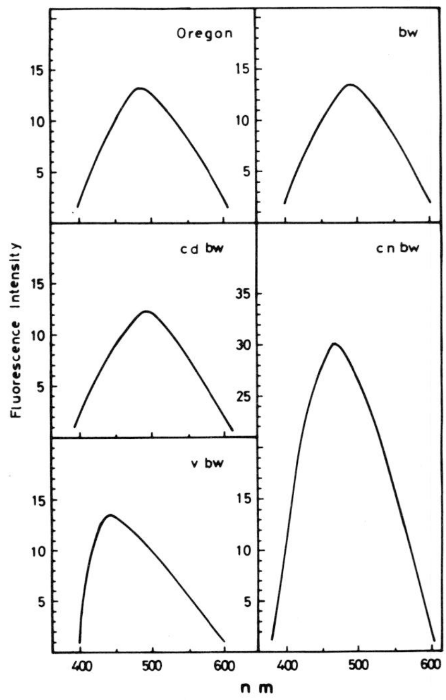

Fig. 1 Fluorescence spectra of the granules in Malpighian tubes.

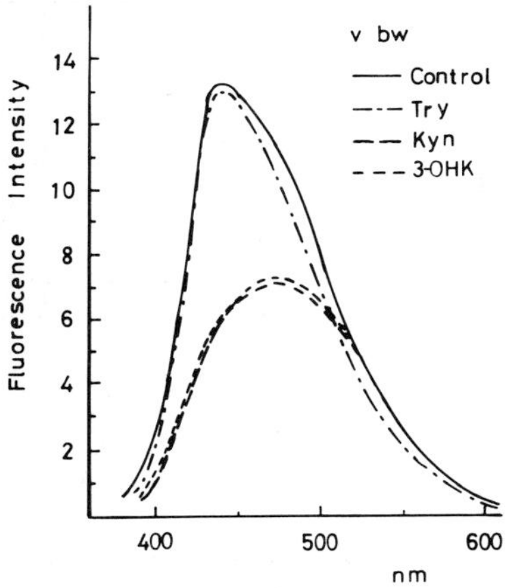

Eig. 2 Fluorescence changes of the granules on feeding tryptophan metabolites to v bw.

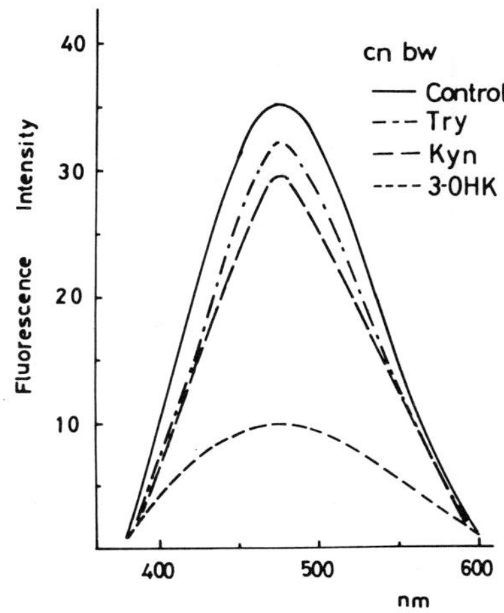

Fig. 3 Fluorescence changes of the granules on feeding tryptophan metabolites to cn bw.

Table 1 Experimental feeding of tryptophan metabolites on the size of the granules in the Malpighian tubes

Strain	Control	Tryptophan	Kynurenine	3-hydroxykynurenine
Oregon	++	++	++	++
v	+	+	++	++
v bw	+	+	++	++
cn	+	+	+	++
cn bw	+	+	+	++
st	-	-	-	-
st bw	-	-	-	-
w	-	-	-	-

in v bw larvae after feeding tryptophan, or in cn bw larvae after feeding tryptophan or kynurenine. Next, isolated Malphighian tubules of v bw larvae and cn bw larvae were placed in Ringer solution (pH 6.8) containing 10 mM tryptophan, 10 mM kynurenine or 5 mM 3-hydroxykynurenine. These excised Malpighian tubules showed similar changes in fluorescence emission as observed in the feeding experiments. The fluorescent materials in Malpiphian tubules are separated by TLC to yield three major components, pteridines, tryptophan and some of its metabolites, and riboflavine (5). This report shows that in the larvae of D. melanogaster changes in fluorescence and size of the granules in Malpighian tubules are closely connected with intracellular tryptophan metabolism. This seems important with respect to the intracellular site of tryptophan metabolism during formation of the brown pigment in Malpighian tubules, and with the chemical relation of the fluorescent materials to tryptophan metabolism.

References

1. Butenandt, A., Weidel, W., Becker, E.: Naturwiss. 28, 477-448 (1940).
2. Butenandt, A., Weidel, W., Schlossberger, H.: Z. Naturforsch. 4b, 242-244 (1949).
3. Beadle, G.W.: Genetics 22, 587-611 (1937).
4. Ephrussi, B.: Quart. Rev. Biol. 17, 326-338 (1942).
5. Wessing, A., Eichelberg, D.: Z. Naturforsch. 23b, 376-386 (1968).
6. Wessing, A.: Verhandl. Deut. Zool. Ges. Saarbrücken 166-174 (1961).
7. Phillips, J.P., Forrest, H.S.: Biochem. Genetics 4, 489-497 (1970).
8. Beadle, G.W., Ephrussi, B.: Genetics 21, 225-247 (1936).
9. Sullivan, D.F., Sullivan, M.C.: Biochem. Genetics 13, 603-613 (1975).
10. Sullivan, D.T., Bell, A.L., Paton, D.R., Sullivan, M.C.: Biochem. Genetics 18, 1109-1130 (1980).

HORMONAL CONTROL OF PIGMENTATION IN MANDUCA SEXTA

Lynn M. Riddiford and Kiyoshi Hiruma
Department of Zoology, University of Washington
Seattle, Washington 98195, U.S.A.

Introduction

Color polymorphism is common among insects. Most of the changes are mediated by the insect's endocrine system, usually during a molt. Despite a vast literature on the description of these color changes and their hormonal control (1, 2), relatively little is known about the cellular actions of the hormones in regulating the changes. The following paper will discuss the hormonal control of the green-black polymorphism in the tobacco hornworm, *Manduca sexta*, focusing on our present knowledge of how the various hormones act on the epidermis.

Normally *Manduca* larval epidermis produces a transparent larval cuticle and a blue pigment, insecticyanin [a protein with a biliverdin IX_{γ} chromophore (3)] . But when juvenile hormone (JH) is absent during a critical period around the time of head cap apolysis during the larval molt, the new cuticle subsequently melanizes and later the epidermis begins to synthesize ommochromes and turns pink (4, 5, 6).

Cuticular melanization

During a larval molt JH must be present at the time of ecdysteroid release to ensure that a new larval cuticle is produced. In *Manduca* JH is necessary for the first 6 hrs of the ecdysteroid rise (7, 8) to enable the epidermis to make a new larval cuticle. Then it is again important about 18 hours later to prevent cuticular melanization (4, 5, 9). When JH is absent at this later time, the epidermis deposits premelanin granules into the newly forming cuticle about 12 hrs later (10). These

Progress in Tryptophan and Serotonin Research

granules contain active phenoloxidase as measured histochemically by the method of Locke and Krishnan (11). Yet no melanization occurs until about 13 hrs later, 3 hrs before ecdysis.

First to determine if some endocrine cue was necessary at this later time, we injected melanization and red coloration hormone purified from Bombyx mori heads (12), and extracts of pharate adult Manduca corpora cardiaca or of purified eclosion hormone (13). None had any significant effect on the timing of the onset of melanization. By contrast, injection of 0.2 μg or greater of 20-hydroxyecdysone (20HE) up to 4.5 hrs before the onset of melanization delayed it in a dose-dependent manner (10). After that time doses up to 5 μg per larva had no effect on subsequent melanization. The ecdysteroid titer in the larva has fallen from its peak value of 3200 ng/ml (20HE equivalents) about 6 hrs after head cap apolysis to about 280 ng/ml by the end of the ecdysteroid-sensitive period (4.5 hrs before melanization), then declines to about 60 ng/ml at ecdysis.

To determine whether the ecdysteroid was acting directly on the epidermis or was acting on some other processes occurring in the larva at that time, we have cultured the epidermis from larvae allatectomized 5-6 hrs before head cap apolysis in the presence and absence of ecdysteroid. When the epidermis was explanted 7 hrs before the onset of melanization and cultured in Grace's medium lacking β-alanine but containing 0.3mM L-dopa for 24 hrs, 96% of the 222 pieces showed strong melanization like that observed in vivo (14). By contrast, when 50 ng/ml 20HE was present, very little melanization occurred. Importantly, when the epidermis was explanted 3 hrs before the onset of melanization, concentrations of up to 1.0 μg/ml 20HE had no effect on the degree of melanization. Therefore, the concentrations and timing of sensitivity to 20HE are similar to those observed in vivo indicating that ecdysteroid appears to act directly on the epidermis to prevent melanization. Whether it prevents the uptake of dopa (the substrate for melanization) into the epidermis or suppresses the activity of the phenoloxidase in the premelanin granules is presently under investigation.

Ommochrome synthesis

Normally during the larval intermolt period Manduca epidermis synthesizes insecticyanin. Then during the molt the mRNA for insecticyanin disappears from the epidermis only to reappear about 12 hrs after ecdysis to the next larval instar (15, 16). When larvae are allatectomized 5-6 hrs before head cap apolysis during the final larval molt, the epidermis begins to turn pink 10 hrs after ecdysis and by 24 hrs is completely pink (5). The pink pigment is primarily dihydroxanthommatin with 15-20% of an unidentified ommochrome pigment (6) and is synthesized for approximately 2 days in the final larval instar (17). Importantly though, the insecticyanin mRNA reappears in these allatectomized larvae after ecdysis but in much reduced amounts as compared to normal larvae (16). The result is that the insecticyanin content by 40 hrs after ecdysis is about half that seen in normal larvae (17).

The lack of JH during the critical period around the time of head cap apolysis thus also alters epidermal pigmentation. Although the tryptophan content of the hemolymph is the same in allatectomized and intact larvae, more kynurenine is taken up by the epidermis of the allatectomized larva and there converted to 3-hydroxykynurenine and thence to the ommochromes (6, 17). Kynurenine hydroxylase activity in this epidermis was 3.2-fold higher than that from normal larva, but there was little difference in the particulate ommochrome synthetase activity (17). Of interest though was the finding that the addition of 0.5mM dopa enhanced by 50% the ommochrome synthetase activity of the 30,000g supernatant of the epidermal homogenate from the black larva as compared to the normal larva. This finding suggested that there might be a link between ommochrome synthesis and melanization as first suggested by Butenandt et al. (18).

In an attempt to determine whether these results bear any relevance to the cellular events occurring in vivo, we have tried to uncouple melanization and ommochrome synthesis. As seen in Table 1, when larvae are allatectomized 1.5 hrs before head cap apolysis, they barely melanize due to the lack of premelanin granules in the newly formed cuticle. Yet

Table 1. Inhibition of melanization and of ommochrome synthesis in allatectomized larvae

Treatment	Number	Melanization score* ± S.D.	Dihydroxanthommatin content (μg/g)
Allatectomy, time before head cap slips			
≥ 4.5 hr	6	4.5 ± 0.6	480
1.5 hr	7	1.2 ± 0.3	504
6 hr after	10	0	0
Allatectomized 5 hr before HCS; 40 μg αMDH 3 hr before expected melanization	16	1.2 ± 0.4	Trace

* Scoring system to Fain and Riddiford (9).

they produce as much dihydroxanthommatin as do larvae allatectomized 4.5-6 hrs before head cap apolysis. Allatectomy 6 hrs after head cap apolysis resulted in normally pigmented larvae—a transparent cuticle underlain by a blue epidermis. Therefore, the critical period for the effects of JH on pigmentation can be divided into two (Fig. 1): first, the period of sensitivity for determination of the deposition of the premelanin granules; second, shortly thereafter, the period of sensitivity for the determination of ommochrome synthesis. Presumably this period is the time of determination of the later enzymatic activities involved.

The second way that we have tried to block melanization is by injection of the dopa decarboxylase inhibitor αMDH (19). This compound also has been found to inhibit phenoloxidase activity in <u>Corcyra cephalonica</u> (20). When 40 μg were injected 3 hrs before the expected onset of melanization, the larvae showed little melanization and had only a trace of dihydroxanthommatin in their epidermis 2 days later (Table 1). When ^{14}C-tyrosine was injected into larvae given αMDH, label was found to accumulate in the hemolymph, half as tyrosine and half as dopa, and very little was incorporated into cuticular melanin (21). Thus the inhibition of the conversion of dopa to melanin also somehow blocks ommochrome synthesis, indicating that some components of the melanization pathway may be important to the conversion of 3-hydroxykynurenine to dihydroxanthommatin. A careful study of the relevant enzyme activities in these larvae is underway.

Recently we have been able to obtain dihydroxanthommatin synthesis *in vitro*. When epidermis was explanted 3 hrs before the expected onset of melanization to β-alanine-free Grace's medium containing 0.3mM dopa for 24 hrs, all the pieces melanized the cuticle as seen previously (14) and half synthesized ommochrome (N=72 pieces from 12 different larvae). Of those that synthesized the pink pigment, half synthesized approximately the same amount as seen *in vivo* by 24 hrs after ecdysis. The addition of 200 μg/ml kynurenine to the culture medium had little effect on the degree of pigment synthesis by these pieces. Experiments to improve the amount of pigment synthesis *in vitro* are underway. Then we should be able to test the effects of dopa and αMDH on this synthesis directly.

Discussion

As summarized in Figure 1, the hormonal control of pigmentation during a larval molt is complex. During the molt the epidermis responds to ecdysteroid by first setting the program for later cuticle synthesis, then the one for pigmentation. Juvenile hormone is necessary during the first period for production of a new larval cuticle, then during the second one for the formation of a transparent cuticle and the resumption of insecticyanin synthesis as the larva feeds and resumes growth after ecdysis. The absence of JH early in the second critical period causes later deposition of premelanin granules whereas its absence 3-4 hrs later in this period only affects epidermal pigment synthesis causing later ommochrome synthesis. By 24 hrs after PTTH and ecdysteroid release all cellular events determining later pigment synthesis are complete.

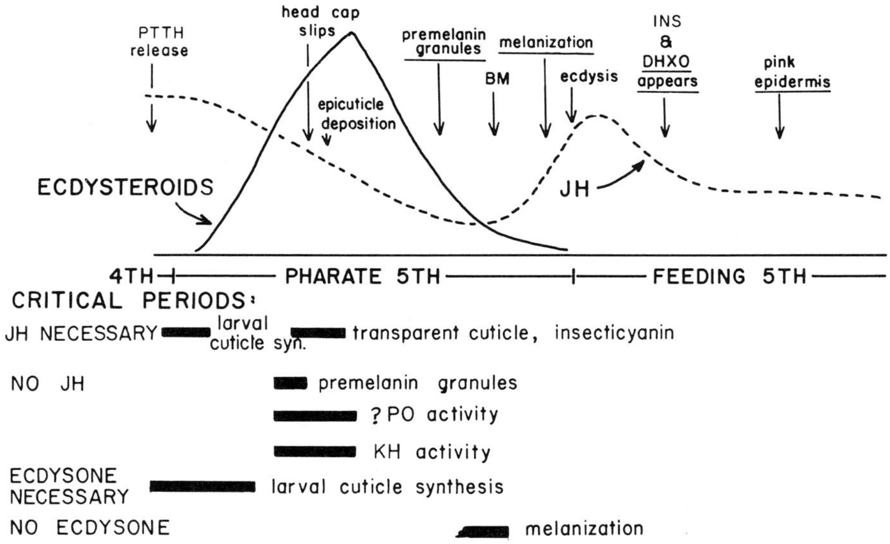

Figure 1. Summary of hormonal control of pigmentation in Manduca during a larval molt. The underlined changes in pigmentation refer to larvae allatectomized 5-6 hrs before slippage of the head cap. Critical periods refer to the times when ecdysteroids or juvenile hormone (JH) are important for determination of morphological features on pigmentation. PTTH, prothoracicotropic hormone; BM, brown mandibles; DHXO, dihydroxanthommatin; INS, insecticyanin; PO, phenoloxidase; KH, kynurenine—3-hydroxylase.

Larval epicuticle begins to be synthesized about the time of the ecdysteroid peak, then endocuticle about 6 hrs later during the falling phase of the ecdysteroid titer (10). After deposition of the premelanin granules, the onset of melanization is then delayed until the ecdysteroid titer has declined still further. By 3 hrs before melanization the epidermis no longer is sensitive to exogenous ecdysteroid (10,14) and also no longer requires exogenous dopa for melanization when explanted (14). Therefore, we favor the hypothesis that ecdysteroid prevents the uptake and possibly also epidermal synthesis of dopa that is the substrate for the melanization reaction in the granules.

In Manduca ommochrome synthesis is linked somehow to melanization. Once the critical period around the time of head cap apolysis is passed, the prevention of melanization by the inhibition of phenoloxidase activity also prevents later ommochrome synthesis. Importantly, however, by removing the corpora allata at a critical time, one can dissociate the two processes and obtain ommochrome synthesis without melanization. These animals lack premelanin granules in their new cuticle, and hence it does not melanize, but the enzymatic activities in the hemolymph and/or epidermis necessary for transport of dopa into the cuticle may be present. A subsequent conversion of dopa to dopaquinone and its reduction back to dopa could then be coupled to ommochrome synthesis as suggested by Butenandt et al. (18).

Other lepidopteran larvae that have similar color polymorphisms during larval life also depend on hormonal events during the molting cycle to determine pigmentation. In the cabbage armyworm Mamestra brassicae JH around the time of head cap apolysis is necessary for xanthommatin synthesis that begins about 12 hrs after ecdysis (22), an action diametrically opposed to its action in the regulation of ommochrome synthesis in Manduca. These findings simply show that different insects use the same hormonal cues to affect pigmentation, but utilize them to control the relevant pigments for their life. Also, melanization which occurs just after ecdysis in Mamestra is controlled differently in that it requires the release of a neurosecretory melanization hormone from the subesophageal ganglion just before ecdysis rather than the fall of ecdysteroid (22). Importantly, however, in Mamestra as in Manduca xanthommatin synthesis requires previous cuticular melanization.

What is now needed is a careful study of the enzymes involved in both ommochrome synthesis and in the melanization reaction and how their activity or synthesis is regulated first by the hormonal milieu during the critical determinative period, then later by an interaction of possible common substrates. Unfortunately not much is known about ommochrome synthetase (23), and phenoloxidases also are very complex and difficult to study (24). Yet the challenge set forth by Butenandt et al. (18) in 1956 of a possible link between these two pigments in biological systems still exists and needs to be tackled.

Acknowledgements

We thank Dr. James W. Truman for comments on the manuscript. The study was supported by grants from NIH and NSF.

References

1. Bückmann, D.: Fortschr. Zool. 22, Heft 2/3, 1-22 (1974).

2. Nijhout, H.F. and Wheeler, D.E.: Q. Rev. Biol. 57, 109-133 (1982).

3. Cherbas, P.T.: Ph.D. Thesis, Harvard University (1973).

4. Truman, J.W., Riddiford, L.M., and Safranek, L.: J. Insect Physiol. 19, 195-204 (1973).

5. Riddiford, L.M. and Hori, M.: In Juvenile Hormone Biochemistry: Action, Agonism, Antagonism (G.E. Pratt and G.T. Brooks, eds.). pp. 241-250. Elsevier/North Holland, Amsterdam (1981).

6. Hori, M. and Riddiford, L.M.: Insect Biochem. 11, 507-513 (1981).

7. Truman, J.W.: J. Exp. Biol. 57, 805-820 (1972).

8. Riddiford, L.M.: Gen. Comp. Endocrin. 34, 438-446 (1978).

9. Fain, M.J., and Riddiford, L.M.: Biol. Bull. 149, 506-521 (1975).

10. Curtis, A.T., Wolfgang, W.J., Hori, M., and Riddiford, L.M.: in preparation.

11. Locke, M. and Krishnan, N: Tiss. Cell 3, 103-126 (1971).

12. Matsumoto, S., Isogai, A., Suzuki, A., Ogura, N., and Sonobe, H.: Insect Biochem. 11, 725-733 (1981).

13. Reynolds, S.R. and Truman, J.W.: In Neurohormonal Techniques in Insects (T.A. Miller, ed.). pp. 196-215. Springer-Verlag, New York (1980).

14. Hiruma, K. and Riddiford, L.M.: in preparation.

15. Riddiford, L.M.: In Proceedings IXth International Symposium on Comparative Endocrinology (B. Lofts, ed.) University of Hong Kong Press, Hong Kong, in press.

16. Riddiford, L.M.: unpublished

17. Hori, M. and Riddiford, L.M.: J. Comp. Physiol. 147B, 1-9 (1982).

18. Butenandt, A., Biekert, E., and Linzen, B.: Hoppe-Seyler's Z. Physiol. Chem. 305, 284-289 (1956).

19. Lipke, H.: In Invertebrate Immunity (K. Maramorosch and R.E. Shope, eds.). pp. 327-336. Academic Press, New York (1975).

20. Raghavan, K.G. and Nadkarni, G.B.: Insect Biochem. 8, 53-57 (1978).

21. Hori, M. and Riddiford, L.M.: submitted.

22. Hiruma, K., Matsumoto, S., Isogai, A., and Suzuki, A.: J. Comp. Physiol. in press.

23. Linzen, B.: In Biochemical and Medical Aspects of Tryptophan Metabolism (O. Hayaishi, Y. Ishimura, and R. Kido, eds.) Elsevier/North Holland, Amsterdam. pp. 215-224 (1980).

24. Brunet, P.C.J.: Insect Biochem. 10, 467-500 (1980).

TRYPTOPHAN METABOLISM AND MORPHOLOGICAL COLOUR CHANGE IN INSECTS

Detlef Bückmann,
Abteilung Allgemeine Zoologie der Universität Ulm
D-7900 Ulm

Vertebrates and insects have both efficiently solved the physiological problems of terrestrial life. However, insects have solved many physiological problems in other ways than vertebrates. As to their tryptophan metabolism, we may ask, why do insects form ommochromes whereas vertebrates do not? There are three possible answers:

1) Insects can form ommochromes because they belong to another main branch of animal evolution, the protostomia, which have evolved somewhat different metabolic pathways. Many protostomia have ommochromes as eye pigments (15, 20).

2) Perhaps insects even must form ommochromes because they cannot cleave the aromatic ring of hydroxy-anthranilic acid. So ommochromes represent the major end products of tryptophan metabolism (15).

3) Possibly insects also must form ommochromes for the sake of their body colouration. Insects are adapted to terrestrial illumination conditions in another way than are the vertebrates. Unlike vertebrates the insects have no special pigment cells for pigment deposition and physiological and morphological colour change. Instead, all normal epidermal cells can form pigments. These pigments can even be different in neighbouring cells, in different developmental stages, and under different biological conditions. Therefrom re-

Progress in Tryptophan and Serotonin Research

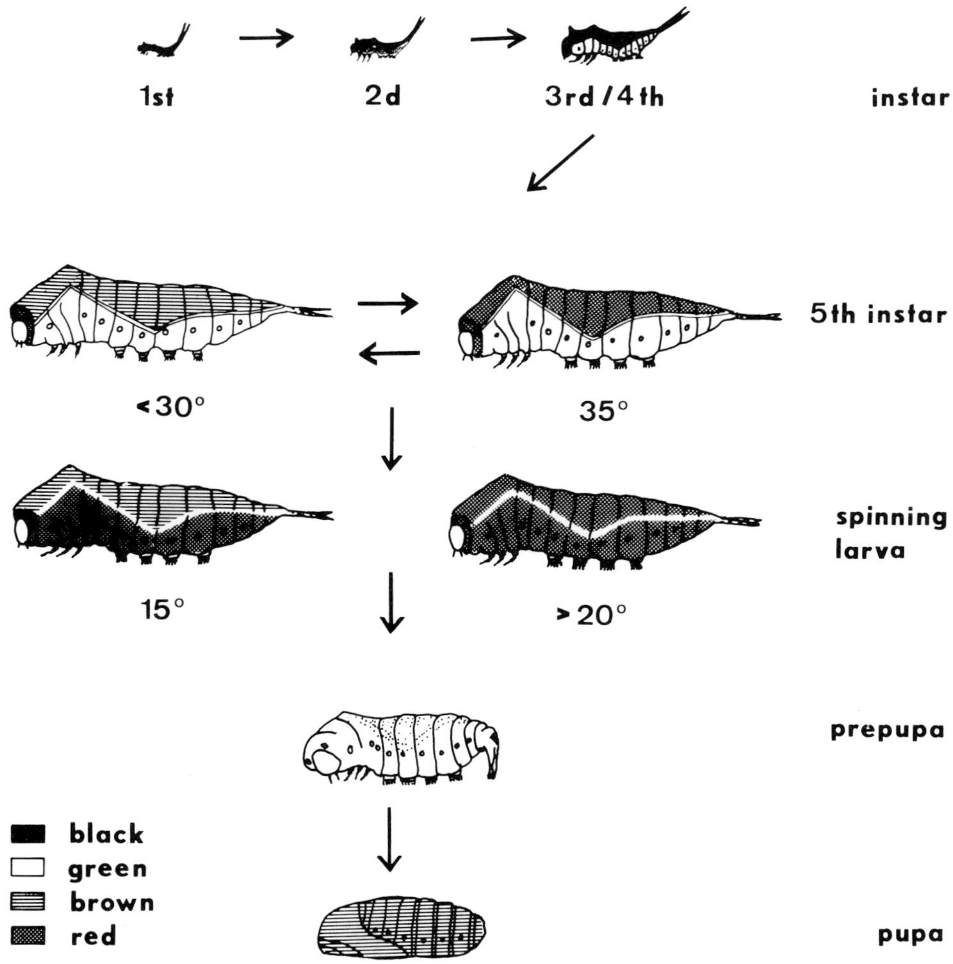

Legend to figure 1:

Fig. 1: Colour patterns and colour changes during the larval development of Cerura vinula L. (Modified after Bückmann, 1974 (6). Black colouration results from cuticular melanin. Red and brown colourations of the larva (not the pupal cuticle) result from epidermal xanthommatin or dihydroxanthommatin respectively. During the 5th larval instar a change of the temperature causes a colour change. The reddening of the brown "saddle patch" can be evoked by high temperature or lack of oxygen. Under normal temperatures (20-25°C) it occurs spontaneously at the spinning stage but can be postponed by low temperature.

sult colour patterns and morphological colour changes. The caterpillar of Cerura vinula may serve as an example for both, colour pattern and colour change (fig. 1). It bears a sharp borderline between ommochrome-containing and -free epidermal cells. This is, at the same time, the borderline between melanin-containing and melanin-free cuticle (5). (These relations between ommochromes and melanin are not yet understood.)

The ommochrome-free epidermal cells are also able to synthesize ommochromes. However, they will form ommochrome at a later developmental stage, causing a colour change when the grown-up larva leaves its food.

Such ommochrome formation will require considerable amounts of tryptophan metabolites. Are these metabolites always available? Or are they even waiting to be removed? What are the relations between colour change and tryptophan balance? In Cerura, ommochromes do also serve as excretory products which are removed via the gut and the Malpighian tubules (3). In other animal groups the deposition of pigments into the integument serves as a form of excretion. Does this apply to integumental ommochrome of insects, too? In other words: Does the morphological colour change serve to excrete tryptophan or can the tryptophan metabolism be changed in order to serve a morphological colour change?

I will present 3 examples: One colour change evidently forced upon the insect merely by its pigment biochemistry, another colour change in which changes of tryptophan metabolism and requirements of body colouration are synchronized by hormonal control, and one case in which a switch in tryptophan metabolism can be evoked by external optical stimulation and causes an adaptive colour change.

As you may know, the brown xanthommatin is a redox pigment,

the reduced form being red. In Cerura vinula this reddening by reduction will occur in the living animal when it is transferred to higher temperature, and may therefore be a response to a higher metabolic rate. It occurs spontaneously when the larva is going to pupate, but can be postponed then by low temperature (fig. 1). The reduction is also evoked by lack of oxygen. If we close some of the tracheal spiracles, only the part of the body thus asphyxiated will redden. Evidently the xanthommatin serves as hydrogen acceptor instead of oxygen. Its respiratory function is, however, probably unimportant. In any case this colour change seems to be a mere result of pigment biochemistry without any adaptive function (5).

However, when the caterpillar is grown up and leaves its food, suddenly the green parts of the integument will form red dihydro-xanthommatin, too. However, those cells already containing brown xanthommatin do not form additional pigment. Their ommochrome is only reduced to its red form. These two reactions are independent of each other. As has been mentioned, in low temperature the brown part will remain brown, while only the green lateral part of the spinning caterpillar will become red by new ommochrome formation (fig.1).

For metabolic reasons the formation of new ommochrome would be desirable if there were a breakdown of proteins rich in tryptophan (- as indeed ommochrome formation can be caused by mere starvation (3) -) or if there were a transformation to other proteins poor in tryptophan. This sounds rather theoretical, but both processes do occur in the spinning larva.

During the breakdown of larval gut and fat body the haemolymph contents of free tryptophan and of free 3-hydroxy-kynurenine increase transiently up to eight-fold. At the same time a cocoon is spun, the material of which is notoriously free of tryptophan. Tryptophan and 3-OH-kynurenine decrease again while ommochromes are being formed (up to 2 mg per g dry weight (16). These are stored as rhodommatin and ommatin D in the fat body and new gut lumen. Later on these pigments are partially removed via the gut.

This is really an effective regulation, because in spite of dramatic variations in the level of individual amino acids the total amino acid content is kept almost constant (11). The resulting amounts of ommochrome cause the young pupa to appear all red inside. Considering this, the reddening of the skin may just be a by-product of excretory ommochrome formation. In this case ommochrome formation might automatically be caused by surplus tryptophan.

However, if we inject tryptophan or 3-hydroxy-kynurenine, the animal will remove this substance without formation of ommochromes. 3-hydroxy-kynurenine increases immediately after injection and decreases again. After injection, tryptophan is evidently transformed to 3-hydroxy-kynurenine which increases temporarily and then decreases again without any ommochrome formation. If, however, we inject the hormone ecdysone, the 3-OH-kynurenine level will increase only slightly, but ommochrome will be formed (7).

This means an insect can remove tryptophan via 3-OH-kynurenine in other ways than by ommochrome formation. Ommochrome production is evoked by a certain hormonal situation only. As part of the pupal ommochrome is removed before the rest is excreted through the gut, ommochromes are not even necessarily metabolic end products in Cerura.

Does the ommochrome formation have any biological adaptive effect at all? Does this striking coloured animal ask for any protective colouration? As a matter of fact the larva of Cerura vinula is well known for the biological adaptations of its colour pattern. The newly hatched larva is all black and resembles the dried furled edges of the poplar leaves. During the following three instars the distribution of cuticular melanin causes a black and green pattern, which looks conspicuous. But this pattern has a "somatolytic" effect. It dissolves the picture of the body among the quivering poplar leaves. In the last instar larva the pattern is the same but the melanisation is somewhat weaker permitting the underlying brown or red epidermal ommochrome to be seen. The green as well as the brown parts of the body appear lighter towards the back. This has been called a "double inverse countershading" (19). In the larva's usual position, hanging among the leaves upside down, the shading by natural illumination from above is compensated by the countershading. The caterpillar's body will not appear corporeal but flat. This may sound unconvincing, but the protective effect has been shown by de Ruiter (19) in experiments with birds feeding on caterpillars.

If the caterpillar is, nevertheless, attacked, it suddenly shows a striking red and black pattern and, at the same time, it sprays a ray of formic acid against the aggressor. These red and black elements are, therefore, interpreted as repellent, a so-called "warning" colouration. When the caterpillar leaves the green leaves in order to spin its cocoon it would look conspicuous on the brown wood. However, at the same time ommochrome formation sets in, and this looks indeed like a protective colouration.

Thus in this case of colour change, the requirements of excretion and of protective colouration are evidently synchronized. How is this brought about?

The colour change occurs invariably at the spinning stage. It is a step of the normal ontogenetic development. In strict zoological terms it may not even be called a "colour change" at all. For terminologic reasons we have proposed to call it an "ontogenetic colour change" (6). The adaptive value of the ontogenetic colour change arises from what looks like a disadvantage at first sight, the rigid and invariable insect instincts. These instincts bring the insect at a certain stage of development reliably into a certain biological situation. For instance, the female lays its eggs always on poplar or related leaves. The young larva will find itself reliably on those leaves. Otherwise it would anyhow die of starvation. No colour adaptation would help it. The same reasoning applies when the spinning larva leaves the green food. Thus in ontogenetic colour change, the synchronization of developmental and behavioural changes guarantee the biological adaptation.

But how is this synchronization achieved? The colour change can only be caused by the hormonal situation which is typical of the developmental stage preceeding pupation: That is the presence of ecdysone and the absence of juvenile hormone (3). Higher doses of exdysone will cause pupation but no colour change (3).

As Prof. Riddiford (18) has shown today, in other lepidopteran species there occur phases of ommochrome formation in other developmental stages (14). They are caused by hormonal conditions typical for each particular developmental stage. One cannot, therefore, ascribe certain pigmentary effects to juvenile hormone or ecdysone. In grown-up Cerura larvae juvenile hormone inhibits ommochrome formation, in feeding Mamestra larvae it does evoke this colouration (13). In Cerura vinula ommochrome formation is caused by ecdysone but later on the same ecdysone evokes the pupal moult (3,4).

There is no rigid connection between a certain chemical type of hormone and a certain biological response. In different cells or different developmental stages, different effects can be evoked by the same hormone and in different animals similar effects can be evoked by different hormones. This phenomenon has been called the "polytropy" of hormones (9). Evidently in the target cells the same hormone receptor complex can be coupled to different final biological responses.

Our last example is the stick insect Carausius morosus. If we blacken the ventral parts of its compound eyes, the world will be looking to the animal as if it were placed in bright illumination on a dark background. Subsequently the insect changes its colour and becomes dark. It can form one hundredfold more ommochrome than a control specimen in which equal areas of the eyes have been blackened but on the upper half (12). The difference is caused only by the different distribution of light stimuli on the eyes. It is an example of the influence of visual information on development.

By the mentioned stimulation this colour change can be evoked at any time, also between moults. But how does the insect adapt its tryptophan supply to its visual situation? Recently we have developed an artificial diet for stick insects. The food has to be applied in form of "artificial leaves" (17). The epidermal ommochrome formation proves to be dependent on the tryptophan content of the diet. The more tryptophan the diet contains, the more ommochrome is formed. However, in every case, those animals, the eyes of which are blackened ventrally, form more ommochrome than the untreated controls. If fed a low tryptophan diet they will behave like living on natural leaves. But in case of higher tryptophan supply the range within which ommochrome formation is varied by the optical conditions remains always restricted (10).

Those stick insects which form more ommochrome will at the same time excrete less kynurenic acid via their excretory organs (1). There is also a correlation of epidermal pterin content to ommochrome level (2) and - similar to the situation in Cerura - a local relation between ommochrome and melanin deposits (12).

That ommochrome formation is not indispensable for the metabolism in Cerura vinula is proved by a probably mutant larva which occurred in our stock. This larva was evidently free of ommochrome but developed normally. Unfortunately no offspring could be obtained from this specimen.

These examples show (1) that ommochrome formation in different insects is a way of excreting tryptophan metabolites, but not the only one and (2) that it can be varied for the purpose of adaptive body colouration though only within a restricted range.

References

1. Berthold, G., Bückmann, D.: J. Comp. Physiol. 100, 347-350 (1975).
2. Berthold, G., Bückmann, D.: J. Comp. Physiol. 111, 33-37 (1976).
3. Bückmann, D.: J. Insect Physiol. 3, 159-189 (1959).
4. Bückmann, D.: Zool. Anz. Suppl. 26, 180-189 (1963).
5. Bückmann, D.: J. Insect Physiol. 11, 1427-1462 (1965).
6. Bückmann, D.: Symp. vergl. Endokr. Mainz, Fortschr. der Zool. 22, 1-22 (1974).
7. Bückmann, D.: Zool. Jb. Physiol. 78, 257-278 (1974).
8. Bückmann, D.: J. Comp. Physiol. 130, 331-336 (1979).
9. Bückmann, D.: Horm. metabol. Res. 15, 211-217 (1983).
10. Bückmann, D.: in preparation
11. Bückmann, D., Willig, A., Linzen, B.: Z. Naturf. 21b, 1184-1195 (1966).
12. Dustmann, J.H.: Z. vergl. Physiol. 49, 28-57 (1964).
13. Hiruma, K., Matsumoto, S., Isogai, A., Suzuki, A.: J. Comp. Physiol., in press.

14. Hori, M., Riddiford, L.: J. Comp. Physiol. 147, 1-9 (1982).

15. Linzen, B.: Adv. Insect Physiol. 10, 117-246 (1974).

16. Linzen, B., Bückmann, D.: Z. Naturforsch. 16b, 6-18 (1961).

17. Maisch, A., Bückmann, D.: in preparation.

18. Riddiford, L., Hori, M.: This volume (1983).

19. de Ruiter, L.: Arch. Neerl. Zool. 11, 1-57 (1955).

20. Summers, K.M., Howells, A.J., Pyliotis, N.A.: Adv. Insect Physiol. 16, 119-166 (1983).

THE STRUCTURE OF PAPILIOCHROME II, THE YELLOW WING PIGMENT OF THE PAPILIONID BUTTERFLIES

Heinz Rembold and Yoshishige Umebachi
Max-Planck-Institut für Biochemie, Martinsried
bei München, West-Germany

Introduction

The yellow wing pigments of the Papilio butterflies are built from tyrosine, ß-alanine and tryptophan as shown by Umebachi (1). The pigments were labelled following injection of anyone of these radioactive precursors into P. xuthus. The pigments of the pale and deep yellow as well as of the reddish brown scales are all related to these precursors. They have been named Papiliochrome II, M, and R, respectively. Papiliochrome II, the main yellow pigment of P. xuthus seems to have the lowest molecular weight of these pigments. It can be separated into the two isomers IIa and IIb by paper chromatography. Both readily decompose to kynurenine and to the dopamine derivative SN-1, when heated in 10^{-3} N HCl for 30 minutes. Using stronger conditions (1 N HCl, 100°C, 2 hrs), the latter is split into ß-alanine and another dopamine derivative (2). SN-1 has finally been shown to have the structure of N-(ß-alanyl)-L-noradrenaline (3).

The present paper gives the linkage between SN-1 and L-kynurenine and as such the structure of Papiliochrome II. The three components, L-noradrenaline, ß-alanine, and L-kynurenine, are contained in Papiliochrome II with a stoichiometry of 1:1:1.

Progress in Tryptophan and Serotonin Research

Materials and Methods

Wings from 300 P. xuthus or demoleus, mostly the male butterflies, were washed twice in a separatory funnel with ether, dried, and then extracted four times with 70 % ethanol (total 1.1 L). The yellow extract was concentrated by use of a rotary evaporator to about 250 ml, filtered, and then applied to a DEAE-cellulose column (3x14 cm). After elution with water the yellow effluent was pooled, applied to an Amberlite CG-50 column (3x11 cm), which was washed with 0.2 % pyridine-acetic acid (pH 5.0), and 0.6 % of the same buffer, respectively. A yellow fraction of the crude Papiliochrome II was then eluted by 23 % pyridine-acetic acid, which was lyophilized, dissolved in a small amount of water and further purified by preparative paper chromotography (Schleicher und Schuell, No. 2668; solvent BAW: upper layer of n-butanol-acetic acid-water, 4:1:5). The band of Papiliochrome II was cut out, eluted with water and, after concentration, was applied to P-cellulose (2.6x10 cm) and eluted with 0.2, 0.6 and finally 2.6 % pyridine-acetic acid (pH 5.5) which eluted Papiliochrome II. The pooled material from the other batches was dried by lyophilisation. A final TLC (cellulose, Merck No. 5716, 20x20 cm, 0.1 mm, solvent BAW) separated the two Papiliochrome II isomers. Yields of IIa: 50,0 mg, of IIb: 38,1 mg, isolated from the wings of 2.500 butterflies.

Results

^{1}H and ^{13}C NMR spectra rule out most of the possible links between SN-1 and kynurenine. In Papiliochrome II, the resonance of carbon 5 is changed, in comparison with N-(ß-alanyl)-L-noradrenaline (see table). The 5-carbon of SN-1 differs in its resonance by more than 15 ppm, whereas all the other signals are shifted by less than 7 ppm. A strong ring current effect of the aromatic protons further demonstrates an interaction of the two aromatic rings.

Table
^{13}C nuclear magnetic resonance data of SN-1, kynurenine and Papiliochrome II

Substance	Position in side chain	ppm	Substance	Corresponding Position in side chains	ppm	Δppm
SN-1	1	46.9	Papilio-chrome II	1	45.0	2
	2	38.0		2	32.0	6
	3	171.6		3	176.8	5
	4	38.5		4	36.0	3
	5	71.2		5	N.O.	>15
kynurenine	1'	174.2		1'	175.4	1
	2'	51.7		2'	51.0	1
	3'	42.0		3'	35.0	7
	4'	202.5		4'	202.5	0

No allocation is possible on the kynurenine side from NMR data. However, field desorption mass spectroscopy yields additional informations. M+1 manifests the molecular weight of 430 for Papiliochrome II. In contrast with kynurenine, which splits into an anilinium ion (m/e 92), no such fragment is formed from Papiliochrome II. Another analytical fragment is m/e 327 for N^{ar}-benzylidenekynurenine. Both these fragments prove that SN-1 and L-kynurenine are linked together through the arylamino group of kynurenine with the 5-carbon atom of SN-1. Papiliochrome II is, according to these spectroscopic data, N^{ar}-[α-(3-aminopropionylaminomethyl)-3,4-dihydroxybenzyl]-L-kynurenine. The two isomeric compounds Papiliochrome IIa and IIb differ, if the molecule model is compared, possibly by rotation of the catechol along the axis C-5 and p-OH to the opposite side. Both isomers are stabilized by the strong ring current.

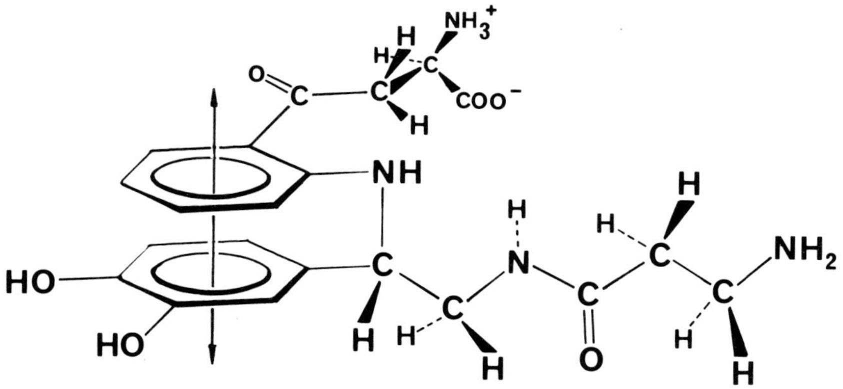

Discussion and Conclusion

The structure of Papiliochrome II can be divided into a peptide backbone with its carboxy end at the L-kynurenine and the amino end at the SN-1 side. Both the aromatic rings are completely exposed from this backbone by a very short bridge formed between the aromatic amino group and the 5-carbon atom. The ring current stabilizes the structure of Papiliochrome II and is responsible for the strong bathochromic shift which ends up with its yellow colour. Protonation of the aromatic amino group easily splits the molecule into two moieties. The new pigment structure of Papiliochrome II may also represent a subunit of the papiliochromes with higher molecular weight. By coupling of more aromatic rings a corridor effect could be induced there which finally explains the red shift of these pigments.

References

1. Umebachi,Y.: Insect Biochem. 5, 73-92 (1975).
2. Umebachi,Y.: Acta vitamin.enzymol.(Milano) 29, 219-222(1975).
3. Rembold,H., Rascher,J., Eder,J., Umebachi,Y.: Z. Naturforsch. 33c, 498-503 (1978).

SOME CHEMICAL PROPERTIES OF PAPILIOCHROME R

Yoshishige Umebachi

Department of Biology, Faculty of Science, Kanazawa University, Kanazawa, Japan

Introduction

Wing-pigments of papilionid butterflies are neither pteridine pigment nor ommochrome. Pigments of the pale yellow, deep yellow, and reddish brown scales of *Papilio* are all related to kynurenine, dopamine, and β-alanine, and have been named Papiliochrome II, M, and R, respectively. Papiliochrome II seems to have the lowest molecular weight among Papiliochromes and has been proved to consist of kynurenine and N-β-alanyldopamine or a derivative (1, 2, 7). On the other hand, the structures of Papiliochromes M and R have remained unsettled, though their chemical properties have already been studied to some extent (3, 4, 5). The present paper deals with further investigations of chemical properties of Papiliochrome R.

Materials and Methods

Reddish brown scales of the anal eye spot in the hind-wings of *Papilio demoleus* and *Papilio machaon* were used.

Components of the scales were fractionated into three fractions: (a) 70 % ethanol-soluble fraction, (b) 4 % HCl-methanol-soluble fraction, and (c) residual scales. In some cases, the residual scales were further treated with 1 N NaOH at room temperature, and the extract was applied to a Dowex 50W x 4 column. After being washed with water, the brown pigment was eluted with 2 N ammonia water (4, 5). The brown pigment fraction thus obtained was lyophilyzed, dissolved in 1 N NaOH, and

Progress in Tryptophan and Serotonin Research

precipitated by adding HCl. The precipitate was found to be a protein-bound brown pigment and is named HCl-ppt fraction.

Kynurenine was determined with Tsuda's reagent (6), and β-alanine was estimated by the DNP-method.

Amino acid analyses were made with a Nihon-Denshi 200A amino acid analyzer.

Incorporation of the following three kinds of ^{14}C-labelled compounds was examined: tryptophan[1-^{14}C], β-alanine[1-^{14}C], and dopamine[1-^{14}C]. Water solutions of these labelled compounds were injected into prepupae of *P. machaon* (0.02 to 0.06 ml, 0.6 to 6 μCi per prepupa).

Results

Autoradiographs were taken of the wings of butterflies (*P. machaon*) which had been injected with ^{14}C-tryptophan, ^{14}C-dopamine, or with ^{14}C-β-alanine during the prepupal stage. The autoradiographs showed that the labelled compounds had all been incorporated into the reddish brown scales. From these labelled scales, the HCl-ppt fraction was prepared and confirmed to be radioactive. Moreover, ^{14}C-tryptophan and ^{14}C-β-alanine were confirmed to be incorporated as ^{14}C-kynurenine and ^{14}C-β-alanine, respectively.

Kynurenine and β-alanine contents of the original scales before fractionation and of each fraction (70 % EtOH, 4 % HCl-MeOH, and residual scales) were determined as shown in Table 1, in which kynurenine and β-alanine contents of each fraction are given as percentage of the total. The table shows that most of the kynurenine and β-alanine in the reddish brown scales are insoluble in 70 % EtOH and 4 % HCl-MeOH, and remain in the residual scales.

The rate of appearance of kynurenine and β-alanine from the residual scales during the hydrolysis in 6 N HCl at 100°C showed

Table 1. Contents and solubilities of kynurenine (upper) and β-alanine (lower)

Species	*P. demoleus*	*P. machaon*
	μg per mg dry weight	
Untreated scales	33.9 73.6	37.3 61.9
Fraction	%	%
70 % EtOH	3.3 3.9	9.5 1.7
4 % HCl-MeOH	6.0 9.9	4.9 5.7
Residual scales	90.6 86.2	85.5 92.5

that kynurenine and β-alanine were more slowly released from the reddish brown scales than from the pale and deep yellow scales.

Amino acid compositions of the residual scales and HCl-ppt fraction are given in Table 2, which shows that the reddish brown pigment contains a large quantity of β-alanine and that the kynurenine content is rather low.

Discussion and Conclusion

From the results of the previous papers (4, 5, 6) and the present paper, Papiliochrome R can be characterized as follows: (a) Papiliochrome R contains kynurenine, dopamine, and β-alanine. (b) In solubilities and absorption spectrum, Papiliochrome R is similar to melanin but does not belong to melanins. The protein-bound brown pigment contains no copper. (c) Papiliochrome R is extracted with 1 N NaOH and precipitated with HCl. Most of the kynurenine is lost during the NaOH treatment. (d) Kynurenine content of the reddish brown scales is rather low in comparison with the pale and deep yellow scales. (e) Both kynurenine and β-alanine are bound more strongly in the reddish brown scales than in the pale and deep yellow

Table 2. Molar ratios of amino acids in the hydrolysates of the residual scales and HCl-ppt fraction from *P. demoleus*

Amino acid	Residual scale	HCl-ppt fraction
Met	3.7*	27.6
Asp	40.7	36.6
Thr	26.2	23.0
Ser	41.9	31.1
Glu	42.8	88.0
Pro	55.1	44.4
Gly	100.0	100.0
Ala	94.7	72.0
Val	111.0	84.4
Ile	32.4	37.7
Leu	46.3	53.0
Tyr	41.0	23.0
Phe	9.5	19.5
β-Ala	197.8	498.1
Kyn	12.3	Trace
Lys	15.5	20.2
His	27.5	64.6
Arg	21.7	——

* Molar ratios are given as the value for 100 of glycine.

scales. From these properties, there is a possibility that Papiliochrome R may be a kind of polymer of N-β-alanyldopamine to which some kynurenine is attached.

References

1. Rembold, H., Rascher, J., Eder, J., Umebachi, Y.: Z. Naturforsch. 33c, 498-503 (1978).
2. Umebachi, U.: Insect Biochem 5, 73-92 (1975).
3. Umebachi, U.: Sci. Rep. Kanazawa Univ. 22, 91-101 (1977).
4. Umebachi, U.: Sci. Rep. Kanazawa Univ. 23, 119-128 (1978).
5. Umebachi, U.: Sci. Rep. Kanazawa Univ. 24, 73-79 (1979).
6. Umebachi, U.: Comp. Biochem. Physiol. (1983) (in press).
7. Umebachi, Y., Yamashita, H.: Comp. Biochem. Physiol. 54B, 55-62 (1976); 56B, 5-8 (1977).

Tryptophan Metabolism in Plants and Microorganisms

METABOLISM OF TRYPTOPHAN IN *PSEUDOMONAS AUREOFACIENS*

Karl-Heinz van Pée, Franz Lingens
Institut für Mikrobiologie der Universität Hohenheim
D-7000 Stuttgart 70

Pseudomonas aureofaciens degrades tryptophan by several different pathways.

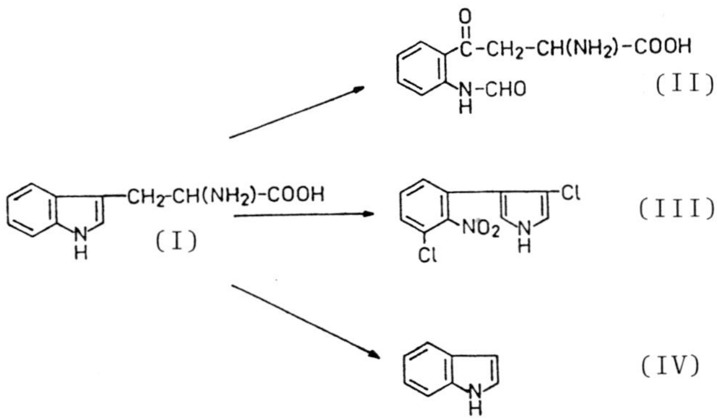

Fig.1 Pathways of tryptophan metabolism in *Pseudomonas aureofaciens*.

Destruction of the indole ring system occurs already in the first step, when tryptophan (I) is metabolized via N-formylkynurenine (II) by the aromatic pathway (1). A different pathway involves step by step degradation of the side chain of tryptophan, before the indole-moiety is attacked (2). Already in an early stage of the biosynthesis of the antifungal antibiotic pyrrolnitrin (III) the indole ring system of tryptophan is converted into a phenylpyrrole ring system (3, Fig.1).

Aromatic pathway

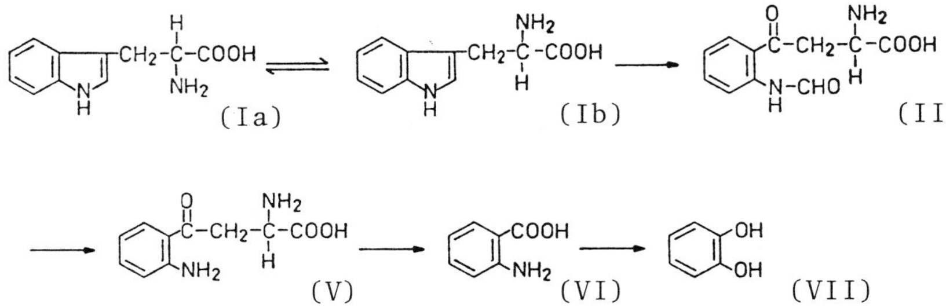

Fig. 2 Conversion of tryptophan by the aromatic pathway

The first step in the aromatic pathway (Fig.2), the formation of L-N-formylkynurenine (II), is catalyzed by the enzyme L-tryptophan 2.3-dioxygenase. This enzyme is spezific for L-tryptophan (Ib) and is induced by its substrate, but not by L-kynurenine (V). The enzymes kynureninase and anthranilate, 1,2-dioxygenase are induced by L-kynurenine (V) and anthranilic acid (VI),as well as by L-tryptophan. D-Tryptophan (Ia) is metabolized by this pathway only after conversion to the L-isomer by a racemase (4).

Degradation of the side chain

The isolation of several 7-chloroindole derivatives,such as 7-chloroindole-3-acetic acid (X),7-chloroindole-3-carboxylic acid (XII) 7-chloroindole (IVa) and 3-chloroanthranilic acid (5,2)suggests the existence of 7-chlorotryptophan as common precursor of these metabolites. In contrast to L-tryptophan, however, 7-chloro-L-tryptophan is not a substrate for L-tryptophan 2,3-dioxygenase and thus can not be metabolized via the aromatic pathway.7-Chloro-L-tryptophan is degraded by a different, constitutive pathway, in which the side chain of tryptophan is removed step by step (2, Fig.3).

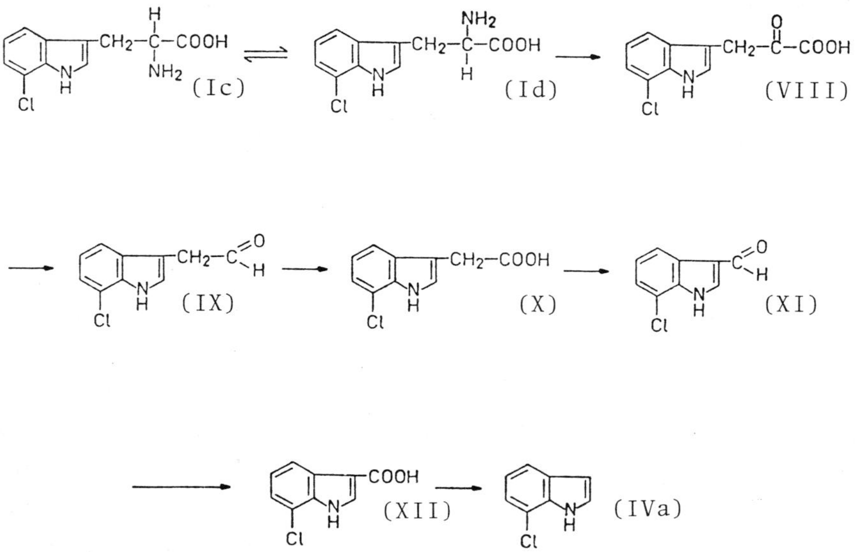

Fig. 3 Proposed pathway of metabolism of the side chain of L-tryptophan and 7-chloro-L-tryptophan in Pseudomonas aureofaciens.

This pathway starts with the transamination of 7-chloro-L-tryptophan (Id) or L-tryptophan (Ib) to yield the corresponding indole-3-pyruvate (VIII). Enzymatical decarboxylation of 7-chloroindole-3-pyruvate (VIII) leads to 7-chloroindole-3-acetaldehyde (IX),which is dehydrogenated to 7-chloroindole-3-acetic acid (X). On the other hand, 7-chloroindole-3-acetic acid can be formed by the spontaneous decomposition of 7-chloroindole-3-pyruvate, too. Wether 7-chloroindole-3-acetic acid (X) is transformed enzymatically to 7-chloroindole-3-carboxaldehyde (XI) as described for several plants (6,7), or whether this aldehyde arises only through decomposition of 7-chloroindole-3-pyruvate (8) is not clear yet. 7-Chloroindole-3-carboxaldehyde (XI) is oxidized enzymatically to 7-chloroindole-3-carboxylic acid (XII), which decarboxylates in aqueous solution to 7-chloroindole (IVa). This metabolite could be transformed to 3-chloroanthranilic acid (9). The isolation of 7-bromoindole-3-acetic acid, 7-bromoindole-3-carboxylic acid and 7-bromoindole suggests,that the pathway for the degradation of the side chain of 7-bromo-L-tryptophan is identical with that of 7-chloro-L-tryptophan and L-tryptophan. As in the case of the aromatic pathway, D-tryptophan can only be metabolized after conversion to the L-isomer.

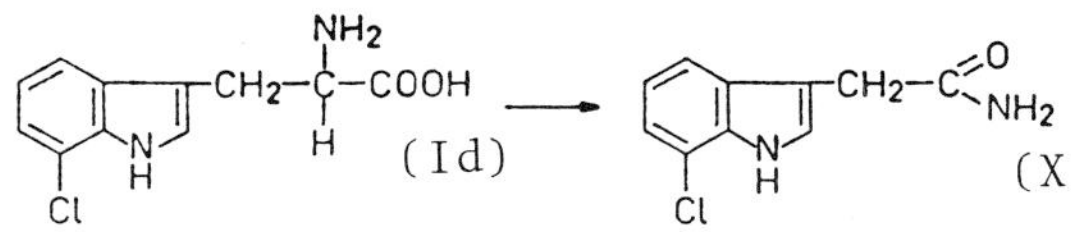

Fig. 4 Conversion of 7-chloro-L-tryptophan to 7-chloroindole-3-acetamide.

An enzyme, that transforms L-tryptophan (Ib) and 7-chloro-L-tryptophan (Id) to the corresponding indole-3-acetamide (XIII) could be detected in crude protein extracts from P.aureofaciens grown in mineral salt medium without addition of L-tryptophan (10, Fig.4).

Biosynthesis of pyrrolnitrin

The starting point for the pathways already mentioned is L-tryptophan. D-Tryptophan has to be converted to the L-isomer before degradation. In contrast to this, the biosynthesis of the secondary metabolite pyrrolnitrin (III) is believed to to start from D - as well as from L-tryptophan (Fig.5). The addition of D-tryptophan 24 h after inoculation increased the production of the antibiotic to 180 % whereas the addition of L-tryptophan decreased the pyrrolnitrin production to 40 % (4), thus indicating D-tryptophan to be the more immediate precursor. Floss and his co-workers (11) demonstrated, by using labeled D- and L-tryptophan, that both isomers are incorporated into pyrrolnitrin with the indole nitrogen giving rise to the nitro group of pyrrolnitrin and the amino nitrogen of tryptophan to the pyrrol nitrogen. The amino group of L-tryptophan however is incorporated much more efficiently. Another difference in the utilisation of the two isomers is the almost complete loss of the α-hydrogen from the side chain of D-tryptophan, but not from the L-isomer. Thus on a pathway from L-tryptophan to pyrrolnitrin the removal of the carboxyl group must occur prior to or simultaneously with the formation of a double bond at the C-2 of tryptophan. This excludes the existence of 2-carboxy-4-(2-amino-3-chlorophenyl)pyrrole (XIV) as an intermediate on a pathway from L-tryptophan. This metabolite, however, could still be an intermediate on a pathway from D-tryptophan to pyrrolnitrin (12). The bromoanalogue of 2-carboxy-4-(2-amino-3-chlorophenyl)pyrrole (XVII) was decarboxylated under resting cell conditions to 4-(2-amino-3-bromophenyl)pyrrole (XVIII, Fig. 6), indicating that the isolated pyrrole-2-carboxylic acids are no dead-end products. Monodechloroaminopyrrolnitrin (XV), the chloride containing decarboxylation product of the pyrrole-2-carboxylic acid, can be accumulated by growing P. aureofaciens in a medium without chloride but in the presence of 7-chlorotryptophan (13). Addition of Cl^- to the medium after accumulation of monodechloroaminopyrrolnitrin (XV) results in the formation of aminopyrrolnitrin (XVI) and pyrrolnitrin (III) from XV (Fig. 5). When grown without chloride and without 7-chlorotryptophan, no phenylpyrrole compound could be detected. This suggests, that the first step in the pyrrolnitrin biosynthesis is a chloride dependent step, e.g. the chlorination of tryptophan to 7-chlorotryptophan. No chloride, however, seems to be necessary for the next steps, the formation of the pyrrole-2-carboxylic acid (XIV) on a pathway from D-tryptophan and the formation of monodechloroaminopyrrolnitrin (XV) on both pathways.

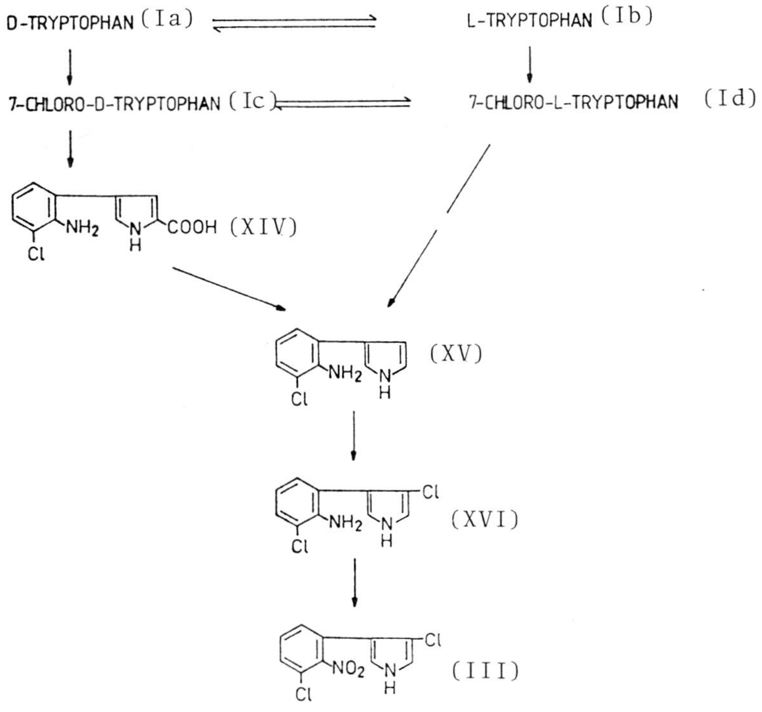

Fig. 5 Hypothetical pathway for the biosynthesis of pyrrolnitrin.

Chlorination in the 3-position of the pyrrole ring leads to aminopyrrolnitrin (XVI). The last step in pyrrolnitrin biosynthesis is the oxidation of the amino to the nitro group (14, Fig.5).

The biosynthesis of brominated pyrrolnitrins (Fig.6) is almost identical with the one described above. With bromide a few additional phenylpyrrole derivatives are formed and the product is 2,3-dibromo-4-(2-nitro-3-bromophenyl)pyrrole(XXI). This is probably due to a higher activity of the bromoperoxi -

dase responsible for the bromination,than of the chloroperoxidase,responsible for the chlorination.

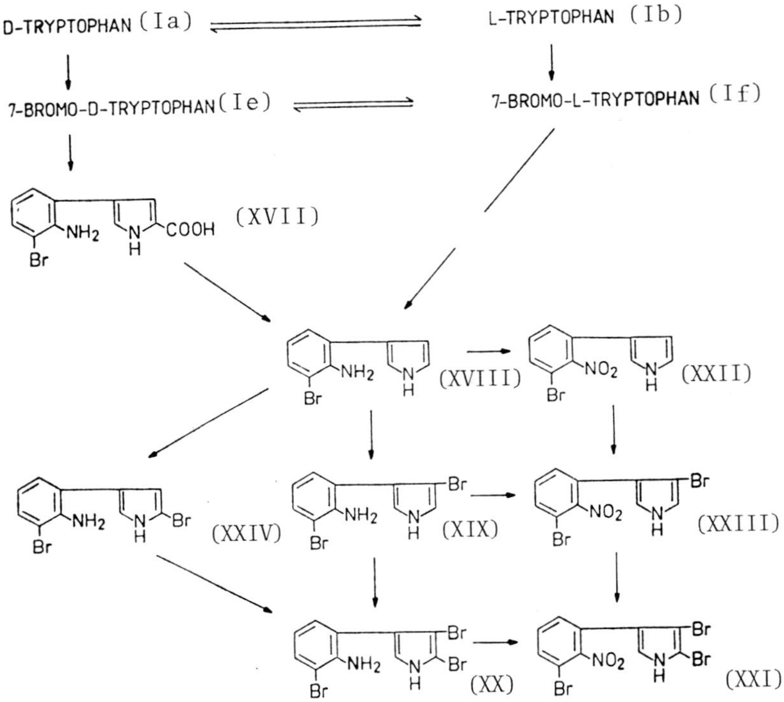

Fig. 6 Hypothetical pathway for the biosynthesis of brominated pyrrolnitrin derivatives

The problem, whether D-or L-tryptophan is the immediate precursor of the antibiotic is still not solved. However, it can be understood now, why the addition of L-tryptophan inhibits the production of pyrrolnitrin whereas D-tryptophan enhances the production of the antibiotic. The reason is the induction of L-tryptophan 2,3-dioxygenase and thus the aromatic pathway by L-tryptophan but not by D-tryptophan, which therefore is available for pyrrolnitrin biosynthesis. D-tryptophan and 7-chlorotryptophan can only be degraded by the " side chain pathway " or serve as precursors for secondary metabolism. The same results, except that no brominated metabolites were isolated could be obtained with Pseudomonas pyrrocinia, another pyrrolnitrin producing Pseudomonad.

References

1. Behrman, E.J., Stella, E.J.: J. Bacteriol. 85, 946-947 (1963).
2. Lübbe, C., van Pée, K.-H., Salcher, O., Lingens, F.: Hoppe-Seyler's Z. Physiol. Chem., in press.
3. Lively, D.H., Gorman, M., Haney, M.E., Mabe, J.A.: Antimicrob. Agents Chemother.: 1966, 462-469 (1967).
4. Salcher, O., Lingens, F.: J. Gen. Microbiol. 121, 465-471 (1980).
5. Salcher, O., Lingens, F.: Tetrahed. Lett. 34, 3101-3102 (1978).
6. BeMiller, J.N., Colilla, W.: Phytochem. **11**, 3393-3402 (1972).
7. Sinha, B.K., Basu, P.S.: Biochem. Physiol. Pflanzen 176, 218-227 (1981).
8. Glombitza, K.-W., Hartmann, T.: Planta (Berl.) 69, 135-149 (1966).
9. Hayaishi, O.: Bacteriol. Rev. 30, 720-731 (1966).
10. Salcher, O., van Pée, K.-H., Lingens, F.: Z. Naturforsch. 356, 340-341 (1980).
11. Chang, C.J., Floss, G.H., Hook, D.J., Mabe, J.A., Manni, P.E., Martin, L.L., Schröder, K., Shieh, T.L.: J. Antibiotics 34, 555-566 (1981).
12. Salcher, O., Lingens, F., Fischer, P.: Tetrahed. Lett. 34, 3097-3100 (1978).
13. van Pée, K.-H., Salcher, O., Lingens, F.: Angew. Chem. Int. Ed. Engl. 19, 828-829 (1980).
14. Hamill , R., Elander, R., Mabe, J., Gorman, M.: Antimicrob. Agents Chemother. 1967, 388-396 (1968).

DOES TRYPTOPHAN SIDE CHAIN OXIDASE FROM PSEUDOMONAS ATCC 29574 CONTAIN A QUINOPROTEIN?

W. van der Graaff, J.A. Duine, J. Frank, Jzn and J.A. Jongejan
Department of Biochemistry
Delft University of Technology
Julianalaan 67, 2628 BC Delft, The Netherlands

Introduction

As is indicated by its name, Tryptophan Side chain Oxidase(TSO) catalyzes the oxidation of the side chain of tryptophan and related indole derivatives.

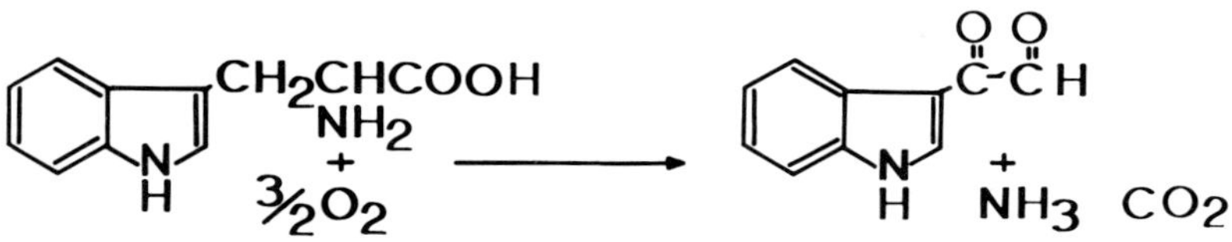

Fig. 1. Oxidation of L-tryptophan catalyzed by TSO.

The enzyme was first described by Roberts and Rosenfeld (1) and Takai et al. (2). In fact it is not a real oxidase but a complex of electron transferring proteins since one of the enzyme forms, TSO-II, consists of a dehydrogenase (detectable with artificial electron acceptors) and a real oxidase which consumes O_2 and oxidizes the dehydrogenase component (3). The dehydrogenase displays a cytochrome-b like absorption spectrum, while no flavin could be extracted from it. However, on reduction and oxidation, typical changes are seen in the 300 - 400 nm region of the absorption spectra. Moreover, an ESR signal was detected with g = 2.00, indicating that still another prosthetic group might be present in the enzyme.

Progress in Tryptophan and Serotonin Research

Quinoproteins

In the past few years we detected a novel class of dehydrogenases: the so-called quinoproteins (4) containing the coenzyme 2,7,9-tricarboxy-1H pyrrolo(2,3-f)-quinoline-4,5-dione (Fig. 2.), abbreviated as pyrrolo-quinoline quinone, PQQ. Quinoprotein dehydrogenases are found in bacteria, widely differing in morphology, physiology and habitat.

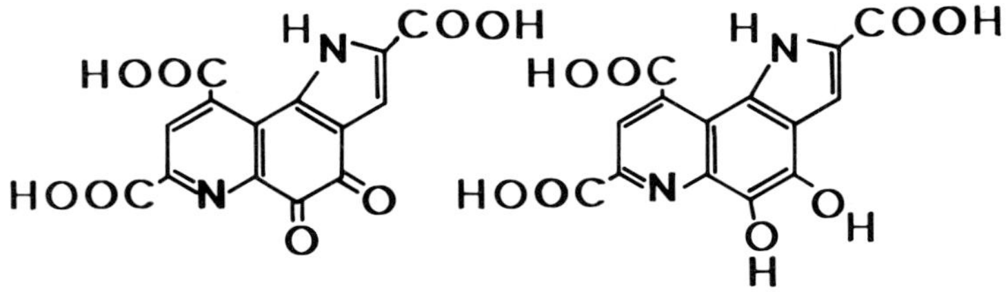

Fig. 2.
The oxidized and reduced forms of pyrrolo-quinoline quinone.

As ESR-spectroscopy revealed that PQQ is present in several dehydrogenases in its free radical state (PQQH·), this prompted us to investigate whether TSO-II dehydrogenase might be a quinoprotein.

Results

Before attempts were made to purify the TSO-II enzyme and the dehydrogenase supposed to be a component of this complex (3), Pseudomonas ATCC 29574 cells were tested (5) for the presence of PQQ. Since the result (1.1 nmole PQQ/g wet cells) clearly indicated that this organism synthesizes PQQ, it was meaningful to develop a purification procedure for the enzymes:

- Fractionation of the crude extract by precipitation with PEG 6000
- Ion-exchange chromatography on DEAE-Sephacel
- Gel-filtration on analytical HPLC columns

The enzymes seem to be firmly bound to the bacterial membranes since good adsorption onto the ion-exchanger was only possible in the presence of a detergent. Enzyme stability greatly increased by addition of polyethyleneglycol to the fractions.
Following the procedure, chromatographically pure TSO-II complex and dehydrogenase were obtained. The absorption spectra are closely similar to those reported in the literature (3).

X-band ESR-spectroscopy of the enzymes showed that both contain a signal with top-top width of 0.9 mT and g = 2.0046 (Fig. 3.).

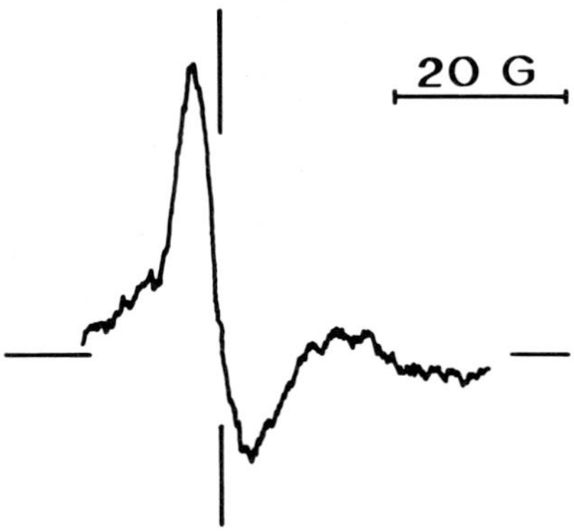

Fig. 3.
X-band ESR signal of purified TSO-II complex, taken at room temperature.

As this suggested the presence of PQQH˙, it was investigated whether the coenzyme is in an extractable form (6). Based on the reported molecular weights it was calculated that TSO-II complex and dehydrogenase contain 10^{-3} and 5.10^{-3} molecules of PQQ per enzyme molecule, respectively.

Conclusions

Pseudomonas ATCC 29574 synthesizes PQQ.
The chromato-raphically pure preparations of TSO-II complex and dehydrogenase show an ESR signal which is similar to that found in other quinoproteins.

Minor amounts of PQQ were found upon extraction of the enzymes. This situation is similar, however, to that found for methylamine dehydrogenase, a quinoprotein in which PQQ is covalently bound to the protein (7).
Therefore it is tentatively assumed that TSO-II contains a quinoprotein dehydrogenase in which PQQ (or a derivative) is in a non-extractable form.

It might be argued that the absorption spectra of the dehydrogenase give no indication for the presence of PQQ. However, in the case of methylamine dehydrogenase from Pseudomonas (8), it is clear that the quinoprotein absorption spectrum is equally obscured by the cytochrome absorption bands. In order to obtain additional evidence, attempts will be carried out to remove the heme prosthetic group from the TSO-II dehydrogenase.

References

1. Roberts, J. and Rosenfeld, H.: J. Biol. Chem., 252, 2640-2647 (1977).
2. Takai, K., Ushiro, H., Noda, Y., Narumiya, S., Tokuyama, T. and Hayaishi, O. : J. Biol. Chem., 252, 2648-2656 (1977).
3. Ushiro, H., Takai, K., Narumiya, S., Ito, S. and Hayaishi, O. : J. Biol. Chem., 254, 11794-11797 (1979).
4. Duine, J. A., and Frank, J. : Trends Bioch. Sci., 6, 278-280 (1981).
5. Duine, J.A., Frank, J. and Jongejan, J.A., Anal. Bioch., (1983) in press.
6. Duine, J.A., Frank, J. and Verwiel, P.E.J. : Eur. J. Bioch. 108, 187-192 (1980).
7. De Beer, R., Duine, J.A., Frank, J. and Large, P.J. : Biochim. Biophys. Acta 622, 370-374 (1980).
8. Durham, D.R. and Perry, J.J.,: J. Bact. 134, 837-843 (1978).

TRYPTOPHAN SIDE CHAIN OXIDASE (TSO)

Structure and Reaction Mechanism

Katsuji Takai, Shintaro Inoue, Flora Zavala, Rudolf Müller, Taihei Hamasaki, Yoshiki Sasai

Department of Medical Chemistry, Kyoto University Faculty of Medicine
Kyoto 606, Japan

Introduction

Tryptophan side chain oxidase (TSO) from *Pseudomonas*, a multienzyme system containing heme(s), catalyzes the dehydrogenation of tryptophan residues (Trp) (1,2,3). It serves as a crystallizable respiratory chain model spanned from polypeptides to O_2, and a newer aspect of tryptophan residue *per se* is also available by analyzing the consequences of oxidation of this electron donor in polypeptides. In the present studies, we present first the organization of two functional units of TSO, then the identification of the primary reaction product as indolyloxazoline bound to polypeptide, and finally the action of TSO on proteins.

Results

Two types of TSO (TSO I/II) have been purified to homogeneity from the same organism (1,2,4,5,6). These were distinguishable in $\underline{M}_r$, $\underline{M}_r$ of subunits, and antigenecity, but appeared to share the similar chromophores (6). Analyses of TSO II, a simpler form, revealed that TSO consisted of two functional units; the *dehydrogenase*, a red/ox system containing protoheme IX, with an apparently high redox potential, that abstracts electrons from Trp but is nonautooxidiable *per se*, and the *oxidase* containing an as yet unidentified heme-like chromophore ($\varepsilon_{m\underline{M}}$ = 100), which mediates the electron transfer from

Progress in Tryptophan and Serotonin Research

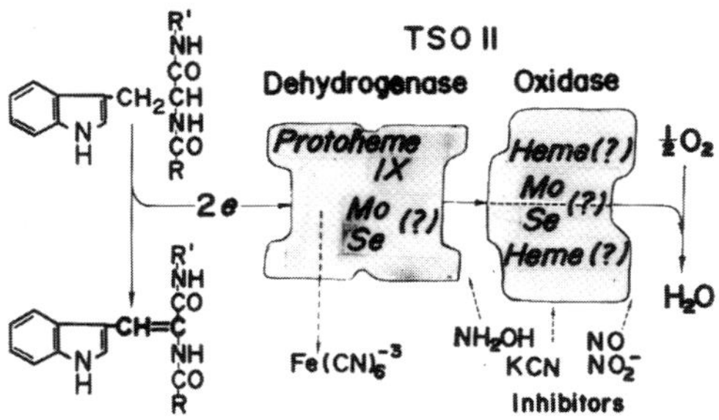

Fig. 1. Structure and function of TSO II —A tentative scheme—

dehydrogenase to O_2 (5,6). Only the latter step was inhibited by KCN, NO, $NaNO_2$, and NH_2OH (6). Se and Mo but not flavins were detectable in TSO II (6). On these bases, a tentative scheme for the structure and function of TSO II is depicted (Fig. 1) (6). By the action of TSO, Trp underwent the pH-dependent dichotomous reaction to give α,β-didehydro (Δ-Trp) and β-hydroxytryptophan (β-OH-Trp) residues, the occurrence of a common precursor being suggested (1,4,7,9). Recently, using a model substrate (Ac-Trp-NH_2), the unstable primary intermediate was isolated, stabilized in an aprotic solvent, and identified as an indolyloxazoline derivative (9). The results implied that Trp deprived of electrons was instantaneously stabilized at the β-carbon atom of Trp by the amide carbonyl oxygen (Fig. 3). In an aqueous medium, this primary product underwent nonenzymatic isomerization to Δ-Trp or hydration to form β-OH-Trp, where the protonation of the imino nitrogen (pK_a = 5.8) of the oxazoline ring prescribed the rate of each

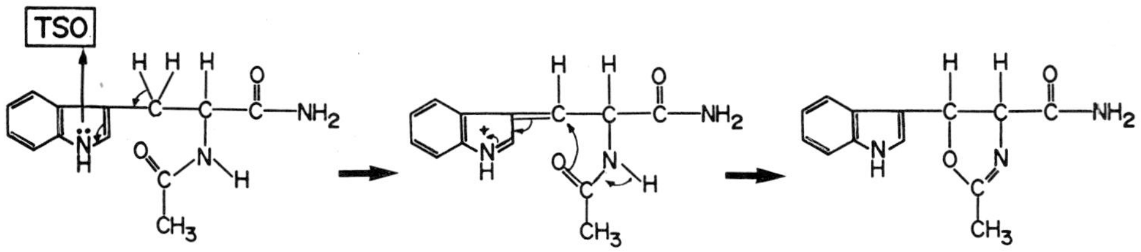

Fig. 2. Possible mechanism of the formation of indolyloxazoline

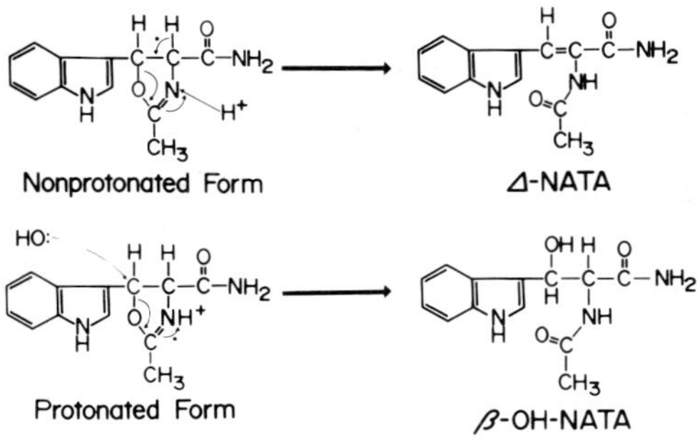

Fig. 3. Mechanism of the pH-dependent dichotomous reaction of the primary product

of the reactions (Fig. 3) (9). Fig. 4 shows the structure of the indolyloxazoline derivative in polypeptide, a new biochemical entity. Noteworthy is that, upon dehydrogenation of Trp, the amide group of the polypeptide reacts with the side chain and, if thiols or other nucleophiles are present, β-substituted adducts are obtained (9). Since the oxidized form of Trp in quest retains the intact indole nucleus, it might be possible that, under certain conditions, Trp in polypeptides may follow a red/ox cycle accompanied by marked structural changes in polypeptides. Search for a reductase activity of the oxidized Trp is underway.

TSO I and TSO II descriminate the site of Trp in polypeptides; N-terminal Trp is a poor substrate of TSO II, C-terminal Trp was equally active for TSO I/II, and Trp inside the polypep-

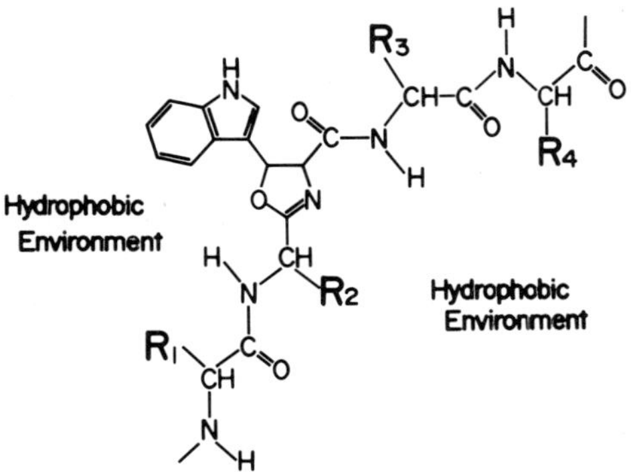

Fig. 4. Indolyloxazoline derivative in polypeptide

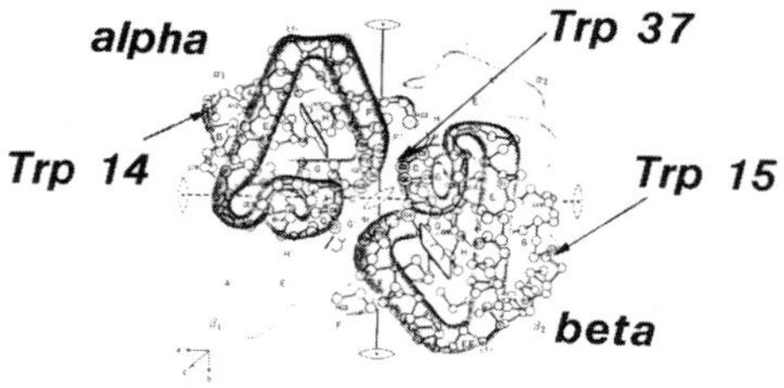

Fig. 5. Trp in hemoglobin

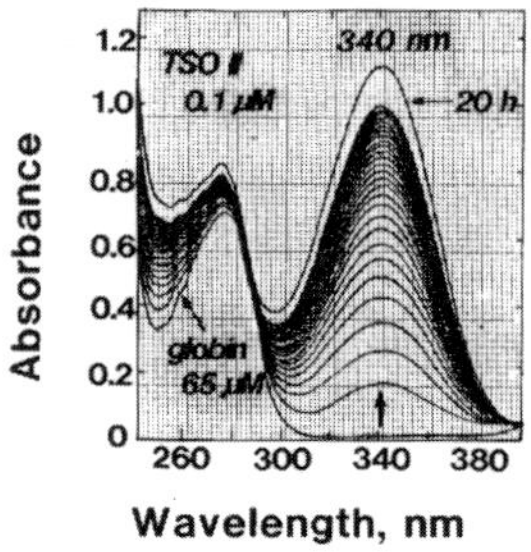

Fig. 6. Spectral changes associated with the modification of Trp-14 of α-globin

tides more or less shows the preference for TSO II (1,4). Upon modification of proteins, TSO II was so far successfully applied. Attempts are underway to modify three Trp of hemoglobin selectively (Fig. 5). Fig. 6 shows the conversion of Trp-14 of α-globin to Δ-Trp, where the reaction was specific as judged by analyses of tryptic fragments (10).

References

1. Takai,K.: Developments in Biochemistry 16; Biochemical and Medical Aspects of Tryptophan Metabolism (Hayaishi, O., *et al.* eds), 103-115, Elsevier/North Holland, Amsterdam (1980)
2. Takai,K.,Ushiro,H.,Noda,Y.,Narumiya,S.,Tokuyama,T.,Hayaishi, O.: J. Biol. Chem. 252, 2648-2656 (1977)
3. Ushiro,H.,Takai,K.,Noda,Y.,Narumiya,S.,Tokuyama,T.,Hayaishi, O.: J. Biol. Chem. 253, 9002-9008 (1978)
4. Ito,S.,Takai,K.,Tokuyama,T.,Hayaishi,O.: J. Biol. Chem.: 256, 7834-7843 (1981)
5. Ushiro,H.,Takai,K.,Narumiya,S.,Ito,S.,Hayaishi,O.: J. Biol. Chem. 254, 11794-11797 (1979)
6. Takai,K.,Müller,R.,Zavala,F.,Inoue,S.,Narumiya,S.,Ito,S., Hayaishi,O.: Oxygenases and Oxygen Metabolism (Nozaki, M., *et al.* eds) 159-172 (1982)
7. Noda,Y.,Takai,K.,Tokuyama,T.,Narumiya,S.,Ushiro,H.,Hayaishi, O.: J. Biol. Chem. 252, 4413-4415 (1977)
8. Noda,Y.,Takai,K.,Tokuyama,T.,Narumiya,S.,Ushiro,H.,Hayaishi, O.: J. Biol. Chem. 253, 4819-4822 (1978)
9. Zavala,F.,Takai,K.,Hayaishi,O.: J. Biol. Chem. 258, 344-351 (1983)
10. Takai,K.,Sasai,Y.,Inoue,S.,Yamazaki,T.,Morimoto,H.: in preparation

CHARACTERIZATION OF THREE TRYPTOPHAN AMINOTRANSFERASES FROM CANDIDA MALTOSA

Rüdiger Bode, Dieter Birnbaum
Ernst-Moritz-Arndt-Universität Greifswald, Sektion Biologie
DDR-2200 Greifswald, G.D.R.

Introduction

The major route of tryptophan metabolism in yeasts is its conversion to indolepyruvate and subsequent degradation to tryptophol and indoleacetate (1, 2, 3). Tryptophan transaminase which is active in the initial step of this metabolic pathway has been detected and partially purified from Candida maltosa EH 15. In this paper we show that three enzymes are involved in this degradation step and some properties of these aminotransferases are described.

Results

To characterize the tryptophan aminotransferase activity, cell free extract of C.maltosa grown on minimal medium (4) was chromatographed on DEAE-cellulose (Fig. 1). Three active fractions were separable upon this chromatography: peak I represents a constitutive aminotransferase, the second fraction (peak II) appears to be regulated by the general control of amino acid biosynthesis, and the aminotransferase of peak III is inducible by tryptophan, phenylalanine, and/or tyrosine (5).
All three enzymes are capable of transaminating prephenate, phenylpyruvate, 4-hydroxy-phenylpyruvate, 2-oxoglutarate, and

Progress in Tryptophan and Serotonin Research

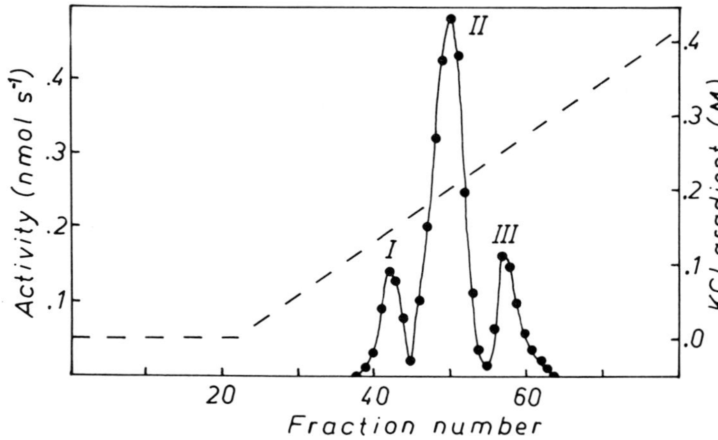

Fig. 1. Fractionation of tryptophan aminotransferase after DEAE-cellulose chromatography. Cell free extract (dissolved in 0.1 M Tris-HCl buffer; pH 8.6) was chromatographed on a DEAE-cellulose column (2 x 17 cm). Elution was carried out by a linear KCl gradient (from 0 mM to 500 mM; total volume 300 ml). Fractions of 4 ml were collected and aliquots were tested for tryptophan aminotransferase activity by the procedure of Gordon and Paleg (6). The reaction mixture contained: 5 mM 2-oxoglutarate, 5 mM L-tryptophan, and 0.1 mM pyridoxal 5'-phosphate.

Table 1. Substrate specificities of tryptophan aminotransferases. Enzyme assay: 5 mM L-tryptophan, 1 mM keto acids, 0.1 mM PLP. Results were expressed as percentage of 2-oxoglutarate-L-tryptophan transamination activity.

Amino acceptor	Aromatic aminotransferase		
	I	II	III
2-Oxoglutarate	100	100	100
Oxaloacetate	50	150	220
Pyruvate	0	0	0
2-Keto-isovalerate	0	0	0
Phenylpyruvate	375	911	1580
4-Hydroxy-phenylpyruvate	275	806	960
Prephenate	400	828	1560

Table 2. Substrate specificities of aromatic aminotransferases. Enzyme assay: 5 mM aromatic amino acids, 5 mM 2-oxoglutarate, 0.1 mM PLP. The method of Granner and Tomkins (7) was used for determination of the activity of phenylalanine and tyrosine aminotransferase. Results were expressed as percentage of 2-oxoglutarate-L-tryptophan transamination activity.

Amino donor	Aromatic aminotransferase		
	I	II	III
L-Tryptophan	100	100	100
D-Tryptophan	0	0	0
DL-6-Methyl-tryptophan	29	43	52
DL-5-Methyl-tryptophan	61	33	32
DL-4-Methyl-tryptophan	48	20	23
DL-6-Fluoro-tryptophan	98	48	54
DL-5-Fluoro-tryptophan	56	43	44
L-Tyrosine	59	50	58
D-Tyrosine	0	0	0
L-Phenylalanine	69	55	64
D-Phenylalanine	0	0	0
DL-p-Fluoro-phenylalanine	26	28	31

oxaloacetate in the presence of L-tryptophan (Tab. 1). We did not obtain any reaction with pyruvate and 2-keto-isovalerate as amino acceptor reactant.

The experimental results about amino donor specificity are shown in Table 2. The aminotransferases are more active in relation to L-tryptophan, than to L-phenylalanine and L-tyrosine, but no activity was found if the D-isomers were present. The purified enzymes were also capable of utilizing various analogues as substrates; no activity was found with other amino acids.

The three aminotransferases possess overlapping specificities for all tested aromatic amino donors and acceptors, and thus are called aromatic aminotransferases. They are also required for normal biosynthesis of L-phenylalanine and L-tyrosine. The apparent K_m-values are reported in Table 3. The Michaelis values for fraction I and III are somewhat lower with the amino

Table 3. Apparent K_m-values (mM) for the aromatic aminotransferases. Values were estimated from abscissa intercepts on double reciprocal plots.

Substrate	Aromatic aminotransferase I	II	III
L-Tryptophan	2.5	3.3	2.0
L-Phenylalanine	0.44	1.1	0.45
L-Tyrosine	0.40	2.0	0.50
Phenylpyruvate	0.08	0.25	0.11
4-Hydroxy-phenylpyruvate	0.05	0.20	0.05
Prephenate	0.08	0.30	0.09
2-Oxoglutarate	5.0	10.0	2.0
Oxaloacetate	3.3	10.0	3.0

donors and acceptors used, than the corresponding values for fraction II.

From these results it is not clear which aminotransferase plays the major role in the degradation of aromatic amino acids in vivo. But aminotransferase III was found to be inducible by the three aromatic amino acids (5), and therefore it is postulated to be mainly involved in utilization of tryptophan as the source of nitrogen.

References

1. Rosazza, J.P., Juhl, R., Davis, D.: Appl. Microbiol. 26, 98-105 (1973).
2. Bode, R., Böttcher, F., Birnbaum, D.: Cell. Mol. Biol. 26, 615-620 (1980).
3. Kradolfer, P., Niederberger, P., Hütter, R.: Arch. Microbiol. 133, 242-248 (1982).
4. Tanaka, A., Ohishi, N., Fukui, S.: J. Ferment. Technol. 45, 617-623 (1967).
5. Bode, R., Birnbaum, D.: Z. Allg. Mikrobiol. in press.
6. Gordon, S.A., Paleg, L.G.: Physiol. Plantarum 10, 39-47 (1957).
7. Granner, D.K., Tomkins, G.M.: Meth. Enzymol. 17A, 633-637 (1970).

ANTHRANILATE HYDROXYLASE AND THE SHORT-CUT PATHWAY FOR THE BIOSYNTHESIS OF NICOTINAMIDE NUCLEOTIDES IN ASPERGILLUS NIGER

Hans-Henning Schott, Karen Zinke, Doris Langenbach-Schmidt
Biochemisches Institut der Universität Giessen, Friedrichstr. 24, D-6300 Giessen

Introduction

Nicotinamide nucleotides are synthesized in microorganisms by two pathways: The O_2- and pyridoxal phosphate-dependent tryptophan-kynurenine pathway and the "anaerobic" aspartate pathway. In addition a third pathway exists which is named the "short-cut pathway", because anthranilate (AA) is directly hydroxylated to 3-hydroxyanthranilate (3HOAA), eliminating eight enzymatic steps.

Two enzymes are associated with the hydroxylation of AA to 3HOAA: 1) anthranilate 3-hydroxylase (1-3) and 2) anthranilate 2,3-hydroxylase (deaminating) which forms - beside the main product 2,3-dihydroxybenzoate (2,3DHB) - 3HOAA as additional product (4-8).

The nature of the anthranilate hydroxylase activity was investigated in Aspergillus niger by purification and characterization of the corresponding enzyme. Also preliminary experiments were carried out to define the possible role of anthranilate 3-hydroxylation and of the short-cut pathway in the metabolism of Aspergillus niger.

Results

Purification. The stabilized anthranilate hydroxylase was isolated from crude extract of Aspergillus niger by protamine

Progress in Tryptophan and Serotonin Research

sulfate and ammonium sulfate precipitations and by chromatographic procedures involving molecular sieve chromatography on Sephacryl S-200 (Pharmacia), adsorption chromatography on hydroxylapatite-Ultrogel (LKB) and chromatofocusing on polybuffer exchanger (Pharmacia). After 2000fold enrichment and an overall yield of 32 %, a specific activity of 15U/mg protein was achieved. The purified enzyme migrates as a single band after sodium dodecylsulfate-polyacrylamide gel electrophoresis or polyacrylamide gel - isoelectric focusing.

Isoelectric point. After chromatofocusing the main fraction of pure anthranilate hydroxylase elutes at pH 5.36.

Product specificity. With the exception of 2,3DHB and 3HOAA, no other direct products of anthranilate hydroxylation were found, in particular no 5-hydroxyanthranilate, salicylate, catechol, 2,4-dihydroxybenzoate or 2,6-dihydroxybenzoate.

Michaelis-Menten constants. The K_m-values for anthranilate and NADPH are 4.5×10^{-6} and 4.1×10^{-5} mol/l (anthranilate 2,3-hydroxylation), resp. 8.2×10^{-6} and 8.3×10^{-5} mol/l (anthranilate 3-hydroxylation).

Molecular weight. The native molecular weight of anthranilate hydroxylase is 45 200 dalton, estimated by molecular sieve chromatography, and 43 200 dalton, measured by sodium dodecylsulfate-polyacrylamide gel electrophoresis.

pH-optimum. Anthranilate hydroxylase shows maximum activity at pH 8.2 in borate-phosphate buffer.

Stability at different pH-values. The stability of anthranilate hydroxylase was tested at pH-values between 4.5 and 10.5 in different buffers (0.01 mol/l). The enzyme is stable from about pH 6 to pH 9.

Inhibitors. From the investigated inhibitors only p-chloromercuribenzoate and Cu^{2+} show a substantial inhibition of anthranilate hydroxylase.

Coenzymes. The homogeneous anthranilate hydroxylase shows the typical spectrum of flavoproteins with absorption maxima at 275, 375 and 450 nm and shoulders at 435 and 475 nm. The flavin was identified as FAD by thin-layer chromatography and high-performance liquid chromatography.

Nature of anthranilate hydroxylase. The described anthranilate hydroxylase from Aspergillus niger is a single enzyme with two activities which catalyses mainly the formation of 2,3DHB from AA, but also the formation of 3HOAA as an additional product to a smaller extent. This was demonstrated as follows:

a) During all purification steps the anthranilate 2,3-hydroxylation activity could not be separated from the anthranilate 3-hydroxylation activity.
b) The proportion of anthranilate 2,3-hydroxylation to anthranilate 3-hydroxylation (about 40:1) was the same in the crude extract, during all purification steps and in the pure enzyme. In addition no major changes in the 40:1 proportion were observed in different buffers, after addition of inhibitors or after storage under oxidizing or reducing conditions.
c) Coinciding enzyme profiles for anthranilate 2,3-hydroxylation and anthranilate 3-hydroxylation were obtained for substrate and cosubstrate specificity, the pH-optimum, the stability at different pH-values, the Michaelis-Menten constants and the reaction to the investigated inhibitors.

The anthranilate hydroxylase under investigation was additionally characterized as a bifunctional (multifunctional) enzyme since the molecular weight determinations demonstrate that it consists only of a single peptide chain.

Possible role of anthranilate 3-hydroxylation in the metabolism. Anthranilate hydroxylase activity (30 mU/g mycel) was induced after the addition of AA (1,5 nmol/l) to the culture medium. Because in this case kynurenine 3-hydroxylase activity cannot be detected, in the presence of AA in the nutrition medium nicotinamide nucleotides can be synthesized only using the short-cut pathway.

References

1. McCullogh, W.G.: Abstr. 131st. Meeting Amer. Chem. Soc. 26c (1957).
2. Nair, P.M., Vaidyanathan, C.S.: Biochim. Biophys. Acta 110, 521-531 (1965).
3. Ueda, T., Otsuka, H., Goda, K., Ishiguro, I., Naito, J., Kotake, Y.: J. Biochem. 84, 687-696 (1978).
4. Sreeleela, N.S., Subba Rao, P.V., Prema Kumar, R., Vaidyanathan, C.S.: J. Biol. Chem. 244,2293-2298 (1969).
5. Langenbach-Schmidt, D.: Dissertation Giessen (1979).
6. Schott, H.-H., Zinke, K., Langenbach-Schmidt, D.: Schriftenreihe Verband der Diplom-Oekotrophologen 2, 239-242 (1980).
7. Schott, H.-H., Langenbach-Schmidt, D., Zinke, K.: J. Chromatogr. 245,85-91 (1982).
8. Zinke, K.: Dissertation Giessen (1983).

BIOSYNTHESIS OF TRYPTOPHAN DERIVED ALKALOIDS IN HIGHER PLANTS: ENZYMATIC FORMATION OF THE SARPAGINE-AJMALINE GROUP

Joachim Stöckigt
Institut für Pharmazeutische Biologie der Universität München
D-8000 München

Indole alkaloids which biogenetically originate from both tryptophan 1 (or its decarboxylation product tryptamine 2) and the monoterpene secologanin 4, comprise a vast and a structurally most diverse group of secondary plant metabolites. From the biogenetic and structural point of view, these alkaloids can be divided into five major classes (1):

Class I: Alkaloids with a non-rearranged secologanin skeleton.

Class II: Alkaloids containing an opened secologanin skeleton.

Class III-V: Alkaloids with a rearranged secologanin skeleton.

Of the class I alkaloids, most interesting groups are the sarpagine and ajmaline group, named for the typical members sarpagine 7 (2) and ajmaline 8 (also called Siddiqui base) (3). Both alkaloids are especially widely distributed in the genus Rauwolfia belonging to the Apocynaceae family. Although there exists a broad variety of alkaloids within both groups (4) and an enormous effort has been made in syntheses (5,6) and degradation studies (7), it is most surprising that only a few biogenetic investigations, exclusivly in vivo feeding experiments, have been conducted during the last 25 years in order to elucidate the biosynthetic formation of these alkaloid groups (8). 1 was recognized early (9) as a precursor of the ß-carboline nucleus of ajmaline 8, after shikimic acid was suggested as an earlier intermediate (10). Accordingly, the C-21 of 8 is not

Progress in Tryptophan and Serotonin Research

derived from a C_1-unit as proven by feeding of labelled N-methyltryptophan, N-methyltryptamine or methionine (11). In addition loganin 3 acts as progenitor of the monoterpenoid part of 8 (12), which in analogy to other monoterpenoid indole alkaloids implies that secologanin 4 is the more direct precursor. In full agreement strictosidine 5, the central precursor of these alkaloids (13) is incorporated into ajmaline 8 (14) whereas in contrast to several proposals (15), 5-carboxycompounds e.g. 5-carboxystrictosidine 6 do not play a role in the elaboration of the sarpagine or ajmaline skeleton (16).

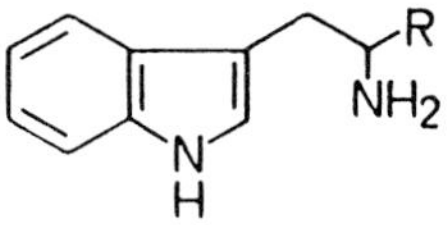

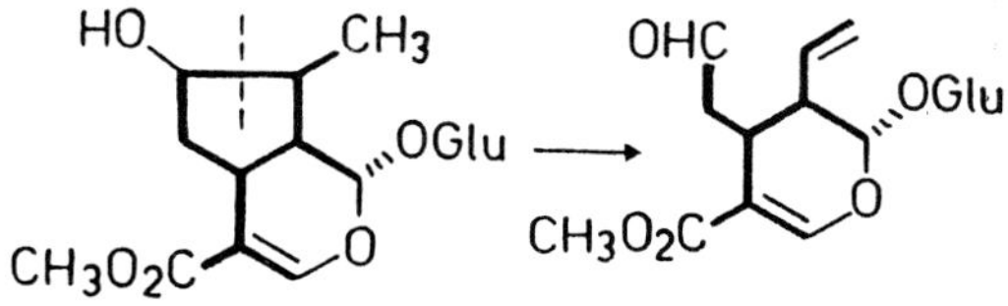

Tryptophan ($R=CO_2H$) 1
Tryptamine (R = H) 2

Loganin 3

Secologanin 4

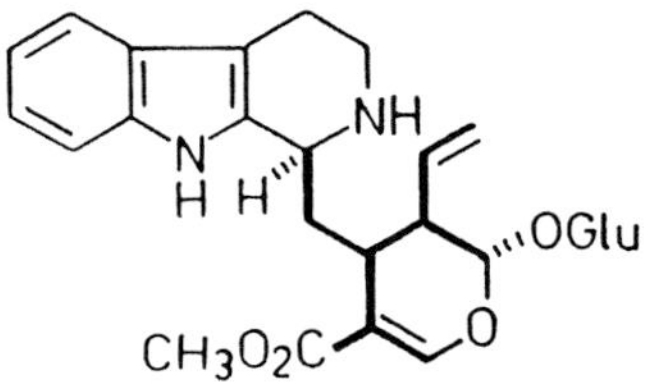

Strictosidine 5

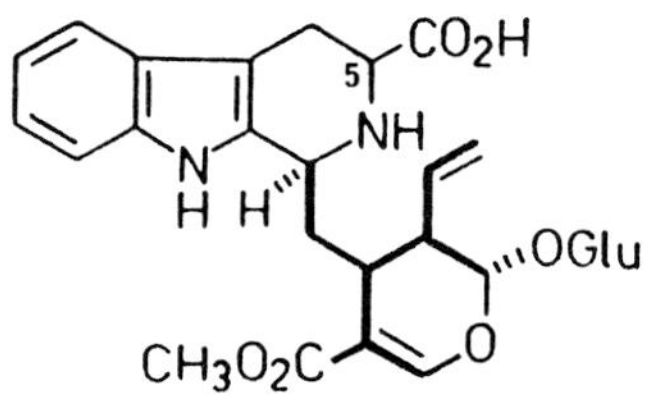

5-Carboxystrictosidine 6

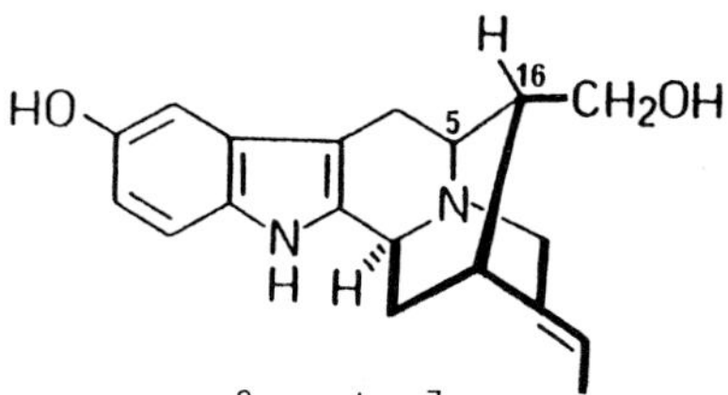

Sarpagine 7

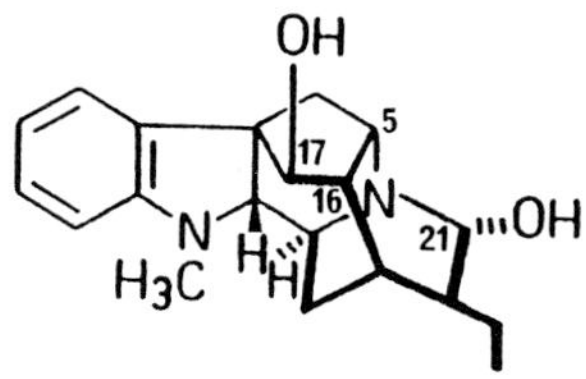

Ajmaline 8

These few experimental findings do not allow description of a comprehensive picture of the biogenesis of 7 and 8 type alkaloids. We therefore applied the methodology developed during the last seven years for the delineation of the heteroyohimbine pathway with Catharanthus roseus cell suspension cultures (17-19) to the study of the biochemical mechanisms of sarpagine/ajmaline formation. This methodology is based on (a) strain-selected cell cultures of R. serpentina specifically capable of synthesizing substantial amounts of sarpagan/ajmalan alkaloids (20) and (b) on isolation and characterization of the enzymes involved in the particular biochemical processes (21). From the structural similarities within the sarpagan/ajmalan group the fundamental mechanisms of their mode of biosynthesis can be formally described by 4 steps:

1. Generation of the characteristic C-5-C-16 bond
2. Conversion of the monoterpenoid C_{10}- into a C_9-unit
3. Formation of the ajmalan-characteristic C-7-C-17 bond
4. "Late" reactions e.g. reduction, deacylation, $N_{(a)}$-methylation, hydroxylation

Whereas of these main features of sarpagine/ajmaline biosynthesis step 2-4 are now basically clarified by our group at the enzymatic level, the first step yet requires experimental verification.

Step 1: Generation of the characteristic C-5-C-16 bond

The mechanism which is commonly accepted in the literature for this step, is the attack of C-16 on an electrophilic iminium species (4,5-dehydrogeissoschizine 9) at C-5. The formation of the N-4-C-5 double bond is assumed to take place by oxidative decarboxylation of a C-5 carboxycompound. This assumption however was previously ruled out by the in vivo incorporation of 5 (14) and the non-involvement of 6 (16) in the biosynthesis of 7 and 8, respectively. A more recent discussion on this particular subject (22) favours the conversion of a well established intermediate of the heteroyohimbine pathway (23), 4,21-dehydrogeissoschizine 9 into 10. Compound 10 would then serve as the immediate precursor for polyneuridine aldehyde 11, bearing the characteristic C-5-C-16 bond.

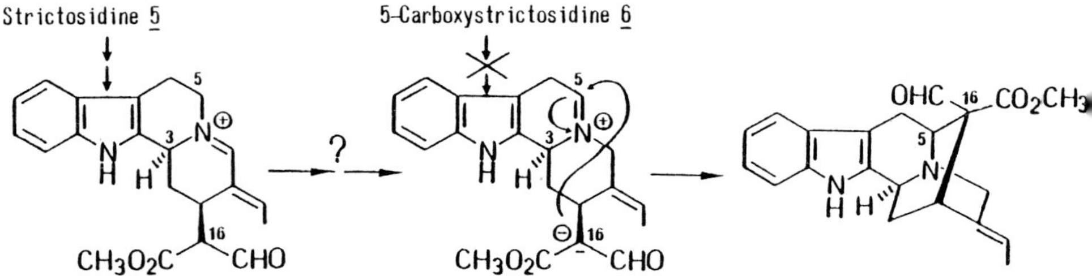

4,21-Dehydrogeissoschizine 9 4,5-Dehydrogeissoschizine 10 Polyneuridine aldehyde 11

Although completely different mechanisms can be postulated for step 1, the biogenetic significance of 11 in the sarpagine/ajmaline biogenesis has been unambigously proven in detail (24).

Step 2: Conversion of the monoterpenoid C_{10}- into the C_9-unit

At the stage of polyneuridine aldehyde 11, a C_1-unit of the monoterpenoid skeleton is removed after hydrolysis of the ester function, which is catalysed by the new plant enzyme polyneuridine aldehyde esterase (PNA-esterase). The direct product is the ß-ketoacid polyneuridine aldehyde acid 12. The acid 12 is highly unstable and cannot be isolated. Therefore we assume that 12 decarboxylates spontaneously, probably still at the enzyme-bound stage. The resulting product is the novel indole alkaloid 16-epivellosimine 13, now bearing the monoterpenoid C_9-unit. The formation of 13 clearly shows that the decarboxylation proceeds with retention of configuration at C-16. In contrast to common hydrolases, PNA-esterase is an outstandingly specific enzyme, accepting exclusively alkaloidal substrates with a C-17 aldehyde group and the sarpagine characteristic C-5-C-16 bond.

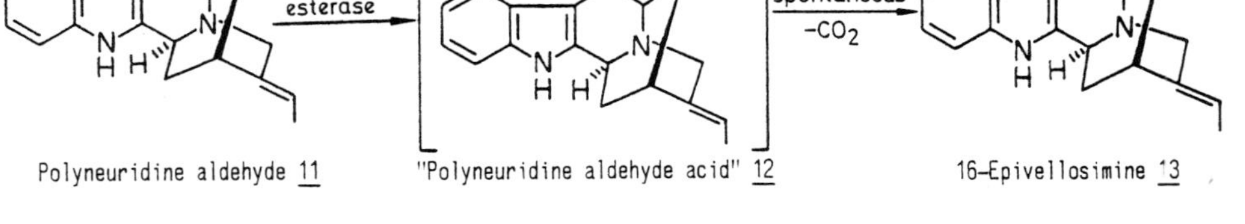

Polyneuridine aldehyde 11 "Polyneuridine aldehyde acid" 12 16-Epivellosimine 13

It is this enzyme which occupies a biogenetic key function because 16-epivellosimine 13 is the necessary substrate for the later enzymes of the sarpagan and ajmalan route. On one hand 13 is a progenitor of 10-deoxysarpagine 20 and on the other hand 13 satisfies the stereochemical requirement for the elaboration of the ajmalan skeleton by generation of the C-7-C-17 bond.

Step 3: Formation of the ajmalan characteristic C-7-C-17 bond

The 16(S) configuration of epivellosimine 13 is a steric prerequisite for the addition of the C-17 carbonyl carbon to C-7. This ring closure can be chemically achieved with analogous compounds under strongly acidic condition (5 N HCl) (25), which might simulate the appropriate biogenetic reaction. Regarding 16-epivellosimine 13, the immediate product formed by the above mechanism would be the indolenine alkaloid deacetylvinorine 14. Indolenines exhibiting a free C-17 OH-group however have neither been isolated from natural sources nor have they been chemically synthesized. They probably are unstable and can be trapped only by reduction of the indolenine double bond or by protecting the OH-group at C-17, e.g. under acetylating conditions (25). Acetylated indolenines like vinorine 15 or vomilenine 16 are however major alkaloids of R. serpentina cell cultures. When 13 was incubated in presence of Rauwolfia enzymes and acetyl coenzyme A at pH 7, vinorine 15 was synthesized. This experiment verifies that at the stage of 13 the ajmalan skeleton is enzymatically formed by a stereochemically controlled addition of C-17 at the nucleophilic C-7 carbon leading to the stable alkaloid vinorine 15 with a ß-orientated acetoxy-group.

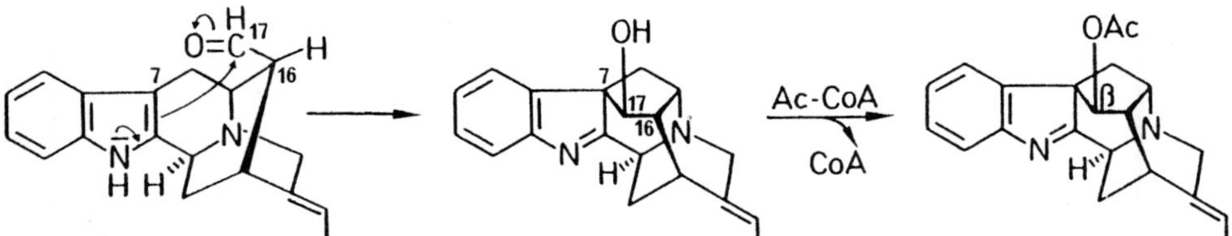

16-Epivellosimine 13 Deacetylvinorine 14 Vinorine 15

Such an indolenine compound then can be expected to serve for the final elaboration of ajmaline 8 catalyzed by "late enzymes" of the ajmalan route.

Step 4: "Late reactions" in the ajmalan route (deacylation, reduction, methylation)

For a cell-free investigation of the late sequence of ajmaline 8 biosynthesis the indolenine alkaloid vomilenine 16 would be the most attractive precursor. Inspection of both structures 16 and 8 resp., demonstrates that at least three enzymes must be involved in the conversion of 16 → 8, a hydrolase (deacylase), a reductase and a methyltransferase. When 16 was incubated in presence of a cell-free extract from Rauwolfia cells and NADPH/S-adenosylmethionine (SAM) as cofactors, the enzymatic formation of 8 was indeed observed by an ajmaline-specific radioimmuno-assay. Therefore 16 is a true biosynthetic progenitor of ajmaline 8 (26). The sequential order of the first two enzymes of this sequence cannot be given at the present time. Both double bonds (1,2 and 19,20) of 16 could be reduced before or after hydrolysis of the acetate function. The involved reductase(s) is a NADPH- and stereospecific enzyme.

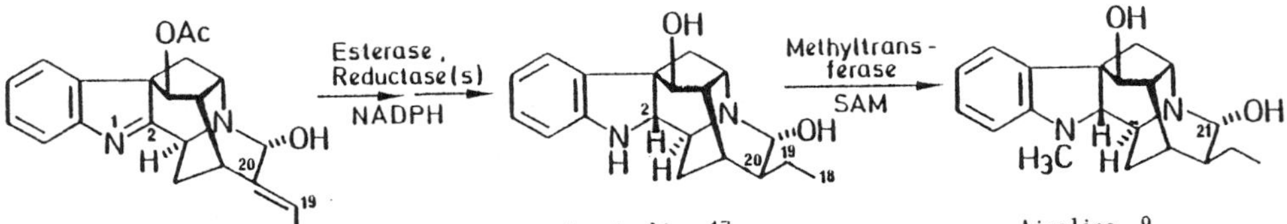

Vomilenine 16 Norajmaline 17 Ajmaline 8

If the SAM was omitted in an effort to intercept the sequence 16 → 8, norajmaline 17 was produced. Isolation and re-incubation of 17 in presence of SAM afforded ajmaline 8, which proves that the methyltransferase is the final enzyme in the whole pathway leading to ajmaline 8.

Step 4: "Late reactions" in the sarpagan route

Concerning the biogenetic route to sarpagan type alkaloids, again 16-epivellosimine 13 plays a crucial role. At pH 7.5 13 isomerizes very easily ($T_{1/2}$= 25 min) forming vellosimine 19 which is a major alkaloid found in Geissospermum vellosii plants (27), Rauwolfia plants (28) and cell cultures (20). Of a large number of alkaloidal aldehydes, 19 is the most efficient substrate for vellosimine reductase (29). This enzyme reduces the aldehyde group of 19 in a NADPH-specific reaction by formation of 10-deoxysarpagine 20.

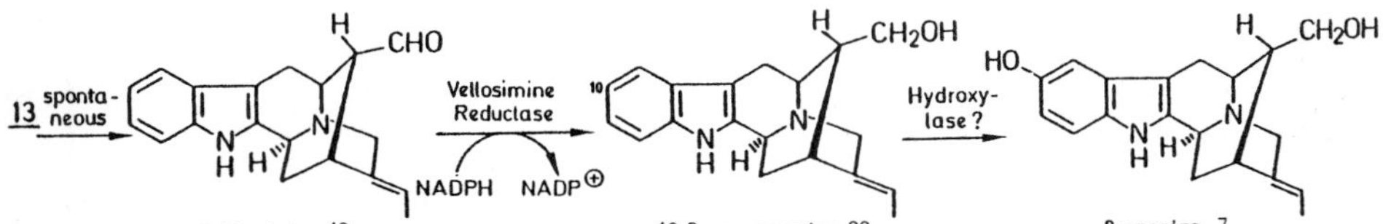

Vellosimine 19 10-Deoxysarpagine 20 Sarpagine 7

We suggest the enzymatic hydroxylation of 20 at C-10 to be the final reaction in the biosynthesis of the Rauwolfia alkaloid sarpagine 7.

The here described enzymatic studies offer for the first time a detailed understanding of the biochemical mechanisms involved in the biogenetic build-up of monoterpenoid indole alkaloids of the sarpagan and ajmalan group. Obviously, both alkaloid groups are biosynthesized by a set of single, soluble enzymes. It seems however, that spontaneous processes of biogenetic importance must be also considered in both routes.

References

1. Kompis,I., Hesse,M., Schmid,H.:Lloydia 34, 269-291 (1971).
2. Stoll,A., Hofmann,A.:Helv.Chim.Acta 36, 1143-1147 (1953).
3. Siddiqui,S., Siddiqui,R.H.:J.Indian Chem.Soc. 8, 667 (1931).
4. Akinloye,B.A., Court,W.E.:J.Ethnopharmacology 4, 99-109 (1981) and literature cited therein.

5. van Tamelen, E.E., Oliver, L.K.: Bioorg. Chem. 5, 309-326 (1976).

6. Cloudsdale, I.S., Kluge, A.F., McClure, N.L.: J. Org. Chem. 47, 919-928 (1982).

7. Bartlett, M.F., Sklar, R., Taylor, W.I., Schlittler, E., Amai, R.L.S., Beak, P., Bringi, N.V., Wenkert, E.: J. Amer. Chem. Soc. 84, 622-630 (1962).

8. Atta-ur-Rahman, Basha, A.: In "Biosynthesis of Indole Alkaloids" (Baldwin, J.E., Goodenough, J.B., Halpern, J., Rowlinson, J.S., Eds.) Claredon Press, Oxford 1983, and cited lit.

9. Leete, E.: J. Amer. Chem. Soc. 82, 6338-6342 (1960).

10. Wenkert, E., Bringi, N.V.: J. Amer. Chem. Soc 81, 1474-1481 (1959).

11. Barton, D.H.R., Kirby, G.W., Prager, R.H., Wilson, E.M.: J. Chem. Soc., 3990-3994 (1965).

12. Battersby, A.R., Kapil, R.S., Martin, J.A., Mo, L.: Chem. Commun., 133-134 (1968).

13. Stöckigt, J., Zenk, M.H.: J. Chem. Soc. Chem. Commun., 646-648 (1977).

14. Nagakura, N., Rüffer, M., Zenk, M.H.: J. Chem. Soc. Perkin Trans. 1, 2308-2312 (1979).

15. van Tamelen, E.E., Haarstad, V.B., Orvis, R.L.: Tetrahedron 24, 687-704 (1968).

16. Stöckigt, J.: Tetrahedron Letters 28, 2615-2618 (1979).

17. Stöckigt, J.: In "Indole and Biogenetically Related Alkaloids" (Phillipson, J.D., Zenk, M.H., Eds.) Academic Press, Inc. London, New York 1980, pp 113-141.

18. Zenk, M.H.: J. Nat. Products 43, 438-451 (1980).

19. Stöckigt, J., Hemscheidt, T., Höfle, G., Heinstein, P., Formacek, V.: Biochemistry, in press.

20. Stöckigt, J., Pfitzner, A., Firl, J.: Plant Cell Reports 1, 36-39 (1981).

21. Pfitzner, A., Stöckigt, J.: Planta medica, in press (and lit. cited therein).

22. Koskinen, A., Lounasmaa, M.: Planta medica 42, 248-249 (1982).

23. Rüffer, M., Kan-Fan, C., Husson, H.-P., Stöckigt, J., Zenk, M.H.: J. Chem. Soc. Chem. Commun., 1016-1018 (1979).

24. Pfitzner, A., Stöckigt, J.: J. Chem. Soc. Chem. Commun., 459-460 (1983).

25. Bartlett, M.F., Lambert, B.F., Werblood, H.M., Taylor, W.I.: J. Amer. Chem. Soc. 85, 475-477 (1963).

26. Stöckigt, J., Pfitzner, A., Keller, P.J.: Tetrahedron Lett. 24, 2485-2486 (1983).

27. Rapoport, H., Moore, R.E.: J. Org. Chem. 27, 2981-2985 (1962).

28. Chatterjee, A., Bandyopadhyay, S.: Ind. J. Chem. 18, 87-88 (1979).

29. Pfitzner, A., Stöckigt, J.: Tetrahedron Lett. 24, 1695-1698 (1983).

ACRIDONE ALKALOID BIOSYNTHESIS IN TISSUE CULTURES OF RUTA GRAVEOLENS

Detlef Gröger and Alfred Baumert
Institute of Plant Biochemistry, Academy of Sciences of the GDR, DDR-4020 Halle (Saale)

Introduction

Anthranilic acid (I) serves as nitrogen-containing precursor for some alkaloid groups e.g. quinoline derivatives, quinazolines and acridones. In general the carboxyl group of I is involved in the cyclization step or the carboxyl group is retained during the course of alkaloid formation. Anthranilic acid is the first intermediate in the tryptophan branch of the aromatic amino acid pathway. Apparently, anthranilic acid directly derived from chorismic acid (with the intact carboxyl group) acts as alkaloid precursor rather than I formed via degradation of tryptophan (1).
About 50 acridone alkaloids have been described which are typical secondary metabolites of the plant family Rutaceae. A number of acridones were isolated (2) from the intact plant and tissue cultures of Ruta graveolens L. = common rue. Rutacridone (VI) possessing a dihydrofuran ring is the progenitor of various acridone-epoxides, which have been found to be phytoalexins (3). The biosynthesis of VI was investigated by in vivo- and in vitro experiments.

Progress in Tryptophan and Serotonin Research

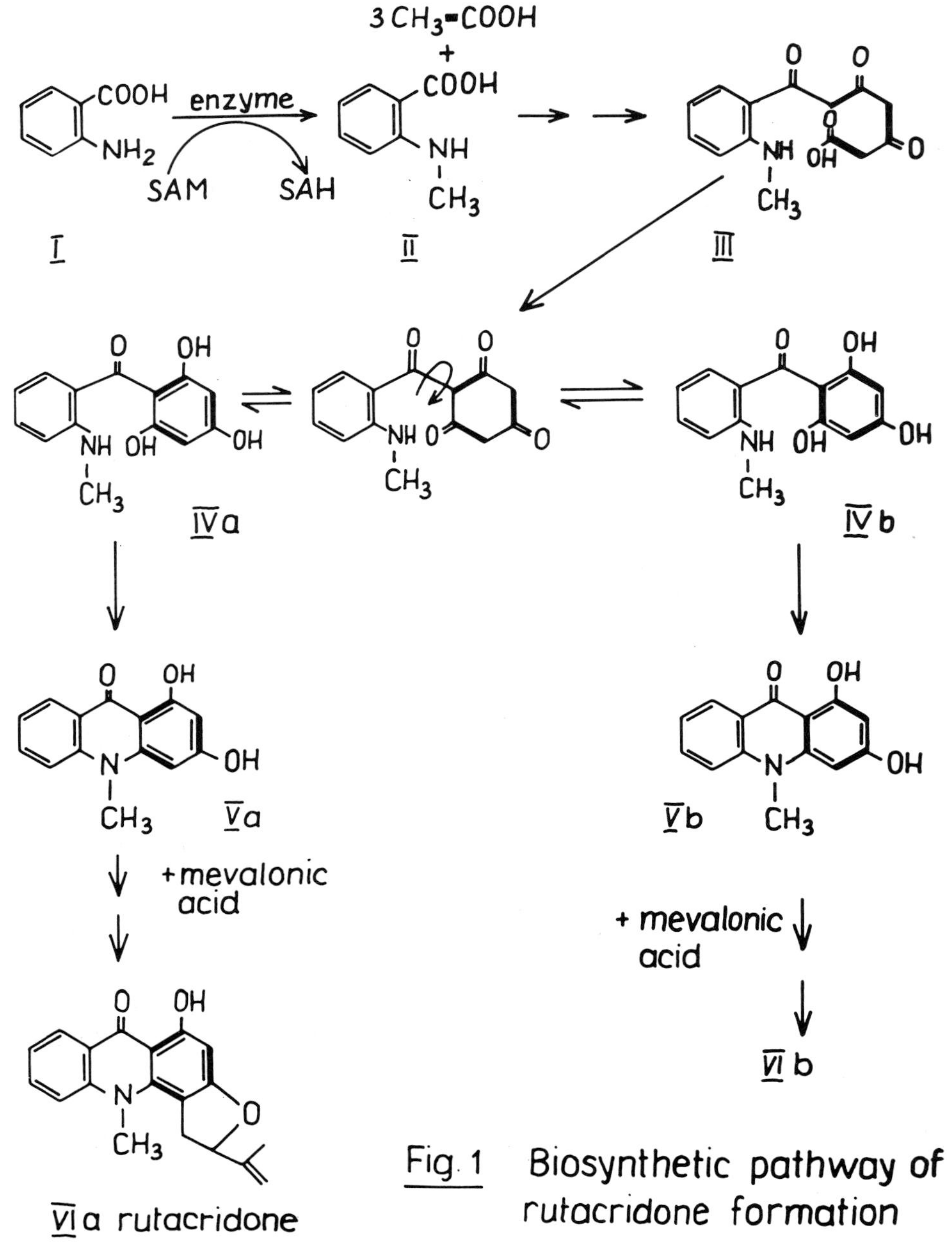

Fig. 1 Biosynthetic pathway of rutacridone formation

Methods

Cell lines of R. graveolens, originally isolated by Dr. I.N. Kuzovkina (Moscow) were used during these studies. Cell lines R-19 and R-20 were cultivated in a modified medium of Murashige and Skoog (4). Calluses and suspension cultures of these particular strains produce as main alkaloid rutacridone (150-200 mg/l alkaloid in liquid media). In vivo experiments were performed using [^{13}C] acetate and various ^{14}C-labeled compounds: [^{14}COOH]anthranilic acid, N-Me-[^{14}COOH]anthranilic acid, [2-^{14}C]acetate and D,L-[2-^{14}C]mevalonic acid. The labeling patterns in VI after feeding of radioactive precursors were determined by degradation to 1-methyl-4-quinolone-3-carboxylic acid according to (5). Cell-free extracts were prepared from lyophilized cells and checked for N-methyltransferase activity. As methyl donor S-adenosyl-L-methionine=SAM was employed.

Results

After feeding of [^{14}COOH] anthranilic acid and degradation of labeled VI all the radioactivity was found in the obtained quinolone derivative. Apparently I is specifically incorporated into VI. Labeling experiments with [2-^{14}C]acetate revealed that a polyketide might be involved in acridone biosynthesis. Surprisingly a specific incorporation of [2-^{14}C] mevalonate into the isopropylidene dihydrofuran part of VI was not observed. Furthermore the specific incorporation rate of mevalonic acid was much lower compared with those obtained with acetate and anthranilic acid (6).
Carbon-13 labeling experiments have shown unequivocally that ring C of rutacridone is acetate-derived. After administration of [1,2-$^{13}C_2$] acetate six carbons were enriched from the incorporation of three intact acetate units. Two diffe-

rent arrangements of acetate units were observed. The rotation of acetate-derived ring C seems to occur at the stage of a N-methylated aminobenzophenone intermediate (7).
The N-Me group of II is derived from L-methionine. N-methylanthranilic acid proved to be an equally good precursor as I for the formation of VI. Furthermore II could be trapped after feeding of anthranilic acid. Cell-free extracts of R. graveolens strain R-20 catalyze the N-methylation of anthranilic acid using SAM as methyl donor. This particular N-methyltransferase was only detectable in acridone synthesizing plant cell cultures. Apparently N-methylation of shikimate-derived anthranilic acid is the first pathway-specific step in the formation of this alkaloid type. The possible pathway of acridone alkaloid biosynthesis is summarized in Fig. 1. The presence of 0.5 mM L-tryptophan in the medium decreased alkaloid formation in cells of strain R-20 by 70 % compared to unsupplemented controls but did not affect growth. Moreover Trp exerts a strong feedback inhibition on anthranilate synthase isolated from R. graveolens cells.

References

1. Gröger, D.: Encyclopedia of Plant Physiology, New Series, Vol. 8, pp 128-159, Springer-Verlag, Berlin Heidelberg, 1980.
2. Szendrei, K., Rozsa, Zs., Reisch, J., Kuzovkina, I.N., Minker, E.: Herba Hungarica 15, 23-29 (1976).
3. Wolters, B., Eilert, U.: Z. Naturforsch. 37c, 575-583 (1982).
4. Murashige, T., Skoog, F.: Physiol. Plant. 15, 475-497 (1962).
5. Gröger, D., Johne, S.: Z. Naturforsch. 23b, 1072-1075 (1968).
6. Baumert, A., Kuzovkina, I.N., Krauss, G., Hieke, M., Gröger, D.: Plant Cell Rep. 1, 168-171 (1982).
7. Zschunke, A., Baumert, A., Gröger, D.: J. Chem. Soc. Chem. Commun. 1982, 1263-1265.

ISOLATION AND BIOLOGICAL ACTIVITIES OF DITERPENE ANTHRANILOYL AND 3-HYDROXYANTHRANILOYL PEPTIDE ESTERS FROM Euphorbia milii.

A. Marston and E. Hecker [+)]

Institut für Biochemie, Deutsches Krebsforschungszentrum,
D-6900 Heidelberg, FRG.

Introduction

Species of the plant families Euphorbiaceae and Thymelaeaceae have given rise to a large number of diterpenoid constituents with interesting biological activities, which include irritation, tumour promotion and cytotoxicity. The tumour promoting substances are esters of tigliane, ingenane and daphnane diterpenes. The most notable of these is the 12-O-tetradecanoyl-13-acetate of phorbol (TPA, see Fig.), which has found widespread use as a model compound in the study of the carcinogenic process [1]. Separation of these active diterpene esters from plant extracts has often proved very awkward and recourse has been taken to multiplicative counter-current distribution techniques for their isolation.

Euphorbia milii ("crown of thorns"), often used as an ornamental plant, is known as a remedy for warts. It contains some remarkable anthraniloyl peptide esters of the diterpene ingenol (milliamines A, B and C) [2] one of which, milliamine A, exhibits considerable irritant activity [3]. In order to probe in detail their biological activities [3], we report here a more extensive investigation of the diterpenoid constituents of E. milii, using a separation procedure guided by an assay for irritation on mouse ear [4]. Irritant activity is read 24 hours after exposure and expressed as Irritant Dose 50 (ID_{50}^{24}) or - more conveniently for purposes of comparison - as Irritancy ($I^{24} = 1/ID_{50}^{24}$).

+) Dedicated to Prof. Dr. Drs. mult. h.c. A. Butenandt on the occasion of his 80th. birthday.

Progress in Tryptophan and Serotonin Research

Results and Discussion

The roots and stems (without leaves) of E. milii collected in Kenya, were extracted with methanol. After partitioning the methanol-solubles between ethyl acetate and water, the organic solvent extract was subjected to a sequence of three Craig counter-current distributions. Final purification of the active sections by preparative TLC and HPLC gave rise to the previously-described milliamines A-C, together with four analogous new ingenol esters, milliamines D-G and, notably, two new 20-deoxy-5-hydroxyphorbol

FIG. STRUCTURES OF MILLIAMINES A-I ISOLATED FROM Euphorbia milii.

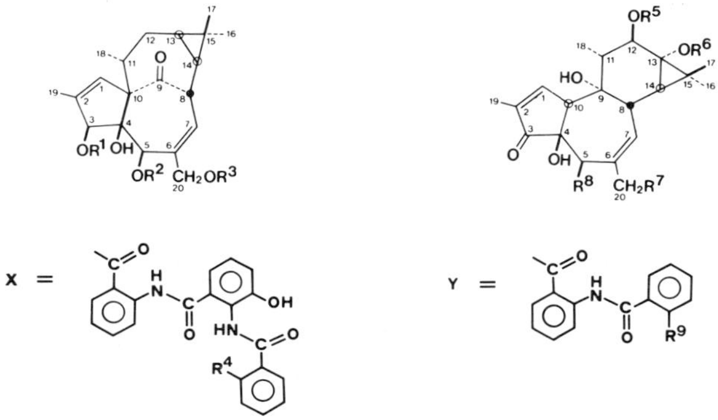

INGENOL:
$R^1=R^2=R^3=H$

3-TI:
$R^1=CO(CH_2)_{12}CH_3$, $R^2=R^3=H$

MILLIAMINE A:
$R^1=X$, $R^2=H$, $R^3=COCH_3$, $R^4=(CH_3)_2N$

MILLIAMINE B:
$R^1=R^2=H$, $R^3=X$, $R^4=(CH_3)_2N$

MILLIAMINE C:
$R^1=X$, $R^2=R^3=H$, $R^4=(CH_3)_2N$

MILLIAMINE D:
$R^1=H$, $R^2=X$, $R^3=COCH_3$, $R^4=(CH_3)_2N$

MILLIAMINE E:
$R^1=X$, $R^2=COCH_3$, $R^3=H$, $R^4=(CH_3)_2$

MILLIAMINE F:
$R^1=R^4=H$, $R^2=X$, $R^3=COCH_3$

MILLIAMINE G:
$R^1=X$, $R^2=R^3=R^4=H$

PHORBOL:
$R^5=R^6=R^8=H$, $R^7=OH$

TPA:
$R^5=CO(CH_2)_{12}CH_3$, $R^6=COCH_3$, $R^7=OH$, $R^8=H$

MILLIAMINE H:
$R^5=Y$, $R^6=COCH_3$, $R^7=H$, $R^8=OH$, $R^9=NHCH_3$

MILLIAMINE I:
$R^5=Y$, $R^6=COCH_3$, $R^7=H$, $R^8=R^9=OH$

esters, milliamines H-I, also possessing anthraniloyl moieties (see Fig.). Milliamines A-E all contain a tripeptide ester group X (see Fig.) consisting of an anthraniloyl moiety, a 3-hydroxyanthraniloyl moiety and a N,N-dimethylanthraniloyl moiety. Milliamines F and G lack the terminal N,N-dimethyl moiety. Milliamines H and I are unusual in that they are strongly fluorescent under long-wave UV light. They too contain an anthraniloyl moiety linked to the diterpene skeleton but, in the case of milliamine H, an amide-linked N-methylanthraniloyl moiety and, in the case of milliamine I, an amide-linked salicyloyl moiety is present.

The milliamines exhibit a range of irritant activities (see Table), the most irritant being milliamine C [5] and the least irritant being milliamines H and I (probably due to the absence of a 20-CH_2OH function in the tigliane moiety). As an irritant, milliamine C is more potent than TPA or

TABLE. IRRITANT ACTIVITIES (ID_{50}^{24}) AND IRRITANCIES (I^{24}) OF MILLIAMINES A-I.

MILLIAMINE	ACYL GROUPS[A)]	ID_{50}^{24} (NMOL/EAR)	IRRITANCY I^{24} ($NMOL^{-1}$)
C	3-X	0.0066	150.0
E	3-X-5-Ac	0.090	11.0
A	3-X-20-Ac	0.040	25.0
D	5-X-20-Ac	0.18	5.6
B	20-X	3.0	0.33
G	3-X	0.015	67.0
F	5-X-20-Ac	0.14	7.1
H	12-Y-13-Ac	8.1	0.12
I	12-Y-13-Ac	> 10.0	< 0.10
TPA	SEE FIG.	0.016	63.0
3-TI	SEE FIG.	0.011	91.0

[A)] X,Y = ANTHRANILOYL GROUPS, Ac = ACETATE

3-O-tetradecanoylingenol (3-TI; see Table). In the standardised two-stage tumour initiation-promotion protocol [4] (initiation on the shaved back skin of mice by dimethylbenz(a)anthracene, followed by treatment with dose p of tumour promoter twice per week), neither milliamine C (p = 5 or 20 nmol/application) nor milliamine A (p = 5 nmol/application) gave rise to papilloma formation. However, milliamine C was strongly hyperplasiogenic (5 nmol dose) on mouse back skin and weakly active as a second-stage [6] tumour promoter. Furthermore, milliamine C is as effective a competitive inhibitor of the binding of (^{3}H) phorbol-12,13-dipropionate to cellular receptors as TPA [7].

With the isolation of milliamines A-I from E. milii, we have now demonstrated for the first time that the irritancy of a Euphorbia species is due to peptide esters of ingenane and tigliane diterpenes. These additionally pose intriguing biosynthetic questions e.g. it remains to be discovered whether the amino-acid moieties of the anthraniloyl peptides arise from metabolism of tryptophan. The fact that milliamine C competitively inhibits the binding of phorbol-12,13-dipropionate to the membrane receptor(s) and yet lacks complete tumour promoting activity means that an important tool has been added to the armoury of compounds necessary for the investigation of multistage carcinogenesis.

References

1. Hecker, E., N.E. Fusenig, W. Kunz, F. Marks, H.W. Thielmann (eds.): Cocarcinogenesis and Biological Effects of Tumor Promoters. Carcinogenesis - a Comprehensive Survey, Vol. 7, Raven Press, New York, 1982.
2. Uemura, D., Y. Hirata; Bull. Chem. Soc. Japan 50, 2005 (1977).
3. Hecker, E., in E.C. Miller et al. (eds.): Naturally Occurring Carcinogens-Mutagens and Modulators of Carcinogenesis, Japan Sci. Soc. Press, Tokyo/Univ. Park Press, Baltimore, p.263 (1979).
4. Hecker, E., in H. Busch (ed.): Methods in Cancer Research, Vol. 6, P. 439, Academic Press, New York, 1971.
5. Marston, A., E. Hecker: Planta medica 47, 141 (1983).
6. Fürstenberger, G., D.L. Berry, B. Sorg and F. Marks, Proc. Natl. Acad. Sci. USA 78, 7722 (1981).
7. Schmidt, R., W. Adolf, A. Marston, H. Roeser, B. Sorg, H. Fujiki, F. Sugimura, R.E. Moore, E. Hecker, Carcinogenesis 4, 77 (1983).

NEW BIOLOGICAL SIGNIFICANCE OF INDOLE-CONTAINING COMPOUNDS AS INITIATORS OR TUMOR PROMOTERS IN CHEMICAL CARCINOGENESIS

Hirota Fujiki, Masami Suganuma, Tomoko Tahira, Mariko Esumi, Minako Nagao, Keiji Wakabayashi, Takashi Sugimura

National Cancer Center Research Institute
Tokyo, Japan

Shozo Takayama

Department of Experimental Pathology, Cancer Institute
Tokyo, Japan

Introduction

The process of chemical carcinogenesis is caused by two qualitatively different factors, initiator and promoter. This two-step concept of chemical carcinogenesis is now widely accepted in tumor formation in skin, liver, bladder and other organs. Since human cancers have been claimed to be caused by environmental factors, we undertook studies to detect initiators as well as tumor promoters in our environment. Several new compounds containing indole nuclei were found to act as initiators or tumor promoters. This paper is concerned with the discovery of strong new initiators, and also of potent new tumor promoters.

Potent new mutagens and carcinogens

Since potent mutagenic activity was found to be produced by pyrolyzing proteins, we subjected various amino acids to pyrolysis. From a tryptophan pyrolysate, we isolated two amino-γ-carboline derivatives, 3-amino-1,4-dimethyl-5*H*-pyrido [4,3-*b*] indole (Trp-P-1) and 3-amino-1-methyl-5*H*-pyrido [4,3-*b*]

Progress in Tryptophan and Serotonin Research

indole (Trp-P-2) (1) (Fig. 1). From pyrolysates of soy-bean globulin, Yoshida et al. obtained two amino-α-carboline derivatives, 2-amino-9H-pyrido[2,3-b]indole (AαC) and 2-amino-3-methyl-9H-pyrido[2,3-b]indole (MeAαC) (2) (Fig. 1). Two amino-γ-carboline derivatives and two amino-α-carboline derivatives are strongly mutagenic. The specific mutagenic activities toward Salmonella typhimurium TA98 of two amino-γ-carboline derivatives were about 10^5 revertants/μg of compounds, which were higher than typical carcinogens such as AF-2, aflatoxin B_1 and 4NQO, and those of two amino-α-carboline derivatives were 10^2 revertants/μg of compound. Trp-P-1, Trp-P-2, AαC and MeAαC were shown to be hepatocarcinogenic *in vivo* in feeding experiments with mice and rats. Moreover, two amino-α-carboline derivatives produced hemangioendotheliosarcomas in brown adipose tissue, which is located in the interscapular region of mice (1).

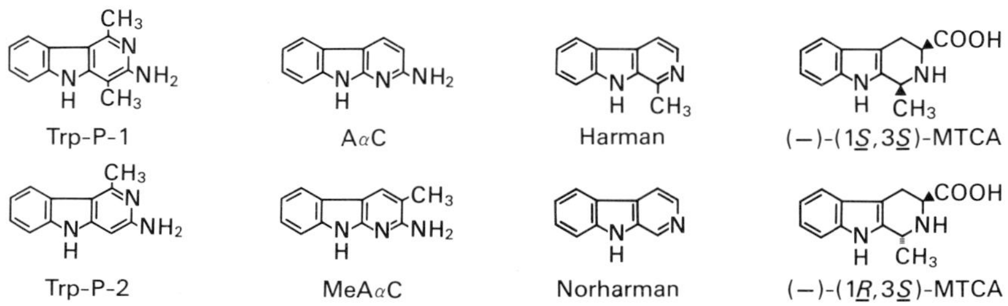

Fig. 1 Structures of new mutagens, comutagens and nitrosable precursors of mutagens

Comutagens

During the isolation of two amino-γ-carboline derivatives, we found that two β-carboline derivatives, harman and norharman (Fig. 1), were also produced by pyrolysis of tryptophan. Harman and norharman are themselves not mutagenic. Aniline, which is not mutagenic, is suspected to produce human bladder tumors. However, it was first demonstrated to be mutagenic in the presence of norharman and S9 mix using TA98 (3). The co-mutagenic action of harman and norharman was shown furthermore with phenylhydroxylamine, nitrosobenzene, *o*-toluidine, yellow OB and N,N-diphenylnitrosamine. Since harman and norharman had been found widely in tabacco tar, Japanese alcoholic drinks such as "Sake" and some mushrooms, it is under investigation whether comutagenic activity may reflect cocarcinogenic potency. At this moment, however, one paper has reported that norharman does not induce carcinogenic activity of aniline when given together in rat experiments (4).

Nitrosable precursors of mutagens

One Japanese condiment, "soy sauce", showed marked mutagenic activity on TA100 without S9 mix after reaction with nitrite. The presence of nitrosable precursors of mutagens called our attention to look for the possible candidates of cancer causes, because N-nitroso compounds are reported to be potent carcinogens. Nitrosable precursors of mutagens were isolated from Japanese soy sauce and identified to be (-)-(1S,3S)-1-methyl-1,2,3,4-tetrahydro-ß-carboline-3-carboxylic acid, (-)-(1S,3S)-MTCA and its stereoisomer (-)-(1R,3S)-MTCA (Fig. 1). About 50% of the total mutagenicity of soy sauce with nitrite depended upon the presence of these two compounds. (-)-(1S,3S)-MTCA and nitrite after 1 hr of incubation at 37°C produced a nitroso substituted compound at position N-2 as the major reaction product. However, this product was not mutagenic. Therefore, the mutagens formed from precursors and nitrite are supposed to be minor products which may have a very high specific mutagenic activity (5).

Potent new tumor promoters, indole alkaloids

Agents that promote the carcinogenic effects in initiated cells are called tumor promoters. Croton oil was introduced by Berenblum at the beginning of the 1940's. The most extensively investigated tumor promoter is 12-O-tetradecanoylphorbol-13-acetate (TPA), which was isolated from croton oil obtained from seeds of Croton tiglium L. (Euphorbiaceae). TPA was found to be as the most active principle in the croton oil (6). The structural elucidation of the most active principle in croton oil was first studied by Hecker when he was in Professor Butenandt's Institute, Max-Planck-Institut für Biochemie. TPA as the only available tumor promoter began the Renaissance of the two-step concept in the development of modern chemical carcinogenesis. We believed that there must be other potent tumor promoters than TPA in our environment, when we were uncertain about the idea that human cancers are not fully understood by the studies of initiators (7). To screen for new tumor promoters other than TPA, we developed a convenient short term test system consisting of three tests; irritation of mouse ear, induction of ornithine decarboxylase (ODC) in mouse skin and adhesion of human promyelocytic leukemia cells (HL-60). Finally, in vivo animal experiments were carried out with a two-step protocol. Among over 300 tested specimens, the indole alkaloids dihydroteleocidin B, teleocidin and lyngbyatoxin A were found to be a new class of tumor promoters on mouse skin (8,9). Dihydroteleocidin B is a hydrogenated derivative of one isomer of teleocidin B (10) (Fig. 2). Teleocidin, which is a mixture of teleocidin A and teleocidin B, was isolated from mycelia of Streptomyces mediocidicus. Teleocidin A, which has a molecular mass of 437 daltons, has two isomers, 14S- and 14R-, and teleocidin B, with a molecular mass of 451 daltons, has four isomers, C(14), C(17)-diastereomers (Fig. 2). Lyngbyatoxin A was isolated from the blue-green alga Lyngbya majuscula by Moore, University of Hawaii (11). Lyngbyatoxin A corresponds to one isomer

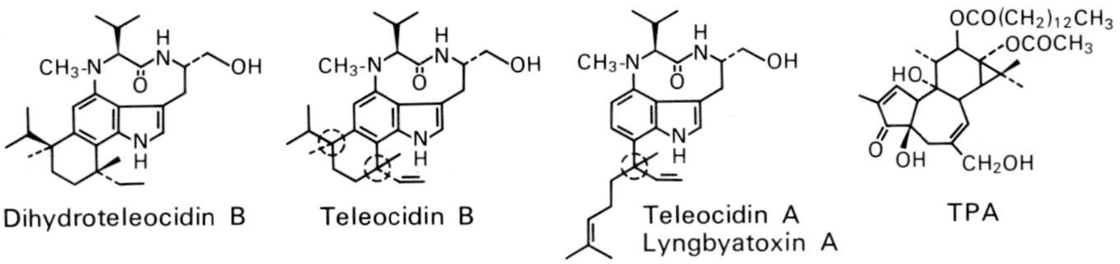

Fig. 2 Chemical structures of indole alkaloids and TPA

of teleocidin A (Fig. 2). As Fig. 2 shows, the chemical structures of these indole alkaloids are entirely different from the structure of TPA.

1) In vitro effects of new tumor promoters

Certain tumor promoters exert many biological effects in vitro in cultured cells, and also in vivo after binding to specific cell surface receptors. Actually some of the in vitro effects are membrane associated. Various biological effects induced by indole alkaloids were compared with those of TPA (Table 1). It is worthwhile to note that dihydroteleocidin B, teleocidin and lyngbyatoxin A induced strong irritation, ODC activity, adhesion and phagocytosis of HL-60 cells, enhancement of aggregation of NL-3 cells (human lymphoblastoid cells) and inhibition of differentiation of Friend erythroleukemia cells, showing similar potencies to that of TPA. Since indole alkaloids are structurally unrelated to TPA, many investigators have attempted to test the effects induced by both indole alkaloids and TPA in their own in vitro or in vivo assay systems. In over 30 biological properties, teleocidin was shown to possess the same high specific activities as to TPA.

2) Two-step carcinogenesis experiment with three indole alkaloids

100μg of 7,12-dimethylbenz(a)anthracene (DMBA) was applied on the skin of the back of 8-week-old female CD-1 mice. One week after initiation, either 2.5μg of dihydroteleocidin B, or 2.5μg of teleocidin, or 3.0μg of lyngbyatoxin A was applied over the initiated back skin twice weekly. For comparison, 2.5μg of TPA was applied by the same procedure as a positive control in each experiment. Fig. 3 shows the tumor formation with DMBA plus dihydroteleocidin B. The percentages of tumor incidence and average number of skin tumors per mouse in week 30 of the groups treated with DMBA plus each of three indole alkaloids are summarized in Table 2. Dihydroteleocidin B, teleocidin and lyngbyatoxin A are very potent tumor promoters like TPA. The results of the group treated with DMBA plus TPA were the average of 3 experiments. Little or no tumors were observed in the groups treated with each of the three indole alkaloids alone or with a single dose of DMBA. All tumors in these groups were histologically examined and classified into squamous cell carcinomas, papillomas, hyperplasia and others (Table 2).

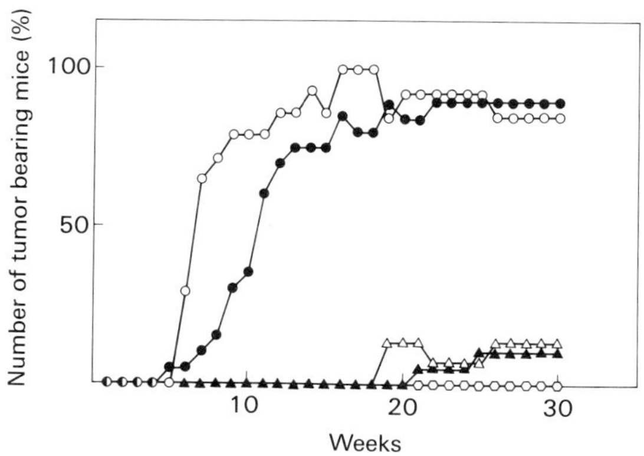

Fig. 3 Effects of dihydroteleocidin B and TPA in tumor formation. DMBA+dihydroteleocidin B (●-●), DMBA+TPA (o-o), dihydroteleocidin B (▲), TPA (△), and DMBA (○).

3) Indole alkaloids and TPA and their bindings to similar membrane receptors

It was of great interest to investigate whether the three indole alkaloids bind to the same receptors to which TPA binds. We applied a method for direct measurement of specific binding of ^{3}H-TPA to a mouse particulate fraction using cold acetone. Dihydroteleocidin B, teleocidin and lyngbyatoxin A inhibited the specific binding of ^{3}H-TPA in a similar way to cold TPA (Table 1). These results are clearly compatible with the inhibition of binding of ^{3}H-PDBu (phorbol-12,13-dibutyrate) to cloned rat embryo fibroblasts (CREF) by these indole alkaloids and TPA, which were produced previously in collaboration with Weinstein's group in Columbia University. Studies on these different systems strongly indicated that the action of these indole alkaloids and TPA may be mediated by the same receptor on cell membranes. The three indole alkaloids and TPA possess both hydrophobic and hydrophilic domains. Studies by computer graphics revealed that these two different classes of tumor promoters have structural similarities both in their hydrophobic and hydrophilic domains (12).

4) Several effects induced more strongly by dihydroteleocidin B than by TPA

Even if we understand that indole alkaloids and TPA exert several biological effects similarly through the same receptor system, several new interesting results with these two classes of tumor promoters were obtained by our collaborators. Kakunaga's group found that dihydroteleocidin B markedly enhanced malignant cell transformation induced by 3 methyl-cholanthrene (13). The effective concentration of dihydroteleocidin B was only one hundredth that of TPA in experiments using cultured mouse cells (A31-1-1). Hayaishi in our laboratory showed that dihydroteleocidin B enhanced the appearance of $CdCl_2$-resistant Chinese hamster lung cells more strongly than TPA did. The over-production of metallothionein I mRNA and amplification of

Table 1 Various effects of indole alkaloids and TPA

	Dihydro-teleocidin B	Teleo-cidin	Lyngbya-toxin A	TPA
Irritant test ID_{50}^{24} nmoles/ear	0.017	0.008	0.011	0.016
Induction of ODC nmoles CO_2/ 5.0μg compound	1.55	1.89	2.05	1.45
Adhesion of HL-60 ED_{50} ng/ml	0.3	4.0	7.0	1.5
Phagocytosis of HL-60 ED_{30} ng/ml	1.4	3.6	2.5	2.5
Aggregation of NL-3 ED_{50} ng/ml	6.5	3.1	2.4	11.2
Inhibition of differentiation of Friend erythroleukemia cells ED_{50} ng/ml	0.2	2.0	0.4	1.0
Inhibition of specific binding of ^{3}H-TPA ED_{50} nM	24.0	4.5	48	4.2

Table 2 Effects of indole alkaloids and TPA on <u>in vivo</u> carcinogenicity experiments on mouse skin

Promoter	Tumor incidence in week 30	Average number of skin tumor per mouse	Squamous cell carcinoma (%)	Papilloma (%)	Hyperplasia (%)	Others (%)
Dihydro-teleocidin B	90.0	4.6	21.0	66.0	6.0	0
Teleocidin	100.0	3.9	13.2	86.8	0	0
Lyngbyatoxin A	87.0	3.7	3.6	96.4	0	0
TPA	92.7	12.0	4.0	94.5	1.0	0.4

the metallothionein I gene were shown in these resistant cells (14). The discovery of a new class of tumor promoters, indole alkaloids, in addition to TPA, gives us a valuable tool to study the biological mechanisms of tumor promotion.

Conclusion

In this paper we indicated the presence of new mutagens, carcinogens and comutagens isolated from the pyrolysates of tryptophan, new nitrosable precursors of mutagens in soy sauce, and also the presence of new tumor promoters isolated from _Streptomyces_ and blue-green alga. They are all indole containing compounds.

Acknowledgements

We are thankful for the support by Grants-in-Aid for Cancer Research from the Ministry of Education, Science and Culture and the Ministry of Health and Welfare of Japan, and the Princess Takamatsu Cancer Research Fund. Work on the isolation of lyngbyatoxin A used in this study was supported by Grant CA 12623-09 to Dr. R. E. Moore at the University of Hawaii from the National Cancer Institute, Health and Human Services, U. S. A.

References

1. Sugimura, T., Kawachi, T., Nagao, M., Yahagi, T., Seino, Y., Okamoto, T., Shudo, K., Kosuge, T., Tsuji, K., Wakabayashi, K., Iitaka, Y., Itai, A.: Proc. Japan. Acad. 53, 58-61 (1977).
2. Yoshida, D., Matsumoto, T., Yoshimura, R., Matsuzaki, T.: Biochem. Biophys. Res. Commun. 83, 915-920 (1978).
3. Nagao, M., Yahagi, T., Honda, M., Seino, Y., Matsushima, T., Sugimura, T.: Proc. Japan. Acad. 53, Ser. B. 34-37 (1977).
4. Hagiwara, A., Arai, M., Hirose, M., Nakanowatari, J., Tsuda, H., Ito, N.: Toxicology Lett. 6, 71-75 (1980).
5. Wakabayashi, K., Ochiai, M., Saito, H., Tsuda, M., Suwa, Y., Nagao, M., Sugimura, T.: Proc. Natl. Acad. Sci. U.S.A. 80, 2912-2916 (1983).
6. Hecker, E.: Naturwissenschaften 54, 282-284 (1967).
7. Sugimura, T.: Cancer 49, 1970-1984 (1982).
8. Fujiki, H., Mori, M., Nakayasu, M., Terada, M., Sugimura, T., Moore, R. E.: Proc. Natl. Acad. Sci. U.S.A. 79, 3872-3876 (1981).
9. Fujiki, H., Suganuma, M., Matsukura, N., Sugimura, T., Takayama, S.: Carcinogenesis 3, 895-898 (1982).
10. Takashima, M., Sakai, H.: Bull. Agr. Chem. Soc. Japan 24, 647-651 (1960).
11. Cardellina, II, J. H., Marner, F. J., Moore, R. E.: Science 204, 193-195 (1979).
12. Horowitz, A. D., Fujiki, H., Weinstein, I. B., Jeffrey, A., Okin, E., Moore, R. E., Sugimura, T.: Cancer Res. 43, 1529-1535 (1983).
13. Hirakawa, T., Kakunaga, T., Fujiki, H., Sugimura, T.: Science 216, 527-529 (1982).
14. Hayashi, K., Fujiki, H., Sugimura, T.: Proc. Jap. Cancer Assoc. 41th Annual Meeting p. 188 (1982).

BIOSYNTHESIS OF SEROTONIN AND AUXIN IN SEEDS OF JUGLANS REGIA

Wolfgang Große

Lehrstuhl für Botanik II der Universität zu Köln
D-5000 Köln 41, Gyrhofstraße 15

Introduction

Serotonin (5-HT) is synthesized in plants and occurs in a wide range of families but little is known about its function in plants. Because of its auxin-like activity it has been proposed that 5-HT may act as a plant growth hormone (1). 5-HT may also protect plants against predation (2) like tannins (3) or alkaloids (4). Now a role for 5-HT as a secondary plant product involved in ammonia detoxification in maturing seeds of walnuts (*Juglans regia* L.) is proposed (5).

Results and discussion

To prevent that the concentration of ammonia increases to toxic levels while water is lost during ripening, the ammonia content of the seeds is reduced by synthesis of Trp and 5-HT (Tab. 1). When the fruits are harvested and stored, 5-HT formation continues up to a level of 5 µmol g fresh weight^{-1} (6).
Ammonia consumption and Trp biosynthesis are performed by the enzyme glutamine synthetase and the constitutive enzymes of the shikimate pathway localized in the plastids of the plant cells (7).

Abbreviations: 5-HT: serotonin = 5-hydroxytryptamine, 5-HO-Trp: 5-hydroxytryptophan, Trp: tryptophan, IAA: indoleacetic acid = auxin, TNH_2: tryptamine, 3-IndAcAld: 3-indoleacetaldehyde

Progress in Tryptophan and Serotonin Research

Tab. 1: Consumption of ammonia and biosynthesis of tryptophan and serotonin by maturing walnut seeds.

day of harvest	ammonia	tryptophan	serotonin	ammonia	water
	µmol / g fresh weight			mM	%
13.8.	7.7	0.3	0	9.5	81
20.8.	5.0	0.8	0	7.4	68
27.8.	3.7	1.3	0	5.8	64
3.9.	2.1	1.5	0	5.1	41
17.9.	1.0	1.6	0.2	4.0	25
30.9.	0.9	1.1	0.7	3.8	24

Trp, normally a substrate for protein and IAA biosynthesis, is transformed to 5-HT in the adult cotyledons (6) when they start

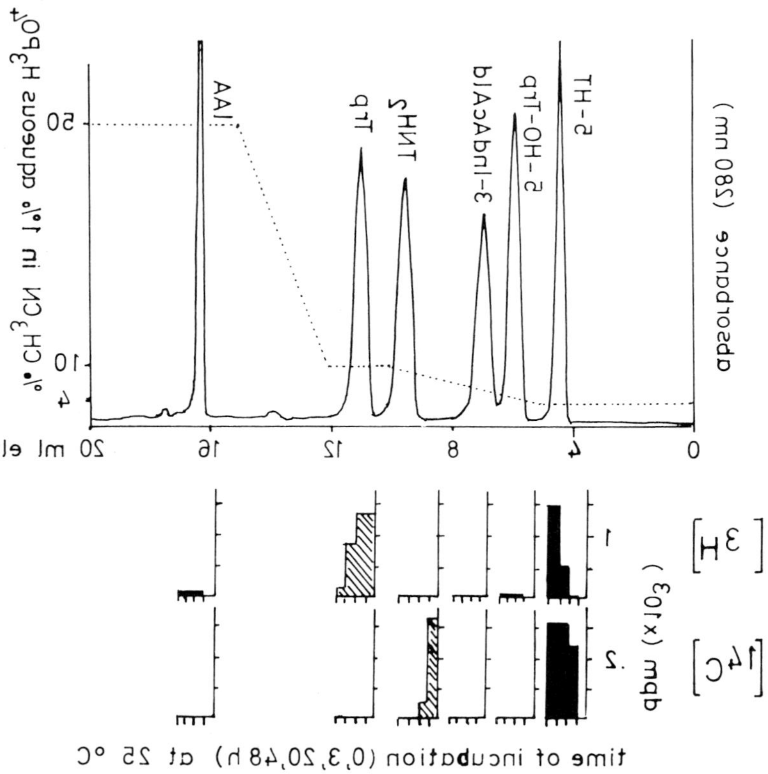

Fig. 1: Reversed-phase (RP-18) HPLC of indole derivatives and there biosynthesis from exogenous [^{3}H]-Trp and [^{14}C]-TNH_2 supplied to aseptically isolated cotyledons from walnuts harvested at 10 th of September.

to function as storage organs (8).
Using [^{3}H]-Trp and [^{14}C]-TNH_2 as tracers, and reversed-phase (RP-18) HPLC for separation of indoles involved in 5-HT and IAA metabolism (Fig. 1), it was shown that the incorporation of [^{3}H]-Trp into 5-HT is slow in contrast to the fast incorporation of [^{14}C]-TNH_2. Therefore, the decarboxylation of Trp to TNH_2 appears to be the first, rate-limiting step in the biosynthetic pathway to 5-HT, catalyzed by the enzyme tryptophan decarboxylase (E.C.4.1.1.27). Subsequently, TNH_2 is transformed to 5-HT by the enzyme tryptamine-5-hydroxylase.
The minute incorporation of [^{3}H]-Trp into 5-HO-Trp demonstrates that TNH_2, but not 5-HO-Trp is the intermediate of 5-HT biosynthesis in walnuts.
The incorporation of [^{3}H]-Trp into the IAA fraction without incorporation of any [^{14}C]-TNH_2 suggests a biosynthesis of IAA in walnuts via 3-indolepyruvic acid involving tryptophan transaminase as a constitutive enzyme in these seeds, but not via TNH_2 (9).
As shown in Fig. 2, cotyledons isolated from young seeds far

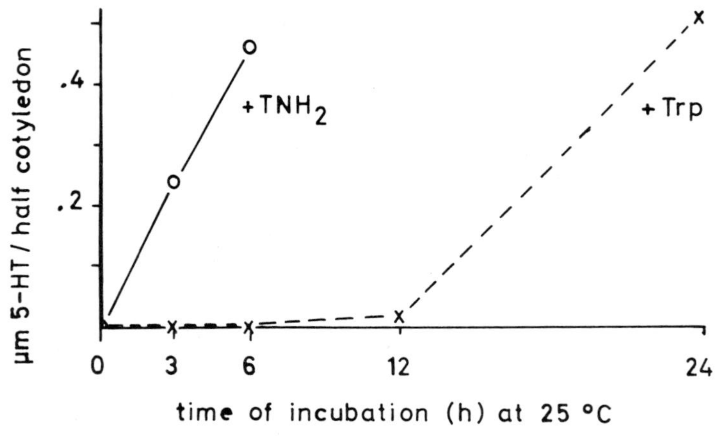

Fig. 2: Biosynthesis of 5-HT from exogenous substrates supplied to aseptically isolated cotyledons from walnuts harvested at 6 th of August.

away from the period of 5-HT accumulation, synthesize 5-HT from exogenous substrates. Therefore tryptamine-5-hydroxylase is a constitutive enzyme in walnut seeds. This observation confirmes also that tryptophan decarboxylase is an adaptive enzyme (10, 11), also expressed in young seeds after abscission. This enzyme appears to be the key controlling element in the 5-HT biosynthetic pathway.

Finally, 5-HT is not distributed throughout the whole cells in mature seeds, but is accumulated in the protein bodies (8); the protein bodies substitute for the vacuoles as storage centres of hydrophilic secondary plant products. In this way 5-HT, like Trp a feedback inhibitor of the anthranilate synthetase (7), is rendered unreactive. Thus it appears that seeds, lacking vacuoles, become able to detoxify ammonia by accumulating 5-HT.

References

1. Niaussat, P., Laborit, H., Dubois, C., Niaussat, M.: C. R. Seances Soc. Biol. Paris 152, 945 - 947 (1958).
2. Schildknecht, H.: Protective substances in arthropods and plants. In: Natural products and the protection of plants. Scripta Varia. Marini-Bettolo, G. B. (ed), Vol. 41 S. 59 - 99. Pontifical Academy of Science, Vatican City 1977.
3. Feeny, P. P.: Ecology 51, 565 (1970).
4. Sturchow, B., Low. I.: Entomol. Exp. Appl. 4, 133 (1961).
5. Große, W.: Phytochemistry 21, 819 - 822 (1982).
6. Bergmann, L., Große, W., Ruppel, H. G.: Planta 94, 47 - 59 (1970).
7. Große, W.: Z. Pflanzenphysiol. 83, 249 - 255 (1977).
8. Große, W., Schmidt, U.: Ber. Deutsch. Bot. Ges. 93, 575 - 585 (1980).
9. Große, W.: Z. Pflanzenphysiol. in press.
10. Große, W., Klapheck, S.: Z. Pflanzenphysiol. 93, 359 - 363 (1979).
11. Klapheck, S., Große, W.: Z. Pflanzenphysiol. 100, 325 - 331 (1980).

BLEPHARISMONE: A CONJUGATION-INDUCING TRYPTOPHAN DERIVATIVE IN THE CILIATE BLEPHARISMA

Akio Miyake

Department of Cell Biology, University of Camerino
62032 Camerino, Italy

Introduction

In conjugation of *Blepharisma japonicum*, a red ciliated protozoon (ca. 300 x 100 μm), cells of complementary mating types I and II stimulate each other by gamones for pair formation (1-3) (Fig.1). Both gamones have been isolated and identified. Gamone of type I cells (gamone 1, blepharmone) is a glycoprotein of 20 kD containing 3 glucosamine,

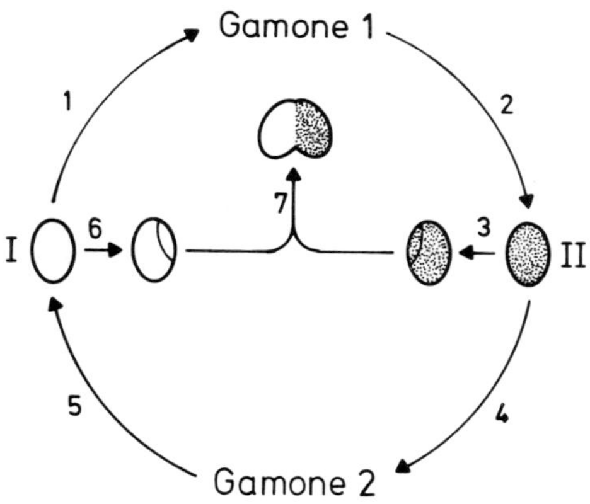

Fig.1. Preconjugant interaction in *Blepharisma japonicum*. Type I cells excrete gamone 1 into the medium (step 1). This gamone acts on type II cells (step 2) and transforms them so that they can unite (step 3) and at the same time induces the production and excretion of gamone 2 (step 4). Gamone 2 acts back onto type I cells (step 5), transforms them so that they can unite (step 6) and also enhances the production and excretion of gamone 1. Transformed cells unite in conjugant pairs upon contact (step 7). From (4).

Progress in Tryptophan and Serotonin Research

3 mannose and about 175 amino acids (5-6). Gamone of type II cells (gamone 2, blepharismone) is calcium-3-(2'-formylamino-5'-hydroxybenzoyl)lactate (7), a tryptophan derivative resembling serotonin (Fig. 2). These two gamones are,

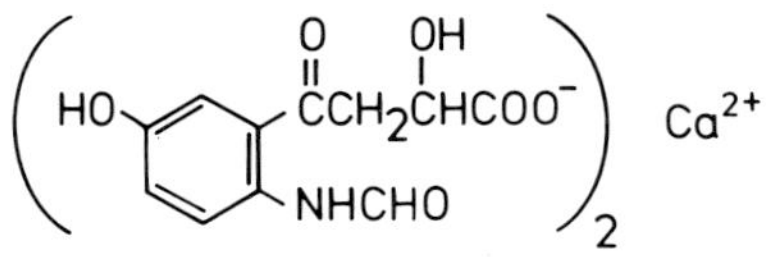

Fig. 2. Blepharismone, gamone 2 of Blepharisma japonicum.

therefore, quite different in the molecular structure, while their biological functions are alike. This molecular asymmetry in a symmetrical system of interacting cells will be discussed from an evolutionary point of view together with possible phylogenetic relationships between gamones in unicellular organisms and hormonal signals in multicellular organisms.

Preconjugants

Well-fed cells neither produce nor respond to gamone. If such cells are deprived of food, they gain capacities to produce and respond to gamone within several hours as they transform into "preconjugants". This transformation is reversible: if preconjugants are fed, they lose both of the acquired capacities. In the following, preconjugants are simply called "cells".

Isolation and Characterization of Blepharismone

Gamone of one mating type can induce pair formation in cells of the other mating type. By means of this cellular response, gamones can be detected and assayed. A unit of activity (U) is defined (2) as the smallest amount of gamone activity that can induce at least one pair in 500-1000 cells suspended in 1 ml SMB, a salt medium for Blepharisma (3). Blepharismone was isolated as follows. Type II cells were suspended (5 ml packed cells per liter) for one day at 25°C in a 9:1 mixture of SMB and cell-free fluid of type I cells. The latter provided gamone 1 which served to induce and to maintain the maximum production of blepharismone by type II cells. The cell-free fluid obtained by removing type II cells by mild centrifugation was used as the starting material for blepharismone isolation. Blepharismone was concentrated in vacuo (<50°C), extracted with ethanol, dried, dissolved in 75% n-propanol and applied to cellulose column chromatography (elution medium: 75% n-propanol). Blepharismone-containing fractions were dried,dissolved in water and chromatographed on paper (developing solvent: 75% n-propanol). Blepharismone activity was localized in a single fraction fluorescing in pale blue under UV irradiation. This fraction was extracted with water. Blepharismone crystallized from water as pale yellow prisms with the activity of 10^6 U/mg. In a typical experiment, 3.68 mg of pure blepharismone was obtained from 48 liters of the cell-free fluid of type II cells mentioned above (7).

Blepharismone was identified (Fig.2) by using X-ray crystallography and other methods (7). It was first called blepharismin(7). But the name was changed to blepharismone to avoid confusion, because the red pigment of Blepharisma, zoopurpurin, was renamed as blepharismin (8) at almost the same time(5). Pure blepharismone dried and kept at < -10°C retains the same activity for years. The stability is probably due to the association of calcium which stabilizes the β-carboxy-β'-hydroxyketone structure (7). If calcium is removed, most of the activity is lost within a few days (9).

Racemic blepharismone was chemically synthesized (10-11). Synthetic blepharismone is approximately half as active as the natural compound probably because one of the enantiometric forms (D) is less biologically active than the other form (L).

Biosynthesis of Blepharismone

If cells of mating type II are treated by gamone 1, blepharismone appears after about 2 hours (2). Even if gamone 1 is removed after 30 minutes of the treatment, blepharismone is produced and the production continues for many hours(4). This result suggests that gamone 1 induces in type II cells an enzyme or an enzyme system participating in blepharismone biosynthesis. Type II cells incubated with ^{14}C-tryptophan produce radioactive blepharismone suggesting that tryptophan is a precursor of blepharismone (3-4).

Induction of Cell Union by Blepharismone

Type I cells treated by blepharismone gain the capacity to unite at the peristomial floor (an area of cell surface near the mouth) and unite in pairs within several hours. Blepharismone induces in type I cells protein synthesis which is essential for this surface change (12). It also induces formation of microtubules emanating in bundles from the internal surface of the peristomial floor. It is assumed that the induced proteins are incorporated into membrane vesicles, transported to the peristomial floor by these microtubules and transform the cell surface so that it can unite with a similarly transformed surface (12-13).

Induction of cell union by blepharismone is competitively inhibited by tryptophan and 5-hydroxytryptophan. L-isomers are more effective than D-isomers in this inhibition. These compounds have no comparable effect on the induction of cell union in type II cells by gamone 1. This suggests that blepharismone acts on type I cells through a specific binding to its receptor. For reviews, see (3, 13).

Chemoattraction by Blepharismone

Blepharismone attracts type I cells. Blepharismone activity as small as or even smaller than 10^{-3} U was effective under the condition used (14). Just as in the induction of cell union, the attraction is inhibited by 5-hydroxytryptophan, suggesting that both chemoattraction and cell union are mediated by the same gamone receptor. However, unlike cell union, chemoattraction does not require the direct participation of protein synthesis (14).

Blepharismone in five species of *Blepharisma*

Gamones of *Blepharisma* have been investigated in five species (*B. japonicum*, *B. stoltei*, *B. americanum*, *B. musculus*, *B. tropicum*). They all contain complementary mating types I and II each excreting gamones 1 and 2 respectively. Gamone 2 of *B. japonicum* induces pair formation in type I cells of all four other species. Conversely, gamone 2 of these four species induces pair formation in type I cells of *B. japonicum*. Gamone 2 of *B. americanum* is not separable from blepharismone by cellulose thin layer chromatography. It appears therfore that gamone 2 in each of these five species is the same molecule, blepharismone. On the other hand, each species appears to have its own specific gamone 1. For review, see (3).

Evolutionary Considerations

All five species of *Blepharisma* mentioned above possess two gamones each of which is excreted by different mating types. However, one may postulate a hypothetical ciliate which has only one gamone and one mating type. In such a ciliate all cells excrete and respond to the same gamone (Fig. 3A). Conjugation then occurs between cells of identical mating type. Nevertheless, such a gamone can play important functions: 1) to control the timing of conjugation through the regulation of gamone production, and 2) to synchronize the acquisition of mating capacity in neighboring cells so that each cell can find a partner. When such a ciliate gains another gamone, a system of complementary mating types may develop (Fig. 3B). Therefore, two gamones in one species can be acquired at different times in the course of evolution. This can partly explain the chemical difference between two gamones in *B. japonicum*.

The modification of protein gamone can be achieved by substituting amino acids, while change in a simple molecule like blepharismone is more difficult requiring new enzymes. This can partly explain the apparently eerie situation that five species of *Blepharisma* use the same molecule, blepharismone, as gamone 2 (see the previous section). The ancestral species of these five species then had blepharismone and gamone 1. New species were later formed, one by one, as gamone 1 molecule was modified (15). This suggests that blepharismone is an old gamone. However, to verify the hypothesis that blepharismone might be as old as or even older than the genus Blepharisma (3, 15), more species must be investigated.

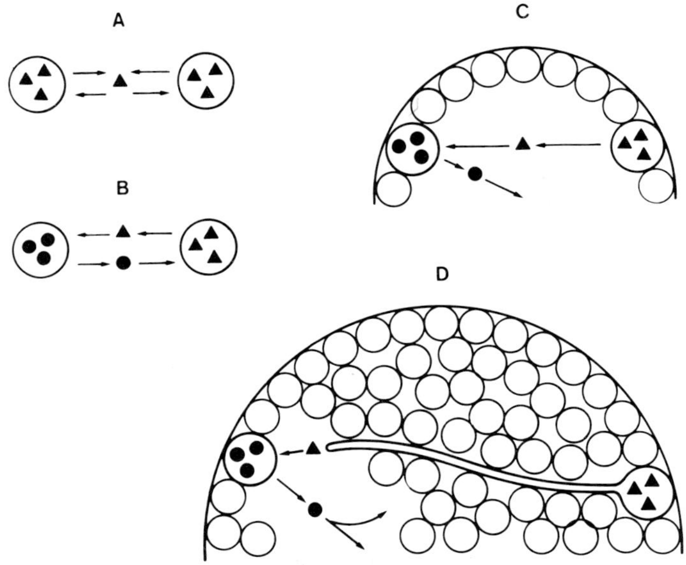

Fig. 3. A hypothetical evolutionary process of communicating cells. Dark circles and triangles: signal molecules. See text.

It may be assumed that ancestors of multicellular organisms were developed from such sexually-reproducinq unicellular
is valid, cells of the first multicellular organisms al-
is valid, cells of first multicellular organisms had already possessed the mechanism of cell-cell communication by chemical signals. They could use this mechanism for cell-cell communication which was needed to build and to maintain their multicellular organization (Fig. 3C). As their size and complexity increased, those which acquired elongated signal-excreting cells would have been selected because of their higher efficiency for specific communication (Fig. 3D), and in this way, primitive nerve cells might have been developed.

The idea to look for the evolutionary origin of cell-cell communication of multicellular organisms in sexual communication of ancestral unicellular organisms is an old one. But the finding that gamone 2 of B. japonicum is a serotonin-like substance makes it more realistic to trace back the origin of neurotransmitters and hormones in multicellular organisms to gamones in ancestral unicellular organisms.

References

1. Miyake, A.: Proc. Jpn. Acad. 44, 837-841 (1968).
2. Miyake, A., Beyer, J.: Exp. Cell Res. 76, 15-24 (1973).
3. Miyake, A.: Sexual Interactions in Eukaryotic Microbes (D.H. O'Day and P.A. Horgen, eds.), pp. 95-129. Academic Press, New York 1981.
4. Miyake, A.: Curr. Top. Dev. Biol. 12, 37-82 (1978).
5. Miyake, A., Beyer, J.: Science 185, 621-623 (1974).
6. Broun, V., Miyake, A.: FEBS Lett. 53, 131-134 (1975).
7. Kubota, T., Tokoroyama, T., Tsukuda, Y., Koyama, H., Miyake, A.: Science 179, 400-402 (1973).
8. Giese, A.C.: Blepharisma (A.C. Giese, ed.), pp. 266-303. Stanford Univ. Press, Stanford, California 1973.
9. Miyake, A.: in Biochemistry and Physiology of Protozoa, Second edition Vol. 4 (M. Levandowsky and S.H. Hutner, eds.), pp. 125-198. Academic Press, New York 1981.
10. Tokoroyama, T., Hori, S., Kubota, T.: Proc. Jpn. Acad. 49, 461-463 (1973).
11. Entzeroth, M., Jaenicke, L.: Z. Naturforsch. 36c, 180--182 (1981).
12. Miyake, A., Honda, H.: Exp. Cell Res. 100, 31-40 (1976).
13. Miyake, A.: in Biochemistry of Differentiation and Morphogenesis (L. Jaenicke, ed.), pp. 211-230. Springer--Verlag, Berlin 1982.
14. Honda, H., Miyake, A.: Nature (London) 257, 678-680 (1975).
15. Miyake, A., Bleyman, L.K.: Genet. Res. 27, 267-275 (1976).

BIOLOGICAL ACTIVITY OF BLEPHARISMONE DERIVATIVES

An Approach Towards Biogenesis of the Hormone

Lothar Jaenicke (+)

Institut für Biochemie, Universität zu Köln
D-5000 Köln 1

Introduction

Pairing of unicellular organisms for sexual reproduction has several particular features that make it a worthwile model of signal-mediated aggregation and morphogenesis in higher, multicellular organisms. A key element in these processes is the so-called "induction" of developmental steps by messengers emitted and received by reciprocal types of cells. In contrast to gamete conjugation in other unicellular organisms in which the two partners fuse totally to form a zygote, in ciliates there is only temporary interaction during which the conjugants exchange nuclear material.

Exhaustive and very elegant studies on the processes during conjugation of the ciliate *Blepharisma japonicum* have been performed by A. Miyake and are summarized in the preceding contribution (1). As usual in ciliates, conjugation in *Blepharisma* is induced by mixing cells of complementary mating type under appropriate conditions. Maturing conjugants immediately start intercommunicating with each other and, after an interval of "pre-conjugant interaction" which is dependent on maturity and nutritional status, they unite into pairs. Pre-conjugant interaction in this species is fluid-mediated, *i.e.* the signal compounds for interaction of the two conjugant types are excreted into the medium from which they may be purified and were characterized as follows (2,3):

(+) To Professor A. Butenandt - a pioneer also in tryptophan research - on occasion of his 80th birthday.

Progress in Tryptophan and Serotonin Research

The two complementary mating types I and II stimulate each other by specific messengers acting on specific receptors and answered by specific biochemical processes thus preparing part of the cell's surface for adhesion, local dissolution and finally, afte nuclear exchange, restoration and individualization. The feedbac cycle of tuning in into conjugation is started by the autonomous excretion by type I cells of gamone 1, or blepharmone, a fairly stable 20 kDa-glycoprotein. In type II cells this compound specifically switches on surface transformation of the mouth region and at the same time excretion of gamone 2, the phenone derivative blepharismone. The latter, in its turn, acts back on type I cells, inducing in them reciprocal surface transformations for conjugation and enhancing excretion of gamone 1. At the height of these processes, the cells unite in conjugant pairs, exchange one of their two germinal nuclei, and separate again. The whole process is completed within a couple of hours. Feedback cycles of this kind might be widely used to control morphogenesis and differentiation (4).

During our studies on cellular messengers for differentiation and morphogenesis (5) we became interested also in the *Blepharisma* system and the biochemical events that lead to the signal compound blepharismone, its mode of action and its functional requirements.

Results and Discussion

Gamone 2 of *Blepharisma japonicum*, or blepharismone, is the calcium salt of 4-(2'-formamido-5'-hydroxyphenyl)-4-oxo-2(S)-hydroxybutyric acid (6). It resembles serotonin chemically as 5-hydroxytryptophan derivative and biologically as specific intercellular messenger.

Racemic calcium blepharismone was first synthesized by Tokoroyama *et al*. (7) starting from acetophenone. However, due to the formation of many isomeric intermediates which have to be removed before condensation with glyoxylic acid, the overall yield is very low. We therefore devised a different approach to the substituted acetophenone which is more versatile, leads immediately to the correct position isomer of any desired acetophenone and thus also to the corresponding derivative or analogue of blepharismone in yields around 40% (8). This path is outlined in Scheme I.

Scheme I

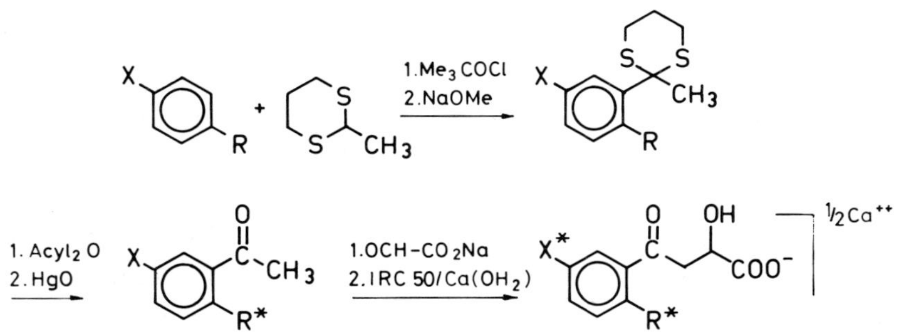

X = H, Hal, Me, MeO, PheCH2O (* HO after hydrogenolysis)
R = H, NH2 (* NHAcyl)

Synthesis of Blepharismones
Acetophenone Route

Starting material is any properly substituted (and appropriately protected) aniline derivative. It is selectively substituted in ortho-position according to Sommelet (9) with acetaldehyde dithioacetal under mildly oxidizing conditions followed by base-mediated 2,3-sigmatropic rearrangement of the intermediate azasulfonium salt (10). In our adaptation of this procedure, the aromatic amine is reacted with 2-metyhl-1,3-dithian (which is easily accessible by BF_3-etherate catalyzed transacetalization of acetaldehyde diethylacetal with 1,3-dimercaptopropanol (11)) at -78°C in an inert solvent and with a small excess of t-butyl hypochlorite and finally rearranged by Na-methylate. In all cases the acylation is strictly ortho-specific and the yield of the corresponding acetophenone thioketal is good. Formylation and hydrolysis of the intermediate gives almost quantitatively the desired 5-substituted 2-formamido acetophenone derivative which is now condensed with Na-glyoxylate following the Tokoroyama synthesis. The yield in the latter step after optimization is about 25%; so that the overall yield is in the range of 10% - an improvement of some 20 to 50-fold (12).

We prepared on this route, besides rac. calcium blepharismone, its analogues with halogen, alkyl, amino, and alkoxy substituents in position 5 as well as compounds lacking the formamido group in position 2 or carrying a higher acylamido residue instead (8, 13). Also in these cases the total yields of the (racemic) compounds are quite satisfactory.

A second effective and still more versatile approach leading at the same time to blepharismone en route of 5-hydroxy tryptophan and also to all putative biological intermediates offers ready access to the gamone and its analogues (14, 15).

This synthesis has its precedent in the work of Gortatowski and Armstrong (16) who synthesized indole lactate from the corresponding gramine derivative and diethyl acetoxymalonate. However, we reduced the required steps by alkylating an indole, substituted in position 5, with racemic methyl-2,3-epoxypropionate using tin tetrachloride in CCl_4 as the ansolvo acid. The resulting racemic methyl-3(5-substituted indolyl) lactates are purified by column chromatography on silica gel (ca. 50% yield), saponified (100% yield) and finally ozonolyzed in methanol at $-78^{o}C$ (17). The primary ozonides are hydrogenolyzed on 5% Pd/C, affording the (N-formamido) phenone derivatives in 50 to 60% yield (18). Thus, racemic 3(5-hydroxy-3-indolyl) lactic acid, a possible biogenetic precursor of blepharismone, is obtained from 5-benzyloxy indole. It is converted to racemic blepharismone in more than 10% yield, calculated for the indole.

Scheme II summarizes the possibilities intrinsic to this concept. As seen, either indole lactates or N-formyl kynurenine analogues may be obtained depending on the starting material and the branching-off of the path.

Scheme II

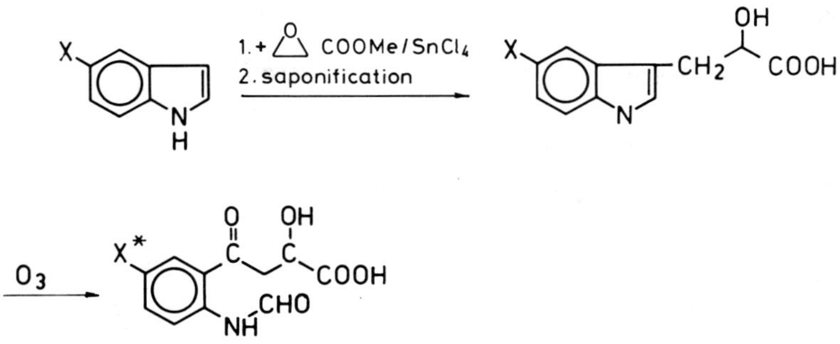

X = H, $PheCH_2O$ (*HO after hydrogenolysis)

Indole Route to Blepharismone and Its Precursors

A modification of this concept, combined with H. Simon's stereospecific electro-enzymatic reduction method (19), finally helped to answer the question of relative activity of the enantiomers of blepharismone: The highly oxygen-sensitive 5-benzyloxy indole-3-pyruvate (in 25% yield from 5-benzyloxy tryptophan by oxidative desamination with pyridine aldehyde (20)) was enantioselectively reduced to R-benzyloxy indole-3-lactate under strictly anaerobic conditions[(+)]. After ozono- and hydrogenolysis of the protected indole derivative in methanol R(+)-blepharismone was isolated as its pale yellow calcium salt and with the correct nmr-spectrum. It is dextrorotatory ($[\alpha]_{RT}^{578}$ = +24,3° (water)) and without biological activity up to 1 µg / ml but does not interfere with the natural hormone or its synthetic racemate (21). Thus, natural blepharismone, active at 10^{-9} g / ml (3), has the S(-)-configuration.

It is obvious and has already been shown indirectly by interference studies with putative intermediates (2) that biogenetically

---

(+) Thanks are due to Prof. H. Simon (TU Munich) for his help.

blepharismone is derived from the amino acid L-tryptophan. This may be attained - with or without preceding hydroxylation in position 5 - either by its transamination to the corresponding indolyl pyruvic acids which are reduced to the indolyl lactic acids as above and then split by indole pyrrolase(s) to the N-formyl-(2-desamino-2-hydroxy) kynurenines, or by first opening the pyrrole ring in this way and then metabolizing the side chain accordingly. This proposal is depicted in <u>Scheme III</u>.

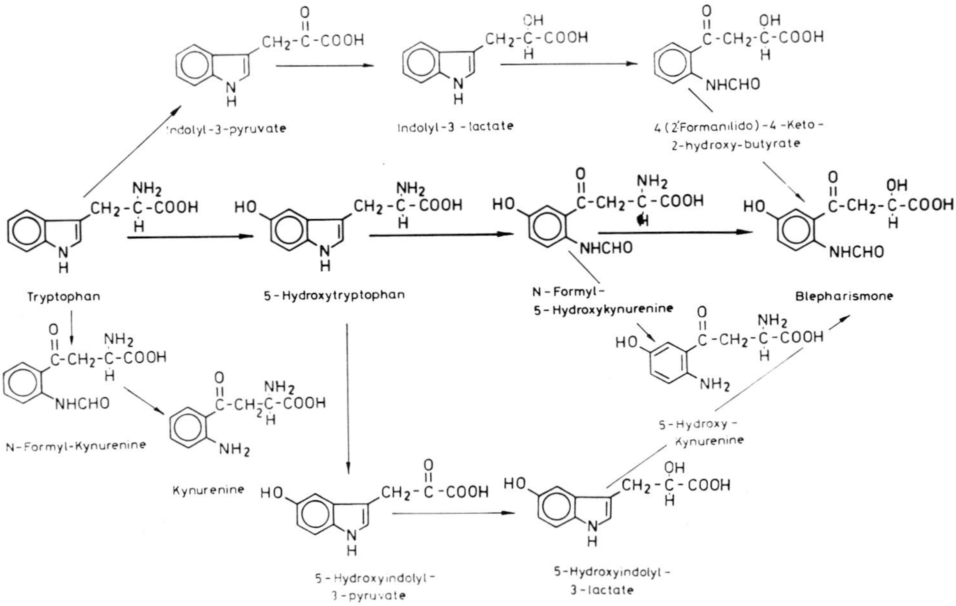

The syntheses described in the foregoing make accessible any of the compounds of this panel. All in all, there are twelve closely related substances, and it was our aim to work out quantitative analyses for this complicated mixture (21). This was achieved by HPLC chromatography using reversed phase columns (Merck Lichrosorb RP18) and tetrabutylammonium sulfate containing phosphate buffers, pH 6.5, without and with methanol as the solvent (22). Such a run is shown in <u>Figure 1</u>. The determination and localization make use of the fluorescence of the indoles and the uv-absorption of all these aromatic or heteroaromatic structures. Fluorescence measurements allow to quantitate as low as 1 ppb

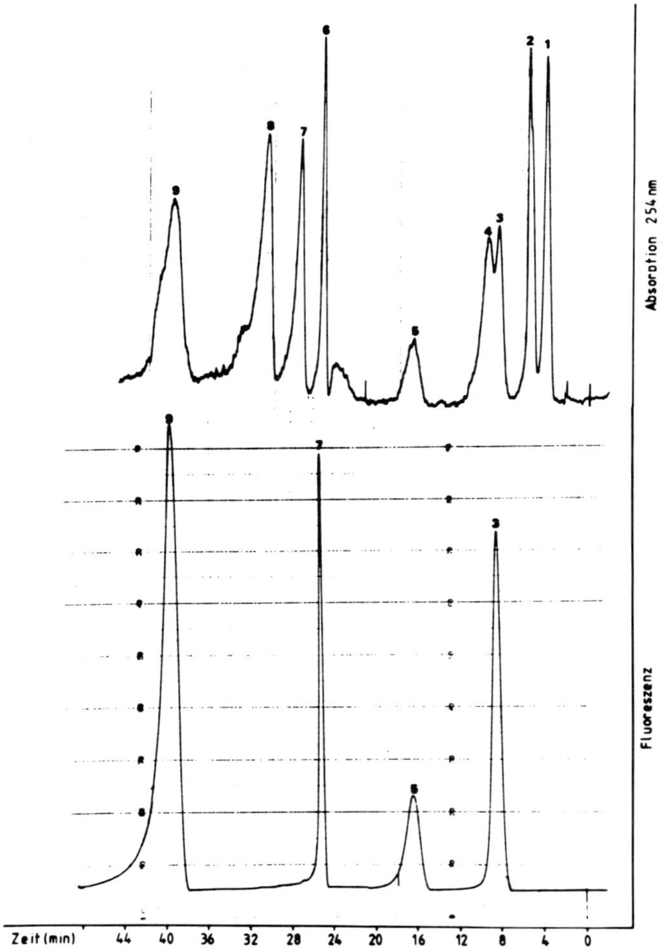

Figure 1

HPLC of indole and 5-hydroxy indole derivatives

Column: Merck Lichrosorb RP 18 (ion pair)
Solvent: 0.01 M phosphate (pH 6.5) + 0.01 M tetrabutylammonium sulfate for amino acids;
after 18 min: same plus methanol 70 : 30 (v/v) for hydroxy acids
Flow rate: 1.6 ml/min
Monitoring: ultraviolet at 254 nm
fluorescence: excitation at 285 nm; emission at 350 nm

(1) 5-OH kynurenine (4.1 min) (2) kynurenine (5.8 min)
(3) 5-OH tryptophan (8.4 min) (4) N-formyl kynurenine (9.6 min)
(5) tryptophan (16.5 min) (6) 5-OH indolyl-3-lactate (25.2 min
(7) blepharismone (27.5 min) (8) desoxyblepharismone (30.6 min)
(9) indolyl-3-lactate (39.6 min)

solutions, whereas uv-determination is about four orders of magnitude less sensitive. However, since in any of the different pathways proposed, there is at least one fluorescent key intermediate, it should be possible to unravel the biogenesis of blepharismone in extracts of induced type II cells.

For the present, however, we tried to tackle this problem indirectly following A. Miyake's demonstration that L-tryptophan as well as 5-hydroxy L-tryptophan are biological inhibitors of blepharismone (2). We used in the same way the structurally related amino acids N-formyl-5-hydroxy-L-kynurenine (A), N-formyl-L-kynurenine (B) and 5-hydroxy-L-kynurenine (C) [(+)] and determined their action on the threshold concentration of blepharismone required for pairing of mature Blepharisma (13). As seen from Figure 2, these compounds have no inducing activity but rather interfere with the gamone the more the closer they are structurally related to it (A > B > C) and desensitize the receptor system. This reacts more strongly to changes in position 2 than in position 5 of the mock-ligands.

[(+)] We thank Prof. O. Hayaishi (Kyoto) for an authentic sample.

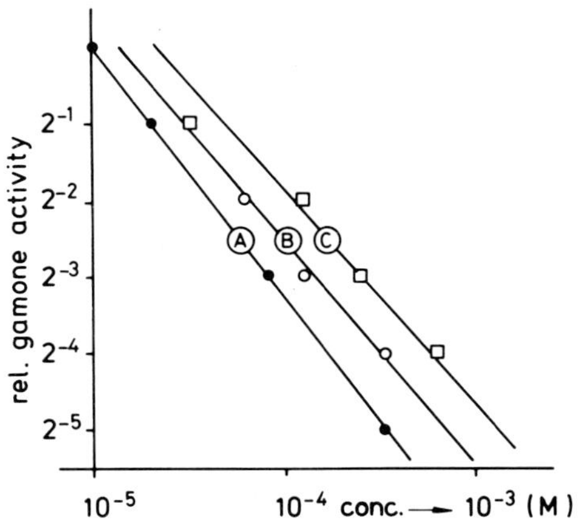

Inactivation of D,L-Blepharismone Activity ($c = 10^{-7}$ M) by Putative Precursors

A = N-Formyl-hydroxy kynurenine
B = N-Formyl kynurenine
C = 5-Hydroxy kynurenine

Figure 2

These two positions seem to be essential for the interaction of the gamone with its postulated membrane receptor. The 2-hydroxy group of the side chain, however, proves essential for the biological action of the hormone, since none of the amino acids structurally related to blepharismone has the potency to induce conjugation of homotypic pairs. From this we deduce that during biogenesis of blepharismone from tryptophan, the transformation of the 2-amino group into the 2-hydroxy group is the final step, preceded by aromatic hydroxylation and pyrrolase action in the sequence:

L-Tryptophan → 5-Hydroxy L-tryptophan → 5-hydroxy-N-formyl L-kynurenine → L-Blepharismone

To check the biological activity of the compounds synthesized we determined the relative threshold concentration for triggering pair formation between type I and type II cells by testing dilution series of each substance and correlating it to synthetic rac. blepharismone as the standard. We found for the analogues in position 5 the following (13):

Except the 5-bromo-, 5-iodo-, and 5-benzyloxy-derivative which have no effect at all, all the other blepharismone analogues are able to induce conjugation in _Blepharisma_, decreasing with increasing hydrophobicity of the substituent in position 5. If the negative logarithm of the relative activities (pA) is plotted against the structural parameter of the lipophilicity (lipophilicity constant π (23)), as in _Figure 3_, a linear dependency is obtained of the type and values indicated. This may mean that blepharismone binds by means of the 5-OH group to a hydrophilic area of the receptor. If the substituent in 5 exceeds a certain size (ca. 2 Å), it can no more be bound properly, and the system is deformed and deactivated. Therefore, analogues with strongly lipophilic and large sized groups have no measurable activity. The opposite effect explains the unexpected high biological activity of desoxy blepharismone which is greater than deduced from the lipophilicity of hydrogen. The small van der Waals diameter and the aromatic character of the compound interfere with the hydrophilic receptor group; apolar association with the receptor

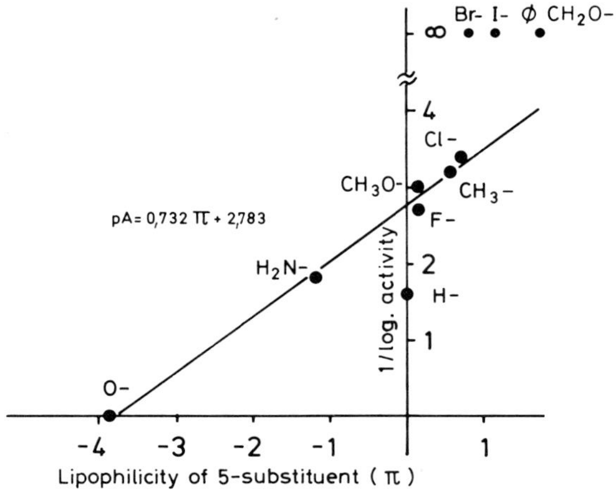

Activity of Blepharismone Analogues as Function of Lipophilicity of the Substituent in 5-Position

Figure 3

of the substance is favoured; consequently, it has a higher biological activity. For binding to the receptor and thus for induction of pairing, the charge distribution of the aromatic system seems to be essential, whereas the ionizing phenolic hydroxyl group is at best another binding increment, but not specific and can be replaced by many other nucleophiles.

On the other hand, Figure 4 shows also that the N-formamido group seems to be absolutely required for the biological function (8). 2-Desformamido blepharismone has less than 0.1% hormone activity; higher concentrations paralyze the excitation transducing system, thus suppressing conjugation. This is counterbalanced by the natural gamone at > 100-fold strength.

In contrast, N-acetyl homologues and analogues of blepharismone are totally inactive. Particularly noteworthy also in this case is the strong desensibilization of the biological system by small amounts of N-acetyl blepharismone which is illustrated in Figure 5. This lets us conclude that the substituent at N-2 of the ligand hormone adds efficiently to the interaction with the receptor; however, it is not the conjugation triggering group.

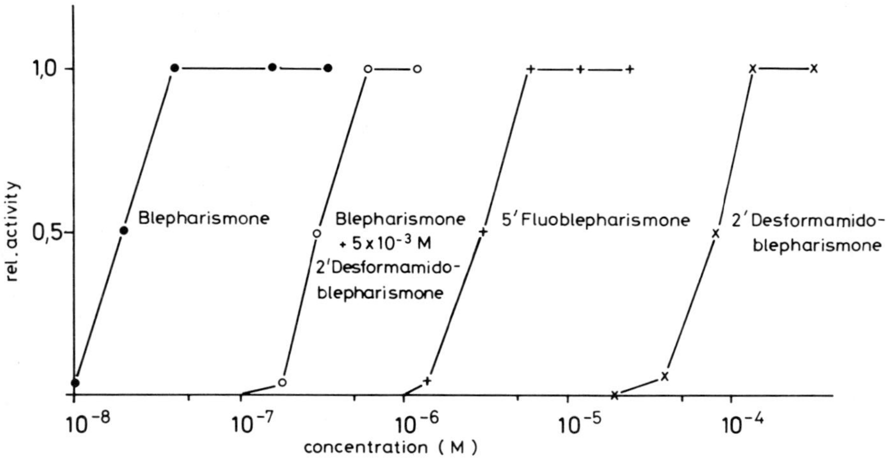

Induction of Conjugation in Blepharisma japonicum
by Blepharismone Analogues

Figure 4

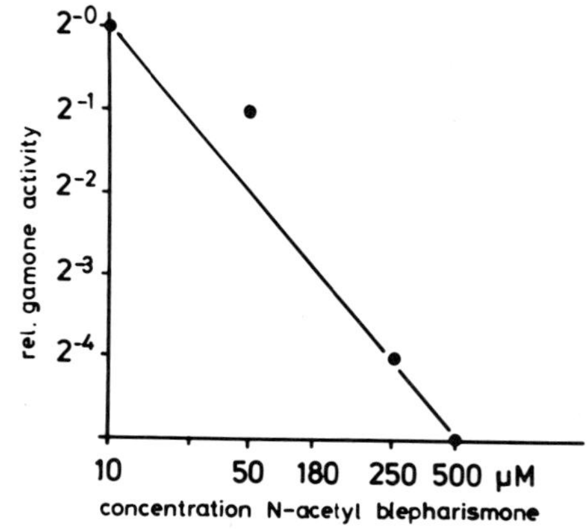

Inhibition of Blepharismone-Induced ($c = 5x10^{-8}$ M L-gamone)
monotypic pairing of Blepharisma japonicum
by N-Acetyl-blepharismone

Figure 5

Thus, we were able, with the synthetic derivative blepharismones, to get some insight into the biosynthesis from L-tryptophan of the hormone and into essential features of its binding and active sites.

References

1. Miyake, A.: preceding paper.
2. Miyake, A.: Curr. Top. Devel. Biol. 12, 37 - 82 (1978).
3. Miyake, A.: in: Sexual interaction in eucaryotic microbes (O'Day, D.H., Hogens, P.A., eds.) Academic Press, New York 1981, pp. 95 - 129.
4. Heckmann, K., Siegel, R.W.: Exp. Cell. Res. 36, 688 - 691 (1964).
5. see: 33rd Mosbach Colloquium: Biochemistry of Differentiation and Morphogenesis (Jaenicke, L., ed.) Springer-Verlag Berlin-Heidelberg-New York 1982.
6. Kubota, T., Tokoroyama, T., Tsukuda, Y., Koyama, B., Miyake, A.: Science 179, 400 - 402 (1973).
7. Tokoroyama, T., Hori, S., Kubota., T.: Proc. Jpn. Acad. 49, 461 - 463 (1973).
8. Entzeroth, M., Jaenicke, L.: Z. Naturforsch. 36c, 180 - 182 (1981).
9. Sommelet, M.: C.R. Acad. Sci. 205, 56 - 58 (1937).
10. Gassman, P.G., Drewes, H.R.: J. Amer. Chem. Soc. 100, 7600 - 7610 (1978).
11. Seebach, D.: Synthesis 1, 17 - 36 (1969).
12. Entzeroth, M.: Diplomarbeit, Köln 1980.
13. Entzeroth, M., Jaenicke, L.: Z. Naturforsch. 37c, 1136 - 1140 (1982).
14. Kunczik, T.: Dissertation, Köln 1981.
15. Entzeroth, M., Kunczik, T., Jaenicke, L.: Liebigs Ann. Chem. 1983, 226 - 230.
16. Gortatowski, M.J., Armstrong, M.D.: J. Org. Chem. 22, 1217 (1957).
17. Witkop, B.: Liebigs Ann. Chem. 556, 103 - 114 (1944).
18. Blumberg, J.: Diplomarbeit, Köln 1982.
19. Simon, H., Günther, H., Bader, J., Tischer, W.: Angew. Chem. 93, 897 - 898 (1981); I.E. 20, 861 - 862 (1981).
20. Ohta, S., Okamoto, M.: Synthesis 9, 756 - 758 (1982).
21. Entzeroth, M.: Dissertation, Köln 1983.
22. Sandberg, G., Andersson, B., Dumbley, A.: J. Chromatogr. 205, 125 - 137 (1981).
23. Seydel, J.K., Schaper, H.J.: Chemische Struktur und Biologische Aktivität. Verlag Chemie, Weinheim 1979, pp. 266 - 288.

Tryptophan in Proteins and Peptides

ADVANCES IN THE CHEMICAL MODIFICATION OF TRYPTOPHAN IN PEPTIDES AND PROTEINS

Angelo Fontana

Institute of Organic Chemistry, Biopolymer Research Centre of CNR, University of Padova, I-35100 Padova, Italy

1. Introduction

Many approaches to the solution of fundamental problems of protein structure and function take advantage of the availability of methods for the selective chemical modification of amino acid side chains of proteins (1,2). The tryptophan residue is an important target for protein modification, due to the high reactivity of the indole nucleus. In this respect, the chemistry of tryptophan in proteins has probably been the object of more intense investigation than that of any other amino acid. Only the sulfur amino acids have received comparable attention from chemists and biochemists.

In the following, a summary will be presented on reagents and techniques used to modify tryptophan developed in the author's laboratory and subsequently extended and applied by other investigators in peptide and protein chemistry studies. A more comprehensive review-article on the chemistry of tryptophan in peptides and proteins has been published (3).

2. Chemical Oxidation

Chemical oxidation of tryptophan has been carried out with peracids, hydrogen peroxide, ozone, periodate and others (3, and references cited therein), but none of these reagents has been shown to be ideal for achieving simple modification of tryptophan residues in proteins, because of the formation of several oxidation products of tryptophan and the concomitant oxidation of several other functional groups of amino acids. N-Bromosuccinimide (NBS) has been extensively used with peptides and proteins to effect oxidation of tryptophan to oxindolylalanine (Oia) (**1**) (4). However, NBS is a very reactive reagent which can bring about modification of several other side chains of amino acids (including peptide bond cleavage). In

Progress in Tryptophan and Serotonin Research

addition, it is experimentally difficult to carry out the reaction under the somewhat critical conditions required for reaction to proceed not beyond the Oia level. Thus, a halogen reagent: substrate ratio of 1:1 is re-

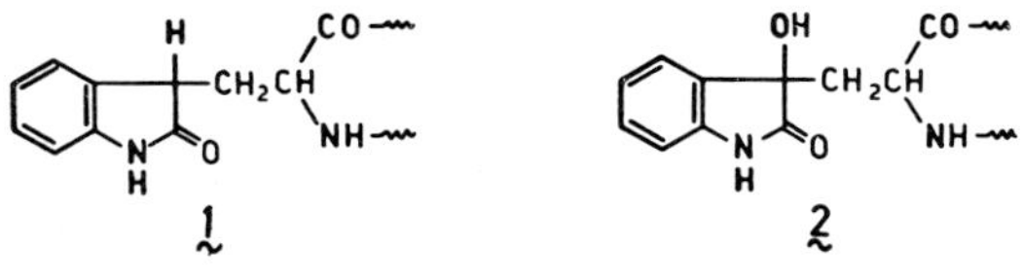

quired, in theory, for the conversion to oxindole (1) and 2:1 for the oxidation to dioxindole (2); with tryptophan in a peptide combination dioxindole formation leads to peptide bond fission (5) (see below). Furthermore, NBS shows such a reactivity that also bromination of the indole nucleus easily occurs and 5-bromo-derivatives are formed (6).

The mechanism of oxidation of 3-alkylindoles to the corresponding oxindoles by positive halogen reagents in the presence of water is shown below (7). 3-Haloindolenine (3) is shown as the initial intermediate, since it is commonly accepted that electrophilic attack at the 3-position of

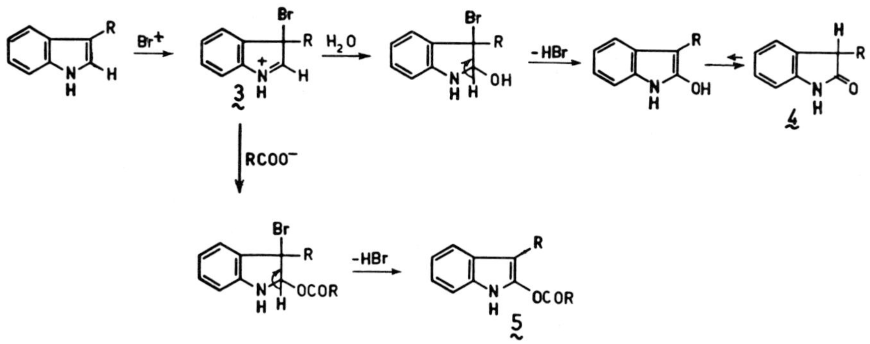

the indole nucleus is favored (8). Oxindole (4) is formed by dehalogenation of 3-haloindolenine through hydrolysis. In the scheme it is shown also that 3-haloindolenine can react at the electrophilic C-2 with a nucleophilic carboxylate, leading to a 2-ester-derivative (5). This could be a side reaction when the usual acetate buffer, pH 4.5, is employed in modification reactions of proteins. This ester formation was shown to occur in the NBS-

-oxidation of lysozyme in acetate buffer (9). A related intramolecular reaction was observed in the oxidation of lysozyme with iodine; in this case the carboxylic group of glutamic acid 35 was covalently bonded to the indole C-2 of tryptophan 108 (10). These results imply that indole-esters can be formed in modification studies of tryptophan by NBS, so that care should be taken to ensure the absence of such esters, if oxindole-derivatives are desired as the end-products of the modification reactions.

The shortcomings of the NBS-modification above discussed, indicate that an alternative method for conversion of tryptophan to oxindole would be desirable. A general procedure for oxidation of 3-substituted indoles to the corresponding 2-oxindoles has been found by Savige and Fontana (11,12). The method, which appears superior to those in current use, consists of treating the indole compound with an acetic acid solution containing dimethylsulfoxide (DMSO) and concentrated HCl. The yields of clean oxidation are high both for simple indoles and for tryptophan peptides and proteins (11,12). The halogenating-oxidizing species likely is the sulfonium ion $(CH_3)_2S^+Cl$ formed by the interaction of DMSO and HCl. This should provide a positive halogen source effecting chlorination to an unstable chloroindolenine which is easily hydrolyzed to the oxindole compound (see above). The method appears suitable for the modification of tryptophan peptides only (12), since the strong acid conditions employed do not appear suitable for proteins and enzymes. Several tryptophan containing peptides (somatostatin, valine-gramicidin A, $ACTH^{1-24}$, LRF) were treated with the DMSO/HCl reagent in acetic acid solution and the corresponding Oia derivatives isolated in over 90% yields after chromatography (12).

The DMSO/HCl procedure oxidizes both tryptophan and methionine, this last being converted to methionine sulfoxide. Suitable procedures to reduce methionine sulfoxide derivatives are available (13), so that it is possible to prepare peptides oxidized at tryptophan only. For example, Gurd and coworkers (14) were able to prepare a glucagon derivative with both tryptophan and methionine oxidized and then, after reduction of the methionine sulfoxide residue with mercaptoacetic acid, to isolate a homogeneous 25-Oia-glucagon.

Oxindolylalanine is quite stable to acid hydrolysis and can be easily quantitated by standard automatic amino acid analysis (11). On the other hand, it is well known that tryptophan is labile under acid or other experimental conditions, so that, for example, it is somewhat difficult to identify tryptophan both using the automatic Edman degradation procedure or the manual dansyl procedure, this last involving the necessity of an acid hydrolysis step (15). Inglis et al. (16) showed that for sequence purposes it is advantageous to oxidize a tryptophan containing polypeptide chain to the corresponding Oia-derivative by the DMSO/HCl procedure, since the Oia residue is easily identified as phenylthiohydantoin-derivative by the Edman procedure.

3. Sulfenylation Reaction

Sulfenyl halides react selectively under acidic conditions with the tryptophan residue in peptides and proteins to form 2-thioether derivatives (17,18). The only other amino acid affected is cysteine, which is converted to a mixed disulfide (19). Since this disulfide can be easily cleaved by thiolysis, it is possible to label peptides and proteins specifically at tryptophan. In principle, all members of the wide class of sulfenyl halides could be used, but the most commonly employed reagents are 2-nitrophenyl compounds, in particular 2-nitrophenylsulfenyl (NPS) chloride. This reagent converts tryptophan to a Trp(NPS)-derivative (6), allowing introduction of a chromophore into a peptide or protein which absorbs in the visible part of the spectrum and thus an easy quantitation of the reaction (20). The bifunctional reagent 2,4-dinitrophenyl-1,5-disul-

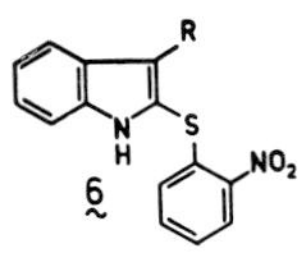

6

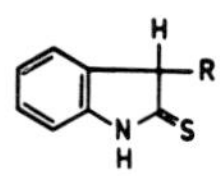

7

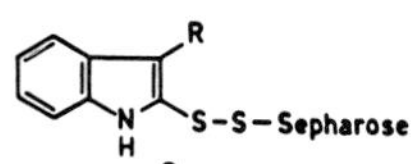

8

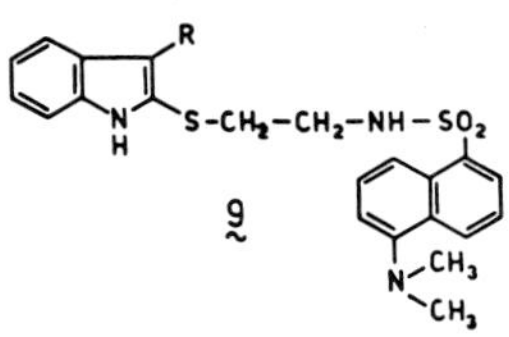

9

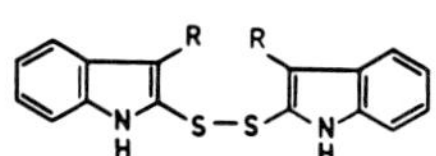

10

fenyl chloride has been used to selectively crosslink peptides between tryptophan units (21,22).

The yellow label covalently bound at tryptophan facilitates identification and separation of tryptophan containing peptides. It has been found that Trp(NPS)-peptides strongly interact with Sephadex LH-20 gel equilibrated with water. For example, the pentapeptide Gly-Ile-Thr-Trp(NPS)-Lys contained in a tryptic digest of $Trp(NPS)^{59}$-apocytochrome C was isolated in a single chromatographic step, since this peptide was selectively adsorbed on a Sephadex LH-20 column in water and subsequently eluted with aqueous acetic acid (A. Fontana, unpublished).

A method for the selective tritiation of tryptophan peptides via the sulfenylation reaction was described by Fromageot and coworkers (23). The procedure consists of removing the sulfenyl label by catalytic hydrogenation in the presence of palladium and tritium gas, leading to derivatives tritiated in the 2-position of the indole nucleus. The method, because of its high specificity, appears of great value for the labelling of biologically important tryptophan derivatives.

Wilchek and Miron (24) described a mild thiolysis method for the quantitative removal of 2,4-dinitrophenyl group from 2-(2,4-dinitrophenylsulfenyl)-tryptophan peptides and proteins to yield thioxindolylalanine (2-thioltryptophan) (7) derivatives. This allows specific incorporation into a peptide or protein of a reactive sulfhydryl group, which can be used in the preparation of additional derivatives of the original tryptophan containing compound. A glucagon affinity column (8) and a fluorescent analog of glucagon, S-(N-dansylaminoethyl)-2-thioltryptophan-glucagon (9) was thus prepared (25). Oxidation of 2-thioltryptophan-glucagon gives the corresponding dimer (10) (26).

The use of phenylsulfenyl chlorides containing an azido ($-N_3$) group for tryptophan modification has been reported by Ramachandran and coworkers (27-30). The particular advantage of reagents such as 2-nitro--4(5)-azidophenylsulfenyl chloride resides in the fact that they allow introduction into a tryptophan-containing peptide of an azido group (11) capable of generating a very reactive nitrene upon photolysis (31). This species, once generated, reacts rapidly in several ways with its immediate environment via insertion reaction (including into an aliphatic C-H bond) allowing covalent binding of the peptide via its tryptophan moiety to, e.g.,

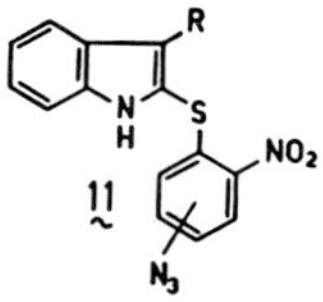

11

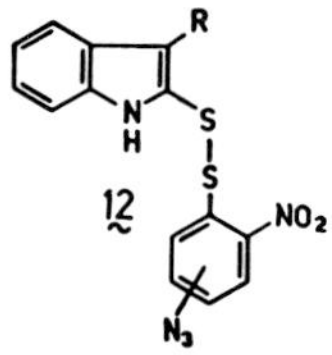

12

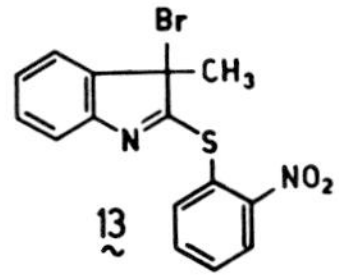

13

its macromolecular receptor. Photoreactive derivatives of the pituitary hormone corticotropin ACTH were prepared and shown to attach covalently to a pituitary protein fraction by photolysis (28,29). Similarly, glucagon was selectively modified at tryptophan-25 with 2-nitro-4-azidophenylsulfenyl chloride and the photolabeled receptor in rat liver plasma membranes was isolated after gel filtration (32). Alternatively, conversion of a tryptophanyl peptide to a 2-thioltryptophan derivative, followed by reaction with an azidophenylsulfenyl chloride, allows preparation of a photoactivable azido-derivative containing the disulfide moiety (12). Upon covalent binding of the peptide to its receptor, reductive methods allow cleavage of the disulfide bond and concomitant splitting of the peptide--receptor complex (30).

4. Fragmentation Reaction

In order to determine the primary structure of a protein, the polypeptide chain is best cleaved by chemical and enzymatic methods into fragments, which can be completely characterized with respect to composition and sequence. Of particular utility are the methods which cleave at relatively rare amino acids (methionine, cysteine, tryptophan), since this would permit production of rather large fragments especially suitable for use in automatic sequencer and for overlapping smaller known sequences (33).

Procedures for effecting specific chemical cleavage of tryptophanyl peptide bonds have been intensively sought since 1960 (34) (see Fontana (35), for a review article). NBS was the first reagent proposed for cleavage at tryptophan (34), but various side reactions of this very reactive reagent with other amino acid side chains, as well as poor yields of peptide bond fission, hampered practical application of the reagent in protein sequence studies. Tribromocresol (36) and N-chlorosuccinimide (37,

38) and active iodine generated with H_2O_2, iodide and peroxidase (39) were subsequently proposed for cleaving peptides and proteins at tryptophan in moderate yields.

Presently, the most often used reagent for protein fragmentation at tryptophan utilizes BNPS-skatole (2-(2-nitrophenylsulfenyl)-3-methyl-3--bromoindolenine) (13) as a source of positive halogen (40). This reagent is highly selective, and yields of cleavage are satisfactory (up to 80%) (cf. 35 for references). Selective cleavage at tryptophan can conveniently be achieved also employing an acetic acid solution containing DMSO and hydrobromic acid (41). Both BNPS-skatole and the DMSO/haloacid reagent cause cleavage of the tryptophanyl peptide bond via a mechanism of oxidative halogenation, most likely involving 3-halooxindole moiety (14) as an

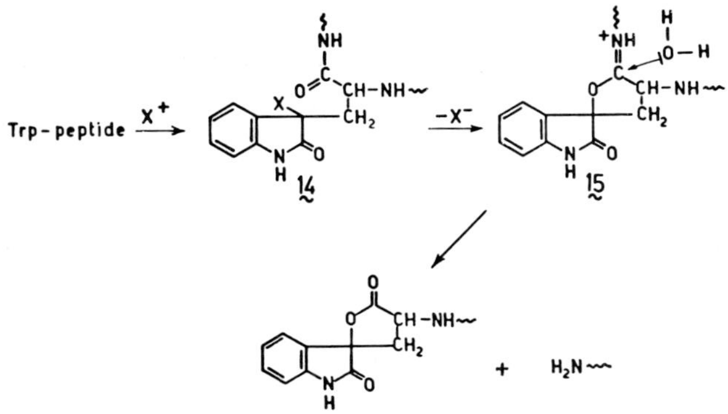

intermediate reactive species leading to an iminolactone (15) of the former tryptophanyl residue which undergoes easy hydrolysis with peptide bond rupture (see ref. 35, for a more detailed discussion of this mechanism).

More recently, two additional methods have been reported to effect cleavage at tryptophan, one involving treatment of a tryptophan containing peptide or protein with a large excess of cyanogen bromide in the presence of heptafluorobutyric acid (48,43), while the other involves treatment with o-iodosobenzoic acid in 4 M guanidine hydrochloride (44). Both these procedures cause cleavage at tryptophan via a mechanism of oxidative halogenation. That cyanogen bromide can act as a halogenating agent was firmly established by isolating brominated derivatives of tryptophan and tyrosine from the reaction mixture (see Fontana et al., these Proceedings). o-Iodoso benzoic acid (16), which appears to be in tautomeric equilibrium with the

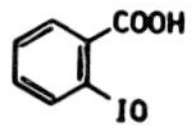

16

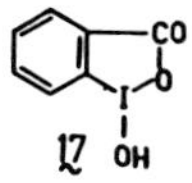

17

cyclic structure 1-hydroxy-1,2-benziodoxolin-3-one (17), in presence of halides (Cl^- when guanidine hydrochloride is added to the reaction medium) and in acid solution causes the formation of halogenating species, including free halogen, resulting from a series of equilibrium reactions. On this basis, Fontana et al. (45) concluded that the modification of other amino acid side chains besides the indule nucleus of tryptophan, as well as cleavage at tyrosine, is an inherent property of the o-iodosobenzoic acid/4 M guanidine hydrochloride reagent and not due to contaminants (e.g. o-iodoxybenzoic acid) present in the commercially available reagent (46).

5. Conclusion

A summary was given on recent advances in methods which can be used to selectively modify tryptophan in peptides and proteins, with particular emphasis given to methods developed over the years in the author's laboratory. As a result of these efforts, some useful procedures for peptide and protein chemistry studies have been developed, as demonstrated by the numerous applications of these procedures already reported in the literature.

References

1. Spande,T.F., Witkop,B., Degani,Y. and Patchornik,A. (1970) Adv. Protein Chem. 24, 97-260.
2. Glazer, A.N. (1970) Ann. Rev. Biochem. 39, 101-127.
3. Fontana,A. and Toniolo,C. (1976) Progr. Chem. Organic Natural Compounds (Herz,W. et al., eds.) Springer-Verlag, Wien-New York, vol. 3, pp. 309--449.
4. Spande,T.F. and Witkop,B. (1967) Methods Enzymol. 11, 498-506.
5. Ramachandran, L.K. and Witkop,B. (1967) Methods Enzymol. 11, 283-299.
6. Lawson,W.B., Patchornik,A. and Witkop,B. (1960) J. Amer. Chem. Soc. 82, 5918-5923.

7. Sundberg,R.J. (1970) in The Chemistry of Indoles, Academic Press, New York, pp. 14-15.

8. Jackson,A.H., Naidoo,B. and Smith,P. (1968) Tetrahedron 24, 6119-6129.

9. Norton,R.S. and Allerhand,A. (1976) Biochemistry 15, 3438-3445.

10. Imoto, T. and Rupley, J.A. (1973) J. Mol. Biol. 80, 657-667.

11. Savige,W.E. and Fontana,A. (1977) Methods Enzymol. 47, 442-453.

12. Savige,W.E. and Fontana,A. (1979) Int. J. Peptide Protein Res. 15, 285-297.

13. Savige,W.E. and Fontana,A. (1977) Methods Enzymol. 47, 453-459.

14. Coolican,S.A., Jones,B.N., England,R.D., Flanders,K.C., Condit,J.D. and Gurd,R.S. (1982) Biochemistry 21, 4974-4981.

15. Hartley,B.S. (1970) Biochem.J. 119, 805-822.

16. Inglis,A.S., Rivett,D.E. and McMahon,D.T.W. (1979) FEBS Lett. 104, 115-118.

17. Scoffone,E., Fontana,A. and Rocchi,R. (1968) Biochemistry 7, 971-979.

18. Fontana,A. and Scoffone,E. (1972) Methods Enzymol. 25, 482-494.

19. Fontana,A., Scoffone,E. and Benassi,C.A. (1968) Biochemistry 7, 980--986.

20. Boccù,E., Veronese,F.M., Fontana,A. and Benassi,C.A. (1970) Eur. J. Biochem. 13, 188-192.

21. Veronese, F.M., Boccù, E. and Fontana,A. (1970) Int. J. Protein Res. 2, 67-74.

22. Epand,R.M. and Cote,T.E. (1976) Biochim. Biophys. Acta 453, 365-373.

23. Marche,P., Girma,J.-P., Morgat,J.-L., Fromageot,P., Ghelis, C., Dubrasquet,M. and Bonfils,S. (1975) Eur. J. Biochem. 50, 375-382.

24. Wilchek,M. and Miron,T. (1972) Biochem. Biophys. Res. Commun. 47, 1015--1020.

25. Wright,D.E. and Rodbell,M. (1981) in Peptides. Synthesis, Structure and Function. Proc. 7th Amer. Peptide Symp. (Rich,D.M. and Gross,E., eds.) Pierce Chem. Co., Rockford, Ill., USA, pp. 135-138.

26. Wright,D.E. and Rodbell,M. (1980) J. Biol. Chem. 255, 10884-10887.

27. Canova-Davis,E. and Ramachandran, J. (1980) Biochemistry 19, 3275-3281.

28. Ramachandran,J., Muramoto,R., Kenez-Keri,M., Keri,G. and Buckley,D.I. (1980) Proc. Natl. Acad. Sci. USA 77, 3967-3970.

29. Ramachandran,J., Hagman,J. and Muramoto,K. (1981) J. Biol. Chem. 256, 11424-11427.

30. Muramoto,K. and Ramachandran,J. (1981) Biochemistry 19, 3280-3286; ibid. 20, 3376-3380; ibid. 20, 3380-3385.

31. Bayley, H. and Knowles,J.R. (1977) Methods Enzymol. 46, 69-114.

32. Demoliou-Mason,C. and Epand,R.M. (1982) Biochemistry 21, 1989-1996; ibid. 21, 1996-2004.

33. Fontana,A. and Gross, E. (1983) in Handbook of Peptide and Protein Chemistry (Darbre,A. and Waterfield,M., eds.) Heyden and Son Lim., London, in press.

34. Patchornik,A., Lawson,L.B., Gross,E. and Witkop,B. (1960) J. Amer. Chem. Soc. 82, 5923-5927.

35. Fontana,A. (1980) in Biochemical and Medical Aspects of Tryptophan Metabolism (Hayaishi, O. et al., eds.) Elsevier/North-Holland Biomedical Press, Amsterdam-New York, pp. 59-72.

36. Burstein,Y. and Patchornik,A. (1972) Biochemistry 11, 4641-4650.

37. Shechter,Y., Patchornik,A. and Burstein,Y. (1976) Biochemistry 15, 5071-5075.

38. Lischwe,M.A. and Sung,M.T. (1977) J. Biol. Chem. 252, 4976-4980.

39. Alexander, N.M. (1974) J. Biol. Chem. 249, 1946-1952.

40. Fontana, A. (1972) Methods Enzymol. 25, 419-423.

41. Savige, W.E. and Fontana, A. (1977) Methods Enzymol. 47, 459-469.

42. Ozols,J. and Gerard,C. (1977) Proc. Natl. Acad. Sci. USA 74, 3725-3729.

43. Ozols,J. and Gerard,C. (1977) J. Biol. Chem. 252, 5986-5989.

44. Mahoney, W.C. and Hermodson, M.A. (1979) Biochemistry 18, 3810-3814.

45. Fontana,A., Dalzoppo,D., Grandi,C. and Zambonin,M. (1981) Biochemistry 20, 6997-7004.

46. Mahoney,W.C., Smith,P.K. and Hermodson,M.A. (1981) Biochemistry 20, 443-448.

PEPTIDE BOND CLEAVAGE AT TRYPTOPHAN WITH CYANOGEN BROMIDE MECHANISM OF THE REACTION

Angelo Fontana, Claudio Grandi, Daniele Dalzoppo and Marcello Zambonin
Institute of Organic Chemistry, Biopolymer Research Centre of CNR, University of Padova, I-35100 Padova, Italy

Introduction. Non-enzymatic methods of cleavage of selected peptide bonds in polypeptides are important tools in both amino acid sequence determination and structure-function studies of proteins (1). Selective high-yield chemical procedures to cleave peptide bonds at methionine, tryptophan and cysteine are presently available (2). Cleavage directed at tryptophan can be performed with several reagents (e.g. BNPS-skatole) which have in common to effect cleavage via a mechanism of oxidative halogenation of the indole nucleus of tryptophan (see Fontana (3), for a discussion of these procedures and mechanism of reactions). The use of cyanogen bromide (CNBr) as a protein reagent is best known as a near quantitative and highly selective procedure to cleave methionyl peptide bonds (4). However, cleavage at tryptophan in moderate yields (~20%) has been observed with β-lactoglobulin (5) and glutamate dehydrogenase (6). Subsequently, Ozols and Gerard (7,8) reported that yields of cleavage can be highly improved by using a large excess of CNBr in heptafluorobutyric acid (HFBA).

Since the chemical mechanism of tryptophanyl peptide bond cleavage by CNBr has not been defined in previous studies, in this report we wish to present data demonstrating that an excess CNBr causes peptide bond fission at tryptophan via a mechanism of oxidative halogenation.

Results. The experiments were performed with a mixture constituted by 1 ml HFBA, 1 ml 90% formic acid and 0.72 g CNBr (7,8); this mixture herewith is called the CNBr/HFBA reagent. It should be noted that these experimental conditions imply a huge excess of CNBr (over 10,000 equivalents over amino acid or peptide to be reacted). When all common amino acids were reacted for 20 hours at room temperature with CNBr/HFBA, it was found by automatic amino acid analysis (Table I) that tryptophan, tyrosine, cysteine and methionine were completely modified by the reagent, while all other amino acids were recovered quantitatively. Methionine was oxidized to methionine sulfoxide and cystine, in part, to cysteic acid.

The modification of free tryptophan with the CNBr/HFBA reagent gave oxidation products of the amino acid, i.e. oxindolylalanine and dioxindolylalanine (diastereoisomers). These compounds were identified by automatic amino acid analysis (3,9). From the reaction mixture of N-acetyl-L-tryptophan reacted with CNBr/HFBA, a preparative chromatography on a Sephadex LH-20 column eluted with water allowed to separate and identify acetylated derivatives of oxindolylalanine (λ max 250 and 280 nm), dioxindolylalanine (λ max 254 and 288 nm) and corresponding lactone (λ max 254 and 300 nm). In addition, the N-acetyl-5-bromo-dioxindolylalanine (λ max 260 and 288) and its lactone (λ max 260 and 310 nm) were clearly identified in the reaction mixture. Tryptophan oxidation products were identified by comparison of their spectral characteristics with those of corresponding 3-substituted oxindoles reported in the literature (10). To test for the presence of bromine, a drop of a solution of the compounds was applied to a copper wire and volatilized in a flame; a green color indicated the presence of a copper halide. The CNBr/HFBA mixture gave variable yields of oxidation and bromination of tryptophan by slight changes of the experimental conditions (reaction carried out with a preformed mixture of the CNBr/HFBA reagent or by successive addition of the components of this reagent, different batches of CNBr).

TABLE I: Recovery of Amino Acids after Incubation with CNBr/HFBA [a].

Amino Acid	% Recovery [b]	Amino Acid	% Recovery [b]
Cysteic acid [c]	61	Valine	99
Meth. sulfoxide [d]	99	Methionine	0
Aspartic acid	103	Isoleucine	100
Threonine	103	Leucine	100
Serine	101	Tyrosine	0
Glutamic acid	100	Phenylalanine	101
Proline	99	Lysine	95
Glycine	99	Histidine [e]	131
Alanine	100	Arginine	97
Hemicystine	0	Tryptophan	0

[a] To a solution of all common amino acids (0.5 µmol each), as used for calibration of the automatic amino acid analyzer, with the exception of cysteine, in 1.2 ml of 90% HCOOH, 1 ml of HFBA and 0.72 g of CNBr were added under stirring. The mixture was held in the dark for 24 hours and then diluted with 10 ml of water and evaporated in vacuo at room temperature. The residue was taken into pH 2.2 citrate buffer and an aliquot of the solution directly analyzed with a Jeol Model JLC-6AH amino acid analyzer (single column). [b] Recoveries of amino acids are expressed on the basis of 100% recovery of alanine. [c] Oxidation product of cystine. [d] Oxidation product of methionine. [e] 3-Bromotyrosine was eluted at the same position of the peak of histidine.

Tyrosine was reacted separately with the CNBr/HFBA reagent. The reaction products were separated by column chromatography on a Sephadex LH-20 column eluted with water and shown to be 3-bromo- and 3,5-dibromotyrosine. These compounds were also identified on the amino acid analyzer and were also obtained reacting tyrosine with N-bromosuccinimide (3,11).

As shown in Table II, the CNBr/HFBA reagent cleaves tryptophanyl peptide bonds in quite high yields (up to 79%). From the data of Table II, it appears that CNBr/HFBA cleaves also at tyrosine, although in low yields. Of note, CNBr cleaves the model peptide Gly-Met-Gly in 80% yield.

Conclusions. The results reported in this study indicate that the CNBr/HFBA reagent acts as an oxidative-halogenating agent. Thus, the reagent causes modifications of amino acids similar to those observed with other reagents of the positive halogen type previously proposed for cleavage of the tryptophanyl peptide bond (12-14). In addition, brominated derivatives of tryptophan and tyrosine have been clearly identified among the products in the reaction mixtures. To explain the halogenating power of the CNBr/HFBA reagent, it seems possible to propose that a nucleophile (side chain of an amino acid) can approach the bromine end of the pseudohalogen CNBr to form HCN and a bromo-derivative, which eventually rearranges to other derivatives (Nu:⟶Br-C≡N:⟶H^+). Alternatively, some free bromine might form by an interaction between CNBr and Br^-, the latter being formed by hydrolysis of CNBr ($Br:^-$⟶Br-C≡N:⟶H^+). These could be side reactions of CNBr in respect to the normal reactivity of CNBr acting as a cyanylating agent (formation of S-cyano-sulfonium derivative of methionine)(4), which become relevant when the reagent is used in huge excess, as in the CNBr/HFBA mixture. On this basis, the mechanism of the peptide bond fission at tryptophan should occur by a mechanism of oxidative halogenation of the indole nucleus involving several halogenation-dehalogenation steps, as previously described for other halogenating agents (BNPS-skatole, dimethylsulfoxide/ haloacid) (3,12-14). Yields of cleavage at tryptophan are quite high (Table II) and similar to those obtained with other reagents earlier proposed to cleave at tryptophan (1,2,12-14).

For the first time, in this report it is shown that CNBr can cleave the tyrosyl peptide bond, although in low yields (Table II). This is in

TABLE II: Cleavage of Peptides by CNBr/HFBA [a].

Peptide	Cleavage Product	Cleavage (%) [b]
Gly-Trp-Gly	Gly	78
Leu-Trp-Leu	Leu	79
Gly-Pro-Trp-Leu	Leu	78
Phe-Val-Gln-Trp-Leu	Leu	64
Leu-Tyr-Leu	Leu	20
Val-Tyr-Val	Val	9
Gly-Met-Gly	Gly	80

[a] Cleavage reactions were performed dissolving the peptide (1 μmol) in 1 ml of the CNBr/HFBA mixture and allowing the reaction to proceed in the dark for 20 hours. All peptides were cleaved separately. [b] Yields of cleavage were determined by automatic amino acid analysis of the released amino acid.

line with the proposal that CNBr/HFBA acts as halogenating reagent , since it is known that cleavage at tyrosine by halogenating agents involves a rapid ortho halogenation, followed by a slower consumption of a third mole of reagent to afford an intermediate iminolactone, which leads to peptide bond fission by hydrolysis (11).

The results obtained with methionine and the peptide Gly-Met-Gly appear to be contradictory, since the free amino acid is oxidized to the sulfoxide derivative (Table I) and the model peptide is cleaved at methionine in high yields (Table II). In fact, the peptide bond at methionine sulfoxide is expected to be resistant to the action of CNBr (4). This would indicate that methionine in a peptide combination is being S-cyanylated by CNBr/HFBA and converted to the homoserine-iminolactone intermediate, which leads to peptide bond fission (4). However, more work is needed in order to clarify this alternative fate (oxidation or cleavage) of methionine.

References

1. Spande, T.F., Witkop, B., Degani, Y. and Patchornik, A. (1970) Adv. Protein Chem. 24, 97-260.

2. Fontana, A. and Gross, E. (1983) in Handbook of Peptide and Protein Chemistry (Darbre, A. and Waterfield, M., eds.), Heyden-Wiley, London, in press.

3. Fontana, A. (1980) in Biochemical and Medical Aspects of Tryptophan Metabolism (Hayaishi, O., Ishimura, Y. and Kido, R., eds.) Elsevier/ North-Holland Biomedical Press, Amsterdam, pp. 59-72.

4. Gross, E. (1967) Methods Enzymol. 11, 238-255.

5. Braunitzer, G. and Aschauer, M.J. (1975) Hoppe Seyler's Z. Physiol. Chemie 356, 473-474.

6. Wootton, J.C., Chambers, G.K., Holder, A.A., Baron A.J., Taylor, J.G., Fincham, J.R.S., Blumenthal, K.M., Moon, K. and Smith, E.L. (1974) Proc. Natl. Acad. Sci. USA 71, 4361-4365.

7. Ozols, J. and Gerard, C. (1977) Proc. Natl. Acad. Sci. USA 74, 3725-3729.

8. Ozols, J. and Gerard, G. (1977) J. Biol. Chem. 252, 5986-5989.

9. Savige, W.E. and Fontana, A. (1977) Methods Enzymol. 47, 442-453.

10. Hinman, R.L. and Bauman, C.P. (1964) J. Org. Chem. 29, 2431-2437.

11. Ramachandran, L.R. and Witkop, B. (1967) Methods Enzymol. 11, 283-308.

12. Omenn, G.S., Fontana, A. and Anfinsen, C.B. (1974) J. Biol. Chem. 245, 1895-1902.

13. Burstein, Y. and Patchornik, A. (1972) Biochemistry 11, 4641-4650.

14. Fontana, A., Dalzoppo, D., Grandi, C., and Zambonin, M. (1981) Biochemistry 20, 6997-7004.

TRYPTHOPHAN, A PITFALL FOR THE PEPTIDE CHEMIST

Luis Moroder and Erich Wünsch

Max-Planck-Institute of Biochemistry, Department of Peptide Chemistry, 8033 Martinsried, Federal Republic of Germany

Dedicated to Prof. A. Butenandt on the occasion of his 80th birthday.

Remarkable progress has been achieved in peptide chemistry over the last years, nevertheless the synthesis of tryptophan-peptides still represents a formidable challenge. In fact the indole function of the tryptophan-side chain is characterized by a high reactivity towards electrophiles, which are frequently employed as reagents or generated as intermediates in many already classical procedures of peptide synthesis. Consequently a series of electrophilic substitutions at the indole ring can occur, the rate of which was found to depend strongly on the chemical environment of the tryptophan residue.
Great attention has been paid in the last years to study in detail part of these side reactions and to find efficient bypasses.

Oxidative degradation of tryptophan

Even if a great number of protecting groups of differentiated selectivity has been developed in peptide chemistry, all the strategies of synthesis both in solution and on solid support imply the use of at least one acidolytic deprotection step to assure a maximum of orthogonal protection. On the other part, indole and indole derivatives are known to be sensitive to acids. Thus, treatment of tryptophan-peptides with strong acids, e.g. with haloacids in acetic or trifluoroacetic acid, with hydrogen fluoride or trifluoroacetic acid, leads to oxidative degradation with production of complex, usually colored mixtures, consisting of components which are not yet

Progress in Tryptophan and Serotonin Research

studied in detail [1,2]. Addition of diethyl phosphite or phenols to free the acid mixtures of halogens [3] as well as operating in inert atmosphere such as under argon [4] or adding antioxidants, preferentially thiols [5,7], constitute efficient methods to prevent or at least minimize the oxidative destruction of tryptophan residues. Additionally deblocking with organic sulfonic acids was found to be usually harmless [8,9].

Alkylation of tryptophan

While above mentioned oxidative degradation had been recognized already in the early times of peptide chemistry, a second even more serious side reaction of the tryptophan moiety, i. e. alkylation of the indole ring resulting from the acidolytic removal of protecting groups, has been overlooked for years. First indications [10-12] in this context prompted us to study this side reaction in detail. By a careful analytical investigation of side-products arising in the synthesis of gastrin-peptides as well as using model compounds we finally succeeded, in collaboration with Kisfaludy's laboratory, to elucidate the alkylation products of the indole ring generated by acidolytic cleavage of protecting groups on tert-butanol basis*) [13-20]. The N^{in}-tert-butyl derivative was found to constitute the main side product. Additionally mono-C-tert-butyl, poly-C-tert-butyl and poly-N,C-tert-butyl derivatives are produced, whereby the extent of such alkylation and the distribution pattern of the various substitutions at the indole function were found to strongly depend on the acids used (trifluoroacetic acid or hydrogen fluoride) and on time and temperature. To minimize this serious side reaction the use of scavengers, e. g. indole derivatives, ethers and thioethers as well as a series of thiols and their mixtures, has been carefully examined by extensive comparative studies [18-20,22]. All efforts to entirely suppress indole-alkylation failed, even if 1,2-ethane-

*) The acidolytic removal of other protecting groups, e. g. of the p-methoxybenzyloxycarbonyl group was found to produce a similar pattern of indole-alkylated derivatives [21].

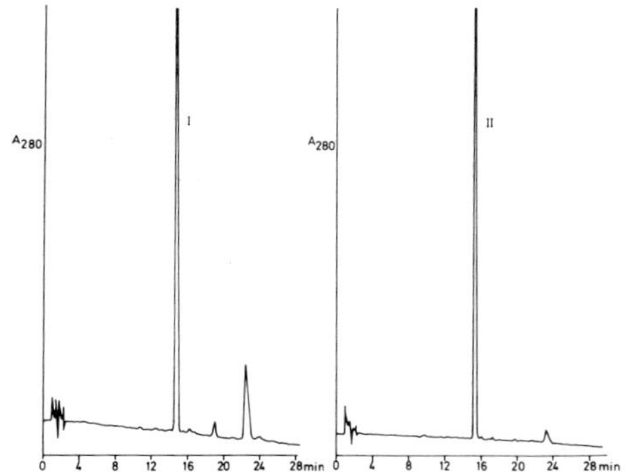

Fig.1. Hplc of crude 1-des-tryptophan, 11-leucine-minigastrinI upon deprotection with I) trifluoroacetic acid/anisole, 10:1 and II) trifluoroacetic acid/anisole/1,2-ethanedithiol, 10:1:0.25. Chromatographic conditions: µ-bondapak C18 (30x0.4 cm); eluents: A) 0.1M sodium phosphate, pH 5.4; B) acetonitrile/eluent A, 45:55; linear gradient of 35 % B to 80 % B in 30 min; flow rate: 1.7 ml/min.

dithiol, dimethylsulfide or their combination were found to efficiently reduce the rate of this side reaction as shown in Fig. 1.

Fortunately various analytical methods, e. g. ^{1}H-nmr, tlc, hplc, are sufficiently sensitive to permit detection and quantitative monitoring of the extent of such side reaction at least for lower molecular weight peptides, thus remarkably facilitating the purification processes.

Considering the findings related to the sulfenylation reaction discussed below [23], it was not surprising at all that the rate of alkylation of the indole function strongly depends on the chemical environment of the tryptophan residue,too. During the synthesis of gastrins and gastrin analogs [14,24] as well as gastrin related peptides [25] tert-butyl-tryptophan derivatives have been detected in extents up to 15 % even applying efficient scavengers, whereas no alkylation of the tryptophan residue was observed upon acidolytic removal of tert-butanol and 1-adamantol derived protecting groups from fully protected

somatostatins [26-27]. Consequently, this sequence-dependent reactivity to electrophiles strongly questions the general efficacy of acidolytic deprotection procedures devised on the basis of model studies with simple tryptophan-compounds as its N^{α}-tert-butyloxycarbonyl derivative.
Besides the use of various,more or less efficient scavengers to prevent alkylation of tryptophan residues, a reaction which undoubtedly becomes critical in repetitive techniques, the N^{in}-formyl protection has been proposed [28-29]. Even if it is effective in suppressing alkylation of the indole function, this protecting group is not the ideal one because of its lability to nucleophiles, its partial removal by hydrogenolysis (a repetive step often employed for the synthesis of peptides in solution), and because of time-dependent saturation of the double bond at positions 2'-3' of the indole group observed to occur during catalytic hydrogenation, too [30]. Additionally, the final removal of the N^{in}-formyl group by nucleophiles or aqueous bases is not free of problems. A progress in this context has, however, recently been achieved using low concentrations of hydrogen fluoride in a hydrogen fluoride-dimethylsulfide-thiol deprotection mixture [31].Of the other N^{in}-protecting groups proposed so far (mainly benzene sulfonyl derivatives [32]) unfortunately none was found to be of general application, partly because of their lability under the normal conditions of peptide synthesis and partly because of the drastic conditions needed for their final cleavage.

Sulfenylation of tryptophan

Among the various amino protecting groups used in peptide synthesis, the 2-nitrophenylsulfenyl group has achieved wide popularity. It is preferentially cleaved by haloacids in organic solvents with concomitant production of sulfenyl halides which, in turn, in the presence of tryptophan residues, react quantitatively with the indole function to produce the related 2'-thioether derivatives [33]. Kinetic studies of this sulfenylation reaction with various indole and tryptophan derivatives as well as tryptophan-peptides clearly revealed its topochemical dependency, but also the possibility to use methylindoles as scavengers [23]. Addition of 2- or 3-methylindole in larger excesses allows a clean and fast cleavage of the sulfenamides by HCl or HBr without any damage to the tryptophan side chain. Thus, alternative procedures, e.g. via thiols or phosphines are superfluous particularly taking into account the slow and partly non quantitative cleavage produced by these nucleophiles.

The fast reaction of sulfenyl halides with indoles, however, prevents the application of a series of new and elegant methods concerning both the removal of S-cysteine protecting groups and the activation of the cysteine-thiol function for a selective synthesis of symmetrical and particularly, of unsymmetrical cystine-peptides. Various commonly used thiol protecting groups are smoothly removed with sulfenyl halides [34-37]; the resulting disulfides are then reduced to the corresponding free thiol derivatives. If tryptophan is present, the thioether formation becomes a critically competing side reaction. Similarly, the activation of cysteine- (or S-protected cysteine-) peptides with sulfenyl halides [38-39] (e.g. with alkoxycarbonylsulfenyl chloride according to Kamber [38]) for a selective pairing of cysteine-peptides as successfully applied in the synthesis of human insulin [40], is not devoid of side reactions in the case of tryptophan-peptides. For identical reasons even the method of Hiskey [41] based on the activation of one cysteine component as sulfenyl thiocyanate as well as the sulfenyl-(diacyl)amide procedure with intermediate formation of S-bromo-cysteine derivatives [42], cannot be used. In this context it is noteworthy that intra- and intermolecular oxidation of cysteine- (or particular S-protected cysteine-) peptides by means of iodine which occurs via the intermediate S-iodo-derivative, inexorably leads in the case of tryptophan-peptides to the production of tryptathione derivatives. Under conditions of intramolecular oxidation the ring size of the tryptathione-peptide decidedly influences the extent of this

side reaction as well documented by an extensive study of Sieber et al. [43].
Consequently, at least for the synthesis of unsymmetrical tryptophan containing cystine-peptides a new method had to be devised. We succeeded in this aim with the sulfenohydrazide procedure [44-45]. It is based on the activation of one cysteine component with azodicarboxylic acid derivatives followed by the reaction of the resulting S-activated sulfenohydrazide with a second cysteine component to yield the unsymmetrical cystine-peptide. In both reaction steps the indole function of tryptophan was found to be inert.
To conclude, even if procedures have been improved for the synthesis of tryptophan-peptides, a protecting group for the indole function capable of reducing its high reactivity towards electrophiles by electron-withdrawal, stable under the normal conditions of peptide synthesis and smoothly removed under mild conditions, has still to be developed.

References

1. Biossonnas, R.A., Guttmann, St., Huguenin, R.L., Jaquenoud, P.-A., Sandrin, Ed.: Helv.Chim.Acta 41, 1867-1882 (1958).
2. Theodoropulos, D., Fruton, J.S.: Biochemistry 1, 993-937 (1962).
3. Wünsch, E.: Collect. Czech.Chem.Commun. 24, 60 (1959).
4. Wünsch, E.: Z. Naturforsch. 22b, 1269-1276 (1976).
5. Blake, J., Wang, K.T., Li, C.H.: Biochemistry 11, 438-441 (1972).
6. Li, C.H., Yamashiro, D.: J. Am. Chem. Soc. 92, 7608-7609 (1959).
7. Sharp, J.J., Robinson, A.B., Kamen, M.D.: J. Am. Chem. Soc. 95, 6097-6108 (1973).
8. Loffet, A., Dremier, C.: Experientia 27, 1003 (1971).
9. Suzuki, K., Endo, N., Nitta, K., Sasaki, Y.: Chem. Pharm. Bull. 26, 2198-2204 (1978).
10. Sieber, P.: Peptides 1968 (E. Bricas, ed.) North-Holland, Amsterdam, 1968, p. 236.

11. Alakhov, Y.B., Kiryushkin, A.A., Lipkin, V.: Chem. Commun. 406-407 (1970).

12. Wünsch, E., Jaeger, E., Deffner, M., Scharf, R.: Hoppe-Seyler's Z. Physiol.Chem. 353, 1716-1720 (1972).

13. Wünsch, E. Jaeger, E., Kisfaludy, L., Löw, M.: Angew. Chem. 89, 330-331, Int. Edn. Engl. 16, 317-318 (1977).

14. Jaeger, E., Thamm, P., Schmidt, J., Knof, S., Moroder, L., Wünsch, E.: Hoppe-Seyler's Z. Physiol. Chem. 359, 155-164 (1978).

15. Jaeger, E., Thamm, P., Knof, S., Wünsch, E.: Hoppe-Seyler's Z. Physiol. Chem. 359, 1629-1636 (1978).

16. Jaeger, E., Thamm, P., Knof, S., Wünsch, E., Löw, M., Kisfaludy, L.: Hoppe-Seyler's Z. Physiol. Chem. 359, 1617-1628 (1978).

17. Löw, M., Kisfaludy, L., Jaeger, E., Thamm, P., Knof, S., Wünsch, E.: Hoppe-Seyler's Z. Physiol. Chem. 359, 1637-1642 (1978).

18. Löw, M., Kisfaludy, L., Sohár, P.: Hoppe-Seyler's Z. Physiol. Chem. 359, 1643-1651 (1978).

19. Chino, N., Masui, Y., Sakakibara, S.: Peptide Chemistry 1977 (Shiba, T., Ed.) Protein Research Foundation, Osaka 1978, pp. 27-32.

20. Masui, Y., Chino, N., Sakakibara, S., Tanaka, T., Kita, H.: Peptide Chemistry 1978 (Izumiya, N., ed) Protein Research Foundation, Osaka, 1979, pp. 189-194.

21. Ogania, H., Sasaki, T., Yajima, H.: Peptide Chemistry 1978 (Izumiya, N., Ed.) Protein Research Foundation, Osaka, 1979, pp. 21-24.

22. Lundt, B.F., Johansen, N.L., Vølund, A., Markussen, J.: Int. J. Peptide Protein Res. 12, 258-268 (1978).

23. Wünsch, E., Fontana, A., Drees, F.: Z. Naturforsch. 22b, 607-609 (1967).

24. Moroder, L., Göhring, W., Nyfeler, R., Scharf, R., Thamm, P., Wendlberger, G.: Hoppe-Seyler's Z. Physiol. Chem. 364, 157-171 (1983).

25. Moroder, L., Wilschowitz, L., Gemeiner, M., Göhring, W., Knof, S., Scharf, R., Thamm, P., Gardner, J.D., Solomon, T., Wünsch, E.: Hoppe-Seyler's Z. Physiol. Chem. 362, 929-242 (1981).

26. Moroder, L., Gemeiner, M., Göhring, W., Jaeger, E., Thamm, P., Wünsch, E.: Biopolymers 20, 17-37 (1981).

27. Moroder, L., Gemeiner, M., Göhring, W., Jaeger, E., Musiol, J., Scharf, R., Stocker H., Wünsch, E., Pradayrol, L., Vaysse, N., Ribet, A.: Hoppe-Seyler's Z. Physiol. Chem. 362, 697-716 (1981).

28. Previero, A., Coletti-Previero, M.A., Cavadore, J.C.: Biochim. Biophys. Acta 147, 453-461 (1967).

29. Ohno, M., Tsukumoto, S., Izumyia, N.: Chem. Commun. 663-664 (1972).

30. Löw, M., Kisfaludy, L.: Hoppe-Seyler's Z. Physiol. Chem. 360, 13-18 (1979).

31. Heath, W.F., Tam, J.P., Merrifield, R.B.: Chem. Commun. 896-897 (1982).

32. Fukuda, T., Wakimasu, M., Kobayashi, S., Fujino, M.: Chem. Pharm. Bull. 30, 2825-2835 (1982).

33. Scoffone, E., Fontana, A., Marchiori, F., Benassi, C.A.: Peptides 1966 (Beyermann, H., ed.) North-Holland, Amsterdam, 1967, pp. 18-23.

34. Guarneri, M., Benassi, G.A., Ferroni, R., Guggi, A., Tomatis, R., Rocchi, R.: Gazz. Chim. Ital. 101, 375-386 (1971).

35. Moroder, L., Marchiori, F., Borin, G., Scoffone, E.: Biopolymers, 12, 493-505 (1973).

36. Pastuszak, J.J., Chimiak, A.: Roszniki Chem. 51, 1567-1569 (1977).

37. Castell, J.V., Tun-Kyi, A.: Helv. Chim. Acta 62, 2507-2510 (1979).

38. Kamber, B.: Helv. Chim. Acta 56, 1370-1381 (1973).

39. Matsueda, R., Kimura, T., Kaiser, E.T., Matsueda, G.R.: Chem. Lett. 737-740 (1981).

40. Sieber, P., Kamber, B., Hartmann, A., Jöhl, A., Riniker, B., Rittel, W.: Helv. Chim. Acta 57, 2617-2621 (1974).

41. Hiskey, R.C., Ward, B.F.: J. Org. Chem. 35, 1118-1121 (1970).

42. Harpp, D.N., Back, T.G.: J. Org. Chem. 36, 3828-3829 (1971).

43. Sieber, P., Kamber, B., Riniker, B., Rittel, W.: Helv. Chim. Acta 63, 2358 - 2363 (1980).

44. Wünsch, E., Romani, S.: Hoppe-Seyler's Z. Physiol. Chem. 363, 449 - 453 (1982).

45. Wünsch, E., Romani, S., Moroder, L.: Peptides 1982 (Blaha, K., ed.), 1983, in press.

PHOTOCHEMICAL AND RADIOCHEMICAL REACTIONS OF TRYPTOPHAN : MODIFICATION OF A SINGLE-STRAND BINDING PROTEIN FROM BACTERIOPHAGE T4.

Jean-Jacques Toulmé

Laboratoire de Biophysique - INSERM U.201, ERA 951 du CNRS, Muséum National d'Histoire Naturelle, 61 Rue Buffon
F. 75005 PARIS - FRANCE

Introduction

Modification of proteins is a strategy to elucidate the role played by a given type of amino acid in many fundamental problems : protein structure, enzyme catalysis, macromolecular interactions, etc... Photochemical and radiochemical methods have been developed to selectively modify amino acids. UV irradiation of proteins results in unspecific modification of the residues which absorb the incident light or which are altered via indirect processes (energy transfer, electron transfer...). Similarly γ radiolysis of aqueous solutions of proteins leads to the unselective attack of several residues (tryptophan, tyrosine, methionine, cysteine, histidine...) by radicals formed by radiolysis of water. Therefore specific radio- or photochemical modifications need a "secondary" reagent which will be responsible for the selectivity. For instance γ radiolysis of an aqueous solution of halide X^- gives rise to $X_2^{\overline{\cdot}}$ radicals which react preferentially with one type of amino-acid depending on the environmental conditions (1). It has also been shown that UV irradiation of tryptophan in presence of trichloroethanol leads to the specific modification of fluorescent tryptophyl residues (C. Ghiron, personal communication, 2).

Stacking interactions between tryptophan and nucleic acid bases provide oligopeptides such as lysyl-tryptophyl-α lysine with a specificity of binding to single-stranded polynucleotides (3,4). Proteins which bind preferentially - and in general cooperatively - to single-stranded nucleic acids have been recently isolated from both eukaryotic and prokaryotic cells. These proteins-termed single-strand binding (SSB) proteins-play

Progress in Tryptophan and Serotonin Research

an essential role in the cell ; for example the protein coded for by gene 32 from bacteriophage T4 (gp 32) is involved in phage DNA replication, recombination, and repair of UV damages (4, 5). From studies with model peptides we may think that stacking of nucleic acid bases with aromatic amino acids convey to SSB proteins part of their specificity of binding. We have studied the affinity of DNA for the gp 32 after γ radiolysis in the presence of SCN^- or UV irradiation in the presence of trichloroethanol.

Material and Methods

1) Material : *E. coli* DNA was purchased from Sigma.[^{32}P]*E. coli* DNA was prepared by nick translation (6). Gp 32 was prepared from *E. coli B* cells infected with an amA292 strain of T4 according to previously described procedures (7, 8).

2) Irradiations : UV irradiations were performed with a HBO 250 W mercury lamp. In order to avoid light absorption by other chromophores than tryptophan the protein solutions were irradiated through a 1 cm cuvette containing toluene. In these conditions the incident dose was 10^{-7} einsteins s^{-1}. Samples containing 50 µM gp 32 in the presence of 0.1 M trichloroethanol were diluted 100 fold before any fluorescence measurements.

Steady state γ irradiations were carried out with a ^{60}Co source which delivered 1.3 Gy s^{-1}. Before irradiation the sample (200 µl) of gp 32 (3.5 µM) in a 1 mM phosphate buffer (pH 6.2) containing 10 mM KSCN was saturated with N_2O.

3) Spectroscopic measurements : Absorption and fluorescence spectra were recorded on a UVIKON 820 spectrophotometer and on a FICA 55 differential spectrofluorimeter, respectively. The concentration of gp 32 solutions was determined using a molar extinction coefficient of 37,000 M^{-1} cm^{-1} at 280 nm.

The tryptophan content of the native and irradiated gp 32 was determined according to a fluorimetric method (9). The native gp 32 contains five tryptophyl residues.

4) Filter binding assay : Sonicated, heat-denatured *E. coli* [^{32}P] DNA mixed with gp 32 in 20 mM Tris, HCl (pH 8.0), 130 mM NaCl, 5 % glycerol 2 mM 2-mercaptoethanol, 1 mM EDTA (buffer A) was filtered at a flow rate of 1 ml mn^{-1} through nitrocellulose filters (Schleicher and Schüll BA85) prewashed with 10 ml of buffer A containing 1 % DMSO. These filters were then washed with 1 ml of buffer A, dried and counted.

Results and Discussion

1) Radiochemical modification of gp 32 : Steady-state radiolysis of water leads to the formation of e^-_{aq} and $OH^{\cdot}$ as major species. Their specificity of reaction towards amino acids is very low. In presence of N_2O, solvated electrons are converted into $OH^{\cdot}$ radicals. Secondarily these radicals can act on X^- ions (X^- stands for I^-, Br^- or SCN^-) according to the following reactions :

$$OH^{\cdot} + X^- \rightarrow XOH^{\bar{\cdot}} \rightleftarrows OH^- + X^{\cdot} \quad ; \quad X^{\cdot} + X^- \rightarrow X_2^{\bar{\cdot}}$$

These radicals $X_2^{\bar{\cdot}}$ react with amino acids with a selectivity which depends on the pH of the medium. The rate constant for the reaction of $SCN_2^{\bar{\cdot}}$ at pH 6.0 decreases in the following order : Trp > Cys >> Met > Tyr >> His. The reaction rate for oxidizing cysteine is at least one order of magnitude lower than that with tryptophan (1). The reaction of $SCN_2^{\bar{\cdot}}$ or $Br_2^{\bar{\cdot}}$ radicals with tryptophan derivatives gives rise to the formation of an oxidation product whose structure is unknown.

The γ radiolysis of gp 32 in N_2O saturated solution containing SCN^- at pH 6.2 results in i) a decrease of its fluorescence emission and ii) a modification of its absorption spectrum. The difference absorption spectrum between irradiated and unirradiated gp 32 exhibits a negative band centered at 280 nm and two positive bands at 290 nm and 240 nm whose intensities increase with the irradiation time (Fig. 1). The modification of absorption and emission spectra of the irradiated gp 32 is accompanied by the decrease of its tryptophan content. Moreover a new fluorescence emission peaking at 420 nm can be correlated with the formation of a radioproduct (Toulmé, unpublished result).

γ radiolysis of gp 32 bound to heat-denatured DNA also leads to a decrease in the fluorescence of the protein. However the number of intact tryptophans in irradiated samples is larger in the complex than in the free protein (3 and 1 residual tryptophan for samples irradiated with a dose of 320 Gy respectively). This suggests that the presence of DNA protects tryptophan against the oxidation by SCN_2^- radicals either by steric hindrance or by electrostatic repulsion between phosphate groups and radical anions. Unexpectedly we have found that some cysteine residues of gp 32 can be attacked upon γ radiolysis in the same conditions. This modification could be due to a particular environment of cysteine residues i.e. the presence in their vicinity of positively charged groups (lysine side chain for instance) which could increase the reaction rate constant of -SH with radical anions.

2) Photochemical modification of gp 32 in the presence of trichloroethanol :

Upon UV irradiation trichloroethanol reacts with tryptophan derivatives in their first excited singlet-state, leading probably to a photoadduct on the indole ring whose structure is unknown (C. Ghiron, personal communication). UV irradiation of N-acetyltryptophanamide (NATA) in the presence of trichloroethanol results in the formation of a photoproduct whose absorption spectrum is shifted to longer wavelengths with respect to NATA. This photoproduct emits a fluorescence centered at about 500 nm (2).

UV irradiation of gp 32 in solutions containing 0.1 M trichloroethanol leads to modifications of its absorption spectrum very similar to those obtained with γ irradiated protein (Fig. 1) : the decrease in the absorbance at 280 nm is accompanied by an increase at long ($\lambda > 290$ nm) and short wavelengths ($\lambda < 260$ nm). Similarly to what has been observed with NATA, the fluorescence emission at 340 nm of gp 32 is decreased upon UV irradiation in presence of trichloroethanol whereas a new emission appears at longer wavelength (Fig. 2). These modifications are clearly related to the photodestruction of tryptophan residue(s). The existence of two isosbestic points at 265 and 292 nm and of one isoemissive point at 410 nm suggests that under our experimental conditions only one type of photoproduct is formed which is probably the same as that one obtained with NATA.

The tryptophan content of native and irradiated gp 32 was determined. As

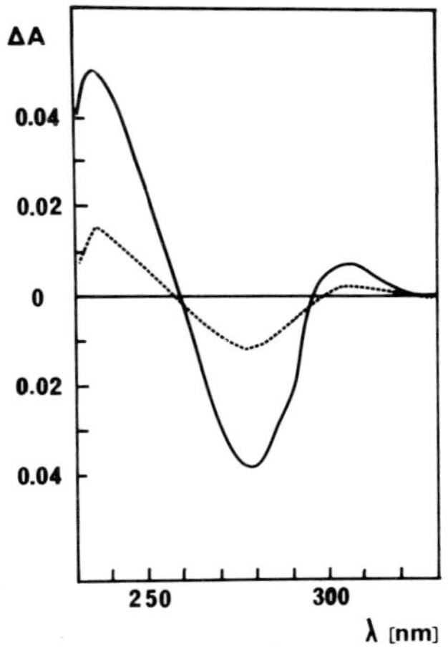

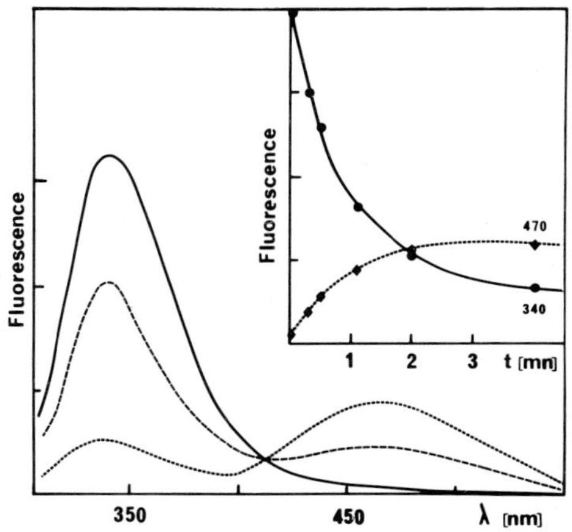

Figure 1 : *Difference absorption spectrum between 3.5 μM native and γ irradiated gp 32 with a dose of 40 Gy (...) or 120 Gy (——). γ irradiation was performed as described in Material and Methods (optical path length 1 cm).*

Figure 2 : *Fluorescence emission spectrum of native (——) and UV irradiated gp 32 during 30 s (---) and 240 s (····), in presence of 0.1 M trichloroethanol.*
INSET : Variation of the fluorescence emission at 340 nm (—●—) and 470 nm (·◆·) versus the time of irradiation (excitation wavelength : 280 nm).

expected from fluorescence and absorption measurements the number of tryptophyl residues decreases upon UV irradiation in the presence of trichloroethanol. But, two intact tryptophans per protein molecule are still present even though the irradiated protein has lost almost all its fluorescence emission. We have already shown that a singlet-singlet energy transfer from tryptophan to tryptophan-trichloroethanol photoproduct could be responsible for the rapid deactivation of unmodified excited tryptophan(s). This process can prevent their reaction with trichloroethanol molecules(10).

3) Binding of photochemically or radiochemically modified gp 32 to DNA :

Previous studies have shown that at least one tryptophyl residue is located in the binding site of gp 32 and is in close contact with nucleic acid bases (4, Toulmé and Le Doan, unpublished results). Whether this residue interacts with polynucleotide constituents is not clear yet. If such

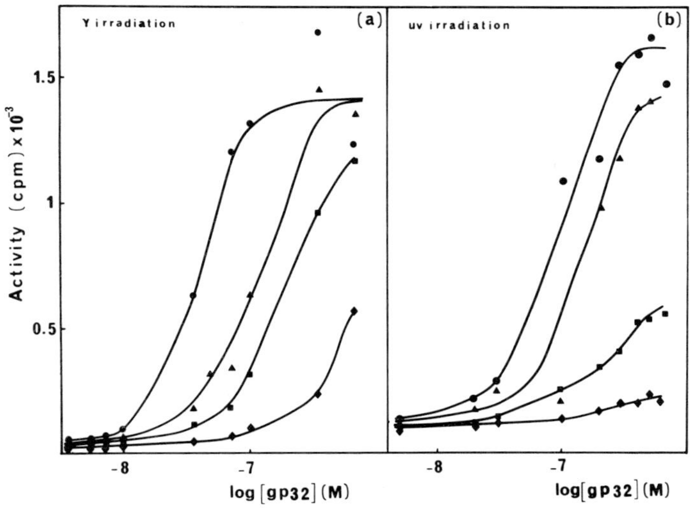

Figure 3 : *Filter retention assay of E. coli [^{32}P] DNA by a)γirradiated gp 32 with a dose of 0 (●), 40 (▲), 80 (■) and 270 Gy (◆). b) UV irradiated gp 32 with a dose of 0 (●), 3 x 10^{-6} (▲), 1.2 x 10^{-5} (■), or 4.2 x 10^{-5} einsteins (◆). Irradiations and retention assays were performed as described under Material and Methods.*

interactions exist and if they are crucial to the formation of gp 32 nucleic acid complexes we expect a decrease in the affinity of DNA for protein molecules whose tryptophyl residues have been modified.

We have investigated the binding of γ or UV irradiated gp 32 to heat-denatured DNA by retention assay on nitrocellulose filters. Proteins are adsorbed onto nitrocellulose whereas DNA molecules flow through the filter. Interactions between protein and DNA promote the retention of the DNA ; the amount of DNA sticking to the filter is a measure of the affinity of the two macromolecules. As shown on Figure 3 retention assays with native gp 32 leads to a sigmoïdal curve which is due to its cooperative mode of binding to single stranded nucleic acids. The amount of DNA retained on the filter for a given concentration of gp 32 decreases with the duration of irradiation for both γ and UV irradiated proteins, thus indicating a loss of affinity of the modified proteins as compared to the native one. With a protein UV irradiated at a dose of about 4 x 10^{-5} einsteins no radioactivity remains bound to the filter. However in this case a slight decrease in the peptidic band of the circular dichroism spectrum of the protein indicates a conformational change. Such a variation was not detected with proteins irradiated at lower doses (< 2 x 10^{-5} einsteins). It was also shown that the photochemically modified protein was not able to compete with native gp 32 for binding to heat-denatured DNA nor did it induce any hyperchromism of poly(A) (10). This suggests that the modification of three out of five tryptophyl residues is responsible for the inactivation of the protein. It is therefore possible that one of these residues is involved in the association of gp 32 with nucleic acids.

Similar results were obtained with γ irradiated gp 32. Although no conformational change was detected by circular dichroism, the modification of cysteine residues upon γ irradiation in the presence of SCN^- ions prevents any clear conclusion to be drawn about the role played by tryptophyl residues in the binding. However the modification of cysteine residues only leads to a slight decrease of affinity of gp 32 as compared to a protein molecule whose both tryptophan and cysteine residues have been modified (Toulmé, unpublished results).

Conclusion

Upon UV irradiation of gp 32 in presence of trichloroethanol three out of five of its tryptophyl residues can be modified. γ irradiation of the protein performed at pH 6.2 in N_2O saturated solutions containing SCN^- ions preferentially leads to the modification of tryptophan. Moreover binding to DNA prevents two tryptophyl residues from reacting with $SCN_2^{\bar{\cdot}}$ radical anions. Both photochemically and radiochemically modified proteins have lost their affinity for DNA. Although modification of cysteine upon γ radiolysis makes the situation less clear, these results suggest that at least one tryptophan residue of the gp 32 could interact with nucleic acid constituents and stabilize the complexes. This conclusion agrees with previous results (4). The above results show both the interest and the limits of the photochemical and radiochemical methods presented. Several points have to be checked before concluding about the involvement of tryptophan residues in some processes: i) are the residues to be modified accessible to the reacting agent ? From this point of view trichloroethanol is particularly convenient because it is a dynamic quencher of the indole fluorescence (11). ii) Is there any side reaction of the agent with other amino acids ? Neither modification of tyrosine nor cysteine have been reported to occur upon UV irradiation of proteins in presence of trichloroethanol (2,10). But we have observed an attack of cysteine residues by SCN_2^- radicals undoubtedly due to an environmental effect. iii) Is the protein conformation conserved upon modification ? This is probably the crucial point. Circular dichroism allowed us to check that no major conformational change occurred upon modification. However this technique cannot reveal local changes induced by the modifying agent.

Acknowledgements : I wish to thank Pr. C. Hélène and R. Santus for fruitful discussions. I am indebted to Mrs R. Debet and N. Loreau for their skillful technical assistance. I am grateful to Drs. T. Le Doan and J. Casas-Finet who have taken a part in this work.

References

1. Adams, G.E., Aldrich, J.E., Bisby, R.M., Cundall, R.B., Redpath, J.L. and Wilson R.L. : Radiat. Res. 49, 278-289 (1972).

2. Privat, J.P. and Charlier M. : Eur. J. Biochem. 84, 79-85 (1978).

3. Brun, F., Toulmé, J.J. and Hélène, C. : Biochemistry 14, 558-563 (1975).

4. Hélène, C., Toulmé, J.J. and Montenay-Garestier, T. in Topics in Nucleic Acid Structures (S. Neidle Ed.) McMillan Press, part 2, 229-285 (1982).

5. Coleman, J.E. and Oakley, J.L. : C.R.C. Critical Rev. Biochem. 7 247-289 (1980).

6. Rigby, P.W.J., Dieckmann, M., Rhodes, C. and Berg, P. : J. Mol. Biol. 113, 237-251 (1977).

7. Alberts, B.M. and Frey, L. : Nature 227, 1313-1318 (1970).

8. Toulmé, J.J. and Hélène, C. : Biochim. Biophys. Acta 606, 95-104 (1980).

9. Sasaki, T., Abrams, B. and Horelker, B.L. : Anal.Biochem. 65, 396-404 (1975).

10. Toulmé, J.J.,Le Doan, T. and Hélène, C. : submitted to Biochemistry.

11. Eftink M.R. and Ghiron, C.A. : J. Phys. Chem. 80, 486-493 (1976).

X-RAY STRUCTURE ANALYSES OF TRYPTOPHAN DERIVATIVES AS MODELS FOR STEREOCHEMICAL ASPECTS OF THE TOXIC PEPTIDES FROM AMANITA MUSHROOMS

Alfred Gieren, Viktor Lamm, Manfred Sturm
Max-Planck-Institut für Biochemie,
Abteilung Strukturforschung I, D-8033 Martinsried.

Introduction

In cooperation with Th. Wieland et al. we have performed the X-ray structure analyses of the tryptophan derivatives 1 (1) and 2 $(CH_3)_2CO$ (2). This work was done in order to elucidate stereochemical features of the poisonous bicyclic oligopeptides - phallotoxins and amatoxins - of Amanita phalloides.

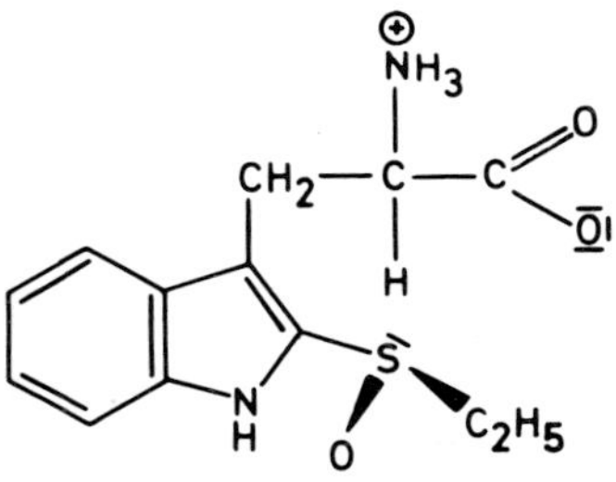
1

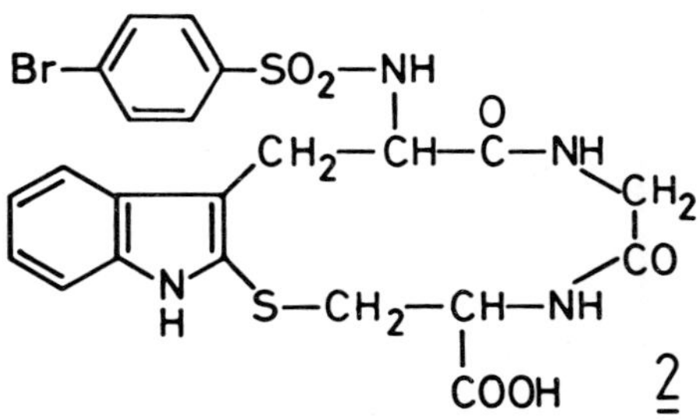
2

Results

The configuration at the sulfur of the tryptophan sulfoxide moiety in phalloidin sulfoxide and amatoxins is essential for the poisonous activity. In the case of the mirror imaged configuration at this sulfur the poisonous activity is dramatically reduced. By comparison of the ORD curves of 1 and the toxic phalloidin sulfoxide and amatoxins it could be concluded, that the configuration at the chiral center has to be the same.

The molecular structure of 1 is represented in Fig.1. The R-configuration at the SO sulfur of the model compound has also to be assigned to the poisons. In the meantime this result was confirmed by an X-ray structure analysis of -amanitin (3).

The structure of the model compound 2 was performed in order

Progress in Tryptophan and Serotonin Research

to get an unambiguous proof for the helicity of the tryptophan-thioether bridge, i.e. the conformation at the S-C bond, connecting the sulfur with the indole ring system. A "free" rotation around this bond is not possible, because it is part of a cyclic system. As shown by Fig.2, at least four different conformations have to be discussed in principle. The helicity of the thioether bridge in 2 $(CH_3)_2CO$ is found as an M_N-helicity. From the fact that 2 and analogous model compounds exhibit a CD spectrum with almost the same but mirror imaged Cotton effects as phallotoxins, to the phallotoxins a P_N-helical thioether structure has to be ascribed. Together with ^{1}H-NMR results and potential energy calculations it was possible, by the

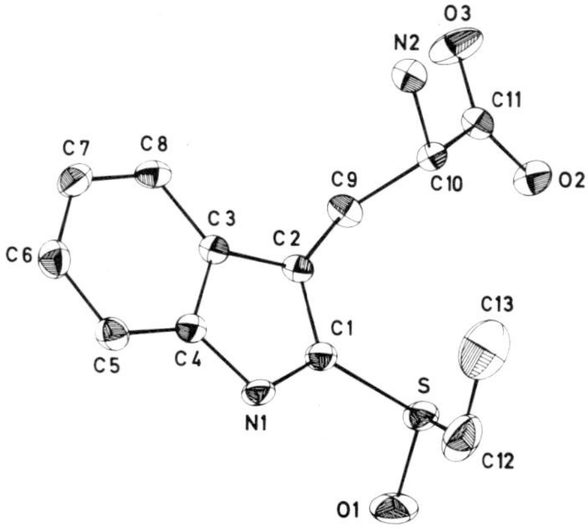

Fig.1: Molecular structure of (R)-2(ind)-ethylsulfinyl-L-tryptophan 1. The atoms are given in form of their thermal ellipsoids; hydrogens are omitted.

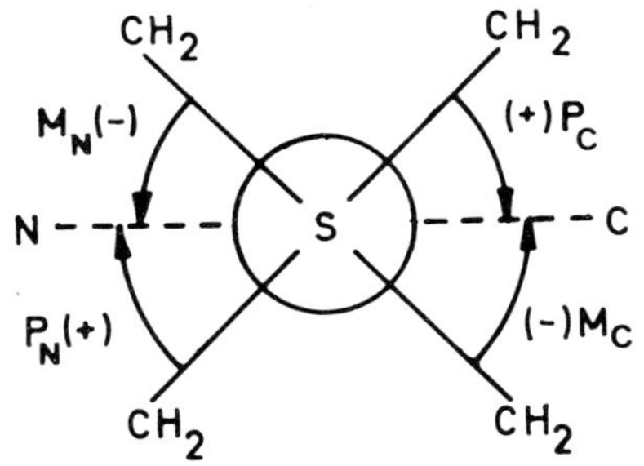

Fig.2: Possible conformations at the S-C(indole) bond in 2.

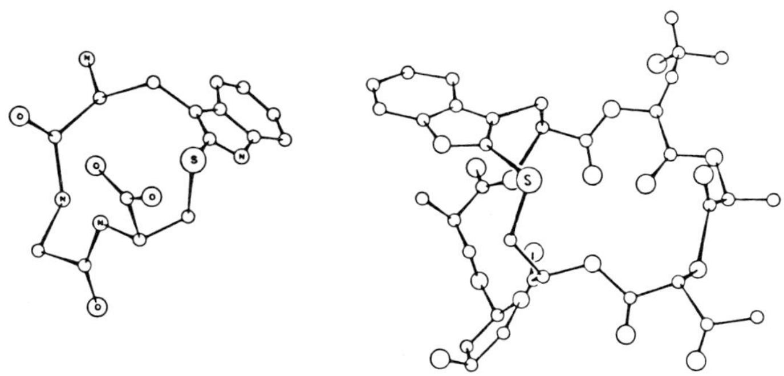

Fig.3: Models of the spatial structure of 2 (M_N-helicity) and of phalloin (P_N-helicity). In the model of 2 the N-p-bromobenzenesulfonyl residue is omitted.

Table 1: Atomic coordinates and thermal parameters of 1 ($P2_12_12_1$; a = 14.534(10), b = 15.605(10), c = 5.743(5) Å; Z=4). The anisotropic thermal parameters are defined by

$$T = \exp[-0.25(B_{11}h^2a^{*2} + B_{22}k^2b^{*2} + B_{33}l^2c^{*2} + 2B_{12}hka^*b^* + 2B_{13}hla^*c^* + 2B_{23}klb^*c^*)].$$

Atom	x	y	z	B_{11}	B_{22}	B_{33}	B_{12}	B_{13}	B_{23}
S	0.5716(1)	0.2703(1)	0.7950(2)	3.32(4)	1.61(3)	2.46(4)	0.35(3)	-1.44(4)	-0.45(3)
O1	0.5166(2)	0.3441(2)	0.8970(7)	5.3 (2)	2.4 (1)	4.5 (2)	1.4 (1)	-1.8 (1)	-2.1 (1)
O2	0.6697(2)	0.0400(2)	1.1977(5)	3.3 (1)	2.1 (1)	1.1 (1)	-0.1 (1)	-0.6 (1)	-0.1 (1)
O3	0.6246(3)	-0.0882(2)	1.0736(5)	6.7 (2)	1.5 (1)	2.2 (1)	-1.0 (1)	-1.4 (1)	0.8 (1)
N1	0.4531(2)	0.2445(2)	0.4301(6)	2.8 (1)	1.1 (1)	2.4 (1)	0.0 (1)	-1.1 (1)	0.1 (1)
N2	0.6315(2)	-0.0376(2)	0.6222(5)	2.2 (1)	2.1 (1)	1.0 (1)	0.2 (1)	0.1 (1)	-0.1 (1)
C1	0.4996(2)	0.2092(2)	0.6161(7)	2.3 (1)	1.4 (1)	1.9 (1)	0.1 (1)	-0.6 (1)	-0.1 (1)
C2	0.4869(2)	0.1216(2)	0.6250(6)	1.8 (1)	1.4 (1)	1.5 (1)	0.2 (1)	0.1 (1)	0.0 (1)
C3	0.4276(2)	0.1020(2)	0.4323(6)	1.6 (1)	1.5 (1)	1.3 (1)	0.1 (1)	0.0 (1)	0.0 (1)
C4	0.4074(2)	0.1800(2)	0.3170(7)	1.8 (1)	1.5 (1)	1.8 (1)	0.1 (1)	-0.2 (1)	0.0 (1)
C5	0.3500(3)	0.1846(2)	0.1226(7)	2.4 (1)	1.9 (1)	1.8 (1)	0.3 (1)	-0.4 (1)	0.3 (1)
C6	0.3105(3)	0.1093(3)	0.0465(7)	2.1 (1)	2.7 (1)	2.1 (1)	0.0 (1)	-0.4 (1)	-0.1 (1)
C7	0.3282(3)	0.0311(2)	0.1627(8)	2.2 (1)	2.0 (1)	2.5 (2)	-0.4 (1)	0.1 (1)	-0.6 (1)
C8	0.3859(2)	0.0265(2)	0.3506(7)	2.3 (1)	1.3 (1)	2.2 (1)	0.0 (1)	0.2 (1)	0.2 (1)
C9	0.5196(2)	0.0629(2)	0.8140(7)	2.0 (1)	1.8 (1)	1.4 (1)	0.2 (1)	0.3 (1)	0.5 (1)
C10	0.6190(2)	0.0287(2)	0.8065(6)	1.8 (1)	1.3 (1)	0.8 (1)	-0.1 (1)	-0.1 (1)	0.1 (1)
C11	0.6399(3)	-0.0107(2)	1.0458(6)	1.9 (1)	1.8 (1)	1.0 (1)	0.2 (1)	0.0 (1)	0.3 (1)
C12	0.6411(4)	0.3147(3)	0.5605(10)	3.9 (2)	3.0 (2)	4.5 (2)	-1.5 (2)	-1.4 (2)	1.1 (2)
C13	0.6941(4)	0.2454(4)	0.4338(12)	4.6 (2)	5.7 (3)	4.2 (3)	-1.8 (2)	0.9 (2)	0.3 (2)

Atom	x	y	z	B(Å²)	Atom	x	y	z	B(Å²)
H1	0.451(3)	0.293(3)	0.388(10)	4.2(11)	H9	0.509(3)	0.099(3)	0.979(8)	2.8(9)
H2	0.589(3)	-0.075(3)	0.627(8)	2.1(8)	H19	0.481(3)	0.024(3)	0.848(9)	3.2(9)
H3	0.633(3)	-0.009(3)	0.468(9)	3.2(10)	H10	0.666(2)	0.073(2)	0.779(7)	1.6(7)
H4	0.686(3)	-0.057(3)	0.641(9)	2.8(9)	H12	0.684(3)	0.354(3)	0.665(10)	4.3(11)
H5	0.342(3)	0.235(3)	0.039(8)	3.1(9)	H22	0.578(4)	0.350(3)	0.442(10)	5.5(13)
H6	0.268(3)	0.110(3)	-0.104(9)	3.0(9)	H13	0.753(5)	0.259(5)	0.357(16)	10.6(22)
H7	0.298(3)	-0.013(3)	0.129(11)	4.7(12)	H23	0.656(5)	0.213(5)	0.340(15)	9.1(19)
H8	0.396(2)	-0.020(2)	0.415(7)	1.4(7)	H33	0.733(4)	0.208(4)	0.557(11)	5.3(13)

knowledge of the chirality of the thioether bridge, to propose the full three dimensional model (2) of the phallotoxins, given in Fig.3.

A comparison of selected bond lengths in 1 and 2 shows no noteworthy differences for the tryptophan fragment. The bond lengths indicate a significant -interaction between the indole ring system and the sulfoxide and thioether residue, respectively.

Same as for 2, in the crystal structure of 1 hydrogen bridges are a dominant factor. There are formed four crystallographically not equivalent intermolecular hydrogen bridges, with all the H-atoms of the ammonium group and the N-H group of the pyrrole ring as donors and the oxygens of the carboxylate and the S-O group as acceptors. As can be seen in Fig.4 one oxygen (O2) of the carboxylate group acts as an acceptor for two hydrogen atoms. The atomic positions and the thermal parameters are given in Table 1.

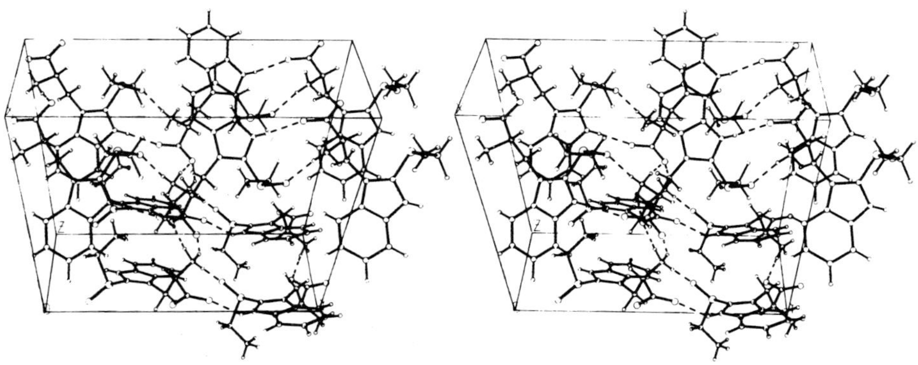

Fig.4: Stereoscopic view of the crystal packing of 1.

References

1. Wieland, Th., Jordan de Urries, M.P., Indest, H., Faulstich, H., Gieren, A., Sturm, M., Hoppe, W.: Liebigs Ann. Chem. 1974, 1570-1579.
2. Wieland, Th., Beijer, B., Seeliger, A., Dabrowski, J., Zanotti, G., Tonelli, A.E., Gieren, A., Dederer, B., Lamm, V., Hädicke, E.: Liebigs Ann. Chem. 1981, 2318-2334.
3. Kostansek, E.C., Lipscomb, W.N., Yocum, R.R., Thiessen, W.E.: J. Am. Chem. Soc. 99, 1273-1274 (1977).

Author Index

Subject Index

ABBREVATIONS:

BH_4:	tetrahydrobiopterin
BNPS-skatol:	2-(2-nitrophenylsulfenyl)-3-methyl-3-bromoindolenine
CSF:	cerebrospinal fluid
DS:	derivative spectrophotometry
EC:	electrochemical detection
F:	fluorometric detection
FANFT:	N-(4-(5-nitro-2-furyl)-2-thiazolyl)-formamide
GC/MS:	gas chromatography/mass spectrometry
GLC:	gas liquid chromatography
HPLC:	high performance liquid chromatography
HPLC/EC:	high performance liquid chromatography/ electrochemical detection
HPTLC:	high performance thin layer chromatography
IDO:	indoleamine 2,3-dioxygenase
LC:	liquid chromatography
LCEC:	liquid chromatography/electrochemical detection
LNAA:	large neutral amino acids
LPS:	lipopolysaccharides
MFO:	mixed function oxygenase
MS:	mass spectrometry
OMA:	optical multichannel analyzer
OPT:	o-phthalaldehyde
PKO:	phenylketonuria
SIM:	selected ion monitoring
TSO:	tryptophan side chaine oxidase